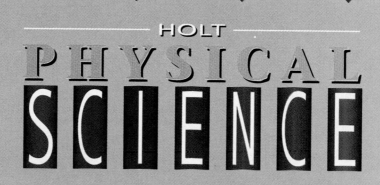

HOLT
PHYSICAL
SCIENCE

Annotated Teacher's Edition

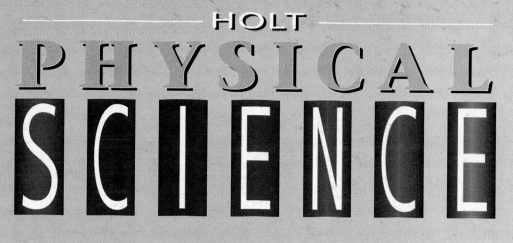

HOLT
PHYSICAL
SCIENCE

Annotated Teacher's Edition

Mapi M. Cuevas
Professor, Department of Natural Sciences
Santa Fe Community College
Gainesville, Florida

William G. Lamb
Physics Teacher
Oregon Episcopal School
Portland, Oregon

SENIOR EDITORIAL ADVISOR
Curriculum and Multicultural Education

John E. Evans, Jr.
Science Education Specialist
Philadelphia, Pennsylvania

HOLT, RINEHART AND WINSTON
Austin • *New York* • *Orlando* • *Chicago* • *Atlanta* • *San Francisco* • *Boston* • *Dallas* • *Toronto* • *London*

Mapi Martinez Cuevas

Dr. Cuevas holds a B.S. in chemistry from the University of Puerto Rico and a Ph.D. in organic chemistry from the University of Florida. Dr. Cuevas is currently a professor in the Natural Sciences Department at Santa Fe Community College in Gainesville, Florida, teaching both chemistry and physical science. She has taught chemistry and physical science at the high school level. In addition to her teaching, Dr. Cuevas has been involved in the development of science curricula, and in the organization of science fairs.

William G. Lamb

Dr. Lamb holds a B.S. in chemistry from Millsaps College and an M.A. and Ph.D. in science education from the University of Texas at Austin. He has also completed all the coursework for a Ph.D. in organic chemistry with a minor in air pollution engineering. Dr. Lamb teaches chemistry, physics, and the impact of science and technology on society at the Oregon Episcopal School in Portland, Oregon, where he holds the Winningstad Chair in the Physical Sciences. He has taught science and been involved with curriculum development at the elementary, secondary, and university levels. In addition, he has served on the advisory board of *The Science Teacher* and as one of the writers for the American Chemical Society's *ChemSource* project. Dr. Lamb is on leave from the Oregon Episcopal School during 1992–1994 to serve as coordinator for the Apprenticeships in Science and Engineering Program, sponsored by the Saturday Academy of the Oregon Graduate Institute of Science and Technology.

John E. Evans, Jr.

Dr. Evans holds an Ed.D. in Education and a Secondary Principal Certification from Temple University, Philadelphia, Pennsylvania. He is currently a Science Program Advisor to the National Science Foundation, Washington, D.C., and was site coordinator for Project 2061 in Philadelphia. He is a member of the Division of Multicultural Science Education Committee of the National Science Teachers Association. He has previously received the Award of Achievement in Science Education, School District of Philadelphia.

Cover Design: Didona Design Associates

Cover: Fireworks. Photo by Gabe Palmer/The Stock Market

Printed in the United States of America   ISBN 0-03-097539-5

4 5 6 7   041   97 96

Acknowledgments

Content Advisors

Iva Brown, Ph.D.
University of Southern Mississippi
Department of Science Education
Hattiesburg, Mississippi

Robert Fronk, Ph.D.
Science Education Department
Florida Institute of Technology
Melbourne, Florida

Theophilus F. Leapheart, Ph.D.
Chemist
Midland, Michigan

Walt Tegge
Chief of Visitor Services
National Historic Oregon Trail
U.S. Department of the Interior
Bureau of Land Management
Baker City, Oregon

Curriculum Advisors

Barbara Durham
Physical Science Teacher
Roswell High School
Roswell, Georgia

Debbie Gozzard
Physical Science Teacher
Lynnhaven Middle School
Virginia Beach, Virginia

Charles Kish
Science Teacher
Saratoga Springs
   Junior High School
Saratoga Springs, New York

Gustavo Loret-de-Mola, Ed.D.
Science Supervisor, Dade County
Fort Lauderdale, Florida

Jim Pulley
Science Teacher
Oak Park High School
North Kansas City, Missouri

Rajee Thyagarajan
Physics Teacher
Health Careers High School
San Antonio, Texas

Brenda Walker
Science Teacher
Tahlequah Middle School
Tahlequah, Oklahoma

Reading/Literature Advisors

Philip E. Bishop, Ph.D.
Professor, Department of
   Humanities
Valencia Community College
Orlando, Florida

Patricia S. Bowers, Ph.D.
Associate Director
Center for Mathematics and
   Science Education
University of North Carolina
Chapel Hill, North Carolina

ANNOTATED TEACHER'S EDITION

Contents

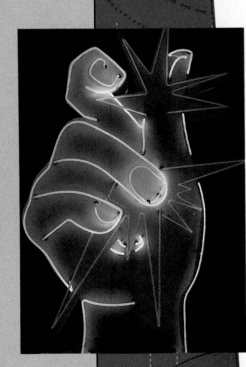

CONTENTS

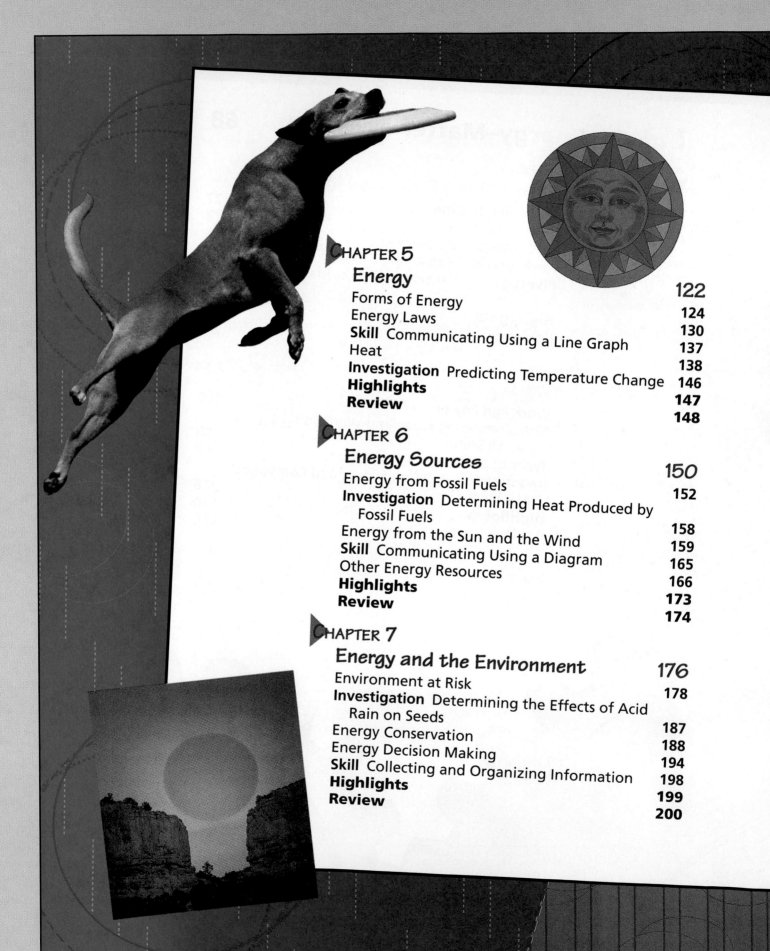

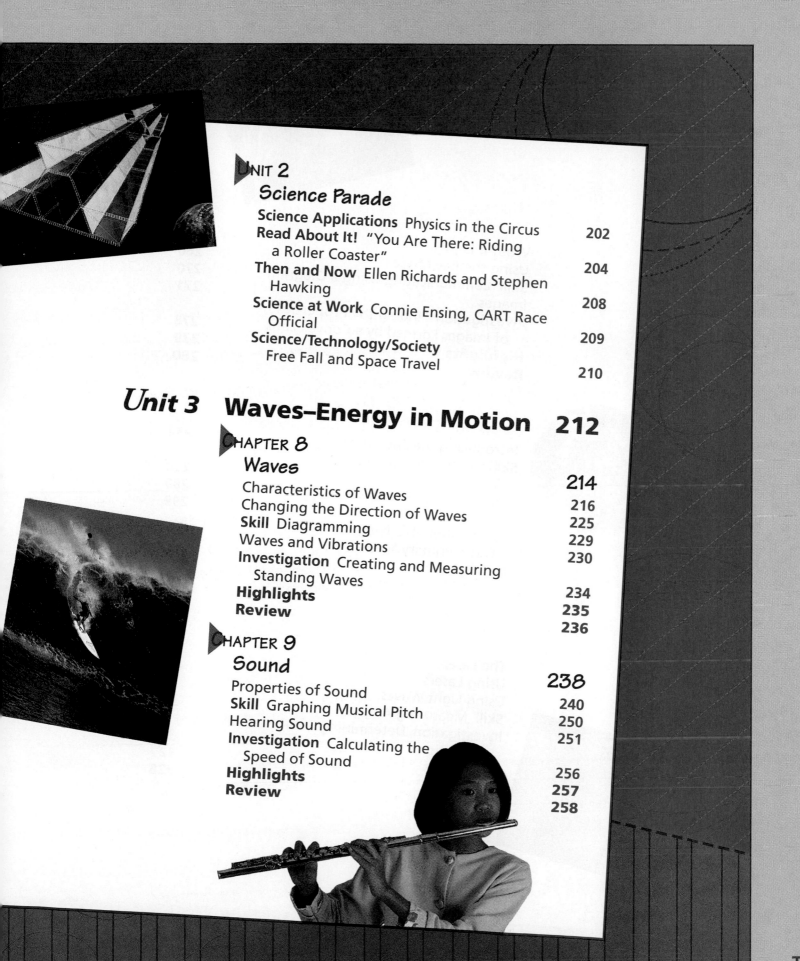

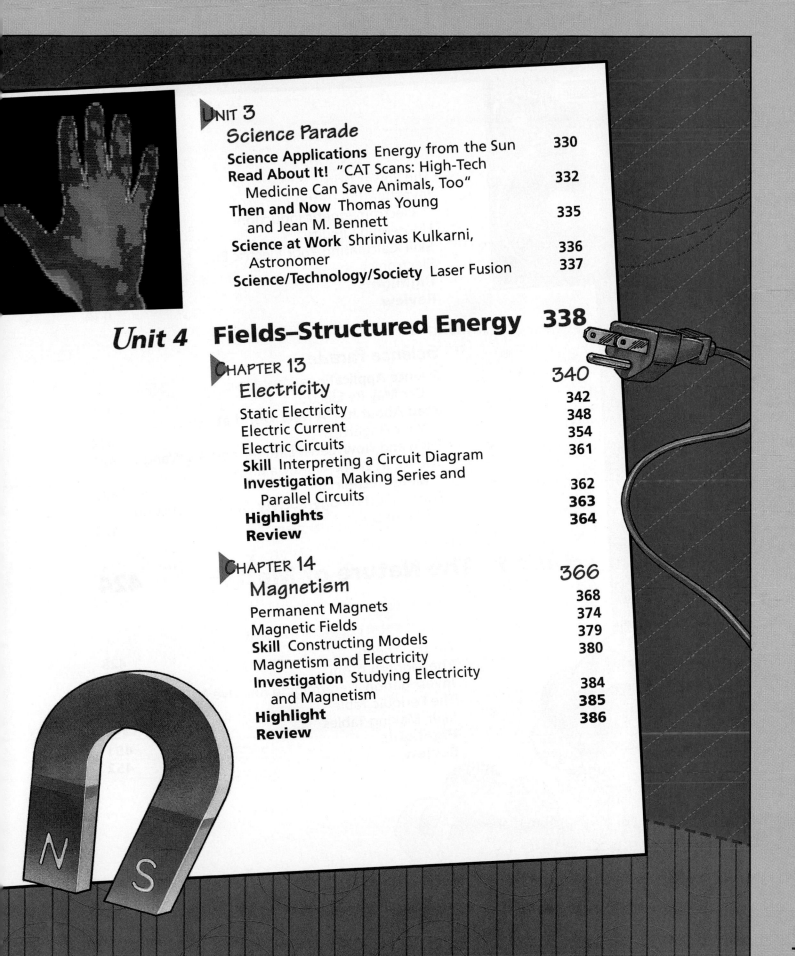

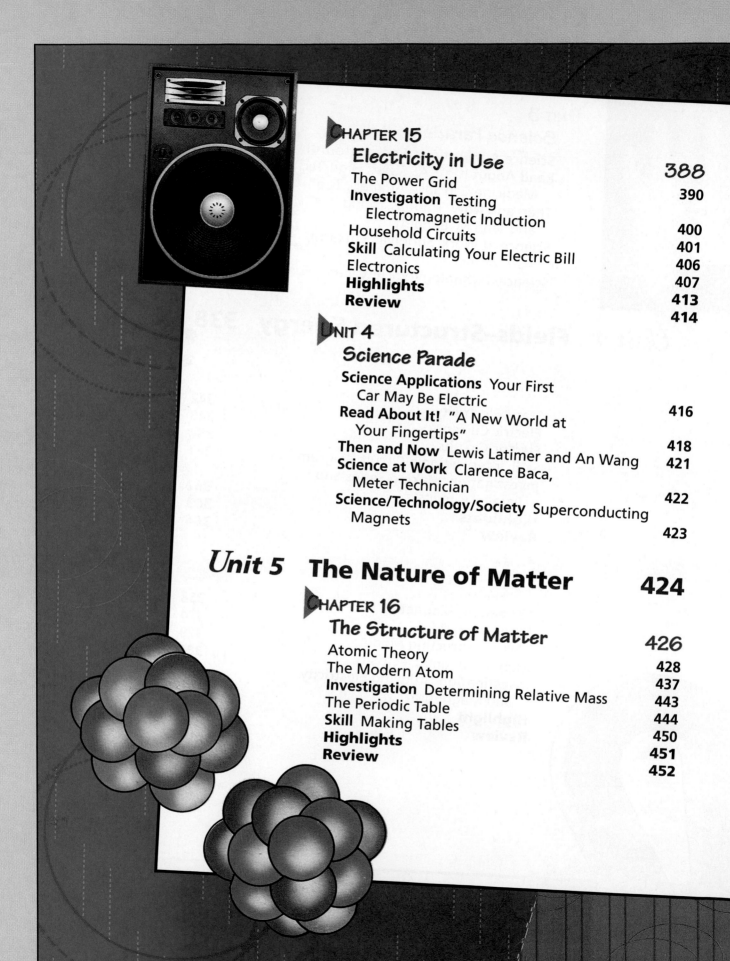

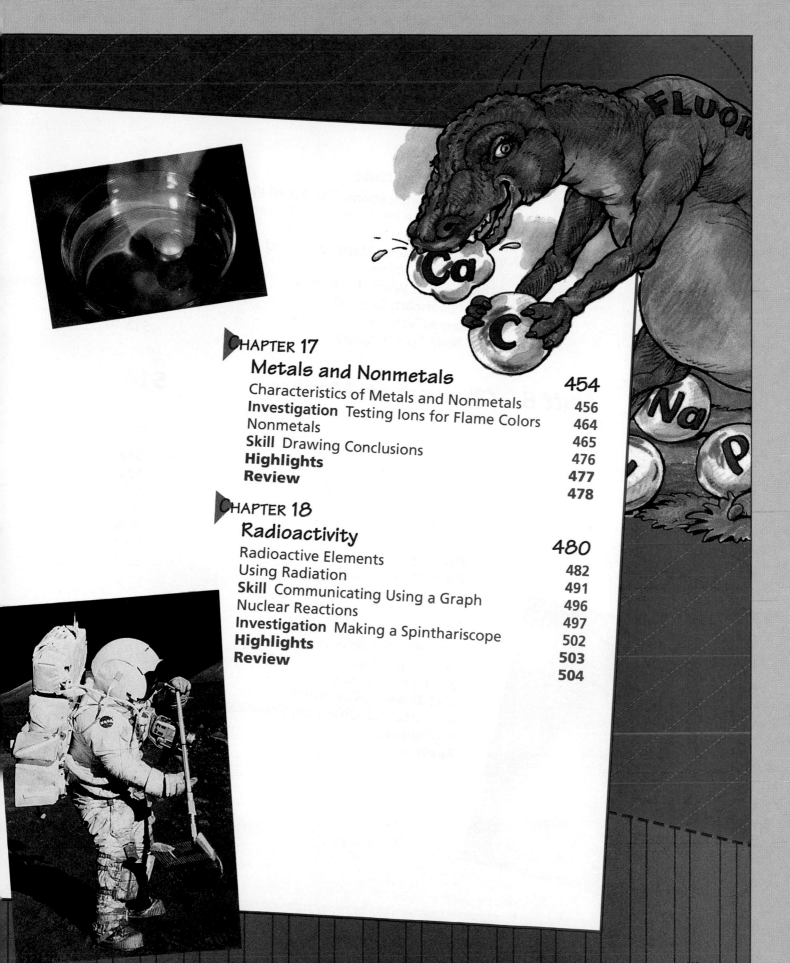

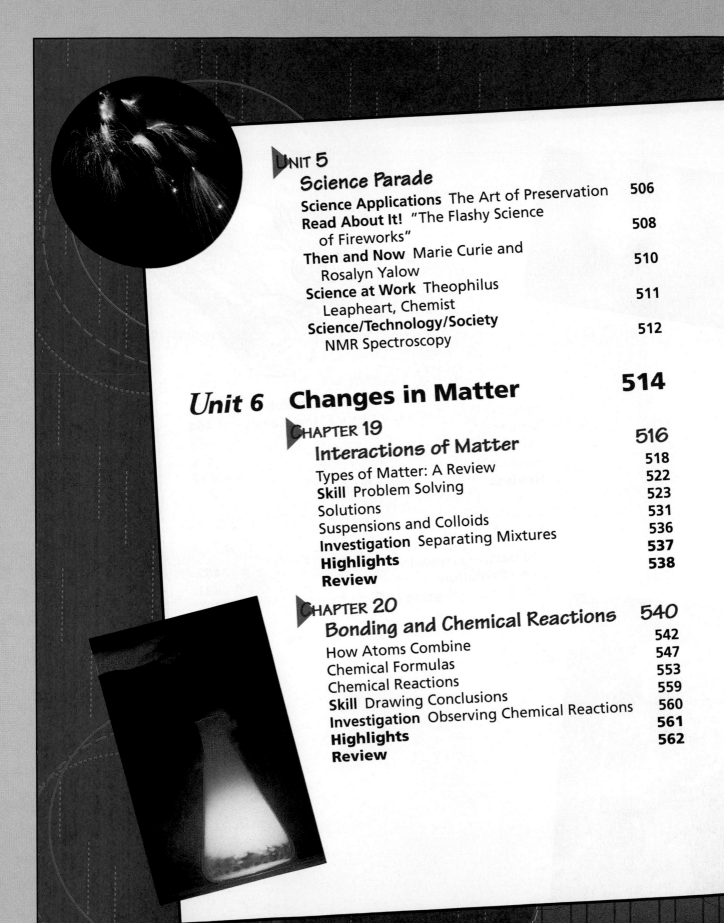

REFERENCE SECTION

SKILL

INVESTIGATION

ACTIVITY

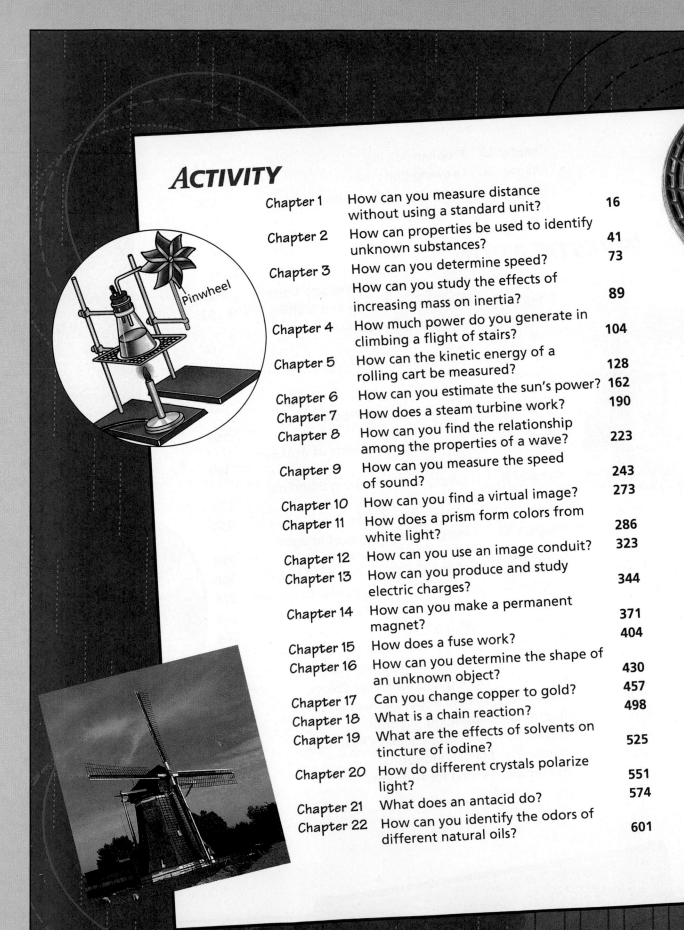

Pinwheel

DISCOVER BY

INTRODUCING
HOLT PHYSICAL SCIENCE

Science is a field of study that is constantly changing. Discoveries are announced with regular and increasing frequency. These advances in science require today's students to prepare themselves to make informed decisions on such questions as preferred medical treatment, genetic engineering, government regulations, and environmental pollution. Success in everyday living depends on the students developing an interest in and understanding of their world.

The general organization of HOLT PHYSICAL SCIENCE proceeds from a survey of the methods of science to an exploration of dynamics, forms of energy, matter, and changes in matter. Unit 1 focuses on the methods of science and the fundamental properties of matter. Unit 2 presents the various forces that affect the position and motion of matter and studies the impact of energy consumption on the environment.

Unit 3 focuses on the characteristics of waves—sound waves, light waves, and electromagnetic waves—and discusses the uses of lasers and fiber optics. Unit 4 is devoted to the study of electricity and magnetism and the continually expanding role of electricity in society. The nature and structure of matter is investigated in Unit 5,

including the characteristics of metals, the characteristics of nonmetals, and radioactivity. Unit 6 discusses the changes in matter and includes an overview of organic chemistry and biochemistry.

The text has been carefully written to provide the students with a readable presentation of science topics. The colorful pages have been thoughtfully designed to integrate explanatory graphics that complement the text.

HOLT PHYSICAL SCIENCE offers intriguing reinforcement of scientific concepts through the use of literature. For example, Chapter 4 relates the daily life of Oregon Trail pioneers to the study of work and machines. Chapter 15 recounts the havoc wreaked upon Florida by Hurricane Andrew and relates it to the generation and use of electricity. Chapter 17 ties an aspect of popular culture, comic book characters, to the study of metals and nonmetals. Interwoven throughout HOLT PHYSICAL SCIENCE, literature focuses and extends the presentation of scientific concepts.

To achieve a balanced presentation, HOLT PHYSICAL SCIENCE integrates all areas of the curriculum, including language arts, mathematics, social studies, and life and earth science. As it investigates energy, the text presents an in-depth discussion of human impact on the environment. The text discusses sound as it describes the numerous roles of sound in our world. These are just a few examples of how HOLT PHYSICAL

SCIENCE integrates the curriculum areas. Through this integration, the text enables the students to understand how science and other curricular areas are related. In addition, the use of real-world examples continually ties the study of science to the world of the student, exploring and relating the concepts to day-to-day personal models. These practical applications of the science concepts allow the students to incorporate what they learn into their daily lives.

From the presentation, text, and graphics in the Pupil's Edition to the carefully structured teaching materials, HOLT PHYSICAL SCIENCE is a readable, teachable program that explores the fascinating topics of physical science.

BECOMING FAMILIAR WITH THE TEXTBOOK

HOLT PHYSICAL SCIENCE is an easy-to-use program designed to aid students in learning and teachers in teaching. For the student, each part of the textbook is an important resource designed to facilitate learning. It is important that you familiarize the students with the learning resources in the textbook.

Unit/Chapter Organization

Introduce the students to the program by asking them to thumb through a unit of the textbook. The consistency of unit and chapter organization makes it easy for the students to find the features in the textbook. Discuss the unit opener photograph and explore the information in the text box. Then have the students skim the chapter summaries, pointing out the range of material that will be covered in the unit. Explain that the *Science Parade* feature covers a variety of subjects related to the unit content, from science applications, reading selections, and information about well-known scientists and careers to science/technology/society issues.

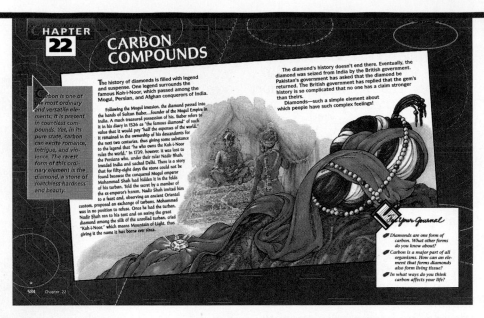

Next, have the students look at the chapter opener and read the introduction. The graphic or literary introductions are designed to provide motivation for the students. The discussion and preview can also provide an informal assessment of the background knowledge that the students bring to each unit and chapter. This assessment will help you fine-tune your planning for the chapters. The *For Your Journal* is a writing connection that introduces the students to a systematic method for recording information and assessing their familiarity with a topic.

Chapter Features

Look through the first chapter with the class, locating such features as the introduction, section objectives, and headings and subheadings. Discuss various methods of organizing information that can be used as the students read and study the chapter. If you encourage them to outline material, the students may use the chapter headings and subheadings as a basis for the development of a detailed outline of chapter material.

While reviewing how the textbook is organized, alert the students to the regular features of the text. Point out the *Discover By* feature, which involves a brief experiment, research, or hands-on activity. Call attention to the *Ask Yourself* questions that provide opportunities for the students to recall what they have just read. The *Section Review and Application* summarizes the content of the section with two forms of questions: Reading Critically and Thinking Critically. The *Skill* focuses on one or more

process skills, such as classifying, interpreting, and communicating. The *Activity* is a short laboratory experience about a topic in the chapter. The *Investigation* provides a more involved inquiry that carefully follows traditional scientific methods.

Chapter Summary and Review

The *Highlights* page begins with *The Big Idea,* which reviews the theme of the chapter and connects the chapter ideas to help the students summarize the important concepts of the chapter. *Connecting Ideas* is a graphic presentation of the concepts/ideas of the chapter. These ideas are presented as a concept map, a table, a chart, an illustration, or a flow chart. *For Your Journal* reviews the students' journal entry from the chapter opener and gives the students the opportunity to revise their original ideas and add new information to their original answers.

The *Chapter Review* summarizes the learnings and provides opportunities for assessing the students' progress. The review also includes *Discovery Through Reading*—a short summary of an interesting book or magazine article related to the chapter theme.

Unit Closure

The *Science Parade* at the end of each unit is a feature of up-to-date information on unit topics and ties the study of science to people, places, events, and issues. The section includes:

◆ Science Applications—topical features related to the unit theme and concepts

◆ Read About It!—articles, book excerpts, or stories related to unit topics

◆ Then and Now—two biographies, one historical and one modern, of people who have done work in areas related to unit topics

◆ Science at Work—information on a career in a field related to physical science, highlighting a real person engaged in that career

◆ Science/Technology/Society—information on developments that have caused or may cause tremendous advances in the physical sciences and that affect everyday life

The Reference Section

While helping the students familiarize themselves with the textbook, do not overlook the *Reference Section*, which begins on page 624. The *Reference Section* includes Safety Guidelines for the classroom and the laboratory, and Laboratory Procedures with information on SI measurement and conversion tables and laboratory methods. The Periodic Table, Electric

Circuit Symbols, and Physical Science Equations are presented. The *Reference Section* also contains an illustrated Glossary and a cross-referenced Index.

Using the Annotated Teacher's Edition

The *Annotated Teacher's Edition* is organized to provide all the information you need to teach HOLT PHYSICAL SCIENCE. It features comprehensive strategies for teaching units, chapters, and chapter sections. The textual material presents teaching strategies, teacher demonstrations, background information, skills activities, answers to all questions, and other useful information.

The basic design of the *Annotated Teacher's Edition* includes reduced Pupil's Edition pages bordered by margin columns, which provide a wealth of information for teachers of all experience levels. All pertinent teacher information is conveniently located for easy use. The notes pertaining to a Pupil's Edition page will be found in the margins adjacent to the page. The margin notes include a complete plan for teaching each section and subsection and for tailoring your teaching style to the ability levels of your students.

Unit Presentation

The *Annotated Teacher's Edition* includes two interleaved pages before each unit. These pages include a Unit Overview and Unit Resources: Print Media for Teachers, Print Media for Students, and Electronic Media. Each unit begins with a discussion activity, which offers suggestions for using the unit opener to set the stage for the chapters that follow. It also provides a journal idea to prompt writing about the topic and investigating prior knowledge.

Chapter Organizational Information

Immediately preceding each chapter are two interleaved pages with charts for Planning the Chapter and Chapter Materials. Planning the Chapter includes all chapter sections, chapter features, and program resources that supplement each section, all with appropriate page references and level designations. Chapter Materials lists all materials necessary to

perform *Investigations, Activities, Discover By* and *Skill activities*, and teacher demonstrations. Below the chart are suggestions for Advance Preparation, Field Trips, or Outside Speakers.

Chapter Teaching Information

The Chapter Opener

Each chapter opener begins with the *Chapter Theme*, which describes the major theme of the chapter, lists other thematically related chapters, and includes any supporting themes. The *Chapter Motivating Activity* is provided to introduce the students to a major concept in the chapter. *For Your Journal* helps explore the students' background and prompts them to record information about the topic. The side columns include a *Multicultural Connection* and *Meeting Special Needs* to help tailor the lesson to the students' abilities. The *About the Literature/Photograph/ Art/Author* gives interesting background information, which you may wish to share with the students.

Section Lessons

Every section has a predictable lesson format that provides section information and a lesson cycle. Section information includes:

◆ Overview Chart that identifies process skills, attitudes, terms, media, and resources for the section

◆ Focus that provides a short overview of section content

The lesson cycle for each section motivates, reinforces, extends, and reviews to provide a complete instructional plan.

1. *Motivating Activity* focuses the students' attention on the lesson and sets the stage for learning.

2. *Teaching Strategies* develop lesson content through reinforcement and extension. Each strategy uses one or more process skills to foster greater understanding and to develop the students' thinking and communicating abilities.

3. *Guided Practice* provides strategies to ensure that the students have understood the section content.

4. *Independent Practice* allows the students to work on their own as they review the section content and concepts.

5. *Evaluation* provides a questioning strategy to encourage the students to demonstrate their understanding of the concepts through application.

6. *Reteaching* offers an additional strategy for students who need extra help to acquire basic concepts presented in the section.

7. *Extension* provides ideas for exploring beyond the content of the section through application, research, and other instructional methodologies.

8. *Closure* suggests individual or group activities that help the students summarize section concepts.

The lesson cycle is accompanied by a wealth of additional notes in the margins of the lessons, including *Science Background, Did You Know?, Multicultural Connection, Integration* (with cross-curricular concepts), *The Nature of Science, Demonstration, Laser Disc* information, *Reinforcing*

Themes, Science/Technology/Society, and *Meeting Special Needs.* The margin areas also contain the answers to any text or activity questions. These answers are printed in red to distinguish them from other notes. In addition, the margin columns have any instructional notes that are needed for the special features in the Pupil's Edition, such as *Discover By, Skill, Investigation,* and *Activity.*

Interspersed throughout the margins are opportunities for assessment: *Ongoing Assessment, Performance Assessment,* and *Portfolio Assessment.* These three types of evaluation are informal tools to help gauge student progress.

Chapter Highlights and Review

The *Chapter Highlights* page provides a suggestion for reinforcing the unit theme, notes reflecting the types of revisions that the students may make in their journal entries, and answers for the *Connecting Ideas* graphic. The *Chapter Review* provides answers to the review questions.

Using Ancillary Materials

A complete program of supplementary materials accompanies HOLT PHYSICAL SCIENCE. Many of the materials can be found in the Teaching Resources, which are organized into six unit booklets and contain the following Blackline Masters.

- Study and Review Guide—pages to reinforce vocabulary and concepts for each section and a chapter review
- Record Sheets for Textbook Investigations—pages for recording data and conclusions for the textbook *Investigations*
- Transparency Worksheets—blackline illustrations of each transparency with accompanying teaching strategies
- Laboratory Investigations—two investigations keyed to each chapter of the textbook and designed to help the students further apply lesson concepts
- Reading Skills—pages designed to improve the students' reading strategies
- Connecting Other Disciplines—pages that integrate physical science with other content areas of the curriculum
- Extending Science Concepts—pages that explore related science topics
- Thinking Critically—pages that focus on science topics while requiring the use of higher-order thinking skills
- Chapter and Unit Tests—materials to evaluate each chapter and unit

Additional laboratory notes, materials lists, other laboratory information, and answer keys are also provided.

Other supplementary materials include Instructional Transparencies, Classroom Reference Posters, and Classroom Instructional Posters. There are 50

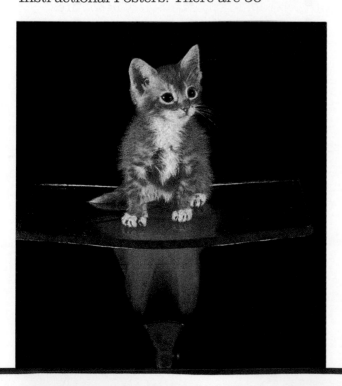

Instructional Transparencies with accompanying worksheets. The Classroom Reference and Classroom Instructional Posters are designed to supplement unit instruction. A separate Test Generator available for Macintosh® and IBM® computers provides a complete testing program that enables teachers to test precisely the content they want by choosing questions from a comprehensive bank of test items.

HOLT PHYSICAL SCIENCE is complemented and supplemented by the *Science Discovery* videodisc program, which consists of still images and motion footage. Specific frames are referenced in the chart preceding each chapter and at point of use in the margin notes, and are indicated by

Scheduling

The teacher can arrive at an appropriate course schedule by considering the ability level of the students; their previous science experience; student interest; teaching style; local science-related resources; and local, county, and state requirements.

While HOLT PHYSICAL SCIENCE is designed with the intention that all the main topics be covered during the typical one-year physical-science course, the individual teacher can adapt the way in which he or she uses the textbook to the needs of the local teaching situation. No teacher should feel compelled to teach every chapter or to give equal emphasis to all chapters of the textbook. The teacher and the school system should dictate what constitutes appropriate course content.

The pacing chart beginning on page T28 is designed to assist teachers in making decisions regarding scheduling, course content, and emphasis. The recommended coverage of a particular chapter or section for a basic, average, or honors course is indicated by a ■. The number of recommended class sessions to be devoted to each chapter and section is also indicated. The days indicated in the pacing chart allow for laboratory work related to the pertinent section. Time for chapter and unit tests is also considered. Use of the Pacing Chart may be supplemented by the Planning the Chapter Chart found on the interleaved pages immediately preceding each chapter.

PACING

Unit	Chapter	Section Number and Title	Basic Course	(days)	Average Course	(days)	Honors Course	(days)
1 INTRODUCTION TO PHYSICAL SCIENCE				29		13		14
	1	**Methods of Science**		12		6		7
		1 Asking Questions and Solving Problems	■	7	■	3	■	3
		2 Measurements	■	4	■	2	■	3
		Chapter Review and Test	■	1	■	1	■	1
	2	**Fundamental Properties of Matter**		16		6		6
		1 Matter	■	5	■	1.5	■	1.5
		2 Characteristics of Matter	■	5	■	1.5	■	1.5
		3 A Model of Matter		5	■	2	■	2
		Chapter Review and Test	■	1	■	1	■	1
		Science Parade and Unit Test	■	1	■	1	■	1
2 ENERGY-MATTER IN MOTION				62		37		37
	3	**Acceleration, Force, and Motion**		16		7		7
		1 Describing Motion	■	7	■	2	■	2
		2 Forces and Motion	■	4	■	2	■	2
		3 Predicting and Explaining Motion	■	4	■	2	■	2
		Chapter Review and Test	■	1	■	1	■	1
	4	**Work**		12		7		7
		1 Work and Power	■	4	■	3	■	3
		2 Types of Machines	■	7	■	3	■	3
		Chapter Review and Test	■	1	■	1	■	1
	5	**Energy**		13		8		8
		1 Forms of Energy	■	5	■	3	■	3
		2 Energy Laws	■	2	■	1	■	1
		3 Heat	■	5	■	3	■	3
		Chapter Review and Test	■	1	■	1		1
	6	**Energy Sources**		10		9		9
		1 Energy from Fossil Fuels	■	3	■	3	■	3
		2 Energy from the Sun and Wind	■	3	■	2	■	2
		3 Other Energy Resources	■	3	■	3	■	3
		Chapter Review and Test	■	1	■	1	■	1

CHART

P A C I N G

Unit	Chapter	Section Number and Title	Basic Course (Days)		Average Course (Days)		Honors Course (Days)
4 FIELDS–STRUCTURED ENERGY			18		27		28
	13	**Electricity**	7		7		7
		1 Static Electricity	0	■	2	■	2
		2 Electric Current	■ 3	■	2	■	2
		3 Electric Circuits	■ 3	■	2	■	2
		Chapter Review and Test	■ 1	■	1	■	1
	14	**Magnetism**	5		10		11
		1 Permanent Magnets	■ 3	■	2	■	2
		2 Magnetic Fields	■ 1	■	2	■	3
		3 Magnetism and Electricity	0	■	5	■	5
		Chapter Review and Test	■ 1	■	1	■	1
	15	**Electricity in Use**	5		9		9
		1 The Power Grid	■ 2	■	2	■	2
		2 Household Circuits	0	■	2	■	2
		3 Electronics	■ 2	■	4	■	4
		Chapter Review and Test	■ 1	■	1	■	1
		Science Parade and Unit Test	■ 1	■	1	■	1
5 THE NATURE OF MATTER			12		21		20
	16	**The Structure of Matter**	6		7		6
		1 Atomic Theory	0	■	1	■	1
		2 The Modern Atom	■ 5	■	2	■	1
		3 The Periodic Table	0	■	3	■	3
		Chapter Review and Test	■ 1	■	1	■	1
	17	**Metals and Nonmetals**	5		6		6
		1 Characteristics of Metals and Nonmetals	■ 2	■	3	■	3
		2 Nonmetals	■ 2	■	2	■	2
		Chapter Review and Test	■ 1	■	1	■	1

CHART

THEMATIC
Science

The word science has its origin in the Latin word *scientia,* which means knowledge. Through years of observation, scientists have amassed a body of knowledge that is quite vast.

Due to the sheer magnitude of knowledge, scientists and educators have had to devise schemes for categorizing this large volume of scientific information. Some state and local curriculum frameworks commonly call for the organization and teaching of science into the separate subject matter categories, or disciplines of life, earth, and physical science. These categories are often subdivided into narrower disciplines. While this categorization is a convenient and seemingly logical method, it fails to connect science knowledge in a comprehensible and interrelated fashion.

In an attempt to integrate and link the disciplines of science, educators and learning theorists have proposed a system of themes for organizing science curricula. Themes span all the disciplines of science. They help make sense out of thousands of disparate science facts, leading to improved student understanding and achievement.

Teaching Themes

The content within HOLT PHYSICAL SCIENCE is organized into major and supporting themes. Each chapter has one of five major themes as a focus as well as a supporting theme. It is important to note, however, that other themes are possible. Feel free to introduce other themes that you consider appropriate.

As they learn the chapter content, help the students understand the relationship of the chapter concepts to the major and supporting themes. The chapter review feature entitled "Reviewing Themes" provides the students with an opportunity to demonstrate their understanding of the various themes.

The HOLT PHYSICAL SCIENCE teacher's edition provides strategies for reinforcing themes in the margin notes.

Definition of Themes

Systems and Structures: The physical world consists of systems and structures at micro and macro levels. Structures interact within systems, and interactions occur between and among systems and structures.

Energy: Interaction within and among systems requires energy. In living things, energy is needed for maintenance, growth, and reproduction. In the physical world, energy can take a variety of forms including wave energy, sound energy, and light energy.

Environmental Interactions: Interconnections exist among life forms as well as between life forms and their abiotic environments. Life forms (including humans) depend both on other life forms and on the physical environment. Interactions also exist between various parts of the inanimate world. Changes in the environment, including the influences of society, pollution, and even natural disasters have a ripple effect that may not be immediately apparent.

Changes Over Time: The natural world is characterized by patterns of change. Change can occur in cycles, in a steady linear progression, or in irregular patterns. Over time, some of these cycles exhibit overall stability in the midst of change.

Technology: Technological advancements have provided scientists additional tools to expand their knowledge of the world's structures and systems. Conversely, scientific discoveries have led to the development of new technologies.

■■ **Major Theme**   ■■ **Supporting Theme**

Chapter	1	2	3	4	5	6	7	8	9	10	11	12	13	14	15	16	17	18	19	20	21	22
Systems and Structures					Major		Major			Major			Major			Major	Major					
Energy			Supporting		Major	Major			Major			Major		Major						Major	Major	
Environmental Interactions	Supporting									Major		Supporting						Major				
Changes Over Time																		Major				
Technology						Supporting		Supporting						Supporting		Major						Supporting

T32

MULTI Cultural
INSTRUCTION

The nation's classrooms are a reflection of the cultural diversity of our people. Increased recognition of cultural diversity and the opportunities it provides has led educators to incorporate multicultural approaches into the curriculum. As a dynamic, ongoing process, multicultural education helps the students gain a greater understanding of themselves, their heritage, and their values as well as a respect for others. By identifying the contributions and strategies of people from diverse cultures, educators can raise the students' consciousness about the contribution of each individual to a group, to the school community, and to society as a whole. Teachers can promote self-esteem, impart a feeling of pride in who we are, and achieve a sense of belonging for each member of a group.

A multicultural approach to education gives each student equal educational opportunity by recognizing differences in cultural heritage and in learning styles.

The students who enter the classroom bring with them widely diverse cultural backgrounds. Such diversity leads to rich and varied experiences that affect the way each individual approaches learning. The teacher can use this diversity to enrich the learning experiences of all students.

As content is taught and strategies are selected, the teacher should keep in mind that the students bring values into the classroom that are a reflection of their heritage and community. The students need an opportunity to share their knowledge and experiences. By using discussion techniques for eliciting prior knowledge and experience, the teacher can detect misconceptions and mold the learning to each student's experiential base. Achievement is greater when the methods and topics connect with who the students are and what they already know. Learning theory suggests that everyone learns new things by finding ways to connect them to what he or she knows and has experienced.

Positive role models from diverse cultural backgrounds are part of the physical science curriculum. The curriculum recognizes that approaches to concept development vary from culture to culture. It provides for the use of different strategies to foster understanding, and it also provides opportunities for interactive learning activities including cooperative grouping strategies.

Teachers are encouraged to cultivate a classroom environment that promotes alternative opinions and different interpretations of conceptual information when appropriate. Using current news events to advance the contributions of people from various cultures is also encouraged.

In short, multicultural education supports enhanced self-concept, values the uniqueness of each individual, and promotes respect for others. Teachers should take appropriate steps to advance multicultural education whenever possible. It is the goal of HOLT PHYSICAL SCIENCE to develop multicultural educational opportunities through content presentation, graphics, and activities.

Our nation's future depends on educating all Americans. Ethnic and gender stereotypes have long served as obstacles to the scientific literacy of all students. A central goal of education must be equal educational opportunities for all. Multicultural education contributes to the fulfillment of this goal. Only when you acknowledge and respect the differences in the students can you begin to achieve educational equity.

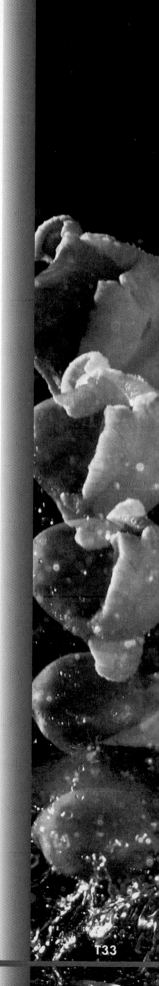

MEETING *Special Needs*

The science class offers a variety of students important information that will help them function and succeed in an increasingly scientific and technological society. Guidance counselors, special-education teachers, teachers of the gifted, and the school nurse may be consulted to help work out the best learning environment as well as realistic goals for each student. The following recommendations apply to mainstreamed students with specific types of learning problems.

Mainstreamed Students

Learning Disabled

Allow the student access to the special-education instructor as needed. Where feasible, make use of teaching helps available from such companies as National Teaching Aids, NASCO, and others. These helps include kits, models, and rubber stamps to reproduce diagrams.

- Give simple, clear directions.
- Encourage repeated efforts.
- Allow for group work during oral assignments.
- Provide for simplified rephrasing of concepts, tests, and reviews.
- Give oral examinations.
- Make certain that easy-to-read science reference material is available.

- Provide a daily, unvarying routine so that your expectations are clear.
- Make certain that instructions are understood before the student starts to work.
- Seat the student where distractions are minimized.
- Allow the student to express ideas with drawings or models if the disability permits. Dyslexic students will be able to develop drawings of their own; the dysgraphic students will be more successful atattaching labels to figures that have been prepared for them.
- Within reasonable expectations, evaluate the student's grasp of the concept, not of spelling, punctuation, or intricate sentence structure.
- Provide spelling lists that the student may use when writing. Accept the use of printing when cursive writing presents too great a challenge.

Hearing Impaired

Allow the student access to your lecture notes if he or she has serious difficulty hearing.

- Avoid speaking while facing the chalkboard.
- Provide seating where the student can hear best or

lip-read most easily.

- Allow for group work during oral assignments.
- Obtain close-captioned films for the hearing impaired.
- Rephrase instructions. Some sounds may be heard better than others.

Visually Impaired

Check with the special-education instructor or with local or state organizations regarding the availability of large-print textbooks. Seat the student close to the chalkboard.

- Stand facing the windows to avoid putting a demonstration into shadow.
- Encourage the handling of materials before or after a demonstration.

- Assign a student to make copies of notes.
- Use verbal cues rather than facial expressions or nods.
- Provide a sighted student guide.
- Provide high-contrast copies of worksheets or chalkboard diagrams.

Health Impaired

The school nurse and special-education teacher can acquaint you with each student's limits.

- Become acquainted with the various symptoms of any health emergencies that might occur.
- Obtain training in first-aid procedures and CPR to be used in the event of a health emergency.

Orthopedically Impaired

Arrange for seating that is comfortable for the student.

- Provide rest breaks.
- Cover the student's desk with felt so that materials do not slip.

Second-Language Support

Many school systems and districts have teachers whose main responsibility is to instruct the limited-English-proficient (LEP) student. Work with this teacher as a resource person in designing lesson plans for the LEP student. If your school system or district has no such resource person, the following suggestions may be of help in dealing with students whose first language is not English. HOLT PHYSICAL SCIENCE offers Spanish language materials for several of its components that can be used by those students whose first language is Spanish.

- Use pictures, diagrams, models, and other props as often as possible.

- Read the captions of illustrations aloud to the student, simplifying the language if necessary.
- Use videotapes or films with accompanying audio to help clarify concepts.
- Prepare read-along tapes for each lesson.
- Provide a word list of key terms (in addition to the new science terms) that may come up often in the discussion of a particular topic.
- Allow the LEP student to work with a partner or in a group whenever feasible.
- Simplify language in descriptions and discussions whenever possible.
- Speak slowly and enunciate clearly. Restate sentences and phrases if necessary.
- Provide for oral testing with oral or pictorial responses from the student. Exercise patience while waiting for a response.

At-Risk Students

At-risk students are those who, for any number of reasons, are liable to perform poorly and who have a high

probability of dropping out. HOLT PHYSICAL SCIENCE is engaging and interesting throughout, appealing to all students. The clear, easy-to-read prose and straightforward, attractive graphics reduce the potential for the students to grow bored. The style of HOLT PHYSICAL SCIENCE is intentionally friendly and unintimidating.

Gifted Students

There are many definitions for gifted students. In the past, definitions focused solely on I.Q. Current definitions identify gifted students as those who are exceptional in an ability area compared with others of the same age. The giftedness includes creative aspects and superior performance in some recognized area.

- Present lessons that accommodate the students' ability to learn faster and remember more.
- Encourage participation in a variety of learning experiences to broaden the students' areas of interest.
- Design curriculum to motivate and enhance the learning process.
- Design independent projects and comprehensive problem-solving tasks.
- Allow student choices to provide options for class activities.
- Avoid comparisons among and between the students in order to develop positive attitudes within the students and toward others.
- Encourage the students to develop leadership skills.

COOPERATIVE LEARNING

Allowing the students to share their learning with one another is a strategy that has been used since the day of the one-room schoolhouse. Throughout every lesson in HOLT PHYSICAL SCIENCE, activities have been developed that are designed to provide worthwhile cooperative learning experiences. These activities can easily be identified by the colorful Cooperative Learning symbol.

Cooperative Learning Groups

Grouping the students for cooperative learning gives them the opportunity to work toward both group and individual goals. The basic elements of a cooperative learning group include positive interdependence, face-to-face interaction, individual accountability, and interpersonal and small-group cooperative skills.

Positive Interdependence

The students must work with and depend on each other to achieve the group goal. Positive interdependence may be achieved through a division of labor, resources, and roles in the group. Rewards for the group, such as the same grade for every member of the group, may be included.

Face-to-Face Interaction

In cooperative learning, it is important that all group members provide support, encouragement, and help to each other. The members should be encouraged to share and discuss their ideas. For cooperative learning to be successful, the students must interact with each other.

Individual Accountability

Individual accountability means that each group member is responsible for knowing the assigned material. In addition, each group member is responsible for the learning of other group members.

Interpersonal and Small-Group Cooperative Skills

A basic ingredient of cooperative learning is teaching the students the skills that are necessary for effective collaboration. The teacher needs to specify interpersonal and small-group cooperative skills such as staying with the group, taking turns, looking at the person who is talking, and checking to make sure that other members understand and agree with the answers. It is important for the students to learn what a cooperative relationship should look and sound like. It is also important to allow the students the time to analyze (process) how well their groups cooperate.

It is necessary to provide the students with opportunities to work as members of both small and large groups to encourage sharing, acceptance of responsibility, and decision making. Successfully guiding cooperative learning activities in the classroom requires a complete understanding of cooperative learning strategies. For further information, you may wish to refer to *Circles of Learning: Cooperation in the Classroom,* Third Edition, (Johnson, Johnson, and Holubec, 1991).

The Teacher's Role in Cooperative Learning

In cooperative learning situations, the teacher should identify the group goal; decide on the group size, group makeup, room arrangement, and materials needed for the activity; and sometimes assign specific roles to group members. The teacher will also need to explain the task, structure the activity, monitor the interpersonal and small-group cooperative skills, and evaluate the product as well as the cooperative skills.

Identify the Group Goal

The students need to understand what is expected of them. Identify the group goal, whether it be to learn specific objectives or to create a product such as a chart, report, or booklet. In addition, you should identify and explain the specific cooperative skills for each activity.

Decide Group Structure and Arrangement

For each activity, you will need to make some basic decisions. These decisions include group size, group makeup, room arrangement, materials needed for the activity, and specific role assignments for each group.

Group Size

Cooperative learning groups may consist of two to six members. If you are using cooperative learning activities for the first time, you may find it easier to keep groups to two or three members. Smaller groups require fewer cooperative skills of the students than do larger groups.

Group Makeup
Use heterogeneous grouping of members when possible. Heterogeneous groups include students with high, average, and low ability levels. This type of grouping encourages greater diversity of thinking. Use homogeneous grouping (grouping of students with similar levels of ability) when acquisition of specific skills is the goal.

Room Arrangement
Have each cooperative learning group sit in a circle so that every member can see every other member.

Materials
Encourage interdependence by providing only one set of the materials to the group. By doing this, you can help the members learn to work together successfully. You may also give different materials to each group member. The completion of the task then depends on how well the members work together.

Role Assignment
For some cooperative learning activities, you may want to assign a *recorder*, who writes down the group's decisions and edits reports; a *summarizer-checker*, who makes sure all group members understand the material; a *research-runner*, who communicates with other groups and gets materials; a *reader*, who reads the directions and questions; and an *observer*, who keeps a record of how well the students perform the required cooperative skills.

Explain the Task
For each activity, make sure that all group members understand the task. To help the students understand, you may want to follow these suggestions:

- Give clear and specific instructions.
- Explain lesson objectives.
- Help the students see the relationship between the concepts and their past experiences and prior learning.
- Define concepts and provide models of what the students must do to finish a task.
- Ask questions to be sure that the students understand the task.

Structure Positive Goal Interdependence
In cooperative learning, it is essential that the students work together. You should stress the "sink or swim together" relationship that is necessary to achieve the group goal. To facilitate positive goal interdependence, you can structure activities in the following ways:

- Have the group produce one product (answer or solution), which each group member signs. Each member must know the reason for the agreed-upon product or answer and be able to explain it.
- Give a group grade. You may evaluate members individually, but reward the group based on the total achievement of the group.
- Evaluate often by giving practice tests, by randomly choosing group members to explain answers or to read papers, and by having members edit each other's work.
- Encourage groups to help other groups by rewarding the entire class if a certain criterion is met.

Monitor Cooperative Skills
As the teacher, you have an important role in cooperative learning. While the students are working together, you can pick up vital information on the task, because in a cooperative setting learning takes place "out loud." You can actually hear the learning while it is occurring. This is also a good opportunity to observe how well the students work together, especially on the skills that you specified before the activity started. You can count and record the number of times appropriate skills occur and provide the group with feedback. Student observers may also be used to record some cooperative skills. It is important that you evaluate the product that each group creates. You may also want to evaluate the group on how well the members worked together.

INTEGRATING
OTHER DISCIPLINES

HOLT PHYSICAL SCIENCE presents a complete physical science curriculum that integrates science with other disciplines. Science is a part of the students' daily lives, and as such, it is an integral part of the learning process. The Pupil's Edition pages explore connections between areas of science, integrating earth science and life science topics. The Pupil's Edition also contains connections across the curriculum with integration of literature and related language skills (reading, writing, speaking, and listening), mathematics, history, geography, social studies, health, art, music, and physical education.

Cross-curricular integration is a regular feature of the *Annotated Teacher's Edition*; margin copy contains teaching tips that weave science concepts into the fabric of the students' knowledge of other curriculum areas. The integrated learning approach is also found in the components of the program. *Connecting Other Disciplines* are specially developed activities that focus on a chapter topic and a related curriculum area. The *Reading Skills, Extending Science Concepts,* and *Thinking Critically* activities also involve cross-curricular content.

Literature and Language Skills
The organization and presentation of topics tie into real-world events as the students read selections from literature. The science content covering people, places, and events recounts famous milestones in scientific study. Integration of language happens as the students discuss, debate, explore, and record information in their journals. The journal activities help the students gather information, record questions, make predictions, draw conclusions, and summarize data for daily lesson activities. The journal provides a tool for learning; keeping a journal is an effective technique for all areas of study. Throughout the program, emphasis is placed on discovery and the inquiry approach, which provides a natural integration of listening and speaking skills.

Mathematics
Science is closely allied to the field of mathematics. Mathematical measurements and amounts are commonly used for gathering, analyzing, and presenting data. Scientists use mathematics to design experiments, to report their findings accurately, and to make predictions on scientific investigations.

Social Studies
History, geography, and other aspects of the social studies curriculum are a natural cross-curricular tie. As the students study the events of science, they recognize the developments over time and how such developments have affected history. The science aspects of many historical events provide for an understanding of possible future developments. Social scientists also use scientific methods to explore current topics, gather data, report findings, and make predictions.

Making Connections
When the students study concepts, it is important for them to understand how the topic relates to their world. Learning isolated data without understanding the intricacies and scope of the content will hamper the students from acquiring true knowledge. Every lesson should be presented without curriculum boundaries, and the students should be encouraged and motivated to explore across all curricular areas.

SCIENCE TECHNOLOGY SOCIETY

The problems of dwindling energy resources and the search for alternate energy supplies are addressed almost daily in newspapers and on television. The impact of computers and related technology on our daily lives has been significant. Social and ethical issues are constantly raised as technology progresses at a rapid pace in the fields of medicine, energy production, and genetic engineering.

To prepare for the future, to understand the role of technology, and to make informed decisions about the social implications of new scientific technologies, the students first need a strong foundation in the basic principles and processes of science. They must develop an accurate image of the nature of science and the usefulness of science in solving problems. Finally, it is also important for the students to gain confidence in their ability to identify science-related social issues and to use their own scientific knowledge to resolve these problems.

Teachers can play a vital role in helping the students gain the knowledge and skills necessary to make responsible decisions about social issues related to science and technology. The sections that follow provide suggestions to help teachers in this role, using features from HOLT PHYSICAL SCIENCE in conjunction with a variety of classroom strategies and outside resources.

Strategies Using Holt Physical Science

HOLT PHYSICAL SCIENCE provides the students with opportunities to analyze science-related and technology-related issues. A discussion of these issues can provide excellent additional opportunities for the development of thinking skills.

"How do scientists solve problems?" and "How does matter change from one form to another?" are two questions often asked by students. Both of these questions are answered in chapters found in Unit 1: Introduction to Physical Science.

Chapter 6 presents basic information about energy and the tradeoffs associated with both existing and alternative energy sources. This chapter includes discussions of the process of energy production and allows the students the opportunity to make recommendations about future energy consumption.

The chapters in Unit 3: Waves – Energy in Motion provide an overview of waves, sound, light and lenses, the spectrum, and uses for light. The structures of these systems are discussed in detail appropriate to the physical-science student. In addition, the technology involved with the production of lasers, optical fibers, and laser applications is highlighted.

The study of HOLT PHYSICAL SCIENCE introduces the students to such topics as electronics, robotics, and synthetic compounds. A discussion of these topics can provide excel-

lent learning opportunities, especially for the development of higher-level thinking skills. The students may also be encouraged to gather information from references. Once information has been gathered, a positive learning environment can be created if the students are taught to respect the right of others to express their views.

These are just a few examples of how physical science is related to the lives of the students. You will find many opportunities to encourage the students to think about how science affects their lives.

Other Classroom Strategies

One popular way to introduce social issues into the science classroom is to have students bring in articles on science and technology from newspapers and magazines. Class time should be set aside on a regular basis to discuss the articles and the students' views. The teacher's role is vital in these discussions. However, the teacher must remember not to make judgments or to reject or praise student responses, but rather to accept all possible answers, ideas, and positions.

Debates are another way to approach the issues of science, technology, and society. For this strategy to be useful, however, the students must do more than repeat the opinions of others. They must be given adequate time and direction to research the issues, collect data, formulate their own opinions, and support their positions.

PROMOTING POSITIVE ATTITUDES

A student's attitude about science will have a strong effect on his or her achievement and future use of scientific knowledge and skills. Your teaching approach and expectations will affect a student's attitude toward science.

For far too many years, the roles of women and people of many cultures in science and mathematics have not been emphasized. Too often, many students have not been able to identify role models in these fields. If this nation is to maximize its human resources and compete effectively in a highly scientific and technical global economy, all schools must promote scientific literacy for *all* students.

Teachers can play an influential role in building positive attitudes toward science by viewing instruction from a three-dimensional perspective—concept development, behavior development, and attitude development. Failure to develop any one dimension is likely to lead to difficulty in developing the other perspectives. It is important to encourage positive attitudes in your program.

Fostering Positive Attitudes

Beyond helping the students take a positive, can-do approach to science, you must base attitude development on critical thinking, decision making, and problem solving. Attitude development also involves fostering scientific values.

Helping the students believe that they can succeed in science includes offering words of encouragement and praise. You should look for positive behavior and success, and acknowledge it. You should create opportunities for success for all students. These opportunities can often be offered effectively in cooperative group settings. You should also be aware that during discussions those with the lowest self-esteem will be slower to respond and to volunteer an opinion. Allow extra time in your questioning sessions and ask open-ended questions that carry less risk of being wrong.

Critical thinking and a positive attitude toward science are closely aligned. Critical thinking is often defined as thinking that considers other points of view. Critical thinkers are better decision makers and problem solvers because they consider all the alternatives. A positive attitude helps the students entertain counterarguments, avoid bias, and consider how others perceive and understand a situation. You can foster such an attitude by creating a classroom environment that openly accepts differing points of view and encourages independent thinking. In addition, asking questions that lead the students to analyze, evaluate, and apply information will contribute to their growth as critical thinkers.

HOLT PHYSICAL SCIENCE encourages the development of positive attitudes by focusing on the following attitudes throughout the program:

◆ caring for the environment
◆ cooperativeness
◆ creativity
◆ curiosity
◆ enthusiasm for science and scientific endeavor
◆ honesty
◆ initiative and persistence
◆ openness to new ideas
◆ precision
◆ skepticism

Benefits of Positive Attitudes

Appreciation and respect for science are also part of building the proper attitude for science. Scientific knowledge and science skills can be both creative and problem-solving tools. When used in a positive manner, they can benefit the planet and all living things that inhabit it.

Experience supports the premise that a positive attitude leads to positive thoughts that then lead to positive actions. In science, this is an important consideration. Many decisions and actions can have far-reaching consequences. All questions are best considered when approached with a positive attitude and the use of critical-thinking skills.

ASSESSMENT
ASSESSMENT

With greater emphasis being placed on higher-order thinking and active learning, traditional assessment with its emphasis on fact recall is no longer compatible with curriculum goals. This is particularly true in science, wherein an enormous explosion of information has rendered sheer knowledge of facts to be of questionable value. Instead, emphasis must be placed on understanding broad concepts and big ideas and using information to solve problems. The students need to know how to retrieve, interpret, and use information effectively. Assessment, then, must determine if the students are acquiring these skills.

Role of Assessment

Assessment serves several purposes. It determines the value that the students place on information and tasks for which there is accountability. Therefore, you should design instruction with attention to the assessment that accompanies it. Assessment should do more than diagnose a student's ability to demonstrate knowledge or perform tasks. It also should provide an opportunity to measure student attitudes, prior knowledge, social interaction skills, learning styles, and communication skills. Used properly, assessment itself can and should be a learning tool. HOLT PHYSICAL SCIENCE offers an effective assessment program by providing a variety of assessment strategies.

Types of Assessment

Current assessment strategies place emphasis on higher-order thinking, demonstration of understanding, ability to communicate, and task performance.

Performance Assessment

This method of testing requires the students to perform a task or series of tasks rather than choose an answer from a ready-made list of responses. These tests can be in the form of open-ended or response items in which the students are expected to respond to some issue or observation. They can be short or extended tasks that the students have to complete. Or they can consist of a portfolio of representative student work.

Portfolio Assessment

Suggestions in the *Annotated Teacher's Edition* provide many opportunities for accumulating activities and assignments for the students' science portfolio. *Portfolio Assessment* designations provide helpful hints for saving documents that demonstrate the students' growth in understanding science concepts and problem-solving techniques.

Performance Assessment

The *Ask Yourself* questions in the Pupil's Edition provide opportunities for the students to pause briefly to summarize and review their understanding of the content. These brief assessments, in coordination with the *Section Review and Application, Highlights Page,* and *Chapter Review,* promote an ongoing evaluation system to help the students track their progress in understanding the content and science processes.

Checklists

In addition to performance-based assessment, student attitude and student demonstration of scientific process should also be assessed. An observational checklist is a good vehicle for these types of assessment. Samples of such checklists are given below. You may use these as models and expand them to meet your needs.

Assessing Scientific Attitudes

ATTITUDE	POOR	GOOD	EXCELLENT
Shows respect for living things	—	—	—
Displays enthusiasm for science	—	—	—
Cooperates and actively participates in small-group projects	—	—	—

Assessing Scientific Processes

ATTITUDE	POOR	GOOD	EXCELLENT
Analyzes data	—	—	—
Observes accurately	—	—	—
Records observations carefully	—	—	—

THE ROLE OF
INQUIRY

HOLT PHYSICAL SCIENCE incorporates an inquiry approach to learning within the context of specific content presentations. Each feature of the program is designed to allow the students the time to explore, analyze, and assess information. From the format of the Pupil's Edition to the **Annotated Teacher's Edition,** the materials are organized to provide inquiry-based activities.

The **Investigation** found in each chapter of HOLT PHYSICAL SCIENCE applies chapter content within the context of a scientific method. You are encouraged to augment the textbook Investigations with those found in the *Laboratory Investigations,* which accompany the Pupil's Edition. Pre-Lab strategies encourage students to form hypotheses and make predictions. The students are offered Post-Lab strategies, which prompt them to reflect on their original hypotheses and predictions and relate new information and conclusions gathered in the *Investigation.*

There are one or more **Activities** in each chapter of HOLT PHYSICAL SCIENCE. These activities apply new concepts by having the students use basic laboratory skills. Each *Activity* allows the student to explore a specific question related to the section.

Discover features are found throughout each chapter of HOLT PHYSICAL SCIENCE. The *Discovers* are short activities that apply new concepts in the chapter to everyday life. The *Discovers* are labeled *Discover By Doing*, *Researching*, *Observing, Calculating*, *Writing*, or *Problem Solving*, and may require laboratory work, library research, or manipulative skills. They may be performed in the classroom or at home. Most *Discovers* require simple, easy-to-obtain materials.

Skill activities enable the student to explore and manipulate the environment, to hypothesize and reason from data, to formulate explanations, and to communicate these explanations to others. In fact, process skills are prerequisites to success not only in science but also in other subjects and in the community and workplace as well.

The acquisition of process skills is a key goal in HOLT PHYSICAL SCIENCE. To accomplish this goal, the program provides numerous opportunities for skills development through engaging narrative style and thought-provoking questions. The common process skills in the Pupil's Edition and in the *Annotated Teacher's Edition* in HOLT PHYSICAL SCIENCE include:

> Analyzing
> Applying
> Classifying/Ordering
> Communicating
> Comparing
> Constructing/Interpreting Models
> Evaluating
> Experimenting
> Expressing Ideas Effectively
> Formulating Hypotheses
> Generating Ideas
> Generating Questions
> Identifying/Controlling Variables
> Inferring
> Interpreting Data
> Measuring
> Observing
> Predicting
> Recognizing Time/Space Relations
> Solving Problems/Making Decisions
> Synthesizing

As they develop and use these skills, the students must also estimate, sequence, describe, record, and draw conclusions. As a result, the students not only acquire basic scientific information but also develop the process skills needed to understand and interpret scientific information.

Field Trip GUIDELINES

ield trips are a desirable part of any science program. Field trips serve to remind the students that the real world exists outside the classroom. It is only in the field, for example, that organisms and physical phenomena can be observed and studied under natural conditions. Whether you go to a museum, a local industry, or an outdoor site, a truly successful field trip should add a new dimension to the students' grasp of science. It should make classroom work more meaningful and more enjoyable. Three things are required of the teacher in order to make any field trip an enriching experience: taking proper safety precautions, setting and evaluating goals, and planning the trip carefully.

Establishing Safety Guidelines

Some precautions in advance of the field trip may eliminate potential safety problems. The following suggestions will help ensure a safe and enriching trip.

◆ When planning a trip to an outdoor site, visit the site in advance and note any potential safety hazards. Discuss necessary safety measures with the students in advance. Include warnings of any hazards discovered during your advance visit.

◆ When planning a trip to a museum, local laboratory, or government agency, visit the location and meet with institution staff prior to the trip. Ask the staff about any precautions the class should be aware of.

◆ Arrange for additional adult supervision when necessary.

◆ If any part of the field trip is to take place on private property, get written permission in advance from the landowner.

◆ Make certain that any consent forms required from parents or guardians or school officials are drawn up, signed, and filed in advance.

◆ Caution students to dress in a manner that will keep them comfortable, warm, and dry.

◆ Pack a basic first-aid kit. Depending on the type of trip, area, and season, consider including an insect repellent. On any trip to a large body of water, be sure to include an adult skilled in water safety and CPR.

◆ Caution the students to report any injury immediately to an adult supervisor.

Setting and Evaluating Goals

Assessing the value of a trip means first setting goals so that you have some standard against which to measure student accomplishments. Well-defined goals must then be conveyed to the students to encourage them to regard the field trip as an important learning experience. Use the following guidelines to set and evaluate realistic goals.

◆ Make sure that the students understand how the field trip relates to the appropriate textbook topic.

◆ Provide the students in advance with an outline of what they will experience on the field trip. If the trip involves an on-site guide, such as a museum curator, try to get an advance summary of the material he or she plans to cover.

◆ If an on-site guide is involved, encourage the students to ask questions.

◆ Give the students a short list of questions to be answered by them after completing the field trip. This will give the students a concrete goal to work toward and will provide you with a means of evaluating the field trip.

◆ Follow up the field trip with a class discussion.

◆ Send personal or student-written thank-you letters to on-site guides or company representatives for their help in planning and conducting the field trip.

It would be beneficial for you to develop a checklist for implementing the planning and organization of field trips. Your checklist can be developed using the guidelines and suggestions provided.

SAFETY GUIDELINES

Many states have school laboratory regulations covering such topics as eye and body protection, storage of combustible materials, fire protection, and the availability of first-aid supplies. Check with your state department of education. Even in the absence of local regulations, the following safety precautions should be routine procedure in any laboratory. You can shape and reinforce proper student attitudes toward laboratory safety by setting a good example.

Equipment

◆ Microscopes, hot plates, and other electric equipment should be kept in good working order. Three-prong plugs must always be plugged into compatible outlets, or an adapter must be used. Never remove the grounding prong.

◆ Before each laboratory session, examine glassware for cracks and ask the students to do the same. Never use cracked or chipped glassware. Glassware used for heating should be made of heat-resistant material.

◆ Broken glassware should be swept up immediately, never picked up with the fingers. Broken glassware should be disposed of in a container

specifically denoted for this purpose. If such a container is not available, broken glass should be adequately wrapped in paper and the paper secured. Alert the maintenance staff that the package contains broken glass.

◆ Never permit the students to operate an autoclave or a pressure cooker for sterilization. Conduct the sterilization yourself.

◆ Mechanical equipment should be set up according to instructions. The students should read all instructions for use before handling mechanical equipment.

Materials

◆ Volatile liquids such as alcohol should be used only in ventilated areas and never near an open flame. To heat such substances, warm them on a hot plate or in a water bath.

◆ For the correct procedures to be used by the students when handling or heating chemicals, please see pages 626 and 627 in the Reference Section.

◆ Used chemicals should be disposed of in a manner that does not pollute the environment. Check on school and state guidelines for proper chemical disposal.

Safety Equipment

Safety equipment commonly found in school laboratories may include fire extinguishers, fire blankets, sand buckets, eyewash fountains, emergency showers, safety goggles, laboratory aprons, gloves, thermal mitts, tongs, respirators, and a first-aid kit.

◆ Note the location of each piece of safety equipment in your laboratory, learn to use it correctly, and teach the students to do the same.

◆ Learn how to use the first-aid equipment. Call the school nurse or a physician immediately in case of injury.

◆ Post the telephone number of a poison-control center in your area.

◆ When teaching large classes, you may wish to set up materials stations in several parts of the room. The students can obtain chemicals and other supplies there, thus minimizing the distance they must carry supplies to their work areas.

◆ Chemicals should be stored in a well-ventilated area that is kept locked at all times. Flammable chemicals should be stored in a fire-resistant cabinet. Never store such chemicals in a refrigerator, unless the refrigerator is specifically marked explosion proof.

◆ Chemical storage areas should be kept clean, orderly, and well lighted.

the students to procedures that require special care.

◆ Do not permit the students to work in the laboratory without supervision.

◆ Do not allow the students to conduct unauthorized experiments.

◆ When an investigation has been completed, insist that the students clean their work areas; wash and store all materials and equipment; and turn off all water, gas, and electric appliances.

General Information

◆ Always perform an experiment yourself before asking the students to try it. Cautionary statements, which are printed in boldface type, are included in the procedures given for all laboratory investigations. Refer to those statements when preparing for the experiments. In addition, safety symbols alert

Safety Symbols

The following safety symbols will appear whenever a procedure requires extra caution.

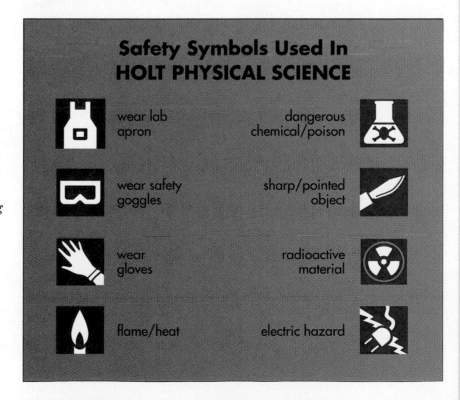

Safety Symbols Used In
HOLT PHYSICAL SCIENCE

wear lab apron

dangerous chemical/poison

wear safety goggles

sharp/pointed object

wear gloves

radioactive material

flame/heat

electric hazard

MATERIALS

The following materials list has been compiled to help the teacher order supplies for the textbook Investigations, Activities, and Skill activities. The items are keyed to the Investigation (I), Activity (A), and Skill (S) in each chapter in which the item is needed. The numeral following the letter indicates the chapter number. Additional information regarding amounts and preparation of materials may be found on the interleaved pages immediately preceding each chapter.

Apparatus and Equipment	Investigations, Activities, and Skills
anvil	S17
balance	S1, I1, A5, I16
boiling chip	A7, I19
Bunsen burner	A7, A12
buret clamps	I2, A13
cart	S5
clamp	I6, I19
clamp, utility	A7
compass	I14
compass needle	A14
copper tongs	A17
corks	I19
cylinders, graduated, 50 mL	I5
dynamics cart	A3, I3
filter	A22
filter paper	I21
strips	A22
forceps	A17
funnel	I21

galvanometer	I15
gloves	I7
glucose test strips	I22
hand lens	I18
hemicylindrical plastic dish	I12
hot plate	I2, A17, I21
image conduit	A12
iron ring	I19
key or knife switches	I13
laboratory apron	I1, A7, I7, A12, I19, I20, A21, I21, I22
laboratory burner	A17, I17, I19
lens, convex	I10
litmus paper	I7
blue	A21
red	A21
magnets	
bar	I15
permanent	A14
medicine droppers	I19, I21
meter stick	A3, I3, A4, A5, S5, A6, A8, I8, A9, A10
metric ruler	S8, I12
microscope (light)	A20
mortar and pestle	I21
plastic tub	S19
polarizing film, 3 cm^2	A20
prism	A11
protractor	S8, A10, I12, S12
pulley with table clamp	I3
ring stand	I2, I6, A7, A13, I9
ring with wire gauze	A7
rolling cart	A5
rubber stopper	S1
ruler	S1, S4, S5, S12
safety goggles	I2, A7, I7, A12, A15, I19, I20, A21, I21, I22
scissors	I4, A20
scoop	I19
solenoid	I15
spring scale	S4
standard water-column apparatus	I9

stoppers	I22
2-hole	A7, I19
stopwatch	I2, A3, I3, A4, A5, A8, I8, A9
test tube rack	I17, I20, I21, I22
thermometer	I5, I6, I9
in 1-hole stopper	A6
in notched corks	I2
triangular file	A12
tuning fork	I9
wire	S17
1.5 m	I14
wire clips	S17
wire connectors	I14
wire gauze	I19

wire leads	
insulated	I15
w/alligator clips	I13
wire loops	I17

Glassware	Investigations, Activities, and Skills
beakers	A21, I19
50 mL	I21
150 mL	I19
400 mL	A17
500 mL	I2
bottles with tops, small	A2
color filters (red, green, blue, cyan, yellow, magenta)	I11
colored glass	A10
dropper bottles with solutions	A20, I20, A21
droppers	I20

flasks	
250 mL	A7
bent glass nozzle	A7
glass tube	A7, I19
graduates	
10 mL	I21, I22
100 mL	S1, I1
graduated cylinder	S19, I20
microscope slides	A20
mirrors	S22
small	S10
Petri dishes with lids	I7
stirring rod	I2, I5, I21, I22

test tubes	I17, I19, I20, I21, I22
large	I2

Local Supply	Investigations, Activities, and Skills
acetate strip	A13
activated charcoal	I19
alcohol	A6
aluminum foil	I7
antacid tablet	A21
assorted objects	S16
battery	I14
6-volt lantern	I13
and holder	S17
beads	I16
blocks	I3
board	A5
books	S4, A5
thin	S5

Biological Supplies

Investigations, Activities, and Skills

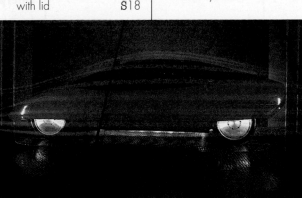

Resources

Electronic Media Suppliers

Agency for Instructional Television
111 West 17th Street
Box A
Bloomington, IN 47402

AIMS Media
9710 DeSoto Avenue
Chatsworth, CA 91311

American School Publishers
P.O.Box 408
Hightstown, NJ 08520

Barr Films
12801 Schabarum Avenue
Irwindale, CA 91706

Bergwall Productions, Incorporated
P.O. Box 2400
Chadds Ford, PA 19317

BFA Educational Media
468 Park Avenue South
New York, NY 10016

Biolearning Systems
Route 106
Jericho, NY 11753

Broderbund Software Direct
P.O. Box 6125
Novato, CA 94948

Carolina Biological Supply Company
2700 York Road
Burlington, NC 27215

Center For Humanities, Incorporated
Communications Park
90 South Bedford Road
P.O. Box 1000
Mt. Kisco, NY 10549

Central Scientific Company
3300 Cenco Parkway
Franklin Park, IL 60131

Charles Clark Company
170 Keyland Court
Bohemia, NY 11716

Churchill Films
12210 Nebraska Avenue
Los Angeles, CA 90025

Clearvue, Incorporated
6465 North Avondale
Chicago, IL 60631

Conduit
The University of Iowa
Oakdale Campus
Iowa City, IA 52242

Connecticut Valley Biological Supply Company
82 Valley Road
P.O. Box 326
Southampton, MA 01073

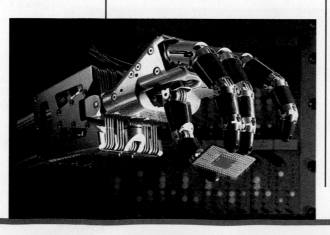

Coronet/MTI Film and Video
108 Wilmot Road
Deerfield, IL 60015

CRM/McGraw-Hill Films
2233 Faraday Avenue
Suite F
Carlsbad, CA 92008

Cross Educational Software
504 East Kentucky Avenue
P.O. Box 1536
Ruston, LA 71270

Datatech Software Systems, Incorporated
19312 East Eldorado Drive
Aurora, CO 80013

Direct Cinema, Ltd.
P.O. Box 10003
Santa Monica, CA 90410

Diversified Educational Enterprises, Incorporated
725 Main Street
Lafayette, IN 47901

Educational Activities, Incorporated
1937 Grand Avenue
Baldwin, NY 11510

Educational Images Ltd.
P.O. Box 3456
West Side Station
Elmira, NY 14905

Educational Services, Incorporated
1725 K Street NW
Suite 408
Washington, DC 20006

Edutech
1927 Culver Road
Rochester, NY 14609

Encyclopaedia Britannica Educational Corporation
310 South Michigan Avenue
Chicago, IL 60604

Films for the Humanities & Sciences
P.O. Box 2053
Princeton, NJ 08543

Fisher Scientific Company
Educational Materials Division
4901 West LeMoyne Avenue
Chicago, IL 60651

Focus Educational Media, Incorporated
485 South Broadway
Suite 12
Hicksville, NY 11801

Frey Scientific Company
905 Hickory Lane
Mansfield, OH 44905

Geoscience Resources
1238 Anthony Road
P.O. Box 2096
Burlington, NC 27216

Human Relations Media
175 Tompkins Avenue
Pleasantville, NY 10570

Instructional Video
P.O. Box 21
Maumee, OH 43537

International Film Bureau
332 South Michigan Avenue
Chicago, IL 60604

Kemtec Educational Corporation
9889 Crescent Park Drive
West Chester, OH 45069

Kons Scientific Company, Incorporated
P.O. Box 3
Germantown, WI 53022

Merlan Scientific
247 Armstrong Avenue
Georgetown, Ontario
Canada L7G 4X6

Micro Learningware
Rural Route #1
P.O. Box 162
Amboy, MN 56010

Modern Talking Picture Service, Incorporated
5000 Park Street North
St. Petersburg, FL 33709

Nasco International, Incorporated
901 Janesville Avenue
Fort Atkinson, WI 53538

Nasco West
P.O. Box 3837
Modesto, CA 95352

National Geographic Society
Educational Services
Department 91
1145 17th Street NW
Washington, DC 20036

National Teaching Aids, Incorporated
1845 Highland Avenue
New Hyde Park, NY 11040

Phillips Petroleum Company
Advertising Department
4th and Keeler Avenue
6-A1 Phillips Building
Bartlesville, OK 74004

Sargent-Welch Scientific Company
911 Commerce Court
Buffalo Grove, IL 60089

Scholastic Software
P.O. Box 7502
Jefferson City, MO 65102

Science Kit and Boreal Laboratories
777 East Park Drive
Tonawanda, NY 14150

Sunburst Communications
P.O. Box 660002
Scotts Valley, CA 95067

Time-Life
1450 East Parham Road
Richmond, VA 23280

Videodiscovery, Incorporated
1515 Dexter Avenue North
Suite 600
Seattle, WA 98109

Resources

Walt Disney Educational Media Company
108 Wilmot Road
Deerfield, IL 60015

Ward's Natural Science Establishment, Incorporated
P.O. Box 92912
Rochester, NY 14692

West Wind Productions
P.O. Box 3532
Boulder, CO 80307

Directory of Organizations

American Association for the Advancement of Science
1333 H Street NW
Washington, DC 20005

American Association of Physics Teachers
5112 Berwyn Road
College Park, MD 20740

American Cancer Society
1599 Clifton Road NE
Atlanta, GA 30329

American Chemical Society
1155 16th Street NW
Washington, DC 20036

American Heart Association
The National Center
7320 Greenville Avenue
Dallas, TX 75231

American Institute of Biological Sciences
730 11th Street NW
Washington, DC 20001

American Institute of Chemists
7315 Wisconsin Avenue
Bethesda, MD 20814

American Institute of Professional Geologists
7828 Vance Drive
Suite 103
Arvada, CO 80003

American Lung Association
1740 Broadway
New York, NY 10019

American Medical Association
Education and Research
Foundation
515 North State Street
Chicago, IL 60610

Centers for Disease Control
1600 Clifton Road NE
Atlanta, GA 30333

Institute for Chemical Education
University of Wisconsin
Department of Chemistry
1101 University Avenue
Madison, WI 53706

National Academy of Sciences
Office of News and
Public Information
2101 Constitution Avenue NW
Washington, DC 20418

National Association of Biology Teachers
11250 Roger Bacon Drive
Number 19
Reston, VA 22090

National Health Council
1730 M Street NW
Suite 500
Washington, DC 20036

National Science Foundation
1800 G Street NW
Room 520
Washington, DC 20550

National Science Teachers Association
1742 Connecticut Avenue NW
Washington, DC 20009

Society of Independent Professional Earth Scientists
4925 Greenville Avenue
Suite 170
Dallas, TX 75206

Superintendent of Documents
U.S. Government Printing Office
710 North Capitol Street NW
Washington, DC 20401

U.S. Environmental Protection Agency
Office of Communications
and Public Affairs
401 M Street SW
Washington, DC 20460

U.S. Public Health Service
Department of Health
and Human Services
200 Independence Avenue SW
Washington, DC 20201

Laboratory Suppliers

Carolina Biological Supply Company
2700 York Road
Burlington, NC 27215

Central Scientific Company
3300 Cenco Parkway
Franklin Park, IL 60131

Connecticut Valley Biological Supply Company
82 Valley Road
P.O. Box 326
Southampton, MA 01073

Damon Industries
DCM Instructional Systems
82 Wilson Way
Westwood, MA 02090

Difco Laboratories, Incorporated
P.O. Box 331085
Detroit, MI 48232

Eastman Kodak Company
BLDG 701 LRP
343 State Street
Rochester, NY 14652

Edmund Scientific Company
101 East Gloucester Pike
Barrington, NJ 08007

Fisher Scientific Company
Educational Materials Division
4901 West LeMoyne Avenue
Chicago, IL 60651

Forestry Suppliers, Incorporated
205 West Rankin Street
P.O. Box 8397
Jackson, MS 39284

Frey Scientific Company
905 Hickory Lane
Mansfield, OH 44905

Hach Company
P.O. Box 389
Loveland, CO 80539

Harvard Apparatus Company
22 Pleasant Street
Natick, MA 01760

Hubbard Scientific Company
3101 Iris Avenue
Suite 215
Boulder, CO 80301

La Pine Scientific Company
13636 Western Avenue
Blue Island, IL 60406

Leica
Instrument Division
P.O. Box 123
Buffalo, NY 14240

Nasco International, Incorporated
901 Janesville Avenue
Fort Atkinson, WI 53538

Nasco West
P.O. Box 3837
Modesto, CA 95352

Sargent-Welch Scientific Company
911 Commerce Court
Buffalo Grove, IL 60089

Science Kit and Boreal Laboratories
777 East Park Drive
Tonawanda, NY 14150

Swift Instruments, Incorporated
P.O. Box 562
San Jose, CA 95106

Triarch, Incorporated
P.O. Box 98
Ripon, WI 54971

U.S. Geological Survey
12201 Sunrise Valley Drive
Reston, VA 22092

Van Waters & Rogers
5353 Jillson Street
Los Angeles, CA 90040

Ward's Natural Science Establishment, Incorporated
P.O. Box 92912
Rochester, NY 14692

Wilkens-Anderson Company
4525 West Division Street
Chicago, IL 60651

HOLT

PHYSICAL

SCIENCE

Annotated Teacher's Edition

About Holt Physical Science

Share the Wonder

Why does ice melt? How much work do you do when you ride a bike? How does our use of energy affect the environment? These are just a few examples of questions asked by physical scientists over the years. The answers to these and many of your own questions about matter and energy can be found in HOLT PHYSICAL SCIENCE.

Explore the Nature of Science

Have you ever seen lightning flash in the sky? Have you changed a tire on a car and washed your hands to get them clean? If so, you have experienced the study of physical science. The study of matter and energy and how they react is what physical science is all about. Science is knowledge gained by observing, experimenting, and thinking. When you study physical science, you use your knowledge to understand how and why things happen the way they do. HOLT PHYSICAL SCIENCE will help you add to your knowledge through reading, discussion, and activities.

Work Like a Scientist

Much of science is based on observations of events that occur in your daily life. From these observations, you can form questions and develop experiments to expand on your observations. During your study of physical science, you will be working and thinking like a scientist. You will develop and improve certain skills, such as your observation skills. With new information, you will try to answer questions, to think about what you have observed, and to form conclusions. You will also learn how to communicate information to others, as scientists do.

Keep a Journal

One way to organize your ideas is to keep a journal and make frequent entries in it. Like a scientist, you will have a chance to write down your ideas and then, after doing activities and reading about scientific discoveries, go back and revise your journal entries. Remember, it's OK to change your mind after you have additional information. Scientists do this all the time!

Make Discoveries

Physical science is an ongoing process of discovery—asking and answering questions about the interaction of matter and energy. If you choose a career in the physical sciences, you may find answers to questions of students and scientists who lived before you. Through curiosity and imagination, physical scientists are able to create a never-ending list of questions about the world in which we live.

Prepare yourself for discovery. Through your studies, you will learn many interesting and exciting things. Allow yourself to explore and to discover the ideas, the information, the challenges, and the beauty within the pages of this book. HOLT PHYSICAL SCIENCE can help you understand the world around you. You will begin to ask new questions and to find new answers. You will discover that learning about physical science is important and fun.

UNIT 1

INTRODUCTION TO PHYSICAL SCIENCE

UNIT OVERVIEW

Unit 1 introduces physical science and presents the basic methods scientists use to ask questions, solve problems, and measure quantities. The basic makeup and characteristics of matter are also presented to provide a solid base for further study.

Chapter 1: Methods of Science, page 2

This chapter provides a description of the methods scientists use to define, explore, and solve problems. Also discussed is the International System of Units (SI) used in science and accepted throughout the world.

Chapter 2: Fundamental Properties of Matter, page 30

This chapter presents properties of the four states of matter and the categories in which matter is organized. Models of matter are introduced that explain and predict many kinds of phenomena. Also discussed is the motion of molecules as matter changes from one state to another.

Science Parade, pages 56–67

The articles in this unit's magazine relate the methods of science and properties of matter to space exploration and advancements in biodegradable plastics. The work of a sculptor is described in the career feature. The biographical sketches focus on a philosopher and a theoretical physicist.

UNIT RESOURCES

PRINT MEDIA FOR TEACHERS

Chapter 1

Bosak, Susan V. *Science Is. . . : A Source Book of Fascinating Facts, Projects, and Activities, 2nd ed.* Scholastic, 1992. This resource contains a collection of hands-on activities, puzzles, and challenges.

Freedman, David H. "Gravity Grabs Antimatter." *Discover* 13 (January 1992): 51. Richard Hughes of Los Alamos National Laboratory offers strong evidence that there doesn't seem to be any gravitational distinction between matter and antimatter.

Peterson, Ivars. "'Baked Alaska' Cooked Up in Liquid Helium." *Science News* 142 (July 18, 1992): 38–39. This article discusses the multiple uses of a new technology. High-energy electrons are created by passing cosmic-ray engineered muons through superfluid helium. Significant amounts of energy—which has the capacity to expand and result in a cold, growing superfluid helium—are deposited.

Chapter 2

Ferris, Timothy, ed. *The World Treasury of Physics, Astronomy, and Mathematics.* Little, Brown, Co., 1991. This collection includes more than sixty pieces by well-known authors of science works.

PRINT MEDIA FOR STUDENTS

Chapter 1

Flaste, Richard, ed. *The New York Times Book of Science Literacy: What Everyone Needs to Know from Newton to the Knuckleball.* Times Books/Random House, 1991. This book describes little-known and widely-known scientific facts in a fun and engaging way.

Gardner, Robert. *Famous Experiments You Can Do.* Franklin Watts, 1990. Experiments from Archimedes to Faraday are presented with clear and easy instructions, using simple equipment.

Chapter 2

Berger, Melvin. *Solids, Liquids and Gases: From Superconductors to the Ozone Layer.* Putnam, 1990. This book discusses three of the states of matter. Timely topics, such as acid rain, superconductors, and the ozone layer, are included.

Foley, Elizabeth. "Inside Ice." *Dolphin Log* (March 1990): 14–15. Properties of ice and varieties of ice are discussed. An experiment is included for further illustration.

Stwertka, Eve and Albert Stwertka. *A Chilling Story: How Things Cool Down.* Silver Burdett, 1991. This book discusses how different types of matter cool down.

ELECTRONIC MEDIA

Chapter 1

 Discovering the Scientific Method: Snigs . . . Flirks . . . Blorgs. Computer Software. Focus Media, Inc. The students use problem-solving exercises to learn what a hypothesis is, analyze observations, organize and evaluate data, and make generalizations from conclusions.

 Measurements: Length, Mass, and Volume. Computer Software. Focus Media, Inc. This program provides an introduction or a review of skills involved in measuring length, mass, and volume.

 Meter, Liter, and Gram. Videocassette. BFA Educational Media. 13 min. Length, volume, and mass are presented in SI units.

 Scientific Measurement. Film. Barr Films. 18 min. This film presents a variety of measurements including time, distance, speed, volume, mass weight, density, and temperature.

 The Scientific Method. Film. Barr Films. 23 min. This film traces the scientific method in classroom experimentation as well as in professional research and development.

Chapter 2

 All About Matter. Videocassette; four-part series. Focus Educational Media. 14 min. each. "Physical or Chemical: What Kind of Change?" "Elements: The Skeleton of Matter," "Compounds: A Special Combination," and "Mixtures: Separate, Yet Together" present information on matter through real-life situations.

 The Invisible World. Videodisc. National Geographic. This fascinating videodisc looks at such things as water crystallizing into ice and the movement of a single atom.

 Motion of a Molecule. Film. BFA Educational Media. 12 min. A hot-air balloon is used to study Brownian Motion and other examples of molecular movement.

 Plasma: The Fourth State of Matter. Film or videocassette. BFA Educational Media. 10 min. Plasma is explored by examining the stars, Earth's ionosphere, neon signs, and other examples.

UNIT 1 INTRODUCTION TO PHYSICAL SCIENCE

"**W**ait, did you pack your toothbrush? Will they have toothpaste? What about your science project?" The questions from your father go on and on. Gee, you think to yourself, it's only a science field trip. People have been visiting and living in space for years.

Although the above paragraph sounds like a fantasy, humans may inhabit space in the near future. When they do, they will make and study new materials that have never been made on Earth.

CHAPTERS

1 *Methods of Science* 2

Scientists solve problems by observing, experimenting, and drawing conclusions. The detective Sherlock Holmes solves mysteries in much the same way.

2 *Fundamental Properties of Matter* 30

The changing properties of water, a familiar form of matter, can turn treacherous and unpredictable on a frigid glacier.

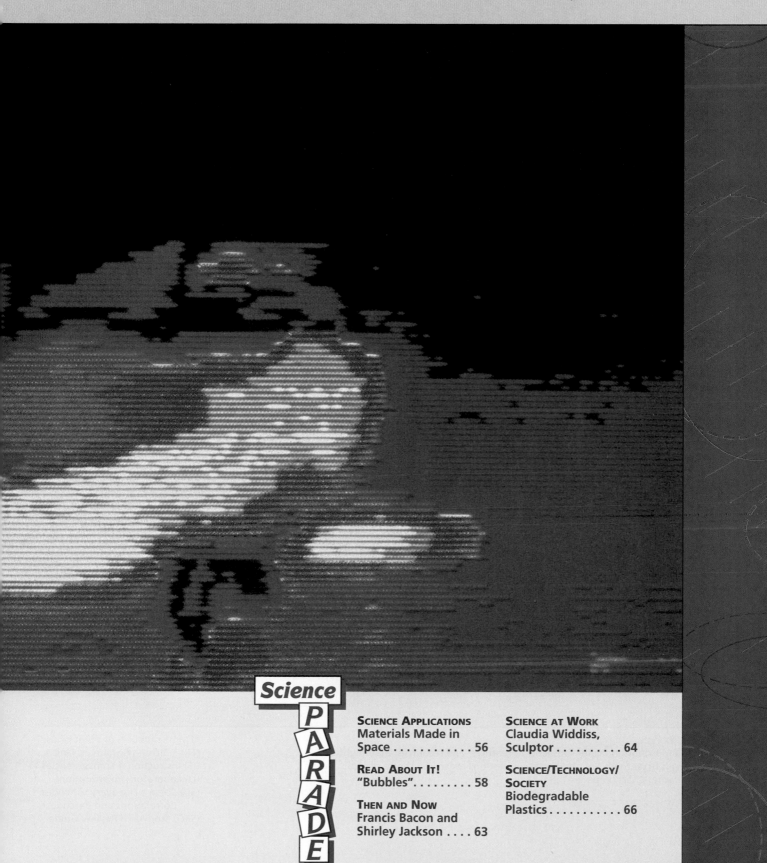

Journal Activity Extend the discussion by asking the students to hypothesize how materials made in space might differ from those made on Earth and how space-manufactured products could change their lives. You might also bring in the following Unit topics: How do scientists solve problems? How is the way scientists solve problems similar to the way a detective solves mysteries? How do the properties of matter change? Finally, ask the students to keep a record in their journals of any questions they might have about these topics. After the students have completed the Unit, encourage them to look again at their questions. A follow-up discussion might help the students realize how much they have learned.

Science PARADE

CHAPTER 1

METHODS OF SCIENCE

PLANNING THE CHAPTER

Chapter Sections	Page	Chapter Features	Page	Program Resources	Source
Chapter Opener	2	**For Your Journal**	3		
Section 1: ASKING QUESTIONS AND SOLVING PROBLEMS	4	Discover by Problem Solving **(B)**	8	*Science Discovery**	SD
• Defining the Problem **(B)**	4	Discover by Writing **(B)**	13	Reading Skills:	
• Suggesting an Answer **(B)**	5	Section 1 Review and Application	13	Finding the	
• Experiments Tell You What You Need to Know **(B)**	5	Skill: Measuring Correctly **(B)**	14	Main Idea **(B)**	TR
• Performing Experiments **(B)**	7			Study and Review Guide, Section 1 **(B)**	TR, SRG
• Drawing Conclusions **(B)**	8				
• Communicating Results **(B)**	11				
Section 2: MEASUREMENTS	15	Activity: How can you measure distance without using a standard unit? **(B)**	16	*Science Discovery**	SD
• The Invention of Measurement **(B)**	15			Investigation 1.1: Using Measurements **(A)**	TR, LI
• The International System of Units—SI **(B)**	16	Discover by Writing **(A)**	18	Extending Science Concepts: Using a Vernier Caliper **(H)**	TR
• Measuring Quantities in SI **(B)**	19	Discover by Doing **(B)**	20		
• Accuracy and Precision in Measurement **(A)**	23	Discover by Problem Solving **(A)**	25	Investigation 1.2: Determining the Density of Solids **(A)**	TR, LI
		Discover by Calculating **(A)**	25	Connecting Other Disciplines: Science and Mathematics, Identifying and Using Significant Digits **(H)**	
• Derived Quantities: Area and Density **(A)**	24	Section 2 Review and Application	25		
		Investigation: Finding Mass, Volume, and Density **(A)**	26		TR
				Thinking Critically **(A)**	TR
				SI Units and Symbols **(A)**	IT
				SI Prefixes **(A)**	IT
				Measurement **(A)**	IT
				Record Sheets for Textbook Investigations **(A)**	TR
				Study and Review Guide, Section 2	TR, SRG
Chapter 1 HIGHLIGHTS	27	The Big Idea	27	Study and Review Guide, Chapter 1 Review **(B)**	TR, SRG
Chapter 1 Review	28	For Your Journal	27		
		Connecting Ideas	27	Chapter 1 Test	TR
				Test Generator	

B = Basic   **A** = Average   **H** = Honors
The coding Basic, Average, and Honors indicates subsections, features, and resources that might be appropriate for different levels of learners. For additional suggestions regarding choice of topic and depth of coverage, see the Pacing Chart on pages T28–T31.

*Frame numbers at point of use
(TR) Teaching Resources, Unit 1
(IT) Instructional Transparencies
(LI) Laboratory Investigations
(SD) *Science Discovery* Videodisc Correlations and Barcodes
(SRG) Study and Review Guide

CHAPTER MATERIALS

Title	Page	Materials
Teacher Demonstration	6	athletic shoe(s)
Discover by Problem Solving	8	(per individual) journal
Discover by Writing	13	(per individual) journal
Skill: Measuring Correctly	14	(per individual or pair) 100-mL graduate, water, rubber stopper, balance, ruler
Activity: How can you measure distance without using a standard unit?	16	(per class) hallway with beginning and ending points marked so that everyone can measure the same distance
Discover by Writing	18	(per individual) journal
Teacher Demonstration	20	different shaped containers of similar volume (3), graduate, water
Discover by Doing	20	(per pair) water, graduate, small rock or marble
Teacher Demonstration	21	objects similar in size with very different masses (2)
Discover by Problem Solving	25	(per group of 3 or 4) objects of different densities (2)
Discover by Calculating	25	(per individual) journal
Investigation: Finding Mass, Volume, and Density	26	(per group of 3 or 4) pennies (10), balance, 100-mL graduate

ADVANCE PREPARATION

For the *Investigation* on page 26, ask the students to bring pennies to class with mint dates between 1955 to the present.

TEACHING SUGGESTIONS

Field Trip
If possible, take the students on a field trip to a local museum that has science or technology exhibits. Before visiting the museum, tell the students that they will have to choose and report on one exhibit that deals with measurement or problem solving. Have them bring their journals to take notes about the exhibit. Students' reports might be in the form of a drawing, a model, or a written description.

Outside Speaker
Invite a pharmacist to class to explain the importance of measurements in pharmaceutical professions. Ask the pharmacist to show the students measuring devices that are used to dispense medications accurately.

CHAPTER THEME—*SYSTEMS AND STRUCTURES*

In Chapter 1 the students will learn how scientists solve problems in a structured way, known as a scientific method. One step of a scientific method involves the collection of data, which often requires the taking of measurements. The students will read about the International System of Units (SI) for measuring various quantities, such as length, volume, mass, and density. A supporting theme in this chapter is **Environmental Interactions.**

◆ **MULTICULTURAL CONNECTION**

Scientific discoveries have been made by people from all parts of the world. Assign a different country to each student. Have the students work independently to research the contributions made by one scientist from that country. You might use the students' written work as part of a bulletin board.

▲ **MEETING SPECIAL NEEDS**

Second Language Support

Discuss with limited-English-proficient students any mystery or detective stories they have read in their native language(s). Elicit a description of the mystery and how it was solved by asking the students simple questions about the story. Then ask them to judge if the detective solved the mystery or crime "scientifically."

CHAPTER 1

METHODS OF SCIENCE

"Elementary, my dear Watson!" the famous detective Sherlock Holmes would cry. Sherlock Holmes often amazed his friend Dr. Watson with his powers of observation and reasoning. In one famous story, the fictional detective sizes up a visitor named Wilson.

▶ **Y**ou may not realize it, but you solve problems every day. Usually, you solve these problems using a scientific method. Scientists use these methods to study nature, synthesize new chemicals, and create new technologies.

"Beyond the obvious facts that he has at some time done manual labour, that he has been in China, and that he has done a considerable amount of writing lately, I can deduce nothing else."

. . . "How, in the name of good fortune, did you know all that, Mr. Holmes?" [Mr. Wilson] asked. "How did you know, for example, that I did manual labour?"

"Your hands, my dear sir. Your right hand is quite a size larger than your left. You have worked with it, and the muscles are more developed."

CHAPTER MOTIVATING ACTIVITY

Cooperative Learning Have the students form small groups. Tell each group to write a short mystery story that includes a solution to the mystery. Then have a representative from each group read the story aloud to the class without revealing the solution. Lead the class in exploring possible methods for solving the mystery. After all possibilities have been discussed, ask the representative to describe how the mystery was solved.

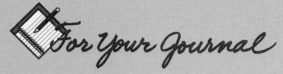

For Your Journal

The students' responses to the journal questions will reflect their understanding of problem solving in science and in everyday life. (The students will have a variety of responses to the journal questions. Accept all answers, and tell the students that they will refer to their journal entries at the end of the chapter.)

"But the writing?"

"What else can be indicated by that right cuff so very shiney for five inches, and the left one with the smooth patch near the elbow where you rest it upon the desk?"

"Well, but China?"

"The fish which you have tattooed immediately above your right wrist could only have been done in China. When, in addition, I see a Chinese coin hanging from your watch-chain, the matter becomes even more simple..."

Mr. Jabez Wilson laughed heavily. "Well, I never!" said he. "I thought at first you had done something clever, but I see that there was nothing in it after all."

> from *The Adventures of Sherlock Holmes*
> by Sir Arthur Conan Doyle

Of course, there was something in Holmes' deductions, a scientific method. While Sherlock Holmes uses his methods to solve crimes, scientists use observation and reasoning to understand the universe.

ABOUT THE AUTHOR

Sir Arthur Conan Doyle—the creator of Sherlock Holmes—was born in Scotland in 1859. Doyle began his professional life as a physician. However, his lack of success in the medical field led him to take up writing. Doyle authored 56 short stories and three novels about the fictional detective. In 1893 he wrote a story in which Holmes was killed. Bowing to public outcry, Doyle wrote a subsequent story in which Holmes "returned to life." Doyle wrote his last Sherlock Holmes novel, *The Valley of Fear*, in 1915.

For Your Journal

- What methods do you use when solving a problem?
- How does measurement help in solving problems?
- How do scientists solve problems and answer questions?

FOCUS

This section identifies and describes the terms and steps used in a scientific method. Examples are given of ways in which scientists report their findings.

MOTIVATING ACTIVITY

Have the students bring to class advertisements that describe various brands of athletic shoes. Display the advertisements, and then work with the students to identify information provided in the advertisements that would help them in choosing an athletic shoe.

PROCESS SKILLS
- Communicating
- Measuring • Comparing
- Interpreting Data

POSITIVE ATTITUDES
- Precision • Enthusiasm for science and scientific endeavor • Skepticism

TERMS
- conclusion • hypothesis
- law • theory

PRINT MEDIA
The New York Times Book of Science Literacy: What Everyone Needs to Know from Newton to the Knuckleball by Richard Flaste (see p. xxib)

ELECTRONIC MEDIA
Discovering the Scientific Method: Snigs . . . Flirks . . . Blorgs Microcomputer Software (see p. xxib)

SCIENCE DISCOVERY
- Decision model
- Observation tests
- Observation and inference
- Quantitative vs. qualitative • Scientific method
- Thinking circle

BLACKLINE MASTERS
Reading Skills
Study and Review Guide

Asking Questions and Solving Problems

SECTION 1

Objectives

Describe the steps of a scientific method.

Define and **distinguish** among the terms hypothesis, law, and theory.

Explain how variables are used in scientific experiments.

Have you ever had a problem that you knew you had to solve? Of course you have. Scientists are experts in solving problems. In fact, they have developed specific methods to solve problems. To understand how this works, let's set up a common problem that you or a friend might have.

Suppose you need to buy a new pair of athletic shoes. How do you choose the right kind for your feet and for the activities that you do?

Defining the Problem

You have a problem. Your shoes are worn out, and you need a new pair. Knowing you need new shoes is the first step in solving your problem.

When scientists work on a problem or question, they usually use a method that has several steps. These steps usually include defining the problem, stating a hypothesis, performing experiments, drawing conclusions, communicating results, and suggesting an answer. These steps are often called a *scientific method*. However, it is important to remember that the steps in a scientific method are not always used in the same order.

Usually, however, asking questions and solving problems cannot begin until a question or problem has been identified. This is called *defining the problem*. Your problem is to find the best pair of athletic shoes for your needs.

Figure 1–1. Choosing the right athletic shoe can be a problem.

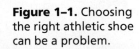

● **Process Skill:**
Generating Ideas

Have the students suggest what they would do if they were on a hike and had to cross a small river, about which they knew nothing. (Answers might include wading or swimming across the river.) Then help the students examine the steps of a scientific method. Point out how these steps could also be used to solve nonscientific problems. (The steps outline a good way to examine a problem, try out solutions, look at the probable results, and try again if the solution is not acceptable.)

● **Process Skill:**
Formulating a Hypothesis

Prepare a "mystery box" by constructing walls in a simple geometric design inside an empty shoe box. Place a marble inside the box, and secure the lid with tape. Ask the students to manipulate the box to hypothesize what the design is. Have them record their hypotheses by drawing the inside of the box on paper. Encourage the students to compare their hypotheses before opening the box.

▼ **ASK YOURSELF**

Why is defining the problem an important part of any scientific method?

Suggesting an Answer

Sometimes scientists experiment in the laboratory or observe nature without any clear idea of what they are looking for. This method is most commonly used when scientists are interested in a subject about which almost nothing is known. More often, however, when scientific questions are being investigated, scientists have an idea of what they expect an experiment to show. Why do scientists have ideas about what to expect? The reason is because they research the problem and gather as much information as possible before beginning to experiment.

Scientists record what they think the answer, or solution, will be before they do an experiment. This educated guess about what might happen is called a **hypothesis**.

▼ **ASK YOURSELF**

What is your hypothesis for the question "Which shoes are the best for my feet and for the activities that I do?"

Experiments Tell You What You Need to Know

A hypothesis is usually tested in one of two ways—observation of nature or laboratory experimentation. Observations collected in a natural setting instead of in the laboratory are called *field studies*. Since you can't observe the action of athletic shoes in nature, you need to experiment to find the answer to your problem.

Whenever you conduct an experiment, you have to decide how to deal with variables. A *variable* is any part of an experiment that can be changed. There are often many variables in an experiment and in field studies. The trick is to test only one variable at a time. That way you will know whether that variable is the one that makes the difference.

Figure 1–2. These scientists are experimenting in a laboratory to gather new information.

● **Process Skill:**
Identifying Variables

Discuss with the students other variables that can influence a person's purchase of athletic shoes. Tie the discussion in with Dr. Nigg's comment that a shoe's features are more important than its price and brand name.

Demonstration

Bring one or more athletic shoes to class to demonstrate the variables of cushioning, control, and torsion as described in the article.

LASER DISC

462–467

Decision model

Some of the variables related to athletic shoes are cushioning, control, and torsion. Of course, you should also consider how they fit! Read the following article to find out more about these variables.

Sneaker Science

How to Choose Your Shoes

from Current Science

Dr. Benno Nigg is a sneaker specialist at the University of Calgary in Canada. Here are some of the features he says any good athletic shoe should have.

■ *Cushioning.* Most sports involve a lot of jumping, and it's important to remember that when you drop through the air, you carry a great deal of momentum. And the bigger you are, the greater is your momentum. When you land, the ground takes that momentum away from you by applying an opposing force to your feet.

If the sole of a shoe is not well cushioned, the opposing force pushing against your feet can injure the delicate cartilage—tissues that lines your joints—the places where your bones meet. "Injured cartilage is the beginning of arthritis, a disease of the joints," says Dr. Nigg.

A Cushioned Landing

Downward Force

Well-Cushioned Sole

Opposing Force

■ *Control.* Dr. Nigg says that the sole of any sneaker should be shaped to help you land naturally. If the heel is too bulky, you may land on the outside of your foot and experience a twisting type of force called *torque.* "If I twist your arm, it hurts because I've created torque. This type of force can break a bone," he says.

"It's important that the shape of a shoe be as close as possible to the shape of the foot. Shoes with big bulky heels and shoes that look like machines can be dangerous because they don't let you land naturally," says Dr. Nigg.

Out-of-Control Landing

Torque

Bulky Heel

■ *Torsion.* Finally, Dr. Nigg says that good running shoes should always allow the front part of the foot to move independently of the rear of the foot. This kind of movement is called *torsion.* "Sometimes when your feet land on the ground, the front part of your foot will rotate. If your shoe has a rigid sole, it will force the rear of your foot to rotate too, and you could sprain your ankle," says Dr. Nigg.

To test for torsion, Dr. Nigg suggests holding one hand on the front part of the shoe and the other hand on the heel. "If you can twist the shoe, then the sole is not rigid and the shoe will allow for torsion," he says.

"These features are more important," says Dr. Nigg, "than whether a shoe costs $100 or has the latest gimmick."

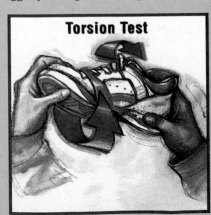

Torsion Test

● **Process Skill:**
Identifying Variables

Tell the class that two students were given denim jeans to wear to test for durability. Ask what factors would have to be controlled to make the test valid.

(Examples include the gender of the students, the length of time the jeans were worn, the size and age of the students, and the activities the students engaged in while wearing the jeans.)

● **Process Skill:** *Evaluating*

Have the students study the chart. Ask them why it might be helpful to test each brand of athletic shoe more than once. (Multiple trials will help make the data more reliable.)

Scientists must keep all variables the same except for the one being tested. This procedure is called *controlling variables*. This is the only way to make the experiment a fair test. After an experiment is completed, the results are carefully recorded.

Even though you can't observe athletic shoes in nature, there are problems that can be solved this way. Scientists often test a hypothesis by making observations of nature to see how certain variables seem to be related to the problem. Scientists measure all the variables thought to be important in the experiment.

 ASK YOURSELF

Describe a situation in which it would be better to observe nature rather than experiment in the laboratory. Explain your choice.

Figure 1–3. These scientists are using collected data to create a map of the moon. They can then analyze the map to further their understanding of the moon's surface.

Performing Experiments

In all kinds of research, measurements of variables must be made and recorded. In science, information such as measurements and observations is called *data*.

What kind of data could you collect about athletic shoes, and how could you collect it? First you would have to set up a controlled experiment. Let's see. Suppose you decided to test five types of shoes that you thought would be good choices. For our purposes we will call them Brands A through E. The variables you want to test are cushioning, control, torsion, and—of course—fit. The first thing you do is make a table in which to record your data as you collect it. It might look something like this one.

	Brand A	Brand B	Brand C	Brand D	Brand E
Cushioning					
Control					
Torsion					
Fit					

ONGOING ASSESSMENT
ASK YOURSELF

Situations that would be better to observe in nature include: natural behavior of living species; environmental problems; and processes or phenomena that occur on a large scale, such as changes in rocks, weather patterns, and population dynamics.

 LASER DISC
472

Observation and inference

473–476
Scientific method

● **Process Skills:**
Interpreting Data, Analyzing

Work with the students to create a hypothetical problem in order to draw conclusions based upon an analysis of the data. Copy these tables on the chalkboard:

Homeroom Grade 11-B	
Left-handed	**Right-handed**
Girls 3	9
Boys 5	11

Homeroom Grade 9-A	
Left-handed	**Right-handed**
Girls 1	10
Boys 4	13

Ask volunteers to draw one conclusion based upon the data. (Conclusions might include that more boys than girls are left-handed and more right-handed girls are in Grade 9-A than in Grade 11–B.)

DISCOVER BY
Problem Solving

After the students complete the Discover by Problem Solving, have them place their lists in their science portfolios.

Now, you need to decide how you are going to collect your data. One way would be to go to an athletic shoe store and try on one brand, testing for all the variables. Another way would be to go to the store and try on each shoe and test for one variable at a time. Each method has advantages and disadvantages.

 **DISCOVER BY Problem Solving** _____

List the advantages and disadvantages of different methods of data gathering. Choose the method you would use to test athletic shoes, and explain why you think it would be a fair test.

Once data is collected, it is analyzed. Data can be analyzed in a number of ways. It can be used to calculate new quantities that describe the variable or to make charts and graphs. Usually data from laboratory experiments is much easier to analyze and interpret than data from field studies because the variables in an experiment are easier to control.

Scientists face similar problems when they are experimenting in the laboratory. There are many different methods for collecting data. Sometimes the method chosen will have an effect on the results, and sometimes it doesn't make any difference at all. Regardless, the choice has to be made and recorded. Then if people want to look at the research, they can repeat the experiment.

ONGOING ASSESSMENT
▼ **ASK YOURSELF**

Recording the method used to collect data allows others to evaluate and repeat the experiment.

▼ **ASK YOURSELF**
Why is recording the method of data collection important?

Drawing Conclusions

Once you have collected your data, you have to analyze it and make a decision about which shoes you are going to buy. In scientific terms, you have to draw a conclusion. A **conclusion** is a judgment based on the analysis of data from experiments or field studies.

You need to draw a conclusion about your athletic shoes. The conclusion will prove or disprove your hypothesis. Remember, hypotheses are often incorrect. However, an incorrect hypothesis can lead to new discoveries and new information as readily as a correct hypothesis.

LASER DISC
477

Thinking circle

▶ **8** CHAPTER 1

Which brand of shoes is best for you? Your conclusion should be based on the data you have collected. However, choosing shoes isn't a totally scientific matter. Your opinion about style and color will also affect your final choice. But, since you've approached the problem in a scientific manner, you won't be choosing your shoes based only on a brand name!

Figure 1–4. You draw a conclusion when you choose which shoes to buy.

Scientific conclusions may lead to the formulation of theories or laws. Probably the testing of athletic shoes won't lead to any new scientific ideas, but that doesn't mean your research is unimportant. Very little scientific research actually leads to new theories or laws. What is the difference between a scientific law and a scientific theory? ①

Scientific Laws   In science, a **law** is a summary of many experimental results and observations. A law only describes what happens, not why it happens. Often scientific laws are summarized as mathematical equations. One example is the laws of motion—you will learn more about them in Chapter 3.

Scientific Theories   An explanation of why things work the way they do is called a **theory**, or a model. A theory explains the results of many different kinds of experiments, observations, and occurrences. A theory also predicts the results of experiments that have not yet been done.

● Process Skill: *Evaluating*

In their journals, have the students provide examples of sentences that use the words *law*, *model*, and *theory*. Discuss whether each word carries its commonly accepted general meaning or the more technical, scientific meaning. Because nearly all examples will use the words in a general sense, stress how important the difference in the general meaning and the scientific, technical meaning is.

● Process Skills: *Comparing, Interpreting Models*

Have the students compare the theories illustrated in Figure 1–6. Ask the students what evidence could have led scientists before Copernicus' time to believe Earth was the center of the solar system. (The sun seems to go around Earth—it rises in the east and sets in the west. The stars, particularly the constellations, seem to revolve around Earth each year.)

THE NATURE OF SCIENCE

The matching jigsaw shapes of the continents led German meteorologist Alfred Wegener to suggest in 1912 that the continents were once joined in a single land mass and slowly drifted to their present positions. However, Wegener could not explain how the continents moved, thus his idea was a hypothesis, not a theory. The accumulation of different kinds of evidence over many years eventually led scientists to form a theory that explains continental movement. This theory, called plate tectonics, explains that continents move because Earth's crust is broken into large sections, or plates, that float on molten rock.

Figure 1–5. A detective who claims to have a "theory" to explain a crime, in a scientific sense has only a hypothesis.

INTEGRATION–
Language Arts

Some scientists use the words *model* and *theory* interchangeably, while others argue that the two terms have slightly different scientific meanings. Before revealing this fact, have the students check several dictionaries to determine what the difference is, if any. Then explain that a model and a theory will be considered to be the same in this book.

Many people confuse the words *hypothesis, law,* and *theory.* Therefore, it may help to review their meanings. A hypothesis is an educated guess based on accurate and relevant information. A hypothesis states what a scientist thinks will happen in a particular experiment or field study. A law is a summary of experimental results and observations that tells what will happen but does not explain why. A theory is an explanation for a group of related results, observations, and occurrences that can be used to predict what will happen in new situations.

The meaning of *theory* in science is often misunderstood because many people use the term to mean an opinion. For example, you might hear a detective on television say something such as "I have a theory to explain Marsha's disappearance." The detective is making an educated guess. A scientist would call this guess a hypothesis.

To a scientist, a theory is much more than a guess. A theory ties together and explains many different kinds of observations. A theory also applies to situations that have not yet happened. If the detective had a scientific theory, he or she would be able to explain why all disappearances occur. The detective would also be able to predict disappearances that had not yet occurred.

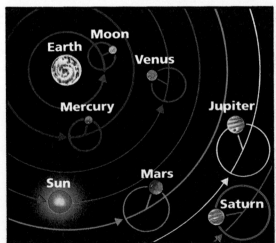 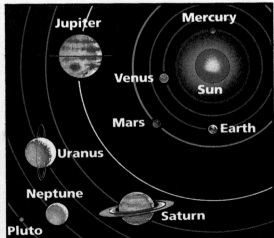

Figure 1–6. Scientific theories may change as new data is found. For example, it was once thought that Earth was the center of the solar system (left). Today we know that the sun is the center (right).

Under certain circumstances, scientific theories may be changed or replaced. Sometimes new observations are made that were not predicted and cannot be explained by a currently accepted scientific theory. Scientists often argue with one another over the meaning of data that does not fit an accepted theory. The scientists who report the new data are sometimes accused of doing poor laboratory work. Other scientists repeat the experiments to verify the new data. If they report the same data, the accepted theory is revised or replaced. Not only must the new theory explain the new data, but it must also explain everything the old theory did.

Scientific theories are never changed without discussion and rechecking of data and experiments. A scientific theory is never replaced unless the theory replacing it is clearly better. You can see that scientists take their theories very seriously. Whenever a scientist uses the word *theory,* he or she means a lot more than just an opinion.

▼ **ASK YOURSELF**

What are the differences among a hypothesis, a law, and a theory?

Communicating Results

Remember the data you collected and the conclusion you drew about your new athletic shoes? Once you have chosen your shoes, you need to communicate your choice. First, you will need to tell the salesperson at the store which shoes you prefer. Then your friends and classmates will see your new shoes. If you wish to, you can even tell people why you chose your shoes.

The final step in any problem-solving method is to communicate results and conclusions. For scientists, this communication occurs in several ways. Scientists may write articles that

Figure 1–7. Scientists usually share their findings and discoveries with other scientists.

ONGOING ASSESSMENT
▼ **ASK YOURSELF**

A hypothesis is an educated guess about what might happen. A law is a summary of experimental results and observations that explains events that happen. A law can be used to make predictions. A theory is an explanation of many different kinds of experimental results, observations, and occurrences that explains why events occur. A theory can also be used to make predictions.

GUIDED PRACTICE

Distribute news or science magazines and newspapers to the students, and ask them to identify any theories they find. Encourage the students to validate their findings by applying them to the meaning of the word *theory* given in the text.

INDEPENDENT PRACTICE

Have the students provide written answers to the Section Review. The students might describe in their journals some common conclusions people make every day and the basis for making those conclusions. An example would be that it will rain because the sky is cloudy.

EVALUATION

Write these statements on the chalkboard. Have the students decide whether the statements are theories and give reasons for their answers. (Neither is a theory because it does not explain why things happen.)

1. Acceleration is the rate of change of speed with respect to time.

2. A small amount of matter can be converted into an enormous amount of energy.

① Communicating informally allows scientists to learn about the latest experimental results without waiting for them to be published or made public.

MEETING SPECIAL NEEDS

Gifted

Suggest that the students write a science fiction story in their journals in which they describe what life would be like if light objects fell a lot faster than heavy objects.

SCIENCE BACKGROUND

The existing theory of falling objects states that heavy and light objects fall at the same rate in a vacuum. The Fischbach paper depended, in part, on the reanalysis of a paper written by Hungarian scientist Baron Roland Eotvos, published in 1922.

Figure 1–8. Ephraim Fischbach reported evidence that lighter objects fall slightly faster than heavier objects in a vacuum.

are published in a scientific journal or magazine. They may write books that report the results of research. Or, scientists may give talks at a meeting of a scientific society. Often, they may do two or even all three of these things.

Scientists also talk to one another about their research. This informal communication network is often as important as articles, books, seminars, and meetings. Why do you think this ① might be so?

As an example of a scientist investigating a problem, consider American physicist Ephraim Fischbach (EE fruhm FIHSH bahk). In 1985 Fischbach first discovered evidence indicating that in a vacuum (a space from which all matter has been removed), lighter objects actually fall a little faster than heavier objects.

Fischbach and his colleagues reported their findings in a scientific journal. When other scientists read about Fischbach's work, there was a great deal of disagreement. Disagreement like this is very common in science. Only after many experiments and much discussion do scientists agree.

Some scientists didn't believe that Fischbach's evidence was strong enough to change the existing theory. These and other scientists conducted experiments to test Fischbach's conclusions. Between 1985 and 1992, literally hundreds of experiments were conducted all over the world. Because most of these experiments did not agree with Fischbach's reports, the theory of gravity was not changed.

You can use scientific methods to solve many problems that occur in your own life. Try the next activity to get a little practice.

RETEACHING

 Cooperative Learning Have the students work in small groups to prepare 10 questions and answers about information in this section. Use the questions and answers in a "Quiz Bowl," having each group compete as a team.

EXTENSION

 Suggest that the students use library references to find out how Belgian scientist Jan Baptiste van Helmont tested his hypothesis about plant growth. Tell the students to outline the steps of a scientific method that van

Helmont used. Have them place their outlines in their science portfolios.

CLOSURE

 In their journals, have the students prepare a list of the terms used in this section and their scientific meanings.

DISCOVER BY *Writing*

Suppose that you are walking home from school and you see a burned log across the sidewalk. Your problem is to determine how the log got there. Write a hypothesis and design an experiment to test your hypothesis. Draw conclusions from your imaginary data. Communicate your results to your classmates. Remember that disagreement is common in the scientific community.

Figure 1–9. Everyday mysteries can be solved using a scientific method.

▶ ASK YOURSELF

Why is disagreement in the scientific community an important part of a scientific method?

SECTION 1 *REVIEW AND APPLICATION*

Reading Critically

1. What are the steps in a scientific method?
2. Why must scientists control variables in their experiments?

Thinking Critically

3. Propose an experiment to answer the question "Which brand of battery lasts the longest?"
4. Define a problem from everyday life, and state a hypothesis that proposes an answer to it. Should this hypothesis be tested in the laboratory or in the field? Explain what difficulties you might encounter in testing the hypothesis.

DISCOVER BY *Writing*

Before the students write their hypothesis, stress that they should consider the evidence available and all reasonable explanations.

ONGOING ASSESSMENT
▼ ASK YOURSELF

Disagreement is important because it leads to more experimentation and new ideas.

SECTION 1 REVIEW AND APPLICATION

Reading Critically

1. The steps in a scientific method are defining the problem, stating a hypothesis, performing experiments, drawing conclusions, communicating results, and suggesting an answer.

2. Scientists must control variables in their experiments to make the data reliable. The variable that caused a change cannot be identified if several variables are changed at the same time.

Thinking Critically

3. Responses should include an experiment that ensures that the proper variables are controlled. Important variables include type of battery, previous use of battery, and load on battery.

4. Hypotheses should clearly reflect the defined problem. The students should give reasons the hypothesis should be tested in the laboratory or in the field. The students should analyze the anticipated difficulties with classmates and brainstorm possible solutions.

Measuring Correctly

Process Skills:
Communicating, Measuring, Interpreting Data

Grouping: Individual or pairs

Objectives
● **Measure** correctly.
● **Compare** measurements.
● **Evaluate** the precision of various measurements.

Discussion
A phenomenon related to the parallax effect is right- and left-eyedness. Ask the students to look at a particular spot several meters away. Have them close one eye and cover the spot with their thumb. Then have the students close the open eye and open the closed eye without moving their thumb. Ask the students to verify that their thumb appears to jump because of the change in angle. The closer the thumb is to the eyes, the bigger the jump appears to be.

▶ Application
1. From above, the volume reading will be consistently

★ PERFORMANCE ASSESSMENT

Have individual students use a centimeter ruler to measure two other items of their choice in the classroom. Evaluate their results by having them repeat their measurement procedure for you.

 After the students have completed the *Skill*, ask them to place their tables and answers in their Science portfolios.

---

lower than from eye level. From below, the reading will be consistently higher than from eye level.

2. The patterns are consistent in that the readings appear to be on the side opposite the side the eye is on. This is the result of parallax.

3. Make sure your eye is on a straight line with the scale to be read.

4. Your eye position should be at the level of the meniscus (when measuring volume) or at the level of a pointer (when measuring mass).

✳ Using What You Have Learned
The amount of milk will be slightly less than what is specified in the recipe. It could make the muffins too dry.

SKILL *Measuring Correctly*

▶ MATERIALS
● graduate, 100 mL ● water ● rubber stopper ● balance ● ruler

▼ PROCEDURE

1. Make a table like the one shown.

TABLE 1: MEASUREMENTS			
	Volume Readings	Mass Readings	Length Readings
Eye level			
Below or left			
Above or right			

2. Fill the graduate about half full of water. Notice that the surface of the water is curved. This curve is called the *meniscus*. To read the volume, look at the water with your eye (a) at the level of the meniscus, (b) below the meniscus, and (c) above the meniscus. Record your readings in the table.

3. Place the rubber stopper on the balance. Look at the pointer with your eye (a) directly lined up with the zero point on the balance, (b) either below or to the left of the zero point, and (c) either above or to the right of the zero point.

(The position depends on the type of balance your school has.) Record your readings in the table.

4. Measure and record the width of this textbook with a centimeter ruler. Read the ruler with your eye (a) directly over the mark, (b) to the left of the mark, and (c) to the right of the mark. Record your measurements.

▶ APPLICATION
1. How did the readings vary as you moved your eye? Why do you think they changed?
2. What patterns do you notice in these variations? Explain why you think these patterns occur.
3. In making measurements, the change that you noticed is called the parallax effect. What can you do to avoid this effect?
4. What does this activity tell you about how you should make measurements?

✳ *Using What You Have Learned*
You are making a batch of muffins for your family. Your little sister is helping you. You see that she is measuring the milk while looking down at the measuring cup. How could this affect your muffins?

FOCUS

Section 2 discusses the International System of Units—SI—as the standard measuring system for scientists. Accuracy and precision in measurement are emphasized as the students become familiar with prefixes that identify different quantities.

MOTIVATING ACTIVITY

Have the students bring to class packaged food items that are nonperishable. On the chalkboard, list the units of measurement printed on the packages. Ask volunteers to distinguish between SI and customary units of measurements on the list. Then initiate a discussion of these two systems of measurement. Tell the students they will learn about SI in this section.

PROCESS SKILLS
- Measuring • Comparing
- Interpreting Data

POSITIVE ATTITUDES
- Precision • Enthusiasm for science and scientific endeavor

TERMS
- density • mass • standard units • volume

PRINT MEDIA
Famous Experiments You Can Do by Robert Gardner (see p. xixb)

ELECTRONIC MEDIA
Meter, Liter, and Gram BFA Educational Media (see p. xixb)

SCIENCE DISCOVERY
- Balance, simple
- Density, hydrometer • How to use a triple-beam balance • Metric conversions
- Metric measurements
- Volume, comparison

BLACKLINE MASTERS
Extending Science Concepts Laboratory Investigations 1.1, 1.2 Connecting Other Disciplines Thinking Critically Study and Review Guide

Measurements

SECTION 2

W hen you go to try on athletic shoes, what is the first thing the salesperson asks you? "What's your size?" If you don't know your size, the salesperson will measure your feet. Then, he or she will bring out several pairs of shoes in your size. What would happen if there were no standard sizes? Each shoe would have to be custom-made!

Shoe sizes are not the only things you measure. You measure your height, your weight, your clothes size, the temperature—it goes on and on. Without measurement, your life would be very different.

Objectives

Identify SI units for commonly measured quantities.

Distinguish between accuracy and precision.

Calculate density.

The Invention of Measurement

No one knows when measurement was first invented, although archaeologists (scientists who study ancient civilizations) can give us a hint about the age of measurement. They have discovered a bone in central Africa that is about 10 000 years old. This bone is important because it has a series of equally spaced marks on it. An analysis of the marks indicates that the bone could have been used as a calendar—to measure time between the phases of the moon. Since early humans were either hunters or farmers, we can hypothesize that they measured time to predict the best hunting or planting seasons.

It is likely that neighboring groups of early humans used the same units of measurement. When everyone agrees to use the same units, they are called **standard units**. It does not really matter what the standard is as long as everyone agrees to use it. Many different types of standard units have been used over the centuries. They have been as varied as the length of a barleycorn and the distance to a star. You can discover more about standard units by doing the next activity.

Figure 1–10. The Ishango bone, shown here, is approximately 10 000 years old. This bone is thought to have been used by early people to measure time.

MULTICULTURAL CONNECTION

Several ancient civilizations developed measurement systems based on units that were the length of certain parts of a person's body. A unit called the cubit was the length of a person's forearm from the elbow to the tip of the middle finger. Archaeologists have found the cubit cut on wooden rods and stone slabs made in Egypt as early as 3000 B.C.

Process Skill: Measuring

Grouping: Individuals

Hints

The longer the distance to be measured, the greater the discrepancy will be among the students' measurements.

▶ **Application**

1. Answers include miscounting, not following the directions, not walking along a designated line, and having feet of different sizes.

2. Measuring three times and calculating the average made the data more reliable.

3. If a standard measuring unit were used, measurements could be compared scientifically.

PERFORMANCE ASSESSMENT

Cooperative Learning

Ask the students to brainstorm other units of measurement that might be used to measure the length of the hall. Example: count the number of tiles in a line from one end of the hall to the other. Ask them to describe their units of measure and determine whether their units are more accurate than the "feet" in the activity. Evaluate their choice of units of measurement for appropriateness.

ONGOING ASSESSMENT
▼ ASK YOURSELF

A standard unit is a unit of measurement everyone in a community or other organization agrees to use.

TEACHING STRATEGIES

● **Process Skill:** *Inferring*

Ask the students to speculate on how a bone might have been useful in measuring standard lengths in earlier cultures. (It is rigid; it would not stretch or shrink; it would not disintegrate for a long time.)

ACTIVITY

How can you measure distance without using a standard unit?

MATERIALS

A hallway with beginning and ending points marked so that everyone can measure the same distance

PROCEDURE

1. Make your own measurement of the chosen distance. Measure the distance in "feet" by placing the heel of one foot against the wall on one side of the hall and then touching the heel of the second foot to the toe of the first foot. Do this along the distance, counting the number of steps you take.

2. Repeat step 1 two times, and average your results. Record your average number of "feet" in the chart your teacher has drawn on the chalkboard.

3. When all students have recorded their measurements on the chalkboard, use the results to answer the Application questions.

APPLICATION

1. What are two reasons that the distance measured by different students produced different numbers of "feet"?

2. Why did you measure three times and calculate the average?

3. How does this activity show the need for standard units?

 ASK YOURSELF

What is a standard unit?

The International System of Units—SI

For thousands of years, small groups of people used similar measures. However, people in different lands used different systems. As travel became more common and trade more important, people found that the different systems of measurement were a problem. This was as true in science as in any other area. For example, when a scientist in Egypt wanted to verify the experiments done by a scientist in China, all the measures had to be converted. Often, this resulted in delays and errors. Something had to be done to standardize measurement!

The problem was solved when a standard system was adopted by scientists all over the world. Today, the International System of Units, or SI, is the standard measuring system for all scientists. The system is called SI from the French name, *le Système International d'Unités.* Using the same standards of measurement makes it easier for scientists to communicate with one another.

SI has units for measuring all types of quantities. The unit is written after the number to show what the number refers to. All measurements must include both a number and a unit of measurement. For example, if someone wants to know your weight and you say 125, the person does not know whether your weight is 125 pounds, 125 newtons, or 125 medium-sized elephants! When recording a number in science, you must always include the unit; otherwise, your measurements and calculations will have no meaning. Common SI units and symbols appear in Table 1–1.

Table 1-1: SI Units and Symbols

Length	
kilometer	km
meter	m
centimeter	cm
millimeter	mm
micrometer	μm

Area	
square kilometer	km²
hectare	ha
square meter	m²
square centimeter	cm²

Mass	
megagram	Mg
kilogram	kg
gram	g
milligram	mg
microgram	μg

Volume of Solids	
cubic meter	m³
cubic centimeter	cm³

Volume of Liquids and Gases	
kiloliter	kL
liter	L
milliliter	mL

Acceleration	
meter per second squared	m/s²

Speed or Velocity	
meter per second	m/s
kilometer per hour	km/h

Frequency	
megahertz	MHz
kilohertz	kHz
hertz	Hz

Density	
kilogram per cubic meter	kg/m³
gram per liter	g/L
gram per cubic centimeter	g/cm³

Force	
kilonewton	kN
newton	N

Energy or Work	
joule	J
kilowatt hour	kWh

Power	
kilowatt	kW
watt	W

Temperature	
kelvin	K
degrees Celsius	°C

● **Process Skill:** *Analyzing*

Have the students explain some standard unit equivalents with which they are familiar. (Examples: 12 in. = 1 ft.; 3 ft. = 1 yd.; 5280 ft. = 1 mi.; 16 oz. = 1 lb.; 60 min. = 1 hr.) Point out the lack of consistency in the conversion factors. Then help the students analyze the organization of the SI units. The students should recognize that SI is based on multiples of 10.

● **Process Skill:** *Communicating*

Encourage the students to become familiar with the prefixes in Table 1–2, since they will encounter them often in the text.

INTEGRATION–*Mathematics*

Ask the students how the exponents of 10 in Table 1–2 are related to the numbers on the left of the equals signs. (The exponent tells the number of places to move the decimal point. For a positive exponent, the decimal is moved to the right; for a negative exponent, the decimal is moved to the left.

$$10^6 = 1\ 000\ 000$$
$$10^{-6} = 0.000\ 001)$$

⬧ **Did You Know?**
Thomas Jefferson first suggested that the United States adopt a system like SI when he was Secretary of State during the 1790s.

DISCOVER BY *Writing*

You might post the students' puns on a bulletin board and encourage the students to add to the list as they think of more puns throughout the year. For examples of puns, see page 61 of *A Random Walk in Science* (Weber, R.L., ed., Crane, Russak & Co., Inc., 1973).

ONGOING ASSESSMENT
▼ **ASK YOURSELF**

Using SI makes it easier for scientists to communicate with each other.

SI works by combining prefixes and base units. Each base unit can be used with different prefixes to define smaller or larger quantities. Each prefix stands for a number by which the base unit is multiplied. For example, the prefix *kilo-* stands for 1000. Therefore, one kilometer (km) is equal to 1000 meters (m). Likewise, the prefix *milli-* stands for one thousandth, or 0.001. Therefore, one milligram (mg) is 0.001 grams (g). Table 1–2 lists common SI prefixes.

Table 1-2: SI Prefixes

Multiplication Factor			Prefix	Symbol	Pronunciation	Term
1 000 000 000 000 000 000	=	10^{18}	exa	E	EHKS uh	one quintillion
1 000 000 000 000 000	=	10^{15}	peta	P	PEHT uh	one quadrillion
1 000 000 000 000	=	10^{12}	tera	T	TEHR uh	one trillion
1 000 000 000	=	10^{9}	giga	G	JIHG uh	one billion
1 000 000	=	10^{6}	mega	M	MEHG uh	one million
1 000	=	10^{3}	kilo	k	KIHL uh	one thousand
100	=	10^{2}	hecto	h	HEHK toh	one hundred
10	=	10^{1}	deka	da	DEHK uh	ten
0.1	=	10^{-1}	deci	d	DEHS uh	one tenth
0.01	=	10^{-2}	centi	c	CEHN tuh	one hundredth
0.001	=	10^{-3}	milli	m	MIHL uh	one thousandth
0.000 001	=	10^{-6}	micro	μ	MY kroh	one millionth
0.000 000 001	=	10^{-9}	nano	n	NAN oh	one billionth
0.000 000 000 001	=	10^{-12}	pico	p	PEEK oh	one trillionth
0.000 000 000 000 001	=	10^{-15}	femto	f	FEHM toh	one quadrillionth
0.000 000 000 000 000 001	=	10^{-18}	atto	a	AT oh	one quintillionth

DISCOVER BY *Writing*

Use Table 1–2 of SI prefixes to help you think of as many prefix puns as you can in five minutes. Here are two examples:

● 1 000 000 000 lings is a form of laughing (*giga*ling or giggling)

● 0.01 rella is a fairy tale heroine (*centi*rella or Cinderella)

Write your ideas in your journal. Share your favorites with your classmates. ✎

▼ **ASK YOURSELF**
What are the advantages of all scientists' using SI?

Measuring Quantities in SI

Any quantity you wish to measure can be measured in SI units. Distance, volume, mass, weight, and temperature are some of the quantities you will be measuring using SI units.

Length   Have you ever bought a new pair of jeans that were too long? What did you do about it? You probably measured (or had someone else measure) the excess length, cut a little of the material off, and rehemmed them. You probably used a measuring tape to make sure they were the same length all the way around.

The most common method for measuring length is to use a meter stick, a ruler, or a measuring tape. You simply put one end of the measuring device at the beginning of what is being measured and read where the end of what is being measured crosses the measuring device.

Figure 1–11. This frame represents a volume of one cubic meter (1 m³).

Length and distance in SI are measured in meters. One meter is about the same as the length of a baseball bat or the height of a doorknob from the floor. Units are created by attaching SI prefixes to the word *meter*. A kilometer, or 1000 m, is about the same distance as 10 soccer fields or 10 football fields. A centimeter, or 0.01 m, is about the same distance as the thickness of your little finger.

INTEGRATION– Mathematics

Remind the students that the volume of a rectangular solid can be determined by using the formula $V = l \times w \times h$. For example, if the length, width, and height of a rectangular solid are each 1 meter, then the volume is:

1 meter × 1 meter × 1 meter = 1 cubic meter = 1 m³.

LASER DISC

450

Metric measurements

447–449

Metric conversions

MULTICULTURAL CONNECTION

In ancient Greece, the "foot" was the most important unit for measuring length. The foot has been widely used in other cultures as well, although its size varies somewhat. For example, the Japanese foot, or *shaku*, is approximately 30.3 cm long; the Nicaraguan *atercia*, 27.9 cm; the Iranian *charac*, 26 cm; and the Anglo-American *foot*, 30.5 cm.

TEACHING STRATEGIES, continued

● **Process Skills:**
Comparing, Predicting

Display 10 objects and ask the students to estimate the mass of each object. Have the students record their estimates on a sheet of paper. Then have volunteers use a balance to determine the actual mass of each object. Finally, have the students compare their estimates with the actual masses.

● **Process Skills:**
Comparing, Analyzing

 Ask the students to examine boxes of dry cereal either at home or in a supermarket. Have the students record the mass listed on three equal-sized boxes of different types of cereal. Encourage the students to record their information in their journals and speculate about why the masses were different.

LASER DISC

455

Volume, comparison

Demonstration

Show the students three different-shaped containers of similar volume. Tell the students that they can determine which container is the largest without using a mathematical formula. Then pour water from a graduated cylinder into one of the containers until it is full. Have a volunteer record the volume of water in each container. Then have the students add the volumes of water to determine the total volume of the container. Finally, have volunteers repeat the process using the other two containers. Ask the students to conclude which container is the largest. (The container that held the greatest volume of water.)

DISCOVER BY *Doing*

 After the students complete the Discover by Doing, you might have them place their data in their science portfolios.

Figure 1–12. In a graduate, the measurement should be read from the middle of the meniscus.

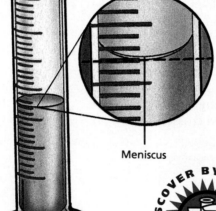

Meniscus

Volume   Quick! Your dad is making breakfast on Saturday morning and is calling to you to measure a cup of milk for the pancake batter. What do you do? You get a measuring cup, pour milk into the cup up to the 1-cup line, and give the milk to your dad.

You measure the volume of liquids all the time. The **volume** of an object is how much space it takes up. A common instrument for measuring liquid volume is the graduate, or graduated cylinder. Like a measuring cup, a graduate has marks on the side that indicate volume. When you read liquid volume, you should follow the same steps you did in the Skill activity on page 14. You can also measure the volume of solids in a graduate. Try it for yourself.

DISCOVER BY *Doing*

Put some water into a graduate, and record the volume. Place a small rock or marble into the graduate, and record the new volume. Subtract the first volume from the second volume. This is the volume of the solid. ✐

The method you just used in the activity is called *displacement* because the solid displaces, or moves, some of the water. The volume of the water that the solid displaced, equals the volume of the solid.

In SI the standard unit of volume is the cubic meter (m^3), which is very large. For many everyday and scientific activities, the cubic centimeter (cm^3) is more useful. A cubic centimeter is about the same volume as about 20 drops of a liquid. A soft-drink can holds about 300 cm^3. However, many types of soft drinks are sold in liter (L) bottles. A liter equals 1000 cm^3. For small volumes, you may use a unit called the milliliter (mL), or 0.001 L. A milliliter and a cubic centimeter are identical measures of volume.

Mass   Suppose you are on the wrestling team. You have an important match today, and you must make sure you have "made weight." That means that you need to check and see if you are within the limits for your weight class. You step on the scale and find that you are right on target! Neither your weight nor your mass has changed.

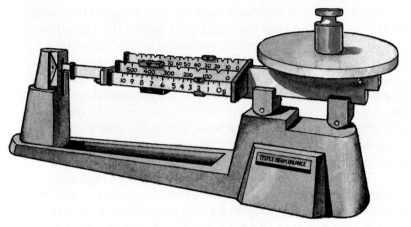

Figure 1–13. A balance, such as this triple-beam balance, is used to measure mass.

Another quantity you measure is mass. **Mass** is the measure of the amount of matter in an object. To measure mass in the laboratory, you use an instrument called a *balance.* Instructions for using a balance may be found on page 630 in the Reference Section.

Wait! Weren't you trying to "make weight" for your wrestling match? Nobody said you needed to "make mass." People often confuse mass with weight, but the quantities are not the same. The mass of an object is the amount of matter in it; mass is a constant value. That means it stays the same no matter where you are. *Weight,* however, is a measure of the force of gravity pulling on the mass of an object. The weight of an object changes depending on the amount of gravitational force acting on it. This is why objects weigh less on the moon than they do on Earth. Because the moon has less gravity, there is less force pulling on the object, making the object lighter. Even though the weight changes, the amount of matter in the object stays the same. Whether on Earth or on the moon, the mass of an object remains constant.

Figure 1–14. All of these objects have approximately the same mass even though they are different sizes.

● **Process Skill:** *Inferring*

Point out that for everyday use and for their laboratory work, the students will use degrees Celsius for measuring temperature. Ask the students why measuring temperature might be important in a laboratory. (Many substances react differently at different temperatures; temperature is a variable that must be controlled in many experiments.)

REINFORCING THEMES
Systems and Structures

Point out to the students that until recently, temperature was measured in the metric system in degrees centigrade. Today, temperature is measured in degrees Celsius, which is the same as centigrade. The term *centigrade* means that the scale is based on 100 divisions. The Celsius unit was named for the scientist who developed it, Anders Celsius, a Swedish astronomer. The kelvin unit for measuring temperature was named for British physicist and inventor, Lord Kelvin, who first conceived of the kelvin thermometer.

✧ Did You Know?

If any matter ever reached 0 K (zero kelvin), all molecular motion would stop and the material would contain no heat energy. This temperature is called absolute zero.

ONGOING ASSESSMENT
▼ ASK YOURSELF

The height of the desk might be measured in meters or centimeters, using a meter stick or a measuring tape.

Well, since you obviously won't be wrestling on the moon, you could consider your mass and your weight the same. For everyday purposes, we consider mass and weight to be the same, although this is only really true at sea level. As you move away from the center of the earth, the pull of gravity decreases. The farther away you are, the less you weigh. How does this fact affect your weight from day to day? If you live in the mountains and you have to wrestle at sea level, you better check your weight carefully. It could go up, even though your mass doesn't change.

The standard SI unit for mass is the kilogram (kg). A 1-L bottle full of water has a mass just a little larger than 1 kg. For many everyday and scientific activities, however, the gram is a more convenient unit to use for mass. A medium-sized paper clip has a mass of about 1 g.

Temperature
A quantity that you must often measure in the science laboratory is temperature. One of the instruments most commonly used to measure temperature is the thermometer.

The kelvin (K) is the SI unit for measuring temperature. Water freezes at 273 K and boils at 373 K. Average body temperature is 310 K.

Kelvin thermometers are very large and cumbersome and are not practical for most common or laboratory uses. Therefore, degrees Celsius (°C) is often used for temperature measurement. To change kelvins to degrees Celsius, subtract 273 from the temperature in kelvins. Water freezes at 0°C and boils at 100°C. Average body temperature is 37°C.

Figure 1–15. A thermometer is used to measure temperature. What is the temperature shown on this thermometer?

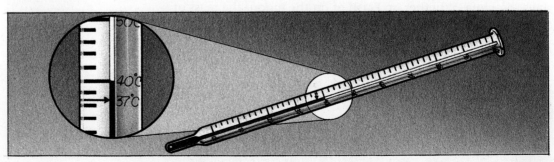

▼ ASK YOURSELF
What unit would you use to measure the height of your desk? What instrument would you use to measure it?

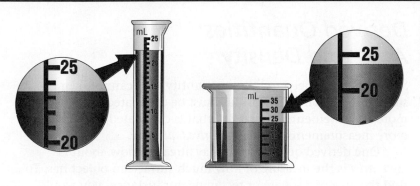

Figure 1–16. These two graduates contain the same volume of liquid. The graduate on the left provides a more precise measurement.

Accuracy and Precision in Measurement

When you measure anything, there are two things you must consider. The first of these is *accuracy.* That is, whether the measured value is the same as the real value. For example, suppose the known mass of an object is 1 kg. If you measure the object's mass and get 1 kg, your measurement is accurate. If you measure its mass and get 2 kg, your measurement is inaccurate.

A second important factor in measuring is *precision,* or the exactness of a measurement. Precision is limited by the smallest division on the scale of the measuring tool. You always estimate to one decimal point beyond the measure marked on the instrument. For example, suppose you used the fat graduate shown in Figure 1–16 to measure volume. Each line on the graduate represents five milliliters (mL). So your measurement could only be precise to 0.5 mL (0.0005 L). The thinner graduate measures to the nearest mL, and it would be precise to 0.1 mL (0.0001 L). The thinner graduate, therefore, gives the more precise measurement.

So, accuracy tells you how good a job you did measuring the liquid. Precision tells you how exact the measuring tool is at showing you the measurement. Accuracy has more to do with the person who is measuring. Precision has more to do with the tool itself.

Both volumes in Figure 1–16 have the same accuracy because they give about the same number. However, the measurement from the thinner graduate is more precise.

Figure 1–17. This person's height is being measured with two different measuring tapes. Which tape gives the more precise reading? Why? ①

▼ **ASK YOURSELF**
What is the difference between accuracy and precision?

In setting up problems on the chalkboard, follow the format used in the text. When you substitute values, express the quantity as a number with units to emphasize that the units are a necessary part of the problem. Then perform the operations to generate the final solution. When writing the steps on the chalkboard, align the equal signs so that the progression from step to step is clear. Encourage the students to use the same orderly and systematic approach when they solve mathematical problems.

Have the students review the equation for calculating the density of an object. Ask a volunteer to state the units for mass and volume (mass: gram; volume: cm^3). Remind the students that 1 cm^3 is equal to 1 mL.

 Have the students provide written answers to the Section Review. Have the students draw a sketch of a room in their home in their journals. Then have them measure the floor area and total volume of the room using metric units.

SCIENCE BACKGROUND

The densities of lead (11.4 g/cm^3) and aluminum (3.7 g/cm^3) are greater than the density of water (1.0 g/cm^3). Therefore, lead and aluminum do not float in water. However, since the density of ice (0.9 g/cm^3) is less than that of water, ice floats in water. Objects that float in water do not float on the very top. They sink until the mass of the water they displace is the same as their mass. This is known as Archimedes' Principle.

 LASER DISC
314

Density, hydrometer

Derived Quantities: Area and Density

Sometimes you need to know a quantity that cannot be measured directly. These quantities must be calculated from two or more measurements. A quantity that is calculated from two or more measurements is called a *derived quantity*.

One derived quantity you may already know about is area. *Area* is the measure of how much surface an object has. To find the area of a square or rectangle, for instance, you would measure the length of two adjacent sides and multiply them. Because area is calculated from two directly measured quantities— the length and the width—area is a derived quantity.

Another property that is derived from measurements is density. **Density** is the ratio of the mass of an object to its volume. Density is often used to help determine what an object is made of. The formula for calculating density is written as follows:

$$\text{density} = \frac{\text{mass}}{\text{volume}} \quad \text{OR} \quad d = \frac{m}{V}$$

Suppose your teacher has given you two objects and has asked you to determine which is more dense without using any equipment. What do you do? Test your ideas in the next activity.

Figure 1–18. By using a balance, you can compare the densities of substances. Since the smaller volume has a greater mass, it must have a higher density.

EVALUATION

Ask the students to calculate the densities of objects with the following masses and volumes:

Object A: 450 g; 30 mL
(15 g/cm^3)

Object B: 2.5 kg; 50 cm³
(50 g/cm^3)

Object C: 550 g; 2 L
(0.275 g/cm^3)

RETEACHING

Provide the students with a list of metric units of measurement, such as millimeter, cubic centimeter, grams, liter, and centigrade. Help them identify the type of quantity that is measured by each unit.

EXTENSION

Interested students might interview a person whose occupation requires taking many measurements, such as a carpenter, meat processor, architect, or engineer. Have them prepare a list of questions prior to the interview, record the person's responses, and report their findings to the class.

CLOSURE

Have the students prepare an outline of the major concepts in this section. (Identify SI units, and determine their measured quantities; distinguish between accuracy and precision in measurement; apply the formula for finding the density of materials.)

DISCOVER BY *Problem Solving*

Your teacher will give you two objects. Determine which is more dense without using any special equipment. Write your method and your results in your journal.

Now you know that you can estimate differences in density. You can also use more precise densities to help identify certain substances.

DISCOVER BY *Calculating*

Imagine that you are an explorer who has discovered a block of something that looks like gold. You measure the volume as 1000 cm³. You use a balance and find that the block has a mass of 19 320 g. Use the formula to find the density of the block. Write your calculations in your journal.

You look up the density of pure gold in a reference book and find that gold has a density of 19.32 g/cm³. Is your block gold? While chances are good that it is, you can't base your answer on density alone. In order to be sure, you would have to do some other tests. You will learn about some of these tests in other chapters.

 ASK YOURSELF

How is a derived quantity different from a measured quantity?

SECTION 2 *REVIEW AND APPLICATION*

Reading Critically

1. Why are standard units of measurement important?

2. Name an instrument used for measuring each of the following quantities: length, mass, liquid volume, and temperature.

Thinking Critically

3. Aluminum has a density of 2.70 g/cm³. What is the mass of a piece of aluminum that has a volume of 100.0 cm³?

4. The width of a table, measured with a ruler that is marked in centimeters, is 65.4 cm (0.654 m). The width of the same table, measured with a ruler marked in millimeters is 652.7 mm (0.6527 m). Which measurement is more precise? Why? Which measurement is more accurate? Why?

DISCOVER BY *Problem Solving*

This activity will help the students assimilate the fact that density is equivalent to the mass in a given volume. Use very diverse items such as a sponge and a marble, or a plastic-foam peanut and a lead sinker.

DISCOVER BY *Calculating*

Using the formula d = m/V, the density of the block is

$$\frac{19 \ 320 \text{ g}}{1000 \text{ cm}^3} = 19.32 \text{ g/cm}^3.$$

ONGOING ASSESSMENT
ASK YOURSELF

A derived quantity is calculated from measurements; a measurement is directly observed.

SECTION 2 REVIEW AND APPLICATION

Reading Critically

1. A standard unit is one that everyone uses. Standard units are important if anything that has to be measured is to be shared between people.

2. Length can be measured with a ruler, mass with a balance, liquid volume with a graduate, and temperature with a thermometer.

Thinking Critically

3. 270 g

4. The measurement 652.7 mm is more precise because millimeters are smaller units of measurement than centimeters. Without knowing the actual width of the table, you cannot tell which measurement is more accurate.

Process Skills: Comparing, Measuring, Interpreting Data

Grouping: groups of 3 or 4

Objectives

● **Use** a balance and a graduate
● **Calculate** density

Pre-lab

Ask the students to predict if a group of pennies will all have the same density.

Hints

The dates on each group's pennies should span at least 30 years. Be sure that each group has some post-1982 pennies. Make sure the pennies will fit in the graduates.

Prepare a chart for the chalkboard similar to the one shown. Include masses by 0.1 g from 2.0 g to 4.0 g.

Mass of Penny (to 0.1 g)	No. of Pennies with this Mass
2.0	
2.1	

Analyses and Conclusions

1. Penny masses are distributed in two groups. The average for each group should be taken as the mass for that group.

2. Yes, the pennies are made of different substances. That is the only conclusion that accounts for different densities, because any wear would remove volume as well as mass or vice versa.

3. If the students do not suggest penny age themselves, suggest it as an option. Older pennies are more dense.

4. The pennies have essentially the same volume but different densities. Therefore, the pennies have different masses.

▶ Application

You could separate brass from gold by calculating the density of each coin.

✳ Discover More

The density of pennies changed in 1980 when the federal government decided it would be less expensive to mint zinc pennies with a copper coating.

Post-lab

Have the students look at their pre-lab predictions. Ask them to revise their predictions based on what they have learned.

INVESTIGATION

Finding Mass, Volume, and Density

▶ MATERIALS

● pennies (10) ● balance ● graduate, 100-mL

▼ PROCEDURE

PART A

1. Make a table like the one shown.

TABLE 1: MASS OF PENNIES	
Penny Number	**Mass of Penny in Grams**
1	
2	

2. Put the pennies on a piece of paper. Write a number on the paper under each penny, and keep them in this order throughout Part A of this investigation.

3. Use a balance to find the mass of each penny in grams as precisely as possible. Round off the values to the nearest 0.1 g. Record the mass of each penny in your table.

4. Your teacher will have a table on the chalkboard that is marked in 0.1 g units. Make a mark in the correct column of the table for each penny your group measures.

5. Compare the masses of the pennies for each group. Are there any noticeable patterns based on mass?

PART B

6. Make two piles of pennies based on the class data. Put each pile on a sheet of paper, and mark the piles "Heavier" and "Lighter." If you cannot decide into which pile a particular penny goes, leave it out.

7. Measure and record the mass of each pile. Write the mass on the paper you are using to identify the pile.

8. Fill a graduate half full of water, and determine the volume as precisely as possible. Record the volume of the water.

9. Carefully put one pile of pennies into the graduate. Measure and record the volume again.

10. Determine the volume of the pennies by displacement. Record the volume.

11. Calculate and record the density of the pile by dividing the volume of the pennies into the total mass of the pennies.

12. Record the density of the pennies on the piece of paper that identifies the pile.

13. Repeat steps 8 through 12 for the other pile of pennies.

14. Record the density of each pile of pennies in the table on the chalkboard.

▶ ANALYSES AND CONCLUSIONS

1. When the pennies were massed, two groups were identified. What was the average mass of each group?

2. Do the measured densities reported by all laboratory groups show two groups of pennies? Explain.

3. Are there any clues that would allow you to separate the pennies into the same two groups without massing them? Explain.

4. How is it possible for the pennies to have different masses?

▶ APPLICATION

You are given 50 coins, some of which are gold. The others are brass. Since the coins look alike, how would you separate them?

✳ *Discover More*

Once you have identified the difference between the heavier and lighter pennies, find out when and why the change occurred.

CHAPTER 1 HIGHLIGHTS

The Big Idea— Systems and Structures

Lead the students to understand that scientists use certain steps, known as a scientific method, to solve problems. Often, scientists need to make measurements. They uniformly use SI in their collection of data.

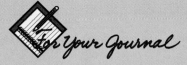

For Your Journal

The students should recognize that using the steps of a scientific method would make their problem solving more scientific.

CHAPTER 1 HIGHLIGHTS

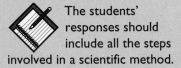

CONNECTING IDEAS

The students' responses should include all the steps involved in a scientific method.

The Big Idea

Science is using methods that involve observing, thinking, and formulating and reformulating ideas about the things around us. Science is using special instruments, solving problems, and trying out new methods of operations. Science is the constant search for information. Scientists must respect evidence that supports new ideas and think carefully about those ideas. They must design fair tests to see whether their ideas are correct.

Measurement is an important part of science. The use of standard units allows scientists to communicate their data to one another easily. Units of measure in SI enable scientists to describe things that are microscopically small and astronomically large.

For Your Journal

Look back at the solution to your problem that you entered into your journal at the beginning of the chapter. Reevaluate how your solution was like a scientific method. How could you have made your problem solving more scientific?

Connecting Ideas

This flowchart shows how scientists might use a scientific method. Remember that measurement is an important part of collecting data. Copy the chart into your journal. Use it to show how you would solve a problem in your everyday life. As information is collected, the problem may have to be redefined, new experiments designed, or further observations made. Scientific methods are not one-way streets.

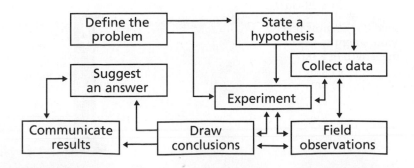

CHAPTER

1

REVIEW

Understanding Vocabulary

Match each term with the correct definition.

1. explanation for a group of related results, observations, and occurrences
2. ratio of the mass of an object to its volume
3. educated guess about what might happen
4. summary of experimental results that can be used to predict what will happen and why they happen the way they do
5. amount of matter in an object
6. judgment based on the analysis of data

a. conclusion (8)
b. density (24)
c. hypothesis (5)
d. law (9)
e. mass (21)
f. theory (9)

Understanding Concepts

MULTIPLE CHOICE

7. In an experiment, controlled variables are
   a) allowed to change.
   b) limited in number.
   c) kept the same.
   d) difficult to change.

8. A person says, "I think that a steel ball bearing and a marble will hit the ground at the same time if I drop them from the same height at the same time." That person is stating a
   a) hypothesis.   b) law.
   c) theory.   d) model.

9. A rock has a volume of 5.43 cm^3 and a mass of 17.86 g. Its density is
   a) 0.30 cm^3/g.   b) 0.30 g/cm^3
   c) 3.29 g/cm^3.   d) 3.29 cm^3/g.

10. Which would be the most precise unit of measurement to use to measure the length of a car?
    a) cm^3 b) mm
    c) cm   d) km

SHORT ANSWER

11. Why are degrees Celsius commonly used in the laboratory rather than kelvin temperatures?

12. What is the difference between an object's mass and its weight?

13. Explain the differences among a hypothesis, a conclusion, a theory, and a law.

Interpreting Graphics

14. The same volume of water was added to each graduated cylinder. What is the volume of the marble in the second cylinder?

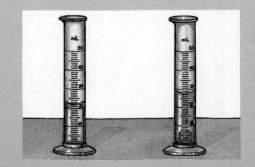

Reviewing Themes

15. The two cubes have different densities.

16. A measuring device can be made more precise by decreasing the size of its units of measurement.

Thinking Critically

17. The block with the greater volume is three times less dense than the other block.

18. The variable being tested is darkness (or lack of sunlight). Students' experimental designs will vary. Variables to be controlled include the types of plants used, where they are located, the amount of water they receive, and the surrounding temperature of the plants.

19. Criteria for evaluating answers should include thought, insight, unusual examples, and explanations that are sensible.

20. The density of the snowball can be increased by packing the snow closer together, thereby decreasing its volume, or by adding another, more dense substance to the middle and increasing its mass.

21. Possible answers include that other scientists might be working on similar experiments and will want to compare the results; to modify the experiments to see whether they will work under different conditions; to evaluate the reporting scientists' reasoning; to provide additional evidence to support the reporting scientists' ideas.

22. The watch is the most accurate for telling the time of day. The stopwatch is the most accurate for telling how much time has passed. The stopwatch is the most precise because it measures in the smallest unit (0.1 s).

23. The wristwatch is the most practical because it tells you the time of day (the stopwatch doesn't), and it is portable (the sun dial isn't). The wristwatch is both accurate enough and precise enough to provide you with the time information you need.

Reviewing Themes

15. *Systems and Structures*
How can two cubes of equal volume have different masses?

16. *Systems and Structures*
How can a measuring device be made more precise?

Thinking Critically

17. Two blocks of wood have the same mass, but the volume of one of the blocks is three times greater than the other. Compare the densities of the two blocks.

18. Jack wishes to determine the effect of darkness on the leaf color of plants. Design an experiment for Jack that controls the necessary variables. Which variable is being tested?

19. How might your life be different if there were no standard system of measurement?

20. How could you increase the density of a snowball without adding more snow to it?

21. Why is it important for scientists to communicate the results of their research to other scientists?

22. Look at the three timepieces shown below. Which is the most accurate? Which is the most precise? Explain your answers.

23. Which of the timepieces is the most practical? Explain why you think so. Include accuracy and precision in your answer.

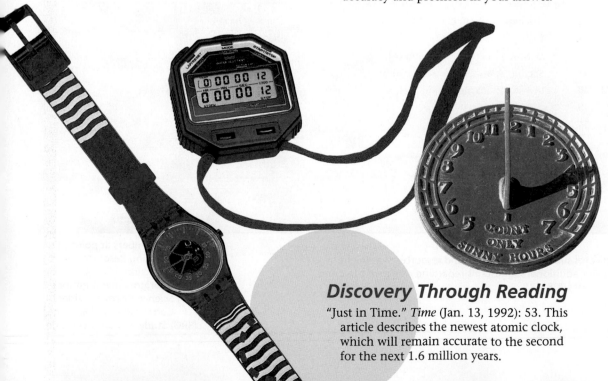

Discovery Through Reading

"Just in Time." *Time* (Jan. 13, 1992): 53. This article describes the newest atomic clock, which will remain accurate to the second for the next 1.6 million years.

FUNDAMENTAL PROPERTIES OF MATTER

PLANNING THE CHAPTER

Chapter Sections	Page	Chapter Features	Page	Program Resources	Source
CHAPTER OPENER	30	**For Your Journal**	31		
Section 1: MATTER	32	Discover by Observing **(A)**	33	*Science Discovery**	SD
• States of Matter **(B)**	32	Discover by Doing **(B)**	33	Extending Science Concepts:	
• Pure Substances **(A)**	34	Discover by Doing **(A)**	35	Determining Solubility **(A)**	TR
• Mixtures **(A)**	36	Discover by Researching **(B)**	37	Investigation 2.1: Using Solubility	
		Section 1 Review and Application	38	to Solve a Mystery **(A)**	TR, LI
		Skill: Hypothesizing **(A)**	39	Study and Review Guide,	
				Section 1 **(B)**	TR, SRG
Section 2: CHARACTERISTICS OF MATTER	40	Activity: How can properties be used to identify unknown substances? **(A)**	41	*Science Discovery**	SD
• Properties of Matter **(B)**	40			Reading Skills: Solving Problems **(B)**	TR
• It's a Matter of Change **(B)**	43	Discover by Doing **(A)**	44	Connecting Other Disciplines: Science and Architecture, Studying Models of Buildings **(A)**	TR
		Section 2 Review and Application	45	Investigation 2.2: Using Models in Science **(A)**	TR, LI
				Thinking Critically **(B)**	TR
				Physical Properties of Matter **(A)**	IT
				Study and Review Guide, Section 2 **(B)**	TR, SRG
Section 3: A MODEL OF MATTER	46	Discover by Researching **(A)**	48	*Science Discovery**	SD
• Models in Science **(A)**	46	Discover by Writing **(B)**	49	Phases of Matter **(B)**	IT
• Moving Particles as a Model of Matter **(H)**	47	Discover by Problem Solving **(A)**	49	Record Sheets for Textbook Investigations **(A)**	TR
• Molecules and Physical Properties **(A)**	48	Section 3 Review and Application	51	Study and Review Guide, Section 3 **(B)**	TR, SRG
• Molecules and Physical Changes **(A)**	49	Investigation: Measuring Melting and Freezing Points **(A)**	52		
• A Change of Phase **(B)**	51				
Chapter 2 HIGHLIGHTS	53	The Big Idea	53	Study and Review Guide, Chapter 2 Review **(B)**	TR, SRG
Chapter 2 Review	54	For Your Journal	53		
		Connecting Ideas	53	Chapter 2 Test	TR
				Test Generator	

B = Basic **A** = Average **H** = Honors
The coding Basic, Average, and Honors indicates subsections, features, and resources that might be
appropriate for different levels of learners. For additional suggestions regarding choice of topic and
depth of coverage, see the Pacing Chart on pages T28–T31.

*Frame numbers at point of use
(TR) Teaching Resources, Unit 1
(IT) Instructional Transparencies
(LI) Laboratory Investigations
(SD) *Science Discovery* Videodisc
     Correlations and Barcodes
(SRG) Study and Review Guide

CHAPTER MATERIALS

Title	Page	Materials
Discover by Observing	33	(per individual) journal, bottle of perfume
Discover by Doing	33	(per individual) journal
Discover by Doing	35	(per group of 3 or 4) paper cups (3), pieces steel wool (3), egg, vinegar (20 mL), water
Teacher Demonstration	36	sugar, salt, alcohol, filter
Discover by Researching	37	(per individual) journal
Skill: Hypothesizing	39	(per group of 3 or 4) materials will vary according to hypotheses tested
Activity: How can properties be used to identify unknown substances?	41	(per group of 3 or 4) small bottles with tops (4), unknown liquids (4), paper cups (4)
Discover by Doing	44	(per individual) plastic graduate, freezer, beaker, small sealable plastic bag
Teacher Demonstration	47	large beaker, water, chalk dust, hand lens
Discover by Researching	48	(per individual) journal
Discover by Writing	49	(per individual) journal
Discover by Problem Solving	49	(per individual) balloon, freezer, tape measure
Investigation: Measuring Freezing and Melting Points	52	(per group of 3 or 4) safety goggles, laboratory apron, ring stand, buret clamps (2), 500-mL beaker, hot plate, large test tube, moth flakes, thermometers in notched corks (2), stirring rod, stopwatch, graph paper

TEACHING SUGGESTIONS

Field Trip
Plan a neighborhood walking tour to locate, observe, and record information about different types of matter. Have the students identify the type of matter and note properties of each type. In the classroom the students can share and discuss their observations.

Outside Speaker
Consider the resources in your community. If there are any oil companies, mining companies, or water treatment plants, you might have someone speak about separation techniques in that industry.

CHAPTER THEME—*SYSTEMS AND STRUCTURES*

Chapter 2 links the structure of matter and its four phases to the kinetic molecular model of matter. The concepts of the atom and its motion are integral to future discussions of chemical reactions, heat energy, and radioactivity. A supporting theme in this chapter is **Environmental Interactions.**

MULTICULTURAL CONNECTION

Mt. Everest—the highest mountain in the world—rises about 8.9 km above sea level. It was named for Sir George Everest, a British surveyor general of India. The mountain is called Sagarmatha by Nepalese and Chomolungma by Tibetans.

The scaling of Mt. Everest by Sir Edmund Hillary of New Zealand and Tenzing Norgay of Nepal in 1953 began a lifetime of friendship between Hillary and the Sherpa people. He and his expedition members devoted much of their time between climbs to the construction of municipal buildings for the people of Nepal, such as schools, an area hospital, an air strip, and sanitary water systems.

MEETING SPECIAL NEEDS

Second Language Support

Use as many illustrations as possible while discussing phase changes, thermal expansion, and the kinetic molecular model. Encourage limited-English-proficient students to draw and label these illustrations in their journals.

CHAPTER

2

FUNDAMENTAL PROPERTIES OF MATTER

Water has unique properties that make it both useful and dangerous. Not even scientists study water's properties as carefully as a climber picking a path through the debris of last week's avalanche.

"**W**e started traversing the glacier sometime after midnight, before the sun could soften the snow. But somehow we have underestimated the distance to the col, or the weight of the packs, or the sheer effort of moving at 18,000 feet. All this slows us down, and now the sun moves across the sky toward noon faster than we are heading toward the col. I prefer traveling across the glacier at night: the headlamp illuminates just those next ten feet of space in front of you. Daylight illuminates too much: the

CHAPTER MOTIVATING ACTIVITY

Place a sample of iron-fortified cereal in a 1 L beaker, and crush it into small pieces. Add enough distilled water to cover the cereal. Then drop in a magnetic stir bar, and place on the stirrer. After a few minutes, have the students examine the stir bar.

Iron should have collected on the magnet. You might extend this activity by pointing out to the students the characteristic properties, mixtures, separation techniques, product labels, elements, and compounds used in this demonstration.

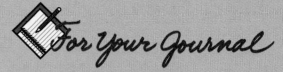

For Your Journal

The journal questions will help you identify any misconceptions the students might have about the properties of matter. In addition, these questions will help the students realize that they do possess knowledge about properties of matter. (The students will have a variety of responses to the journal questions. Accept all answers and tell the students that at the end of the chapter they will have a chance to refer back to the questions.)

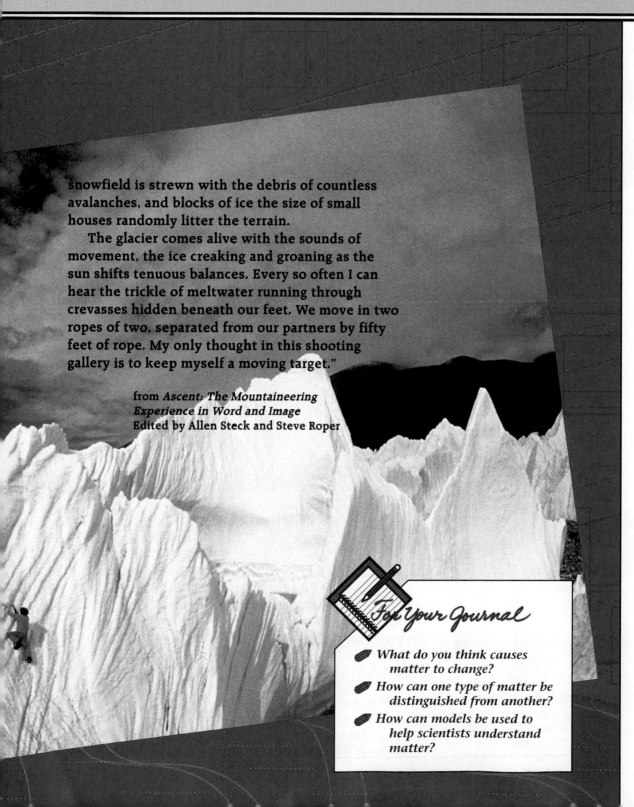

snowfield is strewn with the debris of countless avalanches, and blocks of ice the size of small houses randomly litter the terrain.

The glacier comes alive with the sounds of movement, the ice creaking and groaning as the sun shifts tenuous balances. Every so often I can hear the trickle of meltwater running through crevasses hidden beneath our feet. We move in two ropes of two, separated from our partners by fifty feet of rope. My only thought in this shooting gallery is to keep myself a moving target."

from *Ascent: The Mountaineering Experience in Word and Image*
Edited by Allen Steck and Steve Roper

For Your Journal

- What do you think causes matter to change?
- How can one type of matter be distinguished from another?
- How can models be used to help scientists understand matter?

ABOUT THE PHOTOGRAPH

Mountain climbing—also called mountaineering or Alpinism—is the sport of climbing to the summit of a mountain or hiking in mountainous areas. While mountain climbing can be a rewarding, exciting experience for both men and women, proper training, good judgment, experience, and skill are essential. Accidents are relatively rare and happen primarily to people who lack adequate experience, training, and the proper equipment.

Mountain climbers should always climb with a partner or in a group. Climbers must be aware of the dangers of bad weather and avalanches in some areas. They should be familiar with cold-weather gear and, depending upon the trip, camping procedures.

Climbers should be skilled in a safety procedure called belaying—a method of stopping a fall in case a climber slips while ascending or descending a slope. In belaying, the leader goes first while the companion(s) release the rope from a secure position. While advancing, the leader secures the rope to fixed locations, such as a projecting rock. This procedure provides security until the leader reaches a ledge and can help the other climbers advance.

FOCUS

This section describes and includes examples of the four states of matter: solid, liquid, gas, and plasma. In addition, this section also notes that matter can be classified by its state, purity, and properties.

MOTIVATING ACTIVITY

Cooperative Learning Play a version of "Twenty Questions" with the students. Ask a volunteer to think of, but not name, a simple solid or liquid, such as milk or bread. Then have the students take turns asking questions that can be answered yes or no to determine whether the word is thought of as a solid or a liquid. Repeat the procedure as time permits. Conclude by pointing out to the students that in this section they will learn about types of matter and their characteristics.

PROCESS SKILLS
- Formulating Hypotheses
- Experimenting

POSITIVE ATTITUDES
- Curiosity • Initiative and persistence
- Cooperativeness

TERMS
- pure substance • element
- compounds • mixture

PRINT MEDIA
Solids, Liquids and Gases: From Superconductors to the Ozone Layer by Melvin Berger (See p. xxib)

ELECTRONIC MEDIA
All About Matter— Elements: The Skeleton of Matter; Compounds: A Special Combination; Mixtures: Separate, Yet Together Focus Educational Media, Inc. (see p. xxib)

SCIENCE DISCOVERY
- Compounds, summary
- Desalination plant
- Evaporation; lab setup • Water cycle • Water properties

BLACKLINE MASTERS
Extending Science Concepts
Laboratory Investigation 2.1
Study and Review Guide

SECTION 1

Matter

Objectives

Describe the ways matter can be classified.

Distinguish between elements and compounds.

Propose a method for separating the components of a mixture.

Melting icicles drip onto the frozen ground. Grass and leaves glisten with dew on a cool morning. In a junkyard, automobiles silently rust. Overhead, the night sky is illuminated by exploding fireworks. These are just a few examples of matter changing.

You see many more changes in matter every day. The study of the ways in which matter changes is an important part of science. Learning to predict how matter will change has helped scientists produce useful products, such as nylon, plastics, and medicines. Changes in matter will help produce new products in the future.

Figure 2–1. This wulfenite crystal is one example of a solid.

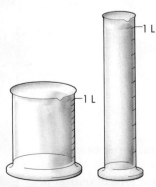

States of Matter

Matter can be classified by its state, purity, and properties. The glacier and the climbers you just read about are all made of matter. Even the air you breathe is made of matter. Matter is anything that has mass and takes up space. It is divided into four states, or phases: *solid, liquid, gas,* and *plasma.*

A solid is anything that has both a definite volume and a definite shape. A rock is a solid, as is a baseball bat and a crystal of salt. In most cases, solid matter can be picked up and carried around without having to place it in a special container.

A liquid is anything that has a definite volume but no definite shape. A key property of liquids is that they flow and can be poured. Liquids take the shape of any container into which they are poured. Look at the photograph, and you will see that if you pour a liquid into a short, wide container, such as a beaker, it will form a short, wide shape. If you pour the same amount of liquid into a tall, narrow container, it will have a tall, narrow shape.

A gas has no definite volume and no definite shape. A gas always takes both the volume and the shape of any container into which it is placed. Try the next activity to find this out for yourself.

Figure 2–2. The liter of liquid shown here has a different shape depending on the graduate in which it is placed.

1 L

1 L

● **Process Skill:**
Classifying/Ordering

Make a list on the chalkboard of the four states of matter and their characteristics. Ask the students how matter differs from energy. (Matter is anything that has mass and takes up space. Energy is anything that causes change in matter.)

● **Process Skills:**
Classifying/Ordering,
Comparing

Have the class brainstorm to create lists of solids, liquids, and gases. Encourage the students to concentrate on both geometric shape and size for solids. You might have the students examine salt crystals through a

hand lens. Point out that powders can be poured, and have the students determine in which category powders would be placed. Compare different liquids with a variety of surface tensions, such as a drop of alcohol and a drop of mercury. Discussion of gases could cover diffusion, as in perfume or

smoke, and pressure of air in tires or of bottled gases (which are liquified under pressure).

DISCOVER BY *Observing*

Your teacher is going to place an open container of perfume in the classroom. In your journal, write down when you first smell the perfume. Does the odor of perfume grow stronger or weaker compared with when you first began to smell it? What does this tell you about gases?

On the earth, a sample of gas would spread throughout the atmosphere. In fact, the atmosphere itself is a mixture of gases that is held near the earth by gravity.

Throughout your day you come into contact with solids, liquids, and gases. This probably isn't much of a surprise to you. However, did you ever stop to consider how often you eat or drink different states of matter?

DISCOVER BY *Doing*

Make a list of all the foods and beverages served in the school cafeteria at lunch today. Then, make a table classifying these foods as solids, liquids, gases, or combinations of these states of matter. Explain how you determined the category for each food. Compare your results with a classmate's. How do your results differ? How are they the same?

DISCOVER BY *Observing*

The smell should grow weaker as the molecules diffuse. How long it takes to smell the perfume indicates the speed of the gas molecules. Note: Some individuals are allergic to perfumes. Allow these students to forgo this activity.

INTEGRATION–
Mathematics

The particles that make up matter are so tiny that a single bubble from a soft drink may contain about 100 000 000 000 000 000 (1×10^{17}) carbon dioxide molecules.

DISCOVER BY *Doing*

PORTFOLIO ASSESSMENT

Examples of foods and beverages might include bread, milk, cheese, pasta, meat, soft drinks, juice, salads, and cookies. Discuss possible combinations, such as pizza (solid topping, liquid sauce), gelatin (liquid in a solid), and soft drinks (gas in a liquid). Have the students place their lists in their science portfolios.

● **Process Skill:**
Classifying/Ordering

If necessary, further clarify the difference between elements and compounds. Use names of some common compounds such as water, salt, sugar, and carbon dioxide, and discuss the ele-ments that combined to form each. Stress to the students that compounds cannot be sepa-rated into their component elements by physical means. Mention some chemical means of separation for comparison, such as electrolysis for water and reaction with acid for sugar.

SCIENCE TECHNOLOGY SOCIETY Working with other scientists, Nobel Prize-winning American scientist Irving Langmuir discovered that some substances reach the plasma state at temperatures between 2800°C and 56 000°C.

Nuclear fusion occurs in plasmas and might be a future energy source. Ask the students why gas in fluorescent tubes or a lightning bolt may be examples of plasmas. (There is no definite volume or shape, and each consists of elec-trically charged particles of high energy.)

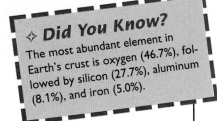

✧ Did You Know?
The most abundant element in Earth's crust is oxygen (46.7%), fol-lowed by silicon (27.7%), aluminum (8.1%), and iron (5.0%).

ONGOING ASSESSMENT
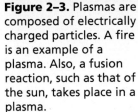 **ASK YOURSELF**

A plasma has no definite volume or shape.

MEETING SPECIAL NEEDS

Mainstreamed

Have the students draw lines on a large sheet of paper so they have four sections and label the sec-tions "Solids," "Liquids," "Gases," and "Plasmas." Then have the stu-dents find and cut out pictures from periodicals that are exam-ples of each type of matter.

Figure 2–3. Plasmas are composed of electrically charged particles. A fire is an example of a plasma. Also, a fusion reaction, such as that of the sun, takes place in a plasma.

The fourth state of matter, plasma, has no definite volume or shape and is composed of electrically charged particles. These charged particles are generally associated with large amounts of energy. A flame is one example of a plasma. The sun is also a plasma.

 ASK YOURSELF
How is a plasma like a gas?

Pure Substances

The second way matter is classified is by its purity. If matter is a **pure substance**, it contains only one kind of molecule. Mole-cules are small particles that make up matter. They are made up of even smaller particles, called *atoms.*

It's Elementary Aluminum, copper, and charcoal are each made of only one kind of atom. A pure substance that is made of only one kind of atom is called an **element.** Copper wire is made up of only copper atoms, and charcoal is made up of only carbon atoms.

If you looked for pure elements in nature, you would be lucky to find more than five or six. Why is this? It is not be-cause there are few elements to look for. There are more than 100 elements, and 92 of them can be found in nature. So, you see, there are plenty of elements to find.

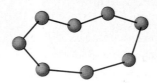

Figure 2–4. Iodine (left) and sulfur (right) are examples of elements. The drawings show their chemical makeup and structure. Each element is made up of only one type of atom.

Elements—Compounded Daily

The reason it is hard to find pure elements in nature is that they usually react to form compounds. **Compounds** are pure substances whose molecules are made of two or more different kinds of elements joined together. You can see examples of elements reacting to form compounds by trying the next activity.

 ᴅɪscovᴇʀ ʙʏ *Doing*

You will need three pieces of steel wool, three cups, an egg, vinegar, and water. Label the cups "Egg," "Vinegar," and "Water." Put a piece of steel wool in each cup. Then add the egg, vinegar, or water. Wait until the next day, and remove the steel wool from the cups. Rinse off the steel wool. Then record what you observe. 🐚

In the activity, you found that elements react when exposed to some chemicals. Molecules of elements and compounds are often represented by colored balls. Each color represents a specific kind of atom. When different kinds of atoms (colored balls) are combined, they make compounds.

Water is a common example of a compound. Water is a transparent, colorless liquid at room temperature. However, water is made of the elements hydrogen and oxygen, both of which are gases at room temperature. As you can see, compounds do not necessarily have the same characteristics as the elements from which they are made.

● **Process Skills:** *Generating Questions, Generating Ideas*

Point out to the students that the search for a greater understanding of matter goes at least as far back as the sixth century B.C., when a Greek named Thales theorized that all substances came from water and would eventually revert to water. Two hundred years later, Aristotle announced his theory that all matter was composed of four elements—fire, air, water, and earth—and that each of these elements was thought to have two of four basic properties of hot, cold, wet, and dry.

● **Process Skill:** *Comparing*

Not until the seventeenth century did Robert Boyle, a British pioneer in the use of scientific method, disprove Aristotle's theory and provide the basic definition of an element. Soon after, a French scientist, Antoine Lavoisier, set up the basic standard for testing whether or not a substance would fit Boyle's definition. Encourage the students to do research to compare Aristotle's theories to Boyle's theories. Ask the students to place their research in their science portfolios.

Demonstration

Demonstrate how sugar and salt can be separated by putting the mixture in alcohol. (Hint: If you prefer, sand can be used in place of sugar.) The salt will dissolve, but the sugar will not. Filter out the sugar, and then evaporate the alcohol to recover the salt. Follow up by asking the students why sugar and salt cannot be separated by putting the mixture in water and then evaporating the water. (Both sugar and salt dissolve in water. They would still be mixed after evaporating the water.)

Oxygen

Hydrogen

Carbon

H₂O

$C_{12}H_{22}O_{11}$

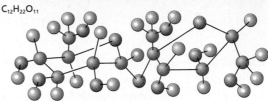

Figure 2–5. Water (left) and sucrose (right) are examples of compounds. Water is a compound made of hydrogen and oxygen atoms. Sucrose is a compound made of hydrogen, oxygen, and carbon atoms. In this photograph, coloring has been added.

One important characteristic of compounds is that they cannot be separated into the atoms they are made of by physical methods. The atoms in a compound are joined chemically. Water, for example, can be frozen or boiled, but the ice or water vapor will still be water. Freezing or boiling cannot separate water into hydrogen and oxygen atoms or molecules. Other physical methods such as straining, filtering, passing the water through a magnetic field, or dissolving it in alcohol cannot separate water into hydrogen and oxygen atoms or molecules.

ONGOING ASSESSMENT
▼ ASK YOURSELF

The two categories of pure substances are elements and compounds.

▼ ASK YOURSELF
What are the two categories of pure substances?

Mixtures

Most matter is not pure but is made of a combination of two or more types of substances. These combinations of matter are called mixtures. A **mixture** is a combination of two or more different kinds of molecules. The key characteristic of a mixture is that each kind of molecule in the mixture keeps its own identity. This means that if you take a mixture apart, the molecules of the substances will be the same as they were before you put them together.

GUIDED PRACTICE

Have the students write a definition in their own words for each vocabulary term and related science term in this section. Encourage the students to also provide at least one example for each definition.

INDEPENDENT PRACTICE

 Have the students provide written answers to the Section Review. In their journals, have the students write brief clues that could be used to identify a substance as a solid, liquid, gas, or plasma.

EVALUATION

 Cooperative Learning Have the students form small groups. Each group should compile a list of pure substances that are in a mixture. Then have the groups trade lists and write a hypothesis stating how they might separate the components in the mixtures listed.

Many mixtures occur in nature. For example, air is a mixture of gases made up mostly of nitrogen and oxygen with small amounts of other gases. The human body is a complex group of mixtures. Your body is mostly water mixed with proteins, fats, minerals, and carbohydrates. A pizza can also be thought of as a mixture. This mixture includes dough, sauce, cheese, and toppings.

In some mixtures, however, it is difficult to see the different substances, or parts. For example, it is difficult to see the sugar in a mixture of sugar and salt because both solids are white crystals.

One way to identify a mixture is to separate it into its different substances. Unlike compounds, mixtures can be separated into their parts using physical methods. For example, you could use a fork to separate the parts of a pizza. The method does not change the characteristics of the parts. Not all mixtures are as easy to work with as the pizza. Some physical methods of separating parts of a mixture are shown in the table below. These are just a few ideas, and some methods work better than others, depending on the type of mixture.

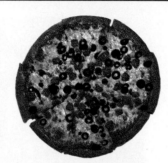

Table 2-1:	Some Physical Methods of Separating Parts of a Mixture
Boil away (evaporate) water leaving solids
Collect metallic parts with magnet
Pick out solid parts by hand or with tweezers
Separate solids from liquids by filtering or straining
Dissolve in water or other liquid

Try the next activity. You'll have a chance to discover more about mixtures, substances, compounds, and elements.

 **Researching** _____

 Look at the labels on items that can be purchased at a grocery store or a drugstore. Use the information to find two examples of mixtures and two examples of pure substances. Use resources in your school library to find out whether the pure substances are compounds or elements. Record your results. ✎

MEETING SPECIAL NEEDS

Mainstreamed

 When discussing separation techniques, you might perform some of them for the students and name the equipment used. Have students draw the equipment in their journal or list each the equipment needed.

 SCIENCE TECHNOLOGY SOCIETY Reverse osmosis is one method used to desalinate sea water. The process places very high pressure on the sea water, forcing it against a semipermeable membrane. Membranes currently used are made of cellulose acetate or nylon. They allow pure water, but no salts, to pass through. Activated carbon filters are installed in front of the membrane to remove large particles first. The process is fairly slow, but the level of water purity is very high.

LASER DISC

2517, 2518, 2519, 2520

Desalination plant

320, 321, 322, 323, 324, 325

Evaporation, lab setup

DISCOVER BY *Researching*

Students' examples of mixtures and pure substances might include cake mix (a mixture), salt (a pure substance), and soft drinks (a mixture).

RETEACHING

 Have the students create a flowchart in their journals beginning with pure substances divided into elements and compounds. Each of those units should then be divided into solids, liquids, and gases.

EXTENSION

 Have the students research a specific molecule and determine its structure. Molecules such as waxes, DNA, saturated fat, or chocolate will reveal long, complicated structures. Ask the students to place their research in their science portfolios.

CLOSURE

Cooperative Learning Assign a portion of Section 1 to each of four groups of students. Direct the groups to prepare an oral summary of its subsection and present it to the other groups.

ONGOING ASSESSMENT
▼ ASK YOURSELF

Examples might include granola-type cereals, fruit or vegetable salads, or cookies. A mixture can be identified by separating it into its different substances.

SECTION 1 REVIEW AND APPLICATION

Reading Critically

1. All matter has mass and takes up space.

2. The four states of matter are solids, liquids, gases, and plasmas. Solids have a definite shape and volume. Liquids have a definite volume but no definite shape. Gases have no definite volume and no definite shape. Plasmas are composed of electrically charged particles and have no definite volume or shape.

Thinking Critically

3. The molecule on the right represents an element because all the atoms in it are the same. The molecule on the left represents a compound because it is made up of two different kinds of atoms.

4. Since the elements found in the earth are mostly found as compounds or as mixtures of compounds, separation techniques are vital to obtain raw materials. Examples include the separation of metal ores and crude oil.

Figure 2–6. This reverse osmosis plant uses high pressure and filters to separate salt from sea water.

▼ ASK YOURSELF

List three examples of mixtures. How do you know they are mixtures?

SECTION 1   REVIEW AND APPLICATION

Reading Critically

1. What are two properties common to all matter?

2. Name the four states of matter, and describe how they differ from one another.

Thinking Critically

3. Pictured here are models of two molecules of different substances. Which one is an element? Which one is a compound? Explain how you can tell the difference.

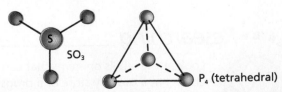

SO_3

P_4 (tetrahedral)

4. The ability to separate mixtures is important for modern society. Explain this statement, and give at least two examples.

SKILL

Hypothesizing

Process Skills:
Formulating Hypotheses, Experimenting

Grouping: Groups of 3 or 4

Objectives
- **Formulate** a Hypothesis.
- **Experiment** to test the hypothesis.

Discussion

The choice of materials needed for experimenting will be determined by the hypothesis. Emphasize the importance of first writing the hypothesis and then, while testing, making careful observations. If the students are at first unsuccessful, they might write another hypothesis and continue experimenting. Discuss with the students how they know that the final products are the same as the substances in the original mixture. Is there a way they can test them? One could test the magnetic property of iron. For some substances, boiling point and melting point could be tested. This discussion could lead into Section 2, where characteristic properties are introduced.

▶ Application

Students might say that they learned one or more ways to separate mixtures, depending upon the materials in the mixture. Accept all reasonable responses the students can justify.

✳ Using What You Have Learned

To write a hypothesis, first determine the problem to be solved. Think of a solution, and propose it in the form of an if/then statement. Then test the hypothesis.

SKILL Hypothesizing

▶ MATERIALS
materials will vary according to hypotheses tested

A *hypothesis* is a statement that describes what you think has happened or will happen. A hypothesis usually begins with the word *if* or *when* and often has the word *then* in the middle.

Here is an example of the use of a hypothesis. Sarah accidentally mixed salt and pepper as she was filling the shakers. She thought she could separate the mixture by filtering the salt out of the pepper. This was her hypothesis: "If I use a strainer to filter the mixture, then I can separate salt from pepper." To test her hypothesis, Sarah got a strainer and slowly poured the mixture into it.

▼ PROCEDURE

1. Here is a situation for which you can write a hypothesis. Suppose you have a mixture of iron filings, salt, and sand. How can you separate them? Write a hypothesis that explains your method. Be sure to include which substance you will separate and what method you will use. (The table on page 37 may be helpful.)

2. Now test your hypothesis. Make a list of the equipment you will need, and give it to your teacher. After you have set up your equipment, try your method. Keep careful records of what you observe.

3. Is your hypothesis correct? One thing to remember is that it is okay if your hypothesis is incorrect. In fact, most scientists make many more incorrect hypotheses than they do correct ones. And they can learn as much from incorrect hypotheses as they can from correct ones. Why do you think this is?

▶ APPLICATION
What did you learn from this activity about separating mixtures?

✳ *Using What You Have Learned*
Suppose you were going to teach someone else to write a hypothesis. What would your instructions be?

★ PERFORMANCE ASSESSMENT

Cooperative Learning

 Ask each group of students to share a real-life example of a hypothesis they developed. Each group should describe the question to be answered or problem to be solved and how they arrived at their hypothesis. Ask each group to also explain whether or not the hypothesis was correct. Evaluate each choice of question or problem and hypothesis for appropriateness.

 Have the students place their hypotheses, observations, and answers from this Skill in their science portfolios.

FOCUS

This section provides a description of a property, and compares physical and chemical properties. A list of physical properties is defined, and the phase changes of condensation, evaporation, and sublimation are described.

MOTIVATING ACTIVITY

Display a wooden block, a glass of water, and an inflated balloon. Initiate a discussion about the characteristics unique to each item. For example, the block is solid; it has a definite shape; it cannot be poured, as can the milk. However, both the milk and the air inside the balloon can take on different shapes, depending upon their containers. Then explain that these are characteristic properties of the items. Conclude by telling the students that they will learn more about characteristic properties of matter in this section.

PROCESS SKILLS
• Predicting • Classifying

POSITIVE ATTITUDES
• Precision • Enthusiasm for science

TERMS
• characteristic property
• physical property • physical change • sublimation

PRINT MEDIA
"Inside Ice" from *Dolphin Log* magazine by Elizabeth Foley (see p. xxib)

ELECTRONIC MEDIA
The Invisible World National Geographic (see p. xixb)

SCIENCE DISCOVERY
• Boiling water; in beaker
• Condensation
• Evaporation and boiling

BLACKLINE MASTERS
Reading Skills
Connecting Other Disciplines
Laboratory Investigation 2.2
Thinking Critically
Study and Review Guide

SCIENCE BACKGROUND

In 1990 NASA launched a satellite carrying 24 canisters packed with barium, calcium, lithium, and strontium. The chemicals will be released into the magnetosphere over the course of a several years' study. These elements have unique characteristics that will help scientists trace the forces involved in magnetic storms and auroras.

SECTION 2

Characteristics of Matter

Objectives

Identify a characteristic property of matter.

Distinguish among melting point, freezing point, and boiling point.

Describe the processes of evaporation, condensation, and sublimation.

Suppose you're watching a game between the Chicago Bulls and the Portland Trailblazers. How can you tell the two teams apart? How can you tell in what city they are playing? You can tell by their uniforms, of course! Substances can be identified in a similar way.

Properties of Matter

A trait that identifies a substance is called a *property*. The properties of a substance stay the same, even if its shape or volume changes. Color, boiling point, odor, density, and chemical composition are all examples of properties.

You can use properties to identify many things. For example, when the Chicago Bulls and the Portland Trailblazers are playing basketball, you can use the property of the color of the uniforms to distinguish between the two teams and to determine which team is playing at home. Most teams wear white at home; so if the Bulls are in red uniforms, that means they are playing in Portland.

Figure 2–7. Each player's uniform helps you identify the players during a game, just as properties of matter allow you to identify different substances.

TEACHING STRATEGIES

● **Process Skill:**
Classifying/Ordering

Have the students begin a list of characteristic properties in their journals. Encourage the students to add properties during the study of this section. You might extend this activity by asking the students to work in pairs or small groups to brainstorm additional characteristic properties.

● **Process Skill:** *Predicting*

Have the students predict what other properties might be tested in the Activity, if smell and taste were not enough to identify all four.

How can properties be used to identify unknown substances?

Process Skills: Predicting, Classifying

Grouping: Groups of 3 or 4

Hints

The four unknown liquids should be water, dilute white vinegar, sugar water, and salt water. Make sure that the containers used for the four solutions are chemically and biologically safe.

Have a large bottle of water and cups available for the students to rinse their mouths. All used liquids should be disposed of in a sink. Used cups should be disposed of safely. Excuse a student from this activity if he or she has a contagious ailment.

▶ **Application**

Accuracy of the students' predictions will vary, based on their observations. Students might note that because the liquids look similar, other tests are required to identify them.

The color of the uniform is a property that can change. However, the name of the team also appears on the uniform. The name of the team is a property that you can use to identify the team, regardless of the color of the uniform. A property that always stays the same is called a **characteristic property** because it is characteristic of a particular kind of matter. How can you use properties to identify unknown substances? Try the next activity to find out.

ACTIVITY

How can properties be used to identify unknown substances?

MATERIALS
bottles with tops (4), unknown liquids (4), paper cups (4)

PROCEDURE
1. Make a chart like the one shown.

TABLE 1: IDENTIFYING LIQUIDS			
Liquid	Prediction One	Prediction Two	Final Prediction
A			
B			
C			
D			

2. Without removing the tops of the bottles, examine the liquids and predict what they are. Record the predictions in your chart.

3. CAUTION: Never put your nose directly over a container. The safe way to smell anything in the laboratory is to open the container and wave the fumes toward you. Take the top off each bottle, and smell the contents. Record what you now think each liquid is.

4. Each liquid is safe to taste. Taste each liquid by pouring a small amount of liquid into a paper cup. Put the liquid on your tongue, and swish it around in your mouth. Spit the liquid back into the cup, and rinse out your mouth with water. Record your final predictions in your chart.

5. Repeat step 4 for each liquid, using a new cup each time.

6. Your teacher will identify each liquid for you. Compare this information with your predictions.

APPLICATION
Were your predictions accurate? Explain why or why not.

PERFORMANCE ASSESSMENT
Cooperative Learning

 After the completion of this Activity, have each group of students brainstorm a list of other properties that might be used to identify unknown substances. (Examples include texture, color, density, and temperature.) Evaluate the lists of properties for appropriateness.

● **Process Skill: Applying**

Have the students think about the density of water in its different phases, and ask them to write their responses in their journals. Is ice more dense or less dense than water? (less dense) Is it common for elements to be less dense in their solid phase? (No, water is unusual in that it is opposite to most pure substances.) What would happen to lakes, ponds, and other bodies of water if ice were more dense than liquid water? (They would freeze from the bottom up; most fish would die.)

✧ Did You Know?

The conductivity or nonconductivity of a substance is another important physical property. Silicon crystals, which are nonconductors, are "doped" with small amounts of conductors, such as germanium or arsenic, to form computer chips. The same idea is being applied to plastics, which are nonconductors; iodine is being added as the conductor. Possible products are stealth coatings for aircraft and thin, flexible batteries.

INTEGRATION–
Astronomy

Air is a mixture of several gases. Oxygen is only one of the gases that comprise air. All of the gases that make up the air condense at different temperatures. Some gases can be found as clouds near the surfaces of very cold planets, such as Jupiter, Saturn, Uranus, and Neptune. The clouds on Earth are clouds of water droplets and water vapor.

Figure 2–8. In addition to his uniform, this player can be identified by his individual playing style. This trait can be considered a characteristic property.

Every substance has two kinds of properties—*physical properties* and *chemical properties*. A chemical property is a trait that relates to a change in the composition of a substance.

Basketball players also have their own characteristic properties. Each name is a characteristic property of that player. Players may also be identified by their special style of play. Either of these is a characteristic property of that player. These properties do not change, regardless of the uniform the player is wearing.

A **physical property** of matter is one that can be observed or measured without changing the composition of the substance. For example, a physical property of water is that it freezes at 0°C. Freezing water does not change its chemical composition—it is still water. Physical properties are often used to identify substances. There are hundreds of physical properties; some of the most common ones are listed in Table 2–2.

Table 2-2 Some Physical Properties of Matter

Property	Definition
Boiling point	the temperature at which a substance char.ges from a liquid to a gas
Condensation point	the temperature at which a substance changes from a gas to a liquid; same temperature as boiling point
Density	the mass of a specific volume of a substance
Freezing point	the temperature at which a substance changes from a liquid to a solid; same temperature as melting point
Melting point	the temperature at which a substance changes from a solid to a liquid
Resistance	the opposition a substance has to the flow of electrical current
Solubility	the degree to which a substance will dissolve in a given amount of another substance, such as water

In many cases, physical properties depend not only on the substance itself but also on the environment of the substance. For example, the temperature at which a substance boils depends upon the atmospheric pressure. Because the atmospheric pressure is lower at high altitudes, water boils at a lower temperature on mountaintops than it does at sea level. You'll learn more about physical properties in Section 3.

▼ **ASK YOURSELF**
Think of a favorite athlete or musician. What characteristic properties can you use to describe that person?

It's a Matter of Change

When you stretch a rubber band or crush a cube of ice, you are causing physical changes. Spreading margarine on toast, bending the tab on a pop-top can, and squeezing a tube of toothpaste are other examples of physical changes. A **physical change** does not produce a new kind of molecule. In each of the previous examples, the molecules remained the same. Only certain characteristics of the matter, such as shape and volume, were changed. Try the next activity to learn more about physical changes.

GUIDED PRACTICE

Cooperative Learning Have the students work in small groups. Choose a substance, such as copper, and have the students list as many characteristic properties of the substance as they can. You might make available a chemical handbook with some tables marked for reference.

INDEPENDENT PRACTICE

Have the students provide written answers to the Section Review. In their journals, have students make a list of liquids for which they would like to know the freezing points, melting points, and boiling points. Then have the students refer to science books and other resources to find the information.

EVALUATION

PORTFOLIO ASSESSMENT Have the students write a description of the physical changes that occur when a person makes a cup of hot tea. Have the students place their descriptions in their science portfolios.

DISCOVER BY *Doing*

Measurements before and after should be the same. Students' second measurements may differ slightly from their first measurements due to poor technique.

 LASER DISC

48655, 48656, 48867

Boiling water; in beaker

2358

Evaporation and boiling

1450

Condensation

MULTICULTURAL CONNECTION

In 1912 African-American inventor Garrett A. Morgan (1877–1963) applied for a patent for his Morgan Safety Hood—a breathing device. This invention later became known as the gas mask. Morgan invented the gas mask to promote safety for workers who were exposed to hazardous gases. In 1914, at the Second International Exposition of Safety and Sanitation in New York City, Morgan received the First Grand Prize for his invention. Soon fire departments throughout the nation began using the Morgan Safety Hood to protect them from poisonous gases and suffocating smoke.

Figure 2–9. When a substance such as water evaporates, freezes, and condenses, it undergoes phase changes. Even though the phase of the water changes, its composition remains the same.

DISCOVER BY *Doing*

Measure a small amount of water, and place it in a sealable plastic bag. Put the bag in a freezer, and leave it there until the water freezes. Then remove the ice from the bag, and place it in a beaker. Cover the beaker, and let the ice melt. Then remeasure the liquid. How does your second measurement compare with your first? Explain your findings. ✎

Freezing, Melting, Boiling—Make Up Your Mind!

Freezing is a common physical change. Melting and boiling are also physical changes. These changes all involve a change in phase, so they are called *phase changes*.

When a liquid freezes, it changes from a liquid to a solid. The temperature at which this occurs is called the *freezing point*. Water, for example, turns to ice at its freezing point, 0°C. An increase in heat causes a change in phase. For example, if ice is heated, it changes back into a liquid. The temperature at which a substance changes from a solid to a liquid is called its *melting point*. For water, the melting point is 0°C. If the water is then boiled, it changes from a liquid phase to a gaseous phase. The temperature at which this happens is called its *boiling point*. Water's boiling point is 100°C.

More Changes of Phase

Additional phase changes include condensation, evaporation, and sublimation. Condensation is the phase change from a gas to a liquid. This change is the opposite of boiling. You have probably seen condensation

RETEACHING

Have the students list the three phase changes—freezing point, melting point, and boiling point—as three column heads. Under each head, help the students recall and list the phase that is undergoing change and to which phase it changes.

EXTENSION

If plastic 10-mL graduated cylinders are available, have the students repeat the Discover by Doing activity on page 44, using exact measurements. (The use of a graduate also eliminates water loss when transferring water to the bag and back to the graduate.) Have the students place 9 mL of water in a graduated cylinder and freeze it. Then have the students take a measurement at these points: after the water has frozen; after the water has melted. Finally, have the students compare their findings with those when they first completed the activity.

CLOSURE

Review the list of characteristic properties that the students have been working on in their journal. Encourage volunteers to share and discuss their lists with classmates.

many times in the form of water droplets on the outside of a glass holding a cold drink. The cold surface of the glass causes water vapor in the air to condense into water drops on the glass.

Evaporation is the opposite of condensation. In evaporation, a liquid changes to a gas and moves into the surrounding atmosphere. A wet sponge gradually dries out, and a fogged window clears by evaporation.

Sublimation is a phase change in which a solid changes directly into a gas. Dry ice is an example of a substance that sublimes. When it does so, the solid dry ice changes to a smoke-like gas without ever melting into a liquid. For this reason, dry ice is often used to simulate fog in television and theatrical productions.

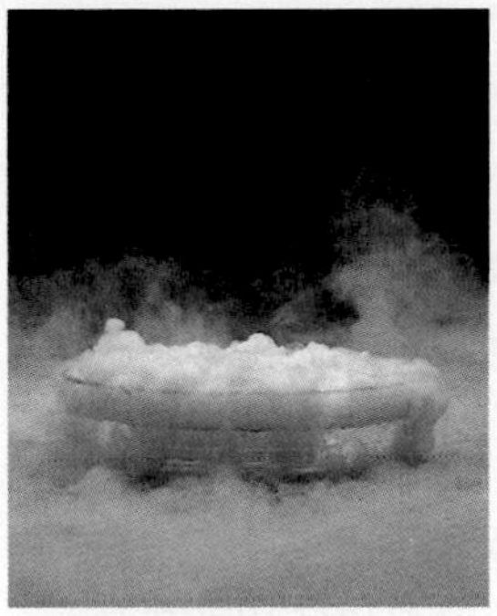

Figure 2–10. Dry ice is an example of a substance that sublimes at room temperature.

Phase changes are related to temperature. Hydrogen, for example, freezes at −259°C and boils at −252°C. It is the only substance that freezes and boils at these particular temperatures. If you have a colorless liquid that boils at 100°C and freezes at 0°C at sea level, you can be reasonably certain the substance is water. The temperature at which a substance changes phase is a characteristic property.

These phase changes are examples of physical changes because whether the water is solid, liquid, or gas, it is still water. Its physical properties, such as volume, density, hardness, visibility, have changed, but it is chemically the same—water.

ASK YOURSELF

Describe the physical changes that occur when you make a cup of hot chocolate.

SECTION 2 *REVIEW AND APPLICATION*

Reading Critically

1. Define the term *property,* and give three examples of physical properties.
2. Explain why the temperature at which a substance changes phase is considered a characteristic property.

Thinking Critically

3. Water boils at a lower temperature on mountaintops than it does at sea level. How would this property of water affect your cooking time if you were trying to make hard-boiled eggs on the top of Mt. Everest?
4. After heavy snowfalls, the snow is often removed from roads and put in piles. As time passes, these piles become smaller, even when the temperature stays below freezing. What do you think causes the snow piles to become smaller?

ONGOING ASSESSMENT

ASK YOURSELF

Answers will depend on the method used to make hot chocolate. Example: A solid dry powder (cocoa) is dissolved in a liquid (milk). As the liquid is heated, steam rises from the surface. The liquid may reach the boiling point, where bubbles of gas emerge from the liquid.

SECTION 2 REVIEW AND APPLICATION

Reading Critically

1. A property is a quality that can be used to identify a type of matter. Physical properties include boiling point, freezing point, density, color, and odor.

2. The temperature at which a substance changes phases varies from substance to substance. Therefore, it is a characteristic property.

Thinking Critically

3. The eggs would have to be cooked for a longer time due to the water's lower boiling temperature.

4. Snow can sublime, changing from a solid directly to a gas.

FOCUS

This section discusses models, including the kinetic molecular model of matter and how models can explain and predict phenomena. Section 3 also describes changes in substances, based on the behavior of their molecules.

MOTIVATING ACTIVITY

Use a shoe box with a lid to construct a black box for the students to examine. Tape small blocks or pieces of cardboard to the box's inner walls to change the configuration, thus creating an obstacle course. Add a small ball or marble, and tape the lid securely onto the box. Then have the students work in groups, examine the sealed box, and draw in their journals how they think the inside of the box looks. Conclude by pointing out to the students that this is a "model" activity of what it is like for scientists to learn about something that they cannot see, such as the atom.

PROCESS SKILLS
- Measuring • Observing
- Experimenting

POSITIVE ATTITUDES
- Cooperativeness • Precision

TERM
- thermal expansion

PRINT MEDIA
A Chilling Story: How Things Cool Down by Eve Stwertka and Albert Stwertka (see p. xxib)

ELECTRONIC MEDIA
Motion of a Molecule BFA Educational Media (see p. xxib)

SCIENCE DISCOVERY
- Melting and boiling points, table
- Melting point
- Rubber ball in liquid nitrogen
- Thermometer, lab setup

BLACKLINE MASTERS
Study and Review Guide

MEETING SPECIAL NEEDS

Gifted

Have the students do research on the theories of John Dalton or Robert Brown. In their journals, have the students state how the ideas of Dalton or Brown contributed to the development of the kinetic molecular model of matter.

SECTION 3

Objectives

Explain three features of a scientific model.

Apply the kinetic molecular model of matter to explain thermal expansion.

Analyze phase changes of matter using the kinetic molecular model.

Figure 2–11. Models are also useful to scientists.

Figure 2–12. Models are often used to represent ideas or phenomena that are either invisible or too large or small to see. This wave machine is used as a model to represent the movement of sound waves.

A Model of Matter

Y ou have probably seen or used many kinds of models. Perhaps you have built model ships or planes, or you have seen a model train. Models are an enjoyable hobby for people of all ages.

In science a model is used to represent an unfamiliar idea or a thing that is either invisible or too large or too small to see. One type of model is a three-dimensional model, such as a model airplane or boat. A second type of model is a diagram or a description. The purpose of a model is to provide a description of an unfamiliar idea or thing using as many familiar ideas as possible. This makes it easier for people to understand.

Models in Science

One example of a scientific model is the wave model of sound. This model compares sound waves to the waves in water. As you know, sound is invisible, which makes it difficult to study. The wave model allows scientists to learn more about sound by studying the visible waves produced in water.

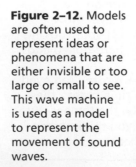

A scientific model has three features.

1. It explains and ties together different phenomena.
2. It is simple to understand and use.
3. It can be used to predict natural occurrences and the results of experiments that have not yet been conducted.

The history of science shows that models, like theories, evolve. Even good models are subject to change. Progress in science constantly produces observations that were not predicted by current models and theories. As this happens, the current models and theories must be changed to reflect new information.

 ASK YOURSELF

How can models help scientists understand new ideas?

Moving Particles as a Model of Matter

Many models have been proposed to describe how matter behaves under various circumstances. The currently accepted model is called the *kinetic molecular model of matter*. This model is based on the idea that all matter is made up of particles and that these particles are in constant, disordered motion.

The idea that matter might be made up of particles is not new. It was first mentioned in the writings of Leucippus (loo SIHP puhs), a Greek philosopher born about 490 B.C. The idea was more fully developed by one of his students, Democritus. The idea that matter is made of particles continued with the work of the British scientist John Dalton (1766–1844).

Another British scientist, Robert Brown (1773–1858), extended the idea of matter and molecules when he used a microscope to observe the movement of particles suspended in drops of water. He proposed that these particles were being hit by even smaller moving particles that we now call molecules. The zigzag motion of small particles suspended in a liquid or gas was named *Brownian motion* after Brown. You can see Brownian motion by observing the dust specks in a beam of light.

Scientists found that many other phenomena could be explained by assuming that matter was made up of tiny particles in random motion. These observations led to the development of the kinetic molecular model of matter.

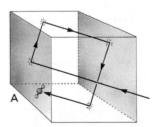

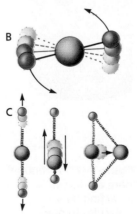

Figure 2–13. Diagram A shows how a gas molecule moves in a closed container. Diagram B shows how the molecule rotates as it moves. Diagram C shows how the bonds holding the atoms of the molecule bend and stretch.

● **Process Skill:**
Constructing/Interpreting Models

Cooperative Learning Use volunteers as individual models of molecules. Tell the students that their bodies and fists represent atoms and that their arms rep-

resent bonds. Use the student "molecules" to illustrate solids, liquids, and gases. Have the students stand close together in a pattern and hold hands, so that they can just stretch but not move from place to place. Then have the students release their grips and begin to move past each other. Finally, have the

"molecules" boil by moving more rapidly, until all the students can spread out in the room.

SCIENCE BACKGROUND

In 1985 at Rice University in Houston, Texas, scientists discovered a new structure for solid carbon. This new structure—called Buckminsterfullerene and nicknamed "Buckyballs"—is based on a soccer-ball shape, similar to Buckminster Fuller's geodesic domes. Since then, scientists have worked to develop practical applications for this new branch of carbon chemistry.

ONGOING ASSESSMENT
ASK YOURSELF

All matter is made of molecules; all the molecules are in constant motion.

INTEGRATION—
Earth Science

It is not high temperatures that make deserts. Instead, it is the lack of available water. The Arctic and Antarctica are considered deserts because water there is in the form of ice and is not available to people or animals.

LASER DISC
44966, 44967, 45956

Rubber ball in liquid nitrogen

Figure 2–14. The molecules of a gas are not attached to one another. When placed in a closed container, the gas molecules move until they fill the volume of the container.

Figure 2–15. Although there are forces attracting liquid molecules to one another, the molecules can slip past each other, as if playing musical chairs.

DISCOVER BY *Researching*

Go to your school library or to the public library, and read about the ideas of Democritus that were very similar to modern scientific ideas. Find out why his ideas were not accepted during his lifetime. Record your findings in your journal.

▶ ASK YOURSELF

On what two characteristics of molecules is the kinetic molecular model of matter based?

Molecules and Physical Properties

Several physical properties can be explained by the kinetic molecular model. For instance, density is the mass of a substance in a given volume. Density tells us how close molecules are to each other. If the molecules of a substance are close together, density is high because there are more molecules in a certain amount of space (volume).

The phases of matter can also be explained by the motion and attachments of molecules. In the gas phase of a substance, molecules are not attached to each other in any way. Each molecule is free and moves about very quickly. The speed of a typical gas molecule is several hundred kilometers per hour. The individual molecules keep moving, often colliding with one another, until they bump into the container walls. This is why they fill any container in which they are put. The distances between the molecules are very large compared with the sizes of the molecules.

The liquid phase is much more dense than the gas phase and the distances between molecules are not large compared to the size of the molecules. The molecules bump into one another much more frequently and the attractive forces between them play a much larger role.

Liquids take the shape of any container into which they are poured. This is because the molecules are free to move past each other. The molecules are close enough together that attractive forces between them keep liquids from changing volume as they are poured from container to container.

In the solid phase, the molecules are also close together. But they have a very limited motion with respect to one another because they are bonded together in a three-dimensional pattern. Therefore, the solid keeps its shape as well as its volume. In solids, the most common molecular motions are the bending and stretching of bonds.

DISCOVER BY *Writing*

In your journal, diagram the molecular arrangements of a solid, a liquid, and a gas. Label the parts of the diagram, and note the relative motions of the molecules.

ASK YOURSELF

How does the movement of molecules in a solid compare with the movement of molecules in a gas?

Molecules and Physical Changes

One commonly observed physical change occurs when the volume of a substance increases as the temperature increases. This physical change is called **thermal expansion.** Most solids, liquids, and gases exhibit thermal expansion.

Hot and Cold   The kinetic molecular model can help you see what causes thermal expansion. Increasing the temperature of a substance increases the motion of its molecules. In solids, in which stretching and bending are the most common forms of motion, faster stretching and bending pushes the molecules farther apart. This occurs even though the molecules remain firmly attached to one another.

DISCOVER BY *Problem Solving*

Fill a balloon with air. Then measure its circumference. Put the balloon in a freezer for 15 minutes. Remeasure the balloon. Are the two measurements different? If so, how do you explain this?

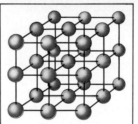

Figure 2–16. All of the molecules of solids are arranged in a regular crystalline pattern and have limited motion. The molecules in these gems, for example, do not move as much as the molecules of liquids and gases.

GUIDED PRACTICE

 Have the students copy in their journals the models of gas, liquid, and solid on pages 47, 48, and 49 and write a brief explanation comparing and contrasting the models and the movements of the three types of matter. You might extend this activity by encouraging the students to share their explanations with classmates and revise their explanations based on classmates' feedback.

INDEPENDENT PRACTICE

Have the students provide written answers to the Section Review. In their journals, have the students draw a picture of the kinetic molecular model. Their drawings should show molecules in solid, liquid, and gaseous states.

EVALUATION

Place enough table-tennis balls in a shallow box so that they nearly fill the box. First move the box very slowly from side to side. Next move the box faster from side to side. Finally move the box vigorously from side to side—some of the balls may leave the box. Ask the students what state of matter each demonstration represents.

Demonstration

Light sticks work by a chemiluminescent reaction in which one compound in a glass vial reacts with another compound inside the stick. The two chemicals are brought into contact by breaking the glass vial and shaking the light stick.

Prepare three beakers, one with ice water, one with room temperature water, and one with hot water. Break and shake three light sticks, and put one in each of the beakers. Have the students observe which stick gives off the most light. (The light stick in the hot water gives off the most light. The heat of the water increases the movement of the molecules in the light stick. Light is given off every time the molecules in the light stick react. Therefore, more light indicates a faster reaction, and thus faster-moving molecules.)

ONGOING ASSESSMENT
▼ ASK YOURSELF

Tires are manufactured to have the proper inflation when they are "hot," or have been driven on for a period of time. As a tire moves along the pavement, the pressure in the tire increases with increasing temperature. Since the manufacturer sets the inflation pressure for cool tires, checking them when warm would give a false reading.

Figure 2–17. Understanding thermal expansion of solids enables scientists and engineers to plan solutions to the problems this phenomenon might cause.

The expansion joint in the bridge shown here provides space so the bridge material can expand on hot days without buckling. The joint also allows the bridge to contract on cold days. The space between the joints is very important. In order for the joints to prevent the walkway from buckling, they must be far enough apart to allow for expansion. Knowing how much the pavement will expand enables engineers to construct joints with the proper separation.

Reaching a Boiling Point

Have you ever filled a pot with soup and put it on the stove to heat? If you filled it too full, you may have seen it spill over. You might say it "boiled over." Careful observation shows, however, that the soup spills before it starts to boil. This is because the soup expands as its temperature is increased.

When liquids are heated, the individual molecules begin to move farther apart. The distance between molecules increases because they are moving faster and farther with respect to each other.

When gases are heated, the gas molecules move faster and farther, causing the gas to expand. In a closed container, the molecules strike the walls of their container with greater force. For this reason, tire manufacturers recommend that tire pressure be checked only when the tires are cool. If you checked tire pressure in tires when they were hot, the pressure would appear too high and you would let out some air. Then when the tires cooled, they would be underinflated.

▼ ASK YOURSELF

Use the kinetic molecular model of matter to describe why you should check tire pressure when tires are cool.

RETEACHING

 Cooperative Learning Have the students work in pairs to prepare a list of phases and phase-change terms in one column and then describe the motion of each in a second column.

EXTENSION

 Have the students think of other solid structures that require an expansion device. Then ask them to think of liquid systems that allow for expansion and, finally, gas systems that allow for expansion. Ask the students to place their ideas in their science portfolios.

(Examples include sidewalk joints, building foundations, overflow space in car radiator tanks, and hot-air balloons or tires.) Note: Tire manufacturers give suggested inflation rates for cold tires, allowing in the guideline for the increase of pressure when hot. Heat and expansion are discussed in more detail in Chapter 16.

CLOSURE

Have the students again play a version of "Twenty Questions" (as they did in the Section 1 Motivating Activity on page 32). However, this time their questions should be more closely related to the concepts they have learned in this chapter.

A Change of Phase

Phase changes can also be explained by the kinetic molecular model. When a solid gets hot enough, its molecules stretch and bend so much that groups of molecules break away from the solid. When the entire solid has separated into free-moving molecules, the material is no longer a solid—it is a liquid. The substance has melted.

Freezing is the reverse of melting. If the molecules in a liquid slow down enough, they are able to attach to each other and hold on. Once all the molecules are attached to each other, the substance is no longer a liquid—it is a solid.

During boiling, individual molecules separate from the liquid and move off independently. When this happens, the liquid becomes a gas. When gases condense into liquids, the individual molecules slow down enough that the weak attractive forces are once again able to stick them to each other. Once this happens, the substance is a liquid with molecules bonded together rather than a gas with independent molecules.

Figure 2–18. This lump of gallium easily melts from a solid to a liquid. During this phase change, the connected molecules separate into individual molecules in close association.

Figure 2–19. Liquid oxygen evaporates very quickly at room temperature. As this substance boils, its molecules gain enough energy to move off as individual gas molecules.

 **ASK YOURSELF**

What happens to the molecules when a solid changes into a liquid?

SECTION 3 REVIEW AND APPLICATION

Reading Critically

1. Name the three features of a scientific model.
2. Use the kinetic molecular model to explain the differences among solids, liquids, and gases.
3. Explain how the density of substances can be related to the kinetic molecular model of matter.

Thinking Critically

4. Dry ice is the solid phase of carbon dioxide. As you read in Section 2, dry ice does not melt. Instead, it sublimes. Use what you have learned from the kinetic molecular model to explain how this happens.
5. Why do musicians "warm up" their instruments before they tune them?

ONGOING ASSESSMENT
ASK YOURSELF

The bonds between molecules stretch so much that molecules break off from the solid to form a liquid.

SECTION 3 REVIEW AND APPLICATION

Reading Critically

1. A scientific model explains and ties together different phenomena, is simple to understand, and can be used to predict phenomena.

2. Answers should focus on molecular structure. For example, in solids, the molecules are attached in a three-dimensional pattern that does not allow much molecular movement.

3. Density of a substance changes when its phase changes. Because of molecular motion, the gas of a substance is less dense than either the liquid or solid of the same substance.

Thinking Critically

4. The bonds that hold the molecules of dry ice together in a solid phase are not very strong. These molecules can easily be bumped out of the solid form directly to gas by the energy of atmosphere molecules hitting the surface of the dry ice or its own thermal energy.

5. Since the density—as well as the volume of the instrument—helps to determine pitch, an instrument must be warmed to performance temperature so that the instrument's molecules and the air molecules inside it expand. If not "warmed up," the instrument could change pitch during a performance.

Process Skills: Measuring, Observing, Experimenting

Grouping: Groups of 3 or 4

Objectives:
● **Measure** the temperature of a liquid as it cools and solidifies.
● **Observe** the pattern of temperature change.
● **Communicate** observations in graph form.

Pre-Lab
Ask the students what freezing and melting points are. Have them use water as an example for their explanation.

Hints
As students complete step 8, you might wish to share this information. To keep the points on the graph separate, put a circle around the temperatures of the moth flakes and a triangle around the water temperatures.

After students complete question 2 of the Analyses and Conclusions, you may wish to extend the activity by asking them the following question: Is there any relationship between the plateau temperature and the time that (a) solid flakes were first observed or (b) all the moth flakes had turned to a solid? (Yes. The plateau began when the crystals appeared and ended when they stopped forming.)

▶ Analyses and Conclusions
1. The water and moth flakes cooled at different rates. The moth-flake curve has a definite plateau that occurred because the moth flakes solidified. Water shows no plateau region because it did not solidify.

2. Yes. Answers will depend on what type of moth flakes were used. Naphthalene solidifies at 80.6°C; paradichlorobenzene solidifies at 53.1°C.

▶ Application
Accept all reasonable answers. Freezers must be set at a certain temperature to keep foods frozen and to prevent the growth of bacteria. Some foods, such as ice cream and gelatin, melt at room temperature and should not be allowed to sit too long before serving.

✳ Discover More
Results will depend on the type and consistency of the juices tested. Thinner juices with higher water contents should melt more quickly than thicker juices with lower water contents.

Post-Lab
Students should realize that the temperature at which freezing and melting occur is the same. Have them create a graph, similar to the one they made in the investigation, for water.

INVESTIGATION

Measuring Freezing and Melting Points

▶ MATERIALS 🔲 🔳
● safety goggles ● laboratory apron ● ring stand ● buret clamps (2)
● 500-mL beaker ● hot plate ● large test tube ● moth flakes
● thermometers in notched corks (2) ● stirring rod ● stopwatch ●
graph paper

▼ PROCEDURE

CAUTION: Put on safety goggles and a laboratory apron.

1. Work in a group, and set up the apparatus as shown.
2. Fill the beaker about 2/3 full of water, and place it on the hot plate. Put the moth flakes in the test tube and place it in the water.
3. Heat the water until the moth flakes have melted.
4. When the moth flakes have melted, place a thermometer in the test tube. Place another thermometer in the water, as shown in the picture. **CAUTION: Do not at-**tach the clamp directly to the thermometer. Stir the water in the beaker with a stirring rod.
5. Make a table like the one shown. One group member should read the stopwatch. A second member should read the thermometers. A third member should record the information. If there are four members, two of them can read the thermometers.
6. When everyone is ready, the timer says "start" and turns off the hot plate. The recorder should write down the starting time. After the start, the timer will say "reading" every 30 seconds for 20 minutes. The readers look at their thermometers; the recorder asks for the water temperature and then the moth flake temperature.
7. Watch the moth flakes closely. Note the time when solid flakes first appear and when no more flakes are forming from the liquid.
8. Record your data on a graph with temperature on the vertical axis and time on the horizontal axis.

TABLE 1: TEMPERATURE READINGS		
Time	Water Temperature	Moth Flake Temperature

▶ ANALYSES AND CONCLUSIONS
1. Using your data, determine whether the water and the moth flakes cooled at the same rates or at different rates. Explain.
2. Is there a section on your graph that shows a period when the temperature of the moth flakes stayed the same? What was the temperature during this flat section (plateau)?

▶ APPLICATION
Why might knowing the freezing points or melting points of foods be helpful to people who prepare meals?

✳ Discover More
Test the melting points of different frozen fruit bars to see whether different juices affect the melting points.

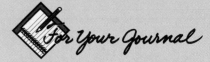

CHAPTER 2 HIGHLIGHTS

The Big Idea

All matter is made up of atoms, has mass, and takes up space. Matter can be classified according to its phase, its purity, and its properties. Scientists use models to show their ideas about the structure of matter and the interaction of its particles. As scientists gain new information from experiments, they may revise their models. Scientists today use the kinetic molecular model of matter to explain how matter behaves.

For Your Journal

Look back at the ideas you wrote in your journal at the beginning of the chapter. How have your ideas changed? Revise your journal entry to show what you have learned about matter.

Connecting Ideas

Read the following poem. Which phases of water are mentioned? Explain how the information in the poem does or does not match what you know about physical changes. Record your answers in your journal.

Walk on a Winter's Day
by Lee Noble

A walk by the lake
on a cold winter's day
the tip of my nose
chilled by biting frosted wind.
Each breath comes as puffs of clouds
that linger for a moment and quickly disappear.
Once shimmering lake no longer rippling
a mass of quiet stillness waiting for the touch of the sun.
Hanging from trees once green with leaves
glittering icy ornaments poised to quietly slip away.
A flurry of soft and gentle flakes
hurriedly drifting and twirling
cover me briefly in a blanket of down
which dissolves and leaves no trace.
The chill of the air, the bite of the wind
the elements on my face
like old friends that come for a visit
and then must go away.

ANSWERS

Understanding Vocabulary

1. Answers should include that only mixtures—not pure substances—can be separated by physical means. More than one characteristic property should be checked before the distinction is made. Students might list dissolving and evaporating, filtering, color, density, and melting point.

2. (a) iron: element; iron oxide: compound

(b) chlorine: element; sodium chloride: compound

(c) copper nitrate: compound; copper: element

Understanding Concepts

Multiple Choice

3. a

4. a

5. c

Short Answer

6. The kinetic molecular model explains and ties together the phenomena of phase change and thermal expansion. It is simple, based on the concept of single atoms and random motion. It can be used to predict how matter may operate, such as gas pressure, expansion and contraction of solids, and condensation on various surfaces.

7. In a solid, the molecules are bonded closely to each other and the only movement is bending, stretching, and vibration. In a liquid, molecules are not as tightly connected and can slide past each other. Bending, stretching, and

vibrating still occur, but groups of molecules can move apart from one another. In a gas, the individual molecules move far apart, and their random, straight-line movement is more rapid than in a liquid.

8. Elements are composed of molecules containing one type of atom, while compounds have molecules containing at least two types of atoms.

Interpreting Graphics

9. The liquid is boiling. The graph plateaus at the boiling point, as the temperature does not continue to rise but remains constant during boiling.

10. The liquid molecules are being supplied with enough energy to move farther apart from each other and eventually become independent.

CHAPTER
2
REVIEW

Understanding Vocabulary

1. Make a list of the ways you could distinguish between a mixture (36) and a pure substance (34).

2. For each pair of terms, use the definitions of *element* (34) and *compound* (35) to help you determine which is an element and which is a compound.
a) iron, iron oxide
b) chlorine, sodium chloride
c) copper nitrate, copper

Understanding Concepts

MULTIPLE CHOICE

3. When a substance freezes, it changes from the liquid phase to the solid phase because
a) molecules stick together.
b) groups of molecules break up into individual molecules.
c) individual molecules stick together and then break apart.
d) molecules slide past each other in disordered motion.

4. Thermal expansion
a) is caused by atoms and molecules moving farther apart.
b) is only observed in liquids and gases.
c) is caused by atoms and molecules getting larger.
d) always happens to a gas when it is cooled.

5. Which of the following is *not* a characteristic property of a substance?
a) phase at room temperature
b) color at room temperature
c) volume at room temperature
d) density at room temperature

SHORT ANSWER

6. How does the kinetic molecular model of matter illustrate the three features of a scientific model?

7. Describe and distinguish between the behavior of molecules in each of the three phases: solid, liquid, and gas.

8. Explain the difference between elements and compounds.

Interpreting Graphics

9. What physical change is taking place when the ethylene glycol is at 198°C? How can you tell?

10. Describe what is happening to the ethylene glycol molecules at 198°C.

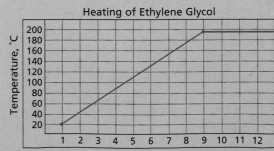

Heating of Ethylene Glycol

Reviewing Themes

11. Temperature (heat energy) causes molecular motion. As the temperature rises, the molecules surrounding the thermometer move faster and collide with the glass wall more often. As a result, the glass molecules vibrate and bend faster and bump into the mercury molecules inside the tube. Thus, the mercury molecules increase their speed and move further from each other—up the capillary tube.

12. Students may find such substances as folic acid, glycerine, sugar, potassium sorbate, vinegar (acetic acid), sorbic acid, cellulose, iron, calcium, salt, sodium citrate, calcium carbonate, and oxalic acid.

Thinking Critically

13. Accept any method students can justify. One method is to skim off the wood chips that will be floating on the liquid. Next, distill the water and methyl alcohol, condensing them in order of their boiling points—methyl alcohol first, then water. This leaves the sand/salt mixture behind. Then dissolve the salt in water, and pour the salt/water/sand mixture into a filter to separate the sand. Finally, evaporate the water to collect the salt.

14. Dry air is composed of approximately 78 percent nitrogen, 21 percent oxygen, and 1 percent argon, with factional amounts of carbon dioxide, neon, and hydrogen.

15. The comparative densities of the sand and gold made panning feasible. A flat pan was used, and a mixture from the stream was swished around with water In the tilted pan. The lighter sand washed out, and the heavier gold remained.

16. Fire is heat and light energy and does not have constant, measurable properties that solids, liquids, and gases have.

17. Density would not be the best property to use. A gas can fill whatever volume it is placed in, which affects its density. However, if the gas' temperature and pressure are specified, then density can be used.

18. A flight simulator shows the pilot what might happen during an actual flight. That is, it shows the results of the pilot's actions, thereby tying different phenomena together. It is simple to use and understand, and it predicts occurrences that have not yet happened but could happen.

Reviewing Themes

11. *Systems and Structures*
A mercury thermometer measures the thermal expansion of liquid mercury as it moves through a very thin tube. Explain how this happens, using the kinetic molecular model of matter.

12. *Systems and Structures*
A pure substance contains only one kind of molecule. Check the labels on foods at home. List some of the pure substances you find.

Thinking Critically

13. Propose a method for separating a mixture containing salt, sand, water, methyl alcohol, and wood chips.

14. The air in the earth's atmosphere is a mixture. Research the gaseous components of the earth's atmosphere. Which is the most abundant gas in the mixture?

15. In the early history of the United States, people would search out sandy stream beds in which small particles of gold were mixed with the sand. The separation process used was called "panning." What characteristic property of the two substances made panning possible?

16. The idea that all matter can be classified in terms of basic elements has been accepted for a long time. In one ancient theory, there were four "elements": earth, air, fire, and water. How does fire differ from the other three "elements"?

17. One characteristic property of a substance is density. Would this be a good property to choose when identifying a gas? Explain.

18. The pictures here show the outside of a flight simulator and what the screen inside looks like. Flight simulators like this one are used to train pilots before they actually fly a plane. Using what you learned in this chapter, explain why a simulator could be considered a scientific model.

Discovery Through Reading

Newton, David. *Consumer Chemistry Projects for Young Scientists*. Franklin Watts, 1991. This illustrated book offers a variety of at-home projects, including several that focus on the properties of matter.

Science PARADE

Discussion

● **Process Skill:** *Comparing*

Ask the students to compare the photographs of the polystyrene beads made on Earth with those made in space. (The size and shape of the beads made in space are much more uniform.)

● **Process Skills:**
Communicating, Inferring

Students are probably familiar with the fact that if oil and water are mixed together, they will separate and the oil will float on the water. Point out to the students that this occurs because oil is less dense than water. Ask the students to speculate what would happen in space if oil and water were mixed together. (They would not separate.)

Science PARADE

MATERIALS MADE IN SPACE

"MADE IN THE USA." You've seen that manufacturing tag many times but not given much thought to it. However, you'd probably look twice at a tag that read "MADE IN OUTER SPACE." You might grin and think that this sounds like a not-very-clever idea from a science fiction book. You can stop grinning. The tag may not exist yet, but new materials are being produced in outer space.

IT CAME FROM OUTER SPACE

Materials with extraordinary—and valuable—properties can be produced in space. For example, polystyrene (tough plastic material) spheres of very uniform size can be manufactured in space. Scientists use these perfectly uniform plastic spheres to calibrate scientific instruments back on Earth. The space balls help measure everything from dust to face powder. The first batch of space-made polystyrene spheres was valued at about $23,000 per gram!

Comparison of polystyrene beads made in space (left) with those made on Earth (right)

Journal Activity
After completing the Unit and reading and discussing the *Science Applications*, ask the students to consult the list of questions they recorded in their journals from the discussion at the beginning of the Unit. (You might have the students turn back to the Unit Opener on pages xxii–1.) Lead a discussion during which the students can verbalize what they have learned. You might also have the students record their new ideas in their journals.

EXTENSION

As a follow-up to the class discussion, you might have the students prepare reports on their ideas for new kinds of space-made products and how these products could be used. Encourage the students to share their reports with classmates.

HOW DOES YOUR CRYSTAL GROW?

The microgravity (near weightlessness) of space is an ideal environment for growing crystals. When crystals grown on Earth reach a certain size, their own weight causes them to break. In microgravity, scientists can grow larger and more complex crystals. On the space shuttle, astronauts baked crystals in individual furnaces that reach temperatures of more than 500°C.

The materials for space crystals range from metals like cadmium to proteins. Some vital space research focuses on growing crystals of the protein interferon, a drug used to treat AIDS, and other AIDS-related proteins. Such space-grown crystals are returned to pharmaceutical labs on Earth, where scientists study them further. Space-made medicines may someday become part of the battle against AIDS.

One space-grown protein may make computers remember more. Scientists extracted a light-sensitive protein from a type of bacterium growing in salt marshes. On the space shuttle, astronaut scientists injected the protein into a polymer carrier.

Protein grown in free fall

The space-manufactured protein may someday be installed into computer memories. Computer engineers think it might make computer memory capacity 10 to 30 times greater. And think of it: from a salt marsh to the space shuttle to a computer memory chip. Quite a journey for a humble protein! ◆

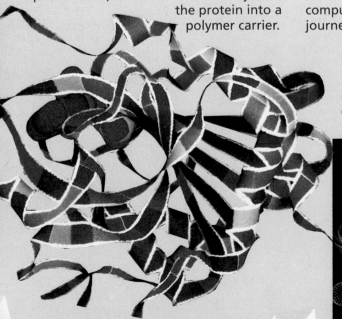

Computer-generated images of proteins

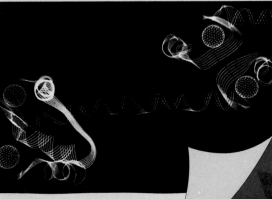

Background Information

All liquids hold together at their surface with a kind of rubbery skin, called surface tension. A soap bubble is filled with air and surrounded by a thin "skin" of water with a layer of soap on both the inside and outside surfaces. Soapy water makes larger, longer-lasting bubbles than plain water does because the soap weakens the surface tension of the water, allowing the surface to be stretched into different shapes. Adding glycerine to a soap bubble solution will also coat the inside and outside surfaces of the bubble. The glycerine helps keep the water in the bubble from evaporating and bursting the bubble.

READ ABOUT IT!

Bubbles

from Kids Discover

Bubbles Everywhere

Bubbles are everywhere. There are bubbles in your bathtub. Bubbles in your soft drink. Bubbles in your shirt. Wait a minute. Bubbles in your shirt? It's true. If you are wearing a cotton shirt made by a certain manufacturer, there may be air bubbles in your shirt—put there to reduce shrinkage when you dry the shirt in a clothes dryer.

What exactly is a bubble? In the purest sense, a bubble is a ball of air or gas in a liquid or with a thin skin of liquid around it. Soap bubbles and carbonated beverages fit this definition. But in recent years, people have given the name of bubble to many things that don't fit this classic definition. The expanded definition of bubble is something like "anything that is shaped like a sphere and hollow inside and that looks and acts like a bubble."

Kids of all ages love to have fun with bubbles. But there is a serious side to bubbles. Scientists and inventors are always studying bubbles, looking for new ideas that might help people. Among other things, they have tried nicotine bubbles to help people stop smoking.

Professor Bubbles, Richard Faverty, makes a gigantic soap bubble that lasts about 30 seconds.

Discussion

● **Process Skills:** *Expressing Ideas Effectively, Formulating Hypotheses*

Ask the students to relate what they learned about the properties of matter in Chapter 2 to how carbonated beverages react. Ask volunteers to explain how the properties of liquids and gases affect carbonated beverages. (Gases can be compressed by increasing pressure, but liquids cannot be compressed.)

● **Process Skills:** *Analyzing, Generating Ideas*

Ask the students to speculate how information about the changing composition of Earth's atmosphere could give scientists clues about the past history of other planets, such as Mars. (Students might note that if the composition of Earth's atmosphere has changed greatly over millions of years, then possibly the atmosphere of Mars has also changed. This could support the idea that, at one time, the Martian atmosphere might have supported some forms of life.)

Bubbles with Gas

Whether you call it pop, soda, tonic, soft drink, or carbonated beverage, there's one thing it's got to have—bubbles. Whenever you reach for a carbonated beverage, the sparkle will come from the same source—carbon dioxide. Carbon dioxide (CO_2) is a colorless, odorless, tasteless gas. When it's trapped with water in a closed container (or underground), the carbon dioxide dissolves in the water. Opening the container reduces the pressure on the water. When the pressure is reduced, the gas expands, and out come those tiny bubbles.

By around 1820, bottled water became popular. Soon flavorings—ginger and lemon—and mineral salts were added.

How do they get the tiny bubbles in the pop? Syrup or concentrate and water are put into a pressurized vessel where the atmosphere is carbon dioxide. This gas is absorbed by the liquid. Later, as the liquid is passing through a pipe (approximately three inches around), very fine bubbles of carbon dioxide are injected into it through a sparger. The sparger's job is to break up large bubbles into tiny bubbles so they can be more easily absorbed into the liquid.

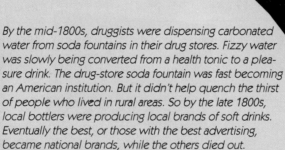

By the mid-1800s, druggists were dispensing carbonated water from soda fountains in their drug stores. Fizzy water was slowly being converted from a health tonic to a pleasure drink. The drug-store soda fountain was fast becoming an American institution. But it didn't help quench the thirst of people who lived in rural areas. So by the late 1800s, local bottlers were producing local brands of soft drinks. Eventually the best, or those with the best advertising, became national brands, while the others died out.

In 1772, Joseph Priestley, nick-named "the father of the soft drinks industry," suggested that with the help of a pump, water would be impregnated with fixed air, thus making carbonated water.

Ask the students to generate ideas and questions about the role bubbles play in scuba diving. (In addition to nitrogen bubbles causing decompression sickness, commonly called the "bends," air bubbles are expelled as a diver exhales. If a diver rises to the surface too quickly while holding his or her breath, expanding air bubbles can rupture sacs in the lungs. In addition, air bubbles that escape into blood vessels can travel to the heart or brain. Eventually, the air bubbles will get stuck in the small blood vessels and shut off the blood supply to that part of the body.)

Have the students mix equal volumes of water, liquid dish detergent, and glycerine to make a bubble solution. Show them how to use heavy wire—such as wire clothes hangers—to make large bubble frames of different shapes. Each bubble frame should have a handle (the neck of the clothes hanger). Then use a large pan or dishpan to hold the bubble-making solution. Encourage the students to experiment to see what shapes and sizes of bubbles they can make.

For centuries fashionable people "took the waters" at health spas in Europe and the United States. Many of the spas were located at places where water (containing dissolved carbon dioxide) mysteriously bubbled out of the ground. This water was believed to cure various diseases and generally to invigorate those who bathed in it and drank it. Before the spa resort was in operation at Saratoga Springs in upstate New York, George Washington visited Saratoga as a guest of the Schyler family. He tried the water and was so impressed that he tried to buy land near the springs, but it had already been purchased.

Saratoga Springs spa park today

In the 17th century, the first marketed soft drinks appeared. They were a mixture of water and lemon juice sweetened with honey. Spa water was first bottled in the late 1700s. So people who couldn't travel to a spa could still benefit from the waters.

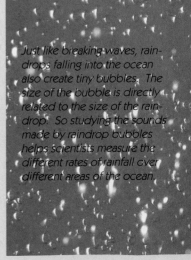

Nitrogen bubbles in the blood cause a dangerous, even deadly, condition called the bends, which results when divers dive too deep, stay down too long, or return too quickly from a deep dive.

Just like breaking waves, raindrops falling into the ocean also create tiny bubbles. The size of the bubble is directly related to the size of the raindrop. So studying the sounds made by raindrop bubbles helps scientists measure the different rates of rainfall over different areas of the ocean.

Background Information

Chewing gum starts with a gum base of chicle and other substances. The base is cooked in huge containers, flavorings are added, and the mixture is kneaded and formed into gum.

Discussion

● **Process Skills:**
Communicating, Formulating Hypotheses

Ask the students to formulate a hypothesis on the best way to get bubble gum off clothes. If possible, let the students test their hypotheses. (The best way to get bubble gum off a piece of clothing is to put it in a plastic bag and put it in the freezer. When the gum hardens, peel it off the clothing. Or, rubbing the clothing with an ice cube will freeze the gum. Then the gum can be peeled off the clothing. Accept reasonable responses students can justify.)

● **Process Skills:**
Communicating

Students might want to do research to discover more about baseball cards. Encourage the students to also research other sports trading cards such as football cards, basketball cards, hockey cards, or automobile racing cards.

Bubble Gum Fun

Most kids love it. Most parents hate it. Most doctors say it can't hurt you. But some people have dislocated their jaws chewing it. It's bubble gum, of course. And it is over 65 years old.

Actually, bubble gum is one of the newer things that people have tried chewing. The ancient Greeks chewed the gum of the mastic tree, and over a thousand years ago the Mayans of Central America chewed chicle, a white gummy substance from the sapodilla tree. In the mid-1800s, a Mexican general brought some chicle to the United States and gave some of it to Thomas Adams, an inventor and businessman. When he and his father added sugar and flavoring for taste, the chewing gum industry began. But it wasn't until 1928 that Walter Diemar, working at the Fleer chewing gum company, came up with the right ingredients to make a gum that was elastic, moist, and not too sticky to peel off faces once a bubble had popped.

In 1928, when Walter Diemar put together the right ingredients for bubble gum, he had only pink food coloring at hand, and that's why bubble gum has been mainly pink ever since.

Those who have to clean up discarded bubble gum complain bitterly about it. But others have found constructive uses for it. Some dentists and speech therapists recommend chewing bubble gum to strengthen jaws. A Columbia University study showed that chewing gum makes people more relaxed. One scientist even claims that liquid bubble gum can be used as an insecticide. When plant-eating bugs eat it, their jaws stick together and they can no longer chew plant leaves.

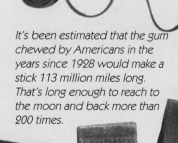

It's been estimated that the gum chewed by Americans in the years since 1928 would make a stick 113 million miles long. That's long enough to reach to the moon and back more than 200 times.

Some card companies no longer include bubble gum with their sports cards because collectors were complaining that the bubble gum was staining the cards!

The new generation of bubble gum includes such exotic flavors as bananaberry split and watermelon. There's sugar-free bubble gum and even bubble gum in the middle of a lollipop!

Background Information

Eiffel Plasterer, the bubbleologist mentioned on this page, began performing at science fairs in 1932 and continued until his eighties. He was a physics teacher and vaudeville performer. One of his favorite projects was creating long-lasting bubbles from a special soap solution and storing the bubbles in sealed jars.

Students who want to know more about bubbles could read one of the following books.

Hendrickson, Robert. *The Great American Chewing Gum Book.* Chilton, 1976.

Pearl, Louis. *Sudman's Bubbleology Guide.* Tangent Toy Co., 1984.

Tchudi, Stephen N. *Soda Poppery: The History of Soft Drinks in America.* Scribner's, 1986.

Zubrowski, Bernie. *Bubbles.* Little Brown, 1979.

Bubbles in the Laboratory

Some bubbles are just for fun. But scientists are finding that certain bubbles may help answer some age-old questions about our universe while they raise some new ones. Air bubbles trapped in amber may hold the key to such ancient mysteries as why the dinosaurs became extinct. Bubbles locked in Antarctic ice may help us understand the world we live in by showing the relationship between carbon dioxide in the atmosphere and temperature. And bubbles in outer space raise some questions about the structure of the universe. ◆

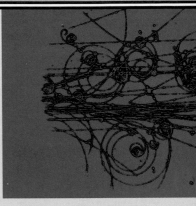

There are countless theories about why the dinosaurs became extinct. Here's one of them. Air bubbles trapped in amber that fossilized 80 million years ago suggest that the air at that time contained 50 percent more oxygen than today's air. If that is true, the dinosaurs may have slowly suffocated as the oxygen in the air was gradually reduced.

A bubble chamber helps in the study of the tiny particles that are the building blocks of matter and the forces that bind them together. In a bubble chamber, liquid is heated high above its boiling point and kept under high pressure to prevent boiling. When the pressure is suddenly reduced, charged particles speeding through the liquid create strings of bubbles. High-speed photographs of the bubbles make possible precise measurements of the particles and their activities.

Eiffel Plasterer, a bubbleologist, sealed many bubbles so they could last longer. How long did his longest bubble last? (340 days!)

When a wave, driven by the wind, breaks at sea, it traps a large amount of air in the water. The trapped air quickly becomes millions of tiny bubbles, which radiate sound as they move in the water. Scientists use sound-detecting equipment to track the bubbles and learn more about ocean currents and waves as they move across the ocean.

THEN AND NOW

Francis Bacon (1561-1626)

Englishman Francis Bacon was a politician and lawyer who lived during the time of Queen Elizabeth I and King James I. Francis Bacon was also a philosopher who wrote about learning from observation and experimentation. The methods of experimental science had been formulated earlier by scientists such as Galileo. Bacon, however, was the first person to write about those methods in a way that convinced others that observation and experimentation were useful ways to learn things.

Bacon's main interest was in discovering practical applications of science that would make life better for the people of the world. In fact, he worked with many skilled artisans from all over Europe to improve methods of metal working and mining. He encouraged artisans to use his methods of problem solving to develop new techniques for use in industry.

Bacon published his ideas, and his work soon attracted widespread interest. In one sense, Francis Bacon's writings helped launch the scientific age in which we now live. ◆

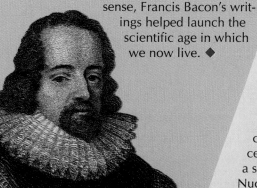

Shirley Jackson (1946-)

Shirley Jackson has always been interested in physics. After graduating as valedictorian of her high-school class, she was admitted to the Massachusetts Institute of Technology (MIT). Four years later she received her bachelor's degree in physics. Jackson decided to continue her education at MIT and became the first African American woman to receive a Ph.D. from that institution. She continues to be involved with MIT as a member of its educational council.

Jackson's first area of research was in the field of high-energy physics. During the early years of her career, she worked in this field at the National Accelerator Laboratory in Batavia, Illinois. Later she was a scientific associate for the European Organization for Nuclear Research in Geneva, Switzerland, and a lecturer at the NATO Advanced Study Institute in Antwerp, Belgium. Now employed at AT&T Bell Laboratories in Murray Hill, New Jersey, Jackson's current research is in low-energy physics. She has written numerous articles on such topics as nutrino reactions and channeling in metals and semiconductors. ◆

Background Information

Rocks are natural, solid materials made of one or more minerals. Scientists classify rocks into three basic types. Sedimentary rocks form from sediments (small bits of soil, older rocks, shells) that settle out of water, ice, or air. Over millions of years, the sediments become pressed together, to form rocks.

Metamorphic rocks are formed when existing rocks are changed by heat, pressure, or chemical reactions. For example, marble is a strong, hard metamorphic rock that can be highly polished. Because of these properties, marble is often used in buildings and monuments.

Sandstone and limestone are sedimentary rocks that can be used as building materials or for sculpting. Limestone varies in color from white to gray to black and in texture from fine-grained to coarse-grained. Marble is metamorphic rock formed from limestone. Pure marble is white, but mineral impurities add colors and form colorful streaks in many varieties of marble. The attractiveness of many marbles make them ideal for use as building materials and for sculpting. Blocks of marble

CLAUDIA WIDDISS
Sculptor

Claudia Widdiss knows about the properties of matter. She uses the properties of stone daily to choose what she will create. Widdiss is a sculptor.

Widdiss begins this limestone sculpture from a block of stone that was discarded by another artist. The block weighs about 181 kg. Widdiss makes a few scale models based on her roughed out block of stone. Scale models are a way of sketching in three dimensions.

Widdiss carves the stone with a hammer and a pointed chisel. Her hammer and chisel are designed much like those used in prehistoric times, except her tools are made of steel instead of stone. It takes three days for her to rough out the basic shape of the sculpture. The basic shape must be enhanced by the natural properties and characteristics of the stone.

used by Michelangelo came from the famous quarries in Carrara, Italy.

Discussion

● **Process Skills:** *Applying, Inferring*

After the students have read the feature about Claudia Widdiss, ask them why it is important for her to understand the properties of stone. (The shape she sculpts must be enhanced by the properties and characteristics of the block of stone she is working with.)

Then ask the students to point out the protective items Widdiss wears. Ask them to also explain the purpose of each item. (Ear coverings protect her hearing from the loud noises produced by the air hammer.

The mask protects her lungs by filtering out tiny particles of limestone dust created during the sculpting process. Gloves protect her hands from rough or sharp edges of the stone. The goggles protect her eyes from dust and pieces of rock.)

After working by hand with hammers and chisels, Widdiss begins more detailed work on the stone. She uses an air hammer, which is activated by compressed air. The air hammer presses a chisel against the stone and carves it into the exact shape she wants. Widdiss carves all around the stone and works on the sculpture from many different positions and angles. The tools that she uses can be very dangerous. That is why she wears protective clothing, such as a mask, gloves, and ear coverings.

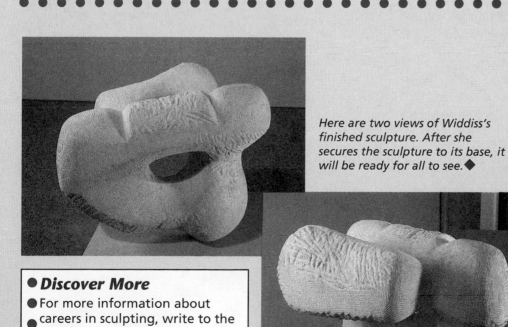

Here are two views of Widdiss's finished sculpture. After she secures the sculpture to its base, it will be ready for all to see. ◆

● **Discover More**
● For more information about
● careers in sculpting, write to the
● Sculptors' Guild
● 110 Greene Street
● New York, NY 10012

Discussion

● **Process Skills:** *Analyzing, Evaluating*

Have the students read the article on biodegradable plastics and speculate how mass production of biodegradable plastics could benefit farmers.

(For those farmers who use plastic on their crops to prevent weed growth, biodegradable plastics could be used instead. Farmers would not have to pay to have the biodegradable plastics removed; they would dissolve. This would save farmers money.)

● **Process Skills:** *Solving Problems/Making Decisions, Classifying/Ordering*

Ask the students to identify the ways they use plastics in their daily lives. Then have them classify the uses into those that could use quickly-dissolving plastics and those that would require slowly-dissolving plastics.

(Quickly-dissolving plastics might include those used for shopping bags, dry cleaning bags, or packaging dry foods or clothing. Slowly-dissolving plastics might include those used for food and drink containers, trash bags, or moist food containers for foods with a short shelf life.)

Biodegradable Plastics

Scientists and manufacturers once thought plastics were the perfect material. Plastics made from petroleum could be shaped into virtually any form—from garbage bags to automobile bodies. Even better, synthetic plastics lasted forever. Unfortunately, plastic's durability has turned out to be a problem.

The Plastic Problem

Synthetic plastics consist of long, tangled carbon polymers. Polymers are giant molecules made by bonding many small molecules together. They resist being broken down by natural forces. Plastic bags, diapers, and bottles are clogging landfills. In response, scientists are developing the technology to produce biodegradable plastics. These natural plastics dissolve under contact with organisms in the earth and water.

Food for Bacteria

To achieve biodegradability, scientists are trying to develop plastics with a special set of properties. First, the material must be an inviting food for naturally occurring bacteria and enzymes. Most biodegradable plastics begin with starch molecules made of carbon, hydrogen, and oxygen. Second, the material must hold together well enough to serve as a plastic. This means that the starch molecules must form long, twisted polymers, like the carbon polymers in synthetic plastic. Finally, natural plastics must degrade into substances harmless to human life and the environment.

It looks and feels like polystyrene, but it isn't! ECO-FOAM is a packaging material based on cornstarch.

Research labs are working on several ways to manufacture biodegradable plastics. In one method, scientists begin with powdery starch from corn or potatoes. As scientists heat the starch to 177°C, the starch molecules form twisted polymers much like synthetic plastics. In another method, scientists grow bacteria that manufacture their own plastic polymer, a polyester similar to what is used in fabrics. The bacteria make and store the polyester for future consumption, much as the human body stores fat. Scientists kill the bacteria and use the naturally produced polymer to make plastics.

A Dissolving Act

Some biodegradable plastics have already reached consumers. Physicians can stitch up internal organs with plastic sutures that eventually dissolve on their own. Some mail-order companies already cushion their shipments with a starch-based packaging material. The packaging is made of 95 percent starch and 5 percent polyvinyl alcohol. Within seconds after getting wet, the plastic packaging dissolves. Another biodegradable plastic used in hospitals holds dirty laundry. Workers simply toss the bags into the washing machine, and the bags dissolve as the clothes are laundered. After this plastic has dissolved, the alcohol in it attracts bacteria in the water. The bacteria attack the alcohol and use the starch as food.

Not So Fast!

Scientists have a harder time making biodegradable plastics that dissolve slowly. No one wants a disposable diaper that dissolves quickly on contact with water!

Some farmers now cover their fields with synthetic plastic to prevent weed growth. However, they must pay to remove the oil-based plastic, which ends up in a landfill. Chemists hope to manufacture a starch-based plastic sheeting to serve as a crop mulch. With a slow-dissolving biodegradable plastic mulch, farmers could let the plastic dissolve harmlessly in their fields.

The Search Continues

Biodegradable plastics are not an easy answer to the problem of plastic waste. The first "biodegradable" plastic bags were a mixture of starch and synthetic plastics that did not fully degrade. Scientists warned that such products might degrade into toxic compounds. Also, biodegradable materials must come in contact with soil and water. Science and technology continue the search for a plastic that will serve consumers' needs without harming the environment. ◆

Even starch-based plastics will not degrade if crammed into a landfill.

ENERGY—MATTER IN MOTION

UNIT OVERVIEW

Unit 2 presents the various forces that affect the position and motion of matter and the energy sources that make this all possible. The concepts of speed, velocity, acceleration, inertia, momentum, and gravity are discussed in detail along with simple and complex machines. Also discussed are measurements of work and power, the laws and forms of energy, pollution, energy conservation, and energy decision-making.

Chapter 3: Acceleration, Force, and Motion, page 70

This chapter defines speed, velocity, acceleration, and motion and describes how they are interrelated. Also discussed are the forces that affect both stationary and moving objects, mass and weight, and the laws of motion and their applications.

Chapter 4: Work, page 98

In this chapter the concepts of work and power are explained. The chapter describes the six types of simple machines as well as compound machines. Mechanical advantage is calculated, and the efficiency of machines is also discussed.

Chapter 5: Energy, page 122

This chapter introduces the concept of energy, various forms of energy, and the laws that describe the behavior of energy. Heat energy, the physical properties related to heat, and the movement of heat are discussed.

Chapter 6: Energy Sources, page 150

This chapter identifies the different types of fossil fuels, their physical characteristics, uses, and locations. Nonrenewable and renewable resources are compared and contrasted. The potential for the uses of solar energy, geothermal energy, and water power as sources of energy is discussed.

Chapter 7: Energy and the Environment, page 176

In this chapter environmental risks of air pollution, acid rain, and nuclear wastes are discussed, as are the advantages and the methods of conserving energy supplies. The arguments for and against storing nuclear waste underground are considered. Methods of energy conservation are discussed. This chapter also includes a discussion of the politics and economics of obtaining energy.

Science Parade, pages 202–211

The articles in this unit's magazine relate the fundamental laws of physics to the antics in a circus, to riding a roller coaster, and to space travel. The work and responsibilities of an automobile race official are described in the career feature. The biographical sketches focus on an environmental chemist and a physicist.

UNIT RESOURCES

PRINT MEDIA FOR TEACHERS

Chapter 3

Brown, Stuart F. "World's Fastest Sailboat." *Popular Science* 238 (January 1991): 56–61. Force, speed, and velocity are included in this discussion of a trifoiler trimaran hydrofoil. A discussion of the boat's design is included.

Chapter 4

Harley, D.W., Jr. "Track and Field Work: The Physics Behind the Slam Dunk and the Shot Put." *The Science Teacher* 57 (September 1990): 58–60. This article gives suggestions on how sports can be used to help demonstrate force and motion.

Chapter 5

Fiorillo, Cheryl M. "Lightweight Block, Heavyweight Insulator." *Popular Science* 236 (May 1990): 106–107. This article focuses on energy-saving polystyrene aggregate insulator blocks that are lightweight and easy to install.

Chapter 6

Perrin, Noel. *Solo: Life with an Electric Car.* Norton, 1992. This is an engaging and humorous account of the author's cross-country trip in an electric car. A capsule history of the electric car in the United States is also included.

Chapter 7

Svitil, Kathy. "Holey War." *Discover* 14 (January 1993): 75–76. Over the past ten years, ozone depletion has intensified and the hole now crosses the tip of South America.

Skurzynski, Gloria. *Robots: Your High-Tech World.* Bradbury, 1990. (NSTA Outstanding Science Trade Book) This book discusses the practical and future uses of robotics in medicine, science, and industry.

PRINT MEDIA FOR STUDENTS

Chapter 3

Gardner, Robert. *Science and Sports.* Watts, 1988. (NSTA Outstanding Science Trade Book) The relationship between science (speed, acceleration, and the laws of motion) and sports is explained.

Hawkes, Nigel. *Vehicles: The Most Extraordinary Achievements Since the Invention of the Wheel.* Macmillan, 1991. This book provides a fascinating look at landmarks in transportation technology.

Hinton, Ed. ". . . And Lived to Tell About It." *Sports Illustrated* 76 (May 18, 1992): 18–23. This informative article discusses the advancements made in protecting race-car drivers from serious injury and death.

Chapter 4

Sletto, Jacqueline Wiora. "Birch Bark Work." *Native Peoples* (Spring 1991): 21–24. The Anishinabe (also known as the Ojibwa or Chippewa) of the upper Midwest and Canada use special tools to cut bark to make birch bark containers.

Taylor, Barbara. *Get It in Gear! The Science of Movement.* Random House, 1991. This exciting book focuses on experiments with machines, energy, and gravity.

Chapter 5

Ardley, Neil. *Heat.* Discovery, 1992. This brief and straightforward book explains and describes aspects of heat.

Cobb, Vicki. *Why Doesn't the Sun Burn Out? And Other Not Such Dumb Questions About Energy.* Lodestar Books, 1990. This book presents nine questions on different kinds of energy such as heat, kinetic, and chemical energy, and their relation to matter.

Sauvain, Philip. *Motion.* Discovery, 1992. This book covers a wide range from the motion of simple machines to pendulums.

Chapter 6

Pringle, Laurence. *Nuclear Energy: Troubled Past, Uncertain Future.* Macmillan, 1989. (NSTA Outstanding Science Trade Book) A comprehensive study of the nuclear energy industry with attention to new developments, safety issues, and global consequences.

Scott, David. "Harnessing Sea Breezes." *Popular Science* 238 (March 1991): 22–24. The future of wind power to generate electricity is discussed. Wind-power sites around the world are highlighted.

Twist, Clint. *Fossil Fuels.* Franklin Watts, 1990. This informative book discusses various fossil fuels, including coal, crude oil, and natural gas, and explains how they are converted to provide energy.

Chapter 7

Brune, Jeffrey. "Lost Horizons." *Discover* 12 (January 1991): 69. Air pollution in several cities is discussed. A computer-enhanced photo illustrates how Denver would look without the haze of air pollution.

Gardner, Robert. *Experimenting with Energy Conservation.* Franklin Watts, 1992. This book is an interesting collection of at-home experiments that focus on energy conservation.

Gay, Kathlyn. *Global Garbage: Exporting Trash and Toxic Waste.* Franklin Watts, 1992. This book discusses the consequences of disposing of trash and toxic waste "somewhere else." But where?

ELECTRONIC MEDIA

Chapter 3

Falling Bodies and Projectile Motion. Film or videocassette. Barr Films. 12 min. Gravity is explored through laboratory experiments and real-life examples.

Force & Motion: Newton's 3 Laws. Film or videocassette. Barr Films. 12 min. Newton's laws of motion are demonstrated through experiments and demonstrations.

Motion and Energy: Physical Simulations. Computer Software. Focus Media, Inc. Simulations illustrate different types of motion.

Chapter 4

Machines, Work, and Energy. Computer Software. Educational Activities, Inc. This program is an introduction to the workings of the lever, inclined plane, wedge, screw, wheel and axle, and pulley.

Simple Machines—Using Mechanical Advantage. Film. Barr Films. 18 min. This film discusses the mechanical advantage of simple machines and complex machines and explains how the modern world depends upon machines.

Chapter 5

Heat, Energy and Temperature. Computer Software. Focus Media, Inc. This tutorial program examines the relationship between heat and temperature while studying conduction, convection, and radiation.

Matter and Energy. Film or videocassette. Coronet Film and Video. 13 min. The law of conservation of matter and energy is discussed.

The Forms of Energy. Videocassette. Focus Educational Media, Inc. 30 min. Different forms of energy, such as potential, kinetic, mechanical, heat, electrical, light, and nuclear are demonstrated.

Chapter 6

The Energy Connection. Film or videocassette. Barr Films. 19 min. The wise use of energy in our everyday lives is explored through the discussion of pollution and dwindling energy supplies.

Energy Conversion. Film or videocassette. BFA Educational Media. 12 min. A hydropower station is the background for a demonstration of the conversion of mechanical energy into electrical energy.

Nuclear Energy: The Question Before Us. Videocassette. National Geographic. This program explains nuclear energy and presents its advantages and disadvantages.

Chapter 7

Energy Seekers. Film. Coronet Film and Video. 12 min. From windmills to tar sands and garbage, this film surveys present and future alternative energy sources.

Help Save Planet Earth. Videocassette. UNI Distribution. 70 min. Celebrities make appearances in this video to encourage environmental awareness. Tips help viewers conserve energy and resources.

Sim City. Computer Software. Brøderbund. Create your own city—or use one of the cities in the program—and participate in its growth. Among the difficult issues to deal with include pollution, power supplies, traffic congestion, and housing costs.

Discussion The high-wire artist and the roller coaster in the background are examples of types of energy. Ask the students who have ridden a roller coaster to describe how the ride starts (being pulled up a steep slope) and what keeps the roller coaster going. (a combination of being pulled up slopes and acceleration) Then discuss with the students what skill is most important for the high-wire artist. (balance) Ask the students what other indication of other motions they see in the photograph.

UNIT 2

ENERGY —
Matter in Motion

CHAPTERS

Journal Activity You can extend the discussion by asking the students what motions are Involved in sports they play. You also might bring in the following Unit topics: How does gravity affect circus performances? Which laws of motion are illustrated by a roller coaster? How much work does a trapeze artist do? Have the students keep a record in their journals of any questions they might have about the topic of physics in the circus. After the students have completed the Unit, encourage them to look again at their questions. A follow-up discussion might help the students realize how much they have learned.

"**L**adies and gentlemen and children of all ages . . . Please direct your attention above the center ring . . ."

Just about everyone loves a circus. You may wonder how the high-wire performer balances on that thin wire, how the acrobats land from great heights without getting hurt, or how a trapeze artist does somersaults in midair. All of these things depend on forces, motion, and energy.

Science PARADE

CHAPTER 3

ACCELERATION, FORCE, AND MOTION

PLANNING THE CHAPTER

B = Basic   A = Average   H = Honors
The coding Basic, Average, and Honors indicates subsections, features, and resources that might be appropriate for different levels of learners. For additional suggestions regarding choice of topic and depth of coverage, see the Pacing Chart on pages T28–T31.

*Frame numbers at point of use
(TR) Teaching Resources, Unit 2
(IT) Instructional Transparencies
(LI) Laboratory Investigations
(SD) *Science Discovery* Videodisc Correlations and Barcodes
(SRG) Study and Review Guide

CHAPTER MATERIALS

Title	Page	Materials
Discover by Calculating	72	(per individual) journal
Activity: How can you determine speed?	73	(per group of 3 or 4) meter stick, toy car, stopwatch, masking tape
Discover by Doing	76	(per individual) journal
Teacher Demonstration	76	marble, ruler, tape
Discover by Problem Solving	77	(per individual) journal
Discover by Doing	81	(per group of 3 or 4) screw eye, spring scale, wooden block, weight, water
Discover by Doing	84	(per pair) slide, glass marble, steel ball bearing, stopwatch
Skill: Interpreting Diagrams	86	(per individual) pencil, paper
Activity: How can you study the effects of increasing mass on inertia?	89	(per group of 3 or 4) dynamics cart, string, springs (2), 1-kg masses (6), stopwatch
Discover by Doing	92	(per pair) spring scales (2), 1-kg mass
Discover by Writing	93	(per individual) journal
Investigation: Calculating Force, Mass and Acceleration	94	(per pair) pulley with table clamp, dynamics cart, masking tape, string, can, sand, standard masses, blocks about equal to cart in mass (4), meter stick, stopwatch or watch with digital numbers

ADVANCE PREPARATION

Have the students bring in toy cars to be sure of an adequate supply for the *Activity* on page 73. Wooden blocks and screw eyes are needed for the *Discover by Doing* on page 81.

TEACHING SUGGESTIONS

Field Trip
A field trip to a local factory can provide a resource for learning how forces and the laws of motion are manipulated in manufacturing. For example, the students can see how friction is lessened, gravity is used, and action-reaction and other forces are depended upon as a product is shaped. Such a trip could provide excellent closure to the chapter study, if prior communications with the business concern have assured that your guide will address the chapter's major concepts at work.

Outside Speaker
A race car driver, bicycle racer, or stunt driver would make an excellent choice to speak to your class about speed, acceleration, and momentum, as well as about effects of gravity and friction upon vehicle movement.

CHAPTER 3

ACCELERATION, FORCE, AND MOTION

CHAPTER THEME—*SYSTEMS AND STRUCTURES*

Chapter 3 will help the students understand forces and how they cause objects to move. Understanding the forces that cause movement—and the factors that can enhance or slow it—can have a profound effect on our everyday lives. From vehicle design to jogging, the study of motion helps us explain the how and why. It also allows the design of ever faster and safer means of transportation. A supporting theme in this chapter is **Energy**.

CHAPTER 3

ACCELERATION, FORCE, and MOTION

"You just wait," his mother said. "We'll build a better sled than Ed Sines has. Now get me a pencil and a piece of paper."

"You goin' to build a sled out of paper?" Orville asked in amazement.

"Just wait," she repeated. . . . The outline of a sled began to appear on the paper. As she drew it she talked. "You see, Ed's sled is about four feet long. I've seen it often enough. We'll make this one five feet long. Now, Ed's sled is about a foot off the ground, isn't it?"

Orville nodded, his eyes never leaving the drawing that was taking shape. It was beginning to look like a sled now, but not like the sleds the other boys had.

Scientists and engineers pursue the goals of faster, farther, and higher, in everything from automobiles to aeronautics. With wind tunnels and computer simulations, they carefully calculate the factors of acceleration, force, and motion.

Bring to class several small toys such as a glider, a car, and a parachute. After asking the students to predict the pattern of motion each toy will have, demonstrate each one. Then ask volunteers to describe precisely each object's motion. You might write the students' descriptions on the chalkboard. Were the students' predictions borne out by the demonstrations? Ask the students to discuss factors they think might have caused the toys to move as they did. Conclude by emphasizing the need for understanding the concepts of velocity, acceleration, friction, and gravity. These demonstrations can provide a reference point as chapter study progresses.

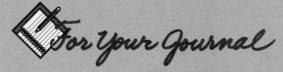

For Your Journal

The journal questions will help you identify any misconceptions the students might have about acceleration, force, and motion. In addition, these questions will help the students realize that they possess knowledge about these concepts. (Students will have a variety of responses to the journal questions. Accept all answers, and tell the students that at the end of the chapter they will have a chance to refer back to the questions.)

"You've made it too low," Will said.

"You want a sled that's faster than Ed's sled, don't you?" His mother smiled. "Well, Ed's sled is at least a foot high. Our sled will be lower—closer to the ground. It won't meet so much wind resistance."

"Wind resistance?" It was the first time Wilbur had ever heard the expression. He looked blankly at his mother.

"Remember the blizzard last week?" she asked. "Remember when you went out to the woodshed and the wind was so strong you could hardly walk to the shed? I told you to lean over, and on the next trip to the wood-shed you did. When you came back with an armful of wood you laughed and said, 'Mother, I leaned 'way forward and got under the wind.' You were closer to the ground and you were able to lessen the wind resistance. Now, the closer to the ground our sled is, the less wind resistance there will be, and the faster it will go."

"Wind resistance . . . wind resistance," Wilbur repeated, and maybe the airplane was born in that moment. Certainly neither Will nor Orville Wright ever forgot that first lesson in speed.

from *The Wright Brothers: Pioneers of American Aviation* by Quentin Reynolds

ABOUT THE LITERATURE

After reading about the death of pioneer glider Otto Lilienthal in 1896, the Wright brothers became interested in flying. They soon began serious study and obtained all the scientific knowledge of aeronautics then available.

In 1900 they tested their first glider near Kitty Hawk, North Carolina. They returned the next year with a larger glider and demonstrated how they could control sideways balance by presenting the tips of the wings at different angles to the wind.

The brothers built a third glider in 1902. This glider had aerodynamic qualities far in advance of any earlier one. On December 17, 1903, Wilbur and Orville Wright made the world's first flight in a power-driven, heavier-than-air machine. During its first flight, the plane flew 37 meters and remained airborne for 12 seconds. It reached a speed of about 48 km/h.

For Your Journal

- What do the laws of motion tell us about how things move?
- How are objects that move swiftly through the air and the water shaped?
- How can understanding how things move help you in your everyday life?

FOCUS

This section considers the components of motion: distance, time, and direction. Speed, velocity, and acceleration are defined and used in calculations to describe motion. Momentum is introduced to convey the idea of the force of a certain mass in motion.

MOTIVATING ACTIVITY

Ask the students to name what is important about their motions in a race. (They might suggest how good their start is, how fast they go, how quickly they can get to top speed, whether they follow a curved or straight track, and how many of the factors that slow them down they can control.) Then ask the students to name ways they think these aspects could relate to a scientific study of motion. Finally, help the students draft a general statement about the importance of understanding speed, velocity, acceleration, and momentum.

PROCESS SKILLS
- Measuring • Solving Problems • Observing
- Analyzing

POSITIVE ATTITUDES
- Cooperativeness • Curiosity
- Precision

TERMS
- speed • velocity
- acceleration • momentum

PRINT MEDIA
Vehicles: The Most Extraordinary Achievements Since the Invention of the Wheel by Nigel Hawkes (see p. 67b)

ELECTRONIC MEDIA
Motion and Energy: Physical Simulations Focus Media, Inc. (see p. 67c)

SCIENCE DISCOVERY
- Acceleration; collision
- Acceleration; falling body • Average speed; animation • Odometer on car

BLACKLINE MASTERS
Thinking Critically
Connecting Other Disciplines
Reading Skills
Laboratory Investigation 3.2
Study and Review Guide

DISCOVER BY *Calculating*

The average speed is 5 blocks (m) divided by 10 minutes, or 0.5 block per minute. Ask the students to put their calculations in their science portfolios.

SECTION 1

Describing Motion

Objectives

Define the terms speed, velocity, acceleration, *and* momentum.

Describe the advantages and disadvantages of momentum an athlete encounters.

Apply the concepts of speed, velocity, acceleration, and momentum when describing motion.

Imagine that you are an Olympic athlete about to compete in a bobsled event. You gaze at the icy, steep course and take a deep breath. The starting bell rings. Grasping the sled, you and your partner push and run alongside it. After a running start, you both jump on and start your first run down the treacherous course. You've made the fastest start of the day! Now, if you and your partner can coax every bit of speed from the sled, you may have a chance to win the gold!

You Were Going How Fast?

In describing both bobsleds and other moving objects, the term *speed* is used. How fast an object moves from one place to another in a specific time is its **speed.** To calculate speed, you divide the distance traveled by the total travel time.

$$\text{average speed} = \frac{\text{total distance}}{\text{total time}} \quad \text{OR} \quad v = \frac{d}{t}$$

DISCOVER BY *Calculating*

Suppose you ride your bicycle to the store. If the store is 5 blocks away, and you take 10 minutes to get there, what average speed were you traveling? Explain your calculations in your journal.

Figure 3–1. Bobsleds can travel at speeds of up to 145 km/h.

To assess the students' ability to differentiate between average speed and constant speed, pose the following problem: The distance from New York City to Orlando, Florida, is 1760 km. If the speed limit is 88 km/h, how long will the trip take? (20 hours) A family makes the trip in 24 hours of actual driving time. How can you account for the difference? (They could not travel at a constant speed of 88 km/h; they stopped along the way.) What was the family's average speed? (73.3 km/h)

ACTIVITY

How can you determine the speed of a toy car?

Process Skills: Measuring, Solving Problems, Observing, Analyzing

Grouping: Groups of 3 or 4

Hints

You might describe each job to the students before beginning the activity: driver—releases car; starter—signals start of each trial; timer—times trial from starting line to finish line; flagger—signals when car crosses finish line; recorder—writes down data for each trial.

▶ Application

1. One way is to let the car travel a longer distance, for example, 10 times farther. Doing so reduces the percentage of uncertainty because there is one more significant figure in the measured quantity.

2. Speed can't be measured directly. Rather, it is based on the relationship between two directly measured quantities— time and distance.

Of course, you know from your bicycling experiences that speed changes when you pedal fast to avoid a snarling dog or when you stop to fix a flat tire. Speed calculated by dividing distance by time is really your average speed. In the next activity you'll have a chance to calculate average speed.

DISTANCE   ÷   TIME   =   SPEED

Figure 3–2. Speed is calculated by dividing distance by time taken to travel that distance.

ACTIVITY

How can you determine speed?

MATERIALS
meter stick, toy car, stopwatch, masking tape

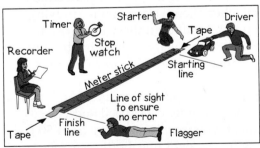

PROCEDURE
1. Lab partners should divide up the following jobs: driver, starter, timer, flagger, and recorder.
2. Set up the materials as shown. Tape the meter stick to the floor or table in an area where the car can move freely. You will use the meter stick to measure the distance the car travels.
3. The driver should place the car so that its nose aligns with the front end of the meter stick. On a signal from the starter, the driver should release

the car so that it travels next to the meter stick. The timer should start the stopwatch at the signal, too.
4. The flagger should signal the instant the nose of the car passes the end of the meter stick. The timer should stop the stopwatch on the flagger's signal.
5. The recorder should record the trial number (first, second, third, and so on), the distance traveled in meters, and the elapsed time in seconds.
6. Repeat steps 3 through 5 to make at least three trials.
7. Calculate the car's average speed for each trial. First average the total distance traveled and the total time elapsed for all trials. Then, divide the distance by the time. Record the average speed on the chalkboard.

APPLICATION
1. How might the distance traveled by the toy car be measured with greater precision?
2. Why is speed considered a derived measure rather than a direct measure?

After the completion of this Activity, have individual students explain how they arrived at their calculations. Check the student's calculations for accuracy.

 After the students have completed this Activity, have them place their observations and answers in their science portfolios.

TEACHING STRATEGIES, continued

● **Process Skills:** *Identifying Variables, Comparing*

Cooperative Learning Give groups of students a graphic illustration of the difference between speed and velocity and the importance of direction. To emphasize the point, ask them to compare the starting and ending points for two cars that travel 2 hours at 80 km/h. Car A travels on a 4-km oval track, while Car B travels on a straight road.

● **Process Skills:** *Expressing Ideas Effectively, Analyzing, Applying*

Assess the students' grasp of the difference between speed and velocity by having them list instances when a hurdler's velocity might change and instances when his or her speed changes. Write the instances on the board; point out velocity changes that do not involve speed change. (Both speed and velocity change from 0 to *x* meters/second at the race's start. Direction, and therefore velocity, changes several times during each leap. Speed probably decreased at the start of each leap and increased as soon as running resumed.)

ONGOING ASSESSMENT
ASK YOURSELF

Responses should show the need to know total distance traveled and total time elapsed during the trip, and they should accurately use the formula $v = d/t$.

① Their velocity changes constantly because their direction is constantly changing as they move around the curve.

MEETING SPECIAL NEEDS

Mainstreamed

Help mainstreamed students make a map of the route they take from home to school on graph paper. Then help the students determine the number of blocks or kilometers traveled and calculate the average speed for trips to school when walking, riding a bike, or riding in a car. The students can estimate if they have not timed these means of movement, or you can suggest average rates, such as 4.5 km/h for walking.

ONGOING ASSESSMENT
ASK YOURSELF

If the cyclist travels the same speed on both the track and the straight road, the velocities will still differ. Both speed and velocity are constant on a straightaway. Direction, and therefore velocity, is constantly changing on a curved track. The speed and velocity of the mountain bike cyclists differs yet from the other cyclists.

▶ **74** CHAPTER 3

Figure 3–3. The cars on this stretch of highway are constantly changing velocity, even when their speed remains constant. Explain why. ①

ASK YOURSELF

Suppose that the speedometer on your bike doesn't compute average speed. How could you find your average speed?

It's Not the Speed . . .

Speed tells us how fast an object is moving. Another word used to describe motion is *velocity*. **Velocity** is the speed of an object in a particular direction. It is easy to confuse speed and velocity because they are both based on the same measurements—distance traveled in a certain amount of time. The difference between speed and velocity is that speed can be motion in any direction, but velocity identifies a certain direction.

Imagine that you are at the roller rink, skating at a steady pace, smoothly following the curve of the rink. Putting one foot in front of the other, you're pushing just hard enough to keep going. Suddenly a friend skates up to you and yells, "Your velocity is changing! Your velocity is changing!" For a moment, you think your friend has lost his mind. You are puzzled, though, because you know you are covering the same distance each minute. Then you remember science class; you're studying speed and velocity. You recall that even though you are traveling at a constant speed, your velocity is changing because you are constantly changing direction to follow the curve of the rink.

Here is another example. If you were to travel in a car at a constant 50 km/h in a straight line, your speed and velocity would both be the same. If you were to speed up from 50 km/h to 70 km/h while going in a straight line, both your speed and velocity would change by the same amount. If you were to travel at a constant 50 km/h in a big circle, your speed would stay the same but your velocity would continually change. In order to have constant velocity, both speed and direction must stay the same.

ASK YOURSELF

A velodrome is a curved, steeply banked track designed for bicycle racing. Explain how the speed and velocity of a cyclist on this track would differ from that of a cyclist on a flat, straight stretch of road. Now, consider the cyclists in a mountain bike race. Their race course consists of rugged terrain that includes mountains and streams. How would their speed and velocity compare to cyclists riding on a velodrome and cyclists riding on a flat, straight road?

Figure 3–4. The parachute on the space shuttle (left) helps bring it to a stop. The dragster's parachute (bottom) also opens to help bring it to a stop.

Slowing Down and Speeding Up

Picture yourself sitting in a dragster, revving the engine, and waiting for the starting light. The track is short and straight; the race will be over in a few seconds. The green starting light flashes, you stomp on the accelerator, and the noise of the squealing tires drowns out the cheers of your crew. As you speed up, your body is pushed back into the car seat. You are feeling acceleration. Both you and the car keep accelerating as the car crosses the finish line. Then you stop the car by using the brakes and releasing a parachute that flares out the back. As the car slows down, you stop accelerating, right? Wrong!

Any time your velocity changes, whether by speeding up, slowing down, or changing directions, you are accelerating. The rate at which velocity changes is called **acceleration.** So when your dragster slows down, its velocity changes and, thus, it accelerates. Your body tends to keep moving forward as the dragster slows. Your seat belt keeps you from traveling forward.

Figure 3–5. These cars are maintaining a fairly constant speed. However, these cars are also accelerating because their direction is changing.

✧ **Did You Know?**

Objects falling through air accelerate until air resistance equals the downward force of gravity; then the objects' velocity is constant. Objects with large surface areas and low weights, such as feathers or a tissue, fall relatively slowly because they quickly reach terminal velocity. This explains why a small insect can fall great distances and walk away unharmed. It falls at a terminal velocity of only a few cm/s.

GUIDED PRACTICE

Have the students compute the speed of several objects in both km per second and km per hour. The huge difference in numerical values should help the students recognize the importance of the units of measure in their answers.

INDEPENDENT PRACTICE

Have the students provide written answers to the Section Review. In their journals, have the students draw and label a diagram that compares the momentum of a child and an adult traveling at the same velocity.

EVALUATION

PORTFOLIO ASSESSMENT Have the students write a paragraph comparing the speed, velocity, and momentum of these two model airplanes and explaining the differences: Plane A weighs 400 g. It flies in a circular pattern and covers 10 m/sec. Plane B weighs 350 g. It flies in a straight line and covers 10 m/sec. Ask the students to place their paragraphs in their science portfolios. (Both planes have the same speed. Plane A's velocity is continually changing because it keeps changing direction, while B's velocity is constant. Plane A has more momentum because it has greater mass.)

LASER DISC

1710

Acceleration, collision

1709

Acceleration, falling body

26607, 26608, 26951

Average speed; animation

2659

Odometer on car

ONGOING ASSESSMENT
▼ ASK YOURSELF

Acceleration shows change in velocity over time. Any acceleration means that either direction or speed (or both) has changed; velocity is speed in a certain direction. The rate of acceleration is slower if the change occurs over a longer time span.

Demonstration

To demonstrate the relationship between velocity and acceleration, roll a marble down the groove in a ruler that has been marked with tape at 0, 10, 15, and 20 cm. Tilt the ruler so that the marble rolls just to the bottom at constant velocity when given a small start. The students should watch and listen for the clicks the marble makes as it hits the tape. Then increase the tilt of the ruler, and have the students predict what will happen to the clicks they hear. (The time interval between clicks will become shorter for every increase in velocity of the marble.)

Figure 3–6. The momentum of this football player moves him forward even though he is being tackled around the legs.

Now imagine that you are trying to qualify for the Indianapolis 500. Only those drivers with the fastest qualifying times get into the race. You drive around the oval track as fast as the car, and your nerves, allows. Remember, because acceleration also applies to changes in direction, you are accelerating even when you come into and get out of a curve. In fact, you are accelerating even if your speed is constant. Because acceleration depends on velocity, the only time you are not accelerating is when both your speed and direction are constant. In the next activity you'll have the opportunity to discover how often you accelerate throughout the day!

DISCOVER BY *Doing*

Today as you are walking from class to class or from class to lunch, notice how you accelerate by speeding up, by slowing down, or by changing direction. Record the results in your journal. Share your results with a classmate. How do they compare? ✐

▼ ASK YOURSELF

How does acceleration affect velocity?

Momentum

Think about all the speeds, velocities, and accelerations that occur during a football game. The players are moving all the time. What affects their motion?

Imagine that you are watching a football game on television. One sports announcer exclaims, "Unbelievable! Number 36 was hit at the three-yard line, but he was just too heavy to stop. His momentum carried him into the end zone for a touchdown!" The other sports announcer says, "It's a good thing Number 28 wasn't carrying the ball. He weighs 40 pounds less than Number 36 and wouldn't have had the momentum to score."

RETEACHING

Graphing may help the students visualize the relationship between velocity and acceleration. Create graphs on the chalkboard or an overhead transparency. Place time in seconds on the x axis and velocity in meters/second on the y axis. Then illustrate how acceleration is large (steep line on the graph) if the change in velocity is large or if the change takes place very quickly. (Note: The line described will be straight if acceleration is constant.) Contrast with another graph showing no acceleration (straight top line at zero) if velocity is constant.

EXTENSION

Supersonic jet pilots and astronauts must cope with enormous changes in acceleration. Have the students do research and report on design features that allow jets and spacecraft to accelerate rapidly. They might also report on equipment and measures used to lessen the physical strain on the pilots.

CLOSURE

Cooperative Learning Have the students work in small groups to summarize what they learned in this section. Encourage the groups to share their summaries with other groups and use the toy vehicles from the Chapter Motivating Activity to illustrate concepts.

Both of the sports announcers are using a physical science term here. In physical science, **momentum** is an object's mass multiplied by its velocity. Momentum, like velocity and acceleration, has definite direction. Unlike acceleration and velocity, the momentum is related to the mass of the moving object.

Problem Solving

Suppose you are playing football on the defensive team. Which sort of ball carrier would you rather face? (a) a small, slow runner (b) a small, fast runner (c) a large, slow runner or (d) a large, fast runner. In your journal, write down your answer. Then explain your choice.

If objects have the same velocity, it will be harder to make the more massive object change velocity. If objects have the same mass, it is harder to change the faster object's velocity.

This applies to people, cars, bobsleds, bicycles, or any other moving object. For example, a small car can corner more easily than a large car (assuming the tires, brakes, and so forth are the same). Because the small car has a lot less mass, it has less momentum at the same velocity. It is easier to accelerate the object with less momentum. Since cornering requires acceleration, the car with less momentum corners better.

► ASK YOURSELF
Think of a sport other than football, and then describe how momentum affects the game.

Figure 3–7. The bicycle and the truck are traveling at the same velocity. Which has more momentum?

DISCOVER BY
Problem Solving

(a) The small, slow runner has the least mass and speed and, therefore, the least momentum, and would require the least force to tackle.

ONGOING ASSESSMENT
► ASK YOURSELF

Responses should take into account the mass of the participants and the object they manipulate, the speeds attained by the people and the object, and the force necessary to place the person in a scoring position.

SECTION 1 REVIEW AND APPLICATION

Reading Critically

1. Speed and velocity both measure distance traveled over a certain amount of time. However, velocity is measured in a particular direction; speed is not.

2. When you slow and stop a car, you change its velocity and it accelerates.

Thinking Critically

3. The weight, or mass, of the car determines its momentum when it is in motion. Greater mass gives a car greater momentum, which makes it harder to accelerate or turn a corner.

4. Velocity changes whenever direction changes; cars change direction often. A meter that changed with every change in direction would be too confusing.

SECTION 1 REVIEW AND APPLICATION

Reading Critically
1. How are speed and velocity alike? How are they different?
2. How does stopping a car make it accelerate?

Thinking Critically
3. When engineers design cars, they are very concerned with the overall weight of the car. Why is this?
4. Why do you think cars have speedometers instead of velocity meters?

FOCUS

Section 2 focuses on balanced and unbalanced forces and their measurement. Motion is described as the result of unbalanced forces. Ways the various forms of friction oppose the motion of objects are discussed. Gravity is defined as a force of attraction between objects. The effects of mass and distance on gravitational pull are explored.

MOTIVATING ACTIVITY

Ask the students to name some forces that are acting on them as they sit in class and as they move through the day. (They might mention gravity, friction, and wind, among others.) Then have them study Figures 3–8 and 3–9 to try to explain the forces that have caused the illustrated movements.

(Gravity helps the soapbox derby racers go downhill; muscles exert a pushing force, and feet push against the ground.)

PROCESS SKILLS
- Communicating
- Interpreting Data
- Recognizing Time/Space Relations

POSITIVE ATTITUDES
- Cooperativeness
- Curiosity

TERMS
- force • newton
- friction • weight

PRINT MEDIA
"... And Lived to Tell About It" from *Sports Illustrated* Magazine by Ed Hinton (see p. 67b)

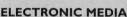

ELECTRONIC MEDIA
Falling Bodies and Projectile Motion Barr Films (see p. 67c)

SCIENCE DISCOVERY
- Car, aerodynamic
- Gravity, direction of • Newton; unit of force

BLACKLINE MASTERS
Extending Science Concepts Laboratory Investigation 3.1 Study and Review Guide

✧ Did You Know?
Everything in the universe is in motion. Even as you sit reading this page, you are moving very rapidly because Earth is rotating on its axis. You are also moving with Earth as it revolves around the sun. In addition, the sun, Earth, and the rest of the planets in our solar system are involved in the general rotation of our galaxy within the universe.

SECTION 2

Forces and Motion

Objectives

Define *the terms* force, friction, *and* gravity.

Explain *how forces can cause changes in motion and how mass and distance affect gravity.*

Compare *friction, drag, and air resistance.*

If you want to do more than describe motion, you must understand what affects motion. What starts it? What stops it? What must you do to make the motion of an object like a bobsled change? The answer is simple. It just takes a push or a pull!

Pushes, Pulls, and Motion

A push or a pull is required to change motion—even in bobsledding. When the bell rings, the bobsled racers *push* the sled from the starting line and jump on as it increases velocity. After that, gravity *pulls* the bobsled downhill, causing the bobsled and its passengers to accelerate. The bobsled accelerates until the brakes and the opposing motion between the sled's runners and the ice *push back* enough against the forward motion to stop the acceleration.

Figure 3–8. This soapbox derby racer is accelerating down the slope as a result of the force of gravity.

Throughout the day you experience pushes and pulls. When you walk down the street, your feet push on the street, moving you forward. When you write a homework assignment, you push on the pencil, making it move across the paper. Every time there is a change in the motion of anything, a force has made that change in motion happen. A **force** is a push or a pull.

● **Process Skills:** *Applying, Communicating*

Ask a volunteer to explain some of the unbalanced forces that allow the boy in Figure 3–9 to push the tricycle up the hill. (The boy's push is greater than the push exerted by gravity and the mass of the tricycle.)

● **Process Skill:** *Applying*

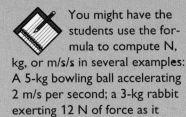

You might have the students use the formula to compute N, kg, or m/s/s in several examples: A 5-kg bowling ball accelerating 2 m/s per second; a 3-kg rabbit exerting 12 N of force as it hops away; 20 N of force making a wagon roll 2 m/s/s. Ask the students to perform their calculations in their journals.

In SI, force is measured in newtons (N), named after the British scientist Sir Isaac Newton. One **newton** is the force required to cause a 1-kg mass to accelerate 1 meter per second each second. If you applied a force of 2 N to a 1-kg brick, it would accelerate by 2 m/s each second. If you only applied 1/2 N to the 1-kg brick, it would accelerate by 1/2 m/s each second.

Newtons = kilograms × meter/second/second

Figure 3–9. Because of the different masses of these two objects, different amounts of force are required to move them.

How Much Acceleration?

The amount of acceleration depends on both the amount of the force used and the mass of the object being pushed or pulled. If you apply a small force to a massive object like a railroad car, you will cause a tiny acceleration. If you apply a large force to a small object like a golf ball or a pencil, you will cause a large acceleration. If you apply a tiny force to a tiny object or a large force to a large object, you will cause a medium acceleration.

Whenever a force is applied to an object, the object's motion is changed. When you kick a soccer ball, you change its motion by applying a force. The kick is a push. When you push a bobsled, you change its motion from no motion at all to moving forward. When you drop a fork in the school cafeteria, the earth pulls the fork down. The earth's pull, gravity, is a force that changes the fork's motion downward. The fork also pulls at the earth, but the earth's acceleration is much smaller.

THE NATURE OF SCIENCE

Sir Isaac Newton, considered to have been one of the greatest scientific minds in history, had to overcome many disadvantages in order to succeed. Newton's father, an illiterate farmer, died three months before Newton was born. Newton himself was born prematurely and was expected to die at birth. When he was four, Newton's mother remarried and left him in the care of his grandmother. When he was in his early teens, his stepfather died, and Newton was forced to leave his studies and work on his mother's farm. However, he was ill-suited to farming, and his mother finally gave in to the pleas of his teacher that Newton return to school and enter Cambridge University.

During his last year at the university, bubonic plague broke out in England; Newton returned to his mother's farm. It was here that he performed many of his outstanding experiments on light and optics. Eventually, when he returned to Cambridge as a full professor, Newton developed the first working reflecting astronomical telescope and published his famous papers on motion and optics.

● **Process Skills:** *Comparing, Expressing Ideas Effectively*

Assess the students' understanding of balanced and unbalanced forces by having them use Figure 3–10 to explain when an added force will cause an object to move, accelerate, or remain at constant velocity.

● **Process Skill:** *Analyzing*

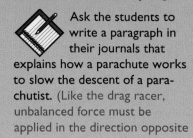

Ask the students to write a paragraph in their journals that explains how a parachute works to slow the descent of a parachutist. (Like the drag racer, unbalanced force must be applied in the direction opposite the parachutist's motion—in this case, downward motion. A parachute is released from the parachutist's pack. The force of the air against the parachute quickly opposes the downward motion of the parachutist. The force of the air against the parachute is unbalanced; the parachutist slows down.)

◢ MEETING SPECIAL NEEDS

Second Language Support

Point out to the students that many words in the English language have multiple meanings. Pair limited-English-proficient students with fluent English speakers. Have the students use dictionaries, etymologies, and other sources to determine how the scientific meanings relate to these uses: "friction" between people; the "gravity" of a situation; someone's being a "force" within his or her community; a "weighty" problem; something that's a "drag"; "accelerated" learning.

ONGOING ASSESSMENT
◣ ASK YOURSELF

If the force is balanced by another force, there will be no movement. However, if the force is unbalanced, the object will move.

Balancing Act One force can cancel the effects of another force. For example, you can hold a fork without dropping it. When you do, you are pushing up with exactly the same force as gravity is pulling down. If a force does not change the motion of an object to which it is applied, the force must have been balanced by another force, because there is no movement when opposing forces are balanced. Only when forces are not balanced does the object accelerate.

Figure 3–10. When a football player is successfully tackled, the player's forward motion is stopped by a force produced by the tackler.

Think about some situations in which objects accelerate—the drag racer, for example. In order for a drag racer to stop quickly, a large, unbalanced force must be applied in the direction opposite the dragster's motion. To apply such a force, a parachute is released from the back of the dragster. The force of the air against the parachute quickly opposes the forward motion of the dragster. The force of the air against the parachute is unbalanced, and the dragster slows down.

Figure 3–11. The result of forces on an object

Forces on a stationary object	Result	Forces on a moving object	Result
Balanced	Object remains stationary	Balanced	Object remains at constant velocity
Unbalanced	Object accelerates	Unbalanced	Object accelerates

◣ ASK YOURSELF
What happens to an object that has a force applied to it?

Friction

Imagine that you are pushing a box loaded with books across the floor. As you push, the floor seems to oppose the motion of the box. What's wrong? Nothing!

You are experiencing **friction,** the force that opposes motion between two surfaces that are touching. In order for you to slide the box of books across the floor, you have to overcome friction.

The amount of friction between two surfaces depends on how hard the surfaces are forced together, the materials that make up the surfaces, and the smoothness of the surfaces. The more books you load into the box, the harder it is to push across the floor. If the floor is covered with rough carpet instead of smooth wood, the box is harder to push. The two surfaces "grab" at each other more. Try the next activity to find this out for yourself.

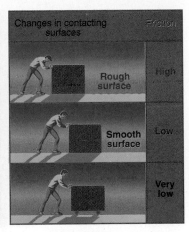

Figure 3–12. The person in this illustration must overcome friction to move this box.

 **BY** *Doing*

Attach a screw eye and a spring scale to one end of a wooden block as shown. Use this device to experiment with friction. First, drag the block along a sidewalk, and note the amount of pull on the spring scale. Then, place a heavy weight on top of the block, and pull it across the same surface. Finally, run water on the sidewalk as you pull the block and weight. Also try pulling the block along a smooth surface like a table top or a waxed floor. What conclusions about friction can you draw from your experiment? ✏

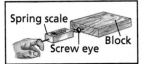

Spring scale
Block
Screw eye

If you placed the box of books on wheels, you could push it across the room much more easily because there would be only a small amount of friction between the wheels and the floor and the wheels and their axles. Friction is not completely eliminated, however. The rolling box would begin to slow down as soon as you stopped pushing it.

TEACHING STRATEGIES, continued

● **Process Skills: *Synthesizing, Expressing Ideas Effectively***

Assess the students' understanding of the concept of gravity by having them explain why the planets stay in orbit around the sun. (Although they are very far apart, each planet is under the influence of the sun's gravity. This attractive force keeps the planets in elliptical orbits around the sun. If gravity suddenly were to disappear, the planets would travel in straight lines tangential to their orbits and leave the solar system.)

● **Process Skill: *Comparing***

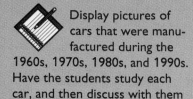

Display pictures of cars that were manufactured during the 1960s, 1970s, 1980s, and 1990s. Have the students study each car, and then discuss with them how car design has changed over the past 40 years to reduce air resistance. Conclude by asking the students to speculate in their journals how improved design contributes to improved fuel economy.

MULTICULTURAL CONNECTION

The invention of the wheel allowed humans greater control over the flow and direction of many forces. The earliest known wheels were constructed in Mesopotamia around 3500 B.C. and are believed to have been fashioned after the potter's wheel. These early wheels were disk-shaped and wooden, and were secured on a round axle by wooden pins. The wheel evolved as sections were carved out of the disk. By about 2000 B.C., radial spokes had been developed. They allowed wheeled carts to replace the sled, advancing civilization tremendously. Animals such as oxen and horses could be better used; most types of work became easier.

REINFORCING THEMES
Systems and Structures

PORTFOLIO ASSESSMENT

Encourage interested students to describe and sketch aerodynamic systems and structures they have observed. Ask the students to put their sketches in their science portfolios.

Figure 3–13. As these boats speed across the water, they encounter both drag and air resistance.

What a Drag! Friction exists with liquids and gases as well as solids. Imagine that you are piloting a motor boat through the water. Wind and the spray of water blow past you. At top speed, you are skimming through the water at 50 km/h.

As the boat moves, its lower surface passes through the water and its upper surface passes through air. There is no direct contact between solid surfaces. However, the boat still encounters friction. Even though the boat's motor continues to apply a force, the boat stops accelerating when it reaches a speed of 50 km/h. And when the force of the motor stops pushing, the boat quickly comes to a stop.

Both the water and the air offer an opposing force against the moving boat. Friction between a solid surface and a liquid or a gas is sometimes called *drag*. When the gas is air, however, the friction is usually called *air resistance*.

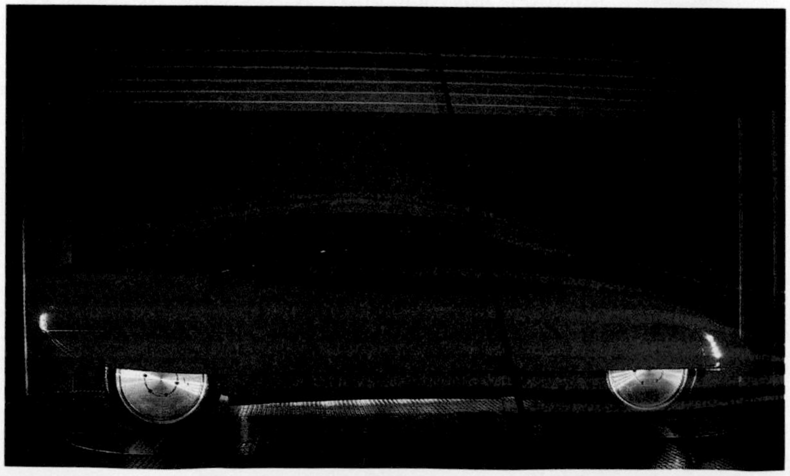

Figure 3–14. In order to achieve high speeds, automobiles must encounter little air resistance. Automobile designers use wind tunnels like the one shown here to check the air resistance of their designs.

How Do They Go So Fast? The top velocity of speed skiers is 224 km/h. They can go this fast because they minimize air resistance. The friction between the skis and snow is very small. The main drag on speed is the air in front of the skiers. Skiers can cut air resistance by decreasing the amount of body exposed to the air. That's why they crouch while skiing. In what other ways can skiers decrease drag to increase speed? ①

Poles held close to body

Aerodynamically designed helmet

Tight-fitting, smooth-surface clothing

Buckles on boots align with surface

Figure 3–15. A downhill skier's clothing and equipment minimize air resistance.

▼ **ASK YOURSELF**

What factors determine the amount of friction between two solid surfaces?

Gravity

As you read at the beginning of this section, gravity is the force that pulls bobsledders and skiers down a mountain. In the seventeenth century, Sir Isaac Newton discovered that all objects with mass attract each other. In fact, he determined that gravity is a fundamental force of nature.

When you hold an object and then let it go, the object accelerates toward the earth because of the force of gravity pulling it. When you throw a baseball, it eventually hits the

GUIDED PRACTICE

Have the students apply the formula N = kg × m/s/s to a given mass and acceleration. Then ask volunteers to explain what will happen to the numbers in the formula if the force pushes the mass across a rough surface, across water, or through air.

(N, or the amount of force used, will increase as the surface over which the object is pushed becomes rougher. N will decrease when the object is pushed across water or through air, as compared to a solid surface.)

INDEPENDENT PRACTICE

Have the students provide written answers to the Section Review. In their journals, have the students draw a picture of a spring scale weighing an object and write a brief explanation of what is being measured in the process.

EVALUATION

After completing the Discover by Doing activity on p. 84, ask the students to summarize their understanding of the relationship of gravity to acceleration of dropped objects. (The gravitational pull of Earth is so much greater than the resistance of the air that the object falls to Earth.)

SCIENCE BACKGROUND

The interferometer was invented in 1881 by American physicist Albert Michelson. The device acted to split a beam of light in two. The separated beams traveled along different paths and then were rejoined at some point. The splitting of light beams in an interferometer is now being used in the development of a new branch of physics, gravitational astronomy. Two large-scale interferometers will be built in the United States, one in Louisiana and one in Washington.

✧ Did You Know?

Gravitational acceleration remains almost constant regardless of an object's mass, providing air resistance is kept to a minimum. It is about 9.8 m/s/s.

DISCOVER BY *Doing*

The marble and the ball bearing should accelerate at the same rate, since they present roughly the same air resistance (having the same shape).

Figure 3–16. Gravitational force, represented by the arrows, varies with the mass of the objects and their distance from one another.

ground because the force of gravity pulls it down. Houses must have support beams to oppose the force of gravity that would otherwise pull the roof down. The moon circles the earth, and the earth circles the sun instead of heading off in another direction because the force of gravity keeps them in orbital paths. Gravity is everywhere in the universe because objects that have mass are present throughout the universe.

Falling Down   The force we call *gravity* is described by the universal law of gravitation, or the law of gravity. The law of gravity says that the strength of an object's gravitational force depends on its mass. The more mass an object has, the more gravitational force it exerts on other objects.

When you drop a fork in the kitchen, the fork falls straight down because the earth has a very large mass and a very large gravity to match. The refrigerator is also pulling on the fork, but you do not see the fork fall toward the refrigerator because, compared to the earth, the refrigerator has a very small mass and thus a very small gravitational pull. Sensitive instruments can measure the gravity pull between the refrigerator and the fork, however.

So the force of gravity between two objects depends on both of their masses—the bigger the mass, the bigger the force. The law of gravity also states that the farther apart two objects are, the weaker the gravitational force that pulls them together.

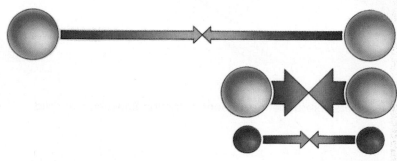

DISCOVER BY Doing

Use a slide to observe the acceleration of objects toward the earth caused by gravity. Go to the top of the slide, and release, at exactly the same time, a large glass marble and a steel ball bearing of the same size. Have someone at the bottom of the slide note when each ball reaches the bottom. What conclusions can you draw about their acceleration? ✎

How Heavy Is It? Any discussion of gravity requires a discussion of the meaning of the words *mass* and *weight*. Many people confuse those two words because they are so closely related. Mass is the amount of matter in an object. You may recall that SI units of mass include the kilogram (kg) and the gram (g). **Weight** is the force of gravity pulling one object toward the center of another object. Usually you think about gravity pulling an object toward the center of the earth. Because weight is a force, weight is measured in newtons (N).

You have probably heard people say they are trying to "lose weight." What they really mean is that they are trying to lose mass, because they want their bodies to have less matter. Of course, as long as they stay on Earth, they will lose weight when they lose matter. What if you were a space traveler? Since weight identifies the gravitational pull of objects, someone who weighs 600 N on Earth would weigh only 100 N on the moon. The moon has only one-sixth the gravitational pull of the earth and so attracts with only one-sixth the force. Although you would weigh less on the moon, the amount of matter would still be the same in both locations. To lose mass, moving to another planet isn't the answer!

◢ ASK YOURSELF
What is the difference between mass and weight?

Figure 3–17. Astronaut Edwin (Buzz) Aldrin weighed less while he was on the moon than when he was on Earth. His mass, however, was unchanged. Why?

SECTION 2 REVIEW AND APPLICATION

Reading Critically

1. What is the relationship between force and acceleration?
2. Explain why friction is a force.
3. What is gravity?
4. Give two examples of objects that are accelerating but are not speeding up. Explain your answer.

Thinking Critically

5. Two identical cars are traveling at identical speeds on identical roads. The only difference is that the second car is traveling in a rainstorm. Which car requires the greater force to move it forward? Explain your answer.
6. Imagine that the sun suddenly collapses into a black hole, with all of its original mass compressed into the volume of a marble or a bead. If the center of the black hole is in the same place as the current center of the sun, how will that affect the earth's orbit around the sun?

Process Skills:
Communicating, Interpreting Data, Recognizing Time/Space Relations

Grouping: Individuals

Objectives:
● **Construct** and **interpret** data from diagrams.
● **Communicate** in visual forms through diagrams.

Discussion

Stress the importance of drawing an arrow to show the force exerted by the spring on the weight. Make sure that the students also draw the size of the arrow relative to the size of the force of gravity. Ask why it might be more useful to explain the forces at work here in a diagram than in written words. (Students' answers should note that visual presentation is more efficient and clear in this case.)

▶ **Application**

The weight on the spring is always accelerating, even when it is momentarily not moving at either the top or the bottom. At these points, the weight's change in direction changes the weight's velocity. The weight, thus, is accelerating.

★ **PERFORMANCE ASSESSMENT**

After the completion of this Skill, have individual students find examples of diagrams in chapters they have already studied. Check the student's comprehension by having him or her interpret the diagram in his or her own words.

▶ **86** CHAPTER 3

✳ **Using What You Have Learned**

1. When the weight is speeding up, both acceleration and velocity are in the same direction. When it is slowing down, acceleration and velocity are in opposite directions.

2. Acceleration is change in velocity in a certain amount of time. Changing velocity by making it slower is acceleration in the direction opposite to the motion of the object.

3. Acceleration is always in the direction of the unbalanced force, as stated in the second law of motion.

4. No. The brakes cause an acceleration of the car in the opposite direction of its forward motion, causing the car to slow down.

 PORTFOLIO ASSESSMENT

After the students have completed this Skill, ask them to place their diagrams and answers in their science portfolios.

SKILL *Interpreting Diagrams*

▶ **MATERIALS**
● paper ● pencil

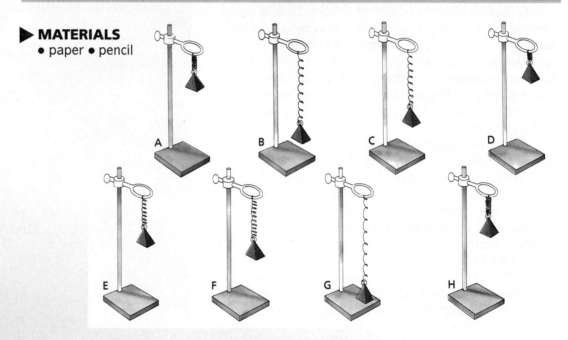

▼ **PROCEDURE**

1. Sketch the diagrams shown. If they are not in the correct sequence, reorder them as you do this step.
2. Show the direction in which the weight is moving with an arrow. If the weight is not moving, do not add an arrow.
3. Draw an arrow to show the force exerted by the spring on the weight. Make the length of your arrow represent the size of the spring's force relative to the size of

the force of gravity. For example, if the spring's force is much greater than the force of gravity, the arrow you draw for it should be longer than the gravity-force arrow in the diagram.
4. Determine the direction in which the unbalanced force is acting, and show it in your diagram.
5. Indicate the direction of acceleration.

▶ **APPLICATION**
Does the weight on the spring ever move without any acceleration or unbalanced force acting on it? Explain why.

✳ ***Using What You Have Learned***
1. When are the acceleration and the direction of motion the same? When are they different?

2. Explain how it is possible to have acceleration in one direction and motion in the other.
3. Is acceleration always in the same direction as unbalanced force? Why or why not?
4. Is it wrong to say that the brakes of a car cause an acceleration opposite to the direction in which the car is moving?

FOCUS

This section discusses the three laws of motion. In explaining these laws, the terms *inertia*, *action force*, and *reaction force* are defined. In addition, the principle of conservation of momentum is introduced: momentum can be transferred from one object to another but cannot change in total amount.

MOTIVATING ACTIVITY

Bring to class two toy cars and several metal washers. Place the washers on top of one of the cars, and set the cars on the floor or a table top so they face one another. Ask the students to predict what will happen if the two cars have a head-on collision. (The washers will continue to move forward though the cars stop each other.)

Then push each car so they collide head-on so the students can determine if their predictions are accurate. Conclude by discussing with the students the forces they think are at work here.

Predicting and Explaining Motion

SECTION 3

An Olympic bobsled race is an exciting place to observe motion, but you also observe objects in motion every day. In the late 1600s, Sir Isaac Newton formulated three laws that describe motion. These three laws identify some universal characteristics of objects, both in motion and at rest. The laws of motion can help you understand why objects move the way they do.

Objectives

Use the three laws of motion to explain how things move.

Explain the terms inertia, action force, and reaction force.

Apply the law of conservation of momentum.

The First Law of Motion

Have you ever been in a car that had to stop suddenly? If so, you probably remember that your body kept traveling forward, even after the car had stopped. Your seat belt provided a balancing force that stopped you from continuing forward. You experienced one aspect of the first law of motion.

This law states that every object maintains constant velocity unless acted on by an unbalanced force. Unless an object receives an unbalanced push or pull, it will just keep doing what it is already doing. In general terms, this law says that objects at rest stay at rest and objects in motion with a constant speed stay in motion unless they receive an unbalanced push or pull, to change the objects' velocity.

Figure 3–18. In a car crash, the driver and passengers keep moving forward until they are stopped by another force. If seat belts are not worn, this stopping force may be the car's windshield.

PROCESS SKILLS
- Measuring
- Interpreting Data
- Communicating

POSITIVE ATTITUDES
- Cooperativeness
- Precision

TERMS
- inertia • law of conservation of momentum

PRINT MEDIA
Science and Sports
by Robert Gardner
(see p. 67b)

ELECTRONIC MEDIA
Force and Motion: Newton's 3 Laws
Barr Films
(see p. 67c)

SCIENCE DISCOVERY
- Surface and motion; friction
- Rocket propulsion

BLACKLINE MASTERS
Study and Review Guide

THE NATURE OF SCIENCE

In his famous work *Principia*, Newton made a brilliant observation: "Nature is essentially simple. Therefore, scientists ought not to introduce more hypotheses than are needed to explain observed facts." He formulated elegantly simple laws of motion that explained the most fundamental phenomena of the universe. Newton's vision allowed him to comprehend the relationships among force, mass, and motion.

In the twentieth century, Newton's profound observation would be echoed by another great scientific mind. Albert Einstein said, "Most of the fundamental ideas of science are essentially simple, and may, as a rule, be expressed in a language comprehensible to everyone."

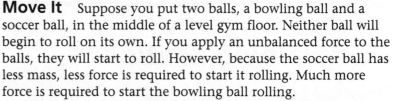
Figure 3–19. Because of the inertia of the objects on this table, the table setting remains intact momentarily as the table is quickly slid from underneath it.

Move It Suppose you put two balls, a bowling ball and a soccer ball, in the middle of a level gym floor. Neither ball will begin to roll on its own. If you apply an unbalanced force to the balls, they will start to roll. However, because the soccer ball has less mass, less force is required to start it rolling. Much more force is required to start the bowling ball rolling.

Now suppose that a friend of yours has started both balls rolling toward you so that they are both traveling at the same velocity. Which ball will be harder for you to stop? The bowling ball, of course! The bowling ball is harder to start and harder to stop because of its greater mass.

The first law of motion also states that objects with greater mass require more force to change their velocity. The resistance any object has to a change in velocity is called **inertia** (in UHR shuh). This is why it is much harder to start and stop the bowling ball than the soccer ball. Because the bowling ball has greater mass, more force is required to start or stop it. It has more inertia than the soccer ball. This property makes small cars easier to handle than big ones and small football players easier to stop than big ones.

Weight Loss

Recall that inertia depends on the mass, not the weight of an object. Objects have inertia regardless of where they are. Inertia even applies to objects in space. In space, massive objects such as support beams will seem to have no weight even though they have large mass. However, once a beam is put in motion, slowing it down or stopping it will require a very large force. The inertia of the beam will be proportional to its mass, not its apparent weight. Try the next activity to learn more about inertia.

Figure 3–20. Astronauts rescuing satellite

ACTIVITY

How can you study the effects of increasing mass on inertia?

MATERIALS

dynamics cart with string tied to both ends, springs (2), string, 1-kg masses (6), stopwatch

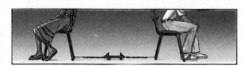

PROCEDURE

1. Set up the dynamics cart as shown.
2. Make a table like the one shown.

TABLE 1: MASS AND INERTIA		
Trial	Mass of the Cart	Added Mass (kg)

3. When the cart is not moving between the two supports, it is at its equilibrium position. Displace it from its equilibrium position by pulling or pushing it toward either support.

4. After moving the cart from its equilibrium position, let it go and measure how many seconds it takes to repeat its back-and-forth motion 10 times. Each back-and-forth motion is called a period. Divide the time for 10 periods by 10 to get the average period, or time, for one back-and-forth motion. Do three trials, and average the results. Record the results in your table.
5. Add a 1-kg mass to the cart, and repeat steps 3 and 4.
6. Repeat the activity with 2 and then 3, 1-kg masses added to the cart.

APPLICATION

1. Construct a line graph with the period versus the added mass, putting the total added mass on the horizontal axis and the period on the vertical axis.
2. Use the first law of motion to explain the relationship shown in your graph.

◥ ASK YOURSELF

You have a research paper due next week but haven't started it. Your English teacher told you that you needed to overcome your inertia. Explain why he used this term.

● **Process Skills:** *Solving Problems, Applying*

Review with the students the formula for representing a force (N = kg × m/s/s). Explain that another way of stating the formula is Force = mass × acceleration or $F = m \times a$. Then ask a volunteer to show how the second law is merely another way of expressing this formula. Finally, help the students calculate the rate of acceleration for the snowballs shown in Figure 3–21, using these assumptions: one snowball has a mass of 5 kg; the other snowball has a mass of 20 kg; 60 N of force are applied to each. (60/5 = 12 m/s/s; 60/20 = 3 m/s/s)

SCIENCE TECHNOLOGY SOCIETY Experiments with the ram accelerator show promise of being able to accelerate projectiles to hypervelocity—up to 10 km/sec. So far, scientists have been able to achieve speeds of 2.6 km/sec. Operating something like a supersonic ramjet aircraft engine, the ram accelerator forces the projectile down a launch tube filled with a fuel/oxidizer mix. The mix is compressed against the walls when the projectile moves through it and ignited behind the projectile to provide thrust. Fins keep the projectile centered.

This huge force acts on the small mass of the projectile to achieve extremely rapid acceleration. Ram accelerator potential, in addition to its efficiency and relatively low cost, might make it useful for hypervelocity research and launch of small payloads into low Earth orbit.

SCIENCE BACKGROUND

Point out to the students that Figure 3–21 illustrates the second law of motion. The two snowballs on the left have the same acceleration. However, because the larger snowball has more mass, it must be pushed with greater force than the smaller snowball. When the same amount of force is applied to the two snowballs on the right, they accelerate at different rates.

Figure 3–21. Rolling snowballs illustrates the second law of motion.

The Second Law of Motion

The best way to understand the second law of motion is to imagine that you are rolling snow into a ball. When the ball is small, you can roll it with little force. You have little trouble accelerating the ball from a standstill. However, as the ball gathers more snow and increases in mass, more force is needed to roll it.

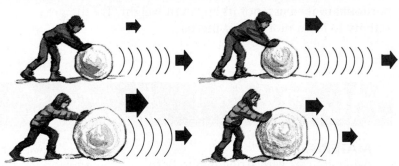

You have to apply more force to the large snowball than to the small snowball to produce the same acceleration. If you applied the same force to each of the snowballs, the small one would accelerate more quickly than the large one. When unbalanced force increases, acceleration will increase. When mass increases, acceleration is inversely proportional to its mass.

The second law of motion describes the relationship between acceleration, force, and mass. One part of this law states that any time an unbalanced force is applied to an object, the object accelerates. The law further states that the amount of acceleration depends both on the mass of the object and the amount of force applied to it.

When subjected to the same unbalanced force, a smaller mass receives a bigger acceleration than a larger mass. This point is illustrated in the photographs of the Olympic athletes.

Figure 3–22. Because of the different masses of these two objects, a similar amount of force imparts very different accelerations.

● **Process Skills:**
Communicating, Solving Problems/Making Decisions

Cooperative Learning Have the students work in small groups to develop an illustration for this problem: You want to dock your boat, so you pull up alongside the pier. You jump from the boat to the pier. What happens to the boat?

Suggest to the students that at least two drawings are needed and that arrows need to be added to show the directions of forces at work. (The jump to the pier means a force is exerted toward the pier; an equal and opposite force of the boat pushes it away from the pier.)

● **Process Skills:**
Applying, Inferring

Assess understanding of the third law of motion by having volunteers explain the action and reaction forces at work in these examples: blastoff of a rocket (action = force of burning fuel downward, reaction = rocket moving upward); swim-ming in water (action = swim-mer's limbs on water, reaction = water's force on swimmer); picking up a backpack (action = hand's force on backpack, reaction = backpack's downward force against hand); jumping from a diving board (action = jumper's force against board, reaction = board's upward spring against jumper).

The second law of motion applies to many common situations. For example, the mass of a car determines how much force its brakes must apply to make it stop. Accidents sometimes occur when people overload their car and then assume the car will stop as easily as it did when empty. An empty car has less mass than the loaded car, so the braking force produces a greater stopping acceleration. A loaded car has more mass. If braking force and velocity are the same, the stopping acceleration of the loaded car will be much smaller than that of the empty car and it will take more time to stop it. This can be expressed as a mathematical equation.

$$\text{acceleration} = \frac{\text{force}}{\text{mass}} \quad \text{OR} \quad a = \frac{F}{m}$$

In the equation, a is the acceleration of an object, F is the un-balanced force pushing or pulling on the object being accelerated, and m is the mass of the object being accelerated.

 ASK YOURSELF

Why is it impossible to roll a bowling ball and a baseball at the same acceleration by using the same force?

The Third Law of Motion

Imagine that you and a friend are warming up for a hockey game. Suddenly, your friend skates into your path, so you give him a push. He rolls away as you intended, but to your surprise, you find yourself rolling off in the other direction. What happened? You have just experienced the third law of motion!

The third law of motion states that for every force there is an equal and opposite force. The movement of you and your friend shows that it is impossible for you to push something (action) without receiving an equal push in the opposite direction (reaction). These pushes, or forces, are often called *action forces* and *reaction forces*.

The original force is the action force. In this example, the push of your hand on your friend's back is the action force. The resulting opposite and equal force is the reaction force. In this case, the push of his back on your hand is the reaction force, and it is this reaction force that causes you to travel backward, away from the spot where you pushed him.

Figure 3–23. These skaters demonstrate the third law of motion. When one skater pushes the other, they both move in opposite directions.

INTEGRATION– Language Arts

Ask the students what they think the word *conservation* means, and have them write the definition in their journals. (preservation; keeping from changing in total amount) As the students read about the law of conservation of momentum on p. 92, stress that this law applies to the objects in a closed system.

ONGOING ASSESSMENT

▼ **ASK YOURSELF**

Given the same force applied to two different objects, acceleration will be less for the object of greater mass (in this case, the bowling ball).

LASER DISC
1836

Rocket propulsion

GUIDED PRACTICE

Remind the students that a newton (N) is the force that will accelerate a mass of 1 kg at a rate of 1 m/s/s. To assure understanding of the relationship of mass and force to acceleration, have the students compute and graph the rate of acceleration for the following data: a 10-kg object to which 10 N of force are applied, then 50 N, then 100 N, and so on. Using a pen or pencil of another color, have the students plot the same data for a 20-kg object to which the same N forces are applied.

INDEPENDENT PRACTICE

 Have the students provide written answers to the Section Review. In their journals, have the students draw a series of pictures with force arrows showing action-reaction pairs and how they are resolved as movement.

EVALUATION

Using the laws of motion, have the students explain the motion that occurs when a pool cue strikes a pool ball. (Students' answers should reflect understanding of inertia, action and reaction forces, and transfer of momentum.)

DISCOVER BY *Doing*

 Students should complete at least several trials of this Discover by Doing before they formulate a conclusion. Students should infer that each scale should measure 9.8 N for the weight of the 1-kg mass, because the reaction force, measured by the first scale, is equal to the action force, measured by the second scale. The first scale would also show the weight of the second scale and thus register slightly more than 9.8 N total. Ask the students to place their explanations in their science portfolios.

ONGOING ASSESSMENT
▼ ASK YOURSELF

An action force applied to an object makes that object respond with an equal, but opposite, reaction force.

MEETING SPECIAL NEEDS

▲ Gifted

 The second and third laws of motion can be combined to produce the law of conservation of momentum. Have students do research to discover how the application of algebraic reasoning can be used in science. Encourage the students to write their research in their journals.

Figure 3–24. The momentum of the ball before impact is equal to the total momentum of the pins and the ball after impact.

The third law of motion applies to all other forces. You move when you walk because the force that your feet apply to the earth is matched by a reaction force that the earth applies to your feet. Remember that you have a lot less mass than the earth. Because of that, your motion when the earth pushes on you is a lot greater than the earth's motion when you push on it. In fact, the earth's motion when you push it is so small that it cannot even be measured. It is the reaction force that moves you forward. To test the third law, try the next activity.

DISCOVER BY *Doing*

Use two spring scales linked together to suspend a 1-kg mass. What does each scale read? This phenomenon relates directly to the third law of motion. Explain how. ✎

▼ ASK YOURSELF
Explain how action and reaction forces affect an object.

Conservation of Momentum

This is it—the last frame of the game. If you make this strike, you will win the bowling tournament of champions. You pick up the heavy ball, wipe it off, and grip it firmly. You line up at the starting point, hold the ball slightly above waist level, and face the pins. Moving forward, you slide gracefully close to the foul line and release the ball. The crowd watches. No one even whispers. The ball rolls toward the pocket between the head pin and the number three pin. Wham! The ball hits the pins, and they go spinning in all directions. You've made the strike!

The total momentum of the bowling ball and pins remains the same whether the bowler knocks down all ten pins or just a few. The momentum of the ball just before it hits the pins is equal to the combined momentum of the pins and ball just after impact. This example illustrates the **law of conservation of momentum.** This law states that momentum can be transferred from one object to another but cannot change in total amount.

RETEACHING

PORTFOLIO ASSESSMENT

Place two roller skates on the floor about 1 m apart. Have a volunteer gently, but firmly, push one roller skate toward the second skate with enough force so that the two skates collide. Then have two volunteers push both skates toward each other with enough force

so that the skates collide. Finally, have the students explain what they observed in terms of the three laws of motion and the law of conservation of momentum. Encourage the students to place their observations and explanations in their science portfolios.

EXTENSION

Interested students might find library books and magazine articles on changes in vehicle design. Encourage the students to write and illustrate reports explaining how a change in vehicle design relates to one or more of the laws of motion, thus making the car safer, easier to accelerate, or more fuel efficient.

CLOSURE

In their journals, have the students write statements, in their own words, that explain each of the laws of motion discussed in this section.

Testing the Law   The law of conservation of momentum is fairly easy to see on a pool table. If one pool ball is shot directly at a second, identical pool ball, the first one will stop or "stick" when the two collide and the second one will shoot forward at the velocity the first one had just before the collision. Because they are identical, their masses are the same. Since the second ball has a velocity after collision equal to the velocity of the first one before collision, the momentum of the first ball has been transferred to the second ball during the collision.

DISCOVER BY Writing

Imagine you and a friend are skating. Instead of skating at a slow speed, you are both skating fast. How would the reaction between gently pushing someone while skating at a slow speed to forcefully pushing someone while skating at a fast speed compare? Write your comparison in your journal. ✎

Saving Up   Rockets work in outer space because momentum is conserved. When fuel is forced out the back of the rocket, the change in momentum of the fuel is offset by the rocket's momentum change in the opposite direction. The total momentum of the rocket and the fuel remains the same.

▼ ASK YOURSELF

How is momentum conserved when a batter hits a baseball?

Figure 3–25. In a simple water rocket, momentum from moving water is transferred to the rocket to make it rise. In this process, momentum is conserved.

ONGOING ASSESSMENT
▼ ASK YOURSELF

The momentum given to the bat by the batter is transferred to the ball. The bat stops moving, but the ball accelerates, conserving the momentum.

SECTION 3 REVIEW AND APPLICATION

Reading Critically

1. The small rock accelerates faster because it has less mass.

2. The reaction force is the force you feel against your foot from the resisting ball.

3. If the car changes velocity suddenly, the passengers will maintain their velocity until they make contact with their seatbelts or the car (windshield)—the first unbalanced force to act on them.

Thinking Critically

4. The rate of acceleration relies on the ratio of mass to force. A small force acting on a small mass (say 10 N to 1 kg) will have the same effect as a large force acting on a large mass (say 100 N to 10 kg).

5. She can get back to the station by throwing something away from the station. When she does, the object thrown will apply an equal force in the opposite direction, accelerating her towards the station (third law of motion). She will accelerate toward the station during the throw (second law of motion), and she will continue to travel toward the station at a constant velocity until acted upon by another force (first law of motion).

SECTION 3   REVIEW AND APPLICATION

Reading Critically

1. How do the acceleration of a small rock and a large rock compare if they are thrown with the same force?

2. You kick a soccer ball, and the action force sends the ball flying through the air. What is the reaction force?

3. How does inertia relate to the safety of passengers in a car as it comes to a stop?

Thinking Critically

4. How is it possible for an object with a large mass acted on by a large gravitational force to have the same acceleration as an object with a small mass acted on by a small gravitational force?

5. A space station construction worker found herself floating free 100 m from the space station because her safety line somehow became unhooked. Attached to her spacesuit were her unhooked safety line, her tool belt and tools, and her oxygen tank. How could she get back to the space station without calling someone for help? In answering, explain the physics of your proposed method.

Process Skills: Measuring, Interpreting Data, Communicating

Grouping: Pairs

Objectives
- **Calculate** force and acceleration.
- **Communicate** relationships with graphs.

Pre-Lab
Have the students state their ideas about the relationship between force, mass, and acceleration.

▶ Analyses and Conclusions
Since the relationship is $F = ma$ if the mass does not change when the acceleration increases, the force used to move the cart must also increase. Force and acceleration are directly proportional to each other.

▶ Application
The stopping distance of a 10-tonne truck traveling at the same speed is 40 m.

⭐ PERFORMANCE ASSESSMENT

Cooperative Learning

After the completion of this Investigation, have each pair of students develop a problem similar to the problem in the "Application." Have the students write their problem and calculate the answer. Check the students' problem for logical statement of the problem and for the correct answer.

✳ Discover More
The liquid mass makes a difference because the water sloshes in the jars and the wood blocks do not move. It is difficult to keep the cart at constant velocity when the water moves from side to side in the jar.

Post-Lab
Ask the students how their ideas about the relationship between force, mass, and acceleration have changed. Ask them to state their new ideas.

PORTFOLIO ASSESSMENT After the students have completed this Investigation, ask them to place their charts and answers in their science portfolios.

INVESTIGATION

Calculating Force, Mass, and Acceleration

▶ MATERIALS
- pulley with table clamp • dynamics cart • masking tape
- string • can • sand • standard masses • blocks of about the same mass as the dynamics cart (4) • meter stick • stopwatch

▼ PROCEDURE

1. Set up the apparatus and make a chart like the one shown.
2. Put enough sand in the can so that when released the dynamics cart will roll at a constant velocity.
3. Remove the middle strip of masking tape to make one 0.50-m interval.
4. Place the cart in front of the first strip of tape and add a 5-g mass to the can.
5. On the timer's signal, release the cart and measure the time taken to travel the 0.50-m distance. Repeat three times, and average the values. Record the average time.
6. Repeat steps 4 and 5 until 25 g have been added to the can.
7. Add one block to the cart, and repeat step 5.
8. Add 25 g to the can, and repeat steps 5–7. Record the results in your chart.
9. Repeat step 5, adding one

TABLE 1: CALCULATING FORCE, MASS, AND ACCELERATION				
Mass Added to Can (g) (m/s/s)	Blocks Added to Cart	Average Time to Go 0.50 m	Calculated Quantities Force (N)	Acceleration

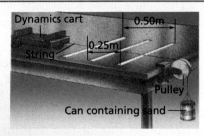
Dynamics cart 0.50m • String 0.25m • Pulley • Can containing sand

block each time until all 4 blocks have been added to the cart.
10. To calculate the unbalanced force, assume 0.01 N for every gram of mass added to the can in addition to the sand added to offset friction.
11. To calculate acceleration, use the following formula

$$a = \frac{d}{t^2}$$

12. Graph the results for steps 4–6 by plotting acceleration on the Y axis and force on the X axis.
13. Graph the results for steps 7–9 by plotting acceleration on the Y axis and cart mass on the X axis.

▶ ANALYSES AND CONCLUSIONS
In steps 5–7 you did not change the mass. What is the relationship between the acceleration and the force used to move the cart?

▶ APPLICATION
A delivery truck with a mass of 5 tonnes, which is traveling at a speed of 15 m/s, requires 20 m to stop. What is the stopping distance of a 10-tonne truck traveling at the same speed?

✳ Discover More
Repeat the Investigation, but use plastic jars of water that weigh about as much as the blocks. How do the results compare?

CHAPTER 3 HIGHLIGHTS

The Big Idea— Systems and Structures

Direct the students to understand that motion is described by its direction, distance, and rate of change in direction. Forces that can act on an object to make it move are friction and gravity. Help the students to see the interconnection among the three laws of motion.

For Your Journal

Students' new lists should reflect an understanding of the laws of motion and of how things move. Encourage the students to think of situations when they either move themselves or must move other objects. Then help them apply their new understanding of motion to these events and speculate how they might do these activities differently. Have the students tell how understanding the way things move could help them in their everyday lives.

CONNECTING IDEAS

Before the students start the Connecting Ideas feature, encourage them to use the science terms they have learned in this chapter. Students' essays should describe the action of the bobsledders from a scientific viewpoint. Essays might include how the laws of motion are demonstrated in what is happening in the pictures.

The Big Idea

Motion can be described, explained, and predicted using a few interconnected ideas. If you can measure the distance an object travels in a certain time, its direction, and how fast its speed or direction is changing, you can describe its motion completely. If you want to know how hard it is to make the object change its motion, you must also measure its mass. To change an object's motion, you must apply a force to it by somehow pushing or pulling it. Common forces include friction and gravity. The three laws of motion explain why things move and why they come to a stop.

For Your Journal

Look back at the ideas you wrote in your journal at the beginning of the chapter. How have your ideas changed? Revise your journal entry to show what you have learned about the laws of motion and how they allow us to predict how things will move.

Connecting Ideas

By looking at the illustration and writing a short essay, you can show how the big ideas of this chapter are related. In your journal, write an essay that describes what is happening or what is about to happen in the pictures.

Understanding Vocabulary

1. (a) Speed is distance traveled over time; velocity is speed in a certain direction. A change in speed always changes velocity, but a change in velocity does not necessarily change speed.

(b) Acceleration is the rate at which the velocity of an object changes, while momentum is the product of an object's velocity and mass (momentum = velocity × mass). Both have definite direction, but momentum is also affected by mass.

(c) A force is any push or pull exerted on matter that attempts to change its motion. For example, friction is a force that resists motion between two touching surfaces. Gravity is the force with which objects attract each other, in proportion to their mass.

(d) Action and reaction forces are always paired. An action force initiates movement of an object, and an equal and opposite reaction force results.

Understanding Concepts

Multiple Choice

2. b

3. c

4. a

5. a

6. d

Short Answer

7. Momentum is equal to mass multiplied by velocity. If a small object has less mass than a large object, it must have a higher

velocity to have the same momentum. If the velocity of the two objects is the same, the small object must have a greater mass to have equal momentum.

8. The arrow travels horizontally at a constant speed because no net force is added to it once it leaves the bow. However, it also travels vertically because gravity exerts a constant force, making it accelerate downward. The net result of the force of gravity acting on the horizontally moving arrow causes its path to curve downward.

Interpreting Graphics

9. Acceleration depends upon force applied and the mass to which it is applied. Given an identical amount of force applied, the tennis ball, with the smaller mass, will travel faster.

10. The third law of motion states that for every force there is an equal and opposite force. Since the tennis ball strikes the wall with a greater speed, it will rebound with greater speed, causing it to roll farther back than the soccer ball.

CHAPTER 3 REVIEW

▶ Understanding Vocabulary

1. Define the terms in each set in a way that explains the relationship of each term to the other.
  a) speed (72), velocity (74)
  b) acceleration (75), momentum (77)
  c) force (78), friction (81), gravity (83)
  d) action force (91), reaction force (91)

▶ Understanding Concepts

MULTIPLE CHOICE

2. If you jog for one hour and travel 10 kilometers, 10 kilometers per hour describes your
  a) momentum.
  b) average speed.
  c) velocity.
  d) speed and velocity.

3. Which of the following objects is *not* accelerating?
  a) a ball being juggled
  b) a horse racing on an oval track
  c) a bowling ball returned to you on a straight, level track
  d) a braking bicycle

4. It would be easier to push a box of fruit up a board and into a truck if you
  a) took some fruit out of the box.
  b) covered the board with rough carpeting.
  c) packed the fruit carefully.
  d) There is no way to know.

5. A person's weight is less on the moon because
  a) the moon has less mass than the earth.
  b) the person's mass is less on the moon.
  c) there is no gravity on the moon.
  d) the moon circles the earth.

6. A loaded delivery truck and a compact car are both traveling 88 kilometers per hour. The truck will
  a) have less inertia than the car.
  b) travel the same distance more quickly than the car.
  c) require less force to travel 100 kilometers per hour.
  d) take longer to stop for a red light.

SHORT ANSWER

7. When could a small object have the same momentum as a large object?

8. Why does an arrow shot from a bow soon start to curve downward?

Interpreting Graphics

9. Look at the figure below. Which ball will travel faster? Explain why.

5N

10. Refer to the figure again. Which ball will roll back further in the opposite direction once it hits the wall? Explain your answer.

Reviewing Themes

11. As it is swung, the baseball bat gains momentum. Its momentum is transferred to the ball when the two objects meet. The more momentum the bat has, the more momentum will be transferred to the ball, and the farther the ball will travel.

12. The energy in the book and the table top do not cause motion while the book lies on it. Gravity is pulling down on the book, but the table top is pushing up with an equal force, so no motion occurs. If the book is pushed off the table, there is no longer an upward push to counter the pull of gravity, and the book falls to the floor. In this case, energy is seen as motion.

Thinking Critically

13. When the brakes are applied, friction acts to oppose the forward motion of the tires. Water or small particles of gravel and dirt on the road can act like ball bearings or a lubricant, reducing the friction needed to stop. Tread provides a channel to remove water and small particles from the tire's surface.

14. The earth's mass—its amount of matter—would remain unchanged. Since gravitational pull is determined by both mass and distance, the greater distance would mean a lesser gravitational pull. (The force of gravitational attraction varies with the inverse of the square of the distance between the attracting objects, so the new gravitational force would be one-fourth as strong.)

15. The second law of motion states that an object accelerates in relationship to its mass and the amount of force applied to it. The marble has less mass than the basketball. However, the force of gravity is proportional to mass. Therefore, objects with small mass have small forces and those with large masses have large forces.

16. As the water drains out, the mass of the tub decreases. Since momentum is equal to mass × velocity, the tub's momentum must decrease even though the velocity remains the same.

17. The wind and rain will increase the air resistance on the boat. The atmospheric turmoil will also cause rough water, increasing the drag caused by the water. The boat will require much greater force (generated by the engine) to maintain the same speed, since friction resists the boat's movement.

18. In all of these excerises, some force to oppose gravity is required, and that is the main force for push-ups and running in place. In running around a track, you are actually pushing the earth behind you as the earth pushes you forward. Air resistance can also be significant on windy days when you are running into the wind. In swimming, in addition to pushing the water backward with your hands, arms, and feet, water drag must be overcome.

Reviewing Themes

11. *Systems and Structures*
A baseball and bat can be thought of as a system. Explain how they interact in accord with the law of conservation of momentum.

12. *Energy*
Energy may do many things; it may cause change or motion. How is energy related to the act of pushing a book off a table top?

Thinking Critically

13. How does the tread on tires help a car to stop quickly? Use your understanding of friction in your explanation.

14. Imagine that the earth is now twice as far from the sun. How would the earth's mass change? How would the gravitational attraction between the earth and sun change?

15. A marble and a basketball dropped at the exact same moment from the exact same height will accelerate at the same rate and reach the ground together. Using what you know about the force of gravity and the second law of motion, explain why this is true.

16. Imagine that a bathtub filled with water is rolling on frictionless wheels in a straight line along a level road. If the stopper pops out and the water drains, the momentum of the tub will lessen even though its velocity stays the same. Why is this true?

17. Why will a boat use more fuel (require more force) to travel 32 kilometers per hour against the wind in a rainstorm than it will on a sunny day with no wind?

18. Exercises such as running, swimming, calisthenics, and aerobics are designed to make you healthier and stronger. When you do these exercises, you are actually moving all or parts of your body in directions that oppose some invisible forces. What are the forces that oppose the following exercises: push-ups, running in place, running around a track, and swimming the length of a pool?

Discovery Through Reading

Macaulay, David. *The Way Things Work.* Houghton Mifflin, 1988. This award-winning book explains how a variety of machines depend on physical forces so they may operate.

CHAPTER 3 **97** ◄

CHAPTER

4

WORK

PLANNING THE CHAPTER

Chapter Sections	Page	Chapter Features	Page	Program Resources	Source
CHAPTER OPENER	98	*For Your Journal*	99		
Section 1: WORK AND POWER	100	Discover by Writing (B)	103	*Science Discovery**	SD
• First Things First (A)	100	Discover by Calculating (A)	103	Investigation 4.1:	
• Work, Work, Work (B)	102	Activity: How much power do		Measuring Power Used to	
• May the Force Be With You (B)	105	you generate in climbing a		Lift Weights (A)	TR, LI
• The Advantage of Machines (H)	106	flight of stairs? (A)	104	Thinking Critically (B)	TR
		Section 1 Review and Application	107	Study and and Review Guide,	
		Skill: Comparing Force and Work		Section 1 (B)	TR, SRG
		Using a Spring Scale (A)	108		
Section 2: TYPES OF MACHINES	109	Discover by Doing (B)	111	*Science Discovery**	SD
• Levers (B)	110	Discover by Calculating (A)	114	Reading Skills:	
• Wheel and Axle (A)	111	Discover by Problem Solving (B)	115	Taking Notes (B)	TR
• The Pulley (A)	112	Discover by Writing (B)	116	Investigation 4.2: Determining	
• I'm Inclined to Use a Ramp (B)	114	Section 2 Review and Application	117	the Mechanical Advantage of	
• Two or More Machines (B)	115	Investigation: Building Simple and		a Simple Machine (A)	TR, LI
		Compound Machines (A)	118	Connecting Other Disciplines:	
				Science and Social Studies,	
				The Effects of the Industrial	
				Revolution on Society (B)	TR
				Extending Science Concepts:	
				Mechanical Advantage of	
				Pulleys (A)	TR
				Levers and Pulleys (A)	IT
				Six Simple Machines (B)	IT
				Record Sheets for Textbook	
				Investigations (A)	TR
				Study and Review Guide,	
				Section 2 (B)	TR, SRG
Chapter 4 HIGHLIGHTS	119	The Big Idea	119	Study and Review Guide,	
Chapter 4 Review	120	For Your Journal	119	Chapter 4 Review (B)	TR, SRG
		Connecting Ideas	119	Chapter 4 Test	TR
				Test Generator	

B = Basic   **A** = Average   **H** = Honors
The coding Basic, Average, Honors indicates subsections, features, and resources that might be appropriate for different levels of learners. For additional suggestions regarding choice of topic and depth of coverage, see the Pacing Chart on pages T28–T31.

*Frame numbers at point of use
(TR) Teaching Resources, Unit 2
(IT) Instructional Transparencies
(LI) Laboratory Investigations
(SD) *Science Discovery* Videodisc
    Correlations and Barcodes
(SRG) Study and Review Guide

CHAPTER MATERIALS

Title	Page	Materials
Teacher Demonstration	100	books
Discover by Writing	103	(per individual) thesaurus, journal
Discover by Calculating	103	(per individual) spring scale, object that can be dragged
Activity: How much power do you generate climbing a flight of stairs?	104	(per pair) stopwatch, meter stick
Skill: Comparing Force and Work Using a Spring Scale	108	(per group of 3 or 4) books (4), ruler, wooden block with an eye hook, spring scale, 20-cm x 30-cm piece of heavy cardboard
Discover by Doing	111	(per individual) meter stick, pencil, 1-kg masses (2)
Discover by Calculating	114	(per individual) book, spring scale, block of wood
Discover by Problem Solving	115	(per individual) journal
Discover by Writing	116	(per individual) journal
Investigation: Building Simple and Compound Machines	118	(per group of 3 or 4) bottle caps, cardboard, craft sticks, empty thread spools, glue, modeling clay, paper, pencils, rubber bands, scissors, shoe boxes, stones, straws, string, tape, any other materials to which you have access

ADVANCE PREPARATION

Have the students begin collecting bottle caps, thread spools, shoe boxes, and other useful materials for the *Investigation* on page 118.

TEACHING SUGGESTIONS

Outside Speaker
Invite a representative of a moving and storage company to speak to the class about the machines used to move furniture. If possible, ask the representative to demonstrate one of the machines or use a photograph or diagram to explain how the machine works.

In this chapter the concepts of work and power are explained, and the interrelationships of the three laws of motion are demonstrated. Mechanical advantage is also described. Chapter 4 also introduces the six types of simple machines and discusses complex machines.

A supporting theme in this chapter is **Energy.**

MULTICULTURAL CONNECTION

Two ancient devices—the Archimedes' screw and the water wheel—are still used on small farms in many third-world countries, such as India and Egypt, and some countries in Central and South America and in Africa. These devices are used to lift water for irrigation.

The Archimedes' screw is believed to have been developed by the Greek inventor Archimedes. The Archimedes' screw consists of a large screw inside a tube. The lower end of the device is placed in the water, and the screw is rotated within the tube to transfer the water gradually to the upper end of the tube.

Historians believe the water wheel was developed in the 100s B.C. At first, it was used mainly to grind grain. Later, it was used for many kinds of mechanical operations. It was a major source of power until the development of the steam engine in the 1700s.

MEETING SPECIAL NEEDS

Second Language Support

Have limited-English-proficient students prepare drawings that illustrate the concepts in the Chapter Motivating Activity. They should show work not being performed (lifting a desk without moving it) and using a machine (in this case a lever) to perform work.

CHAPTER

4 WORK

*E*migrants bound for Oregon in the 1840s would have envied the motorized machines we have today for hauling materials. The hardships of the trek to the West gave them life-and-death lessons in the science of work.

Albert Bierstadt, "The OregonTrail" 1869, 31" x 49", oil on canvas. The Butler Institue of American Art.

These days a moving van easily hauls a family's possessions from Missouri to Oregon in just a few days. Fresh produce speeds across half a continent in a few hours, before the peaches ripen and the melons spoil. A century-and-a-half ago, the same journey required immense physical labor.

Obtain a 4" × 4" board about the same length as the height of your unattached teacher's desk (or a large heavy table) and a long pipe. Ask several volunteers to sit on the desk or table, and then ask another volunteer to lift the desk. The desk should be heavy enough so that it cannot be lifted by muscle effort alone. After a few minutes of effort, have the volunteer sit down. Ask the students if the volunteer did any "work."

Ask a volunteer to lift the desk by pushing on the pipe, using the board as a fulcrum. Show the volunteer how to set it up first. Explain the science definition of work and that a machine makes work easier.

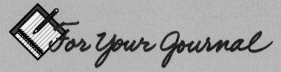

For Your Journal

The journal questions will help you identify any misconceptions students might have about force and work. In addition, these questions will help the students realize that they do possess knowledge about work and machines. (Students will have a variety of responses to the journal questions. Accept all answers, and tell students that at the end of the chapter they will have a chance to refer back to the questions.)

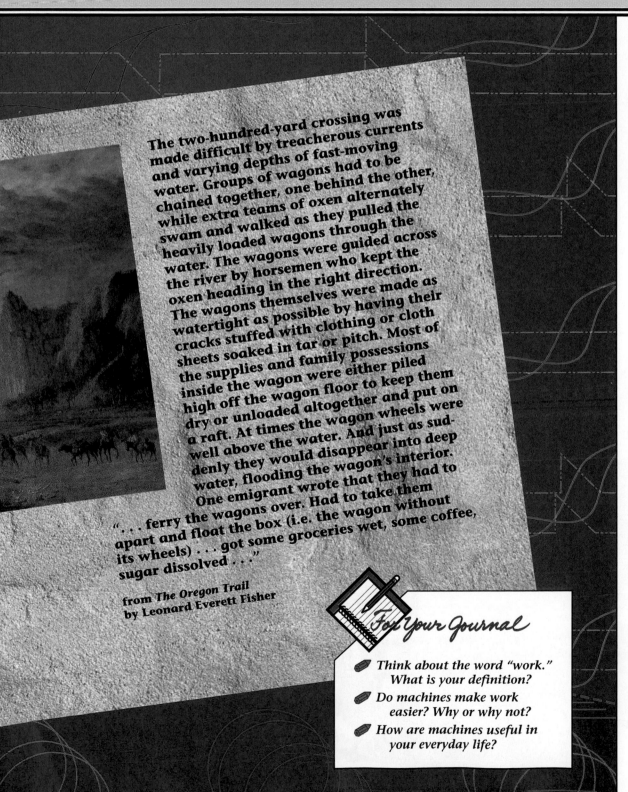

The two-hundred-yard crossing was made difficult by treacherous currents and varying depths of fast-moving water. Groups of wagons had to be chained together, one behind the other, while extra teams of oxen alternately swam and walked as they pulled the heavily loaded wagons through the water. The wagons were guided across the river by horsemen who kept the oxen heading in the right direction. The wagons themselves were made as watertight as possible by having their cracks stuffed with clothing or cloth sheets soaked in tar or pitch. Most of the supplies and family possessions inside the wagon were either piled high off the wagon floor to keep them dry or unloaded altogether and put on a raft. At times the wagon wheels were well above the water. And just as suddenly they would disappear into deep water, flooding the wagon's interior. One emigrant wrote that they had to

". . . ferry the wagons over. Had to take them apart and float the box (i.e. the wagon without its wheels) . . . got some groceries wet, some coffee, sugar dissolved . . ."

from *The Oregon Trail*
by Leonard Everett Fisher

ABOUT THE LITERATURE

Wagon trains spent five arduous months traveling the Oregon Trail from the Missouri River to the far West. Humans and animals had to haul provisions and possessions up mountain ranges and across rivers. Teams of oxen plodded up the sloping pass that took emigrants over the Rocky Mountains. At times, the oxen had to be unhitched and the wagons lowered by ropes and pulleys over bluffs.

Between 1841 and 1850 more than 10 000 people departed from Independence, Missouri, for the frontier. After the discovery of gold in the Blue Mountains of Oregon, thousands more set out over the Oregon Trail.

If you or the students want to know more about the Oregon Trail, consult the following book.

> Fisher, Leonard Everett. *The Oregon Trail.* Holiday House, 1990. The photographs, art, and easy-to-read narrative make this book a valuable reference about the Oregon Trail.

You might also encourage the students to visit or write to the following address for more information about the Oregon Trail.

National Historic Oregon Trail Interpretive Center
Bureau of Land Management
P.O. Box 987
Baker City, OR 97814

For Your Journal

- Think about the word "work." What is your definition?
- Do machines make work easier? Why or why not?
- How are machines useful in your everyday life?

FOCUS

In this section, the concepts of work and power are defined, and the relationships among force, power, and work are explained. The concept of mechanical advantage is also discussed. Equations to calculate power and mechanical advantage are given.

MOTIVATING ACTIVITY

Cooperative Learning Students often have difficulty differentiating between the common usage and scientific definitions of the terms *work* and *power*. Arrange the students in small groups, and instruct them to make a list of examples that illustrate the terms scientifically. Have the groups share their ideas with the class, and have them review their examples during the study of this section to see whether the examples conform to the scientific definitions of the terms.

PROCESS SKILLS
- Experimenting • Measuring
- Observing • Comparing
- Applying

POSITIVE ATTITUDES
- Cooperativeness
- Curiosity

TERMS
- work • machine • power
- mechanical advantage

PRINT MEDIA
Get It in Gear! The Science of Movement by Barbara Taylor (see p. 67b)

ELECTRONIC MEDIA
Machines, Work, and Energy Educational Activities, Inc. (see p. 67c)

SCIENCE DISCOVERY
- Work; input and output • Axe
- Axe; splitting wood

BLACKLINE MASTERS
Laboratory Investigation 4.1
Thinking Critically
Study and Review Guide

Demonstration

To compare work and force, hand a volunteer a few books and ask him or her to carry them across the room and back. Add two or three more books to the volunteer's load, and ask him or her to repeat the procedure. Then ask the students whether the volunteer applied more force during the second trip than during the first trip. (The volunteer applied more force to support the books during the second trip.) How much work was done during each of the volunteer's trips? (Work was done during each trip the books were moved. However, more work was done during the second trip because more force was needed to move the additional books.)

SECTION **1**	# *Work and Power*

Objectives

Define work *and* power.

Explain *how machines can multiply force without multiplying work.*

Use *the concept of mechanical advantage to explain how machines make work easier.*

I magine that it is 1843. You and your family have been traveling by foot, horseback, and wagon for six months. The 3218 kilometer journey along the Oregon Trail was grueling; many people and animals died along the way. But at last you have arrived in the Oregon Territory and are surrounded by vast evergreen forests. Here you will build a new home, a new life.

First Things First

By the time you reach the Oregon Territory, it is already late fall, too late to plant crops or build a home. Therefore, you and your family stay with others who have homes and plenty of food. During the winter months, you listen to their stories about life in the beautiful, but harsh, Willamette Valley.

At last spring arrives! Clearing the land is the first task. Crops need to be planted to provide the food supply for next winter. The need to clear the land is so great that you will put together a temporary shelter and live in that until the crops are planted. The trees that you clear from the land will be used later to make a log cabin and barn.

Figure 4–1. Pioneers lived in half-shacks until they could build permanent housing.

● **Process Skill:** *Recognizing Time/Space Relations*

Ask the students to explain how two people can do the same amount of work using equal force if one person pushes a lightweight object and the other pushes a heavier object.

Encourage the students to place their explanations in the science portfolios. (Since work depends only on force and distance, the work would be the same in both cases. The lighter object would be moving faster at the end of that distance (if no other forces apply)).

● **Process Skill:** *Analyzing*

Some students might equate the phrase *less work* with *easier work*. Point out that machines do not change the amount of work done; they simply change the force, the distance, or both.

Figure 4–2. The work this person is doing while pushing the car is equal to the force she applies to the car multiplied by the distance the car moves.

You take a look at one of the smaller trees and figure you could probably push it over without too much effort. You push and push, but the tree doesn't move. Your hands and arms begin to ache, and you start sweating. Hard work, right? Wrong! You haven't done any work at all. To have done work, your push would have had to move the tree.

In science, *work* does not mean just making an effort. Work means accomplishing something. More precisely, **work** means not only applying a force but also using that force to make something move. In order to do work on the tree, you have to make it move. The work you do on the tree is the force you apply to it multiplied by the distance you make it move. The equation for work is

$$\text{work} = \text{force} \times \text{distance} \quad \text{OR} \quad W = F \times d$$

Pushing on a wagon or a barrel might allow you to do work on those objects, but you probably won't do much work pushing on a tree. To do work on a tree, you must exert a large force. Some of the other pioneers would recommend that you use a machine—an ax. A **machine** is a device that helps you do work by changing the size or direction of a force. Why do you think an ax is considered a machine? ①

Although the pioneers did not have modern-day farm machines to clear the land, they did have machines such as the ax and plow. They either brought these machines on the journey or made them once they arrived at their destination. In the next section, you will learn about different types of machines.

▼ **ASK YOURSELF**

Why is no work being done when you push on the side of a building?

THE NATURE OF SCIENCE

Some of the earliest machines to be invented were the tools used by prehistoric people in gathering and hunting their food. Tools roughly chipped from stone date back about one million years, and stone axes and spearheads dating as far back as the emergence of civilization have been found in some archaeological sites. The inclined plane became the first principle of technology to be put to work in making cutting tools. It might have enabled people to build the pyramids in Egypt in 2600 B.C., using high earth ramps to raise massive stone blocks into position.

 LASER DISC
1854

Work; input and output

1820

Axe

1821

Axe; splitting wood

① An ax is considered a machine because it increases the size and direction of a person's force.

ONGOING ASSESSMENT
▼ **ASK YOURSELF**

Responses should include that although force was applied, the building did not move. Therefore, no work was done.

● **Process Skill:**
Identifying/Controlling Variables

Ask the students whether two people performing the same task can accomplish the same amount of work and generate the same amount of power

using different amounts of force. (If the distances times the forces yield equivalent amounts of work, they can.)

● **Process Skill:** *Comparing*

Power measures the amount of time it takes for work to be done. Power is a ratio that expresses the relationship of work to time. Ask the students to compare a machine, such as a car, with another machine of the same type that has twice as

much power. Have them write a ratio in their journals to express the difference in power between the two machines. (The second machine can do the same amount of work in half the time as the first machine.)

INTEGRATION—
Geography

One obstacle to the emigrants' progress was the Snake River in Idaho, which wagon trains followed for 402 kilometers before crossing into Oregon. The river crossing taxed animals and people, threatening to drench their food stores. Although many people and animals drowned in the swirling waters of the Snake River, others continued the journey on to Oregon or Washington.

① Power is a rate. The runner in front does the work at a faster rate than the other runners. Therefore, he is using more power than the other runners.

Figure 4–3. Although these runners are doing the same amount of work, they do not all have the same power. The runner in front is using more power than the others. Why? ①

✧ Did You Know?
The athletes at the 1968 Olympics suffered due to the thin atmosphere in 2240-m-high Mexico City. However, the thin atmosphere also provided less resistance, allowing many world records to be broken.

MEETING
SPECIAL NEEDS

Second Language Support

Be sure that limited-English-proficient students understand the text's references to the pioneers. If any of the students do not understand, ask a volunteer to explain as clearly as possible that part of American history.

Work, Work, Work

The land has been cleared, and the crops have been planted. Already the days have grown shorter; there is less daylight in which to get chores done. Now it is time to build a log cabin and after that, a barn. To do the work quickly, you need lots of power.

Power, like work, is a word that means one thing in ordinary language and another in the language of science. In everyday language, power is often used to mean "strength." Power has even been used as a synonym for force. In science, however, **power** is the amount of work done in a certain period of time. Power is a rate, like speed. As you read in the last chapter, speed is the distance traveled in a certain period of time. Since power can tell us the time it takes to do a certain amount of work, motors and engines are given power ratings to identify how fast they can do work. In mathematical terms, power is represented as

$$\text{power} = \frac{\text{work}}{\text{time to do work}} \quad \text{OR} \quad P = \frac{W}{t}$$

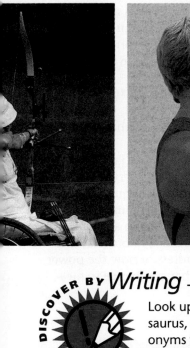

Figure 4–4. In most situations, the force applied in doing work is not constant. In these situations, average force is used to calculate work.

 DISCOVER BY *Writing*

Look up the words *work* and *power* in a thesaurus, and write the commonly accepted synonyms for each of them in your journal. Use a dictionary to look up the meanings of the synonyms with which you are unfamiliar. Tell whether each synonym is also a synonym for *work* and *power* as these words are used in science. ✐

More Power to You   To help you build your log cabin, you could ask your neighbors for assistance. Many people working together supply more power than just one person. They can do the same work at a greater speed, or they can do more work in the same amount of time. Try the next activity to find out the relationship between power and speed.

DISCOVER BY *Calculating*

Connect a spring scale to an object that you can pull across a table. Then slowly pull the object across the table while you take a reading on the spring scale. Repeat this procedure, only this time pull the object quickly. Finally, measure the distance across the table. Use this measurement and the spring scale readings to calculate the work and the power developed when pulling the object. What is the relationship of the speed with which the object is pulled to the amount of power developed? ✐

SECTION 1   **103**

ACTIVITY

How much power do you generate in climbing a flight of stairs?

Process Skills:
Experimenting, Measuring, Observing

Grouping: Pairs

Hints

Caution the students to follow safety rules as they try to rapidly climb the stairs.

▶ **Application**

1. Because the quickest students are often less massive, a slower, more massive student is usually more powerful.

2. No, because power was measured only for a short period of time.

3. (100 kg) × (9.8 m/s) × (1 m) = 980 J of work done in lifting the object. Since that was done in 1 second, (980 J)/(1 s) = 980 W or 0.98 kW.

MEETING SPECIAL NEEDS

Mainstreamed

A student in a wheelchair can measure power for the Activity as follows. Roll the wheelchair at a low constant velocity, and measure the velocity. Stop propelling the wheelchair, and let it coast to a stop, measuring the time required. Calculate the acceleration of the wheelchair as initial velocity divided by time to roll to a stop. Then multiply this acceleration by the mass of the student and the wheelchair to get the force of friction. The force required to maintain a constant velocity is then equal to the frictional force. Finally, multiply that force by the distance the wheelchair is moved at a constant velocity to get force.

TEACHING STRATEGIES, continued

● **Process Skill:** *Solving Problems/Making Decisions*

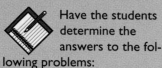

Have the students determine the answers to the following problems:

1. How much work is needed to lift a 10-kg carton a distance of 3 m? (294 J)

2. How many times more work is needed to move the crate twice as much? (2 times, or 588 J)

● **Process Skills:** *Analyzing, Applying*

Although horsepower is not an SI unit, it is often used in measuring the power of engines. Ask the students to name different engines whose capacities are measured in horsepower. (Examples include automotive engines, lawn-mower engines, and aircraft engines.)

Measuring Force and Work In SI, force is measured in *newtons* (N). Recall that a newton is the force required to cause a 1-kg mass to accelerate 1 m per second each second. For instance, if you held a 1 kg mass in your hand, the force pulling down on your arm would be about 9.8 N.

In SI, work is measured in **joules** (J) (JOOLZ). If you lift a weight of one newton one meter from the ground, you do one joule of work. In the next chapter, you will learn that energy is also measured in joules.

The name of the metric unit that measures 1 J of work done each second is the *watt* (W). One watt is a very small amount of power. Therefore, the kilowatt (kW), equal to 1000 W, is more commonly used to measure large amounts of power. You have probably heard the term *kilowatt* used in association with electric power. However, it is important to note that the watt is the standard unit of power, regardless of how the power is produced. In the next activity, you'll have a chance to measure how much power you can generate.

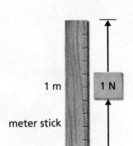

Figure 4–5. One newton lifted 1 m requires 1 J of work.

ACTIVITY

How much power do you generate in climbing a flight of stairs?

MATERIALS
stopwatch, meter stick

PROCEDURE

1. Determine your weight in newtons. If your school has a metric scale that provides mass in kilograms, multiply kilograms by 9.8 m/s/s to get your weight in newtons. If your school has a scale that weighs in pounds, multiply your weight by 4.45 N/lb to get your weight in newtons.

2. One student should use the stopwatch to time how long it takes each student to walk quickly up the stairs. Another student should record the times.

3. Measure the height of one step in meters. Multiply the number of steps by the height of one step to get the total height of the stairway.

4. Multiply your SI weight by the height of the stairs (in meters) to get the work you did in joules. The equation for work is
   work = force × distance OR $W = F \times d$

5. To get your power in watts, divide the work you did by the time (in seconds) that it took you to climb the stairs.

6. Record your name and your power rating on the chalkboard.

APPLICATION

1. Who in your class was the most powerful? Was it the person you expected it to be?

2. Do the power ratings of the students provide any indication of endurance? Explain.

3. How powerful is someone who can lift a 100-kg mass a distance of 1 m in 1 s?

● **Process Skill:** *Comparing*

Point out to the students that crowbars come in varying lengths. Then discuss with the students how the length of a crowbar affects how it increases force.

● **Process Skill:** *Analyzing*

Display a claw of a hammer, and have the students determine how the hammer is actually two machines in one. (The hammer-head is used to transfer the force from your hand to a nail. The claw end is a crowbar.)

● **Process Skills:** *Generating Ideas, Applying*

 Tell the students that bearings, lubrication, and streamlining help reduce friction. Ask the students to jot down their ideas about how these things cannot eliminate friction completely. (If a machine is not continually sup-

plied with fuel or electricity, friction gradually consumes its energy of movement and the machine slows down and stops. This is why a perpetual motion machine can never be invented on Earth.)

▶ **ASK YOURSELF**

An elephant and a mouse raced each other up the stairs. The mouse beat the elephant by a full second but the elephant claimed, "I am more powerful than you are, and this race proves it." How can that possibly be true when the mouse won the race?

May the Force Be With You

An ax helps you cut down trees by changing the size or direction of a force. When a machine is used to change the size of a force, it does so by either multiplying force or multiplying distance. No machine can do both.

The worker in the photograph is using a crowbar to open a crate. The end of the crowbar pushes the top of the crate up with great force. However, the end of the crowbar that is pushing up on the crate moves only a short distance. The other end of the crowbar receives a much smaller force but moves a larger distance.

In this example, not only is the force multiplied, but the direction of the force is also changed. When you push down on the crowbar, the top of the crate is forced up. In many cases, a change in the direction of the force makes work easier.

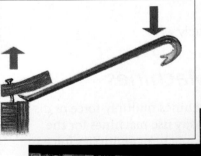

Figure 4–6. This crowbar converts a small force applied over a large distance into a large force applied over a small distance.

ONGOING ASSESSMENT

▶ **ASK YOURSELF**

The elephant has a larger mass than the mouse. Even though it took longer, the work done in moving the larger mass against the pull of gravity is much greater.

SCIENCE BACKGROUND

The joule was named after British physicist James Prescott Joule, whose experiments contributed significantly to the acceptance of the relationship between heat and work. The joule is the metric unit of energy. It is the amount of work equal to the force of 1 newton moving an object a distance of 1 meter.

MEETING SPECIAL NEEDS

Gifted

Have the students compare the work done to lift a 25-kg rock 1 m high on the earth with the work done to do the same on the moon, considering that the moon's gravitational attraction is only 0.16 that of the earth. (Earth: 25 kg × 9.8 N × 1 m = 245 J. Moon: 25 kg × 9.8 N × 0.16 × 1 m = 39.2 J)

SECTION 1 **105** ◀

GUIDED PRACTICE

Review with students the meaning of *ratio*. Explain that a ratio expresses the relationship between two similar quantities in respect to the number of times the first quantity contains the second quantity. Then write this example on the board: one car can go from 0–100 km/h in 30 seconds; another car can reach 100 km/h in 10 seconds. The comparison between the two can be expressed in ratios: 30:10 or 3:1. Finally, encourage the students to think of other examples of ratio, and help them express the ratios on the board.

INDEPENDENT PRACTICE

Have the students provide written answers to the Section Review. In their journals, have the students write out the formulas and define the terms introduced in this section.

EVALUATION

Have the students write a brief description of a task, explaining the components that cause work to be done and describing the power generated in terms of *force*, *work*, and *power*. Have the students place their descriptions in their science portfolios.

INTEGRATION— *Earth Science*

Friction always slows a machine's efficiency on Earth. But in space, no air exists to create friction and slow a spacecraft. A spacecraft could travel forever in space without ever using its engine. Some of the unattended research crafts, like *Voyager*, have already achieved perpetual motion, a feat impossible in Earth's atmosphere.

ONGOING ASSESSMENT
▼ ASK YOURSELF

A machine does not multiply work because of friction. Less work is always received from a machine than what is put into it.

REINFORCING THEMES
Systems and Structures

To emphasize the theme of systems and structures between force and work, draw the following diagram on the board or make it into an overhead transparency. Ask the students to point out the interactions involved in the diagram.

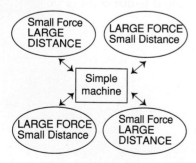

Figure 4–7. Efficiency is increased when friction is reduced. For example, skiers use wax to reduce friction between the skis and the snow.

A machine does not multiply work. Because of friction, which you read about in the last chapter, less work is always received from a machine than is put into it. When distance is increased, force is reduced. Perhaps you've used a screwdriver to open a can of paint. As you move one end of the screwdriver, the other end moves in the opposite direction. Your hand moves much farther than the paint-can lid. Increasing the distance you move decreases the force necessary to open the can. In either case, the work put into the machine is always greater than the work you get out of it.

▼ ASK YOURSELF

Explain why a machine does not multiply work.

The Advantage of Machines

Pioneers used machines because machines multiply force or distance to make work easier. People today use machines for the same reason. The number of times a machine multiplies force is its **mechanical advantage (MA)**. Mechanical advantage is the ratio of the force that comes out of a machine to the force that is put into the same machine. This is expressed as

$$\text{mechanical advantage} = \frac{\text{force output}}{\text{force input}}$$

Some machines, such as a hammer, multiply distance instead of force. As force is increased, distance is reduced. When you use a hammer to pound in a nail, your arm moves a small distance while the hammerhead moves a much larger distance. You have increased the output distance of the hammerhead by reducing the output force. However, because the hammerhead has a large mass, its momentum increases as its speed is increased. An increase in the hammer's momentum increases the amount of force it can apply to the nail.

Discuss this situation with the students, using it to reinforce the concepts taught in this section: A weightlifter pulls up on a barbell, lifts the barbell over his or her head for three seconds, and then lets the barbell drop to the floor. When was the weightlifter doing work? (only when lifting the barbell) When was no work done by the weightlifter? (when the barbell is overhead and when it is dropped to the floor)

 Have students research and write a report about James Watt. In addition to defining horsepower as a unit of power, Watt also invented the micrometer in 1772; the sun-and-planet or epicycloidal gear in 1781, and the centrifugal governor, which is a sensor.

 Cooperative Learning Have the students work in small groups to write a summary of the section. Then ask each group to share its summary with classmates.

Figure 4–8. Without machines, it would probably take the collective force of hundreds of people to move this large stone.

Machines make work easier by changing the amount of force, distance, direction, or speed. However, machines never reduce the amount of work. They actually make more work by adding friction. Because of friction, no machine is 100 percent efficient. You can never get as much work out of a machine as you put in because some of the work must always go to overcome friction. Lubricants can reduce friction and increase efficiency.

▼ **ASK YOURSELF**
What is mechanical advantage?

SECTION 1 *REVIEW AND APPLICATION*

Reading Critically
1. In terms of science, how do work and power differ?
2. Explain why a machine can't be 100 percent efficient.

Thinking Critically
3. How might your life be different if there were no machines?
4. It requires just as much work to stop a pole-vaulter landing in a foam-filled pit as it does to stop a pole-vaulter landing on asphalt. However, a pole-vaulter landing on asphalt is subjected to much more power. Explain why this is true.

SCIENCE BACKGROUND

Cooperative Learning
 Point out that some machines actually depend on friction to be effective. Then challenge the students to think of and list as many machines as possible that depend on friction. (Lists should include car tires, which need friction to grip the road; brakes and clutches in cars, which use friction to work.)

ONGOING ASSESSMENT
▼ **ASK YOURSELF**

Mechanical advantage is the number of times a machine multiplies force.

SECTION 1 REVIEW AND APPLICATION

Reading Critically
1. Work is the product of a force and the distance an object moves in the direction of the applied force. Power is the rate at which work is done.

2. Because of friction, less work is always received from a machine than is put into it.

Thinking Critically
3. Without machines, everything we do would take longer.

4. Power is equal to work divided by time. When a pole-vaulter hits the asphalt, he or she is stopped in a lot less time than if he or she lands in a foam-filled pit. This makes the power larger, even though the amount of work done is the same.

SKILL

Comparing Force and Work Using a Spring Scale

 After the students have completed the Skill, ask them to place their tables and answers in their science portfolios.

Process Skills: Comparing, Measuring, Applying

Grouping: Groups of 3 or 4

Objectives

● **Organize** measured data in table to show relationships between observations.
● **Apply** concepts of force and work to data.

Discussion

Point out to the students that tables are used in science to organize and classify experimental data. Check students' tables to make sure that they have labeled the columns properly and are filling in the information in the correct spaces.

▶ Application

It takes more force to lift the block than to pull the block up the ramp. Theoretically, both require the same amount of work, but friction on the ramp reduces efficiency and so slightly more work is required using the ramp.

✳ Using What You Have Learned

You would tell the person that force differs from work. Force can be measured with a spring scale. Work equals force × distance the object was moved.

★ PERFORMANCE ASSESSMENT

Cooperative Learning

Have each group of students repeat the procedure in the Skill using a different-sized wooden block. Have the students compare their results. Check their calculations for accuracy.

SKILL Comparing Force and Work Using a Spring Scale

▶ MATERIALS

● books (4) ● ruler ● wooden block with an eye hook
● spring scale ● 20-cm × 30-cm piece of heavy cardboard

▼ PROCEDURE

1. Make a table like the one shown to record your data.

TABLE 1: COMPARING FORCE AND WORK
Force × Height =
Force × Length =

2. Stack the books on the floor or on a table. Measure how high the stack is in meters.

3. Attach the block of wood to the spring scale. Lift the spring scale and block to the top of the stack. Record the amount of force needed in newtons.

4. Calculate the work you did by multiplying the force by the height of the books. Record your calculations.

5. Now place the cardboard against the books to make a ramp. Attach the block to the spring scale again. Place the block at the bottom of the ramp.

6. Pull the block up the ramp. Record the amount of force needed.

7. Calculate the work you did this time by multiplying the force by the length of the ramp. Record your calculations.

▶ APPLICATION

Look at your data. Which takes more force—lifting a block or pulling it up the ramp? Which takes more work? Explain.

✳ Using What You Have Learned

Suppose you were going to teach someone to compare force with work using a spring scale. What directions would you give?

Types of Machines

SECTION 2

The logs for your new home have been prepared. You
used an ax to chop notches close to the ends. The notches will hold the logs to each other when they are fitted
together to form the sides of the cabin. The sides of the
log cabin will be about 2.4 meters high. You cannot lift
the heavy logs without help, so you've gathered your
neighbors together for a house-raising. However, neither
the force you can apply with your muscles nor that of
your neighbors will be enough to accomplish your goals.
What is the solution?

One of your neighbors who has assisted at many
house-raisings takes charge. He and your other neighbors
use long pieces of wood to roll a log into place at the
bottom of the cabin. Then they use poles and rope to
roll the log up to the top of the cabin. Construction
takes place before your eyes! What did your neighbors
know that you didn't? They knew how to use the
resources around them as machines. The machines they
used are **simple machines,** machines that have only one
or two parts. There are six simple machines: the lever,
the wheel and axle, the pulley, the inclined plane, the
wedge, and the screw. Which of these machines might
be used at a house-raising? ①

Objectives

Name and **describe**
the six types of simple
machines.

Evaluate the
mechanical
advantage of simple
machines.

Design and
construct a
compound machine.

① The students might respond
that all six types of machines
could have been used at a house-
raising. Some of the machines
might be made on-site, such as a
ramp; others might be tools, such
as an ax.

● **Process Skills:**
Experimenting, Observing

Cooperative Learning The concepts dealt with in this section are more easily grasped when the students are given the opportunity to actually work the machines discussed. As each

simple machine is presented, give groups of students machines (pulleys, levers, and so on) with which to test the concepts being taught. Frequent, short, hands-on activities are more effective than longer, more complex activities for conveying the basic mechanical principles discussed in this section.

● **Process Skill:** *Analyzing*

Ask the students where levers are located in their own bodies. What body parts serve as fulcrums? (Legs and arms act as levers; joints act as fulcrums.)

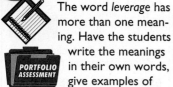

INTEGRATION– *Language Arts*

The word *leverage* has more than one meaning. Have the students write the meanings in their own words, give examples of each meaning, and give their ideas why the same word is used in a mechanical sense and a non-scientific sense. Ask the students to put their examples in their science portfolios. (The meanings are: 1) the action of a lever and the mechanical power resulting from it, and 2) increased means of accomplishing some purpose.)

PORTFOLIO ASSESSMENT

SCIENCE BACKGROUND

Levers are simple machines. Evidence suggests that the lever was in use for thousands of years prior to its law being expressed mathematically by the Greek scientist and mathematician Archimedes. He is credited with the discovery of the principle of the lever. Archimedes is also thought to have defined the principle of the screw. He showed that by making observations and experimenting, it was possible to deduce basic principles that explain why and how many machines work.

Levers

Remember the long pieces of wood the workers put under the log to move it to the bottom of the cabin? They were using the wood as a lever. A **lever** is a bar used for prying or dislodging something. Many common tools, such as crowbars, rakes, and wheelbarrows, are examples of levers.

The part of the lever that you push on is called the *effort arm*. The part that pushes on the object you want to move is called the *resistance arm*. The point on which the lever pivots is called the *fulcrum*. There are three ways to arrange the parts of a lever. The arrangement of the parts determines whether the levers are first class, second class, or third class. The three classes of levers are illustrated in Figure 4–9.

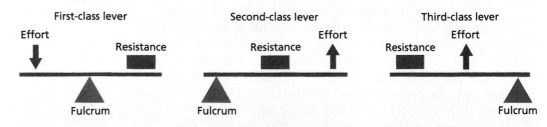

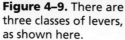

Figure 4–9. There are three classes of levers, as shown here.

One kind of lever you are probably familiar with is the seesaw in a playground. The support for the seesaw is the fulcrum, which allows the seesaw to pivot. The part of the seesaw resting on the fulcrum is the only part that will not go up or down.

Anyone who has been on a seesaw knows that changing how far you sit from the fulcrum of the seesaw changes how hard you must push down to make your partner go up. The farther you are from the fulcrum, the more your force is multiplied. Work becomes easier as you increase the length of the effort arm.

The mechanical advantage of a lever can be calculated by dividing the length of its effort arm by the length of its resistance arm. If you wanted to lift someone on the seesaw who weighed twice as much as you do, you would really need to sit a little farther than twice as far away because the friction requires you to put in a little extra force.

Shovels, oars on a rowboat, balance scales, scissors, and pliers are all examples of levers. Identify the effort arm, the resistance arm, and the fulcrum of all these levers. What other tools can you think of that are levers? Try the next activity to find out more about the mechanical advantage of levers. ①

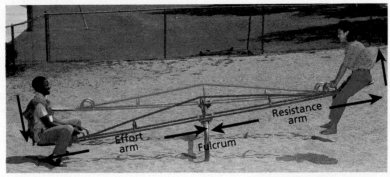

Figure 4–10. A seesaw is actually a first-class lever. The seesaw is supported at its center by a fulcrum.

①

 DISCOVER BY *Doing*

Use a meter stick, a pencil, and two 1-kg masses to observe the mechanical advantage of a lever. The meter stick will act as the lever, the pencil as the fulcrum, and the masses as the effort and resistance forces. Arrange the items as shown. Experiment with the lever by repositioning the fulcrum and one of the masses. What conclusions can you draw about the lever's mechanical advantage? 🖉

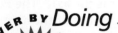

▼ **ASK YOURSELF**

Why does a long crowbar have a bigger mechanical advantage than a short crowbar?

Wheel and Axle

Have you ever made ice cream? There are electric ice-cream makers, and there are also the ice-cream makers that you crank by hand. When you make ice cream "by hand," it seems to take forever until the finished product is ready. It may appear that you are not using a machine at all, but you are. As a matter of fact, the ice-cream maker is an example of a **wheel and axle,** a simple machine consisting of a lever connected to a shaft.

Wheel

Figure 4–11. The crank on this ice-cream maker is a wheel and axle.

Axle

SECTION 2   **111** ◀

● **Process Skill:** *Recognizing Time/Space Relations*

Point out that the wheel operates like a lever, being at a distance from the axle. Then help the students recognize that the fulcrum is at the center of the axle. Finally, ask the students what would be equivalent to an increase in the radius of a

wheel in a lever apparatus. They should use the equation $W = F \times d$ to explain this. (The radius of a wheel is equivalent to the length of the effort arm; therefore distance increases in both cases.)

● **Process Skill:** *Synthesizing*

Have the students make a diagram that illustrates the relationship between force and distance when an axle is used to turn a wheel. Encourage the

students to place their illustrations in their science portfolios. (If an axle is used to turn a wheel, then a large force applied over a small distance is changed into a smaller force applied over a larger distance—the circumference of the wheel.)

THE NATURE OF SCIENCE

The earliest "wheeled" carts were probably no more than flat planks with a series of logs rolling along underneath them. This method of transportation was at best cumbersome, since the logs had to be continually picked up and placed in the front of the plank. While the inventor of the first wheel and axle is unknown, it is known that carts with a wheel on an axle appeared in Sumeria by the year 3500 B.C.

ONGOING ASSESSMENT
▼ ASK YOURSELF

Turning the wheel a small distance multiplies your force on the axle; the wheel moves at a greater speed than the axle.

✧ *Did You Know?*

Tunnels that carry trains are being built under the English Channel between England and France to speed up the trip between the two countries. Besides many smaller tools such as shovels, jackhammers, bulldozers, and cranes, the workers use a giant tunnel boring machine (TBM). Each TBM is about 215 m long and weighs more than 1 million kg. Challenge the students to name which simple machines might be incorporated in the very complex machine needed to drill through the earth. Remind students that these machines are designed to make sure that the walls do not collapse on the workers.

Figure 4–12. This person is using a wheel and axle in the form of a screwdriver to apply a twisting force to a screw.

For something to be considered a wheel and axle, the wheel cannot spin on the axle. The pioneers used a wheel and axle to winch heavy loads of ore up from mines and occasionally to pull ferries across rivers.

Every time you turn a doorknob or a knob on a radio, you are using a wheel and axle. The knobs are the wheels, and the shafts are the axles. When you turn the knob, you are multiplying your force in turning the shaft.

Another example of a wheel and axle is the steering wheel of a car. A small force applied to the large steering wheel is multiplied as the smaller axle turns to create a very large force. In this application of a wheel and axle, the effort arm is the radius of the wheel, and the resistance arm is the radius of the axle. When a car is steered around a curve, the driver must move the steering wheel a large distance to move the axle a short distance. However, the small force on the steering wheel becomes a large force to turn the wheels of the car.

A wheel and axle also has mechanical advantage. Think about the common screwdriver. The handle of the screwdriver is the wheel, and the blade of the screwdriver is the axle. The radius of the wheel is the effort arm, and the radius of the axle is the resistance arm. This explains why it is easier to insert screws with a screwdriver that has a wide handle. The mechanical advantage is greater because the effort arm—the radius of the handle—is bigger than the resistance arm—the radius of the blade.

▼ ASK YOURSELF
How does a wheel and axle make work easier?

The Pulley

Now both your new home and barn are complete, and the crops have been harvested. It is time to store the hand-bundled hay in the upper floor of the barn. What is the most efficient way to get the bundles up there? You could use a pulley. A **pulley** is a simple machine that consists of a wheel that is free to spin on its axle. This is different from the wheel and axle, which must move together. Most pulleys have a groove around the circumference to hold a rope or a cable. Force is applied to the rope that travels in the groove of the pulley. The rope is attached to the object to be moved, in this case, the bundle of hay.

Pulleys can be either fixed or movable. If they are fixed, they are attached to something that does not move, like a building or a tree. Fixed pulleys only spin; they do not move up and

MA=2   MA=3   MA=4   MA=? ①

down with the weight being lifted. Therefore, the mechanical advantage is one. It multiplies whatever force you put in by one. Remember, though, that because of friction it will really be less than one and you will get less force out than you put in.

In most cases, fixed pulleys are used to change the direction of force. They cannot be used by themselves to multiply force. By using a fixed pulley, you can pull down on the rope to cause an upward force on the hay bundles and get them to the upper floor of the barn.

Figure 4–13. The mechanical advantage of a system of pulleys can be determined by counting the number of ropes supporting the resistance of force, or load.

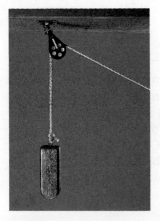

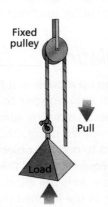 Fixed pulley

Pull

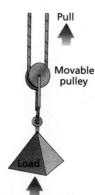 Pull

Movable pulley

Load

Movable pulleys, however, are attached to the object being moved. Unlike fixed pulleys, they move up and down with the weight being lifted. Movable pulleys are used to multiply force. They can multiply force even more when they are used with fixed pulleys in a *block and tackle,* as shown in Figure 4–13. Block and tackles are sometimes used by automobile mechanics to remove heavy engines from cars. By moving the end of the

Figure 4–14. The fixed pulley (left) is used to change the direction of a force. The movable pulley (right) is used to multiply force.

● **Process Skill:** *Comparing*

Ask the students to name examples of fixed pulleys and movable pulleys that they might use or see. If necessary, remind the students that the two types of pulleys are not interchange- able as far as their use. A fixed pulley is only used to change direction; a movable pulley is used to multiply force.

● **Process Skill:** *Experimenting*

Remind the students that work does not change. If force increases, then distance must decrease; if distance increases, then force must decrease. The amount of work changes only if force and distance both change in the same direction. You can reinforce this concept by pro- viding an object on an inclined plane for the students to exper- iment with.

MEETING SPECIAL NEEDS

Mainstreamed

Illustrate the advantage a screw has over a nail in terms of grip- ping or holding ability. Insert a small wood screw into a piece of wood. Then pound a nail into the same piece of wood. Have the students compare the amount of force needed to pull the screw and nail out of the wood using a claw hammer. (The students should note that more force was needed to pull out the screw. Therefore, the screw has greater gripping or holding ability.)

ONGOING ASSESSMENT

ASK YOURSELF

Boat building and repair and con- struction of buildings, escalators, and elevators are some examples of other industries where pulleys would be useful.

① The students might respond that other uses for ramps could include ramps to make buildings wheelchair-accessible and entrance and exit ramps on highways.

DISCOVER BY *Calculating*

PORTFOLIO ASSESSMENT

The students should find that the inclined plane reduces the effort as the slope of the incline decreases. This is, of course, at the expense of distance covered. Have the students place their results in their science portfolios.

rope a long distance with a small force, a mechanic is able to lift the motor a small distance with a large force. This shows again the trade-off between force and distance. You can put a small force over a large distance into a machine and get a large force moving a small distance out of a machine.

ASK YOURSELF
In what industries other than those mentioned might pulleys be useful?

I'm Inclined to Use a Ramp

At the house-raising, your neighbors used a ramp to roll the logs to the top of the cabin. A ramp is really an **inclined plane**, a simple machine con- sisting of a flat, sloping surface. The mechanical advantage of a ramp can be calculated by divid- ing the ramp's length by the ramp's height. This tells you that a longer, more gently sloped inclined plane has a higher mechanical advantage than a short, steep one. How many other uses can you think of for ramps? ①

DISCOVER BY *Calculating*

Prop up a book so it forms an inclined plane. At- tach a spring scale to a block of wood or other object. Using the spring scale, pull the object up the inclined plane. Record the amount of force needed to pull the object. Compare this force with the amount of force required to pull the object up inclined planes at different angles. How does the effort required to pull the object change with the slope of the inclined plane? ✎

In chopping down trees and processing them into boards and beams, you would use many different kinds of tools, in- cluding a wedge. A **wedge** is an inclined plane that is used to push objects apart. The head of an ax is a wedge. Other wedges include the blades of knives, saws, razors, and other cutting tools. Sharpening these tools makes the cutting edge thinner. This increases the mechanical advantage and makes them easier

to use. Nails are also wedges and are often used to fasten boards
together.

Another kind of fastener is the screw. A **screw** is an inclined plane that is wrapped around a cylinder. Screws are used
on many items, such as screw-on jar lids, water faucets, and
some kinds of automobile jacks. The mechanical advantage of
the screw is much like that of the inclined plane. Looking at the
threads on a screw is like looking at many tiny inclined planes.
The more gentle the slope of these tiny ramps, the easier it is
to turn the screw. That means it has a higher mechanical advantage. Remember though, if it takes less force to turn, it will require that you turn it a greater distance.

Figure 4–15. A screw is
basically an inclined
plane wrapped around
a cylinder. This illus-
tration shows the
mechanical advantage
of a screw with gently
sloping threads.

gentle slope threads
high MA
small force required
GREAT DISTANCE

steep slope threads
low MA
LARGE FORCE REQUIRED
small distance

 Problem Solving _____

Most of the examples of machines given in this
section are machines that enable people to exert
more force than their muscles are capable of.
Describe one situation in which it is important for
humans to exert less force than their muscles normally do. Write your description in your journal. ✏

▼ **ASK YOURSELF**

How are an ax blade, a paring knife, and a screw similar?

Two or More Machines

Most machines are made up of combinations of machines.
Compound machines are two or more simple machines put
together to do work. The ax is a compound machine. You just
read that the head of the ax is a wedge. What simple machine ②
is the handle? In the next activity, you'll have a chance to
design your own compound machine.

Cooperative Learning Remind the students that the mechanical advantage of every machine is less than predicted because of friction. Then have the students work in groups to make six columns with the name of each simple machine at the head. Have the students list the sources of friction for each machine. Provide assistance as needed.

Have the students provide written answers to the Section Review. In their journals, have the students list the six types of machines and one or more examples of each to reinforce the concepts.

Have each student work with a partner. One student should name a compound machine. The other student should name the simple machines that are part of the compound machine. Then have the students reverse roles and repeat the procedure.

SCIENCE BACKGROUND

Some bicycles built in the 1800s had wooden wheels and iron tires and were justifiably called "boneshakers." In 1874, H. J. Lawson built what was referred to as the safety bicycle. Its wheels were smaller than the earlier high-wheeler bicycle and positioned the rider closer to the ground. Modern bicycles are based on the design of the safety bicycle.

INTEGRATION– *Mathematics*

Have interested students find out what the correct air pressure should be in a bicycle tire. (Correct air pressure is related to tire size. Most tires have the correct air pressure stamped on the sidewall.) Encourage the students to find out why correct air pressure improves a bicycle's ride. Ask the students to place their research in their science portfolios. (It reduces the amount of friction between the tires and the roadway.)

SCIENCE TECHNOLOGY SOCIETY A robot is an example of a compound machine. A robot is computer-controlled; its movements can be designed to be extremely accurate and precise. The automotive industry is among many industries that use robots to increase efficiency and productivity.

DISCOVER BY *Writing*

Rube Goldberg, a famous cartoonist, was known for his drawings of imaginary machines like the one shown here. Although humorous, his strange and unusual machines employed many types of simple machines. Study the Rube Goldberg drawing shown here, and list all the simple machines you recognize. Try your hand at designing and drawing your own "Rube Goldberg machine." Which simple machines did you use in your machine? How efficient do you think your machine is? ✎

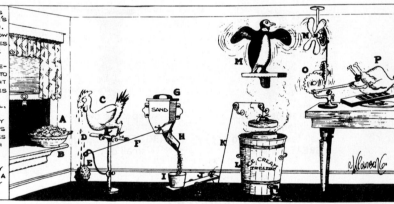

SIMPLE WAY TO CARVE A TURKEY. THIS INVENTION FELL OFF THE PROFESSOR'S HEAD WITH THE REST OF THE DANDRUFF. PUT BOWL OF CHICKEN SALAD(A)ON WINDOW SILL(B) TO COOL. ROOSTER(C)RECOGNIZES HIS WIFE IN SALAD AND IS OVERCOME WITH GRIEF. HIS TEARS(D) SATURATE SPONGE(E), PULLING STRING(F) WHICH RELEASES TRAP DOOR(G)AND ALLOWS SAND TO RUN DOWN TROUGH(H)INTO PAIL(I). WEIGHT RAISES END OF SEE-SAW(J)WHICH MAKES CORD(K) LIFT COVER OF ICE CREAM FREEZER(L). PENGUIN(M)FEELING CHILL, THINKS HE IS AT THE NORTH POLE AND FLAPS WINGS FOR JOY, THEREBY FANNING PROPELLER(N)WHICH REVOLVES AND TURNS COGS(O)WHICH IN TURN CAUSES TURKEY(P) TO SLIDE BACK AND FORTH OVER CABBAGE-CUTTER UNTIL IT IS SLICED TO A FRAZZLE — DON'T GET DISCOURAGED IF THE TURKEY GETS PRETTY WELL MESSED UP. IT'S A CINCH IT WOULD HAVE EVENTUALLY BECOME TURKEY HASH ANYWAY.

Bicycles Built for One or More Bicycles are compound machines consisting of a variety of simple machines. The back wheel of a bicycle is a wheel and axle. Attached to its axle are gears that allow the chain to grip tightly as it turns on one of the various different-sized wheels. The teeth on these gears are actually a series of levers spaced evenly around the outside of each gear. These gears are attached to another gearwheel and axle, which is attached to two pedals. In addition to the gearwheel and axle system, the bicycle also has levers that are used to apply the brakes and to steer the front wheel.

Figure 4–16. The gears on this bicycle are actually small levers attached to a wheel and axle. The brakes consist of two levers that pivot on a single fulcrum.

Winter—At Last   Outside the wind blows, and the piles of snow reach above the windows. Inside your cabin, a fire in the fireplace blazes. Before winter set in, you helped to make the tables and benches you and your family now use. The winter will be long, but there is a lot to do inside the cabin—grinding corn, repairing farm tools, and making candles and clothing. Next spring, you and your neighbors plan to build a schoolhouse. Why not? The resources and machines to do the job are just outside your front door.

▼ **ASK YOURSELF**

Identify a compound machine not mentioned in this chapter, and list the simple machines it contains.

SECTION 2   REVIEW AND APPLICATION

Reading Critically

1. Categorize the six simple machines as either levers or inclined planes.
2. Is a pulley considered to be a wheel and axle? Why or why not?
3. Explain why an ax is a compound machine.

Thinking Critically

4. Give an example of a device that contains one or more of each of the following simple machines: lever, wheel and axle, and wedge.
5. How do you think a tandem bicycle (a bicycle "built for two") might be similar to and different from a single-passenger bicycle?

Objective

● **Understand** the difference between a simple and a compound machine.

Pre-Lab

Have the students name and describe models they have constructed. (Examples include models of airplanes, houses, and cars.) Then tell the students they will construct and compare models in this Investigation.

Analyses and Conclusions

1. Responses will depend upon the machines built. Students' machines might contain any or all of the six simple machines discussed in this section.

2. The machines were combined to work together and to provide a use that simple machines alone could not provide.

3. While mechanical advantages will vary, remind the students that because of friction, mechanical advantage is less than 100 percent.

★ PERFORMANCE ASSESSMENT

Cooperative Learning

Have the students display and demonstrate their machines for classmates. Encourage classmates to ask questions about the machines. Check each group's responses to questions for accuracy.

Application

Although the students' machines will vary, the students should be able to point out each simple machine within their compound machines. The students should also briefly explain how their machines make work easier.

✳ Discover More

Have the students point out the simple machines that make up the compound machines. For example, a radial-arm saw has a wheel and axle, and a drill press has a screw.

Post-Lab

Have volunteers list reasons why models are important in science. The students might point out that models help scientists visualize or manipulate objects that are too small or too large to actually visualize or manipulate.

INVESTIGATION

Building Simple and Compound Machines

▶ MATERIALS

● bottle caps ● cardboard ● craft sticks ● empty thread spools ● glue ● modeling clay ● paper ● pencils ● rubber bands ● scissors ● shoe boxes ● stones ● straws ● string ● tape ● any other materials to which you have access

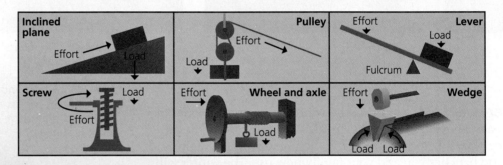

▼ PROCEDURE

1. Make an inclined plane using any of the materials listed above. Compare your inclined plane with those of other groups.

2. Take the inclined plane apart. Now make a pulley, a lever, a screw, and a wheel and axle. Compare your group's simple machines with those of other groups.

3. With the others in your group, decide how to make a compound machine. In your journal, draw a picture of your machine and label the parts. Be creative! Your compound machine can be a machine that already exists or one that you have invented.

4. Build your machine. Then demonstrate to the other groups how your machine makes work easier.

▶ ANALYSES AND CONCLUSIONS

1. How many simple machines are in your compound machine? What are they?

2. Why did you combine the machines in this way?

3. What are the mechanical advantages of your machine?

▶ APPLICATION

Design a compound machine that has all six simple machines in it. Explain what the machine will do and how it will make work easier.

✳ Discover More

If your school has an industrial arts or technology department, ask the teacher to show you some of the large tools, such as a lathe, drill press, and radial arm saw, as well as some of the special-purpose hand tools. Examine each of these compound machines to determine the simple machines that make up their parts.

CHAPTER 4 HIGHLIGHTS

The Big Idea— Systems and Structures

Lead the students to an understanding that work is the application of force to make something move a certain distance. Power is the amount of work done in a certain amount of time. Also help the students to understand that there are six types of simple machines.

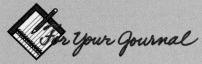

CONNECTING IDEAS

Both machines are compound machines. The tricycle contains several wheels and axles, and screws, which are used to hold the tricycle together. The food grinder is made up of two wheels and axles and two screws.

CHAPTER 4 HIGHLIGHTS

Albert Bierstadt, "The Oregon Trail" 1869, 31" × 49", oil on canvas. The Butler Institute of American Art.

The Big Idea

Work is the application of force to make something move a certain distance. Power is the amount of work you can do in a certain amount of time. Because there is a limit to the force even a very strong human or several humans can exert, people need machines to help them do their work.

There are six types of simple machines. Simple machines are often combined to make many different kinds of compound machines. All machines, whether simple or compound, have a mechanical advantage—the ratio of force that comes out of a machine compared with the force that is put into the machine. Mechanical advantage is always less than 100 percent because of friction.

For Your Journal

Look back at the ideas you wrote in your journal at the beginning of the chapter. How have your ideas about work and machines changed? Revise your journal entry to show what you have learned. Include information on how machines are useful in your everyday life.

Connecting Ideas

Consider the food grinder and the tricycle shown. Identify these two machines as either simple or compound, and name the simple machine or machines that make each of them work.

Understanding Vocabulary

1. (a) Each term deals with the concept of work. Work is the force applied to an object multiplied by the distance the object moves. Power is the amount of work done in a certain period of time.

(b) Each term is related to machines. A simple machine is a device that helps you do work by changing the size or direction of a force. A compound machine is two or more simple machines put together to do work. Mechanical advantage is the ratio of the force that comes out of a machine to the force that is put into it.

(c) Each term names a type of simple machine. A lever is a bar used for prying or dislodging something. A wheel and axle is a lever connected to a shaft. A pulley is a wheel that is free to spin on its axle.

(d) Each term names a type of inclined plane. A ramp is simply an inclined plane. A wedge is an inclined plane that is used to push objects apart. A screw is an inclined plane that is wrapped around a cylinder.

Understanding Concepts

Multiple Choice

2. b

3. d

4. a

5. d

6. a

Short Answer

7. Since work = force × distance, if you change the distance through which the work is done, you must also change the force. If input distance is large compared with output distance, input force must be small compared to output force.

8. The blades of the plow that separated the soil into two piles were a wedge.

Interpreting Graphics

9. The bottle opener is a second-class lever. It has the fulcrum at one end (the point of the opener), the effort arm at the other end (where the opener is held), and the resistance arm in between.

10. The fishing pole is a third-class lever. It has the fulcrum at one end (a person's wrist), the resistance arm at the other end (the end of the pole), and the effort arm in between (a person's hand).

CHAPTER 4

REVIEW

► Understanding Vocabulary

1. For each set of terms, define each term and explain how the terms are related.
  a) work (101), power (102)
  b) simple machine (109), compound machine (115), mechanical advantage (106)
  c) lever (110), wheel and axle (111), pulley (112)
  d) ramp (114), wedge (114), screw (115)

► Understanding Concepts

MULTIPLE CHOICE

2. A screw is an example of what kind of simple machine?
  a) wedge
  b) inclined plane
  c) pulley
  d) wheel and axle

3. Work is
  a) making an effort.
  b) applying force.
  c) making something move.
  d) applying a force and making something move.

4. When a fixed pulley is used, the force and the load
  a) move in opposite directions.
  b) move in the same direction.
  c) do not move.
  d) are always equal.

5. Machines can multiply
  a) only force, but not distance.
  b) only distance, but not force.
  c) neither force nor distance.
  d) force or distance.

6. The name of the metric unit that measures 1 joule of work done each second is the
  a) watt.   **b)** kilowatt.
  c) newton.   **d)** gram.

SHORT ANSWER

7. Briefly explain how levers and inclined planes allow force to be made larger by making the distance through which the force is applied smaller.

8. Explain which part of a pioneer's plow was a wedge.

Interpreting Graphics

9. Look at the art below. What class of lever is the bottle opener? Explain your answer.

10. Look at the art below. What class of lever is the fishing rod? Explain your answer.

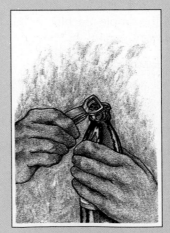

Reviewing Themes

11. *Systems and Structures*
Choose a compound machine, such as scissors. List the simple machines that make up the compound machine and explain how the simple machines work together.

12. *Energy*
Explain how energy relates to machine usage.

Thinking Critically

13. In a wheel and axle, the radius of a wheel is like the effort of a straight lever. The radius of the axle is like the resistance arm. What part of the wheel and axle is like the fulcrum of a straight lever? Explain your answer.

14. Think about machinery commonly used today. If the lever were to be eliminated, how many machines would still be in use? Explain your answer.

15. An elevator is a single pulley lifting machine. Explain in your own words how you think this machine works or draw a diagram showing how you think it works.

16. Explain why a merry-go-round like the one shown here is not considered a wheel and axle.

17. In moving the piano, the movers in the top photograph will use more force but will do slightly less work than the movers in the bottom photograph. Explain why.

Discovery Through Reading

Rosta, Paul. "Elevators that Capture Energy." *Popular Science* 237 (December 1990): 98–99. Millions of elevators waste energy. This article discusses an invention that taps and stores energy to run other systems.

CHAPTER
5

ENERGY

PLANNING THE CHAPTER

B = Basic   A = Average   H = Honors
The coding Basic, Average, and Honors indicates subsections, features, and resources that might be appropriate for different levels of learners. For additional suggestions regarding choice of topic and depth of coverage, see the Pacing Chart on pages T28–T31.

*Frame numbers at point of use
(TR) Teaching Resources, Unit 2
(IT) Instructional Transparencies
(LI) Laboratory Investigations
(SD) *Science Discovery* Videodisc Correlations and Barcodes
(SRG) Study and Review Guide

CHAPTER MATERIALS

Title	Page	Materials
Teacher Demonstration	126	rubber bands of different thicknesses (2), rulers of equal length (2)
Discover by Writing	126	(per individual) journal
Activity: How can the kinetic energy of a rolling cart be measured?	128	(per group of 3 or 4) several books, board, masking tape, meter stick, balance, rolling cart, stopwatch
Teacher Demonstration	130	balls (golf ball, table-tennis ball, basketball, tennis ball, superball), meter stick
Teacher Demonstration	131	mass, string, ringstand, ring or clamp
Discover by Doing	134	(per individual) bicycle pump, deflated bicycle tire
Skill: Communicating Using a Line Graph	137	(per group of 3 or 4) thin books (10–20), stiff cardboard 15 cm x 40 cm, ruler or meter stick, cart, tape, graph paper
Discover by Doing	138	(per group of 3 or 4) bowl, water, thermometer, electric mixer
Teacher Demonstration	141	Florence flasks (3), colored water, glass tubing, one-holed stoppers (3), beakers (3), ice water, boiling water, water at room temperature
Discover by Doing	143	(per individual) safety goggles, laboratory apron, glove (per group of 3 or 4) burner, metal rod, candle, matches
Teacher Demonstration	143	heatproof glass cake pan, books or blocks, hot plate, potassium permanganate, water
Investigation: Predicting Temperature Change	146	(per group of 3 or 4) plastic-foam cups (3), thermometer, 50-mL graduates (2), cold water, hot water, stirring rod

ADVANCE PREPARATION

For the *Discover by Doing* on page 134, all students might not have a bicycle pump with which to inflate a tire. You might ask volunteers to bring in pumps and tires.

TEACHING SUGGESTIONS

Field Trip
A field trip to the local electric energy plant is an excellent addition to your instruction on energy. Nearly all electric energy plants involve multiple conversions of energy forms, with heat as one form. Heat transfer and entropy losses are problems at all energy plants. This field trip is especially valuable if you correspond with the tour leader in advance and discuss with her or him the operation of the plant in terms of energy transfer, the laws of thermodynamics, and heat and temperature. It is especially impressive to students when the energy company representative relates physics concepts and processes to the everyday working of the plant.

Outside Speaker
If you cannot go on a field trip, have a representative from your local energy plant speak to your classes. Again, this will be more effective if you arrange with the speaker to cover the points studied in the chapter.

CHAPTER THEME—*ENERGY*

This chapter will help the students understand that energy is found in many forms and that each of the forms is responsible for shaping our lives. Without energy, technological advancements, such as motorized machines, would not be possible. Without energy, life itself would not be possible. Discussions of potential energy, kinetic energy, and heat are used to develop the greater concept of energy. A supporting theme in this chapter is **Systems and Structures.**

CHAPTER
5 ENERGY

ew things are as dramatic or display as many forms of energy as the launch of a space shuttle. The light, heat, the deafening sound of the engines and rockets, and the movement of the huge orbiter are all forms of energy. The amount of energy released during a launch is tremendous.

CHAPTER MOTIVATING ACTIVITY

Perform the following demonstrations, and ask the students to list the forms of energy involved: turn on a portable radio (chemical energy into electric energy and sound energy); "crack" an instant athletic hot pack (chemical energy to heat energy); have the students rub their hands together (mechanical energy to heat energy); turn on a television or computer (electric energy to light and/or sound energy); demonstrate a radiometer (light energy to mechanical energy); turn the lights in the classroom off and on (electron motion or electric energy to heat and light energy)

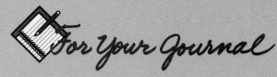

For Your Journal

The journal questions will help you identify any misconceptions the students might have about energy. In addition, these questions will help the students realize that they already possess some knowledge about energy. (Students will have a variety of responses to the journal questions. Accept all answers, and tell the students that at the end of the chapter they will have a chance to refer back to the questions.)

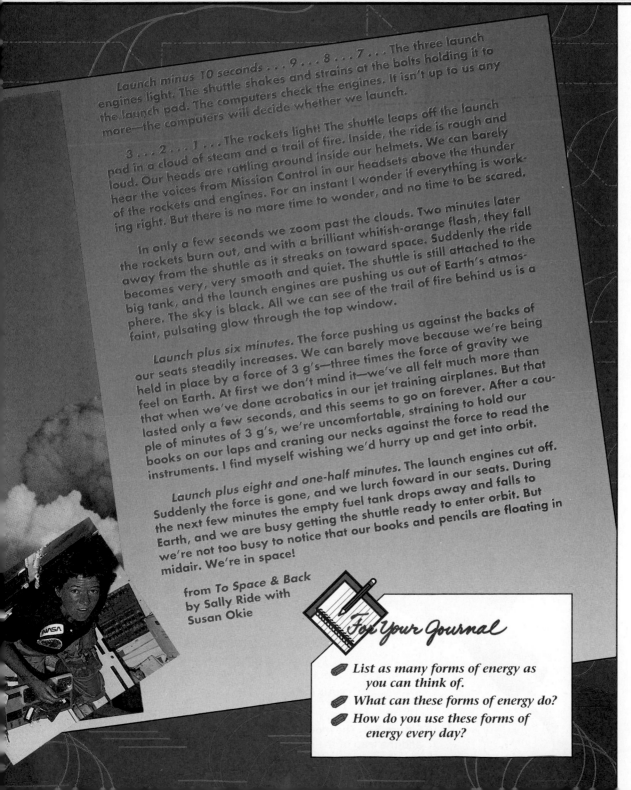

Launch minus 10 seconds . . . 9 . . . 8 . . . 7 . . . The three launch engines light. The shuttle shakes and strains at the bolts holding it to the launch pad. The computers check the engines. It isn't up to us any more—the computers will decide whether we launch.

3 . . . 2 . . . 1 . . . The rockets light! The shuttle leaps off the launch pad in a cloud of steam and a trail of fire. Inside, the ride is rough and loud. Our heads are rattling around inside our helmets. We can barely hear the voices from Mission Control in our headsets above the thunder of the rockets and engines. For an instant I wonder if everything is working right. But there is no more time to wonder, and no time to be scared.

In only a few seconds we zoom past the clouds. Two minutes later the rockets burn out, and with a brilliant whitish-orange flash, they fall away from the shuttle as it streaks on toward space. Suddenly the ride becomes very, very smooth and quiet. The shuttle is still attached to the big tank, and the launch engines are pushing us out of Earth's atmosphere. The sky is black. All we can see of the trail of fire behind us is a faint, pulsating glow through the top window.

Launch plus six minutes. The force pushing us against the backs of our seats steadily increases. We can barely move because we're being held in place by a force of 3 g's—three times the force of gravity we feel on Earth. At first we don't mind it—we've all felt much more than that when we've done acrobatics in our jet training airplanes. But that lasted only a few seconds, and this seems to go on forever. After a couple of minutes of 3 g's, we're uncomfortable, straining to hold our books on our laps and craning our necks against the force to read the instruments. I find myself wishing we'd hurry up and get into orbit.

Launch plus eight and one-half minutes. The launch engines cut off. Suddenly the force is gone, and we lurch forward in our seats. During the next few minutes the empty fuel tank drops away and falls to Earth, and we are busy getting the shuttle ready to enter orbit. But we're not too busy to notice that our books and pencils are floating in midair. We're in space!

from *To Space & Back*
by Sally Ride with
Susan Okie

For Your Journal

- List as many forms of energy as you can think of.
- What can these forms of energy do?
- How do you use these forms of energy every day?

ABOUT THE AUTHOR

Sally Ride, one of the coauthors of this literature piece, is a familiar name to many people because she was the first American woman in space.

Born in 1951 in Los Angeles, California, she earned her Ph.D. in physics at Stanford University, became an astronaut, and in June 1983 became the first American woman in space during a mission of the space shuttle *Challenger*.

During her two shuttle missions, Dr. Ride successfully launched satellites, using a specially designed remote manipulator arm. She later served on the presidential commission investigating the explosion of the space shuttle *Challenger*.

Since Dr. Ride's historic mission, other women have traveled into space. These women include Mary Cleave, Bonnie Dunbar, Anna Fisher, Mae Jemison, Judith Resnik, and Kathy Sullivan. (Judith Resnik was killed in the January 28, 1986, *Challenger* explosion, as was Christa McAuliffe who was to be the first teacher in space.)

FOCUS

In this section, the many forms of energy are discussed. Heat, light, chemical, mechanical, and sound are defined as forms of energy. Potential energy and kinetic energy are introduced, discussed, and compared.

MOTIVATING ACTIVITY

Cooperative Learning Have the students work in cooperative groups to make a poster that illustrates various forms of energy. Students can cut pictures from magazines that show different forms of energy, arrange the pictures on poster board, and write captions that describe the forms of energy. Have the groups display their energy posters in the classroom. You might remind the students that most of the pictures will show more than one kind of energy.

PROCESS SKILLS
• Measuring • Inferring

POSITIVE ATTITUDES
• Curiosity • Openness to new ideas

TERMS
• potential energy • kinetic energy

PRINT MEDIA
Why Doesn't the Sun Burn Out? And Other Not Such Dumb Questions About Energy by Vicki Cobb (see p. 67b)

ELECTRONIC MEDIA
The Forms of Energy Focus Educational Media, Inc. (see p. 67c)

SCIENCE DISCOVERY
• Energy; potential vs. kinetic • kinetic energy; hammering a nail

BLACKLINE MASTERS
Laboratory Investigation 5.1
Study and Review Guide

INTEGRATION–
Biology

The human body shows many examples of the transformation of energy. For example: Chemical energy is released when food is chewed and digested. Chewing is a mechanical process. Digestion is mostly a chemical process. The energy from the food is finally converted into mechanical, chemical, and heat energy when muscles are flexed.

SECTION 1

Forms of Energy

Objectives

Compare potential and kinetic energy.

Calculate the kinetic energy of a moving object.

Relate the amount of kinetic energy and the speed of an object to the force of a collision.

Our universe is rich in different kinds of energy. You know that during a shuttle liftoff, a tremendous amount of energy is released. Sometimes, though, energy can be hard to recognize. Is energy in action when a flag flutters in the breeze? Does energy cause the ice in your soda to melt? Yes. Energy is recognized mostly by what it does.

Energy can be violent, as when thunder claps so loudly the windows shake. But energy can also be gentle, like a dewdrop quietly evaporating on the petal of a rose. You might say energy is an expert at disguise.

Energy Has Many Forms

Suppose you wanted to move your textbook from your backpack to your desk. You couldn't do that without using energy. You would use mechanical energy—supplied by muscle power—to move the book. Energy is found in many forms. Heat is the energy produced by moving atoms and molecules. Light energy travels in waves to reach your eyes. Chemical energy is the energy that binds atoms and molecules together. Your ear receives energy vibrations as sound. Mechanical energy is the energy produced by moving objects. It is this form of energy that you would use to move your book. A Frisbee® flying through the air has mechanical energy. A moving bicycle, flowing water, and a rolling marble also have mechanical energy. What's more, this is only a small sampling of the types of energy that affect your life.

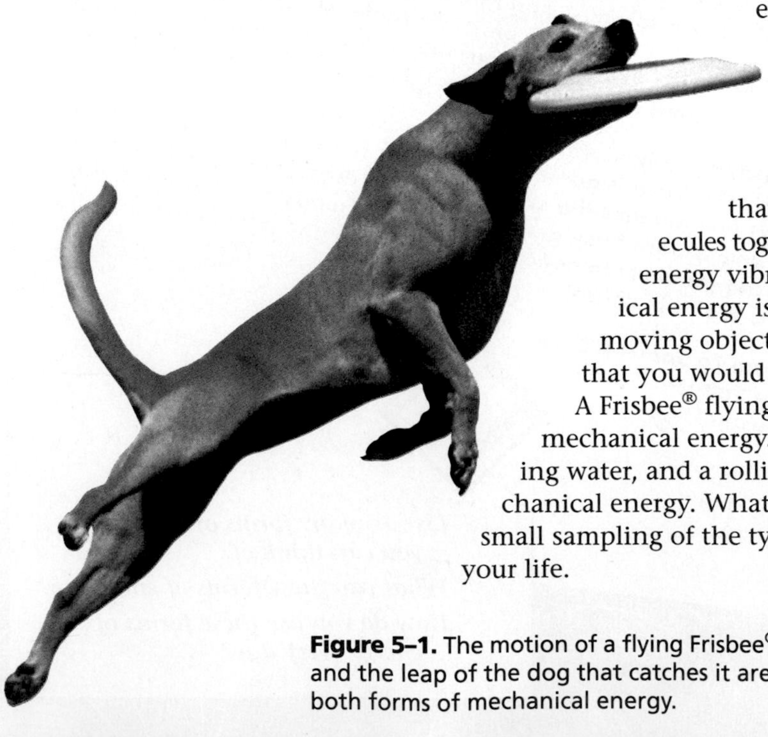

Figure 5–1. The motion of a flying Frisbee® and the leap of the dog that catches it are both forms of mechanical energy.

TEACHING STRATEGIES

● **Process Skills:**
Communicating, Generating Ideas

Point out to the students that energy is all around them. Although they use energy and see the effects of it, the actual energy itself cannot be seen or touched. Ask the students what they think energy is. (Use the

students' responses to form an operational definition similar to "energy is anything that causes change," or "energy is the ability to move or change matter in some way.")

● **Process Skill:** *Inferring*

Tides and winds are forms of energy. Ask what change or movement is caused or brought about by waves and wind. (Answers include moving boats, turning turbines or wheels, changing beaches by moving and depositing sand, weathering, and transporting rocks.)

● **Process Skill:** *Describing*

In their journals, have the students describe and illustrate the forms of energy involved in rolling a strike in bowling. (Answers should reflect an understanding of the forms of energy.)

You know that energy can take many different forms, but what exactly is it? *Energy* is the ability to cause change. Think back to removing the book from your backpack. In that case, the change was in the position of the book and the change was caused by mechanical energy. Let's look at another example.

Suppose you get home from school and there's a note asking you to wash the dishes. Sheesh! You are always having to do the dishes. Muttering under your breath, you collect the dirty dishes and stack them up. You run some hot water and add dishwashing detergent. After scrubbing the dishes clean and rinsing them with warm water, you stack them to dry. In this example, energy was used to cause many changes. How many can you name? ①

Energy is a necessary ingredient for change. The amount of energy used to cause a change is measured in *joules (J)*. You may recall from Chapter 4 that work also is measured in joules. This is because work is a type of mechanical energy.

Figure 5–2. The energy of a moving bowling ball causes dramatic changes when it crashes into the pins.

▶ **ASK YOURSELF**

Energy has many forms, and each of the forms causes change. Give an example of your own that shows this is true.

Stored Energy

Think about a rubber band. Just lying around on a desk, a rubber band cannot force anything to move or do any work. However, if you pull back on the rubber band and hold it, it has stored energy. The energy stored in an object due to its position is called **potential energy.** The energy used to stretch the rubber band becomes stored in the rubber band as potential energy.

If you release the stretched rubber band, it snaps back to its original shape. When this happens, the potential energy in the rubber band changes to motion.

REINFORCING THEMES
Energy

Point out to the students that energy never disappears; it changes form. Then ask the students to look around the classroom and list, in their journals, those objects that have energy because of their position. (Possible answers include water dripping from a faucet, a door on hinges, and pens in a student's shirt pocket.)

① There are many kinds of energy used when washing dishes: the energy used to heat the water, the energy the person uses who washes the dishes, the energy of the reaction of the dishwashing soap with the leftover food on the plates, and the evaporation of the water from the clean dishes. Most students will identify the mechanical energy involved. You might point out that heat energy and chemical energy are also involved, as well as possible electrical energy.

✧ **Did You Know?**
The energy released every second by the three main engines of the space shuttle is 23 times greater than the energy output of the Hoover Dam every day.

ONGOING ASSESSMENT
▶ **ASK YOURSELF**

Responses should include information about the types of energy involved and the changes that result.

● **Process Skill:**
Communicating

Point out to the students that riding a bicycle requires the rider to have energy—the ability to do work. When you ride your bicycle one block, you use about 90 000 J of energy.

Demonstration

Obtain two rubber bands of different thicknesses and two rulers of equal length. Place the rubber bands on a table, and have the students observe their shapes. Stretch each rubber band over a ruler. Pluck each band. Ask how the thicker rubber band can be made to produce the same pitch as the thin one. (To produce the same pitch, stretch the thicker rubber band until both rubber bands have the same thickness or diameter.)

LASER DISC

1869, 1870, 1871, 1872, 1873

Energy; potential vs. kinetic

DISCOVER BY *Writing*

PORTFOLIO ASSESSMENT   After the students have completed the Discover by Writing, have them place their poems or short stories in their science portfolios.

✧ *Did You Know?*
The nutritional energy of a glass of milk is about the same as the kinetic energy of a car traveling at 100 km/hr—about 1 million joules.

The potential energy stored in a stretched rubber band is called *elastic potential energy.* Elastic potential energy can be stored in stretched springs as well as stretched rubber bands. Elastic potential energy can also be stored in compressed springs and bent vaulting poles. In fact, any object that can be forced into a shape that is different from its natural shape can store energy. If the object is able to return to its natural shape, it has elastic potential energy.

DISCOVER BY *Writing*

Imagine that you are the pole shown in the photographs below. Write a poem or a short story in your journal that describes how your energy changes and how you feel as it happens. ✎

Figure 5–3. A pole-vaulter stores energy in the pole as it bends. As the pole straightens, the potential energy is released and the vaulter is launched over the crossbar.

Figure 5–4. As the drop of water falls and as the pitcher moves to throw the ball, potential energy is released.

Another form of potential energy is *gravitational potential energy*. Gravitational potential energy is the energy an object has when it is in an elevated position. A baseball in the air and a drop of water suspended from a faucet both have gravitational potential energy. Anything that can fall or drop has gravitational potential energy.

▼ **ASK YOURSELF**

Describe two examples of potential energy, and explain how they are different.

Energy in Motion

Think back to the rubber band example. Remember that when the rubber band is stretched, it has potential energy. When you let go of the rubber band, it snaps back to its original shape and the potential energy changes to motion. What kind of energy is the motion? ①

The energy an object has due to its mass and its motion is called **kinetic energy.** Anything in motion has kinetic energy. Mathematically, kinetic energy is described as follows:

$$\text{kinetic energy} = \frac{\text{mass} \times \text{speed}^2}{2}$$

Note that this equation uses speed instead of velocity. The amount of kinetic energy in an object is the same no matter what direction it moves. Try the next activity to find out more about kinetic energy.

ACTIVITY

How can the kinetic energy of a rolling cart be measured?

Process Skills: Measuring, Inferring

Grouping: Groups of 3 or 4

Hints

Be sure the board is wide enough to allow several centimeters on each side of the cart. Calculations would be easier if the distance between the starting and finishing lines were in whole numbers of meters. The equations in step 8 can be used because the starting speed is 0 and the acceleration is uniform.

▶ **Application**

1. The kinetic energy at the bottom of the ramp is less than the gravitational potential energy at the top of the ramp for two reasons— some of the energy has been converted to heat, and some of the cart's kinetic energy is in turning wheels, as well as in its forward motion.

2. On the higher hill, you have more gravitational potential energy.

PERFORMANCE ASSESSMENT

Cooperative Learning

Ask each group of students to share a real-life example of kinetic energy. Each group should describe the forms of energy and the types of energy transformations demonstrated. Ask them also to describe their methods and tools of measurement. Evaluate their choice of energy and measurement tools as appropriate.

GUIDED PRACTICE

Have the students note that the speed component in the formula for kinetic energy is squared, that is, multiplied by itself. Have them practice using this formula with one digit numbers ($2^2 = 4$, $3^2 = 9$). Remind the students that to get the value for kinetic energy, the product of the mass of the object and its speed squared must be divided by 2.

INDEPENDENT PRACTICE

 Have the students provide written answers to the Section Review. In their journals, have the students draw and label a picture that shows different forms of energy.

EVALUATION

Have the students write examples of potential energy and kinetic energy on slips of paper. Collect the papers in a box or other container. Then have the students take turns randomly selecting a slip, reading the example aloud, and identifying the example as potential energy or kinetic energy. Encourage the students to justify their answers.

ACTIVITY

How can the kinetic energy of a rolling cart be measured?

MATERIALS

several books, board, masking tape, meter stick, balance, rolling cart, stopwatch

PROCEDURE

1. Make a ramp with the books and board as shown. Make sure the starting line is far enough from the top so the cart can be placed behind the line.

2. Make a table like the one shown.

TABLE 1: KINETIC ENERGY OF A CART					
Length of Ramp	Mass of Cart	Time of Trial			Average
		1	2	3	

3. Measure and record the distance (in meters) between the start and the finish lines. Also measure how much higher the middle of the cart is at the top of the ramp as compared with the cart's middle when it is at the bottom of the ramp.

4. Use the balance to find the mass of the cart in kg. Record the mass.

5. Set the cart so that it is behind the starting line and release it. Time how long it takes for the cart to reach the finish line.

6. Repeat step 5 two more times, and average the results.

7. You already know the kinetic energy of the cart before it begins to roll down the ramp. The cart is not moving, so its speed is zero. Therefore, its kinetic energy is zero. What is the kinetic energy of the cart at the bottom of the ramp? To determine kinetic energy at the bottom, you need to know the cart's mass *(m)*, which you have already measured, and its speed *(v)*.

8. You can find the cart's average speed by dividing the distance traveled down the ramp by the average time taken to travel that distance. You have measured both these quantities. But the cart is accelerating from no speed at the top of the ramp to a speed at the bottom of the ramp that is faster than average. Because it is accelerating smoothly, the speed at the bottom is just twice the average speed.

9. Now that you know the speed at the bottom of the ramp, you can calculate kinetic energy as $1/2\ mv^2$.

APPLICATION

1. Gravitational potential energy is calculated as *mgh*, where *m* is the cart's mass, *g* is the acceleration due to gravity (9.8 m/s/s), and *h* is the height of the ramp. How does the cart's gravitational potential energy at the top of the ramp compare with its kinetic energy at the bottom of the ramp?

2. Suppose you are out riding your bike and you come to the top of a small hill. You coast down and enjoy the ride. Later you come to the top of a large hill. You fly down. Why does your bike go faster on the large hill?

RETEACHING

Have the students create a two-column chart in their journals. Label one column "Potential" and the other "Kinetic." Ask the students to list examples of each type of energy. (Wherever possible, the examples should parallel each other. Lists should include examples, such as a car parked on a hill has potential energy; a car rolling down a hill has kinetic energy.)

EXTENSION

Have the students use library materials to read about Einstein's equation $E = mc^2$. They should find the meaning of each letter and the correct unit for measuring each quantity. Have the students determine the amount of energy that could be produced from a 2-kg mass if $c = 300\,000$ km/s. $(1.8 \times 10^{17}$ J$)$

CLOSURE

Cooperative Learning Have the students work in small groups to write a summary of the section. Then ask each group to read aloud their summary.

Based on what you learned in the activity, try to answer the following question: How would the crashes be different if one car going 2 km/h and another car going 1 km/h each hit a tree? ①

Because kinetic energy increases in proportion to the square of the speed, you know that the car going 2 km/h has more kinetic energy and will hit the tree with more force. How much more force will there be at impact? The kinetic energy of the car going 2 km/h is four times as great as the car going 1 km/h: $2^2 = 4$ and $1^2 = 1$. Any time the speed of a vehicle is doubled, the force of impact is four times greater. Now you know why car accidents at higher speeds are much worse than accidents at lower speeds. If everything else is the same, a crash at 60 km/h causes four times the damage that a crash at 30 km/h would cause.

Now you know that the faster an object travels, the more energy it will have as it hits something. Think about what this means if you want to stop *before* you run into something. The amount of force required to make you stop in a certain distance is equal to your kinetic energy divided by the distance. Using your common sense, you can see that the faster you are moving the greater the force necessary to stop you. It all has to do with energy.

Figure 5–5. The damage from this wreck is severe. When two cars hit each other head-on, the damage is the result of the kinetic energy of both vehicles.

① As the students continue to read, they will discover that the increase in damage is an exponential function of speed. In this case, there would be four times as much damage to the faster moving car.

REINFORCING THEMES
Systems and Structures

Using the charts made in the Reteaching activity above, point out the kinetic and potential aspects of each example of energy given. Lead students to an understanding of the relationship between kinetic and potential energy

ONGOING ASSESSMENT
▼ **ASK YOURSELF**

No, he is four times as hard to tackle.

SECTION 1 REVIEW AND APPLICATION

Reading Critically

1. Responses should include information about the fuel before and after takeoff, as well as information about the position of the shuttle itself.

2. $KE = \dfrac{mv^2}{2} = \dfrac{(50\ kg) \times (6\ m/s)^2}{2} = 900$ J

Thinking Critically

3. The students could compress the spring with the book, remove the book quickly, and allow the spring to jump off the desk. Potential energy was stored in the spring as it was compressed and released as kinetic energy when the book was removed.

4. If the line drive is three times faster, the ball has nine times as much kinetic energy.

▼ **ASK YOURSELF**

A football player runs twice as fast as an opponent. Is he twice as hard to stop?

SECTION 1 *REVIEW AND APPLICATION*

Reading Critically

1. Use the space shuttle as an example to show the difference between potential energy and kinetic energy.

2. A 50-kg gymnast swings off a bar at 6 m/s. What is the gymnast's kinetic energy as she leaves the bar?

Thinking Critically

3. A friend doesn't believe that energy can change forms. You have a spring and a book on your desk. How can you use these things to prove to your friend that energy can change forms?

4. Baseball players who play third base often field bunted balls with their bare throwing hands. A line drive, which travels about three times as fast as a bunt, might break the player's hand if the ball were caught barehanded. Explain why this is true.

FOCUS

In this section, the first and second laws of thermodynamics are discussed. Both are illustrated with common examples, and their applications to everyday situations are presented.

MOTIVATING ACTIVITY

 Have the students look at the word *thermodynamics* and write the definitions in their journals. Ask them to identify the two words that are combined to make this word (*thermo* and *dynamics*). Then ask them to list other familiar words that begin with the prefix *thermo.* (Answers will probably include thermometer and thermostat.) Ask what the word dynamic means to them. (Answers should include "active" or "changing.") Derive an operational definition of thermodynamics. (Change or action due to heat.)

PROCESS SKILLS
• Predicting • Applying

POSITIVE ATTITUDES
• Cooperativeness
• Creativity

TERMS
none

PRINT MEDIA
Motion by Philip Sauvain (see p. 67b)

ELECTRONIC MEDIA
Matter and Energy Coronet Film and Video (see p. 67c)

SCIENCE DISCOVERY
• Bowling ball pendulum
• Refrigerator
• Friction creates heat • Friction; rubbing hands together

BLACKLINE MASTERS
Laboratory Investigation 5.2
Study and Review Guide

Demonstration

Collect a variety of balls: golf, table tennis, basketball, tennis, superballs, and others if possible. Drop each ball from a known height, for example, 2 m. Measure the height of rebound. The height of rebound/ the height of drop = efficiency of ball. Ask the students which ball is most efficient. Have them list the balls in order of efficiency. Ask the students what happened to the remaining energy. (It became waste heat or was used to overcome friction or drag.)

SECTION 2

Energy Laws

Objectives

Illustrate *the first and second laws of thermodynamics by tracing energy flow through a series of events.*

Apply *the first and second laws of thermodynamics to explain energy exchange.*

Explain *how energy conversions result in energy being wasted as heat.*

It's a hot summer day, and you and your friends are outside playing basketball. The group decides to take a break. You run into your house to get something cold to drink. But before you reach for something to drink, you stand in front of the open refrigerator door and enjoy the cool air. If you stood there long enough with the door open, the entire room would become cooler, right? Wrong! You may not know it, but right before your eyes, the laws of energy are at work. Understanding these laws makes most technology possible—even refrigerators. Read on to find out more.

Energy Is Conserved

Figure 5–6. When these children jump on their pogo sticks, they are converting potential energy to kinetic energy and back again.

A pogo stick is basically a pole on which footrests are supported by a strong spring. When you jump on a pogo stick, the spring compresses and changes kinetic energy into potential energy. Then the spring begins to expand and converts potential energy back into kinetic energy. The amount of kinetic energy produced by the spring, plus a small amount of kinetic energy changed into heat energy, is equal to the amount of potential energy stored by the spring. This is an example of the law of conservation of energy.

The law of conservation of energy states that the total amount of energy in the universe always stays the same. Another way to say this is that energy cannot be created or destroyed. It can be changed from one form to another, but the total amount of energy never changes. The law of conservation of energy is also called the *first law of thermodynamics.*

Thermodynamics in Action   The operation of the pogo stick shows how elastic potential energy is converted into kinetic energy and vice versa. However, the pogo stick also illustrates how gravitational potential energy is converted into kinetic energy. As the spring converts its potential energy into kinetic energy, the person and the pogo stick are moved up and into the air. As the person rises, the kinetic energy is being converted back into potential energy. This time the potential energy is in the form of gravitational potential energy. When the person reaches maximum height, all of the kinetic energy has been converted into gravitational potential energy. As the pogo stick moves downward, this potential energy is converted back into kinetic energy.

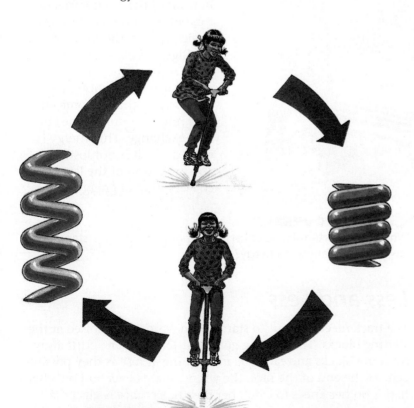

Figure 5–7. Energy is changed from one form to another in cycles.

 LASER DISC
29833–29964

Bowling ball pendulum

 **SCIENCE TECHNOLOGY SOCIETY**   Advertisements for cars try to convince consumers that cars are highly efficient machines. However, cars are highly inefficient. Less than 15 percent of the energy in the gasoline put into a car is used to propel the car down the roadway. More than twice that much energy—33 percent—leaves the exhaust system as waste heat. Another 33 percent of the energy is used to run the cooling system. The remaining energy is lost in friction and running the car's accessories.

● **Process Skills:**
Analyzing, Applying

Point out that a refrigerator needs "breathing room." Ask the students why it might not be a good idea to have a refrigerator next to a stove. (Answers might include that since a refrigerator's heat flows out the back, if the back has

already been heated by a stove, the refrigerator motor will have to work harder to get the heat to flow.)

LASER DISC

2341

Refrigerator

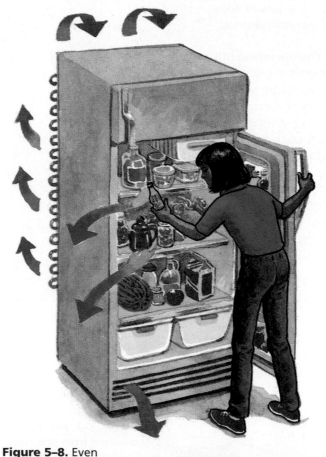

Figure 5–8. Even though you feel cool air when you open a refrigerator, the kitchen does not become cooler. The heat removed from inside the refrigerator is released into the kitchen by the refrigerator's working parts.

ONGOING ASSESSMENT

▼ **ASK YOURSELF**

Responses should include that without this understanding, the scientists and engineers might produce highly inefficient machines. With this understanding, they can provide for ways to channel off waste heat and sometimes even to use this heat productively.

Remember that the total amount of energy always stays the same, but some energy is given off as heat. In the pogo stick example, the person on the pogo stick gives off some heat, and the spring, as it is compressed and released, gives off some heat.

Cooler or Hotter? You are probably still wondering about the open refrigerator and why it can't cool the entire room. The refrigerator works by pumping heat from the inside to the outside of the refrigerator. When you open the door, cool air passes out of the refrigerator and is replaced by warm air from the room. Any heat that goes into the refrigerator is also pumped out of the refrigerator.

You do feel cooler standing in front of the open refrigerator, but the room will not get cooler. In fact, the room will get slightly hotter. This is because some of the refrigerator's parts, such as the motor and compressor, change some energy into heat during the energy exchange. This energy is wasted, but it is not lost. This is an application of the law of conservation of energy.

▼ **ASK YOURSELF**

Why is it important for scientists and engineers to understand the law of conservation of energy?

Less and Less

The track meet is about to start. Seven women are poised at the starting blocks. The starting gun fires! The women sprint away from the blocks and race 100 m, running as fast as they possibly can. At the end of the race, the winner bends over and puts her hands on her knees to catch her breath. Her face is glistening with sweat.

Figure 5–9. Runners use energy to race. This use of energy results in waste heat.

How is this track meet related to the conservation of energy? Any time energy changes from one form to another—such as when potential energy from food changes into motion—some of the energy changes into heat. The runners' bodies heat up, and they sweat to cool down. This is the basis of the *second law of thermodynamics*. Whenever there is an energy conversion, heat is produced that cannot be used to do useful work. The unuseable heat is called *waste heat*. Why do you think it is called "waste"? ①

Applied to machines, the second law of thermodynamics says that no machine is 100 percent efficient. The amount of work a machine does is always less than the amount of energy put into it. Some of the input energy always becomes waste heat, even if all friction is eliminated. Even the most carefully designed system cannot convert all input energy into work. What would happen if you could invent a machine that was 100 percent efficient? ②

You can observe waste heat when you use almost any machine. You can even observe it using a bicycle pump. See for yourself how this happens.

TEACHING STRATEGIES, continued

● **Process Skills:**
Comparing, Identifying

Ask the students to list situations or experiences, other than the bicycle pump, in which heat is generated and not used for useful work. (Possible answers include the heating of a saw blade while sawing and the heating of the casing of any power tool while it is operating.)

MEETING SPECIAL NEEDS

Second Language Support

To carry out the *Discover by Doing,* you might have limited-English-proficient students work with English-fluent students. When they have finished the activity, have them discuss their findings with the English-fluent students.

DISCOVER BY *Doing*

PORTFOLIO ASSESSMENT

After the students have completed the Discover by Doing, you might have them write an explanation of what happened. Students should keep their work in their science portfolios.

LASER DISC

20380–20587

Friction; rubbing hands together

ONGOING ASSESSMENT
▼ ASK YOURSELF

The students might respond that the blood carries waste heat to the capillaries beneath the skin at the surface of the body. Body heat is released through the skin into the environment.

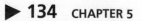
▶ **134** CHAPTER 5

Figure 5–10 Not all of this person's work is being used to pump up the bicycle tire. Some is converted into waste heat.

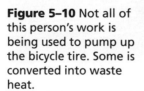

Figure 5–11. Bending a coathanger back and forth produces waste heat.

DISCOVER BY *Doing*

Use a bicycle pump to pump air into a very deflated bicycle tire. Before you start pumping, feel the bicycle tire and the outside of the pump. Make a note of the temperature of each one. After you finish pumping air into the tire, feel the tire and the pump. What happened? Explain why. ✎

You can also observe waste heat by bending wire from a clothes hanger back and forth. As you bend the wire, it becomes warm. In fact, if you bend it enough to break it, it becomes quite hot. The wire is warmed by waste heat. **CAUTION: If you try this, be careful with the broken ends of the hanger.**

▼ ASK YOURSELF
How do people give off waste heat?

GUIDED PRACTICE

Cooperative Learning Have the students work in pairs to solve this problem: Randy carries a bowling ball weighing 70 N up a flight of stairs 20 m high. How much potential energy has the bowling ball gained? (1400 J)

INDEPENDENT PRACTICE

Have the students provide written answers to the Section Review. In their journals, have them state the first and second laws of thermodynamics in their own words.

EVALUATION

Have the students describe and illustrate the forms of energy involved in skiing downhill. (Answers should reflect an understanding of the forms of energy and the laws of thermodynamics.)

Perpetual Motion

Many people throughout history have tried to make perpetual motion machines. A perpetual motion machine is one that makes at least as much energy as it uses. In other words, a perpetual motion machine, once started, would continue operating forever with no loss of motion, no waste heat, no friction, and no additional energy input. No one has ever succeeded in making one of these machines because no one has yet found a way to get around the second law of thermodynamics. Even complex natural systems—such as the solar system—that appear to run without energy input have been found to be running down very slowly.

Figure 5–12. For centuries, inventors have attempted to design perpetual motion machines. Here are two examples. One from ancient Greece and one from a modern museum exhibit.

▼ **ASK YOURSELF**

Why can't there be a perpetual motion machine?

ONGOING ASSESSMENT
▼ **ASK YOURSELF**

The students might reason that since every machine produces some waste heat, it has to be expected that additional energy is required to make up for the energy that goes into making waste heat.

SECTION 2 **135** ◀

RETEACHING

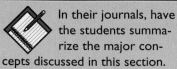

In their journals, have the students summarize the major concepts discussed in this section. (Any conversion of energy from one form to another results in some of the energy changing into heat. Some input energy becomes waste heat.)

EXTENSION

Have the students use reference books and other resources to find out how machines get rid of waste heat. (They should find that many machines have fans and openings to the outside to release the waste heat created when the machines operate.)

CLOSURE

Cooperative Learning Have the students work in small groups to write a summary of the section. Then ask each group to prepare questions that correspond to the summary so that anyone knowing the answers to the questions will understand the main concepts of Section 2.

① The desert rabbit converts its food into energy. The cactus gets its energy from the sun.

ONGOING ASSESSMENT
▼ ASK YOURSELF

The rabbit would likely die from starvation since it converts its food into energy.

SECTION 2 REVIEW AND APPLICATION

Reading Critically

1. At every bounce, the total of the gravitational, elastic potential, kinetic, and heat energy are always the same. The ball finally stops bouncing because the energy has changed to waste heat.

2. Examples include waste heat generated by machines, heat does not flow spontaneously from cold to hot objects, mixtures do not "un-mix" without help.

Thinking Critically

3. Thermodynamics takes into account the heat added to a system as work is done and the energy used to do the work. When analyzed carefully, you can determine mathematically that although energy is changed from one form to another, it is not lost.

4. One potential "energy web" for bicycling might be: light from the sun is converted by plants to chemical potential energy, which is then converted by the cyclist's muscles into mechanical energy, which is used to move the pedals of the bicycle. The total energy is conserved, but energy is given off at each change in the form of waste heat.

Figure 5–13. Animals convert the nutrients from plants or other foods into heat and motion.

Energy and Life

Even living systems obey the second law of thermodynamics. Plants, animals, and all other creatures require a source of energy, such as food. Animals get food from eating plants or eating other animals. Plants make their own food using sunlight, water, and carbon dioxide. Regardless of the type of living system and how it gets its energy, each produces heat and motion as a result of that energy.

Think back to the 100-m race at the track meet. The women get the energy to run the race by eating nutritious foods. When the race is over, they replace their energy by eating again.

Look at the picture of the desert rabbit. Where does it get its energy? Where does the cactus get its energy? ①

The second law of thermodynamics is fairly simple mathematically, but it is difficult to explain how it works—especially when explaining living systems. The explanation here uses only the simplest examples. Like all the other laws of nature, the second law of thermodynamics describes *how* things happen; it does not explain *why* things happen.

 ASK YOURSELF

What would happen to the rabbit if it couldn't find a supply of energy?

SECTION 2 *REVIEW AND APPLICATION*

Reading Critically
1. Relate the action of a bouncing ball to the laws of thermodynamics. Assuming it is not caught, what finally causes the ball to stop bouncing?
2. Give two examples that illustrate the second law of thermodynamics.

Thinking Critically
3. The term *thermodynamics* is formed from the Greek words *therme* meaning "heat" and *dynamis* meaning "power, strength, or movement." How does the meaning of *thermodynamics* apply to energy conservation? How does it apply to energy that is wasted?
4. Trace the forms of energy and their changes in a simple action, such as pedaling a bicycle from one place to another. Be sure to include the input energy. Show where energy is conserved, and describe how some of the energy changes to waste heat at each change in the form of energy.

Objectives
● **Construct** graphs to show the relationships between variables.
● **Interpret** data from graphs.

Discussion
Straight-line graphs are used in science to analyze, interpret, and draw conclusions from recorded data. Ask the students why an increase in one variable on the graph causes an increase in the other variable. (A straight-line graph shows that the two variables are directly proportional to each other.)

▶ **Application**

The higher the ramp, the farther the cart travels. Therefore, the line on the graph moves up and to the right. The students should extend the line of their graphs to predict how far the cart would roll if the ramp were 30 cm high.

✳ **Using What You Have Learned**

Some patterns are easier to understand when the data is shown on a line graph because you can see the relationships between mathematical quantities.

SKILL *Communicating Using a Line Graph*

▶ **MATERIALS**
● thin books (10 to 20) ● stiff cardboard, 15 cm × 40 cm ● ruler or meter stick
● cart ● tape ● graph paper

▼ **PROCEDURE**

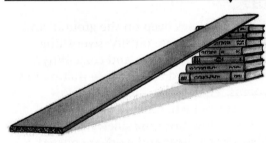

3. Mark points for the distances you measured. Then make a smooth line, crossing as many points as you can.

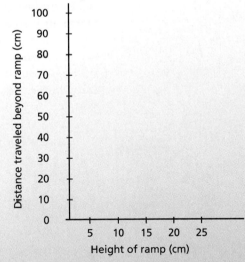

1. Set up a ramp as shown. You will do five trials to gather the data for your graph. Using different numbers of books, make the ramp a different height each time. Record the height of the ramp in your data table. Let the cart roll down the ramp, and then measure and record how far it travels beyond the bottom of the ramp.

2. Use the graph shown as a guide to make the graph of your data.

▶ **APPLICATION**
Look at your graph. What pattern do you observe? How does increasing the height of the ramp affect the distance the cart rolls? Use your graph to predict how far the cart might roll if you started it at a height of 30 cm.

✳ *Using What You Have Learned*
Are some patterns easier to understand when the data is shown on a line graph? Why?

★ **PERFORMANCE ASSESSMENT**

Cooperative Learning

Ask each group of students to develop another set of data that can be plotted on a line graph. Examples include the daily high and low temperatures over a week and the number of students in class each day over a similar period of time. You can check the students' work by examining their line graphs for accuracy.

PROCESS SKILLS
• Predicting • Measuring

POSITIVE ATTITUDES
• Curiosity • Skepticism

TERMS
temperature • kinetic energy
• conduction • conductors
• convection • radiation

PRINT MEDIA
Heat by Neil
Ardley (see p. 67b)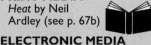

ELECTRONIC MEDIA
*Heat, Energy, and
Temperature*
Microcomputer
Software. Focus Media, Inc.
(see p. 67c)

SCIENCE DISCOVERY
• Expansion with
heat; balloon
• Conduction of
heat; hot frying pan
• Heat conductors

BLACKLINE MASTERS
Connecting Other Disciplines
Reading Skills
Extending Science Concepts
Thinking Critically
Study and Review Guide

① The sun's heat reaches the earth by radiation. Heat from the sun cannot be transferred by conduction or by convection since these two methods of heat transfer depend on matter through which the heat moves.

DISCOVER BY *Doing*

After the students have completed the *Discover by Doing,* you might have them write out their explanations and place them in their science portfolios.

FOCUS

This section provides information on the characteristics of heat as a form of energy, the difference between heat and temperature, the physical properties of heat including thermal expansion, and the ways in which heat is transferred.

MOTIVATING ACTIVITY

Cooperative Learning Have the students work in groups to devise ways to change mechanical energy into heat energy. Have the groups demonstrate their findings to the class. You might remind the students that heat energy also is produced in many chemical reactions.

SECTION 3

Heat

Objectives

Distinguish between heat and temperature.

Describe the properties of thermal expansion.

Infer how heat is transferred when given an example.

When you think of the sun, you probably think of its blazing heat. Considering that the sun is 150 million kilometers away, how does this heat reach the earth? ①

Heat Is Energy

It is a cold winter day. The snow lies deep on the ground, and school has been cancelled. Children are outside swooshing down hills on their sleds. Inside, it is warm and cozy. Why? It's warm because the house is heated by a furnace that pumps warm air into all the rooms. This heat is a form of energy.

Heat is used and is produced when many kinds of work are done. When you shovel the snow from the sidewalk, you give off heat. You probably know that heat and work are related, but what is the connection? Try the next activity, and see for yourself.

DISCOVER BY *Doing*

Fill a bowl with water, and allow the water to come to room temperature. Carefully measure and record the temperature of the water. Then, using an electric mixer set on high speed, beat the water for three minutes. Stop the mixer and recheck the water temperature. Has the temperature changed? Explain why. 🖋

Many scientists have had questions about the relationship between work and heat. In the late 1700s, an American scientist named Benjamin Thompson, better known as Count Rumford, noticed that the barrels of cannons became hot as the holes were drilled in them. At that time, people thought heat was a substance called *caloric* that flowed from object to object. However, Thompson noticed that in drilling a cannon, there was no source of heat. He concluded that the work done in making the holes somehow changed into heat. Therefore, there had to be a connection between heat and work.

Several years later, English physicist James P. Joule concluded that heat and work were different forms of the same

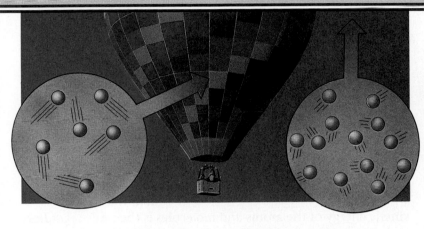

Figure 5–14. When a substance is heated, its molecules move faster and farther apart; it becomes less dense. A hot-air balloon rises because the hot air inside is less dense than the air outside.

thing. Because of his work, the unit of energy in SI is named after him.

Now you know that heat is a form of energy, but what causes it? To help you understand what causes heat, think about this. As an object becomes hotter, its atoms and molecules move faster. As an object becomes colder, its atoms and molecules move slower. So you can say that heat is due to the motion of molecules or atoms. Specifically, the **heat** of an object is the total kinetic energy of the random motion of its atoms and molecules.

Research about heat and work didn't stop with James Joule. Today, scientists are still trying to understand more about heat and how heat can be used to do work. One of these scientists is Annie Easley, who works for NASA. Easley's work has centered on identifying how energy changes from one form to another. She applies what she finds out to improve existing technology. Some of her discoveries have been used to improve the operation of electric-powered vehicles used by NASA.

 ASK YOURSELF

Why are joules used to measure both heat and work?

Temperature and Heat

If someone asked you the difference between temperature and heat, what would you say? Maybe the following statements will help you answer the question.

A large pot of boiling water has more heat than a small pot of boiling water, even though their temperatures are the same. A small pot of boiling water has the same temperature as a large pot of boiling water, even though the amount of water is different.

Benjamin Thompson (Count Rumford)

James Joule

Annie Easley

● **Process Skill:** *Inferring*

Ask the students which they think will melt more ice—a teacup filled with water at 100°C or a bathtub filled with water at the same temperature. Have them justify their answers in their journals. (The students should emphasize the quantities involved. Since both amounts are the same temperature, the bathtub water contains more particles at 100°C, so it has the ability to do more work.)

● **Process Skills:** *Comparing, Inferring*

Ask the students to explain how it is possible for one sample of water to have more heat than a second sample if the first sample has a lower temperature. (The larger the quantity of water involved, the more particles there are present. More particles mean more motion, even if the average motion of each particle is less.)

✧ Did You Know?

Another unit often used to measure heat is the calorie. A calorie is the amount of heat required to raise the temperature of 1 g of water 1°C. One calorie equals 4.186 J. Therefore, a calorie is a larger unit of heat than a joule. Food calories are really kilocalories or 1000 calories. They are abbreviated kc or C.

INTEGRATION– *Mathematics*

Reinforce the students' understanding of temperature and heat by reviewing the concepts of *average* and *total*. The students might relate to the idea of the total points scored on five tests to the average score of the five tests.

SCIENCE BACKGROUND

Count Rumford showed that mechanical energy could be converted into heat, but scientists did not have a theory of matter that could explain how this happened. In the early 1800s John Dalton proposed an atomic theory of matter that suggested that matter was made of moving particles. This particle theory became widely accepted by scientists because it could explain both chemical reactions and heat more easily than the fluid theories. The students will learn more about Dalton's work in Chapter 16.

Temperature = 100°C

Temperature = 100°C

Heat

Heat

Figure 5–15. Temperature is average kinetic energy, and heat is total kinetic energy.

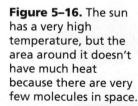

Figure 5–16. The sun has a very high temperature, but the area around it doesn't have much heat because there are very few molecules in space.

Still stumped? The difference between temperature and heat has to do with the movement of atoms and molecules—the kinetic energy—of substances. For example, when a pot of water is boiling, the average kinetic energy of the atoms and molecules is the same regardless of the size of the pot. However, a large pot of boiling water has many more water molecules than a small pot. Therefore, the large pot contains more heat energy. Likewise, you have to add more energy to the molecules and atoms in the large pot to give them the same average kinetic energy as the molecules and atoms in the small pot. That is why it takes a large pot of water longer than a small pot of water to begin boiling, even when the burner is set on high under both. There are more molecules to heat in the large pot.

How Hot Is It? Many people confuse the meanings of temperature and heat, but now you know they're not the same. Temperature is *not* a measure of how much heat a substance contains. **Temperature** is the measure of the average kinetic energy of the moving atoms and molecules of a substance. Compare this definition to the definition for heat—the *total* kinetic energy of the moving atoms and molecules of a substance. To put it more simply, think of temperature as measuring how fast the average atom or molecule in a pot vibrates. Heat measures both how fast an atom or molecule vibrates and also how many molecules and atoms are vibrating.

It's Not the Heat . . . When you think of heat and temperature as the total and the average kinetic energies of a substance, you can understand many seemingly strange things. For example, suppose you were in a spacecraft and you traveled in a path very close to the sun. You would certainly get very hot, right? Not necessarily! If you could shade your spacecraft from the light, you could actually freeze. This is true because, although the temperature near the sun is quite high, there is very little heat. The average kinetic energy of the atoms and molecules near the sun is very large; thus, the temperature is high. However, there are very few atoms and molecules in space. For this reason, the amount of heat near the sun is low.

ASK YOURSELF

How does the number of molecules in a substance affect temperature?

Thermal Expansion

Have you ever noticed that beverage manufacturers do not fill glass bottles to the very top? Why do you think this is? Here's a hint: As a substance becomes hotter, its atoms and molecules move faster and the substance expands.

Now that you've had a chance to think about the question, you have probably figured out the answer. If a bottle were filled to the top with a beverage and then warmed, the liquid would expand and shatter the glass.

You have just discovered one of the physical properties related to heat: thermal expansion. *Thermal expansion* is an increase in the size, or volume, of a substance due to an increase in the motion of its molecules and atoms.

When a substance is heated, the kinetic energy of the atoms increases. This increase in kinetic energy causes the atoms to move farther apart. The farther apart the atoms move, the greater the thermal expansion. Exactly how far apart the atoms move determines the thermal expansion of the particular substance. Thermal expansion is a characteristic property of substances.

Turn Up the Heat Understanding the thermal expansion of substances can be very useful in technology. For example, a thermostat that controls the temperature in a house relies on thermal expansion to turn the heat on and off automatically. The thermostat contains a metal coil made of two different metals fastened together. The metals expand at different rates. When the temperature increases, the outside metal expands more than the inside metal. This action causes the coil to curl up on itself and trip a switch, which turns on the heat.

Figure 5–17. As this metal strip inside the thermostat warms or cools, it changes shape. This change turns the heater circuit on and off.

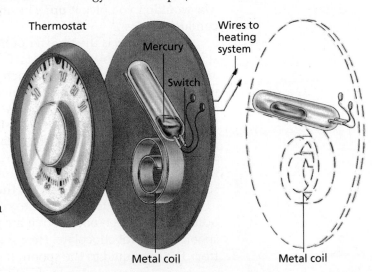

Thermostat

Mercury

Switch

Wires to heating system

Metal coil

Metal coil

● **Process Skill:** *Applying*

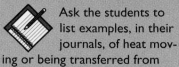

Ask the students to list examples, in their journals, of heat moving or being transferred from place to place. (Answers might include the heating of a room, water heating on a stove, touching a spoon that was in a cup of hot liquid, and the warmth of the air on a summer day.)

Figure 5–18. Expansion joints such as this one allow structures to expand and contract.

Bridging the Gap   Knowledge of thermal expansion is also helpful to engineers. When building a bridge, engineers know that the metal and concrete in the bridge are going to expand when the weather is warm and contract when the weather is cold. To prevent the bridge from buckling and collapsing, expansion joints are inserted at regular intervals to allow the bridge to expand and contract without damage. An expansion joint is shown here.

▼ **ASK YOURSELF**

Why is understanding the thermal expansion of substances useful in technology?

ONGOING ASSESSMENT
▼ **ASK YOURSELF**

Responses should include that without an understanding of the thermal expansion of substances, many useful technological advances would not have occurred, such as the development of jet planes and rockets.

MEETING SPECIAL NEEDS

Second Language Support

Encourage limited-English-proficient students to work with English-fluent students to prepare descriptions of what happens to the molecules during the conduction of heat.

 LASER DISC

30233

Conduction of heat; hot frying pan

2072

Heat conductors

Movement of Heat

You have just made a cup of hot chocolate. You stir it and leave the spoon in the cup. In a short while, the spoon gets warm. This happens because the hot chocolate transfers some of its heat to the spoon. This is just one example of how heat moves. Heat energy moves in three different ways: conduction, convection, and radiation.

Conduction   If you saw a quarter lying on a hot stove, would you use your hand to pick it up? Of course not! But *why* wouldn't you pick it up? The answer has to do with conduction.

Conduction is the transfer of heat energy from one substance to another by direct contact. Conduction occurs when atoms or molecules bump into each other. When the fast-moving molecules in a hot substance hit the slow-moving molecules in a colder substance, the slow-moving molecules speed up and the fast-moving molecules slow down. Note the heat transfer always goes from the hotter substance to the cooler substance. That's because heat energy is transferred from the faster molecules to the slower molecules. As more and more of the slow-moving molecules go faster, the substance heats up.

Because conduction is due to colliding molecules, it can occur only between objects that are touching. Think back to the spoon in the hot chocolate. Heat is transferred by conduction from the hot liquid to the spoon. If you were to hold the cup of

● **Process Skills:** *Inferring, Applying*

Ask the students to explain the transfer of heat in cooking an egg in a pan on a stove. (Answers should include observations such as: heat is transferred from the burning fuel or electric coil of the stove to the pan; heat moves through the pan and is transferred to the egg by conduction.)

● **Process Skills:** *Analyzing, Applying*

Call the students' attention to Figures 5–19 and 5–20. Have the students discuss how heat is conducted or not conducted in each example. Show the students a thermos. Discuss how coffee or soup is kept hot in a thermos. (The inside of a thermos is a vacuum bottle. The air is removed from between the two layers of glass.) Ask the students how this keeps the liquid hot. (The heat cannot move across the vacuum by conduction or convection.)

hot chocolate in your hands, your hands would warm up too. This is because heat is conducted from the hot chocolate to the cup, and then from the cup to your hands.

You can feel the heat from the hot chocolate because the cup allows the conduction of heat. Some materials and substances allow heat to travel more easily through them than others. Materials and substances that readily allow the transfer of heat are called **conductors**. Many metals, such as copper and aluminum, conduct heat well. If the hot chocolate were in a metal cup, the cup would probably be too hot to hold. To find out more about conduction, try the following activity.

DISCOVER BY Doing

CAUTION: Put on safety goggles, a laboratory apron, and a glove before trying this activity. Get a burner, a metal rod, a candle, and some matches. Light the candle and put four spots of wax, equally spaced, on the metal rod. Hold the rod in your gloved hand, and place the end of the rod into the flame. Record how long it takes for each spot of wax to melt. Why does the wax melt in this pattern? How does this example demonstrate conduction?

Some materials limit the amount of heat that passes through them. These materials and substances are called **insulators.** Air is an example of a very good insulator. If the hot chocolate were in a cup made of plastic foam, which has many small air pockets, the cup would transfer less heat and only be warm to the touch.

Look at the picture of the two students in the lab. Both are heating rods in a flame. The girl's rod is glass; the boy's is aluminum. The boy is wearing a glove. Why?

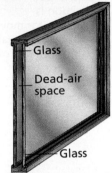

- Glass
- Dead-air space
- Glass

Figure 5–19. This house has special windows that reduce heat loss in the winter. The air space between the panes of glass acts as an insulator.

Demonstration

To demonstrate conduction of heat in liquids, support one end of a heatproof-glass cake pan with books and the other end with a hot plate. Fill the dish about two-thirds full of water. Sprinkle a few crystals of potassium permanganate on the water over the hot plate. Turn on the hot plate. Observe the currents in the liquid, both over the hot plate and from one side of the dish to the other.

Figure 5–20. Study this photograph closely. Are these students using proper safety precautions? Explain your answer.

DISCOVER BY *Doing*

PORTFOLIO ASSESSMENT

After the students have completed the *Discover by Doing*, have them record their observations and place them in their science portfolios.

① to keep from burning his hand

② Yes. They are wearing safety equipment and using caution.

GUIDED PRACTICE

Discuss with the students what would happen in the aquarium in Figure 5–21 if convection currents were not set up. (The water around the heater would get very hot; the water farthest away from the heater would take a long time to be heated by conduction.)

INDEPENDENT PRACTICE

In their journals, have the students compare the three methods by which heat moves from one place to another. Tell the students that in many homes heat is provided by large metal objects called radiators. Have them discuss whether or not the name *radiator* is an appropriate name for these objects. (The heat from the metal moves by convection and radiation, so *radiator* is an appropriate name.)

EVALUATION

Ask the students to explain why lakes rarely freeze solid. (The students' responses should be related to the fact that convection currents in the water maintain the water at temperatures above freezing.)

INTEGRATION–
Geology

Have the students use library materials and other resources to research plate tectonics. This information can extend the concepts of density, conduction, and convection.

MEETING
SPECIAL NEEDS

Gifted

Have the students use information learned about convection currents to write a news report about the movement of weather systems across the country. Encourage the students to refer to weather maps that appear in daily newspapers when gathering their information. Have them put their finished reports in their science portfolios.

MULTICULTURAL
CONNECTION

Amir Faghri at Wright State University in Dayton, Ohio, has discovered a way to heat the hands of workers who must work at below-freezing temperatures. He puts heat pipes in gloves that transfer the heat from the elbows to the colder fingertips.

Each heat pipe is a sealed container that holds a liquid. Heat from the elbow vaporizes the liquid. The heated liquid travels to the cooler end of the pipe where it condenses, leaving its heat to warm the fingers.

Figure 5–21. Convection is responsible for moving heat from the aquarium heater to the rest of the water.

Figure 5–22. The shimmer above this road is caused by convection currents. Winds and storms are caused by convection currents in the earth's atomosphere.

Convection   Look at the picture of the aquarium. It contains tropical fish that must be kept in water that stays at a constant temperature. To maintain this temperature, a small heater is installed in one corner. Because of convection, this small heater can keep the entire aquarium at a constant temperature. **Convection** is the transfer of heat in liquids and gases as groups of molecules move in currents.

The water surrounding the aquarium heater is heated by conduction. However, as the water gets hot, its molecules move faster and take up more space. This action causes the warm water around the heater to become less dense, because the molecules are farther apart. As a result, the water is pushed upward by the cooler, more dense water in the aquarium. Warmer water, like warmer air, rises. As the cooler water takes its place near the heater, it is also heated. Meanwhile, the warm water that was pushed away from the heater begins to lose kinetic energy to the surrounding cooler water. As this cycle continues, a *convection current* is set up due to the changing densities of the water.

Under certain conditions, you can see convection currents. During cold weather, for example, the air over a hot spot, such as a heated building, will rise and seem to shimmer. This shimmer is caused by the different ways light travels through hot air and cold air. Even during the summer, convection currents can be seen as a shimmer above hot roads. Sometimes this shimmer appears from a distance to be water on the road—a *mirage*.

RETEACHING

Have the students create a three-column chart. Label one column "Conduction," the next column "Convection," and the third column "Radiation." Ask the students to list examples of heat transfer in the proper column.

EXTENSION

Air is a good insulator. Discuss with the students why they think the "layering" of clothing is more effective in keeping them warm in cold weather. (The air trapped between layers of clothing provides additional insulation.)

CLOSURE

Cooperative Learning Have the students describe the concepts of heat transfer in their own words in cooperative groups. They should give at least one example of each kind of heat transfer that is not discussed in the section.

A type of convection current is also responsible for the weather. Air that is over warm surfaces near the ground is warmer and less dense than air above. The warm air then rises and the cold air sinks. This forces the bottom layer of the atmosphere, the one nearest the earth, to turn over. This overturning is a type of convection, which creates the weather systems of the world. In a similar way, convection currents also cause some ocean currents and the movement of water in ponds and lakes. This helps to mix the nutrients in the water and, therefore, support more life.

Radiation It is a warm summer day. You are outside, and the sun feels hot on your skin. How does the heat from the sun travel 150 million kilometers through space to Earth? It travels by radiation. **Radiation** is the transfer of energy by electromagnetic waves. The movement of heat by radiation does not require matter. Both conduction and convection depend on molecules to carry heat.

The word *radiation* can also be used to refer to the waves and particles given off by radioactive materials such as uranium and plutonium. Therefore, you must pay careful attention to how this word is used whenever you see it.

There are many examples of heat transfer by radiation. For example, if you warm your hands by a fire in a fireplace, the heat travels to your hands by radiation. For another example of radiation, consider again the cup of hot chocolate. If you place your hand about 1 cm to the side of the cup, you can still feel the warmth from the hot chocolate.

▼ ASK YOURSELF

When you walk barefoot on the street in the summer, the pavement can be so hot that it burns your feet. How is heat transferred from the sun to your feet in this example?

SECTION 3 REVIEW AND APPLICATION

Reading Critically
1. Describe how heat is transferred by convection.
2. Why is thermal expansion considered to be a characteristic property of substances?

Thinking Critically
3. Someone asks you how warm you are, and you tell them your temperature. Explain why just telling them your temperature does not give enough information to answer the question.
4. When entering a burning building, firefighters stay low to the floor. Explain why this is so.

ONGOING ASSESSMENT
▼ ASK YOURSELF

Radiant energy from the sun is transmitted through space by means of electromagnetic waves, also known as radiation. These waves originate from accelerated charged particles. This energy sent out from the sun can be absorbed by another body. The black asphalt is an excellent absorber of heat. It can also emit the heat by radiation to your feet. (Note: The students will learn more about electromagnetic waves in Chapter 11.)

SECTION 3 REVIEW AND APPLICATION

Reading Critically

1. As a liquid, such as water, is heated, its molecules move faster and take up more space. As the molecules move farther apart, the warmer liquid becomes less dense and rises. At this point, denser, cooler water moves in to take its place, causing circulation within the liquid.

2. Thermal expansion is a characteristic property because each substance has its own special, distinctive rate of expansion.

Thinking Critically

3. Unless you have a fever, your temperature is usually around 37°C. However, different parts of your body have different amounts of heat. Heat and temperature must be considered when answering the question.

4. The cooler air is near the floor. The smoke is hot so it rises.

Predicting Temperature Change

Process Skills: Predicting, Measuring

Grouping: Groups of 3 or 4

Objectives
- **Predict** the final temperature of a mixture
- **Observe** patterns of heat transfer
- **Relate** observed patterns of heat transfer to daily life.

Pre-Lab

Have the students break into lab groups and discuss a situation in which predicting the temperature of a liquid might be helpful. Have them record the criteria they would use for such a prediction.

▶ Analyses and Conclusions

1. Responses will depend upon the students' predictions. Differences should be explained based on a change of ideas about heat transfer.

2. The temperature was in between that of the hot and the cold water.

3. The temperatures changed in direct proportion to the amount of cold or hot water added.

4. The final temperature of the mixture is directly proportional to the temperatures of the water added together. By setting up a mathematical proportion, it is possible to predict the final temperature.

$$\frac{(m_1 \times t_1) + (m_2 \times t_2)}{m_3} = t_3$$

▶ Application

You would have to add 40 L of water at 60° C. At this temperature, you would not have to remove any cold water. If you used water at a lower temperature, you would have to increase the amount of water you added.

Post-Lab

Have the students review the criteria they listed before the investigation. Ask them to revise their criteria based on what they now know.

✳ Discover More

Responses will depend on the liquid used. Differences are the result of different specific heats.

⭐ PERFORMANCE ASSESSMENT

Cooperative Learning

Have the groups of students compare their results from the main part of the Investigation and their results from Discover More. Check their work by having the students verbally summarize their findings.

INVESTIGATION

Predicting Temperature Change

▶ MATERIALS
- plastic-foam cups (3) ● thermometer ● graduated cylinders, 50 mL (2)
- cold water ● hot water ● stirring rod

▼ PROCEDURE

1. Copy the table shown.

TABLE 1: TEMPERATURE CHANGES					
Cold Water		**Hot Water**		**Mixture**	
Amount	Temperature	Amount	Temperature	Amount Predicted	Temperature Actual
20 mL		20 mL			
20 mL		40 mL			
40 mL		20 mL			

2. Label the cups "A," "B," and "C."

3. Place 20 mL of cold water in cup A and 20 mL of hot water in cup B.

4. Measure and record the temperature of the water in each cup.

5. Predict what the temperature will be if you mix the water in cups A and B. Record your prediction.

6. Pour the water from each cup into cup C, and stir with the stirring rod.

7. Measure and record the temperature of the mixture.

8. Repeat steps 3–7, but change the amounts of water used to 20 mL of cold water and 40 mL of hot water.

9. Repeat steps 3–7, but change the amounts of water used to 40 mL of cold water and 20 mL of hot water.

▶ ANALYSES AND CONCLUSIONS

1. How did your predictions compare with the actual temperatures? Explain any differences.

2. How did the temperature of the mixture compare with the starting temperatures when equal volumes of hot and cold water were mixed?

3. How did the temperature of the mixture compare with the starting temperatures when unequal volumes of hot and cold water were mixed?

4. How can you predict the final temperature of a mixture of hot and cold water?

▶ APPLICATION

Suppose you want to warm up a child's wading pool, which holds 200 L of water. Now it has 120 L of water at 20° C. You want the pool to be 30° C. How much hot water, at what temperature, would you have to add? Would you have to take out some of the cold water before you added the hot water?

✳ Discover More

Repeat this investigation using a liquid other than water. Does the temperature of the liquid change in the same way as water did? Explain your results.

CHAPTER 5 HIGHLIGHTS

The Big Idea— Energy

Lead the students to an understanding that energy is the ability to change something or to do work. Also help the students to understand that in predictable ways, energy can be transferred from one form to another.

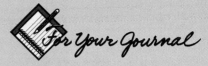

For Your Journal

Students' ideas should reflect an understanding of how energy is used and how it changes form. You might have the students share their ideas with the rest of the class.

CHAPTER 5 *H*IGHLIGHTS

The Big Idea

Energy is the ability to change something or to do work. Energy is involved anytime the position, appearance, or makeup of something changes. All changes involve energy.

Energy is found in many forms and can change from one form to another. Kinetic energy can change to potential energy, chemical energy to heat energy, electrical energy to light energy, and so on. Energy can also be stored and transferred. Energy is transferred when objects move.

Every time energy is transferred, some of the energy becomes waste heat. Heat energy can be used to do work. Heat can be transferred by conduction, convection, or radiation.

For Your Journal

Look back at the ideas you wrote in your journal at the beginning of the chapter. How have your ideas changed? Revise your journal entry to show what you have learned. Be sure to include information on how energy is used and how it changes form.

CONNECTING IDEAS

1. potential
2. convection
3. radiation
4. heat
5. temperature

Connecting Ideas

This concept map shows how the big ideas of this chapter are related. Copy the map into your journal, and fill in the blanks to complete it.

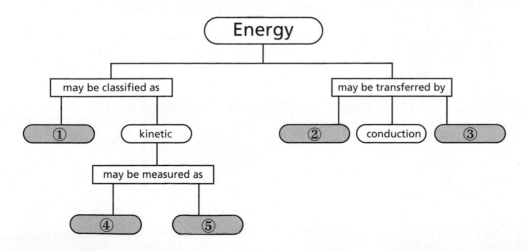

Understanding Vocabulary

1. (a) Both are types of energy. Potential energy is stored energy; kinetic energy is the energy of motion. When potential energy is released, kinetic energy is the result.

(b) Both are measures of kinetic energy. Heat is the total kinetic energy of the atoms in an object; temperature is the average kinetic energy of the atoms in an object.

(c) Both affect heat transfer. A conductor allows heat to travel from one place to another. An insulator prevents the transfer of heat.

(d) All three are methods of heat transfer. Conduction occurs when two solid objects touch each other. Convection occurs as heat is moved in a current through a liquid or a gas. Radiation is the transfer of energy by electromagnetic waves.

Understanding Concepts

Multiple Choice

2. c

3. d (The entire washer will expand and become larger since metal expands in all directions. The hole in the center of the washer will become smaller if it is small compared to the diameter of the washer. If it is a large hole, it will get bigger.)

4. c

5. c

6. b

Short Answer

7. To cool the inside of a refrigerator, the motor and coils must pump heat from the inside of the refrigerator out the back. This heat goes into the room. The amount of heat pumped out of the inside of the refrigerator is the same as the amount of heat pumped out the back. In addition, the motor is not 100 percent efficient, so some of the energy that makes the motor run is also converted into heat. A running refrigerator always heats the room it is in, even if the door is open.

8. A child on a swing is basically a pendulum in which gravitational potential energy is converted to kinetic energy and vice versa. The total amount of energy remains constant, but small amounts of energy are converted to waste heat at each swing. In order to keep swinging, some source of motion energy must be put into the system to replace the motion energy converted to waste heat.

Interpreting Graphics

9. b, because this is the highest hill, therefore gravitational potential energy is greatest

10. a, because it is higher on the hill

CHAPTER 5
REVIEW

▶ Understanding Vocabulary

1. For each set of terms, explain the similarities and differences in their meanings.
- **a)** potential energy (125), kinetic energy (127)
- **b)** heat (139), temperature (140)
- **c)** conductor (143), insulator (143)
- **d)** conduction (142), convection (144), radiation (145)

▶ Understanding Concepts

MULTIPLE CHOICE

2. Which of the following actions does *not* describe potential energy being changed into kinetic energy?
- **a)** releasing a stretched rubber band
- **b)** a child sliding down a slide
- **c)** a spring being compressed
- **d)** an apple falling from a tree

3. A washer is a round, flat piece of metal with a round hole in it. Suppose you heat a washer. The hole in the washer will
- **a)** get larger.
- **b)** get smaller.
- **c)** stay the same size.
- **d)** There is no way to know.

4. Starting at rest, a skateboarder rolls downhill. At the bottom, the speed is 10 m/s. If the rider's mass is 30 kg, what is the kinetic energy of the rider?
- **a)** 300 J
- **b)** 1000 J
- **c)** 1500 J
- **d)** 3000 J

5. Every time energy changes from one form to another, some of the energy always changes into
- **a)** kinetic energy.
- **b)** potential energy.
- **c)** heat energy.
- **d)** mechanical energy.

6. Joules could be used to measure
- **a)** the mass of a car.
- **b)** the energy produced by a car engine.
- **c)** the power produced by a car engine.
- **d)** the speed of a car.

SHORT ANSWER

7. Use the two laws of thermodynamics to briefly explain why it is impossible to cool a room by leaving a refrigerator door open.

8. Explain the energy conversions that apply to a child on a swing. Why does the child need energy input from time to time?

Interpreting Graphics

9. Look at the figure below. Which car has the greatest potential energy? Explain why.

10. Look at the figure below. Which car has the greatest potential energy? Explain why.

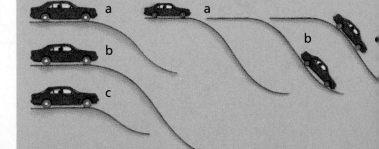

Reviewing Themes

11. *Energy*
During the transfer of energy from one form to another, or from one place to another, waste heat is given off. Explain this occurrence using the laws of thermodynamics.

12. *Systems and Structures*
Using an example other than a pogo stick, explain how objects and people interact with energy to form a repeating cycle.

Thinking Critically

13. The food web is a key idea in biology. Producers (plants) in the food web are eaten by herbivores, which are in turn eaten by carnivores. All three—plants, herbivores, and carnivores—are recycled by decomposers into nutrients for plants. Briefly explain the food web idea in terms of energy conversion. Do not forget to include an energy source(s) for the plants.

14. A greenhouse is a structure made mostly of glass, which allows light in but limits the flow of air in and out. Why does the air in a greenhouse get hot during the daytime?

15. The picture below shows several ways in which heat can be transferred. Identify each transfer, and explain how it occurs.

16. A trick many campers use for rapidly baking potatoes in a bed of coals involves sticking a large nail through the potato, wrapping the potato in foil, and making sure it is surrounded by hot coals. Explain why this method would cook a potato more rapidly than just wrapping it in foil and putting it in an oven at the same temperature as the coals.

17. The popping of corn is a very interesting phenomenon. Popcorn kernels "pop" when they are heated because each kernel contains a small amount of water within its hard shell. Explain in scientific terms why popcorn pops.

Discovery Through Reading

"Can a machine run forever?" *Current Science* 76(August 1991):10. This article tells about a modern attempt at making a perpetual motion machine.

CHAPTER 6

ENERGY SOURCES

PLANNING THE CHAPTER

B = Basic   A = Average   H = Honors
The coding Basic, Average, and Honors indicates subsections, features, and resources that might be appropriate for different levels of learners. For additional suggestions regarding choice of topic and depth of coverage, see the Pacing Chart on pages T28–T31.

*Frame numbers at point of use
(TR) Teaching Resources, Unit 2
(IT) Instructional Transparencies
(LI) Laboratory Investigations
(SD) *Science Discovery* Videodisc Correlations and Barcodes
(SRG) Study and Review Guide

CHAPTER RESOURCES

Title	Page	Materials
Discover by Researching	154	(per individual) journal
Discover by Researching	154	(per individual) journal
Discover by Researching	157	(per individual) journal
Investigation: Determining Heat Produced by Fossil Fuels	158	(per group of 3 or 4) small can, thermometer, ring stand, clamp, charcoal briquette, wire screen, tripod, lighter
Teacher Demonstration	159	cardboard box, thermometer, tape, light bulb
Activity: How can you estimate the sun's power?	162	(per group of 3 or 4) piece of thin metal, 2 cm x 6 cm; graphite suspended in alcohol; meter stick; thermometer in 1-hole stopper; jar and lid with hole; 100 watt light lamp
Discover by Doing	164	(per group of 3 or 4) electric fans (2), ammeter
Skill: Communicating Using a Diagram	165	(per individual) paper and pencil
Discover by Researching	172	(per individual) journal

ADVANCE PREPARATION

For the *Investigation* on page 158, you may want to ask the school cafeteria to save the appropriate sized cans for you ahead of time, or send a note home with the students alerting parents to begin saving the cans.

TEACHING SUGGESTIONS

Field Trip
Plan a field trip to a utility company where students can observe the production of energy. If that is not feasible, plan a field trip to an industry that uses photovoltaic systems to provide electricity.

Outside Speaker
Your local utility company will most likely be able to provide a speaker to discuss the aspects and various methods of production of electricity in your area. The speaker may also be able to discuss the integration of technology into their list of energy producing options. Such technology might include the testing or development of small-scale and large-scale solar power applications.

CHAPTER 6
ENERGY SOURCES

CHAPTER THEME—*ENERGY*

This chapter identifies the different types of fossil fuels and describes how they are used in the production of usable energy. The methods of using these fuels and other energy sources—including hydroelectric and nuclear energy to produce electric energy—are discussed in detail. The potential for the uses of solar energy, geothermal energy, wind energy, ocean tide energy, and ocean thermal energy is also examined. A supporting theme in this chapter is **Technology**.

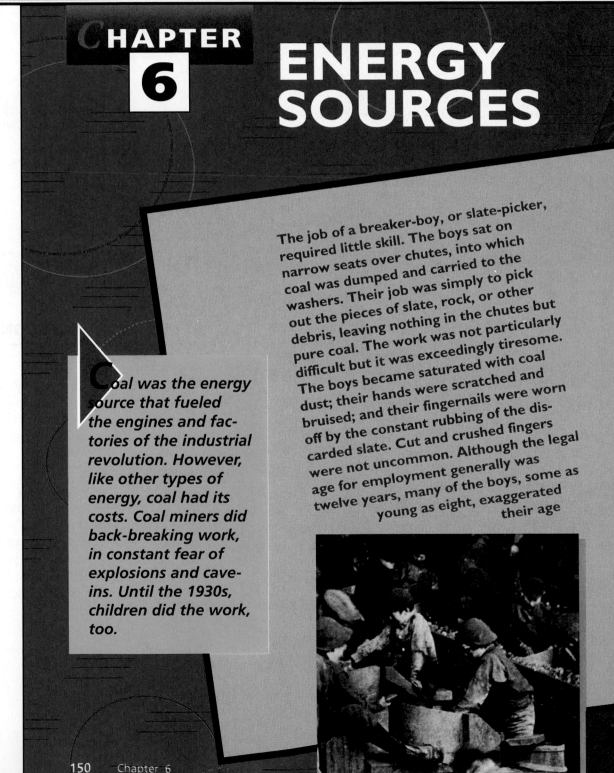

CHAPTER 6
ENERGY SOURCES

Coal was the energy source that fueled the engines and factories of the industrial revolution. However, like other types of energy, coal had its costs. Coal miners did back-breaking work, in constant fear of explosions and cave-ins. Until the 1930s, children did the work, too.

The job of a breaker-boy, or slate-picker, required little skill. The boys sat on narrow seats over chutes, into which coal was dumped and carried to the washers. Their job was simply to pick out the pieces of slate, rock, or other debris, leaving nothing in the chutes but pure coal. The work was not particularly difficult but it was exceedingly tiresome. The boys became saturated with coal dust; their hands were scratched and bruised; and their fingernails were worn off by the constant rubbing of the discarded slate. Cut and crushed fingers were not uncommon. Although the legal age for employment generally was twelve years, many of the boys, some as young as eight, exaggerated their age

Perform the following demonstration: Light an alcohol burner, a kerosene lamp, a Bunsen burner, and a charcoal briquette. Turn on a light bulb and solar calculator. Have the students observe each example and then write the source of fuel for each of them. Extend this demonstration with a discussion of fossil fuels and other types of energy sources. Point out to the students that the light bulb can be traced back to several different sources of energy, depending on the geographic location of your class.

For Your Journal

The journal questions will help the students begin to identify energy resources of their geographic area and distinguish among the sources that are used in their homes. They also will begin to think of the tradeoffs involved with each energy choice they make. (The students will have a variety of responses to the journal questions. Accept all answers, and tell the students that at the end of the chapter they will have an opportunity to refer back to these questions.)

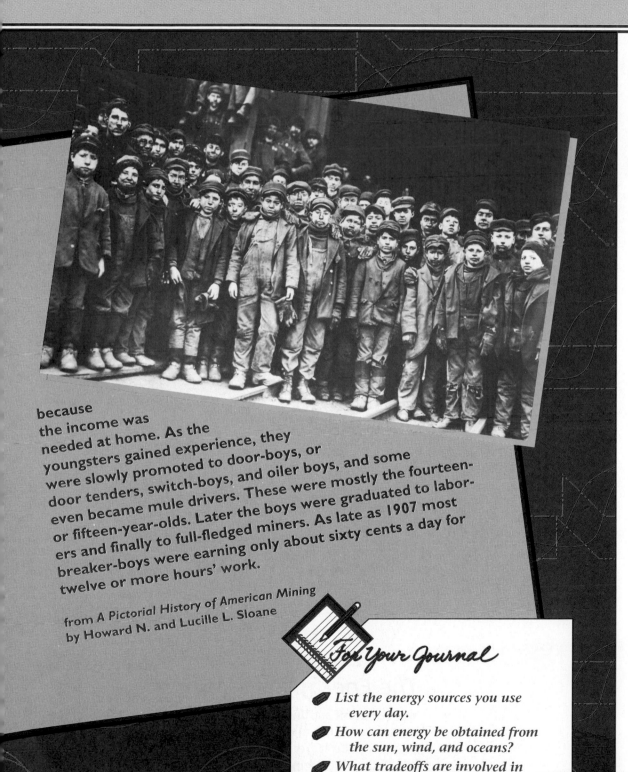

because the income was needed at home. As the youngsters gained experience, they were slowly promoted to door-boys, or door tenders, switch-boys, and oiler boys, and some even became mule drivers. These were mostly the fourteen- or fifteen-year-olds. Later the boys were graduated to laborers and finally to full-fledged miners. As late as 1907 most breaker-boys were earning only about sixty cents a day for twelve or more hours' work.

from A Pictorial History of American Mining by Howard N. and Lucille L. Sloane

ABOUT THE LITERATURE

Health risks to coal miners are numerous and spring largely from air contaminants. The poisonous gases sometimes present are hydrogen sulfide, methane, carbon monoxide, carbon dioxide, and sulfur dioxide. All of these are also odorless, except the two containing sulfur. Early miners carried birds, such as canaries which are sensitive to air contaminants, into the mine with them. If the birds became ill or died, it was an indication that poisonous gas was present. Miners also carried flame safety lamps, which change flame color in the presence of poisonous gases.

A second major hazard is coal dust, which has been found to cause a lung disease known as black lung disease. Finally, some mine dust has explosive properties. Much work has been done by the mining industry to reduce mine hazards.

For Your Journal

- List the energy sources you use every day.
- How can energy be obtained from the sun, wind, and oceans?
- What tradeoffs are involved in using different sources of energy?

PROCESS SKILLS
• Experimenting • Measuring

POSITIVE ATTITUDES
• Caring for the environment
• Curiosity

TERMS
• fossil fuels • nonrenewable resources • refinery

PRINT MEDIA
Fossil Fuels by Clint Twist (see p. 67b)

ELECTRONIC MEDIA
The Energy Connection Barr Films (see p. 67c)

SCIENCE DISCOVERY
• Map of world resources • Oil mining; off shore • Oil rig • Pipeline, oil

BLACKLINE MASTERS
Connecting Other Disciplines
Reading Skills
Study and Review Guide

SCIENCE TECHNOLOGY SOCIETY When the supertanker *Exxon Valdez* spilled more than 41 million liters of crude oil into Alaska's Prince William Sound, the cleanup effort joined biologists, naturalists, chemists, and engineers. High-pressure hot-water jets, rakes, shovels, and paper towels were used to clean contaminated areas. Chemists designed floating barriers made of polypropylene to absorb oil. Nitrogen-phosphorus fertilizer was sprayed on oil-laden shores to stimulate the growth of oil-eating bacteria, which greatly accelerated the cleanup.

FOCUS

This section identifies and discusses three major types of fossil fuels. The students will be asked to compare and contrast the physical characteristics, uses, and locations of coal, petroleum, and natural gas. In addition, the students will learn to identify and define nonrenewable resources.

MOTIVATING ACTIVITY

Cooperative Learning Have the students work in groups and list as many petroleum products they can think of and the different ways these products are used in a home. Then combine each group's list on the board. Point out that the heavy dependence on these products for gasoline, asphalt, and plastics, makes their eventual replacement important since petroleum is nonrenewable.

SECTION 1

Energy from Fossil Fuels

Objectives

Identify *the three major types of fossil fuels.*

Compare *and* ***contrast*** *the physical characteristics, uses, and locations of coal, petroleum, and natural gas.*

Define *and* ***identify*** *nonrenewable resources.*

There you are, standing outside the public library, ready to begin your research paper on energy use. You push on the doors, but they are locked. The library not open on a Saturday? Through the crack in the doors, you smell a musty, closed-up odor. How long has the library been closed? You turn your head and notice the rusted cars that appear anchored to their parking spots. There are no traffic sounds, no blaring radios, and no human voices.

The footsteps of an approaching stranger startle you. "Conduct your research wisely," he cautions. "The energy choices and decisions you make now may affect future generations. Without energy, the future will be as bleak and silent as this landscape." The stranger shuffles away. Suddenly, you throw off the covers and sit up in bed. What a nightmare!

The Energy Research Begins

You've put off starting your research paper for as long as possible. However, the nightmare has given you the motivation to begin work on it. Your topic is to identify current energy sources and to describe them. You are also supposed to make recommendations about energy choices for the future.

TEACHING STRATEGIES

● **Process Skill:** *Classifying/Ordering*

Guide the students in making an information organizer for their journals. Have the students list the fossil fuels from this section in their journals, with three headings for each fossil fuel: Location, Advantages/Disadvan-tages, and Processing. As you discuss each fossil fuel, have the students list information under these headings. Tell the students to leave ample space under each category so they can add information as they learn more in this chapter.

● **Process Skill:** *Comparing*

PORTFOLIO ASSESSMENT

Cooperative Learning Have the students work in groups to research the areas of the United States where the three fossil fuels are found. Provide the students with an outline map of the United States, and have them mark the locations of major fossil fuel deposits. Then have them develop a legend for their maps, explaining the symbols they have used for each fuel. You may wish to have students put their completed maps in their science portfolios.

A Fossil Is a Fuel? You discover that the most commonly used energy source today is supplied by burning three major fossil fuels—coal, petroleum, and natural gas. They are called **fossil fuels** because they are formed from plant and animal material that was buried in sediment millions of years ago. As time passed, bacterial action, heat, and pressure converted these fossils into hydrocarbons—molecules consisting of carbon and hydrogen. When hydrocarbons are burned, they release heat, water, carbon dioxide, and some pollutants.

Not Everything Old Becomes New Again

Fossil fuels take millions of years to form. Because they take so long to form, fossil fuels are considered to be nonrenewable resources. **Nonrenewable resources** are those that cannot be replaced after they are used. You go to the library, check out several books about fossil fuels, and head for home.

ASK YOURSELF

Why are fossil fuels considered nonrenewable resources?

Coal

The first fossil fuel you read about is coal. Coal is a dark brown or black rock that can be burned to release energy. It is used most commonly as a fuel for electric-energy plants and for industrial processes, such as iron smelting, that require high temperatures. Coal is found in layers of sedimentary rock thought to be the sites of old peat bogs that were covered with earth and compressed for millions of years. When the layers in coal are split, fossils of plants and animals are often found.

Coal is found in many areas of the United States in layered deposits called *seams*. Coal is removed by two principal methods, depending on the location of the seam in the earth. The first method, strip mining, is used if the seam is near the surface. *Strip mining* involves stripping away the top layer of soil and then removing the layers of coal underneath. This leaves huge scars on the earth's surface and should be—but most often isn't—followed by reclamation. *Reclamation* is the process of restoring the land to its original fertile condition.

Figure 6–1. This piece of coal contains the fossil of a fern that lived millions of years ago.

Figure 6–2. Peat that has been cut and stacked. Peat can be burned for fuel.

✧ *Did You Know?*

Figures at the beginning of the 1990s indicate that the current world reserves of fossil fuels at present consumption rates are as follows: oil: 50–90 years, coal: 260 years, natural gas: 500+ years, depending on amount of future discoveries.

ONGOING ASSESSMENT
▼ ASK YOURSELF

Since it takes millions of years for fossil fuels to form, they are considered nonrenewable.

SCIENCE BACKGROUND

The primary pollutants in fossil fuels are nitrogen oxides and sulfur oxides. Molecules of these substances are emitted during the burning of the fuel and slowly diffuse into the atmosphere. They settle back on Earth in the form of acid rain, resulting in deforestation and fish kills. These pollutants also cause smog, resulting in respiratory problems. There is also increasing evidence supporting the greenhouse effect, which is caused by the production of excess carbon dioxide. These issues are discussed in Chapter 7.

● **Process Skill:** *Comparing*

Point out that dredging—like strip mining—Is a form of surface mining. Dredging is used where mineral-bearing sand and gravel layers are exceptionally thick. You might encourage interested students to use reference resources to find out more about dredging.

● **Process Skills:** *Analyzing, Comparing*

Discuss different types of pollution caused by the burning of fossil fuels. Have the students list the names of specific pollutants they have heard of. Differentiate between the sulfate and nitrate pollutants in the air, which cause acid rain, and the emission of excess CO_2. Conclude by leading the students to an understanding that this excess is thought to be a major cause of the greenhouse effect, which they will learn more about in Chapter 7.

THE NATURE OF SCIENCE

In 1596 the personal physician of Duke Frederick of Württemberg mentioned that mineral oil distilled from oil shale was good for healing. In 1694 patent number 330 was granted to three English subjects who had found a means of extracting "pitch, tarr, and oyle out of a sort of stone." In 1838 a commercial oil industry began in France, producing oil for lamps. During that century, as the whaling industry began to lose the ability to meet the demand for oil, other means of producing oil became a concern.

Finally, the first commercial oil well in the United States was drilled in 1859 by Edwin L. Drake near Titusville, Pennsylvania. On August 27, 1859, the well began producing about 35 liters of oil per day.

DISCOVER BY *Researching*

Dangers associated with deep mining include cave-ins, gas explosions, and long-term disabilities from exposure to coal dust. Deep mining is used when a mineral deposit lies deep beneath the earth's surface. However, deep mining is more dangerous and more costly than surface mining. Have the students put their research in their science portfolios.

ONGOING ASSESSMENT
▼ ASK YOURSELF

The two processes used to mine coal are strip mining and deep mining.

▶ **154** CHAPTER 6

DISCOVER BY *Researching*

Find information on a strip mine that has been reclaimed. How was the land restored to its original condition? What can the land now be used for? How much did the reclamation cost? You may wish to prepare a poster that shows the results of your research. ✏

Coal located deeper in the earth is removed by *deep mining*, the second method of removing coal. In this process, deep shafts are dug into the earth to reach the coal. The specialized machinery shown in this picture is used to dig tunnels directly into the seam. As the coal is dug away, it is taken to the surface of the mine.

Figure 6–3. In strip mining (left), the top layer of soil is removed to expose the coal. The coal is dug and then carried away in large trucks. In deep mining (right), specialized equipment is used to remove seams of coal from deep beneath the earth's surface.

DISCOVER BY *Researching*

Find information on deep mining. What are the dangers involved in this type of mining? What are the advantages and disadvantages of deep mining? Share your findings with your classmates. ✏

▼ ASK YOURSELF
What are the different methods used to obtain coal from the earth?

Petroleum

The second fossil fuel you research is petroleum. Black gold is one name for petroleum. It is also called crude oil, or simply oil. Petroleum is a dark brown, thick liquid found in underground pools. These pools are located within layers of porous sedimentary rock, such as sandstone. Oil is collected by drilling wells down to the deposits and pumping the oil to the surface. Some oil wells are drilled as deep as seven kilometers into the earth.

Unlike coal, oil cannot be easily used in the same form in which it is taken from the earth. It must first be taken to a **refinery**—a large industrial plant that separates crude oil into products such as gasoline, diesel fuel, heating fuel, petroleum jelly, and asphalt. Oil is separated into various products in a *fractionation tower,* as shown in this drawing.

Figure 6–4. Refineries (left) consist of many fractionation towers. The towers distill crude oil into many useful products, as shown on the right.

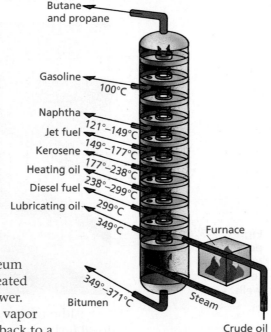

Butane and propane

Gasoline — 100°C

Naphtha —
Jet fuel — 121°–149°C
Kerosene — 149°–177°C
Heating oil — 177°–238°C
Diesel fuel — 238°–299°C
Lubricating oil — 299°C
349°C

Furnace

349°–371°C
Bitumen

Steam

Crude oil

Within a fractionation tower, the petroleum undergoes fractional distillation. The oil is heated into a gas, which rises in the fractionation tower. Since the products contained in the crude oil vapor have different boiling points, they condense back to a liquid state at different temperatures and levels in the tower. Each product is then pumped from its condensation level in the tower to a separate holding tank.

During the distillation process, many petroleum products undergo still another process called *cracking*. In this process, heat and chemicals are used to break apart the most dense molecules. This process is used primarily to produce more gasoline than is possible through fractional distillation alone.

▼ **ASK YOURSELF**

What are the two processes used in refining crude oil?

GUIDED PRACTICE

 Have the students draw an outline of a house in their journals. Then have them label the different places in the house where a fossil fuel or a product of a fossil fuel might be used. (Accept all reasonable answers the students can justify.)

INDEPENDENT PRACTICE

Have the students provide written answers to the Section Review. Then have them reexamine their list of advantages and disadvantages of each fuel in their journal and propose a usage plan for the United States. Ask the students to place their usage plans in their science portfolios.

EVALUATION

Have the students summarize the concepts in this section and choose the fossil fuel they feel is least environmentally damaging. (The students will likely choose natural gas since it is easy to collect and produces little pollution. Accept any choices the students can justify.)

 **SCIENCE TECHNOLOGY SOCIETY**

In 1967 the U.S. government and private industry cooperated in Project Gasbuggy, the first use of a nuclear explosion to produce natural gas. In New Mexico's San Juan Basin, a hydrogen device equal to 26 300 metric tons of TNT was detonated 1292 meters underground. The explosion freed huge gas deposits trapped in rock formations that were too hard for normal drilling to release. The blast created an underground chamber, which filled with more than 2.8 million cubic meters of gas from surrounding rock. Drillers then easily reached the chamber. Two other experimental nuclear explosions—in 1969 and 1973—released trapped natural gas in Colorado's Piceance Creek Basin.

MULTICULTURAL CONNECTION

Natural gas from wells was used in China by 900 B.C. About 1000 A.D., the Chinese first burned coal for fuel.

ONGOING ASSESSMENT
ASK YOURSELF

Natural gas took millions of years to form, as did the other fossil fuels. However, there is a theory that natural gas might be one of the building blocks of Earth, as is oxygen.

▶ **156** CHAPTER 6

Natural Gas

The third fossil fuel is natural gas. The flame on a gas stove or Bunsen burner is created by burning natural gas. Natural gas is a colorless gas consisting mostly of methane (CH_4). Methane is commonly found with petroleum deposits, but it is also found by itself. Like deposits of petroleum, natural gas deposits are reached by drilling wells. Natural gas is an important fuel because it is relatively easy to collect from the earth, and, compared with some other fossil fuels, it produces little pollution when burned. It is collected and transported by pipelines to processing plants and then distributed to consumers. Natural gas is most often used for heating and generating electricity.

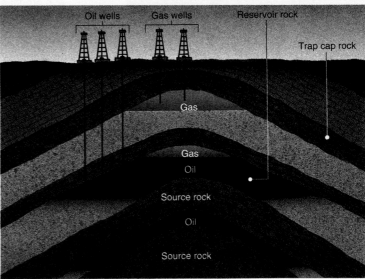

Figure 6–5. Both petroleum and natural gas deposits are reached by drilling wells.

Although natural gas has traditionally been considered a fossil fuel, some scientists have recently challenged this classification. These scientists now hypothesize that natural gas comes from the rocks and dust from which the earth was *originally* made. The facts that some meteorites contain hydrocarbons and that simple hydrocarbons like methane have been discovered in interstellar space are used as evidence to support this hypothesis. One prediction from the hypothesis is that there are huge quantities of natural gas deep in the earth, much deeper than levels where fossils are found.

▼ ASK YOURSELF
Why do some scientists consider natural gas a fossil fuel, while others do not?

RETEACHING

Have the students work independently or in pairs to prepare a poster to display information about one of the fossil fuels. For example, a poster for coal might include deep mining, stripping the earth, or producing electric- ity. One for petroleum might include pictures of products derived from oil. You might display the posters in the classroom and use them as a basis for future discussions on energy sources.

EXTENSION

Encourage interested students to read about the trans-Alaska pipeline. The students might draw a simple map of Alaska, mark the route of the pipeline, and indicate the site of the *Exxon Valdez* oil spill. Have the students place their maps in their science portfolios.

CLOSURE

Cooperative Learning Have the students work in small groups to write a summary of this section. Encourage each group to share its summary with other groups.

No Easy Answers

Whew! Doing research is tiring work. You consult your outline and begin to add more information to it. Now that you've finished reading about the fossil fuels used today, would you recommend that fossil fuels be used as the primary source for the earth's future energy needs? Why or why not? What are your other choices? Complete the activity to get more information.

 **DISCOVER BY Researching**

Plant and animal materials that were recently living are referred to as *biomass*. For example, firewood is a form of biomass. Research what other types of biomass people have used to supply their energy needs. What are the pros and cons of using forms of biomass? Record your research in your journal. Take turns reading your research findings aloud.

After you look over your research notes, you conclude the bottom line is that the supply of fossil fuels is nonrenewable. In addition, the burning of fossil fuels releases pollutants. Future energy choices must include clean, renewable resources. You continue your research—perhaps solar energy and wind energy would be a good place to start.

▼ **ASK YOURSELF**
What are the drawbacks involved in using fossil fuels?

Figure 6–6. The burning firewood is an example of biomass.

SECTION 1 REVIEW AND APPLICATION

Reading Critically
1. What are the similarities and differences among the three fossil fuels?
2. Identify the nonrenewable resources discussed in this section.

Thinking Critically
3. Some people think the United States should use its own fossil fuels rather than purchase such fuels from other countries. Others think we should continue to purchase fossil fuels from other countries. Which do you favor, and why?
4. Many decisions related to energy production and consumption determine how tax money is spent. Briefly discuss both the pros and the cons of the following energy-related tax issue: Interstate highways were strongly subsidized with taxes and railroads were not.

ONGOING ASSESSMENT
▼ **ASK YOURSELF**

Although fossil fuels are relatively easy to collect from the earth, they are nonrenewable, and prices will rise as scarcity increases. Pollution from the burning of the fuel causes health and environmental problems.

SECTION 1 REVIEW AND APPLICATION

Reading Critically

1. Similarities: Fossil fuels were all formed from geologic heat and pressure on organic matter, they are nonrenewable, they are pollutants when used as fuel, and they can be used to produce electricity. Differences: Coal is found and mined only in specific geologic areas. Oil and natural gas can both be drilled for but are not always found together. Natural gas is burned directly to produce heat. Crude oil requires refining before it becomes a fuel.

2. The nonrenewable resources discussed are coal, petroleum, and natural gas.

Thinking Critically

3. Students' answers should demonstrate an understanding of the concepts presented in the section. Accept all reasonable positions the students can justify.

4. Pro: Travel is more flexible with autos and trucks than with trains for both private commerce and government purposes. Con: Train travel is more efficient, especially considering maintenance costs of highways.

Determining Heat Produced by Fossil Fuels

Process Skills:
Experimenting, Measuring

Grouping: Groups of 3 or 4

Objective

● **Determine** the heat per unit mass produced by burning a parafin candle, a fossil-fuel derivative.

Pre-Lab

Have the students list and discuss some ways used to indicate that heat is produced. Examples might include using a thermometer and a change in the state of matter.

Hints

Have the students bring in soup cans from home. The cans should be clean and have the labels removed.

Procedure

6. The increase in water temperature will depend upon the intensity of the heat produced by the burning candle.

7. The amount of wax burned can be found by subtracting the final mass reading from the original.

8. The students should calculate the number of joules of heat energy as follows: 100 g × temperature rise × 4.2 J/g · °C. For example, if the temperature rose 20°, multiply 100 g × 20°C × 4.2 J/g · °C. (8400 J)

9. Joules per gram is found by dividing total joules by number of grams burned.

Analyses and Conclusions

Much of the heat escaped into the atmosphere. Some heat was transferred to the air via conduction and was radiated away as infrared radiation.

▶ Application

90 km/h × 1 L/10.5 km = 8.57 L/h
8.57 L × 800 g/L = 6856 g
6856 g × value from step 7 (J/g) = number of joules produced (J)

✳ Discover More

Natural gas, alcohol, a peanut, or a potato chip will provide a different heat per unit mass than the candle.

Post-Lab

Ask the students to revise their lists to incorporate what they learned about how heat production is indicated.

 PORTFOLIO ASSESSMENT Have the students place their data tables, observations, and answers from the Investigation in their science portfolios.

INVESTIGATION

Determining Heat Produced by Fossil Fuels

▶ MATERIALS

● small can
● thermometer ● ring stand ● clamp ● candle
● balance ● calculator ● lighter
CAUTION: Be careful when you light the candle.

▼ PROCEDURE

1. Make a data table.
2. Set up the equipment as shown. If your candle is not already stuck to the can lid, light the candle and attach it to the lid with a few drops of melted wax.
3. Measure 100 mL (100 g) of water into the can. Put the thermometer in the water, and record the water's temperature.
4. Determine the mass of the candle and can lid.
5. Place the candle under the can of water so that the top of the wick is about 1 cm from the bottom of the can, as shown in the illustration.
6. Light the candle. After 15 minutes, record the water temperature and blow out the candle. How much did the temperature rise?

7. Again determine the mass of the candle. How much wax was burned?
8. Calculate the total number of joules of heat energy

produced (100 g × temperature rise (°C) × 4.2 J/g · °C).
9. Calculate the number of joules produced by 1 g of wax.

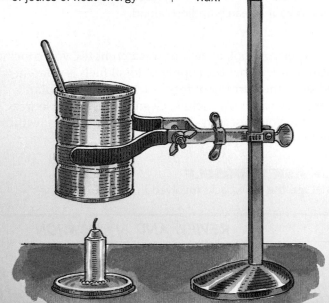

▶ ANALYSES AND CONCLUSIONS

If all the heat produced by burning the candle was not used to heat the water, where did some of the heat go? Explain your answer.

▶ APPLICATION

A small car gets about 10.5 kilometers per liter of fuel when traveling at 90 km/h on a straight, level highway. If each liter of fuel

is 800 g and has the same heat of combustion as candle wax, how many joules of energy are expended each hour while traveling at that speed?

✳ Discover More

Try the experiment again using different fuels, such as a peanut or potato chip. Be sure to use the proper safety precautions.

Energy from the Sun and the Wind

SECTION 2

The media specialist hands you a package of information you requested. A photograph at the beginning of an article shows a sleek racing car. You wanted an article about solar energy, not automobiles! Shrugging your shoulders, you begin to read:

Spectators watched as the car, making only a high-pitched hum, glided down the road. Long and slender, the car looked like a cannelloni on wheels! Obviously, this was no ordinary car. It was the Sunvox I solar-powered race car—winner of the cross-country portion of the Second Annual Sun Day Challenge. The car's driver, Tamara Polewick, a 19-year-old engineering student from Dartmouth College, may have a "sunny" future ahead of her—either as a race car designer or driver.

Objectives

Describe *how solar energy can be used to heat a home.*

Explain *how energy from the sun can be used to produce electricity.*

Compare *and* **contrast** *solar energy and wind energy.*

Figure 6–7. A car of the future

Solar Energy

Putting aside the article, you look at the other materials and continue your research on solar energy. Each day, you learn, the sun bathes the earth with as much energy as would be contained in a trainload of coal more than 2 million kilometers long. Energy from the sun is called **solar energy.** Solar energy is a renewable resource. It travels through space as sunlight and provides the earth with both light and heat. Solar energy drives the winds. Solar energy is responsible for the water cycle (the evaporation and precipitation of water that results in rain and snow) that eventually drives hydroelectric generators.

● **Process Skill:** *Analyzing*

Help the students comprehend different ways in which heat can be moved. Examples include the importance of fans and how a radiator moves heat with water.

● **Process Skills:**
Communicating, Generating Ideas

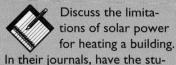

Discuss the limitations of solar power for heating a building. In their journals, have the students speculate how geography plays a role. Then provide information on the average number of sunny days in different parts of the country. Discuss with the students where solar heating might be most beneficial.

Many of the nation's largest utility companies are participating in a project to develop large-scale solar power applications. In addition, manufacturers are trying to develop less expensive, more efficient solar cells. These steps would help make solar panels practical and competitive with other electric-generating sources.

At least one power company has started to use solar panels to operate irrigation pumps and to meet other small energy needs for customers. As the use of solar panels becomes more widespread and costs fall, solar power will become more common.

ONGOING ASSESSMENT

▼ **ASK YOURSELF**

Solar energy is light energy from the sun.

⟡ **Did You Know?**
Seventy-two percent of solar collectors sold are used to heat swimming pools.

MEETING SPECIAL NEEDS

Gifted

Cooperative Learning

Have students work in small groups to research and prepare drawings of home designs that accommodate passive and active solar heating systems.

Figure 6–8. The sun is the earth's ultimate source of energy.

Figure 6–9. In passive solar heating, the heat moves naturally through the home. In the diagram shown here, excess heat from the sun is stored in the masonry wall.

The sun provides the energy necessary for plant growth. Without plants, which support the food chain, there would be no life on Earth and fossil fuels would never have formed. Even coal, petroleum, and natural gas can be thought of as forms of concentrated, stored solar energy.

Life on Earth is dependent on solar energy for its very existence. More sunlight strikes Earth than is needed or is used. This excess is an important energy source and is currently used to meet some energy needs.

 ASK YOURSELF

What is solar energy?

Solar Heating

Humans can harness solar energy for many uses. One major use of solar energy is for heating homes and buildings. Many different systems for converting sunlight to heat are now in operation. Most of the systems use either passive solar heating or active solar heating. These two systems differ in the ways that they collect and transfer heat.

Soaking Up Some Rays Homes that are heated as a result of the simple absorption of solar energy use **passive solar heating.** In this method of heating, no mechanical devices are used to transfer heat from one place to another. The drawing shows an example of a home with a passive solar heating system. In this system, large glass panels trap heat in the home. Natural convection currents carry the heat throughout the rooms. Excess heat is absorbed by a thick concrete floor and wall and is released into the home at night or during cloudy periods.

Concrete wall

Collecting the Rays

More complex solar-heating systems use **active solar heating.** This method of heating uses mechanical devices such as fans to move heat from one place to another within a building. Most active solar-heating systems include *solar collectors,* which are devices that gather sunlight and convert it into heat. The photograph shows a home that has many solar collectors installed on its roof.

Most active solar-heating systems use either air or water as the working fluid to transfer heat. In most cases, pumps or fans move the working fluid through the solar collectors. If the working fluid is air, it can be routed directly from the collectors into the home, where it heats the rooms. When water is used, a *heat exchanger* is necessary to transfer the heat from the water to the air in the rooms. A radiator is an example of a heat exchanger. When hot water flows through the radiator, the fins on the radiator become hot and heat the surrounding air.

In your notebook, you jot down your ideas about solar energy. Perhaps you should recommend that solar energy be the primary energy resource for the future. After all, the sun is expected to shine for billions of years, so it will certainly provide enough nonpolluting energy. But how can you estimate the amount of power the sun produces? You decide to conduct an experiment.

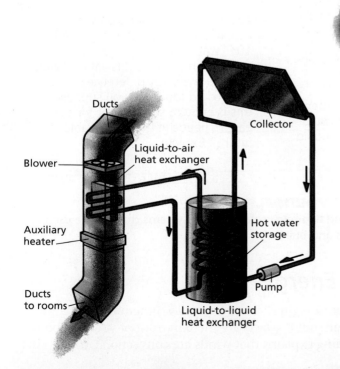

Ducts
Collector
Liquid-to-air heat exchanger
Blower
Auxiliary heater
Hot water storage
Ducts to rooms
Pump
Liquid-to-liquid heat exchanger

Figure 6–10. The solar collectors on this roof collect energy from the sun.

Figure 6–11. This diagram shows the operation of one type of active solar-heating system. This particular system uses water as its working fluid.

ACTIVITY

How can you estimate the sun's power?

Process Skills:
Experimenting, Interpreting Data

Grouping: Groups of 3 or 4

Hints

Have the students read their thermometers approximately every 30 seconds until the temperature has not changed for several minutes. This will be the maximum temperature.

▶ **Application**

1. The students' calculations should come reasonably close to the calculated value for the sun's power (3.40 × 10²⁵ watts). Accept all reasonable explanations for differences that the students can justify.

2. Many factors could cause a difference between the students' results and the accepted value. Among these are cloud cover, latitude of location compared to the latitude of maximum solar power, the angle at which the student held the collector, and lack of efficiency of some collectors.

3. The students will likely feel that there would be ample power for the earth's needs in the foreseeable future. Accept all responses that the students can justify.

 Have the students place their observations, calculations, and answers from this Activity in their science portfolios.

PORTFOLIO ASSESSMENT

TEACHING STRATEGIES, continued

● **Process Skill:** *Analyzing*

Review with the students how a turbine operates. Discuss the change of solar energy to electricity, which does not involve mechanics but rather a photo voltaic cell. Compare that conversion of energy to wind, which uses a turbine, with fossil fuels, which also use a turbine to produce electricity. Discuss with the students the difference in the way wind and fossil fuels operate a turbine.

ACTIVITY

How can you estimate the sun's power?

MATERIALS
piece of thin metal, 2 × 6 cm; graphite suspended in alcohol; meter stick; thermometer in 1-hole stopper; jar and lid with hole; 100-watt lamp

PROCEDURE

1. Carefully prepare your collector as shown in the diagram.

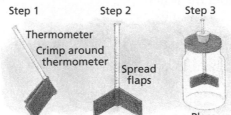

Step 1 — Thermometer / Crimp around thermometer / Metal strip

Step 2 — Spread flaps / Paint these sides with graphite

Step 3 — Place thermometer in jar

2. Place the jar collector outside in the sun. The graphited surfaces of the metal should face the sun. Tilt the jar until the shadow cast by the metal flaps is as large as possible.

3. As soon as it stops rising, record the collector's maximum temperature and return to the classroom.

4. Allow the collector to cool, and then place it 25 cm from a 100-watt bulb. The graphite side of the metal should be facing the bulb.

5. Allow the collector to reach the same maximum temperature it reached while in the sun. If necessary, move the collector closer to or farther from the bulb to achieve the same temperature. If you moved the collector, measure the distance from the bulb to the jar.

6. Using the formula below, calculate the sun's power versus the power of a 100-watt light bulb. The distance from the sun is 1.50 × 10¹¹ m. If you had to move the jar in step 5, replace 0.25 m with your measurement.

$$\frac{\text{power of the sun}}{(1.50 \times 10^{11}\text{ m})^2} = \frac{100\text{ watts}}{(0.25\text{ m})^2}$$

APPLICATION

1. Based on your calculations, what is the power of the sun?

2. The value for the sun's power, based on careful measurements, is 3.40 × 10²⁵ watts. How close did your calculation come to this value? What could explain any difference?

3. If we could tap it, do you suppose this is enough solar power for the earth's needs in the foreseeable future? Explain.

ONGOING ASSESSMENT
▼ **ASK YOURSELF**

Passive solar heating uses no mechanical devices to transfer heat from one place to another. Active solar heating uses mechanical devices, such as fans, to move heat from one location to another within a building.

 ASK YOURSELF

Compare the two basic solar-heating systems that can be used to heat homes and buildings.

Wind Energy

You'd never thought of wind energy as something that comes from the sun, but the facts are there in front of you. The book you're reading explains that winds are convection currents that

GUIDED PRACTICE

Have the students prepare a concept map in their journals, placing solar energy (the sun) at the center. Encourage the students to brainstorm all of the effects solar energy has on the earth. After the students have completed their concept maps, ask them to place their work in their science portfolios. (The students might think of such effects as wind energy; plant growth changing to biomass energy; the creation of fossil fuels; and the creation of precipitation, generating water flow.)

INDEPENDENT PRACTICE

Have the students provide written answers to the Section Review. In their journals, have them list both renewable and nonrenewable energy sources discussed thus far.

EVALUATION

Choose two or more geographic locations that represent climate extremes, such as San Diego, California, and Concord, New Hampshire. Have the students work individually or in pairs and plan the type of renewable energy resources that could be used to provide heat or electricity for a home in each location.

are formed when the sun heats some parts of the atmosphere more than others. Wind can transfer some of its kinetic energy to other objects. Sailboats, for example, use kinetic energy from the wind to propel them forward.

Figure 6–12. This windmill in the Netherlands uses the energy of the wind to pump sea water away from the land.

Wind Turbines The energy of wind can be converted into electricity with **wind turbines.** A typical wind turbine consists of a fanlike turbine attached to an electric generator. When wind strikes the turbine, the wind makes the turbine spin. This spins the generator to produce electricity. In some regions, many wind turbines are grouped together in installations called *wind farms.* A single wind farm may consist of more than 30 wind turbines. Wind farms in the United States are located in California, Vermont, Oregon, Montana, and Hawaii.

Energy for Sale Wind turbines can also be used to generate electricity for individual homes. Many homeowners have installed wind turbines to produce their own electric energy. They can use the electricity from their wind turbines directly, or they can sell it to the electric company serving the area. When energy is sold to the electrical company, it simply flows into the company's electric grid system. Try the next activity to find out more about wind turbines.

Figure 6–13. Many wind turbines grouped together are called wind farms. This wind farm is in California.

MULTICULTURAL CONNECTION

The earliest known windmills date from about 600 A.D., and were built in Persia (Iran). Constructed using a verticle shaft and horizontal sails, which were partially protected from the wind, they were used to grind grain.

THE NATURE OF SCIENCE

The original Middle Eastern windmill design had horizontal blades, but the Europeans changed it and used a side mill. The hollow-post mill was invented in the fifteenth century and was used to drain sea water from low-lying land. English engineer John Smeaton was the first to scientifically investigate the design of the windmill in 1759. He proposed five sails, or blades, instead of four.

The windmill remained one of the main sources of power throughout Europe until the invention of the turbine engine. After World War I, the turbine engine became the main method of power generation. The windmill is still in use today, mainly in peasant economies; more than 1000 were still in use in Portugal in 1965. Although conservationists are promoting a comeback of the windmill, they have not yet met with large-scale commercial success.

 LASER DISC

1779

Wind farm

1786

Windmill, prairie

Cooperative Learning Have the students work in small groups to prepare questions on this section. Then have the groups exchange questions, and after about 10 minutes, have each group read aloud its questions and the answers that have been developed.

PORTFOLIO ASSESSMENT Have interested students work in pairs and invent a design for either a passive or an active solar energy system for a home. The students will need to do research to get ideas. Remind the students to consider factors such as the direction the collectors should face, the location of nearby trees or other structural elements (such as chimneys or roof peaks casting shadows), the amount of tilt of the collectors (dependent on latitude), the reflection from nearby buildings, and the placement of storage systems for heat (rock bins or water tanks). Have the students place their research and designs in their science portfolios.

Cooperative Learning Have the students work in pairs or small groups to design a chart or other graphic that lists the sources, advantages, and disadvantages of each energy source discussed in this section. Encourage the groups to share their efforts with classmates.

✷ DISCOVER BY *Doing*

The students will obtain the most accurate responses if they use a sensitive ammeter.

ONGOING ASSESSMENT
▼ ASK YOURSELF

A wind turbine uses the kinetic energy of the wind to create electricity.

SECTION 2 REVIEW AND APPLICATION

Reading Critically

1. active solar heating system

2. a wind turbine

Thinking Critically

3. Passive solar heat systems do not rely on mechanical systems to transfer heat. Active solar heat systems use mechanical systems, such as pumps and blowers, to transfer heat. The success of a passive solar system depends on the location and design of the home. This type of system can be practical in new homes where precise location can be controlled. Active systems are more practical for existing structures that may not have the best location.

4. Solar energy would be most useful in areas that have a high percentage of sunny days and areas with warm mean temperatures. It would be least useful in cloudier areas of the country.

✷ DISCOVER BY *Doing*

Obtain two electric fans and an ammeter. Place the fans on a table so they face each other and are only a few centimeters apart. Plug in one of the fans, and allow it to blow on the other fan. The wind that the first fan creates should cause the other fan to start spinning, even though it is not plugged in. Connect the ammeter across the prongs of the disconnected plug to measure the current the "wind turbine" produces. ✐

Like all other sources of energy, wind turbines have advantages and disadvantages. Wind energy, like solar energy, is renewable and nonpolluting. But wind turbines work only when the wind is blowing and can be noisy during operation.

Tough Questions, Tough Decisions Energy supply presents challenging questions. In a rough draft of your paper, you've already decided people should not depend completely on fossil fuels. Should you recommend solar energy or wind energy as the primary energy supply?

The advantages and disadvantages of each are many. Weather and climate systems must be very carefully considered. If there are long cloudy periods, solar methods might not capture enough energy to support society's needs. Perhaps the solution is to recommend the use of more than one type of energy source. After all, the healthiest ecosystems are those with the largest number of different species in them. Might that principle apply to human energy systems as well?

▼ ASK YOURSELF
How can wind be converted into electricity?

SECTION 2 *REVIEW AND APPLICATION*

Reading Critically
1. Which type of solar-heating system uses solar collectors?
2. What device converts the kinetic energy of the wind into electricity?

Thinking Critically
3. Compare passive and active solar heating. Describe a situation in which each type of heating system would be more practical.
4. Where do you think solar energy would be most useful? least useful?

Objective
● **Draw** and **label** a diagram as a method of communicating an idea.

▶ **Application**
Students may respond that they could diagram a plan to build a school float, to decorate a room, or to design the parts of a tree or flower.

✳ **Using What You Have Learned**
Students' tips might include drawing the diagram to scale, labeling all important parts, and using arrows to indicate movement. Encourage the students to share their lists with classmates.

SKILL *Communicating Using a Diagram*

▶ **MATERIALS**
● paper ● pencil

Observe the diagram. It shows how a solar hot-water system works. Notice that all the important parts of the diagram are labeled. Arrows are used to show how the water and heat travel through the system. The diagram lets you see at a glance how the system works.

▼ **PROCEDURE**

1. Make a diagram to explain one of these processes: how your home is heated, how a windmill is used to generate electricity, or how a water wheel can be used to generate electricity.
2. To make a good diagram, you need to understand what you want to show. You may have to do some research before you begin. If you do, make some sketches as you read about your topic. Try out different ways of making your diagram.
3. When you finish your diagram, show it to another person. Ask him or her to explain what the diagram shows.

▶ **APPLICATION**
In what other parts of your life might diagramming skills be useful?

✳ **Using What You Have Learned**
Diagrams are important for sharing ideas. Get together with a group, and make a list of tips for making good diagrams. Share your list of tips with other groups.

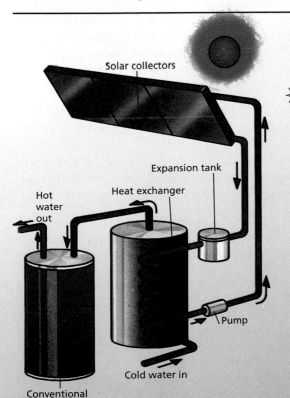

Solar collectors

Expansion tank

Heat exchanger

Hot water out

Pump

Cold water in

Conventional hot water tank

⭐ **PERFORMANCE ASSESSMENT**

Cooperative Learning

Ask each group of students to think of another set of data or idea that could be presented as a diagram. Evaluate each group's choice and plan for appropriateness.

PORTFOLIO ASSESSMENT
Ask the students to place their diagrams, observations, and answers from the Skill in their science portfolios.

FOCUS

This section discusses the use of several energy sources other than fossil fuels, wind energy, and solar energy. Among the sources considered are geothermal energy, hydroelectric energy, ocean tide energy, and nuclear energy.

MOTIVATING ACTIVITY

Display pictures of the sun, petroleum products, and firewood. Challenge the students to determine how the items are related. (Each can be used as a source of energy.) Then display a picture of the ocean and a dam, and repeat the procedure. (Each item is related to water.) Conclude by explaining to the students that they will learn how the ocean and a dam can provide sources of energy.

POSITIVE ATTITUDES
• Caring for the environment
• Enthusiasm for science

TERMS
• geothermal energy
• nuclear reactors

PRINT MEDIA
Nuclear Energy: Troubled Past, Uncertain Future by Laurence Pringle (see p. 67b)

ELECTRONIC MEDIA
Nuclear Energy: The Question Before Us National Geographic (see p. 67c)

SCIENCE DISCOVERY
• Cooling tower
• Geothermal energy

BLACKLINE MASTERS
Laboratory Investigation 6.2
Thinking Critically
Study and Review Guide

MEETING SPECIAL NEEDS

Mainstreamed

Use as many illustrations as possible in this section for mainstreamed students. Pictures of natural geysers, large dams, and nuclear power plants, along with diagrams will aid the students' comprehension. Be sure to point them out and use them to prompt discussion. If a mainstreamed student has a tutor accompanying him or her, make the tutor aware of the key illustrations in this chapter.

SECTION 3

Objectives

Tell how geothermal energy is converted into usable energy.

Compare and **contrast** methods of extracting energy from water.

Explain how energy is converted from one form to another in alternative methods of energy production.

Figure 6–14. Geysers are produced by geothermal energy.

Figure 6–15. Turbine generators (left) are used to produce electricity. The turbines (right) convert steam into rotary motion to turn the generator.

Other Energy Resources

At last, you believe, you are on the right track. In your research paper, you intend to suggest that energy choices for the future should include a combination of fossil fuels, solar energy, wind energy, and other energy resources. What other energy resources are there? You check out a video from the library that may have some useful information.

Geothermal Energy

The first image you see on the video is a geyser. The narrator explains that deep in the earth's core, radioactive nuclei produce heat as they decay. Because the crust prevents most of the heat from escaping, it builds up. This heat is a source of energy, called **geothermal energy.** In some places, geological forces push large masses of melted rock, or *magma,* to within four to six kilometers of the earth's surface, forming geothermal *hot spots.* As water seeps down to these hot spots, the water becomes very hot. Sometimes this hot water forces its way back to the earth's surface, forming geysers or hot springs.

Turbines Powered by Steam In most cases, the fluid that turns generator turbines is steam made by boiling water. The fuel used to boil the water is most often a fossil fuel. In some cases the fuel is a nuclear fuel such as uranium. The steam turbines in electric generating plants work just as well on steam from under the earth's surface as they do on steam from a boiler. Therefore, if steam is released from hot spots, electricity can be generated.

● **Process Skill:** *Comparing*

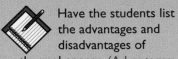

Have the students list the advantages and disadvantages of geothermal energy. (Advantages: Geothermal energy is free [not including trapping and processing costs]. It can be used directly rather than needing to be converted to electricity. Disadvantages: Geothermal energy is only available in a limited number of places. The water in some locations is mildly radioactive.)

Figure 6–16. A geothermal electric plant receives high-pressure steam from a geothermal hot spot located beneath the earth's surface.

In some cases, hot spots do not already contain water and steam. In these situations, water is pumped down to the hot spots to produce the necessary steam to operate the turbines.

Geothermal Electric Plants

Many countries, including the United States, the Commonwealth of Independent States, Italy, Australia, Japan, and Iceland, already have electric plants that operate on geothermal energy. Some countries, such as Iceland, use geothermal energy directly rather than converting it to electricity. Such direct uses include heating homes, buildings, and household water. The hot water from the earth is even used to heat swimming pools.

Like solar or wind energy, geothermal energy is free, if you don't count the cost of trapping and processing it. Unfortunately, it is only available in a limited number of locations. Many of these are not near places where many people wish to live. Also, the water from some hot spots contains high concentrations of dissolved minerals that are slightly radioactive. This radioactive water is less dangerous than spent fuel from a nuclear reactor, but it must be disposed of carefully.

Mining the Earth's Heat

If geothermal energy is to be used as a future energy resource, hot spots must be close enough to the surface to allow inexpensive "mining" of the earth's internal heat. This occurs most commonly where tectonic plates push into each other or pull apart.

 ASK YOURSELF

What produces geothermal energy?

 **LASER DISC**
1140

Geothermal energy

 **MULTICULTURAL CONNECTION**

Geothermal energy has been utilized by many countries around the world. Iceland has more than 100 geothermal wells near its capital, Reykjavik. The Philippines is the world's second largest geothermal producer, behind the United States. Central America has high seismic activity, and power plants already exist in El Salvador and Nicaragua. Costa Rica and Guatemala have plans to build such plants. The United States has geothermal plants operating in California, Montana, and Hawaii.

ONGOING ASSESSMENT
ASK YOURSELF

During decay, radioactive nuclei produce heat as they decay deep within the earth's core.

● **Process Skill:** *Comparing*

Point out to the students that dams have other uses in addition to providing hydroelectric energy. Some dams are used to hold water for irrigation. Other dams are used for flood control, holding back flood water so the water can be released gradually.

● **Process Skill:** *Predicting*

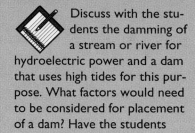

Discuss with the students the damming of a stream or river for hydroelectric power and a dam that uses high tides for this purpose. What factors would need to be considered for placement of a dam? Have the students

write their responses in their journals. (Both dams operate in the same manner—water drops and flows through a turbine. The land to be flooded behind a river dam would have to be considered. The mean high and low tide differential would need to be known for a tidal dam.)

ONGOING ASSESSMENT

▼ **ASK YOURSELF**

Water under pressure (controlled by the dam) is channeled past a turbine. The moving water causes the turbine to rotate and power an electric generator.

The cold-intake pipe in an OTEC plant may extend more than one kilometer below the surface of the ocean. A mini-OTEC ammonia-based plant was built as an experimental prototype in 1979. It operates off Keahole Point on the west coast of Hawaii. At-sea testing and problem solving is still going on.

Hydroelectric Energy

The video continues with footage of a gigantic concrete structure. The narrator identifies the structure as a hydroelectric dam. In areas where deep bodies of water are collected behind dams, the fluid turning the turbine is water falling through a dam. Electricity produced in this fashion is called *hydroelectric energy*. The drawing on this page shows the basic arrangement of a hydroelectric dam.

Figure 6–17. The Grand Coulee Dam in the state of Washington is 165 m tall and 1272 m wide. It is so large it can easily be seen from orbit with a simple telescope! In a hydroelectric dam, water flows from a reservoir through turbines. Each turbine, in turn, operates a generator to produce electricity.

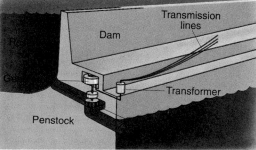

▼ **ASK YOURSELF**

How is electricity produced by a hydroelectric dam?

Ocean Tide Energy

Water flowing down a river can be used to produce hydroelectric energy if that river can be trapped behind a dam. However, this is not the only way to use the energy of moving water. Another method involves the ocean tides, which also represent large masses of moving water. In a few locations on Earth, dams can be used to trap water within a bay area during high tide. Then at low tide, the water is allowed to drain through turbines located in the dams to produce hydroelectric energy.

Water Temperature Energy?

The oceans also represent a source of energy that is not based on water movement. Instead, this energy source is based on water temperature. The water at the surface of the oceans absorbs large amounts of solar energy. As a result, surface water is much warmer than deeper water. You have probably experienced this difference while swimming. You can even feel it in a swimming pool as you dive to the bottom. A process that uses the differences in ocean water temperature to produce electricity is called *ocean thermal energy conversion,* or *OTEC*.

The OTEC System

The OTEC system works much like a refrigerator. It uses warm surface water to heat a special fluid, such as ammonia or Freon, that has a very low boiling point. When surface water heats the fluid, the fluid begins to boil. This boiling action creates pressure that turns turbine generators, producing electricity. Once the fluid is used in this way, it is routed down to a cold layer of the ocean. Here the fluid is cooled and converted back into a liquid, and the cycle begins again.

One of the advantages of the OTEC system is that it is basically pollution free. It uses solar energy to heat the surface of the ocean and can operate 24 hours a day. Unfortunately, OTEC plants are very expensive to build. Also, since they are built offshore, the energy they produce is hard to transport to cities where it will be used. It is possible to convert the energy into microwave radiation, beam it ashore, and then convert it to electricity. But that would add a lot to the cost. Moreover, scientists are not sure how OTEC energy factories will affect the marine life of the area.

Figure 6–18. On the Rance River in France, ocean water is trapped behind a dam at high tide. At low tide, the water is used to operate turbine generators.

Figure 6–19. This illustration shows what an OTEC plant might look like.

▼ ASK YOURSELF
How can tides be used to produce energy?

● **Process Skill:**
Classifying/Ordering

Have the students list all the energy sources in this chapter that produce steam to run a turbine and produce electricity. (Sources should include all of the fossil fuels, biomass, geothermal energy, and nuclear energy.)

● **Process Skill:** *Analyzing*

Ask the students to predict what might happen if control rods were not used to regulate a nuclear chain reaction. (The reaction would continue to accelerate, producing more heat and power, and end in an eventual meltdown.)

● **Process Skill:** *Analyzing*

Cooperative Learning Ask the students to work in small groups to consider and discuss how the discovery of superconducting materials might aid some of the renewable energy sources that have been presented. (Transporting costs would be much cheaper, making geothermal, tidal, ocean thermal, and wind sources more economical.)

SCIENCE TECHNOLOGY SOCIETY One of the most difficult problems to solve relating to nuclear energy is the problem of disposing of spent fuel. The basic requirement is a site's capability to contain and isolate the waste safely until it naturally decays. A spray calciner has been developed that solidifies the waste in glass or ceramic. The Environmental Protection Agency (EPA) has determined that a successful site must isolate the waste for 10 000 years. Nine locations in the United States are currently under consideration.

THE NATURE OF SCIENCE

The natural radiation of a uranium compound was discovered by French scientist Henri Becquerel in 1896. In 1897 Marie Curie, a French chemist, demonstrated that the radiation came from the uranium atom. She and her husband, Pierre, discovered that the element thorium also produced radiation. She then discovered two new radioactive elements, polonium and radium.

In 1939 a German physicist, Otto Hahn, discovered nuclear fission. That same year, Hungarian-born physicist Leo Szilard combined his knowledge of a chain reaction and the new fission development and realized a nuclear bomb was possible. In 1943, J. Robert Oppenheimer headed the Manhattan Project in the United States, which developed the first nuclear bomb. An atomic bomb eventually ended World War II and began the nuclear age.

Figure 6–20. Nuclear reactors, such as this one, use fission reactions to produce electricity.

Figure 6–21. Shown here is a diagram of a typical nuclear reactor. This particular type of reactor is called a *pressurized water reactor.*

Nuclear Energy

The final portion of the video includes a discussion of nuclear energy—energy stored within the nuclei of atoms. Nuclear energy is very concentrated, but it is also very hard to release in a controlled fashion. Complex devices called **nuclear reactors** are used to convert nuclear energy into heat. The heat from nuclear energy is used to boil water into steam. The steam is then used to spin a generator turbine that produces electricity.

Splitting Up All operating nuclear energy plants use nuclear fission. *Fission* means "splitting." In nuclear fission, the nucleus of a uranium or plutonium atom splits when a free neutron crashes into it. When each nucleus is split, it releases heat energy, smaller atoms, and additional neutrons. The additional neutrons go on to split other nuclei to cause a *chain reaction.* This chain reaction must be carefully regulated to keep it under control.

Nuclear reactors are designed in many different ways. A simple diagram of one kind of nuclear reactor is shown on this page. Most reactors, regardless of their design, contain the same basic parts.

The nuclear chain reaction takes place and is controlled within the *reactor core.* The reactor core contains uranium (or plutonium) fuel pellets that have been sealed inside long rods called *fuel rods.* The uranium inside these rods is the fuel for the entire nuclear reactor.

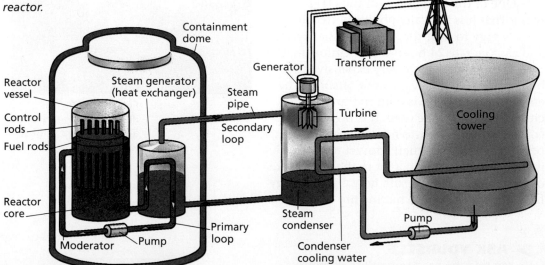

GUIDED PRACTICE

 Have the students write a few paragraphs in their journals supporting or criticizing the use of nuclear energy instead of fossil fuels. Remind the students to use information in this section to back up their positions. Encourage the students to share their efforts with classmates.

INDEPENDENT PRACTICE

 Have the students provide written answers to the Section Review. Compile the answers to question 4 (p. 172) on the board for the students to copy. In their journals, have the students copy the diagram of a geothermal plant. Remind them to label the energy source clearly.

EVALUATION

Have the students write a description of what happens when a technician "powers up" a nuclear reactor. Ask several volunteers to read aloud their descriptions so classmates might discuss them.

Another set of rods is also located inside the reactor core. These rods are called *control rods* because they are designed to control the nuclear chain reaction. The control rods are filled with materials such as boron and cadmium that can absorb free neutrons. When the control rods are lowered into the reactor core, more neutrons are absorbed, so a smaller number of nuclei are split. This, in turn, slows the chain reaction, and less heat is produced. When reactor technicians "power up" or "power down" a nuclear reactor, they are raising or lowering the control rods.

Is There a Moderator in the House?

In addition to fuel rods and control rods, the reactor core also contains a moderator. The moderator is a substance such as water or graphite that can slow down the speed of free neutrons. If neutrons are moving too fast, they will not cause the uranium or plutonium nuclei to split. In order to have many of the neutrons travel slowly enough to cause fission, they must bounce off something that can absorb some of their kinetic energy. Neutrons transfer some of their kinetic energy to the atoms of a moderator in collisions. Most commercial nuclear energy reactors in the United States use water as a moderator.

The water in nuclear reactors also acts as a coolant. It flows around the fuel rods, where it absorbs heat. If left uncooled, the reactor core could get so hot that it would actually melt, causing radiation to escape. This situation is called a *meltdown*. Partial meltdowns have occurred at Chernobyl in the Ukraine and at Three Mile Island in Pennsylvania.

Reactors also contain *shielding* to prevent radiation from escaping the core. Some shielding materials reflect stray neutrons back into the core. Others absorb radiation to protect the structure of the reactor and to keep radiation from escaping. Radiation can cause biological damage to the people who work at the reactor.

The *containment dome,* which covers the entire reactor, is designed to prevent radiation from escaping into the environment even if an accident occurs in the reactor. The Three Mile Island accident caused little damage to the environment because the radiation was held in the reactor by the containment dome. By contrast,

Figure 6–22. New fuel rods are shown here being installed into a reactor core. Once the rods are installed, they will not need to be replaced again for approximately one year.

Figure 6–23. The liquid around the fuel and control rods in a reactor core is the moderator.

INTEGRATION– *Social Studies*

Nuclear energy also has great value because it produces nuclear radiation—high-energy particles and rays. Nuclear radiation is used in medicine, industry, and science. Nuclear energy also powers some ships and submarines. These vessels have a nuclear reactor to create heat for making steam. The steam is used to turn the ship's propellers.

✧ Did You Know?

A nuclear plant uses much less fuel than does a fossil-fuel plant. The fissioning of less than one metric ton of uranium fuel provides about as much heat energy as the burning of 2.7 million metric tons of coal or 1.2 million barrels of oil.

 LASER DISC
2000

Cooling tower

Cooperative Learning Display a large map of the United States and assign groups of students an energy source. Then have each group note on the map areas where its assigned energy source is located. Encourage the students to use reference sources, if necessary.

Have interested students do research to find the location of hydroelectric dams in the United States. The students should list the locations of the dams and note them on an outline map of the United States or the resources map they created earlier in this chapter. Instruct them to design a symbol for the dams and add the new symbol to the map legend. Encourage the students to place their research and maps in their science portfolios.

Cooperative Learning Have the class consider their geographic location to decide which alternate energy sources might work in their area. Place them in small groups and ask them to write a proposal for their idea. Then have each group present its proposal to the class.

ONGOING ASSESSMENT

ASK YOURSELF

Nuclear fission naturally emits heat. Water, the moderator, is heated by this reaction and turns to steam.

SECTION 3 REVIEW AND APPLICATION

Reading Critically

1. Students' choices might include geothermal energy, hydroelectric energy, nuclear energy, or solar energy. Accept all reasonable explanations of the scientific processes.

2. Geothermal energy is free. However, it is only available in a limited number of locations, many of which are not near where people live. In addition, the water from some locations contains high concentrations of dissolved minerals that are slightly radioactive. While this radioactive water is less dangerous than spent fuel from a nuclear reactor, it must still be disposed of carefully.

Thinking Critically

3. Students should discuss an alternative energy source described in this section. Accept all choices that the students can reasonably justify.

4. The source and methods for producing electricity are alike in that they all power a turbine in some manner, usually through the production of steam. They differ in that some methods are more risky or more costly to use than others.

the Chernobyl reactor did not have a containment dome, and the Chernobyl accident released large quantities of radiation into the environment.

The Solution? Should you recommend full implementation of nuclear fission as an energy alternative? Although nuclear fission can produce tremendous amounts of energy, fissionable materials are expensive and not readily available. In addition, there is currently no accepted method for storing or disposing of radioactive waste from a nuclear reactor.

Through all of your research, you've noticed a pattern. There are advantages and disadvantages to each energy resource. You have read over and over again that the success or failure of humanity depends on its energy supply. But the energy supply must fit both the needs of the people and the resources of the planet. As you've discovered, there are many different energy resources available, but the decision to use each one must be made with care.

DISCOVER BY Researching

Some alternative energy sources that have not been discussed include (a) hydrogen fuel from ocean water, (b) solar energy used to operate steam engines to produce electricity, and (c) solar energy collected in space and transmitted to Earth with microwaves. Conduct research on one of these or another alternative energy source, and prepare a poster. In your poster, include the tradeoffs associated with each energy source.

ASK YOURSELF
How do nuclear reactors convert nuclear energy into heat?

SECTION 3   REVIEW AND APPLICATION

Reading Critically
1. Choose one alternative energy source, and explain the scientific process by which it produces useful energy.
2. What are the advantages and disadvantages of geothermal energy?

Thinking Critically
3. Of the alternative energy sources described in this section, which one do you think is the best choice for widespread use? Explain your answer.
4. In what ways are the methods for producing electricity from conventional, nuclear, and alternative energy sources alike? In what ways are they different?

CHAPTER 6 HIGHLIGHTS

The Big Idea—Energy

Be sure the students understand that energy is never "used up" or destroyed. You might practice a series of energy conversions with the students, such as tracing the burning of a fossil fuel to the running of a microwave oven in their home.

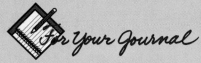

For Your Journal

After the students have completed this chapter, help them review the tradeoffs associated with each energy source discussed. Encourage them to revise their original journal entries as necessary.

CHAPTER 6 *H*IGHLIGHTS

The Big Idea

Heat, light, sound, and electricity—these are all forms of energy. Without energy, everything on Earth would stand still. Nothing moves without it.

Most of the energy people use comes from nonrenewable sources such as fossil fuels. Because fossil fuels are nonrenewable resources, energy alternatives must be developed. Everyone is concerned about energy sources for the future. Scientists are exploring new sources—renewable ones that won't run out. They are also working on new technologies to make energy from these sources continuously available.

For Your Journal

Think about what you have learned in this chapter about types of energy sources. Look back at what you have written in your journal, and revise or add to the entries to reflect your new understanding about energy sources.

CONNECTING IDEAS

Students' completed concept maps should be similar to the one shown below.

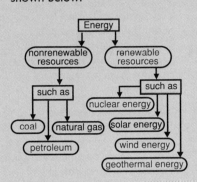

Connecting Ideas

Copy this concept map into your journal. Complete the map by filling in details where needed. Remember that a concept map tells you not only which concepts are hooked together, but also what their relationships are.

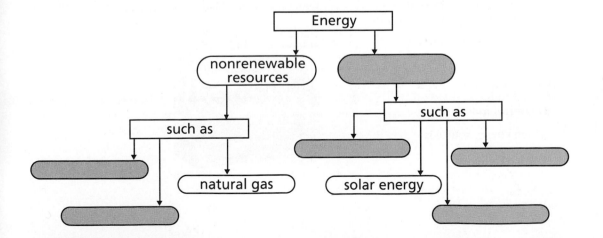

ANSWERS

Understanding Vocabulary

1. (a) type of energy
   (b) method of converting energy
   (c) method of converting energy
   (d) method of converting energy
   (e) type of energy
   (f) method of converting energy

2. A nuclear reactor uses nuclear or atomic energy to produce power. A refinery separates crude oil into gasoline, heating oil, and other products. Both a hydroelectric plant and an OTEC plant use water to produce energy. However, an OTEC plant uses water's thermal energy to produce power.

Understanding Concepts

Multiple Choice

3. c

4. d

5. a

6. a

7. b

Short Answer

8. Coal mining either strips the surface of land or results in large, underground caverns. Hydroelectric plants usually require the damming of a river, which floods the land behind the dam. Students might also mention visual changes caused by oil wells, wind turbines, or wells to tap geothermal energy.

9. The number of available hours of sunlight is less during the winter than during the summer. As a result, a solar collector has less opportunity to operate during the winter.

Interpreting Graphics

10. The steam enters an enclosed portion of the engine, where it becomes compacted and builds up pressure. The increased pressure pushes on the blades, causing them to turn. An exit is provided for the steam so that the increasing pressure does not cause an explosion. However, the steady input of steam keeps the blades turning.

CHAPTER

6

REVIEW

▶ Understanding Vocabulary

1. Classify each term into one of two categories: types of energy or methods of converting energy.
   a) solar energy (159)
   b) wind turbines (163)
   c) passive solar heating (160)
   d) active solar heating (161)
   e) geothermal energy (166)
   f) nuclear reactors (170)

2. Compare and contrast these types of industrial plants: nuclear reactor (170), refinery (155), hydroelectric dam (168), OTEC system (169).

▶ Understanding Concepts

MULTIPLE CHOICE

3. Which term does not relate to coal?
   a) reclamation   **b)** seam
   c) refinery   **d)** strip mine

4. A type of energy production not dependent on location is
   a) tidal.   **b)** solar.
   c) wind.   **d)** nuclear.

5. Which of the following is not a fossil fuel?
   a) solar energy   **b)** coal
   c) natural gas   **d)** petroleum

6. Which part is not found in a nuclear reactor?
   a) fractionation tower
   b) containment dome
   c) fuel rods
   d) control rods

7. Which energy source is nonrenewable?
   a) geothermal
   b) oil shale
   c) solar
   d) wind

SHORT ANSWER

8. Which types of energy collection alter the surface of Earth, and in which ways?

9. Why would solar energy be less efficient during the winter than during the summer?

Interpreting Graphics

10. Follow the path of the steam in this turbine engine. Explain how the blades are turned by the steam.

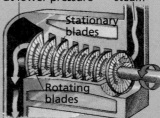

Outgoing steam at lower pressure   Incoming steam

Stationary blades

Rotating blades

Reviewing Themes

11. The sun provides the solar energy that heats the surface of the ocean. This creates a temperature differential between the upper and lower water layers. A floating power plant can use this thermal energy by causing a liquid, such as ammonia (which boils at a low temperature), to create steam to run a turbine. The turbine is connected to a generator, which produces electricity. The electric energy goes to your light bulb, which converts the energy to light. Heat is also produced in this conversion and along the way in the wire.

12. The OTEC system is basically pollution-free. It uses solar energy to heat the surface water of the ocean. The system can operate 24 hours a day. However, OTEC plants are extremely expensive to build, and their energy is difficult to transport to where it would be used. While it is possible to convert the energy into microwave radiation, beam it to land, and then convert it to electricity, it would also be very costly. In addition, scientists are unsure how OTEC energy systems might affect marine life in the surrounding area.

Thinking Critically

13. Earth gives off as much energy as it receives. Some of the energy is reflected and makes Earth visible from space. Most of the energy is emitted back as infrared radiation.

14. The explorers saw natural gas seeping out from underground that had been accidentally set on fire, perhaps when struck by lightning.

15. Pollutants emitted from coal-burning power plants can escape into the atmosphere and be carried hundreds of kilometers by winds and weather. The acidic pollutants could then be dropped on these distant locations when it rains.

16. The homeowner's bill would be reduced $50 each month. It would take 30 months for the savings to equal the cost of installing the solar collectors ($1,500 divided by $50 = 30 months).

However, the time needed for the savings to equal the cost of installation could be affected by a higher- or lower-than-average heating bill due to unseasonably cold or warm weather conditions or a higher-than-average heating bill due to lengthy periods of cloudy weather.

17. Advantages include: energy source would be low cost, energy would be inexhaustible, energy would be available day and night. Disadvantages include: advanced technology would be required and installation would be expensive, microwaves are potentially dangerous, system could be easily disrupted by foreign power.

Reviewing Themes

11. *Energy*
Starting with the sun shining on the ocean, trace the number and types of energy changes that occur during the OTEC system of energy production. Follow the process all the way to a light bulb in your house.

12. *Technology*
What are the advantages and disadvantages of the OTEC system?

Thinking Critically

13. Most of the solar energy hitting Earth becomes thermal energy. However, the average temperature of Earth does not increase with time. What are some possible reasons for this?

14. In 1775 French explorers in the Ohio Valley saw what they called "pillars of fire." Explain what they saw.

15. How could acid rain damage forests and lakes that are hundreds of kilometers from the nearest coal-burning power plant?

16. Suppose it costs $1,500 to install solar collectors on the roof a house. The homeowner's $100 monthly heating bill goes down 50 percent after installing the solar collectors. How long does it take for the savings to equal the cost of installation? Explain your answer.

17. One proposal for using solar energy is to build massive collectors in space. The large solar collectors would then receive energy from the sun as they orbit the earth. The energy would then be transmitted to the earth in the form of waves. What do you think would be the advantages and disadvantages of this proposed system? Do you think this system will ever become a reality? Explain.

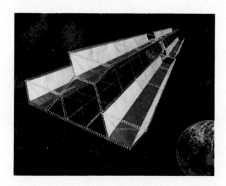

Discovery Through Reading

Haybron, Ron. "Sun Lover." *Discover* 11 (July 1990): 30. Read about a solar dynamic generator planned for the *Freedom* space station.

ENERGY AND THE ENVIRONMENT

PLANNING THE CHAPTER

Chapter Sections	Page	Chapter Features	Page	Program Resources	Source
CHAPTER OPENER	176	**For Your Journal**	177		
Section 1: ENVIRONMENT AT RISK	178	Discover by Doing **(B)**	179	*Science Discovery**	SD
• Air Pollution **(B)**	178	Discover by Doing **(A)**	182	Extending Science Concepts: Determining the pH of Acid Rain **(A)**	TR
• Raindrops Falling on My Head **(B)**	182	Section 1 Review and Application	186	Investigation 7.1: Measuring Air Pollution **(B)**	TR, LI
• Nuclear Energy Is NOT Waste Free **(A)**	185	Investigation: Determining the Effect of Acid Rain on Seeds **(A)**	187	Thinking Critically **(B)**	TR
				Earth's Atmosphere and the Greenhouse Effect **(A)**	IT
				Atmospheric Ozone **(B)**	IT
				Study and Review Guide, Section 1 **(B)**	TR, SRG
Section 2: ENERGY CONSERVATION	188	Activity: How does a steam turbine work? **(A)**	190	*Science Discovery**	SD
• Spending Energy Wisely **(B)**	188	Discover by Writing **(B)**	192	Reading Skills: Making Sequences **(A)**	TR
• Conservation at Home **(A)**	191	Section 2 Review and Application	193	Connecting Other Disciplines: Science and Social Studies, Evaluating Power Plants **(B)**	TR
				Study and Review Guide, Section 2 **(B)**	TR, SRG
Section 3: ENERGY DECISION MAKING	194	Discover by Calculating **(B)**	195	*Science Discovery**	SD
• Energy Systems **(H)**	194	Discover by Writing **(B)**	196	Investigation 7.2: Determining that Acid Can Come From Coal **(A)**	TR, LI
• Energy Economics **(B)**	195	Section 3 Review and Application	197	Record Sheets for Textbook Investigations **(B)**	TR
• Energy Politics **(A)**	197	Skill: Collecting and Organizing Information **(A)**	198	Study and Review Guide, Section 3 **(B)**	TR, SRG
Chapter 7 HIGHLIGHTS	199	The Big Idea	199	Study and Review Guide, Chapter 7 Review **(B)**	TR, SRG
Chapter 7 Review	200	For Your Journal	199	Chapter 7 Test	TR
		Connecting Ideas	199	Test Generator	

B = Basic   **A** = Average   **H** = Honors
The coding Basic, Average, and Honors indicates subsections, features, and resources that might be appropriate for different levels of learners. For additional suggestions regarding choice of topic and depth of coverage, see the Pacing Chart on pages T28–T31.

*Frame numbers at point of use
(TR) Teaching Resources, Unit 2
(IT) Instructional Transparencies
(LI) Laboratory Investigations
(SD) *Science Discovery* Videodisc Correlations and Barcodes
(SRG) Study and Review Guide

CHAPTER MATERIALS

Title	Page	Materials
Discover by Doing	179	(per group of 3 or 4) dark cloth; small, empty aquarium or glass lid; thermometer; plastic wrap or glass lid
Discover by Doing	182	(per pair) tap water, strips of blue and red litmus paper (2 each), vinegar
Teacher Demonstration	182	vinegar (10 mL), bicarbonate of soda (1 g), samples or photographs of limestone, dolomite, or marble affected by acid rain
Teacher Demonstration	185	dominoes, small open boxes
Investigation: Determining the Effect of Acid Rain on Seeds	187	(per pair) safety goggles, laboratory apron, and gloves (for each), Petri dishes with lids, paper towels, solution to simulate acid rain, seeds, tape, aluminum foil, labels, litmus paper
Activity: How does a steam turbine work?	190	(per group of 3 or 4) safety goggles and laboratory apron (for each), ring stand (2), utility clamp (2), ring with wire gauze, 250-mL flask, boiling chip, 2-hole stopper to fit flask with bent glass nozzle and thin glass tube (safety valve), Bunsen burner, toy pinwheel
Discover by Writing	192	(per individual) journal
Discover by Calculating	195	(per individual) journal
Discover by Writing	196	(per individual) journal
Skill: Collecting and Organizing Information	198	(per individual) paper and pencil

ADVANCE PREPARATION

For the *Teacher Demonstration* on page 182, collect samples of limestone, dolomite, or marble affected by acid rain. For the *Investigation* on page 187, collect weed seeds ahead of time.

TEACHING SUGGESTIONS

Field Trip
If possible, visit a place with your students where the effects of acid rain are visible. If this is not possible, visit a museum that has a display on energy and the environment. Have the students prepare a newspaper dedicated to the environmental issues discussed in this chapter. Encourage them to write articles on the issues that most interest them.

Outside Speaker
If you cannot go on a field trip, have a representative from a wildlife preserve or environmental agency in your area speak to the class about environmental problems created by energy use. Provide the speaker with a list of topics discussed in Chapter 7 and the points that the students would most like to have addressed. Have the students prepare questions beforehand to ask the speaker. After the talk, have the students work in groups to discuss what they have learned and to prepare a report.

CHAPTER 7

ENERGY AND THE ENVIRONMENT

John Muir Papers, Holt-Atherton Special Collections, University of the Pacific Libraries, © 1984, Muir-Hanna Trust

The naturalist John Muir spent his life exploring and describing the natural beauty of the American West. He was troubled that people saw in nature only "resources" to be exploited. In the last ten years of his life, Muir battled plans to dam the Hetch Hetchy valley, then a part of Yosemite National Park in California. Muir wrote of the valley's awesome beauty.

The artist . . . wandered day after day along the river and through the groves and gardens, studying the wonderful scenery; and, after making about forty sketches, declared with enthusiasm that although its walls were less sublime in height, in picturesque beauty and charm Hetch Hetchy surpassed even Yosemite.

CHAPTER MOTIVATING ACTIVITY

PORTFOLIO ASSESSMENT

Ask the students to describe their favorite vacation area that they have visited or would like to visit and provide details of what makes the location so appealing. Ask whether this spot has changed over the years and what might be causing the changes. Then explain to the students that in this chapter they will learn how uses of energy impact on the environment. Encourage the students to place their descriptions in their science portfolios.

For Your Journal

The journal questions will help you identify any misconceptions the students might have about the effect of energy use on the environment. In addition, these questions will help the students to realize that they do possess knowledge about energy and the environment. (The students will have a variety of responses to the journal questions. Accept all answers, and tell the students that at the end of the chapter they will have a chance to refer back to the questions.)

John Muir Papers, Holt-Atherton Special Collections, University of the Pacific Libraries, © 1984, Muir-Hanna Trust

That anyone would try to destroy such a place seems incredible; but sad experience shows that there are people good enough and bad enough for anything. The proponents of the dam scheme bring forward a lot of bad arguments to prove that the only righteous thing to do with the people's parks is to destroy them bit by bit as they are able . . . Landscape gardens, places of recreation and worship, are never made beautiful by destroying and burying them.

from *The American Wilderness*
by John Muir

ABOUT THE LITERATURE

John Muir (1838–1914) lost the battle to save the Hetch Hetchy Valley from being flooded. Engineers completed O'Shaugnessy Dam in 1923, flooding what Muir called his "temple" at Hetch Hetchy. Today the dam provides electricity and drinking water to the city of San Francisco. However, Muir's struggle was not in vain. He helped spur the public into an awareness of conservation issues.

Students who want to know more about John Muir could check out and read the following book.

Force, Eden. *John Muir*. Silver Burdett, 1990. This award-winning biography is about John Muir, the naturalist and conservationist responsible for the creation of the U.S. National Park Service.

For Your Journal

- What are our natural resources?
- How does the burning of fossil fuels affect the environment?
- What can you do to reduce the amount of energy you use?

FOCUS

This section discusses the kinds of pollution that result from the use of energy-producing fuels. The dangers to the environment and the problems of dealing effectively with pollution are emphasized.

MOTIVATING ACTIVITY

Cooperative Learning Have the students work in small groups to develop slogans and ideas for posters that show the importance of protecting Earth from air pollution. The students

might find pictures in magazines that show smog in certain cities. Their posters should show the damage along with the cause. Encourage the students to display their posters and refer to them throughout their reading of the chapter.

PROCESS SKILLS
- Experimenting
- Interpreting Data
- Comparing

POSITIVE ATTITUDES
- Curiosity • Caring for the environment • Cooperation
- Skepticism

TERMS
- greenhouse effect • acid rain • nuclear waste

PRINT MEDIA
"Lost Horizons" from *Discover* magazine by Jeffrey Brune (see p. 67b)

ELECTRONIC MEDIA
Help Save Planet Earth UNI Distribution (see p. 67c)

SCIENCE DISCOVERY
- Acid rain; formation and deposition
- Atmosphere; layers • Ozone plot
- Pollution, air

BLACKLINE MASTERS
Extending Science Concepts
Laboratory Investigation 7.1
Thinking Critically
Study and Review Guide

INTEGRATION– Health

Laboratory studies show that people might develop headaches, become dizzy, or show other signs of illness when exposed to 100 parts per million of carbon monoxide (CO). Heavy motor vehicle traffic in many urban areas often creates high levels of CO. Lower levels of this gas are also thought to cause impairment of vision and judgment. In higher altitudes, the harmful effects of CO are multiplied.

SECTION 1

Environment at Risk

Objectives

List and **discuss** the sources of air pollution, acid rain, and nuclear waste.

Compare and **contrast** the arguments for and against storing nuclear waste underground.

Evaluate the pros and cons of various pollution-control methods.

O n everything from billboards to T-shirts you see the slogans: Save the Planet, Save Our Earth, Preserve the Planet, Keep Earth Green, and on and on. What's happening? We have come face to face with our environment—an environment at risk. The issues related to the environment are enormous—air pollution, acid rain, nuclear waste, energy conservation, energy economics, and energy politics. You are probably willing and eager to "Save the Planet," but you need information about the issues. This information will be your survival tool—a tool with which you *can* help the environment.

Air Pollution

Nearly 90 percent of the United States' energy requirements are supplied by fossil fuels. Unfortunately, the use of fossil fuels creates air pollution. The burning of fossil fuels produces nitrogen oxides, unburned and partially burned hydrocarbons, sulfur oxides, carbon oxides, and smoke particles. These pollutants affect the atmosphere in several ways.

Nitrogen oxides and partially burned hydrocarbons react with sunlight to produce *smog*. Smog looks like dirty fog and is a health hazard. The word itself comes from a combination of the words *smoke* and *fog*. In cities such as Los Angeles, high concentrations of smog can even lead to the cancellation of outdoor athletic events and sometimes city residents are advised to stay indoors.

Figure 7–1. Smog, visible in the photograph on the left, is a serious health hazard. The photograph on the right shows the same city on a low-smog day.

TEACHING STRATEGIES

● **Process Skills:**
Classifying/Ordering, Inferring

Discuss with the students the kinds of pollution they have seen. Then have them list the sources of common pollution. (Examples include motor vehicle exhaust, chemicals, and garbage.)

● **Process Skills:**
Inferring, Applying

Review with the students the second law of thermodynamics from Chapter 5. Point out that pollution is a form of entropy, as is waste heat. Then ask the students if a pollution-free

process for eliminating pollution would be possible. (Cleaning up pollution always requires energy from somewhere, resulting in pollution somewhere else.)

● **Process Skills:** *Analyzing, Comparing, Inferring*

Ask the students if they think air pollution existed in the nineteenth century before the gasoline-powered automobile was invented. (Coal and wood burned for industrial and home-heating uses created air pollution in the nineteenth century as well.)

Incomplete combustion of fuels also produces carbon monoxide, a colorless and odorless pollutant that is poisonous to humans and other animals. Carbon monoxide reduces the ability of red blood cells to carry oxygen in the body. It can stick to hemoglobin in the same way oxygen does, blocking the oxygen. Carbon monoxide is not usually a problem because it is produced in low concentrations and is changed in the atmosphere to carbon dioxide. However, when air circulation is poor, such as in closed garages, in long tunnels, or during unusual weather conditions, the effects of carbon monoxide can be deadly.

Reducing air pollution is difficult. For example, devices used to clean polluted air from factory smokestacks often do so by trapping the pollutants in water. However, this causes water pollution. When pollutants are removed from the water, they often become a solid-waste problem. There are no easy solutions. Often, though, one option is better than another. The solid waste from the water and air can be stored in one place and be kept track of. When it is still in the air or water, it can make its way anywhere on the planet and is uncontrolled.

A Hot Time on the Old Planet

In addition to causing air pollution, the burning of fossil fuels causes other problems. Have you noticed how hot it gets inside a parked car on a sunny day with the windows rolled up? Something similar happens on Earth. Carbon dioxide in the earth's atmosphere acts like a glass globe surrounding the planet. The carbon dioxide allows sunlight to pass inward to the earth but reduces the flow of heat energy outward into space. More carbon dioxide in the air traps more heat energy. This phenomenon is called the **greenhouse effect** because the carbon dioxide acts much like the glass of a greenhouse. To find out more about the greenhouse effect, try the next activity.

Figure 7–2. To avoid carbon monoxide poisoning, this mechanic attaches an exhaust hose to the tailpipe of a car. The hose carries exhaust fumes out of the garage.

DISCOVER BY *Doing*

Ask the students what the purpose of the plastic wrap or glass lid is in this activity. (The light rays pass through the plastic or glass into the aquarium, but most of the heat remains trapped inside. Little heat escapes into the atmosphere.) The temperature in students' "greenhouses" should be at a consistently higher temperature than the temperature outside.

PERFORMANCE ASSESSMENT

After the students complete the Discover by Doing, have them demonstrate how the greenhouse effect occurs. You can check a student's understanding of the importance of the greenhouse effect on the planet by asking the student to relate the activity to what happens in the atmosphere. The students should compare the plastic or glass to CO_2 in the atmosphere.

Discover BY *Doing*

Place a dark cloth on the bottom of a small, empty aquarium. Now lay a thermometer on the dark cloth. Put the aquarium in direct sunlight, and cover it completely with plastic wrap or a glass lid. How does the temperature in this "greenhouse" compare with the temperature outside? In your journal, write an explanation for this, including a diagram. ✎

SECTION 1 **179** ◀

● **Process Skills:**
Inferring, Predicting

Discuss with the students how changes in climate due to the greenhouse effect might affect plant and animal life on Earth. Ask the students to record their ideas and revise them as necessary. Encourage the students to place their ideas in their science portfolios. (Some of the species of plants and animals that inhabit colder climates might become extinct. Other species would probably adapt to a warmer climate.)

MULTICULTURAL CONNECTION

Chico Mendes, a Brazilian rubber tapper, formed a union to protest the burning of the rain forests. He tried to convince cattle ranchers that using the rain forest in less destructive ways could preserve it for the future. Mendes' efforts helped reduce some of the air pollution from the burning, but he lost his life in the process.

① Students may respond that if the planet warmed too much the polar icecaps might begin to melt and raise sea levels, causing flooding in coastal areas throughout the world. Other students may recall concepts presented in Chapter 2 and respond that water in the oceans will expand due to the action of the molecules.

LASER DISC
1472

Atmosphere; layers

1492

Ozone plot

INTEGRATION— *Health*

Smokers are perhaps most at risk from carbon monoxide (CO) poisoning. Inhaled cigarette smoke contains CO, which combined with CO from pollution sources to raise the level of CO in a smoker's blood. The first symptom of CO poisoning is drowsiness, which some people think might account for some motor vehicle accidents.

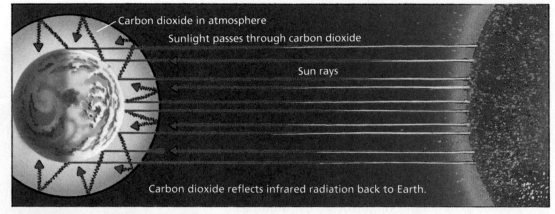

Carbon dioxide in atmosphere
Sunlight passes through carbon dioxide
Sun rays
Carbon dioxide reflects infrared radiation back to Earth.

Figure 7–3. Excess carbon dioxide in the atmosphere might cause global warming due to the greenhouse effect.

Scientists hypothesize that the greenhouse effect may result in an increase in the average temperature of Earth's atmosphere. This heating is called *global warming*. One predicted effect of global warming is a large change in wind and rainfall patterns, which could lead to crop failures. What other changes might occur if the planet warmed too much?

One way to reduce the greenhouse effect and global warming involves reducing fossil-fuel consumption or using alternative fuels that produce fewer pollutants. The major alternative fuels that are now under consideration as the "fuels for the future" include natural gas, propane, ethanol, and methanol. Each fuel has its advantages and disadvantages. Over the next decade, federal Clean Air Act rules will force more vehicles—from school buses to garbage trucks—to burn alternative fuels.

Holes in the Shield
You know that the earth is surrounded by atmosphere, which is divided into distinct layers, as shown in Figure 7–4. Near the top of the stratosphere is a layer called the *ozone layer*. Ozone is a form of oxygen. The ozone layer protects Earth from the sun's ultraviolet radiation by absorbing it. This protection is important because too much ultraviolet radiation can cause problems like skin cancer, impairment of vision, and even death.

There are several ways the ozone is being depleted. For example, nitrogen oxides, produced by high-flying airplanes, can damage the ozone layer by reacting with the ozone. The ozone layer is also damaged by compounds called *chlorofluorocarbons* or *CFCs*. Freon, an example of CFCs, is used in refrigeration and air-conditioning units and in some manufacturing processes. When Freon is released into the air, it works its way up into the atmosphere and destroys the ozone layer. Nitrogen oxides and

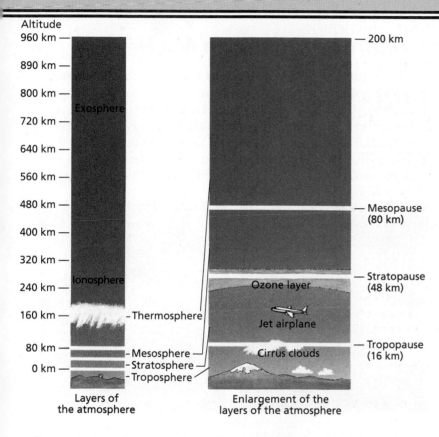

Layers of
the atmosphere

Enlargement of the
layers of the atmosphere

Figure 7–4. Layers of the earth's atmosphere

Freon destroy the ozone layer faster than it can form through natural processes.

How can the production of CFCs be lessened? American manufacturers have been ordered to stop production of all ozone-depleting chemicals by the end of 1995. Scientists are searching for replacements that do not harm the environment. In the meantime, automobile air-conditioner repair shops have begun using a machine that can recycle the refrigerant, so none is vented to the atmosphere. So far, more than 200 000 recycling machines have been sold.

Figure 7–5. As you can see from these photographs, the ozone hole over the Antarctic has grown larger.

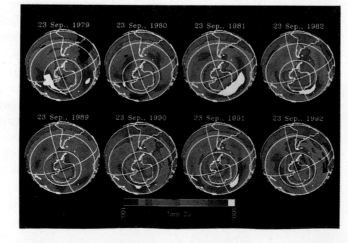

▼ **ASK YOURSELF**
What is air pollution, and how can it be stopped?

● **Process Skill:** *Solving Problems/Making Decisions*

Ask the students to hypothesize why acid rain kills trees and marine life. (Plants and fish thrive in environments that have a certain range of acidity. However, highly acidic soil or water damages or even kills the organisms that live there.)

● **Process Skill:** *Analyzing*

Have the students suppose that a car is left outside. In the morning, it is covered with dew. Throughout the morning, the sun burns off the dew. Then ask the students how acid rain could damage the car's paint even though it didn't rain. (Air pollutants combine with the dew to form an acidic water. The water stands on the paint. When the sun burns off the dew, the pollutants are left behind on the paint.)

Demonstration

Demonstrate what happens when 10 mL of vinegar is added to 1 g of bicarbonate of soda. Ask a volunteer to explain the reaction. (The acid reacts chemically with the base to produce a salt.) Then display samples or photographs of limestone, dolomite, or marble that have reacted to acid rain in a similar way, resulting in the destruction of statues or buildings made of these materials.

DISCOVER BY *Doing*

Point out to the students that when red litmus paper touches an acid, the paper stays red. When red litmus paper touches a base, the paper turns blue. The opposite occurs with blue litmus paper. The rain samples will reflect various levels of acidity. The acidity of the tap water and vinegar solution will depend on the ratio of water to vinegar. Ask the students to put their results in their science portfolios.

★ PERFORMANCE ASSESSMENT

After the students complete the Discover by Doing, have individual students demonstrate how they might test for acid rain.

Figure 7–6. Acid rain can destroy forests (left) and kill fish and other organisms living in lakes (right).

Raindrops Falling on My Head

Rain that contains acid is called **acid rain.** Acid rain is formed when air pollutants (sulfur oxides and nitrogen oxides) combine with oxygen and water vapor in the atmosphere to form weak sulfuric and nitric acids. This acidic water can travel as clouds for hundreds and even thousands of kilometers before falling to the earth as acid rain. You can find out whether acid rain is a serious problem in your neighborhood by doing the next activity.

DISCOVER BY *Doing*

Test the acid content of water. Take a water sample from the tap, and dip one strip of blue litmus paper and one strip of red litmus paper into the sample. Then add a couple of drops of vinegar to the water sample, and repeat the procedure. How would the results compare to those of normal rainwater and acid rain? Collect rain samples and test them. How do they compare? ✐

It's Raining Acid Acid rain affects our environment in many ways. As runoff, it might make lakes and rivers acidic, killing fish and other aquatic organisms. Some lakes become so acidic that not even bacteria can live very well in them. The water in these lakes is very clear because none of the tiny organisms that cause water to look cloudy can survive. Acid rain also reacts with soil and rocks to dissolve ions of aluminum and other metals, which can be harmful to living organisms. When washed into rivers and lakes, these dissolved metals pollute the water.

● **Process Skill:** *Inferring*

Ask the students how the amount of nitrogen oxide in the air can be decreased. (by making car engines smaller and by using pollution-control systems on motor vehicles)

● **Process Skills:** *Analyzing, Communicating*

 Discuss with the students how some nations have been found to have extreme levels of air pollution due, in part, to the operation of seriously outdated

manufacturing facilities. Help the students do research to recognize the relationship between this pollution and acid rain in neighboring countries. Ask the students to record their research and ideas in their journals.

Figure 7–7. Damage to the Statue of Liberty was caused largely by the effects of acid rain.

REINFORCING THEMES
Environmental Interactions

Cooperative Learning Have the students work in pairs or small groups to design a diagram that illustrates the interactions of acid rain. The diagrams should show how acid rain is formed, how it travels, and how it affects the environment. Encourage the students to share their diagrams with classmates.

✧ **Did You Know?**
Motor vehicles manufactured for sale in California must meet especially high pollution-control standards due to that state's levels of air pollution in cities such as Los Angeles. The high levels of air pollution are due, in part, to the state's limited number of public transportation systems and heavy reliance on automobiles.

Acid rain is also harmful to land organisms. Some food crops have been harmed by the effects of acid rain. Acid rain is killing huge forests in Canada, Europe, and the northeastern United States. Much of the acid rain that is destroying Canadian forests comes from polluted air in the United States. Much of the acid rain that is killing Swedish and Norwegian forests comes from Germany and Poland. As you can see, acid rain is a global problem that can be solved only with international cooperation.

Acid Rain Damages Structures   In addition to killing living organisms, acid rain damages buildings, bridges, statues, and other structures. Much of the damage done to the Statue of Liberty was caused by acid rain. The total repair cost for the Statue of Liberty was $30 million. Structures made of limestone or marble react with acid rain in a manner similar to the reaction that occurs between sodium bicarbonate and the acid in vinegar.

The only way to control acid rain is to reduce the emission of sulfur oxides and nitrogen oxides into the air. The main source of sulfur oxides is the burning of coal and other fossil fuels in electric-energy plants. As much as 50 percent of the

MEETING SPECIAL NEEDS

Mainstreamed

Mainstreamed students might find the environment a cause they would like to become involved with. Provide materials, or help the students find materials on wilderness areas that might be endangered by energy use. Discuss with the students what they might do to preserve the areas.

SECTION 1   **183** ◀

TEACHING STRATEGIES, continued

● **Process Skills:**
Analyzing, Communicating

Use the diagram of the catalytic converter to explain to the students how the device works. A catalytic converter contains a finely divided platinum-iridium catalyst into which exhaust gases from the engine are passed together with excess air so that carbon monoxide and hydrocarbons are oxidized into carbon dioxide and water.

LASER DISC

2664

Acid rain; formation and deposition

SCIENCE BACKGROUND

In Donora, Pennsylvania, a "death smog" occurred in 1948. Twenty people died, and almost half the people in the region had some type of eye, nose, or throat irritation. Many people suffered from headaches, nausea, coughing, and breathing difficulties. In London, England, heavy air pollution caused 4000 deaths in 1952 and 1000 deaths in 1956. In December 1973, the air pollution in Belgium's Meuse Valley caused 60 deaths and serious illness in 6000 people.

ONGOING ASSESSMENT
▼ **ASK YOURSELF**

An acidic lake is clear because the organisms that produce some of the cloudiness in normal lakes cannot live there.

smog and 90 percent of the carbon monoxide in the air pours out of automotive tailpipes. Nitrogen oxides are produced by internal combustion engines. When fuel is burned at high temperatures, oxygen reacts with nitrogen in the air as well as with the fuel, producing nitrogen oxides. Making engines smaller and less powerful is one way to reduce nitrogen oxides. But there is something else that is in place now that works, the pollution-control systems on most automobiles.

In addition, in November 1992, new regulations required specific urban areas to implement tough emission testing for cars and light trucks. The simple tailpipe gauge that measures exhaust while engines idle has been replaced with a new high-tech treadmill device. It collects exhaust while the car idles, accelerates, and brakes. The information is fed into computerized equipment that is so sensitive that millions of cars that formerly passed emissions tests will now probably fail. The vehicles that fail must have repairs made and attempt the test again. The new regulations may help reduce pollution from vehicles by 30 percent in many cities.

Figure 7–8. Most passenger vehicles sold in the United States have a catalytic converter—a device used to remove pollutants from automotive exhaust. Catalytic converters, such as this one, help reduce carbon monoxide, nitrogen oxides, and hydrocarbons.

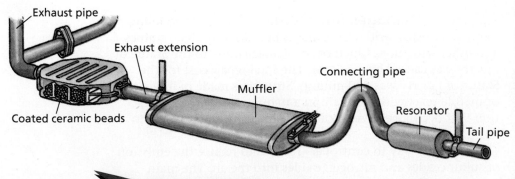

Exhaust pipe

Exhaust extension

Connecting pipe

Muffler

Resonator

Coated ceramic beads

Tail pipe

▼ **ASK YOURSELF**
Why is the water of an acidic lake clear?

GUIDED PRACTICE

Ask the students to name some advantages and disadvantages of using nuclear energy. Have the students place their lists in their science portfolios. (Advantages: Nuclear energy does not add pollutants to the air and generates a large amount of energy from a small quantity of fuel. Disadvantages: Nuclear energy requires huge amounts of water to cool the reactors, creates thermal pollution if the hot water is dumped in rivers and oceans, and requires long periods of time for nuclear waste to become less radioactive.)

INDEPENDENT PRACTICE

Have the students provide written answers to the Section Review. In their journals, have them draw and label a diagram that explains the greenhouse effect.

EVALUATION

Discuss with the students which method they think would be better—reprocessing spent nuclear fuel rods or storing them in underground caverns. (Answers should reflect pros and cons of both methods.)

Nuclear Energy Is NOT Waste Free

A typical 1000-megawatt (MW) nuclear energy plant creates about 25 tons of spent, or used, fuel each year. This fuel is highly radioactive and can have a half-life of hundreds or even thousands of years. That means that the material will be dangerous for extremely long periods of time. Spent fuel and other radioactive products from nuclear reactions are called **nuclear waste.** These wastes are extremely dangerous to most life forms and, if not disposed of properly, can cause great environmental damage.

Even though nuclear energy has been used for decades, no permanent storage facility for nuclear wastes currently exists. The wastes from the first experiments in nuclear energy, which took place in 1919, are still in temporary storage. The nuclear energy industry identifies categories of nuclear wastes. The most critical storage problem relates to spent nuclear fuel.

Currently, spent fuel is stored temporarily in cooling ponds on the grounds of the nuclear power plants that produce it. In this case, "temporary" seems to mean several decades. Some engineers and scientists hypothesize that the spent fuel rods will eventually corrode and crack open, releasing radioactive waste into the environment. Also, the storage ponds at some older nuclear power plants are reaching their capacity. It is vitally important that we find new and improved methods of storing nuclear wastes.

A number of ways of dealing with spent fuel rods have been proposed. One way is to reprocess the rods to extract reusable uranium and plutonium. This reprocessing would lessen, but not eliminate, the radioactivity of the wastes. Disposing of the reprocessing wastes would still be a problem.

Another plan, which is receiving serious consideration, is to store wastes in underground caverns. The wastes would first be incorporated into solid cylinders of glass or ceramic. These cylinders would then be covered with stainless steel shells and

Figure 7–9. Spent fuel from nuclear reactions is currently stored in cooling ponds, as shown here. These ponds are located on the same site as the nuclear reactor.

Demonstration

Use dominoes to model a nuclear chain reaction. Demonstrate the function of control rods by setting small open boxes behind a few of the dominoes in the arrangement.

The same kind of camera that can show thermal pollution (a disadvantage in producing nuclear energy) can also detect heat loss in homes. Home energy consultants use these cameras and other special devices to help homeowners find out how best to conserve energy.

✧ Did You Know?

The first sustainable nuclear chain reaction occurred on December 2, 1942, under the west stands of Stagg Field at the University of Chicago. You may wish to encourage students to research the reason this nuclear reaction occurred.

RETEACHING

Have the students create a three-column chart in their journals with the headings: Ozone Layer, Global Warming, and Acid Rain. Have them list all the terms related to each heading under these subheadings: Causes, Effects, and Possible Solutions. Encourage the stu-

dents to compare their charts with classmates' charts and add to their chart, if necessary. Ask them to quiz one another on the basis of the information in their charts.

EXTENSION

Have the students use reference materials to research half-life. Ask them to relate their findings to the problem of nuclear waste storage. (Half-life is the amount of time required for half the nuclei in a specific quantity of nuclear waste to undergo radioactive decay in which it changes to a less harmful substance.)

CLOSURE

Cooperative Learning Have the students work in small groups to write original questions and answers for this section. Have each group ask its questions of another group.

ONGOING ASSESSMENT
ASK YOURSELF

Spent nuclear fuel rods are stored in cooling ponds on the grounds of the nuclear power plant that produces the waste.

SECTION 1 REVIEW AND APPLICATION

Reading Critically

1. Nitrogen and sulfur oxides are acidic, causing damage to lungs when inhaled and creating acid rain. Carbon dioxide is a greenhouse gas, perhaps causing long-term changes in the climate. Carbon monoxide is a poison.

2. Acid rain makes lakes and streams acidic and dissolves minerals, such as aluminum, that damage organisms.

Thinking Critically

3. Answers should point out that cleaning up pollution usually generates more pollution.

4. Answers should mention international relations and the interdependence of economic and ecological systems.

Figure 7–10. This diagram shows a planned underground facility for permanently storing nuclear waste.

placed along the hallways of underground caverns. When full, each hallway would be sealed off with rock and stone.

This plan may not be foolproof. Critics of the plan maintain that certain circumstances, such as earthquakes, could cause the release of radioactive materials into the environment. The critics also point out that the security of the storage area would have to be maintained for hundreds or thousands of years. Supporters of the plan claim that the chances of wastes escaping from a well-designed storage facility are very low compared with the chances of wastes escaping from temporary storage. As you can see, there is disagreement about how nuclear waste should be handled. Competent scientists and engineers are found on both sides of the technical arguments.

 ASK YOURSELF

Where and in what form is spent nuclear fuel currently stored?

SECTION 1 REVIEW AND APPLICATION

Reading Critically

1. What problems are caused by high concentrations of nitrogen oxides, sulfur oxides, carbon dioxide, and carbon monoxide in the air?

2. What are some ways acid rain damages the environment?

Thinking Critically

3. Provide background information that supports this statement: There are no easy solutions to the pollution problem, only difficult choices.

4. If the United States benefits from the processes that produce acid rain, why should the United States be concerned about forests in another country being damaged by that acid rain?

Objectives
● **Compare** the effect of regular tap water and of "acid rain" on seed germination.
● **Calculate** the germination rates for seeds under both conditions.
● **Determine** whether weed seeds or seeds from food plants are more likely to survive in an area of acid rain.

Pre-Lab
Ask the students to predict what they will find out from this Investigation.

Hints
To make the "acid rain" solution, mix 2 parts 3 M sulfuric acid with 1 part 6 M nitric acid in a very large container. Add water until the pH is 4. A measure of 100 mL of sulfuric acid/nitric acid mixture should make 10 000 mL of "acid rain" solution.

You might give each team "acid rain" samples at different pH levels. The class results can be used to graph results of pH versus germination rate.

In Step 3 of the procedure, tell students to label their dish with the kind of solution, kind of seeds, and date and time of the experiment.

INVESTIGATION

Determining the Effect of Acid Rain on Seeds

▶ **MATERIALS**
● safety goggles
● laboratory apron ● gloves
● petri dishes with lids ● paper towels
● solution to simulate acid rain ● seeds ● tape
● aluminum foil ● labels ● litmus paper

▼ PROCEDURE

CAUTION: Put on safety goggles, a laboratory apron, and gloves.

1. Fold a paper towel to cover the bottom of a petri dish. Then dampen the towel with "acid rain" solution.
2. Count out 25 seeds, and place them on the towel.
3. Cover the seeds with another folded paper towel that has been dampened with "acid rain" solution. Put the top on the petri dish, tape it shut, and wrap the dish with foil to protect the seeds from light.

4. Design a control for your experiment. Have your teacher OK your ideas before you proceed. Then set up the second petri dish.
5. Use litmus paper to determine the pH of the "acid rain" solution and the regular tap water, and record it in your journal.
6. Allow the petri dishes to sit undisturbed for the period of time recommended by your teacher. The precise time period will be determined by the kind of seeds you are using.

7. Unwrap and open the petri dishes. Count the seeds that have germinated and the seeds that have not in each dish.

▶ ANALYSES AND CONCLUSIONS
1. Calculate the germination rate for seeds under both conditions. To calculate the germination rate, divide the number of seeds that have germinated by the total number used. How do they compare?
2. Collect data from others in the class who used the same kinds of seeds you did. Average the results for plants germinated in "acid rain" and plants germinated in tap water. How do the average results compare to your results?

▶ APPLICATION
A farmer in your area complains to the county commission that her lettuce crop has suffered ever since the new coal-fired power plant was built. Based on your results from the germination experiment and your knowledge of pollution from coal-fired power plants, how likely is it that she is right?

✳ Discover More
Collect some weed seeds and test them as well as seeds purchased at a seed store. Based on your new research, are food plants or weeds more likely to survive in an area of acid rainfall?

▶ Analyses and Conclusions
1. The germination rate of seeds dampened with regular tap water should be greater than that of seeds dampened with the "acid rain" solution.

2. The average results should be about the same as the results based on one pair's experiment.

▶ Application
The burning of coal produces sulfur dioxide and nitric oxide. These combine with oxygen and water vapor to form weak sulfuric acid and nitric acid in the atmosphere and fall as acid rain. The lettuce was probably affected by acid rain.

✳ Discover More
Responses will depend on the seeds. The fact that weeds survive so well without any help from farmers seems to indicate that they are more likely to survive acid rain than are food plants.

Post-Lab
Ask volunteers how their original predictions differ from what they learned in this Investigation. Encourage the students to add to the summaries to make them as complete as possible.

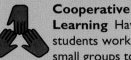

SECTION 2

Energy Conservation

Objectives

State three ways new technology can help people conserve energy, and give an example of each.

Evaluate the positive and negative effects of methods of conserving energy.

Fossil fuels are nonrenewable energy resources. Because coal and oil are also raw materials for drugs, plastics, and many other useful compounds, it would be unwise to burn them up completely. How can we use fossil fuels wisely? What are the options?

Spending Energy Wisely

Using a resource efficiently is called **conservation.** Energy conservation has several advantages. By conserving nonrenewable energy resources, existing supplies will last longer. Because any energy transformation creates waste heat, using less energy will produce less pollution. Overall, reducing the amount of energy used will reduce, but not eliminate, the negative impact of energy use on our environment.

There are only three ways to conserve energy—decrease activities that require energy, perform these activities more efficiently, or use different methods of producing and transforming energy that are more efficient. These options can involve new technology, new processes and procedures, and new behavior.

Figure 7–11. This chart shows the sources of energy in the United States.

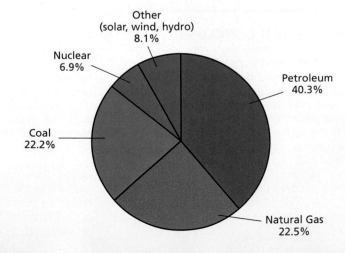

The Sources of Energy
Source: U.S. Dept. of Energy

Other (solar, wind, hydro) 8.1%
Nuclear 6.9%
Petroleum 40.3%
Coal 22.2%
Natural Gas 22.5%

● **Process Skill:** *Solving Problems/Making Decisions*

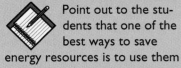

Point out to the students that one of the best ways to save energy resources is to use them wisely. Ask the students to brainstorm ways in which

energy can be used wisely and to write their ideas in their journals. (Plan trips in the car so that a number of errands can be accomplished in one trip; walk or ride a bicycle; use public transportation.)

● **Process Skills:** *Inferring, Analyzing*

In the past 20 years, automobiles have become smaller and lighter in weight, homes are more thoroughly insulated, and advertisements for both cars and homes include phrases such as *fuel efficient* and *energy efficient.* Discuss with the students

what these trends reveal about energy conservation. (People are becoming more aware of the costs of energy, and they realize they can save money by conserving energy.)

Conservation in the Workplace

In industrial settings, using new technology is a common way to increase efficiency. New technology is often applied in two ways—using new materials and using new processes.

Chemists and engineers are constantly trying to produce new materials with useful properties. Some of these materials help to conserve energy by increasing efficiency. For example, synthetic materials have replaced metal panels in some automobile bodies. These panels are much lighter than comparable metal panels. Automobile bodies made with these panels weigh less and can be equipped with smaller, more energy-efficient engines with no loss of performance.

Using new processes can also result in energy savings. A simple example of this can be found in the trucking industry. Many large trucks today are equipped with wind deflectors on their cabs, as shown in the photograph on this page. These simple wind deflectors reduce resistance to air flow around the trucks' trailers. This reduction in resistance increases the fuel mileage of the trucks.

Only a few decades ago, no one had even thought of putting wind deflectors on trucks. Simple ideas that can make a difference are often overlooked. The idea of the wind deflector, it is rumored, was thought up by a high school student for a science project. This idea resulted in greater fuel efficiency in the trucking industry.

Figure 7–12. Wind deflectors on trucks (left) save considerable amounts of fuel. Why do you think some drivers have substituted this netting device (right) for the metal tailgate? ①

How does a steam turbine work?

Process Skills:
Experimenting, Comparing

Grouping: Groups of 3 or 4

Hints

Be sure to have a supply of bent glass nozzles. The end of each nozzle should be constricted to produce high-pressure steam.

Be sure the students understand that this Activity is not truly an example of cogeneration, since the heat is not a byproduct of a manufacturing process.

▶ **Application**

The similarity is that the pinwheel turns because steam is blown on it. Some differences include size, efficiency, enclosure of the turbine to channel the steam over the wheel, number of vanes, and attachments of the turbine to a generator.

PORTFOLIO ASSESSMENT Have the students place their observations and answers from this Activity in their science portfolios.

◇ Did You Know?

Co-generation is also effectively used at sewage treatment plants. Methane gas produced by sewage is used as fuel to heat the plant.

TEACHING STRATEGIES, continued

● **Process Skills:**
Communicating, Expressing Ideas Effectively

Point out that while robots sometimes put people out of work, other people might find new jobs building and maintaining robots. Have the students write in their journals what they would do if they worked in a factory that was planning to introduce robots to do the work they currently perform.

Another example of a simple process that increases efficiency is something called *cogeneration*. This refers to the use of the heat produced in manufacturing to boil water to use in a steam generator. Electricity is generated along with the manufactured product, so it is *co*generated. By producing two useful forms of energy from one fuel source, cogeneration results in fuel savings. You can explore co-generation by doing the next activity.

ACTIVITY

How does a steam turbine work?

MATERIALS

safety goggles; laboratory apron; ring stand (2); utility clamp (2); ring with wire gauze; flask, 250 mL; boiling chip; two-hole stopper to fit flask with bent glass nozzle and thin glass tube (safety valve); Bunsen burner; toy pinwheel

CAUTION: Put on safety goggles and a laboratory apron. Keep them on during this activity.

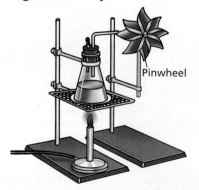
Pinwheel

PROCEDURE

1. Assemble the equipment as shown. Fill the flask about half full of water, and add the boiling chip.
2. Insert the stopper with the nozzle and valve assembly into the flask so that the nozzle points toward your pinwheel.
3. Light the burner and heat the water.
4. Once the water starts boiling, adjust the location of the nozzle so that the turbine (pinwheel) starts to spin.
5. Observe the interaction of steam and turbine as you make the turbine spin. Carefully feel the air near the turbine on the side away from the nozzle.

APPLICATION

How is your steam turbine similar to the steam turbines used in electric-energy plants? How is it different? What could you do to make your steam turbine more efficient?

Robots An example of a technology that involves both new materials and new processes is robotics. Robots are computer-controlled tools. Today, most robots are used in factories to assemble products. Robots can do jobs that are dangerous to humans because they are unaffected by the heat, fumes, and blinding arcs produced in welding. Also, robots do not need

Figure 7–13. Robots and other computer-controlled devices can improve the efficiency of manufacturing. Here, robots are being used to weld car frames.

Technological advancements have made many contributions to energy conservation at home. New refrigerators are nearly twice as energy efficient as models manufactured 10 years ago. Programmable, "set-back" thermostats help reduce heating and cooling costs when no one is home. Many dishwashers now have energy-saving features, such as an air dry cycle. Low-flow shower heads use less water but provide a shower spray equivalent to less efficient shower heads.

good lighting or a comfortable environment. The use of robots results in energy saved on lighting, heating, and air conditioning. A disadvantage of using robots is that people who once did the jobs of the robots have to find work elsewhere. The introduction of new technologies often produces a shifting of jobs.

 ASK YOURSELF

In what ways can technology help conserve energy?

Conservation at Home

As stated earlier, the only ways to conserve energy are to limit activities that require energy, to do these activities more efficiently, or to use different methods of producing and transforming energy that are more efficient. In industry, increasing efficiency is the goal, and this is most often done with new processes and technology. In the home, even small changes in behavior can result in large increases in efficiency. For example, an average dishwasher uses about 25 to 30 L of hot water per load. It has been estimated that if every family that owned a

SECTION 2　**191** ◀

GUIDED PRACTICE

Ask the students to determine whether a car that burns pure gasoline costing $0.33/L and gets 8.4 km/L is more or less efficient than a car that burns gasohol costing $0.28/L and gets 6.3 km/L. (The car burning gasoline gets 2.5 more kilometers per dollar than the one burning gasohol.)

INDEPENDENT PRACTICE

Have the students provide written answers to the Section Review. In their journals, have the students list the behaviors in Table 7–1 (p. 193) that they think their family would benefit from.

EVALUATION

Have the students describe and illustrate how cars have become more energy efficient in the past decade. (Answers should reflect an understanding of energy efficiency, such as increased fuel efficiency due to design aspects and new technologies.)

DISCOVER BY *Writing*

Cooperative Learning After the students have completed the Discover by Writing, have them work in small groups to share some of the energy uses from their lists. Have each group compile a list of the best ways they could change their behaviors to decrease energy use. Evaluate their choices for appropriateness. Encourage the students to keep their lists in their science portfolios.

Figure 7–14. If all Americans turned down their heat by six degrees, we would save 500 000 barrels of oil each day! What benefits could this have to the environment?

dishwasher did one less load of dishes a week, the energy savings would equal the energy necessary to heat about 150 000 homes during the winter.

It is sometimes a struggle to change your behavior. Many people don't even realize what their behavior is. To become aware of your behavior, try the next activity.

DISCOVER BY *Writing*

Write in your journal all the applications of energy you make during the day. Beside each use, identify the energy conversion. For example: Electric iron—electric energy to heat energy. Make a chart of your energy use. Next to each use, mark whether the use is necessary or a convenience. Next to that, suggest at least one way you could change your behavior to decrease the energy used for that activity.

Table 7–1 lists some of the ways that you and your family can decrease your personal energy use and help conserve energy resources. Many suggestions for conserving energy are based on common sense. For example, turning off the television when you are not watching it is a matter of common sense. So is turning out the light when you leave the room, or keeping the doors and windows closed during cold weather. Remember, energy costs money. The more energy you save, the more money you save, and the environment benefits as well.

RETEACHING

PORTFOLIO ASSESSMENT Have the students illustrate Table 7–1 with pictures from magazines or by means of simple figures (or comic strip drawings) that show ways to save energy. Have the students place their sketches in their science portfolios.

EXTENSION

Have the students visit an appliance store, choose one category of appliance, and record the information shown on the yellow "Energy Guide" tag (the estimated cost of using that appliance for one year) on each appliance. Have them make a chart listing several appliances in one category from highest to lowest Energy Guide ratings. Ask them which features might cause the Energy Guide rating to worsen for the category of appliance they chose.

CLOSURE

Cooperative Learning Have the students work in small groups to write an outline of this section. Ask each group to display its outline or write it on the board.

Table 7-1 Some Ways to Conserve Energy

1. In the winter, turn down the heat and put on a sweater to help keep warm. In the summer, use the air conditioner less often.
2. Take short, warm showers instead of long, hot showers. Remember, heating that water takes a lot of energy.
3. Turn off lights and appliances when you are not using them.
4. Avoid letting the ice build up in the freezer compartment of your refrigerator, and keep the heat exchanger coils on the back of the refrigerator clean and free of dust. Both ice and dust act as insulators and make it harder for the refrigerator to transfer heat.
5. Use washers, dryers, and dishwashers only when you have a full load. They use about the same amount of energy whether they are full or not. Hang clothes out to dry on a clothesline when the weather permits.
6. Walk or ride your bike when you need to go somewhere. Remember that cars are among the major energy users in the United States.

ASK YOURSELF

What are several ways of conserving energy?

SECTION 2 REVIEW AND APPLICATION

Reading Critically

1. How can energy conservation help reduce pollution?
2. List three ways new technology can help decrease the amount of energy used in industry, and give a short example of each.
3. How do robots help conserve energy?

Thinking Critically

4. How would putting plastic over the windows and using curtains properly help to decrease home energy consumption during the winter months?
5. Table 7–1 lists six behaviors around the home that will help your family conserve energy. List two others things you could do in your home, and explain how they will help conserve energy.

ONGOING ASSESSMENT
ASK YOURSELF

Answers could include the ways suggested in Table 7–1, among others.

SECTION 2 REVIEW AND APPLICATION

Reading Critically

1. Since using energy creates pollution, conservation decreases pollution by creating less pollution in the first place.

2. When cars are made with new light-weight materials, they operate more efficiently. Using wind deflectors on trucks increases the fuel efficiency of the trucking industry. New materials are used in robots, allowing them to perform more efficiently than humans. Accept all reasonable responses.

3. Robots can work continuously in cold or hot conditions and in poor light. Energy is conserved because little energy must be used for heating, air conditioning, and lighting. Since they can work continuously, robots can also produce more products in less time than can human workers.

Thinking Critically

4. Covering windows hinders convection by stopping the motion of air and blocking radiation. The home retains heat better.

5. Examples include using a hand tool instead of its electrical counterpart and doing several errands at once instead of making separate trips. Accept all reasonable responses.

FOCUS

In Section 3 the students learn about the cost of energy, including the hidden costs. The reasons for the importance of energy sources in international politics are also discussed.

MOTIVATING ACTIVITY

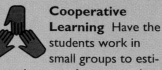

Cooperative Learning Have the students work in small groups to estimate the cost of energy in their home by assigning percentages of the whole monthly energy cost to any or all of these that they have: refrigerator, heating, freezer, water heater, color TV, lighting, dishwasher, dryer, microwave, range, other. Have them compare their estimates with those of other groups and tally the estimates for each use to find the class average.

PROCESS SKILLS
- Classifying/Ordering
- Measuring • Interpreting Data

POSITIVE ATTITUDES
- Caring for the environment
- Openness to new ideas

TERMS
none

PRINT MEDIA
Global Garbage: Exporting Trash and Toxic Waste by Kathlyn Gay (see p. 67b)

ELECTRONIC MEDIA
Energy Seekers Coronet Film and Video (see p. 67c)

SCIENCE DISCOVERY
- Earth Day 1990
- Recycling center

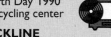

BLACKLINE MASTERS
Laboratory Investigation 7.2
Study and Review Guide

MULTICULTURAL CONNECTION

The neem tree provides firewood and shade in India and many other countries. In addition to these energy benefits, its twigs prevent tooth decay and gum disease. Chemicals from the tree kill athlete's foot and block the growth of ringworm. Its seeds produce a safe pesticide.

The neem tree also offers several advantages to the ecosystem. It is so hardy that it grows even in drought-stricken parts of the world. If planted on a large scale, it could improve declining ecosystems and prevent erosion.

SECTION 3

Energy Decision Making

Objectives

Evaluate the economic factors to be considered in energy decision making.

Discuss three issues in international politics that are related to energy policy.

How much does energy cost? This is a very important question, but it is not a simple one to answer. You have probably heard about new energy systems that are designed to reduce energy use. In many situations, these new systems do use less energy and do save money spent on fuel. However, the benefits are not the total story.

Energy Systems

Jazz, Rock, Country, Rap, the Blues, and Classical—these are all different types of music. Whenever you use a stereo system to listen to your favorite music, you are using electric energy. However, this energy is the end product of many processes that also require energy. In other words, energy was required to make energy available to your stereo.

Energy in the form of coal may have been used at the electric generating plant to drive the turbine generators that produced the electricity. Energy was used to mine the coal and transport it to the energy plant. Energy was also used in manufacturing the equipment necessary to harness and control all the other energy. For example, energy was used to create equipment such as coal-mining machinery, turbine generators, steam boilers, and electric lines. The total energy system includes much more energy than may be obvious when you use the electricity that is finally produced.

Figure 7–15. Digging machine at the Peabody coal strip mine in Kankakee, Illinois

● **Process Skills:** *Inferring, Evaluating*

Discuss with the students some of the hidden costs in producing and using energy. (Make a list on the board as the students offer suggestions. Examples include the cost of cleaning up oil spills, reclaiming land after strip mining, and installing catalytic converters on automobiles.)

● **Process Skills:** *Communicating, Generating Ideas*

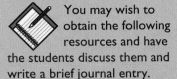 You may wish to obtain the following resources and have the students discuss them and write a brief journal entry.

Brown, Lester R., Christopher Flavin, and Hal Kane. *Vital Signs*

1992: The Trends That Are Shaping Our Future. Norton, 1992. This book is an overview of environmental trends surfacing all over the world.

Luoma, Jon R. "Bungee Jumping in Brazil." *Discover* 14 (January 1993): 78 This article discusses the role of the EPA at the June 1992 Earth Summit.

By thinking of energy use as a system, you can compare the energy output of a particular system to its input and calculate the net energy. For example, to determine the net energy produced by petroleum, you must first determine its energy input. This includes the energy used to find the petroleum, drill for it, pump it into tankers, transport it to refineries, refine it, transport it to users, and deal with its wastes. If a particular energy system receives support from taxes, that must also be considered an energy input, since money also requires energy to be produced. This combined energy input must then be subtracted from the total energy produced by burning the petroleum. In most energy systems, net output is less than 20 percent of the combined inputs.

Any thought about using or saving energy can be much more complex than it may first appear. Much information must be considered before sometimes tough choices must be made.

 ASK YOURSELF

Why is energy use considered a system?

Energy Economics

Before you can determine the benefits of a new energy system, you must figure the total price of a new system and compare the energy savings with the amount of energy used by the old system.

DISCOVER BY Calculating

Suppose your family is considering spending $2,000 on a new solar water heater. The solar hot-water system is designed to cut your previous fuel charge for hot water in half. If your family spends $40 per month on hot water, the new system will reduce this bill to $20 per month. Is this system a wise investment? Explain your answer. How long will it take for the system to pay for itself?

Figure 7–16. A solar hot-water system can help cut energy costs. However, it is also an expensive investment.

Another way of looking at the investment is from a break-even standpoint. At a savings of $240 a year, it will take a little more than eight years for the system to pay for itself, or for you to break even. If your family is planning to live in the house for only a couple of years, the new system will not be worth the investment, unless it adds significant value to the house.

The economics of energy also include some hidden costs. Fossil fuels are relatively inexpensive fuels for producing energy. However, the acid rain caused by fossil fuels has damaged food crops and killed fish. The reduced food supply has caused increases in the prices of certain foods. In addition, many people have required additional medical care as a result of pollution from fossil fuels, driving up the cost of insurance. Paying for this care and the increased insurance premiums are also hidden costs of using fossil fuels.

Figure 7–17. Fossil fuels have hidden costs. The pollution from using fossil fuels can cause health problems that require expensive medical attention.

DISCOVER BY *Writing*

 Imagine that you have the authority to select the type of energy plant to be built in your community. What would be your choice and why? Look in newspapers and magazines for examples of effective advertisements. Then write three advertisements that would help convince the citizens in your community that your choice is the best one.

ASK YOURSELF
Name two hidden costs of the use of fossil fuels.

RETEACHING

Have the students write a letter to a person in Canada telling him or her why they hope that the acid rain problem will be resolved soon. The students should also describe efforts being made to solve the problem. Ask the students to place their letters in their portfolios.

EXTENSION

Have the students research the results of the 1992 Earth Summit in Rio de Janeiro and the official position of the United States at the summit. Encourage them to report and discuss their findings with classmates.

CLOSURE

Have the students write a summary in their journals of the section. Encourage them to share their summaries with their classmates.

Energy Politics

Energy plays an important role in domestic and international politics. One political and economic alliance, called the *Organization for Petroleum Exporting Countries,* or *OPEC,* was formed to help member countries obtain control of their oil resources and to control oil prices. From their point of view, petroleum is their most valuable natural resource. The countries feel that they need to control the production and price of their oil so they can adequately prepare for the future, when their petroleum will be used up. Countries that import oil, on the other hand, feel they are overcharged and captive consumers.

Petroleum is not the only energy resource that affects international politics. Other international issues involve the control of nuclear energy and acid rain. Yet another international issue is the control of carbon dioxide emissions, which may alter climates and produce the greenhouse effect and global warming. These issues can only be confronted with international cooperation.

A step toward the goal of international cooperation was taken in 1992, with the first Earth Summit held in Rio de Janeiro, Brazil. That meeting, designed to address global environmental problems, was attended by a collection of world leaders. Among the issues discussed at the first Earth Summit was the control of carbon dioxide emissions to decrease global warming.

Figure 7–18. The United Nations Conference on Environment and Development—the Earth Summit—was held in June 1992 in Rio de Janeiro.

ASK YOURSELF

Why are acid rain and the control of carbon dioxide emissions considered international issues?

SECTION 3   REVIEW AND APPLICATION

Reading Critically

1. Why was OPEC formed?
2. Describe how the medical costs of treating illnesses resulting from pollution can be considered hidden costs of energy use.

Thinking Critically

3. Develop a plan for diversifying United States energy sources so that the United States no longer has to depend on imported petroleum.
4. Present two arguments in favor of and two arguments against energy independence for the United States.

 LASER DISC

2615

Earth Day 1990

2652

Recycling center

ONGOING ASSESSMENT
ASK YOURSELF

The atmosphere has no boundaries. The acids and CO_2 emissions that are produced in one country are carried in the atmosphere to other countries.

SECTION 3 REVIEW AND APPLICATION

Reading Critically

1. OPEC was formed to help member countries obtain control of their own resources and to control oil prices.

2. Since these costs are directly related to the production and use of energy, they can be considered in calculating the overall cost of the energy.

Thinking Critically

3. Answers should show an understanding of energy sources and their impact on the environment and society. Accept all reasonable responses.

4. By being energy independent, the United States would not be subject to price increases demanded by energy exporting nations and would be improving its own economy by reducing the trade deficit. On the negative side, the United States would be using its own natural resources at a much faster rate, and the additional production of energy would increase pollution.

SECTION 3   **197**

Objectives

● **Estimate** the energy cost per year in a home.
● **Use** information gathered to suggest ways to reduce the amount spent on energy in a year.

Discussion

The data in the table has been compiled and converted to consistent units from a variety of sources. Recent technological advances, such as solid-state electronics and computer-enhanced efficiency, might have lowered the energy use for some appliances below the levels indicated. Ask the students how the use of solar heaters might affect the energy bill. (The bill for heating the house and for hot water might be lowered, but the cost of the solar heating system must be taken into account.)

PERFORMANCE ASSESSMENT

Ask each student to share a real-life example of energy saving. Each student should describe a way to decrease energy consumption. The students should mention how much energy would be saved by carrying out their plan. Evaluate their ways to save energy for feasibility.

Application

Accept reasonable solutions such as using appliances less often or in a more efficient manner. Check students calculations to ensure their accuracy.

* Using What You Have Learned

1. Students' results will vary depending on their family's usage and geographic location. Accept reasonable answers.

2. Students' answers regarding which appliance uses the most energy and least energy in a year will vary according to the family's usage of appliances. Students' calculations should be based on information available in the table.

PORTFOLIO ASSESSMENT

Have the students place their charts, observations, and answers from this Skill in their science portfolios.

SKILL Collecting and Organizing Information

▶ MATERIALS
● paper ● pencil

▼ PROCEDURE

1. Make a chart to collect data on your home electric-energy use. Organize the chart by appliances found in your home. Your chart should also have at least three columns: one column for data, one column for calculations, and one column for the products of your calculations.
2. For a week, survey how often and how long your family uses lights and appliances.
3. From the results of your survey and from the table, calculate how many watts of power are used by each appliance in the course of a year. Show your calculations in your chart.
4. Add the individual values to determine the total power in watts your family uses in a year.
5. Convert your energy total to kilowatt-hours. A kilowatt-hour is 1000 watts, or 1 kilowatt, of energy used for 1 hour. To convert kilowatts to kilowatt-hours, divide the number of kilowatts by 3600 (the number of seconds in an hour).

TABLE 1: ENERGY CONSUMPTION FOR COMMON APPLIANCES

Appliance	Energy Consumption
Automatic dishwasher	4.3 kWh/use
Clothes dryer	3.0 kWh/use
Coffee maker	8.9 kWh/month
Dishwasher with no dry cycle	3.8 kWh/use
Electric clock	1.4 kWh/month
Electric range	
Manually cleaned oven	85 kWh/month
Self-cleaning oven	100 kWh/month
Freezer	
Chest-type	110 kWh/month
Upright, frost-free	210 kWh/month
Hand iron	0.83 kWh/hr
Hot shower (5 minutes)	1.5 kWh/use
Portable electric heater	1.3 kWh/hr
Radio (3 hours/day)	0.08 kWh/hr
Refrigerator	150 kWh/month
Room air conditioner	0.9 kWh/hr
Stereo (3 hours/day)	0.12 kWh/hr
Television (6 hours/day)	
Color	0.21 kWh/hr
Black-and-white	0.05 kWh/hr
Toaster	3.3 kWh/month
Vacuum cleaner	3.9 kWh/month
Warm bath (tub half-full)	1.3 kWh/use
Washing machine	
Hot wash, warm rinse	7.8 kWh/use
Hot wash, cold rinse	6.3 kWh/use
Warm wash, warm rinse	5.5 kWh/use
Warm wash, cold rinse	2.8 kWh/use

▶ APPLICATION

Using your survey, propose ways your family could decrease its energy consumption. In each case, calculate the amount of energy and money that would be saved in a year.

※ Using What You Have Learned

1. Multiply the number of kilowatt-hours in your estimate by the cost of a kilowatt-hour of energy. The result is the estimated cost of the electric energy your family uses in a year.
2. Which appliance in your home uses the most energy in the course of a year? the least?

The Big Idea—Energy

Lead the students to an understanding of the importance of energy to their lives. Help them to realize that although energy production and consumption produce pollution, there are ways to minimize this pollution.

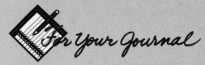

For Your Journal

Students' ideas should reflect an understanding that conserving energy and reducing the use of fossil fuels will decrease the amount of pollution caused by energy production. You might have the students share their new ideas with the rest of the class.

CONNECTING IDEAS

● The students might respond that automobiles produce pollution by burning fuel and warming the environment directly with the heat given off and indirectly by a greenhouse mechanism. All energy use produces pollution and wastes.

● The students should realize that energy is required to produce the machinery that mines the oil that will provide the energy needed to run factories and to produce electricity.

● The students should make the connection with acid rain as an energy pollutant. Acid rain has contributed to the death of fish.

● The students might respond that energy is used to keep the lights, heating, or air conditioning on. However, they should also note that energy is being wasted when no one is in the room and the door is left open.

CHAPTER 7 HIGHLIGHTS

The Big Idea

Energy may be the most important topic you will ever study. Everything you do requires energy, and energy is central to civilization as we know it. Energy production and consumption produces pollution. Even if we solve problems like the greenhouse effect and acid rain, we may never escape creating pollution with our energy activities. The best we can do is to reduce the pollution to a minimum by using alternative types of energy, more efficient technology, and changing our behavior. Solving energy problems will always be politically and economically difficult because energy production and consumption make up a very complex system of interactions.

For Your Journal

Review the questions you answered in your journal before you read the chapter. Then add an entry that shows what you have learned from the chapter. Include ideas about the relationship between energy and the environment.

Connecting Ideas

Below are a series of illustrations relating to this chapter. In your journal list and explain the concepts you can find illustrated.

ANSWERS

Understanding Vocabulary

1. (a) Each is a type of pollution. Acid rain is formed when air pollutants combine with oxygen and water vapor in the atmosphere to form weak sulfuric and nitric acids. Smog is produced when nitrogen oxides and partially burned hydrocarbons in the form of sulfur oxides react with sunlight. Nuclear waste is the spent fuel and other radioactive products from nuclear reactors.

(b) The greenhouse effect is caused when carbon dioxide in Earth's atmosphere acts like a glass globe surrounding the planet. The carbon dioxide allows sunlight to pass inward to Earth but reduces the flow of heat energy outward to space. The greenhouse effect might result in an increase in the average temperature of Earth's atmosphere—a condition called global warming. However, using resources carefully and efficiently to minimize pollution, such as carbon dioxide, is one way to practice conservation.

Understanding Concepts

Multiple Choice

2. a

3. b

4. d

5. c

Short Answer

6. To find the net energy produced by petroleum, you have to include all the energy used to find the petroleum, drill for it, pump it into tankers, transport it to refineries, refine it, transport it to users, and deal with wastes. This combined energy input must be subtracted from the total energy produced by burning the petroleum.

7. 4 years ($120 divided by $30 per year)

Interpreting Graphics

8. People are consuming less energy today than they were ten years ago. One pattern students might mention is the decrease in use of nonrenewable energy sources. It appears that the use of renewable energy sources will increase, while the use of nonrenewable energy sources will decrease.

CHAPTER 7
REVIEW

Understanding Vocabulary

1. Explain how the terms in each set are related.
  a) acid rain (182), smog (178), nuclear waste (185)
  b) greenhouse effect (179), global warming (180), conservation (188)

Understanding Concepts

MULTIPLE CHOICE

2. Which of the following is not a way to conserve energy?
  a) Wash and dry small loads of laundry more often.
  b) Walk or ride your bike when you need to go someplace.
  c) Take short, warm showers instead of long, hot showers.
  d) Turn off lights and appliances when you are not using them.

3. The ozone layer is important because it
  a) increases global warming.
  b) protects Earth from ultraviolet radiation.
  c) protects Earth from smog.
  d) increases the greenhouse effect.

4. An air pollutant that is not the direct result of burning fossil fuels is
  a) carbon dioxide.
  b) sulfur dioxide.
  c) nitrogen oxides.
  d) chlorofluorocarbons (CFCs).

5. Which of the following is not the result of burning fossil fuels?
  a) greenhouse effect
  b) acid rain
  c) nuclear waste
  d) smog

SHORT ANSWER

6. Using petroleum production as an example, explain how the net output of energy is generally much less than the combined energy input.

7. An energy-efficient air conditioner costs $120 more than another model, but is $30 a year cheaper to run. How many years would it take before your savings equaled the additional cost?

Interpreting Graphics

8. This graph shows energy consumption by major sources. What patterns of energy use do you observe? What predictions about future energy use can you make? Explain your predictions.

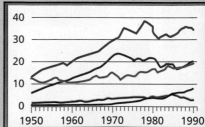

Energy Consumption* 1950–1990
(quadrillion Btu)

*One British thermal unit (Btu) = 1055 joules

Reviewing Themes

9. *Energy*
Earth radiates heat energy into space as a way of getting rid of waste heat. Explain how the greenhouse effect interferes with this process.

10. *Environmental Interactions*
Chlorofluorocarbons have been linked to the destruction of Earth's ozone layer. Use the release of Freon into the atmosphere as an example of how this interaction is harmful.

Thinking Critically

11. Energy has become increasingly important in global politics. Which topics related to energy might you expect to be most important during future international negotiations? Explain your answer.

12. If Earth's average temperature rose a few degrees, less energy would be needed to heat homes. Why, then, is global warming not considered beneficial to the energy problem?

13. A "fuel-saver" thermostat can be programmed to set back the heat or air conditioning for periods of time a homeowner chooses. To be most effective, the time periods must be longer than two hours. Based on this information, when would it be most practical to set back the heat or air conditioning?

14. The problems of the environment are global in aspect. Explain why international cooperation is necessary to solve these problems.

15. For many industrial tasks, robots are reportedly more efficient than humans in assembling a simple product. Offer pros and cons that should be considered when introducing more robots into industry.

Discovery Through Reading

Pack, Janet. *Fueling the Future.* Children's Press, 1992. This illustrated book presents the relationships between issues pertaining to energy and the environment.

Science PARADE

Science PARADE

Background Information

This *Science Applications* is about the part physics plays in circus acts. Point out to the students that anyone who is involved in sports knows how important it is to keep his or her muscles strong. Then discuss with the students the forces a trapeze artist must exert as he or she goes through a routine. Probably most students have swung on a swing at some time and can recall the back-and-forth, up-and-down motion and the split second of no motion.

Discussion

You might go through the list of physics concepts in the last paragraph of *Science Applications* and have the students suggest how each concept is displayed in the photographs.

Physics in the Circus

Acrobats, high wire artists, and trapeze performers all use physics in their acts.

The crowd grows quiet as Carla Hernandez mentally prepares herself for the quadruple somersault attempt. Her brother, Raul, is already in position to catch her. She jumps from the platform and begins to swing, back and forth, building up height and speed. Suddenly, she lets go! Her body is a blur as she flies through the air, spinning like a cannonball. She extends her arms, and Raul is there. He grabs her wrists, and the crowd goes wild.

● **Process Skills:**
Communicating, Expressing
Ideas Effectively

Have the students try to describe the forces they exert to keep a swing going and the forces that are exerted on their bodies as they keep the swing going. (You may wish to have

students consult Chapter 3 to help them with this answer. Students may respond that they use the force of their muscles to keep the swing going or they use an unbalanced force to start the swing moving and to keep it moving against air resistance and friction. Accept other reasonable responses.)

Journal Activity
After completing the Unit and reading and discussing the *Science Applications*, ask the students to consult the list of questions they recorded in their journals from the discussion at the beginning

of the Unit. (You might have the students turn back to the Unit Opener on pages 68–69.) Then lead a discussion during which students can verbalize what they have learned. You might also have the students record their new ideas in their journals.

Trapeze Tricks

As if by some kind of sixth sense, trapeze artists always seem to know just where the trapeze will be as they fly through the air. A physicist, of course, could calculate where the trapeze would be using the laws of motion. However, by the time the calculations were complete, the trapeze would have moved and the physicist would have fallen into the net below.

When they perform, trapeze artists appear to be flying free, making their bodies move at will. What is really happening is that they are pulling against the force of gravity to do their tricks. At the same time, gravity is pulling on the center of their bodies. They practice pushing and pulling their bodies against gravity. The different parts of their bodies also push and pull on each other using muscle power. Whether they realize it or not, trapeze artists take advantage of the laws of physics.

It's Not the Fall . . .

Trapeze artists know that even if they fall into a net, they probably won't be hurt. The physics of this is not difficult to understand. When a performer falls into a net, the net sags. This

slows the fall gradually, instead of stopping it quickly. If the stop were sudden, the force applied to the artist's body would be tremendous, and he or she could be seriously hurt. But with a long, drawn out stop, the force applied to the artist is small. As long as the net is properly placed below the performers and they land on the broad parts of their bodies, falls are not especially dangerous. It is important to know, however, that trapeze artists must stay in superb physical condition. Strains on joints and muscles due to the rapidly changing accelerations of the act could hurt a performer much more than any fall.

Obey the Laws

Once you learn the physics of speed, acceleration, momentum, inertia, work, and mechanical advantage, understanding circus tricks becomes much easier. Figuring out how circus performers take advantage of the laws of physics makes watching them even more fun. Knowing the science behind a trick may even give you insight into just how rigorously circus performers must train in order to accomplish their amazing feats successfully. ◆

Physics helps the "human cannonball" determine where he will land.

Even clowns use physics in their stunts.

Background Information

The Magnum XL-200 opened in May 1989. It is not truly the tallest roller coaster in the world, but it does have the greatest vertical drop—more than 60 m, during which a speed of more than 116 km/h is reached. Cedar Point has nine roller coasters, making its total more than that of any other amusement theme park.

Discussion

● **Process Skills:** *Applying, Interpreting Data*

Ask students to relate what they learned about force, acceleration, and motion in Chapter 3 to explain how a roller coaster works. Then discuss with the students how hill height affects potential energy. (The higher the hill, the greater the potential energy.)

READ ABOUT IT!

You Are There:

RIDING A ROLLER COASTER

Thrills, Chills, and Fun Physics
by Hugh Westrup from *Current Science*

How much of a daredevil are you? Daredevil enough to ride the Magnum XL-200, the world's tallest roller coaster? I did, and now that my hands have stopped shaking, I can tell you all about it.

I rode the Magnum to learn about the science behind the sensations–the exhilaration and the extreme fear–of a roller coaster ride. I wanted to find out how roller coasters achieve some of their heart-stopping effects and still remain safe.

One of the first things I learned was that we are now living in a golden age of roller coasters. Every year several new coasters spring up across the United States, each one attempting to break the previous year's records for the tallest, fastest, and steepest coaster. With names like King Cobra, Great American Scream Machine, and Shock Wave, these new megacoasters are also breaking records for thrills and chills.

Getting to the Point

Surely no roller coaster is scarier than the Magnum XL-200, a hump-backed monster of crosshatched tubular steel that looms over Cedar Point Park in Sandusky, Ohio, on the shore of Lake Erie. So wicked is the Magnum that its designer, Ron Toomer, president of Arrow Dynamics, Inc., refuses to ride on it!

Having braved the Magnum, I'm not sure which was more frightening: the ride itself or standing in line to reach it! The Magnum is so popular that waiting for a ride takes up to an hour. All the while, the blood-curdling shrieks of riders on the Magnum mingle in the air with a continuous recorded message warning away anyone with a heart, back, or blood-pressure problem. Some fun!

But all too soon the wait was over and I found myself quickly ushered into a seat on the 36-passenger train. I took a long, deep breath as a ride operator lowered a padded steel bar over my head and into my lap. An instant later, the train pulled out as the crowd behind me burst into a round of applause. And away we went!

On the Fright Track

Some people who have ridden the Magnum say that the first 30 seconds of the ride may be the most agonizing ones of your life. As the little train leaves the station, it latches on to a moving chain that pulls you slowly up the first hill. On my trip, my riding companion, a Cedar Park tour guide, motioned toward the magnificent view of Lake Erie

● **Process Skills:** *Generating Ideas, Applying*

Remind the students that in Chapter 5 they learned about energy transformation. Ask the students to list the energy

transformation(s) that occur during a roller coaster ride. (Responses should show an understanding of the second law of thermodynamics.)

unfolding before us. She said that on a clear day you can see the skyscrapers of Cleveland, Ohio, 60 miles (96 kilometers) away.

But who could be in a mood for sight-seeing at a time like this? The only sight I could focus on was the alarming gap growing between the train and the ground as we reached 12 stories, then 13, 14, 15 . . . . Was it too late to turn back?

As I learned later, most roller coasters begin with their tallest hill so as to supply each train with a large reserve of potential energy. *Potential energy* is energy that is stored in an object. For the train to achieve this potential energy, a motorized chain pulls the vehicle uphill. After the train reaches the top and starts downward, the potential energy changes to *kinetic energy*, or energy of motion. This kinetic energy provides much of the fuel for the effects that send chills up and down the spine.

As our train rounded the crest of the hill at a height of 20 stories (205 feet, or 62 meters), for a half-second we seemed to hang in mid-air, not going up or down. A small weather-vane at the top turned gently in the breeze.

But then, with a gutwrenching whoosh, the train's potential energy abruptly changed to kinetic energy and we barreled down the back of the hill at a speed of 10 . . . 20 . . . 30 . . . 40 . . . 50 . . . 60 . . . 70 miles per hour! Eeeyyyooowiii!!!

Because the first drop is so long (195 feet, or 60 meters) and the angle of the track so steep (60 degrees), the Magnum accelerates to a speed that broke the record for fastest roller coaster in the world.

As we hurtled downward, the track beneath us seemed to disappear. Like Wile E. Coyote plunging off a desert cliff, we appeared to be in a free-fall, with the ground rushing upward to smash us to bits.

G Whiz

With only a few feet to spare, our fall was suddenly broken as the train bottomed out on the coaster's first curve. Entering the curve, I found myself overcome by a strange, sinking sensation, as if I had just swallowed a cow. Various factors had combined to hold me in my seat with a force of 3.5 g's.

A *g* refers to the combination of forces acting on a moving vehicle and the people in it. The number of g's that a person or roller coaster is subjected to depends upon such factors as the acceleration of the roller coaster, the pull of gravity, and the curvature of the roller coaster tracks.

Earth's gravity pulls all things down with a force of 1 g. As you sit in class, Earth's gravitational attraction is exerting a force of 1 g on you. As I waited in line for a ride or sat in the train at the top of the first hill, the g force on me was 1.

Riders hurtle down the back side of the Magnum XL-200's first hill. As they move into the first curve, they experience large g forces that make them feel

● **Process Skills:**
*Communicating, Classifying,
Constructing/Interpreting
Models*

Ask the students whether a
roller coaster is an example of a
simple or a complex machine.
(complex machine) After the
students have determined that

a roller coaster is a complex
machine, ask them to describe
the simple machines that make
up a roller coaster. (Students
might turn to Chapter 4 to
refresh their memories about
work, simple machines, and
complex machines.)

● **Process Skill:** *Evaluating*

As a follow-up to the class discussion, you might have the students prepare a report on the
principles of physics involved in
their favorite sport or activity.

For example, they could write
about how knowledge of
physics might affect their performance in a particular sport,
such as basketball or soccer.

However, the
number of g's
quickly changed
during my first
plunge. As the roller
coaster accelerated
downhill, the
increasing speed
offset somewhat
the effect of gravi-
ty. The g force
became less than 1
but not quite zero.

As the roller
coaster tracks began
to curve, I felt an
increasing pull on
my body. The force
pressing on me
jumped to 3.5 g's,
making me feel as if
I were 3.5 times
heavier than normal.

*A view from the top of the Magnum XL-200,
205 feet in the air. Riders accelerate to more
than 70 miles (112 kilometers) an hour on
the downhill slide.*

in jet aircraft that take
roller coaster rides in
the sky. Each time a jet
flies over the top of a
"hill," its passengers
become weightless and
float around the cabin
of the plane.

After the Magnum's
second hill, what fol-
lowed was a pretzel-
shaped series of turns
and more out-of-seat
experiences, including
numerous speed
bumps; more breathless
plunges; a hairpin cor-
ner that seemed about
to dump us into the
lake; a sharp dive into a
tunnel that threatened
to cut off our heads;
and a fantastic show of

Later I learned that 3.5 g's is no small force,
considering that space shuttle astronauts
experience 4 g's while blasting off from Cape
Canaveral. At 5 g's, human eyesight dims and
a person loses consciousness.

As our train pulled out of the first curve,
the momentum gained from our precipitous
drop shot us to the top of the second hump,
157 feet (45 meters) in the air. As the car
zoomed toward the top, I found myself on a
sudden crash diet, losing all that "weight" I
had just gained—and more. At the top of the
hump, the number of g's holding me in my
seat dropped to below zero (*negative g's*).
The acceleration of the train overcame the
pull of gravity and I was lifted off my seat. I
was weightless!

Had it not been for the lap bar holding me
down, I would have taken off into space. The
train would have taken off too if not for its
two sets of wheels, one riding above the
track, the other riding below.

The National Aeronautics and Space
Administration trains astronauts for the
weightlessness of space by giving them rides

smoke, strobe lights, and sound effects inside
the tunnel.

Thrills but No Spills

Every jolt and whipcrack corner on the
Magnum reminded me of what designer

An Exciting Journey of Great Gravity

*This diagram shows how the number of
g's varies during the first part of the
Corkscrew, another roller coaster at Cedar
Point. Unlike the Magnum XL-200, the
Corkscrew has a loop-the-loop turn.
During the ride, you feel as if you're lifting
out of your seat when the force on your
body is only one-half normal gravity. You
become weightless when your acceleration
totally offsets the pull of gravity (zero g's)
at the top of a hill. At the top of the loop,
you feel a force of 1 g that is directed
straight up into the air because the car is
upside down.*

Journal Activity Encourage interested students to do research on one of these world record-setting roller coasters and record their research in their journals: the longest roller coaster—The Beast—at Kings Island, near Cincinnati, Ohio; the tallest roller coaster—the Moonsault Scramble—at Fujikyu Highland Park near Kawaguchi, Japan; the tallest looping roller coaster—the Viper—at Six Flags Magic Mountain in Valencia, California.

Ron Toomer once said about making roller coasters: "The aim is to build in every fright imaginable." But as I learned at Cedar Point, a roller coaster is one of the safest forms of transportation. The Magnum XL-200's tracks get a careful inspection every day. And once a year, a team of engineers uses X-ray equipment to examine those parts of the coaster where riders experience excessive positive and negative g forces.

Look at it this way. The odds of getting killed on a roller coaster are about 1 in 140 million, compared with 1 in 60 million on a domestic airliner. Statistically speaking, roller coasters are safer than merry-go-rounds. The few deaths that happen on roller coasters usually result from the antics of riders who disregarded the safety rules.

But surely as roller coasters grow taller and steeper, they must test the outer limits of steel and concrete technology. Not so, say roller coaster designers. Roller coasters could climb much higher with no increase in danger to riders.

The limit to a roller coaster's height is not in the engineering; it's in the economics. Roller coasters are expensive; Cedar Park spent $8 million on the Magnum.

Passengers ride upside down in a double helix turn on the Typhoon roller coaster at Busch Gardens in Tampa, Florida.

The other limit is human endurance. Twenty-story structures like the Magnum may be about as high as roller coaster riders will dare to go. How many thrills can a person take anyway?

Well, speaking for myself, I can only say that when the train came to a final stop, my throat was hoarse from screaming and my knees were like jelly. I was breathing hard. Stepping off the train, I couldn't help but turn to the tour guide and exclaim: "Let's do it again!" ◆

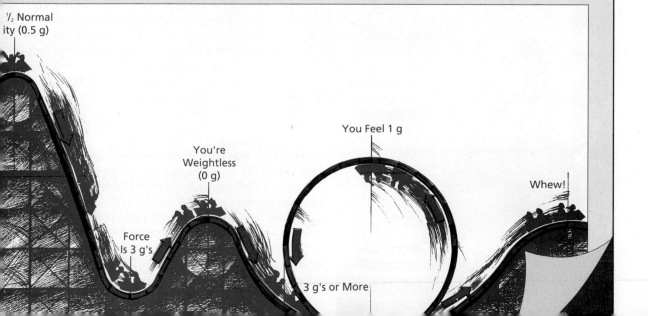

½ Normal
ity (0.5 g)

You're Weightless (0 g)

You Feel 1 g

Whew!

Force Is 3 g's

3 g's or More

Discussion

● **Process Skills:** *Evaluating, Generating Ideas*

After students have read the biography on Ellen Richards, ask them why her work was important in the field of energy use.

Encourage students to do additional research on the life and work of Richards or one of the scientists listed below.

• Lise Meitner (1878–1968)
• Ernest Rutherford (1871–1937)
• Leon Serpollet (1858–1907)
• Karl Benz (1844–1929)

Journal Activity
Students who are interested in Stephen Hawking can read one of the following books and write a journal entry.

Hawking, Stephen. *A Brief History of Time.* Bantam, 1988. In this best-selling book, Hawking describes his theories.

Simon, Sheridan. *Stephen Hawking: Unlocking the Universe.* Dillon Press, 1991. This book tells about Hawking's childhood and career.

White, Michael and John Gribbin. *Stephen Hawking: A Life in Science.* Dutton, 1992. This book tells about Hawking's life.

THEN AND NOW

Ellen Richards
(1842-1911)

Ellen Richards was probably one of the first women to be concerned with water pollution and energy conservation. In the 1870s, she studied the quality of water in Massachusetts rivers and wrote several books about energy conservation in the home.

Richards was born in Boston in 1842. She grew up on a farm in the nearby town of Westford. Although it was unusual for a young woman to receive a formal education, Richards attended Vassar College and Massachusetts Institute of Technology (MIT). In fact, she was the first woman to be accepted at MIT. She received a B.S. degree in chemistry from MIT and an M.A. from Vassar.

Richards was dedicated to improving the home environment. She started the Home Economics Association and also helped to improve opportunities for women in science through the Association of University Women. As a result of her efforts, the Women's Laboratory was established at MIT. The laboratory offered women training in chemistry, biology, and mineralogy. ◆

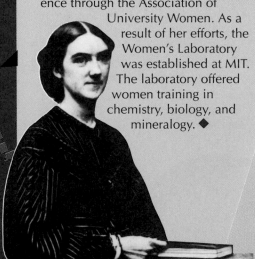

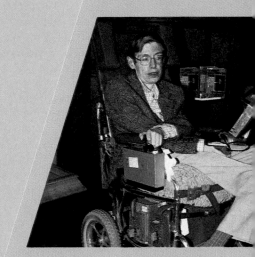

Stephen Hawking
(1942-)

While a student in graduate school, Stephen Hawking was diagnosed with amyotrophic lateral sclerosis (ALS), a disease of the nervous system. Also called *Lou Gehrig's disease*, ALS slowly destroys the nerves that control muscle activity. Since then he has been a wheelchair user and needs help with most physical activities.

ALS has not kept Hawking from scientific research, however. He has made some of the most important discoveries about gravity since Albert Einstein. His work supports the theory that the universe started with a big bang.

Hawking was born in Oxford, England, in 1942. At the age of 20, he received a B.A. degree from University College, part of Oxford University, and went on to receive a Ph.D. from Cambridge University. Hawking is perhaps best known for his work in applying Einstein's general theory of relativity to conditions that might lead to the formation of black holes in space. His book, *A Brief History of Time*, introduces people to his theories. ◆

Discussion

● **Process Skills:** *Evaluating, Expressing Ideas Effectively*

After the students have read the feature on Connie Ensing, discuss why her job is important.

(She and her crew make sure that cars do not have unfair competitive advantages. When Ensing and her crew spot safety problems, they are preventing potential accidents.)

● **Process Skills:** *Interpreting Data, Recognizing Time/Space Relations, Formulating Hypotheses*

Perhaps you or your students have seen video clips of spectacular auto racing collisions. Ask the students to think about what they learned in Chapter 5 about kinetic energy to help

them explain why auto racing collisions are so devastating. (Students learned that kinetic energy increases in proportion to the square of the speed. Any time the speed of a vehicle is doubled, the force of impact is four times greater. The faster an object travels, the more force it has when it hits something.)

SCIENCE AT WORK

CONNIE ENSING
CART Race Official

As the sleek race car downshifts into pit row, Connie Ensing springs into action. During the few seconds that the car is in the pit, she carefully observes the activity of the pit crew that services and fuels the racer. She makes sure that everything is done safely and according to the rules. Ensing is a technical official for CART, **C**hampionship **A**uto **R**acing **T**eams, Inc.

Officials of CART travel to races all over the United States and even to Canada and Australia. At each event, Ensing recruits and trains as many as 25 volunteers who will assist her before, during, and after the race.

Before a race, her team examines each car to certify that it meets all regulations. Designs are checked to make sure they satisfy CART rules. Measurements are taken of each car's length, wheel width, and suspension system. Eight-foot tape measures are used for large measurements; calipers for small measurements. Other tools include levels and plumb bobs. These help the team to determine whether the angle of the rear wing on each race car is correct.

Team members report their findings to Ensing, who then determines whether a car has an unfair advantage or a safety problem. This information is shared with the car's pit crew, so that corrections can be made.

During a race, Ensing and her team watch for oil leaks, mechanical problems, and even driver fatigue. If a pit crew violates safety procedures or maintenance rules, Ensing calls the control tower. If necessary, crews or drivers are warned on the spot. Imagine stepping out in front of a race car to signal its driver! Sometimes Ensing does exactly that.

After a race, Ensing and her team may recheck some of the cars. They do this if they think any cars have been altered during the race. Cars can still be disqualified at this point. The findings of the inspectors sometimes change the outcome of a race.

Ensing enjoys her work as a CART technical official. If you are interested in car racing and like using math and science skills, you, too, might enjoy this kind of work. ◆

Discover More

For more information about a career as a CART official, write to the

Championship Auto Racing Teams, Inc.
390 Enterprise Court
Bloomfield Hills, MI 48013

Background Information

In order for you to get on the NASA *Educational Horizons* newsletter mailing list and to locate the NASA Teacher Resource Center for your geographic area, write to the following address.

NASA Education Division
Educational Publications Branch
Code FEP, Rm 2-J34
300 E Street, SW
Washington, DC 20546

The *Educational Horizons* newsletter tells about the latest educational events and scientific news from NASA Headquarters and Centers. It provides opportunities for teachers to interact with NASA through workshops or seminars, through the acquisition of educational products such as printed information or videos, as well as through television programming and teleconferencing. *Educational Horizons* is published three times a year.

Free Fall and Space Travel

Remember the last movie you saw about travel in space? People probably moved around inside the spacecraft pretty much as they do on Earth. Most movies don't show one common feature of space travel—weightlessness.

FREE FALL

Since most space travel is done relatively close to Earth, astronauts are not truly weightless. What makes space travel unique is that astronauts are actually falling freely. Gravity is not making them push on anything as it would on Earth. Nothing under them provides support, because their spacecraft or space station is also falling freely.

Even though astronauts are not totally weightless, they and everything around them behave in rather strange ways. For example, if an astronaut placed a lead weight in a bucket of water, the weight would not sink. Both the lead weight and the bucket of water would be falling toward Earth at the same speed. The lead weight would float. If an astronaut placed a drop of water in the center of a spacecraft, it would stay there, because the spacecraft and the water drop would fall at the same rate.

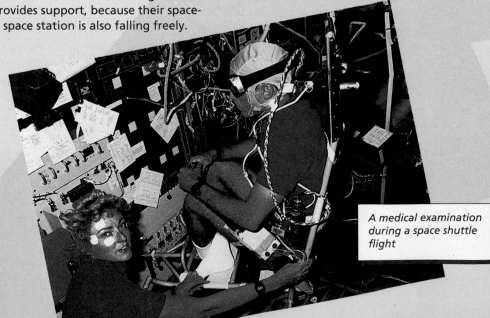

A medical examination during a space shuttle flight

In free fall, astronauts can lift and move heavy objects with ease.

PLUSES AND MINUSES

Free fall has some advantages and some disadvantages. One big disadvantage is that it causes space sickness. Imagine how your stomach would feel as you dropped down the first hill of the Magnum XL-200 roller coaster you read about earlier. Now imagine how your stomach would feel if, instead of reaching the bottom of the hill and starting back up again, you just kept falling. This is what space sickness is like.

The advantages of free fall are many. Weight is a problem for many manufacturing processes on Earth. For example, many metals cannot be mixed because they have different densities. When mixed, these metals separate as oil and water do. In space, metals of different densities stay mixed. This allows for many new kinds of metals to be produced.

MADE IN SPACE

Free fall has other advantages, too. Many kinds of semiconductor crystals that cannot be grown on Earth can be grown in space. Also, very strong but lightweight ball bearings can be made in space by inflating drops of steel like balloons, then allowing them to cool. Because the molten steel and the air bubble are falling at the same rate, the air does not rise through the steel and pop the bubble as it would on Earth.

New metals, semiconductor crystals, and lightweight ball bearings are only three products that could be made in space. There will undoubtedly be many more manufacturing processes that will benefit from a free-fall environment. ◆

WAVES—ENERGY IN MOTION

UNIT OVERVIEW

This unit describes the concept of waves as energy. Energy in the forms of sound and electromagnetic radiation is presented in detail. Light, lenses, and lasers are discussed comprehensively.

UNIT RESOURCES

PRINT MEDIA FOR TEACHERS

CHAPTER 8

"Flukes and Foils." *Discover* 12 (June 1991): 16. The ways in which whales might use wave energy to swim is presented.

CHAPTER 9

Gardner, Robert. *Experimenting with Sound.* Watts, 1991. This book provides experiments for teachers and students to perform to learn about sound and sound waves.

CHAPTER 10

Freedman, David H. "Photon Motel." *Discover* 13 (January 1992): 51. Physicists Azriel Genack and Narciso Garcia built a cage to "catch" photons.

CHAPTER 11

Miller, Judy K. "Ink." *Chem Matters* 11 (February 1993): 8–10. This article discusses the history of ink, pigments, vehicles, and solvent pollution.

CHAPTER 12

Bromberg, Joan Lisa. *The Laser in America: 1950–1970.* The MIT Press, 1991. A history of the laser is presented in an enjoyable and conversational style.

PRINT MEDIA FOR STUDENTS

CHAPTER 8

Walker, Jearl. *The Flying Circus of Physics with Answers.* John Wiley & Sons, 1977. This collection of problems about physics in the real, everyday world includes a section on waves. The questions focus on relevant, fun phenoma involving familiar objects.

Arnosky, Jim. *Near the Sea: A Portfolio of Paintings.* Lothrop, 1990. (NSTA Outstanding Science Trade Book) Through his text and paintings, artist and naturalist Jim Arnosky captures the beauty and rhythms of the sea.

"Wild Waves." *National Geographic World* 193 (September 1991): 16–21. Characteristics of waves are illustrated. How do surfers catch waves? This article tells all about it.

CHAPTER 9

Berger, Melvin. *The Science of Music.* Crowell, 1989. (NSTA Outstanding Science Trade Book) This book describes the relationship between music, sound, and science.

Levinson, Nancy Smiler. *Chuck Yeager: The Man Who Broke the Sound Barrier.* Walker, 1988. (NSTA Outstanding Science Trade Book) Well-written text describes the experiences, goals, and ambitions of the man who first broke the sound barrier.

CHAPTER 10

Feynman, Richard. *QED: The Strange Theory of Light and Matter.* Princeton University Press, 1985. In this brilliant explication of quantum electro-dynamics (QED), Feynman treats the layperson to a mind-expanding tour of the curious world of light and matter.

Grotta, Sally and Daniel. "Digital Photography." *Popular Science* 241 (September 1992): 62–66. This interesting article discusses digital photography.

"Why in the World?" *National Geographic World* 175 (March 1990): 23. Why do people need glasses? This question is answered through photographs and diagrams.

CHAPTER 11

Heiligman, Deborah. "There's A Lot More to Color Than Meets the Eye." *3-2-1 Contact* (November 1991): 16–20. The effects of color on people are colorfully presented. Color puzzles are included at the end of the article.

Hitzeroth, Deborah. *Radar: The Silent Detector.* Lucent Books, 1990. A history of radar and how it has been used in the military is presented. Historic photos add to the appeal of this part of the nonvisible spectrum.

Roman, Mark. "The Little Waves that Could." *Discover* 10 (November 1989): 54–60. A history of microwaves as well as a discussion of present-day uses is outlined.

CHAPTER 12

Asimov, Isaac. *How Did We Find Out About Lasers?* Walker, 1990. (NSTA Outstanding Science Trade Book) A clear explanation of the laser includes a brief history of the scientific investigation of electromagnetic radiation and current uses for lasers.

Gross, Neil. "The New Worlds Lasers Are Conquering." *Business Week* 3169 (July 16, 1990): 160–162. In this article, the author discusses the latest laser applications.

Talbot, Michael. *The Holographic World.* HarperCollins Publishers, 1991. This book discusses the idea that the universe itself may be a giant hologram, constructed, at least in part, by the human mind.

CHAPTER 8

Mechanical Waves. Film or video-cassette. Barr Films. 20 min. The properties of mechanical waves are presented through real-life examples and laboratory experiments. Topics include frequency, wavelength, diffraction, and the Doppler effect.

Standing Waves and the Principle of Superposition. Videocassette. Britannica. 11 min. Computer animation and experiments demonstrate how standing waves are produced.

What Is a Wave? Videocassette. Focus Educational Media, Inc. 16 min. A simple, yet in-depth, description of wave motion is presented.

CHAPTER 9

Sound, Energy, and Wave Motion. Film or videocassette. Coronet Film and Video. 14 min. A jazz trio playing in an open field is used to demonstrate how energy moves from player to instrument to air and then transferred by waves to produce energy changes in the listener's ears.

Sound Waves and Stars: The Doppler Effect. Film or videocassette. BFA Educational Media. 13 min. The Doppler effect is explained by discussing radar waves, light waves, and changes in pitch.

CHAPTER 10

Science Essentials: Light. Videodisc. Britannica. Students explore light through experimenting with how light travels, lenses, and colors.

The Simple Lens. Film or videocassette. BFA Educational Media. 12 min. The ways in which a lens works to organize light are explored.

The World of Light Energy. Videocassette; 5-part series. Focus Media, Inc. 14 min. each. Light energy is explored through several topics.

CHAPTER 11

All About Matter and Energy. Computer Software. Focus Media, Inc. This learning game promotes the development of vocabulary and concepts about matter and energy.

Light and the Electromagnetic Spectrum. Film or videocassette. Coronet Film and Video. 14 min. The relationship between light, gamma rays, X-rays, ultraviolet and infrared rays, radio waves, and light waves is illustrated.

Light, Color and the Visible Spectrum. Film or videocassette. Coronet Film and Video. 13 min. Color is examined by mixing colored lights and filters and by adding and subtracting colors.

CHAPTER 12

Fiber Optics. Videocassette. Britannica. 20 min. This video captures the wonder of how fiber optics works.

Introduction to Holography. Videocassette. Britannica. 17 min. Through laboratory experiments, this video demonstrates types of holograms and illustrates the hologram's three-dimensional character.

The Laser: The Light Fantastic. Film or videocassette. BFA Educational Media. 21 min. How a laser produces light as well as the uses of the laser are explored.

UNIT
3

Discussion The light from the sun is only one kind of wave energy that the students are familiar with. Ask the students what other kinds of waves they have observed and where they have observed them. (Answers might include waves in water, such as a lake, pond, or ocean; in a field of grain on a windy day; in a flag flapping in the breeze.) Ask the students if they are able to see all waves. (No. Most waves are not visible to the human eye. For instance, sound travels in waves and is not normally visible.)

WAVES —
Energy in Motion

CHAPTERS

8 Waves 214

From the toddler splashing in a puddle to the artist painting the chill Atlantic surf, humans find endless fascination in the movements of waves.

9 Sound 238

Both scientists and musicians study the properties of sound. When asked what he had to say to young musicians, trumpeter Wynton Marsalis replied, "Work on your sound."

10 Light and Lenses 260

Jan Vermeer used the camera obscura to produce his paintings. The effect the lenses had on the light changed the way Vermeer painted his subjects.

11 The Spectrum 282

Van Gogh used colors to convey emotion in his paintings. Colors are just one way in which electromagnetic waves affect us.

12 Uses for Light 310

From Star Wars to laser surgery, the laser has had an enormous effect on our imaginations and on our lives.

Journal Activity You can extend the discussion by asking the students to list all the different kinds of waves they can think of. You might also bring in these Unit topics: What characteristics do all waves have in common? What are the properties of sound waves? What parts of the spectrum are not visible? What does color have to do with waves? Ask the students to keep a record in their journals of their questions. After the class has completed the Unit, have the students look again at their questions. A repeat discussion might help the students realize how much they have learned.

Much has been written about the beauty of our nearest star—the sun. You have probably seen radiant sunsets such as the one shown here. However, the sun does much more than just light the day and warm the earth. The sun sends energy to the earth in the form of waves. These are not the kind of waves you see at the beach, but they are not so very different, either.

Science PARADE

WAVES

PLANNING THE CHAPTER

B = Basic   **A** = Average   **H** = Honors
The coding Basic, Average, and Honors indicates subsections, features, and resources that might be appropriate for different levels of learners. For additional suggestions regarding choice of topic and depth of coverage, see the Pacing Chart on pages T28–T31.

*Frame numbers at point of use
(TR) Teaching Resources, Unit 3
(IT) Instructional Transparencies
(LI) Laboratory Investigations
(SD) *Science Discovery* Videodisc Correlations and Barcodes
(SRG) Study and Review Guide

CHAPTER MATERIALS

Title	Page	Materials
Teacher Demonstration	217	Petri dish, water, overhead projector, pencil or thin wooden rod
Discover by Problem Solving	218	(per individual) journal
Discover by Doing	220	(per individual) journal
Discover by Writing	222	(per individual) journal
Activity: How can you find the relationship among the the properties of a wave?	223	(per group of 3 or 4) metal spring about 3 m long, meter stick, stopwatch, chalk
Discover by Doing	226	(per pair) basin of water, journal
Discover by Doing	226	(per group of 3 or 4) small rubber ball, protractor, masking tape
Skill: Diagramming	229	(per group of 3 or 4) pencil, metric ruler, sheet of unlined paper, protractor
Teacher Demonstration	231	tuning forks with slightly different frequencies (2)
Discover by Problem Solving	233	(per individual) journal
Investigation: Creating and Measuring Standing Waves	234	(per group of 3 or 4) rope clothesline about 3 cm, stopwatch, meter stick

TEACHING SUGGESTIONS

Field Trip
Visit your school's music department. Ask the music teacher to demonstrate the various instruments of the orchestra to show how they produce sounds of controlled frequency.

Outside Speaker
Invite an air traffic controller to speak to the class about how waves play a part in his or her occupation. You might also ask the U.S. Army Corps of Engineers to supply a representative to speak to the class about the Corps' efforts to control beach erosion.

CHAPTER 8 WAVES

CHAPTER THEME—ENERGY

This chapter will help the students understand that energy is carried in waves and that a wave is affected by the medium through which it travels. Three types of waves are introduced: transverse, longitudinal, and standing waves. The properties of waves are explored including reflection, refraction, interference, amplitude, wavelength, frequency, and speed. A supporting theme in this chapter is **Environmental Interactions**.

MULTICULTURAL CONNECTION

Katsuda Sezawa spent many years studying tsunamis, beginning in the 1930s. Tsunamis are wide shallow waves that are generated by earthquakes or underwater volcanic activity. These waves may be up to 160 kilometers long and travel as fast as 724 kilometers an hour. Tsunamis have caused great destruction along coastal areas and are quite difficult to predict.

MEETING SPECIAL NEEDS

Second Language Support

Have limited-English-proficient students create a "mini dictionary" of unfamiliar terms in their journals as they study this chapter. The students should write each term in both English and their first language. Encourage the students to add phrases, drawings, or any other notes that will help them become familiar with the terms.

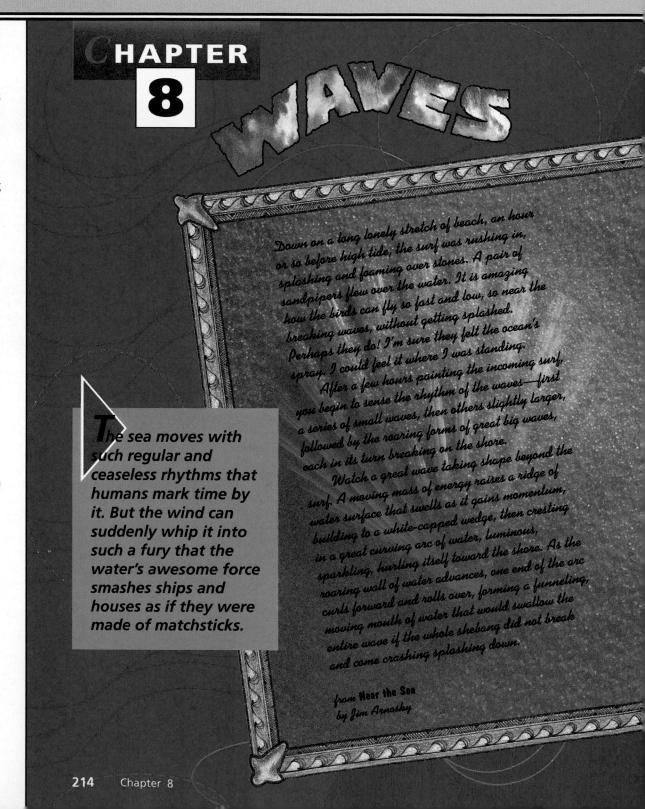

CHAPTER 8 WAVES

The sea moves with such regular and ceaseless rhythms that humans mark time by it. But the wind can suddenly whip it into such a fury that the water's awesome force smashes ships and houses as if they were made of matchsticks.

Down on a long lonely stretch of beach, an hour or so before high tide, the surf was rushing in, splashing and foaming over stones. A pair of sandpipers flew over the water. It is amazing how the birds can fly so fast and low, so near the breaking waves, without getting splashed. Perhaps they do! I'm sure they felt the ocean's spray. I could feel it where I was standing.

After a few hours painting the incoming surf, you begin to sense the rhythm of the waves—first a series of small waves, then others slightly larger, followed by the roaring forms of great big waves, each in its turn breaking on the shore.

Watch a great wave taking shape beyond the surf. A moving mass of energy raises a ridge of water surface that swells as it gains momentum, building to a white-capped wedge, then cresting in a great curving arc of water, luminous, sparkling, hurling itself toward the shore. As the roaring wall of water advances, one end of the arc curls forward and rolls over, forming a funneling, moving mouth of water that would swallow the entire wave if the whole shebang did not break and come crashing splashing down.

from *Near the Sea*
by Jim Arnosky

CHAPTER MOTIVATING ACTIVITY

Show the videodisc *The Puzzle of the Tacoma Narrows Bridge* (Carolina Biological Supply Company). Initiate a discussion about the wave motion of the bridge. Ask the students to study the motion of the bridge carefully. Are there any points on the bridge that do not move? (There is no motion at either pier, at a point halfway between the piers, or along the center line.) Have the students describe the motion of a car on the bridge. Conclude by having the students observe what happens to a point across the roadway when one side of any point on the bridge is moving upward. (When one side of the bridge is moving upward, the opposite side is moving downward.)

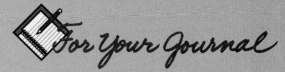
For Your Journal

The journal questions will help you identify any misconceptions that the students might have about waves. These questions will also help the students realize that they do possess knowledge about waves. (Students will have a variety of responses to the journal questions. Accept all answers, and point out that at the end of the chapter they will be able to refer back to the questions.)

Copyright © Jim Arnosky, from NEAR THE SEA: A PORTFOLIO OF PAINTINGS by Jim Arnosky. By permission of Lothrop, Lee & Shepard, a division of William Morrow & Co., Inc.

ABOUT THE AUTHOR

Jim Arnosky was born in 1946 in New York City. He currently lives in Vermont. His book, *Near the Sea,* is a collection of his paintings and writings. Each painting depicts a place on a small Maine island. To economize on weight carried while hiking, Arnosky limited his palette to six colors: red, light yellow, yellow ocher, permanent blue, raw umber, and white. Arnosky has also written and illustrated a number of other books, including *I Was Born in a Tree and Raised by Bees, Outdoors on Foot,* and *Crinkleroot's Animal Tracks and Wildlife Signs.*

For Your Journal

- *Observe water in motion—in a downspout, water fountain, or faucet—and describe its movements.*
- *How are water waves like other kinds of waves?*
- *How can waves carry energy?*

● **Process Skills:** *Comparing, Generating Ideas*

Explain that the example of a wave driving a pump handle is not very practical because the resistance of the pump handle to the motion will be much greater than the energy provided by the rope. Challenge the students to hypothesize why

this is so. (The density of the handle is much greater than that of the rope; therefore, the wave will slow down when it hits the pump handle. The effect is that very little energy gets to the pump.)

● **Process Skill:** *Formulating Hypotheses*

 Cooperative Learning Ask the students to work in small groups and discuss the differences between waves on an ocean and waves on a lake. (Ocean waves are usually larger and contain more

energy.) Then ask the groups to develop a hypothesis as to where ocean waves get their energy. Have the students share their ideas. (The students should hypothesize that ocean waves get their energy from the wind. Ocean waves usually have more energy because the winds blow over them for a longer distance.)

✧ *Did You Know?*

When a storm is forecast, small boats are often sent out to sea so that they can bob harmlessly up and down as waves pass under them. If they stayed in the harbor, they could be damaged from the energy of breaking waves.

SCIENCE BACKGROUND

Pure transverse waves can occur only in solids. Liquids and gases cannot support the disturbance caused by a transverse wave. They do not have the self-restoring abilities of a solid.

DISCOVER BY
Problem Solving

The people in the line correspond to the medium through which a wave travels. The medium does not move (the people are standing still), but the particles in the medium move up and down (the people's hands and arms are moving). The buckets of water correspond to the energy as it moves through the medium.

Figure 8–3. A person is making a transverse wave with a rope to operate a bicycle pump. As the wave moves to the right, the points of the rope move up and down. As a crest reaches the pump, the handle is raised. As a trough reaches the pump, the handle is lowered.

Figure 8–4. The way a bucket brigade gets water to a fire is similar to the way energy travels as a wave.

in the rope that travels down the rope to the other end. The wave consists of a series of crests and troughs that follow each other down the rope. Ms. Evans explains that the wave carries energy through the rope because each point on the rope passes its energy on to the next point.

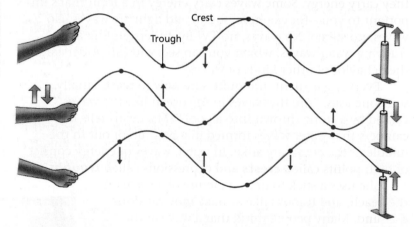

"In a water wave, as in a rope wave, particles move perpendicular to the path of the wave. That means the water moves up and down. This kind of wave is called a **transverse wave.**" Then she shows you a picture of a bucket brigade from her book and has you try to solve an interesting problem.

DISCOVER BY *Problem Solving*

Study the picture that shows how a bucket brigade gets buckets of water to the fire. How is this similar to the way energy travels as a wave? Meet with some other students in a small group to discuss this question.

● **Process Skills:** *Constructing/Interpreting Models, Analyzing*

Have the students draw examples of longitudinal waves on the board. (Waves should include compressions and rarefactions.) Then ask

other students to circle and label each area of compression or rarefaction in the waves. Ask the students to place their drawings in their science portfolios.

● **Process Skills:** *Generating Ideas, Communicating*

Tell the students that the Dutch scientist Christian Huygens (HOY guhnz) invented a mathematical model that describes the behavior of waves. He is also famous for the invention of the

pendulum clock and for his discovery of the rings of Saturn. Have interested students do research on Huygens and have them share their research with classmates.

Side to Side

Side to Side   Ms. Evans turns the page of her book and shows you another kind of wave, a **longitudinal wave.** In this kind of wave, the particles move parallel to the path of the wave. That is, they move back and forth rather than up and down.

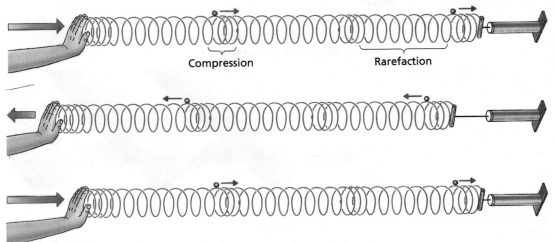

Compression   Rarefaction

Notice that you don't see crests or troughs in this kind of wave. The particles in a longitudinal wave move differently from those in a transverse wave. Every time the person in Figure 8–5 pushes the spring forward, the coils get closer together. The areas where the coils are closer together are areas where the molecules are closer together. These areas are called **compressions.**

When the person pulls her hand back, she separates the coils. In the areas where the coils separate, the molecules are farther apart. These areas are called **rarefactions.** A longitudinal wave consists of a series of compressions and rarefactions following each other down the spring. The compressions and rarefactions move the pump handle in and out.

Remember that in both types of waves, energy is transferred from one place to another. The particles may move up and down perpendicular to the wave's direction as in a transverse wave, or the particles may move back and forth as in a longitudinal wave. The particles don't move down the rope or spring, but the energy does.

Figure 8–5. Here a longitudinal wave created by compressing a coiled spring is used to operate a bicycle pump.

THE NATURE OF SCIENCE

Like most data in oceanography, wave measurements have undergone substantial sophistication during the last 25 years. Until that time, data on wave characteristics was obtained by direct or indirect visual methods involving a variety of sighting methods and photographic techniques.

Electronic wave recorders are now widely used. In addition, photography is among the most practical methods used to measure wavelength. Aerial photographs are commonly used to determine wavelength and examine refraction patterns of waves as they enter shallow water.

MEETING SPECIAL NEEDS

Mainstreamed

Have all the students write in their journals definitions in their own words for *longitudinal wave, compression,* and *rarefaction.* Encourage them to draw diagrams illustrating each term. For mainstreamed students, check their diagrams for accuracy. You might have them work with a classmate. Instruct the students to check each other's labels to make sure the terms are spelled correctly.

ONGOING ASSESSMENT
▼ ASK YOURSELF

Both are ways for energy to be carried through a medium.

▼ **ASK YOURSELF**
What do transverse and longitudinal waves have in common?

SECTION 1   **219** ◀

TEACHING STRATEGIES, continued

● **Process Skill:** *Analyzing*

 Write the lowercase and uppercase Greek letters for lambda on the chalkboard. Explain to the students that Greek letters are used in scientific formulas because scientists need an international means of communicating their ideas to each other.

Then ask the students to name any other Greek letters they know of that are used in either mathematics or science. Examples include pi, delta, beta, gamma, and alpha, among others. Write the letters on the board, and encourage the students to write them in their journals and label them correctly. Have the students research the ways in which the letters are used. For example, Δ is used to represent "change in," so if t = time, Δt would be "change in time."

① As the frequency of the wave increases, the wavelength decreases.

DISCOVER BY *Writing*

 Check the students' diagrams to be sure they have labeled all parts properly. In particular, be sure that they have not reversed the terms *transverse waves* and *longitudinal waves*. Encourage the students to place their drawings in their science portfolios.

SCIENCE BACKGROUND

The SI unit for frequency, the hertz, is named for Heinrich Hertz (1857–1894). Hertz was a German scientist whose work with waves led to the development of modern radio, radar, and television.

SCIENCE TECHNOLOGY SOCIETY Ultrasound, sound waves with frequencies above 20 000 hertz, has many uses. For example, physicians use ultrasound waves to destroy kidney stones and some types of tumors without surgery. The high frequency of the waves "shatter" the kidney stones or tumors.

▶ **222** CHAPTER 8

Figure 8–9. The rope on the bottom is being shaken twice as fast as the rope on the top. What happens to the wavelength of a wave as the frequency increases? ①

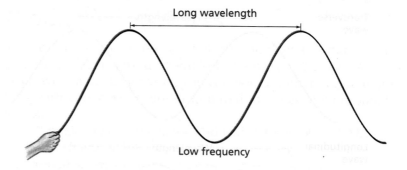

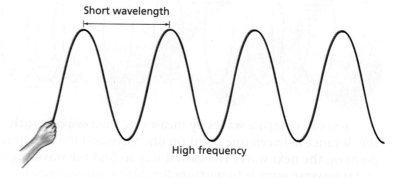

She explains that scientists measure frequency in a unit called *hertz (Hz)*; one hertz is equal to one vibration per second. Then she gives you and your friend a sheet of paper and asks you to draw diagrams.

DISCOVER BY *Writing*

In your journal, make a diagram of two transverse waves, one with a high frequency and one with a low frequency. Label all the parts and properties of the waves. Then diagram two longitudinal waves showing the same parts and properties. ✎

Faster Than a Speeding Bullet "There's one more property of waves you can measure—speed," Ms. Evans says. "How could you find out how fast a wave is moving?" Your friend says, "Maybe I could swim beside the crest of a wave and time the trip." The counselor replies, "You could measure the speed of a wave by following the crest of a transverse wave or the compression of a longitudinal wave and timing the trip, but you'd have to move pretty fast. Here's an easier method."

GUIDED PRACTICE

Cooperative Learning Provide the following problems for the students to solve, either in pairs or small groups.

Imagine a ball floating on a wave. At 10 Hz, the ball will move up and down twice as fast as it would at an original frequency of 5 Hz. How fast would its perpendicular movements be at 2.5 Hz? (The movements would be one-half the original speed.)

INDEPENDENT PRACTICE

Have the students provide written answers to the Section Review. In their journals, have the students write the formula for determining the speed of the wave.

ACTIVITY

How can you find the relationship among the properties of a wave?

Process Skills: Measuring, Inferring

Grouping: Groups of 3 or 4

Hints

Repeat steps 2, 3, and 5 in order to get a more accurate value. Suggest that the students take an average.

▶ **Application**

1. The speed of the wave depends on the frequency (which is given) and the wavelength (which is not given). The students might assume that the wavelength is the same, in which case the speed of the wave with the higher frequency would be greater.

2. Assuming the speed remains constant, the wavelength decreases.

Ms. Evans then explains that scientists measure the speed of a wave mathematically in meters per second (m/s). In any wave, the speed of the wave is equal to the product of the wavelength and the frequency. The counselor then writes an equation.

$$v = \lambda f$$

OR

$$\text{speed} = \text{wavelength} \times \text{frequency}$$

The symbol for wavelength is the Greek letter lambda (λ). In the next activity you'll have a chance to calculate the speed, wavelength, and frequency of a wave.

ACTIVITY

How can you find the relationship among the properties of a wave?

MATERIALS
metal spring about 3 m long, meter stick, stopwatch, chalk

PROCEDURE
1. Place a spring on the floor in a straight line. Measure and record its length in meters.
2. Have one group member shake the spring on the floor from side to side two or three times to produce a 2 Hz or 3 Hz frequency.

3. With a stopwatch, time how long it takes the crest of one wave to travel the entire length of the spring.
4. Determine the speed of the wave. Divide the length of the spring by the time it took a single crest to travel from one end of the spring to the other.

5. With a stopwatch, time how long it takes to produce 10 complete cycles in the spring. Use the formula again to determine frequency.

$$f = \frac{v}{\lambda}$$

Now to find this same quantity by measurement, divide 10 into the total time. Do the two quantities match?

6. Find the wavelength. Use chalk to mark the floor at the top of the first crest. Then mark the top of the second crest. Measure the distance between the marks.
7. Repeat the procedure using a frequency of 4 Hz to 5 Hz.

APPLICATION
1. How does the speed of the wave with the lower frequency compare with the speed of the wave with the higher frequency?
2. What happens to the wavelength when the frequency is increased?

PERFORMANCE ASSESSMENT

Cooperative Learning

At the completion of this Activity, have each group demonstrate their procedure. Check each group's calculations for accuracy.

LASER DISC
1496, 1497

Wave action

11059, 11060–11210

Wave motion

2220

Waves; parts

● **Process Skills:** *Analyzing, Comparing*

Have the students determine what the following situations have in common: a hockey puck bouncing off the side of an ice rink, a light beam bouncing off a shiny surface, and a ball bouncing off a wall. (Each is an example of reflection. The hockey puck, light beam, and ball move in a straight line until they strike a barrier—in these cases, the side of the rink, the shiny surface, and the wall.)

● **Process Skill:** *Inferring*

Discuss with the students how it might be possible for a wave to hit a barrier and not be reflected. (The barrier would be composed of material that absorbs the energy of the wave, such as a shoreline that absorbs the energy of a tidal wave.)

PORTFOLIO ASSESSMENT

Have the students work independently or in pairs and draw diagrams that illustrate angle of incidence and angle of reflection. After you have checked the students' diagrams for accuracy, have them place their work in their science portfolios.

---

① The wave has changed direction because when a wave strikes a barrier, it reflects off it.

DISCOVER BY *Doing*

The waves produced are reflected by the side of the basin.

MULTICULTURAL CONNECTION

Many games are based on a knowledge of reflection. Tennis has conflicting stories explaining its origin. Some historians believe an early form of tennis was played by the ancient Egyptians and Arabs. Years later, the Greeks and Romans copied the game. The game was similar to modern handball in that a single person hit a crude ball against a wall with bare hands.

LASER DISC

10772, 10773–11058

Waves hitting the shore

DISCOVER BY *Doing*

PORTFOLIO ASSESSMENT

Emphasize to the students that prediction is always based on experience (as opposed to a guess). Encourage the students to recall similar experiences before making their predictions. Have the students place their predictions in their science portfolios.

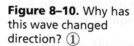

Figure 8–10. Why has this wave changed direction? ①

Later you see Ms. Evans in the dining hall and tell her what you observed about waves against the raft. "Good observation," she says. "Waves move in a straight line until they strike a barrier. The 'bouncing back' of a wave after striking a barrier is called **reflection.** You can see this in a basin of water."

DISCOVER BY *Doing*

Get a basin of water. Let the surface of the water become very calm, and then create a wave with your hand. Watch the wave travel and hit the side of the basin. In your journal, record what happens. 🖋

There is a way to predict how a wave will bounce off a barrier. Try the next activity to see if you can figure it out.

DISCOVER BY *Doing*

Find a hall where you can safely use a small rubber ball. Stand about a meter away from the wall, and then roll the rubber ball to the wall from a variety of angles, such as 25, 45, 75, and 90 degrees. Predict the angle at which it will bounce back. Use a protractor to measure the angle at which you will roll the ball to the wall and the predicted angle of the reflection. Mark the angle with masking tape. How close were your predictions? 🖋

Have the students provide written answers to the Section Review. In their journals, have the students diagram a wave with the following specifications: a wavelength of 3 cm and a wave line 10 cm long.

Have the students describe in their own words the differences between the terms in each set: *reflection, refraction; angle of incidence, angle of reflection.*

Use this analogy to clarify the concept of refraction further: A sled's parallel runners travel more quickly over snow than over a dry roadway. If the sled is moving at an angle, one runner will hit the dry roadway before the other runner. Since the platform of the sled connects the runners, the faster-moving runner (the one on the snow) turns toward the slower-moving runner (the one on the roadway). When the faster-moving runner hits the roadway, the sled moves forward in a new direction, but at a slower speed.

Look at the diagram on this page. Notice both the green and the red lines. The green line has been drawn perpendicular to the barrier. A line perpendicular to a barrier is called the *normal.* The red line shows the direction in which the wave is traveling as it approaches and leaves the barrier. The angle between the normal and the direction in which the wave is traveling is called the *angle of incidence.* Think of it as the angle of the wave as it runs into the barrier. The angle is labeled *i.* The angle between the normal and the direction in which the reflected wave is traveling is called *angle of reflection.* This angle is marked *r.* The angle of incidence always equals the angle of reflection.

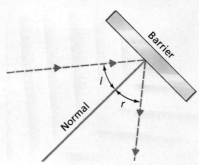

Figure 8–11. When a wave strikes a barrier, it is reflected, as shown here.

▼ ASK YOURSELF

How does the angle of incidence compare with the angle of reflection when a wave reflects from a barrier?

Waves That Bend

The next day before boating class, Ms. Evans tells you that in addition to reflection, there is another way waves change direction. She asks you to observe what happens when a wave created by a motor boat approaches the shore.

You can see that the crests of the wave get closer together as the wave approaches the shore. "Why do you think that is?" she asks. You don't have an answer, so she continues. "The waves get closer together because the lake bottom becomes more shallow. The more shallow the water, the slower the wave travels. When one crest line slows down, the crest behind it, still in slightly deeper water, catches up a little."

She goes on to explain that ocean waves behave the same way. As an ocean wave enters shallow water, the crest lines bend and become nearly parallel to the shore. The direction at which the wave approaches the shore becomes nearly perpendicular.

Figure 8–12. This motorboat creates waves that will eventually approach the shore.

▼ ONGOING ASSESSMENT
▼ ASK YOURSELF

The angle of incidence and the angle of reflection are equal. If the students' activity results did not agree with this concept, suggest some reasons why they might not have agreed.

SCIENCE BACKGROUND

Since the angle of incidence always equals the angle of reflection, ask the students to apply this principle to a carnival funhouse mirror. Why are the images distorted? (Because the glass is distorted, the angle of incidence from the light wave does not equal the angle of reflection. Therefore, the image is not true.) Mention to the students that more information about the structure and properties of light waves will be presented in Chapter 12.

> ✧ **Did You Know?**
> The word *reflection* comes from the Latin *reflectere,* meaning "to bend back." *Refraction* comes from the Latin *refringere,* meaning "to break up."

PROCESS SKILLS
• Measuring • Observing
• Interpreting Data

POSITIVE ATTITUDES
• Openness to new ideas
• Cooperativeness

TERM
• standing waves

PRINT MEDIA
Near the Sea: A Portfolio of Paintings by Jim Arnosky (see p. 211b)

ELECTRONIC MEDIA
Standing Waves and the Principle of Superposition Britannica (see p. 211c)

SCIENCE DISCOVERY
• Waves

BLACKLINE MASTERS
Extending Science Concepts
Study and Review Guide

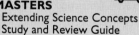

MULTICULTURAL CONNECTION

Japanese prints originated in the early 1600s as illustrations in popular books. Since many people were interested in the illustrations themselves, publishers began to produce the illustrations separately from the books. Printmaking flourished from approximately 1603 to 1867. The middle class arose and prospered in Japanese cities. These people were the chief purchasers of Japanese prints, using them as inexpensive substitutes for paintings.

Most Japanese prints portray scenes from everyday life, the theater, and other popular forms of entertainment. Landscape prints of the Japanese countryside and coastline and the seas surrounding Japan became popular during the early and mid-1800s.

FOCUS

This section introduces the concept of two or more waves passing through a medium at the same time. Constructive and destructive interference are discussed, and the concept of standing waves is introduced. Standing waves and their nodes and antinodes are also highlighted in this section.

MOTIVATING ACTIVITY

Pluck the strings of a guitar or other stringed instrument at various points along their length, and have the students note the different tone of each. Then shorten the length of each string, one at a time, by placing your finger along the neck of the instrument. Have the students note how the tone changes when the length of the string changes. (The shorter the string, the higher the pitch.) This effect is due to the shorter wavelengths of the sound waves; as they are given a shorter distance to move, their frequency becomes higher.

SECTION 3

Waves and Vibrations

Objectives

Describe what happens when two waves arrive at a point at the same time.

Analyze the parts and properties of a standing wave.

In the evening, all the campers and counselors gather around a campfire to talk and sing songs. The moon shines overhead and is reflected in the water. One counselor takes out a guitar and sings a folk song about a ship called the *Edmund Fitzgerald*. This ship sank in huge waves on Lake Superior. One camper says that waves more than 30 m have been recorded on the Great Lakes, where he comes from. Another camper knows about tsunamis—giant sea waves caused by earthquakes or volcanos. She says that in the Ryukyu (ree OO kyoo) Islands of Japan, where her family originally came from, a powerful tsunami once produced breakers 85 m high! That's about the same height as a 25-story building.

Hokusai, 36 Views of Fuji, Great Wave of Kanagawa. Private Collection.

Figure 8–14. This drawing is an artist's idea of what a tsunami looks like.

Interference

After hearing these stories about great waves, you ask Ms. Evans how waves can become so large and powerful. She explains that waves do not often occur in isolation. One wave usually follows another, and not all waves travel at the same speed with the same frequency. If the crests of two waves meet, their energy combines to form a more powerful wave. This process is called *constructive interference.*

After the campfire, you and your friends join Ms. Evans in the dining hall to continue your discussion about waves. She

TEACHING STRATEGIES

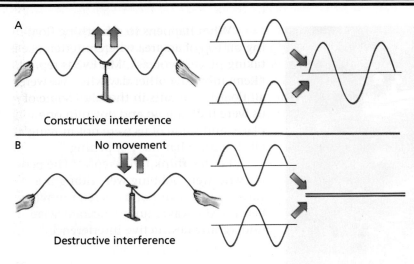

A
Constructive interference

B
No movement
Destructive interference

Figure 8–15. In the top drawing, two waves undergo constructive interference and produce a single, stronger wave. In the bottom drawing, the waves undergo destructive interference and cancel each other out.

has you look at a diagram in her book. Ms. Evans points out that in Part A of the diagram, which shows constructive interference, the bicycle pump handle rises twice as high when the crests of two waves meet as it does when the crest of one wave raised the handle.

Then she points out what happens in Part B of the diagram when the trough of one wave and the crest of another wave meet at the pump handle: the energy of one cancels out the energy of the other. Because the waves cancel each other out, the handle does not move. When two or more waves come together in such a way that a crest meets a trough, the waves undergo *destructive interference*.

Figure 8–16 shows waves created in a ripple tank. In the picture you can see that as two waves meet, constructive interference takes place in some areas while destructive interference takes place in other areas.

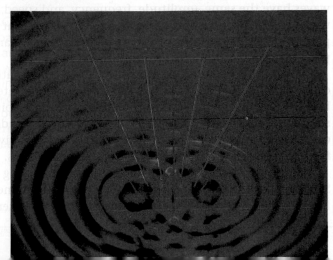

Figure 8–16. Two pointers are vibrating to produce two waves that interfere with one another. The green lines show where the interference is constructive. The red lines show where the interference is destructive.

Creating and Measuring Standing Waves

Process Skills: Measuring, Observing, Interpreting Data

Grouping: Groups of 3 or 4

Objectives
● **Produce** standing waves in a rope.
● **Identify** the frequencies that make a wave stand.

Pre-Lab
Discuss with the students the measurement standards scientists use. (SI units is an example.) Then have volunteers hypothesize how scientists might measure standing waves. (Accept all reasonable hypotheses.)

Analyses and Conclusions
1. Wave speed remains constant while frequency and wavelength vary inversely.

2. Frequencies of standing waves form a harmonic series in which all frequencies stand in simple whole number ratios. This is a direct consequence of the fact that the wavelength is always twice the length of the rope divided by an integer.

PERFORMANCE ASSESSMENT
Cooperative Learning
Have each group repeat the procedure again, this time using cord that is much thinner and lighter in weight. Then have the groups explain why the results differed. (The wave velocity is fixed by the thickness of the rope. The velocity is faster for the thinner cord. The students need less force to start the wave.)

Application
Standing waves in rivers are caused by roughness on the bottom in an area where the depth of the water is equal to the amplitude of a wave. This causes the water in the wave to move up and down without traveling forward.

✳ Discover More
Since the wave velocity is fixed by the thickness of the rope, the velocity would be slower for the thicker rope. Also, the students will find a need for greater force to start the wave.

Post-Lab
Discuss with the students how measuring standing waves differs from measuring other things, such as volume. (SI units can be measured very precisely. The students' waves are measured less precisely.)

INVESTIGATION

Creating and Measuring Standing Waves

▶ MATERIALS
● rope clothesline, about 3 m ● stopwatch ● meter stick

▼ PROCEDURE

1. Prepare a data table like the one shown.

TABLE 1: WAVE CHARACTERISTICS			
Number of Antinodes	Wavelength (m)	Frequency (Hz)	Speed (m/s)
1			
2			
3			

2. Tie one end of the rope to a fixed support, such as a doorknob. Measure the length of the rope.

3. Shake the other end of the rope up and down. Be sure not to swing the rope around. Adjust the frequency until the entire rope is vibrating with just one loop. There is now a node at each end of the rope and an antinode in the middle.

4. Remember that nodes are always one-half of a wavelength apart. Therefore, with one loop in the rope, the rope is one-half wavelength long. What is the wavelength of the standing wave in the rope? Once you know the frequency and the wavelength, you can calculate the speed of the wave.

5. Determine the frequency of the wave. Using a stopwatch, time how long it takes to shake the rope up and down 10 times. This is equal to 10 cycles. Calculate the frequency by dividing 10 into the total time. Record your answer in the data table.

6. Increase the frequency of the wave until there is a node in the middle of the rope. The rope will vibrate with two loops. The rope is now one wavelength long. Determine the frequency and the speed of the wave in the rope. Enter this data in your table.

7. Continue increasing the frequency to get a vibration with three loops in the rope. Now the rope is one-and-one-half wavelengths long. Make the calculations, and fill in your table.

▶ ANALYSES AND CONCLUSIONS
1. What have you found out about the speed of the wave?
2. What have you found out about the frequencies that produce standing waves? Explain your answer.

▶ APPLICATION
Standing waves also occur in rivers, especially in fast-flowing areas called *rapids*. What conditions do you think cause standing waves in rivers?

✳ Discover More
Repeat the procedure with a rope that is much thicker and/or heavier in weight. Make sure it has the same length as the first rope you used. What are your results? If they differ, why do you think this is so?

CHAPTER 8 HIGHLIGHTS

The Big Idea— Energy

Lead the students to an understanding that waves carry energy through a medium. They move the energy along from one point to another. All waves have properties such as amplitude, wavelength, frequency, and speed.

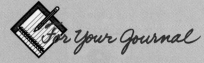

For Your Journal

The students' ideas should reflect an understanding of the different types of waves, how they carry energy, and how they interact. You might encourage the students to share their new ideas with classmates.

CHAPTER 8 HIGHLIGHTS

The Big Idea

Waves carry energy; they deliver their energy from one point to another as they travel. They can travel through various media as either transverse or longitudinal waves. All waves have properties, such as amplitude, wavelength, frequency, and speed, that tell us how much energy they possess.

Waves react with one another predictably to form patterns. Waves may bounce as reflections or bend as refractions. They may combine either to increase their energy or to cancel it. The energy carried by waves may be heard as sound or seen as light. It can even be used to see through objects. The energy of waves is a powerful force in nature.

For Your Journal

Look back at the ideas you wrote in your journal before you began this chapter. Revise your journal entry to reflect what you have learned about waves. Be sure to include information that tells how waves carry energy.

Connecting Ideas

CONNECTING IDEAS

The students' charts should be completed with the following terms:

Transverse waves:
1. trough
2. frequency
3. amplitude
4. wavelength
5. destructive
6. refraction

Longitudinal waves:
7. compression
8. speed
9. wavelength
10. frequency
11. constructive
12. destructive
13. reflection
14. refraction

This chart shows how the big ideas of this chapter are related to both transverse and longitudinal waves. Copy the chart into your journal and complete it.

PROPERTIES OF WAVES					
Type of wave	Zone of most dense molecules	Zone of least dense molecules	Properties that can be measured	Kinds of possible interference	Ways in which direction changes
Transverse	crest	___①___	speed ___②___ ___③___ ___④___	constructive ___⑤___	reflection ___⑥___
Longitudinal	___⑦___	rarefaction	___⑧___ ___⑨___ ___⑩___ amplitude	___⑪___ ___⑫___	___⑬___ ___⑭___

CHAPTER 9

SOUND

PLANNING THE CHAPTER

B = Basic **A** = Average **H** = Honors
The coding Basic, Average, and Honors indicates subsections, features, and resources that might be appropriate for different levels of learners. For additional suggestions regarding choice of topic and depth of coverage, see the Pacing Chart on pages T28–T31.

*Frame numbers at point of use
(TR) Teaching Resources, Unit 3
(IT) Instructional Transparencies
(LI) Laboratory Investigations
(SD) *Science Discovery* Videodisc Correlations and Barcodes
(SRG) Study and Review Guide

CHAPTER MATERIALS

Title	Page	Materials
Discover by Doing	241	(per group of 3 or 4) candle, matches, flat piece of glass, piece of thin stiff wire, tape, tuning fork
Discover by Doing	242	(per pair) flat table, eraser, paper clip
Activity: How can you measure the speed of sound?	243	(per group of 3 or 4) large building with a flat surface in an open space, hammer, wooden board, stopwatch, meter stick
Discover by Observing	245	(per class) microphone, oscilloscope
Skill: Graphing Musical Pitch	250	(per group of 3 or 4) graph paper, pencil
Teacher Demonstration	252	whistle, rubber tubing (1 m)
Discover by Writing	253	(per individual) journal
Discover by Doing	254	(per group of 3 or 4) plastic wrap, tin can, rubber band, sugar, metal tray
Discover by Doing	254	(per individual) journal
Investigation: Calculating the Speed of Sound	256	(per group of 3 or 4) standard water-column apparatus, tuning fork (G = 392 Hz), thermometer

ADVANCE PREPARATION

If you have hearing-impaired students in class, be sure to structure activities that include them. For example, you might put them in charge of illustrating, finding graphics for, and recording the findings for the activities. Also be sure that these students are given preference in any activities for which hearing is not required, if there is not time for all students to complete the activity.

TEACHING SUGGESTIONS

Field Trip
There are several areas in and around the school that you could use for demonstration purposes. For instance, take the class to a band or orchestra rehearsal and have them judge the acoustics of the rooms being used. After checking out the safety regulations, take them to a construction site to hear the decibel levels. Do they think these levels are dangerous to their ears? to the ears of the workers? How do the workers protect themselves?

CHAPTER 9 SOUND

CHAPTER THEME—ENVIRONMENTAL INTERACTIONS

This chapter discusses the compressions and rarefaction of sound waves as they pass through matter. The students will learn that the amplitude of sound waves is measured in decibels and that the pitch of a sound wave is determined by the frequency of its waves. The parts of the human ear and the larynx also are identified and explained. Finally, the students will learn how sounds are produced and that sound waves have many uses, particularly in the field of medicine. A supporting theme in this chapter is **Technology**.

MULTICULTURAL CONNECTION

PORTFOLIO ASSESSMENT

The field of music has drawn performers from a wide variety of cultural backgrounds. In the United States the influence of African-American performers and composers has been particularly strong in the areas of jazz and blues. Musicians both in Europe and the United States have created many rock and pop groups. In addition, the world of music serves to communicate the political concerns of different cultural groups. For instance, the lyrics of folk music often focus on human rights, while modern rock bands or rappers often sing about environmental and social issues. Have the students pick a favorite song with an environmental theme and find out who wrote the song and which musician(s) made it popular. Ask the students to place their research in their science portfolios.

MEETING SPECIAL NEEDS

Second Language Support

Work with limited-English-proficient students to compile a list of objects that make sounds in their school and homes. Ask the students to write their lists in their journals. (Examples include a television, a radio, human voices, and a telephone.)

CHAPTER 9

Sound

*F*ew sounds are as familiar or unique as the human voice. The human voice is so distinctive that police can identify criminal suspects by voiceprint (the electronic record of an individual voice). Telephone company computers may soon identify long-distance callers by their voiceprint instead of a credit card number.

Wynton Marsalis

The sound of an individual voice is a blend of the other voices we have heard all our lives. Family and friends teach us the peculiar ways of speaking that make for regional accents. A simple word like "nice" sounds much different in an Alabama drawl than in the clipped tones of the Midwest. Regional accents may themselves be blends of different ways of speaking, as in the American South, where the speech of African slaves affected the speaking style of whites.

Display objects that make different kinds of sounds. Include objects that are struck, blown, or plucked. Have the students examine and try the different sound makers, and ask them what the objects have in common. Discuss the way in which each object makes its sounds. Then work with the students to prepare a list of words that can be used to describe the sounds made by the objects. Ask the students which words they consider "scientific." Accept any answers the students can justify.

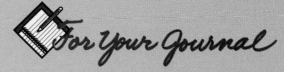

For Your Journal

The students' answers to the journal questions will help you identify what the students already know about sound and what they need to learn. (The students will have a variety of responses to the journal questions. Accept all answers, and tell the students that at the end of the chapter they will have a chance to reread the questions and consider their original answers in light of what they have learned.)

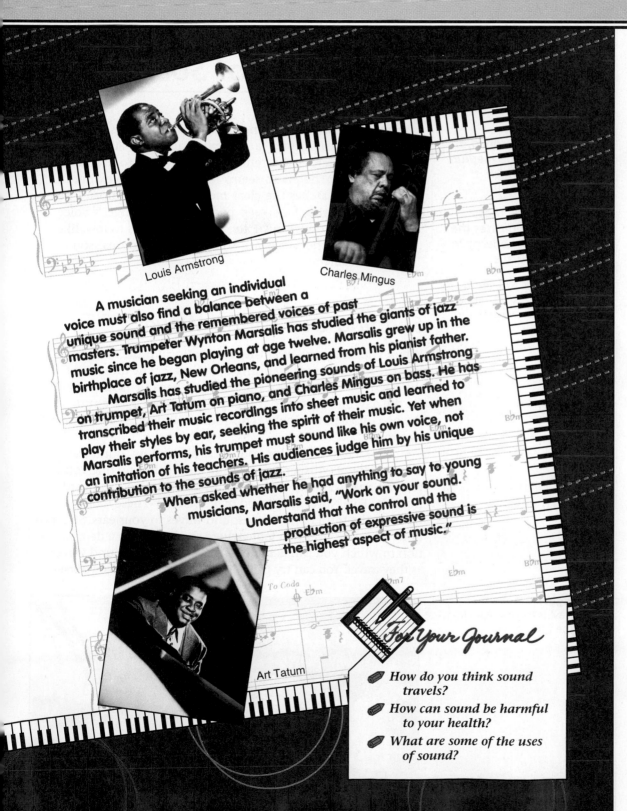

Louis Armstrong

Charles Mingus

A musician seeking an individual voice must also find a balance between a unique sound and the remembered voices of past masters. Trumpeter Wynton Marsalis has studied the giants of jazz music since he began playing at age twelve. Marsalis grew up in the birthplace of jazz, New Orleans, and learned from his pianist father. Marsalis has studied the pioneering sounds of Louis Armstrong on trumpet, Art Tatum on piano, and Charles Mingus on bass. He has transcribed their music recordings into sheet music and learned to play their styles by ear, seeking the spirit of their music. Yet when Marsalis performs, his trumpet must sound like his own voice, not an imitation of his teachers. His audiences judge him by his unique contribution to the sounds of jazz.

When asked whether he had anything to say to young musicians, Marsalis said, "Work on your sound. Understand that the control and the production of expressive sound is the highest aspect of music."

Art Tatum

ABOUT THE MUSIC

Jazz is a unique form of American music, created in the late 1800s by African Americans. It contains elements of many other forms of music, such as gospel songs and spirituals, European classical harmonies, and work songs. Originating in the Deep South, jazz is characterized by improvisation, that is, parts of a work are created spontaneously by the musicians as they play.

Many giants of the American musical scene are and were jazz musicians. Besides those named in the selection, Miles Davis, Dizzy Gillespie, Diane Schur, Chick Corea, Peggy Lee, Oscar Peterson, Charlie Haden, and Keiko Matsui remain noteworthy jazz musicians.

For Your Journal

- *How do you think sound travels?*
- *How can sound be harmful to your health?*
- *What are some of the uses of sound?*

PROPERTIES OF SOUND

PROCESS SKILLS
• Measuring • Evaluating
• Communicating
• Measuring • Comparing
• Interpreting Data

POSITIVE ATTITUDES
• Creativity • Precision

TERMS
• volume • pitch
• noise pollution • decibels

PRINT MEDIA
Chuck Yeager: The Man Who Broke the Sound Barrier by Nancy Smiler Levinson (see p. 211b)

ELECTRONIC MEDIA
Sound, Energy, and Wave Motion Coronet Film and Video (see p. 211c)

SCIENCE DISCOVERY
• Decibel level; table
• Vibration and pitch
• Volumes and tones; lab setup
• Waves; sound

BLACKLINE MASTERS
Connecting Other Disciplines
Reading Skills
Study and Review Guide

INTEGRATION–
Language Arts

Remind the students that sound travels through various media—such as solids, liquids, and gases. Point out to the students that the term *media* has more than one meaning. For example, students may be familiar with the term *media* used in connection with communications—radio, television, and publications. Help the students draw the correlation that just as sound travels through various material media, information travels through various communications media.

FOCUS

This section focuses on how sound waves are produced and the speed with which they travel through air. The distance sound travels is calculated given the speed of sound, the temperature of the air, and the time elapsed. Relationships among volume, amplitude, frequency, and pitch are explored.

MOTIVATING ACTIVITY

Cooperative Learning Have the students work in groups to make a sound-relationship collection. Each picture they draw or find in a magazine should show a relationship between something or someone making a sound and someone or something hearing or detecting the sound.

Examples include a music group playing to an audience; a teacher talking to a student; a prairie dog calling to another to warn of a predator; a person listening to a radio or television. In each relationship, have the students label both the source and the receiver of the sound, as well as the sound itself.

SECTION 1

Properties of Sound

Objectives

Relate the speed of sound to the medium through which it travels.

Recognize that volume indicates the amount of energy in a sound wave.

Identify the cause of noise pollution and its effect on hearing.

Imagine that your science class has been chosen to sit in on a rehearsal of a jazz combo called the Five Notes. The night before the rehearsal, you listen to a tape of their music. The piano seems like a rich, complex net of sound that catches the clear melody of the trumpet. The drum and string bass keep pace while the singer's voice leaps up and down the scale, from high notes to low, like an acrobat. The songs are sometimes humorous, sometimes melancholy.

Sending Out Sound Waves

When your class arrives at the city auditorium the next morning, the musicians are warming up on stage. The director of the science museum, Ms. Mendoza, will be your "sound guide" for the day. She ushers your group into a room adjacent to the box office in the lobby. This room will serve as your "sound classroom."

Ms. Mendoza asks for your ideas about how sound travels. There's a lot of whispering and murmuring, but no one volunteers an answer. She explains that the sounds you just heard from the Five Notes are caused by vibrations. These vibrations travel in the form of invisible sound waves to your ears. She motions everyone over to a table where she has set up a demonstration in which you can actually see the pattern of sound waves as they travel. You can try this sound demonstration for yourself.

Figure 9–1. Sound travels to your ears in the form of waves.

● **Process Skill:** *Inferring*

Remind the students that they cannot see sound coming from a musical instrument or from someone's voice. How do they think sound gets from one place to another? Ask the students to write their responses in their journals. (The students might respond that sound travels through air.)

● **Process Skills:** *Observing, Communicating*

Have the students observe the actions of the prongs of a tuning fork as you strike it. What are such movements called? (vibrations) Point out that all sound is caused by vibrations.

DISCOVER BY *Doing*

CAUTION: Use care around the lit candle. Place a lit candle near the surface of a flat piece of glass until the glass is blackened with soot. Attach a piece of thin, stiff wire with tape to one prong of a tuning fork. Strike the tuning fork so that it vibrates. Let the tip of the wire just touch the blackened area of the glass. What does this show about the motion of the tuning fork?

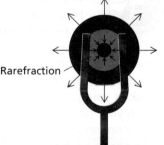

Compression

Rarefraction

Figure 9–2. As this tuning fork vibrates, it creates waves.

How did the pattern of sound waves form when the tuning fork prongs vibrated? As the tuning fork prongs got closer together, they squeezed the air between them to form a dense zone of molecules called a *compression*. As the tuning fork prongs moved apart, they formed a zone in which molecules were less densely packed, called a *rarefaction*. A series of compressions and rarefactions made up the pattern of the sound waves. You may recall reading about compression and rarefaction in the previous chapter.

◤ ASK YOURSELF
How does a tuning fork produce sound waves?

Faster Than the Speed of Sound

As you and your classmates are walking toward the auditorium, one student wants to know how fast sound travels. Before the question can be answered, you see a bolt of lightning flash outside in the sky. Seconds later, you hear the crash of thunder. Ms. Mendoza speaks loudly over the thunder. "Nature just gave us a classic demonstration of the speed of sound," she says. "The lightning and thunder occur at the same time, but it takes longer to hear the thunder because the speed of light is faster than the speed of sound."

Figure 9–3. How far away is the lightning? Count the number of seconds between the lightning and the thunder. Divide by three to determine the approximate distance of the lightning in kilometers.

● **Process Skills:**
Communicating, Comparing

Point out that the word *volume* has two major scientific uses, both connected with the concept of quantity. The volume of a sound is how loud or soft it is, while the volume of amount is how much space is occupied by an object.

● **Process Skill:**
Experimenting

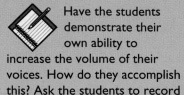 Have the students demonstrate their own ability to increase the volume of their voices. How do they accomplish this? Ask the students to record their accomplishments in their journals. (force more air from their diaphragms, or lungs) What does this cause? (more energy in the sound waves) What is the end result? (a louder voice)

INTEGRATION–
Music

Many students may not know or think about the technical, mathematical, and scientific complexity of sound and music. The following list of words describes the different aspects of complexity of music.

- Dynamics (soft or loud)
- Register (pitch)
- Duration (staccato or sustained)
- Harmony voice (soprano, alto, tenor, bass)
- Notation (written music)
- Intervals (steps between notes)

MULTICULTURAL CONNECTION

Music is one of the oldest forms of human art. Song and language have gone hand-in-hand in the development of human culture. Early instruments consisted of hunting tools or sticks banged into one another, as well as woodwind devices made from bones and thick reeds.

① Sound wave C has the most volume because it has the greatest amplitude.

② As the volume of a sound increases, the amplitude of the sound waves also increases.

ONGOING ASSESSMENT
▼ **ASK YOURSELF**

The more energy a sound wave has, the louder the volume of the sound.

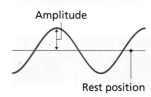

Amplitude

Rest position

(a) small amplitude

Amplitude

Rest position

(b) medium amplitude

Amplitude

Rest position

(c) great amplitude

Figure 9–6. Which sound wave has more volume? How do you know? ①

Volume

Your group is now sitting in the auditorium as the drummer and trumpeter rehearse a passage. Sometimes the sounds seem soft and gentle like a spring breeze. At other times, the sounds seem harsh and angry. In fact, you feel as though the loudness of the sound has so much energy it will hurt your ears. Are you right about this?

The greater the vibrations, the more energy will be carried in the sound wave. The more energy carried in the sound wave, the louder the sound. So when the trumpeter blows hard and creates greater vibrations in the trumpet, the result is a sound wave with greater energy. When the drummer softly taps the snare drum, the result is a sound wave with less energy. Both softness and loudness of sound are referred to as **volume.** The amplitude—that is, the distance a sound wave moves from a rest position—indicates the intensity of volume. The greater the amplitude, the greater the volume.

② You saw the pattern sound waves made in an earlier activity. How does a change in volume affect the pattern of sound waves? Look at the Figure 9–6. It shows three different volumes of sound.

▼ **ASK YOURSELF**

How does the amount of energy in a sound wave affect volume?

Pitch

Later in the morning, the pianist rehearses a solo. She stops and asks the singer if she is playing in a high enough pitch. The vocalist suggests she play in a higher pitch. What does that mean, you wonder? As you listen to the pianist adjust the sound, you infer that the **pitch** of a sound means its highness or lowness. Pitch is not affected by the volume—the softness or loudness of the sound.

During a rehearsal break, your group is invited on stage. You ask the pianist how she increased the pitch on the piano. She invites you and the rest of the group to look under the lid of the grand piano. The pianist points out that shorter strings are attached to keys that play higher notes; longer strings are attached to keys that play lower notes. She tells you that 2000 years ago ancient Greek philosophers knew that when a string was shortened to half its length, the note it produced was twice as high in pitch. The Italian astronomer and physicist Galileo (1564–1642) suggested that the reason this happens is because the frequency doubles.

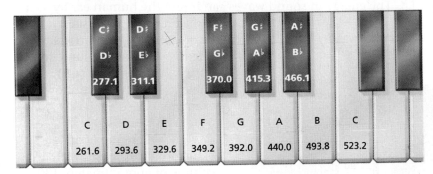

Figure 9–7. Higher pitched notes are created when keys attached to shorter strings are struck. The pitch produced by each of these piano keys is determined by the frequency.

You now understand that piano keys attached to shorter strings have higher pitch. And, according to Galileo, the shorter strings create a higher pitch because they have a higher frequency when struck. This means they vibrate more times per second than longer strings. Try the next activity to find out how pitch affects the pattern of sound waves.

 *Observing* _____

Observe the measurements of an oscilloscope that is connected to a microphone. An oscilloscope is shown here. Observe the wave patterns of five different sounds. How does the pitch change as frequency changes? How does the oscilloscope pattern show the amplitude of the wave? ✎

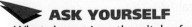 **ASK YOURSELF**
What determines the pitch of a sound?

TEACHING STRATEGIES, continued

● **Process Skills:** *Applying, Inferring*

Encourage the students to offer other examples of robots used in industry, manufacturing, and other enterprises. (Examples might include the use of robots on assembly lines in automobile manufacturing.)

● **Process Skill:** *Comparing*

Have the students work independently or in pairs to research the similarities and differences between ultrasound and x-rays. (The students should note the advantages and disadvantages of each application.)

GUIDED PRACTICE

Cooperative Learning Have the students work in groups to outline the content of this section. Suggest that the students use topic heads to structure their outlines. Then have each group share its outlines with classmates and revise its outlines based on classmates' feedback.

INDEPENDENT PRACTICE

Have the students provide written answers to the Section Review. In their journals, have them write a letter to a newspaper editor on the subject of noise pollution from a local industry or airport.

 A new type of cane for visually impaired people uses ultrasonic waves. The cane sends out waves of about 50 000 hertz. The signals reflect off obstacles and return to a sensor that the person wears. A sounding device in the cane gives off sounds of different frequencies for different distances.

REINFORCING THEMES
Environmental Interactions

Strike a tuning fork, and touch it to a table tennis ball hanging from a string. (The ball will bounce away.) Lead the students to understand that the energy of the sound waves caused the ball to move.

MEETING SPECIAL NEEDS

Mainstreamed

Sound waves are invisible but can, in a sense, be made visible. Strike a tuning fork, and ask mainstreamed students whether they can see it vibrating. (no) Then touch the vibrating fork to the surface of the water in a beaker. Ask the students to describe in their own words what happened. (The vibrations of the tuning fork cause waves in the water.)

Figure 9–11. Some robots are equipped with sonar to detect obstacles in their path. The robot shown here was designed to replace guide dogs used by the visually impaired.

Figure 9–12. An ultrasonic cleaner such as the one shown here uses high-frequency sound waves to clean jewelry.

Sonar is even used in some robots. The sonar enables the robots to "sense" obstacles or to detect moving objects. Computers process the signal from the sonar, allowing the robots to react to the location or movement of objects. In this manner, some robots are able to respond to unexpected situations.

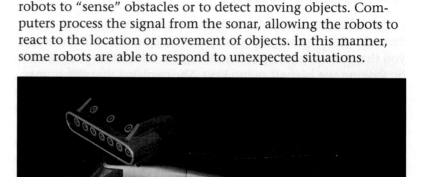

Sparkling Clean Ultrasonic waves have uses in addition to sonar. One common device uses ultrasonic waves to clean jewelry, machine parts, and electronic components. This device, called an *ultrasonic cleaner,* consists of a container that holds a bath of water and a mild detergent. Sound waves of about 50 kHz are then sent through the bath. These sound waves create such intense vibrations in the water that they remove dirt from the items placed in the bath. The major advantage of ultrasonic cleaners is that they are nonabrasive, so they do not scratch the items as they are cleaned.

Using Sound Waves in Medicine Ultrasonic waves used for medical applications are referred to as *ultrasound.* In some cases, ultrasound is used to remove kidney stones without

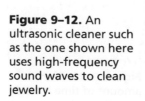

EVALUATION

Have the students write a simple dialogue between two people in which the two speakers show their understanding of the meaning of *sound waves, volume, pitch, noise pollution,* and *ultrasonic waves.* Have the students place their dialogues in their science portfolios.

RETEACHING

Write the section subheads on the board. Using each as a section main idea, have the students take turns telling what each subhead means in terms of sound. For example, for the subhead "Sending Out a Signal," a student might note that sound travels in waves caused by vibrations made by people or objects.

EXTENSION

Have the students use library resources to find out what the frequency range is for different singing voices—soprano, alto, tenor, bass—and what the frequency range is for a woodwind instrument, a percussion instrument, and a stringed instrument.

CLOSURE

Cooperative Learning Have the students work in small groups to create a sound web, filling it in with concepts, definitions, and examples related to the section content.

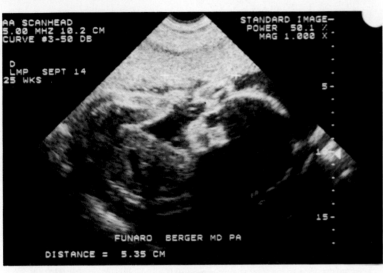

Figure 9–13. This sonogram shows a 25-week-old fetus. Sonograms are used to determine the size, health, and position of a fetus.

surgery. Kidney stones are crystals that form in the kidneys. Ultrasound is used to shatter these crystals without damaging the soft tissues nearby. The tiny fragments can then pass out of the body in the urine.

Ultrasound also provides a way to "see" inside the body. The ultrasonic waves bounce off high-density tissue, such as a tumor. The reflected waves are converted into electrical signals and fed into a computer. The computer uses these signals to create an actual picture, called a *sonogram.* Using this technique, physicians can locate tumors and gallstones. They can also examine developing fetuses inside their mothers.

◢ ASK YOURSELF

Explain two uses of ultrasonic waves.

SECTION 1 *REVIEW AND APPLICATION*

Reading Critically

1. Explain how sound waves travel through air.
2. What two measurements would you make to determine accurately how far away a lightning bolt struck in a thunderstorm?

Thinking Critically

3. Does the volume of a sound increase or decrease as sound waves move further and further from their source? Explain your answer.
4. Tell which of the following sounds is best described as noise pollution. Then explain your answer.
  a. Three people singing different songs at the same time in high pitches.
  b. The sound of an airplane taking off from a runway.

SCIENCE TECHNOLOGY SOCIETY Conventional generators of ultrasound usually produce multiple beams that fan out from their source. Scientists have recently developed improved ultrasound generators that emit single, well-defined beams. These improved beams will prove beneficial to sonar and medical ultrasound imaging.

ONGOING ASSESSMENT
◢ ASK YOURSELF

Ultrasonic waves can be used to clean jewelry and electronic parts and can be used for medical applications.

SECTION 1 REVIEW AND APPLICATION

Reading Critically

1. The vibrations of an object set up zones of compression and rarefaction, creating waves that travel through air and other media.

2. Count the number of seconds between seeing the lightning and hearing the thunder, and then divide by three.

Thinking Critically

3. The volume decreases. For example, the sound of a jackhammer decreases as you move further from it.

4. Choice **b** is an example of noise pollution because the decibel range is so high it can harm human hearing. Choice **a** will simply sound unpleasant but will probably not be a danger to human health.

Process Skills:
Communicating, Measuring, Comparing, Interpreting Data

Grouping: Groups of 3 or 4

Objectives

● **Communicate** data by plotting a graph based upon the frequency of a string at 10-cm intervals.
● **Interpret** the data to show the relationship between the length of the string and the frequency.

Discussion

Graphs are used in science to analyze, interpret, and draw conclusions from recorded data. Ask the students why a decrease in one variable on the graph would cause an increase in the other variable. (A graph in which one variable increases while the other decreases shows that the two variables are inversely proportional to each other.)

Application

1. The graph should show that shortening the string increases the frequency produced by the string.

PERFORMANCE ASSESSMENT

Cooperative Learning

After the completion of this Skill, have each group of students think of another way they could present the data. (Examples include a horizontal or vertical bar graph or a pictograph. Accept all reasonable suggestions.)

2. Find the string length for middle C by reading the length of the string for a frequency of 261.6 Hz. The string should be about 38 cm long.

3. The frequency is approximately 414 Hz.

✳ **Using What You Have Learned**

The thickness and tightness of the string also determine pitch.

PORTFOLIO ASSESSMENT

Have the students place their graphs, observations, and answers from this *Skill* in their science portfolios.

SKILL *Graphing Musical Pitch*

▶ MATERIALS
● graph paper ● pencil

Study the table. It shows the frequency of the sound produced each time a guitar string was plucked. The string was plucked eight times, each time shortened 10 cm by the placement of the musician's finger. The frequency was measured in Hz by an audio oscillator—an electronic device used to identify the pitch of sounds accurately.

TABLE 1: FREQUENCY OF GUITAR STRING	
Length (cm)	Frequency (Hz)
80.0	124
70.0	142
60.0	165
50.0	198
40.0	248
30.0	331
20.0	496
10.0	992

▼ PROCEDURE

1. Use graph paper to create a graph, and mark the horizontal axis in centimeters. Make the scale large enough so that the final reading is near the right side of the paper.
2. Now mark the vertical scale of the graph in Hz. The 1000-Hz mark should be near the top of the page.
3. Plot each of the eight measurements, or data points, on the graph, and connect them with a smooth curve.

▶ APPLICATION

1. What happens to the frequency produced by the string as the string gets shorter?
2. Middle C has a frequency of 261.6 Hz. Use your graph to determine how long the string would have to be to sound a middle C.

3. A♭ and G♯ correspond to 25.3 cm. What frequency is that?

✳ **Using What You Have Learned**
Considering what you know about a guitar, what besides the length of the string determines its pitch?

FOCUS

This section describes how the human ear responds to sounds and how the organs of the voice produce sound. The effect of the motion of a source of sound upon pitch—the Doppler effect—is discussed, as well as the characteristics of sound transmission and methods for improving sound quality.

MOTIVATING ACTIVITY

Cooperative Learning Have the students work in small groups to make collages or posters of pictures they have collected or illustrated that show the ears of various species of animals, including humans. Have the students do research on each ear and then write a brief description of each kind of ear and how this adaptation serves the particular animal. Encourage the students to display their posters in the classroom.

PROCESS SKILLS
• Measuring • Calculating
• Analyzing • Interpreting
Data

POSITIVE ATTITUDES
• Openness to new ideas
• Curiosity

TERMS
• Doppler effect • acoustics
• reverberation • larynx

PRINT MEDIA
The Science of Music
by Melvin Berger
(see p. 211b)

ELECTRONIC MEDIA
Sound Waves and Stars: The Doppler Effect BFA
Educational Media
(see p. 211c)

SCIENCE DISCOVERY
• Ear; human
• Senses; hearing
• Sound to ear

BLACKLINE MASTERS
Laboratory Investigations
9.1, 9.2
Extending Science Concepts
Thinking Critically
Study and Review Guide

Hearing Sound

D uring lunch break, your group sits on benches outside the auditorium. Instead of the rhythms of the Five Notes weaving together, you hear the daily sounds of a busy city street. Ms. Mendoza asks you to use your imagination to discover a rhythm in these street sounds. You close your eyes and listen: a mourning dove repeats soft coos from a tree branch above; car engines hum steadily in a chorus; the traffic light lets out rhythmic clicks as the colors change from green to yellow then red and back to green again. Your ear has learned to hear a new kind of music!

SECTION 2

Objectives

State how the motion of the source of sound affects pitch in the Doppler effect.

Evaluate and **explain** how reverberation influences acoustics.

Locate parts of the human ear, and relate them to the process of hearing.

The Doppler Effect

Just as you step back inside the building, the intense scream of a siren signals a firetruck's approach. The pitch of the siren gets higher and higher as the truck gets closer to where you are standing. Then, when the truck passes by, the pitch of the siren suddenly drops. Why did that happen?

Look at the illustration of the firetruck. You can see sound waves in front of the truck and behind it. The sound waves in front of the truck are packed tightly together as the truck moves closer to the last wave it produced. They have a shorter wavelength and a higher frequency. That's why you hear

Figure 9–14. The sound waves from a siren change pitch due to the Doppler effect.

● **Process Skill:**
Communicating

Reinforce the concept of the Doppler effect by having the students make two drawings. One drawing should show the position length of the sound waves when a car remains in one place with its engine running. The second drawing should show the position length

of the sound waves when a car moves in one direction. (The first drawing should show sound waves moving out evenly in all directions from the standing car. The second drawing should show that sound waves in front of the car are pushed together while sound waves behind the car are spread farther apart.)

● **Process Skills:**
Experimenting, Observing

 Cooperative Learning Work with groups of students to construct one or two experiments they could set up with available objects to test the Doppler effect. For example, the students could blow a whistle or

ring a bell while standing still and then while walking in an empty hall or outside at an appropriate time. Encourage the students to share their experiments and results with classmates.

Demonstration

To demonstrate the Doppler effect, attach a whistle to the end of a 1 m length of flexible rubber tubing so that you can sound the whistle by blowing into the other end of the tube. Then, while blowing into the tube, swing the whistle around in a vertical oval. The students should hear the pitch of the whistle rise and fall.

ONGOING ASSESSMENT
▼ ASK YOURSELF

An ice cream truck equipped with a bell approaches a listener and then drives past. As it approaches, the pitch of the bell increases and then becomes lower as the truck passes. The frequency of the sound waves of the bell increases as the truck approaches and then decreases as it passes. Accept all reasonable responses.

SCIENCE BACKGROUND

In 1842 Christian Johann Doppler, an Austrian physicist, published a paper describing what is now known as the Doppler effect. Three years later a group of scientists decided to test his theory further. They loaded a train with 15 trumpet players and took a position along the track. As the train came by with trumpets blaring, the scientists found that the pitch of the instruments increased and then decreased, just as Doppler had predicted.

the pitch of the siren get higher as the firetruck approaches. Then look at what happens when the truck passes. The sound waves are farther apart from one another because the truck is now moving away from the last wave it sent toward you. That's the reason the pitch of the siren drops as the firetruck passes. A change in the frequency of waves caused by a moving wave source or a moving observer is called the **Doppler effect.**

▼ ASK YOURSELF
Give an example of the Doppler effect, and explain how it occurs.

Acoustics

Everyone noisily reassembles in the auditorium. The voices, laughs, and coughs seem to echo all around you. In fact, the design of an auditorium affects the quality of the sound produced by musicians or speakers on stage. Ms. Mendoza tells you that the city auditorium will soon be renovated to improve its acoustics. **Acoustics**, she explains, is the branch of physics that deals with the transmission of sound. She points out that lighting fixtures, seats, and other decorative details in an auditorium affect the acoustics.

Ms. Mendoza hands out a diagram of an auditorium, showing the way sound waves travel to four different seats. She wants you to determine whether or not there are acoustical problems with reverberation (rih vuhr buh RAY shuhn). **Reverberation** is the combination of many small echoes, occurring very close together. It is caused when reflected sound waves meet at a single point, one right after another. Some reverberation gives music or speech a richness or fullness. Too much reverberation, though, makes speech hard to understand and music fuzzy.

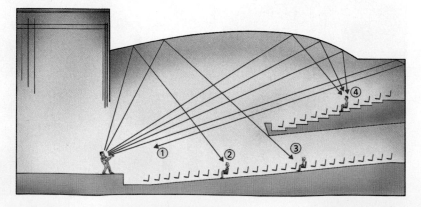
Figure 9–15. This diagram shows how sound reflects from the ceilings and walls of an auditorium.

DISCOVER BY *Writing*

Look at the diagram of the auditorium in Figure
9–15. Follow the sound pathways from the
speaker to the people in the audience marked
"2" through "4". Is this auditorium well designed
or poorly designed? In your journal, write a paragraph that supports your answer. Make sure that you explain
whether or not reverberation would be a problem and why. ✎

ASK YOURSELF

How can reverberation affect the way you hear voices or music in
an auditorium?

How the Human Ear Hears

As you think about the variety of sounds you've heard so far
today, you can't help but wonder how the human ear hears. You
ask Ms. Mendoza and she says, "Actually, your ears don't hear
sounds. Rather, they convert energy from sound waves into
nerve impulses. These impulses travel from your ears through
nerves that lead to your brain, where they are interpreted as
sound."

She passes out an illustration that shows the ear and describes how it works. Then she suggests an activity that simulates a vibrating eardrum. After you look at the illustration of
the human ear, try the activity to see how the eardrum receives
sound.

Figure 9–16.
(1) The outer ear collects sound waves and
passes them through the outer ear until they
strike the eardrum. (2) The eardrum vibrates
when sound waves hit it. The hammer, a small
bone of the middle ear, is connected to the
eardrum. When it vibrates, the anvil and
stirrup amplify the vibrations from the
hammer. (3) Energy from the vibrations pass to
the cochlea in the inner ear. This snail-shaped
tube is filled with liquid and lined with nerve
endings. (4) Vibrations in this liquid move
small hairs that stimulate nerve cells, which
send messages to the brain. The brain inter-
prets these messages as sound.

The Human Ear

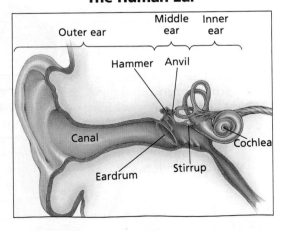

Calculating the Speed of Sound

Process Skills: Measuring, Calculating, Analyzing, Interpreting Data

Grouping: Entire class

Objective
● **Experiment** to determine the speed of sound.

Pre-Lab
Ask the students to describe methods for determining the speed of sound.

Hints
In the course of this *Investigation*, disregard a slight increase in volume you might notice when the length of the column is about 1/8 wavelength. At 1/4 wavelength, the effect is marked and unmistakable.

You might expand on this Investigation by repeating the demonstration at a different frequency to show that the velocity does not depend on frequency.

▶ Analyses and Conclusions
1. The speeds will vary slightly, but they should be between 340 m/s and 350 m/s.

2. You should expect agreement within 5 m/s.

PERFORMANCE ASSESSMENT

Cooperative Learning
Have the students summarize the process they used in this Investigation. Check the students' data and calculations for accuracy.

3. On a warm day, the velocity is greater, so the air column at resonance will be longer.

4. The speed will be slower.

▶ Application
Because the daytime temperature would be higher, the air columns at resonance would be longer, allowing your voice to travel faster during the day.

✳ Discover More
Sound would travel faster outdoors if the outdoor temperature was much warmer than the indoor temperature. Sound would travel more slowly outdoors if the outdoor temperature was much colder than the indoor temperature. Varying air temperature accounts for the varying results.

Post-Lab
Ask the students to summarize what they have learned about calculating the speed of sound.

Have the students place their measurements, observations, and answers in their science portfolios.

INVESTIGATION

*C*alculating the Speed of Sound

▶ MATERIALS
● standard water-column apparatus ● tuning fork, G = 392 Hz
● thermometer

▼ PROCEDURE

1. Put enough water in the metal reservoir of the water column so that the length of the air column can be adjusted.
2. Adjust the height of the water in the reservoir until the air column is about 20 cm long.
3. Strike the tuning fork, and hold it over the open end of the tube.
4. Lower the water in the reservoir until you hear a sudden increase in the volume.
5. Continue to move the reservoir up and down until you are sure you have found the place where the volume is at the maximum.

6. Measure the length of the air column (in meters), and multiply it by four to get the wavelength.
7. Multiply the wavelength by the frequency of the tuning fork to find the speed of the sound. The frequency of a tuning fork is stamped on its handle.
8. Measure the air temperature, and double-check the speed you just calculated using the following procedure. At 0°C, sound travels though air at 331.5 m/s. For every 1°C increase in temperature, the speed of sound increases by 0.6 m/s. Calculate the speed of sound at your room's air temperature.

▶ ANALYSES AND CONCLUSIONS
1. According to your measurements of the air column, what is the speed of sound?
2. How does the speed compare with the speed calculated from the temperature?
3. What would happen to the length of the air column on a warm day? How would you use the information above to support your response?
4. Describe the results of the same experiment performed in a much colder room.

▶ APPLICATION
If you were in the desert with a friend, would your voice travel faster during the day or during the night? Why?

✳ Discover More
If the air temperature outside is substantially different from the air temperature inside, repeat this investigation outdoors. How do the results change if the temperature outdoors is much warmer than it is indoors? much colder? What do you think accounts for the different results?

CHAPTER 9 HIGHLIGHTS

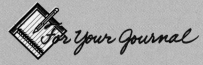

CHAPTER 9 HIGHLIGHTS

The Big Idea

Sound is vibrations—the rapid back-and-forth movement of matter. Sound, like other forms of energy, can travel through matter. The amount of energy in a sound wave determines the intensity of volume. The frequency of a sound wave determines the pitch. Sound travels through some forms of matter faster than through others. The medium through which the sound wave moves and the temperature of the medium affects the speed at which a sound wave travels.

Properties of sound are understood not only by characteristics of the source that produces the sound wave but also by the object that receives the sound wave. The human ear receives sound waves; the human voice produces sound waves.

For Your Journal

Look back at the ideas you wrote in your journal at the beginning of the chapter. How have your ideas changed? Now that you know that sound can be harmful to your health, what will you do to protect your hearing? Add your new ideas about the uses of sound in medicine and industry.

Connecting Ideas

These illustrations show examples of some of the concepts presented in this chapter. In your journal, write a caption that describes each of the three scenes.

Understanding Vocabulary

1. (a) Both terms relate to sound. Volume is the loudness or softness of a given sound, while pitch is the highness or lowness of the same sound.

(b) Both terms relate to volume. Volume is measured in units called decibels. Noise pollution is sound that has a high decibel reading or is loud enough to be harmful to the human ear.

(c) Each term defines sounds that humans cannot hear. Ultrasonic waves are sound waves above the limit of human hearing. Sonar is a navigation system that uses ultrasonic waves to determine water depth. Ultrasound is the medical application of ultrasonic waves.

(d) Both terms are related to the transmission of sound. Acoustics is the branch of physics that studies such phenomena. Reverberation is the combination of many small echoes very close together, the amount of which affects the quality of sound.

Understanding Concepts

Multiple Choice

2. d

3. b

4. c

5. d

Short Answer

6. You are correct. Noise pollution is a question of volume and damage to human hearing, not musical preference. A chain saw has a decibel level that causes harm to the human ear. Marching band music, as long as it does not exceed a safe volume, is not considered noise pollution.

7. Ultrasonic waves are directed toward a specific organ of the body, where they bounce off the organ's tissues. The waves then are converted into electronic signals that produce a "picture" of the organ.

Interpreting Graphics

8. (a) eardrum, hammer, anvil, stirrup

(b) outer ear, canal

CHAPTER 9

REVIEW

Understanding Vocabulary

1. For each set of terms, explain the similarities and differences in their meanings.
   a) volume (244), pitch (245)
   b) noise pollution (246), decibels (246)
   c) ultrasonic waves (247), sonar (247), ultrasound (248)
   d) acoustics (252), reverberation (252)

Understanding Concepts

MULTIPLE CHOICE

2. A sonar device can use the echoes of ultrasonic waves to find the
   a) speed of sound.
   b) temperature of water.
   c) height of any waves.
   d) depth of water.

3. Ultrasonic waves are *not* used to
   a) locate tumors.
   b) improve acoustics.
   c) clean jewelry.
   d) shatter kidney stones.

4. During a thunderstorm you see lightning before you hear thunder because
   a) thunder occurs after lightning.
   b) thunder is farther away than lightning.
   c) light travels faster than sound.
   d) sound travels faster than light.

5. A person with his or her ear to the rail hears an approaching train sooner than someone standing up next to the tracks because
   a) sound travels faster closer to the ground.
   b) the train is within sight.
   c) the first person has better hearing.
   d) sound travels faster through a solid than through a gas.

SHORT ANSWER

6. You complain of the noise pollution caused by chain saws being used to remove trees brought down by a storm. Your sister complains that while the school band does not play excessively loud, it causes noise pollution because it plays too many marching songs. Which of you is correct?

7. Describe how ultrasonic waves are used to "see" inside the body.

Interpreting Graphics

8. Which part or parts of the ear
   a) vibrate when sound waves hit them?
   b) collect sound waves and pass them to the eardrum?

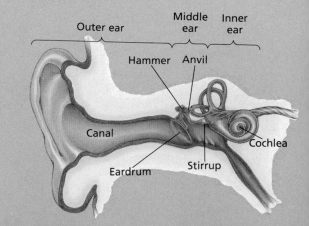

Reviewing Themes

9. *Environmental Interactions*
Explain how each of the following contributes to the sounds you make when you speak: lungs, larynx, vocal cords, mouth, tongue, lips.

10. *Environmental Interactions*
If a tree falls in a forest and no one is there to hear it, does the tree make a sound as it falls? Explain your answer.

Thinking Critically

11. An acoustic engineer might place drapes on the walls of an auditorium. How might the drapes improve the acoustics in the auditorium?

12. A popular science fiction film used this statement in its advertisements: "In space, no one can hear you scream. . . ." Explain the statement in terms of what you have learned about sound waves.

13. How might a robot equipped with sonar be more useful than a guide dog? How might the robot be less useful?

14. Your friend has heard about something called a sonogram, which physicians use to examine a fetus inside the womb. Your friend says, "If it has to do with sound, it must hear and measure the heartbeat." How would you answer your friend?

15. An orchestra is playing in a huge outdoor amphitheater. Thousands of listeners sit on a hillside far from the stage. The amphitheater has an amplification system to increase the volume produced by the orchestra. A special computer slows down the amplified sound of the orchestra by a fraction of a second. Why is this computer used? What would happen if the speed of the amplified sound was not slightly decreased?

16. Police often use radar guns to help them determine the speed of approaching vehicles. These guns send out high-frequency waves that are reflected back from moving objects. How does the Doppler effect relate to the operation of these radar guns?

Discovery Through Reading

"New Devices Wipe Out Noise." 77 *Current Science* (March 13, 1992): 11. This article discusses the development of a "noise-cancellation device."

LIGHT AND LENSES

PLANNING THE CHAPTER

Chapter Sections	Page	Chapter Features	Page	Program Resources	Source
CHAPTER OPENER	260	*For Your Journal*	261		
Section 1: WHAT IS LIGHT?	262	Discover by Doing (B)	263	*Science Discovery**	SD
• The Brightness of Light (B)	262	Discover by Doing (B)	264	Connecting Other	
• The Ray Model of Light (B)	263	Section 1 Review and Application	265	Disciplines: Science and	
• Other Models of Light (B)	264			Mathematics, Measuring	
				with Light (H)	TR
				Reading Skills: Finding the Main	
				Idea and Supporting Details (B)	TR
				Study and Review Guide,	
				Section 1 (B)	TR, SRG
Section 2: USING THE RAY MODEL	266	Discover by Doing (A)	267	*Science Discovery**	SD
		Discover by Observing (B)	268	Investigation 10.1:	
• Bending Light Rays (A)	266	Section 2 Review and Application	269	Changing Image Size	
• Lenses and Light Rays (A)	267	Skill: Communicating Using		with Optics (H)	TR, LI
		a Diagram (A)	270	Study and Review Guide,	
				Section 2 (B)	TR, SRG
Section 3: IMAGES	271	Activity: How can you find a		*Science Discovery**	SD
• Real Images from Lenses (A)	271	virtual image? (A)	273	Investigation 10.2:	
• Not So Real Images (A)	272	Discover by Doing (H)	276	Observing Multiple Images	
• The Human Eye (A)	274	Section 3 Review and Application	277	in Plane Mirrors (A)	TR, LI
• Flat Mirrors (H)	276	Investigation: Finding the		Extending Science Concepts: Types	
		Properties of Images Formed		of Mirrors and Their Images (H)	TR
		by a Convex Lens (A)	278	Thinking Critically (A)	TR
				Light and Lenses	IT
				The Human Eye	IT
				Record Sheets for Textbook	
				Investigations (A)	TR
				Study and Review Guide,	
				Section 3 (B)	TR, SRG
Chapter 10 HIGHLIGHTS	279	The Big Idea	279	Study and Review Guide,	
Chapter 10 Review	280	For Your Journal	279	Chapter 10 Review (B)	TR, SRG
		Connecting Ideas	279	Chapter 10 Test	TR
				Computer Test Bank	

B = Basic   **A** = Average   **H** = Honors
The coding Basic, Average, and Honors indicates subsections, features, and resources that might be appropriate for different levels of learners. For additional suggestions regarding choice of topic and depth of coverage, see the Pacing Chart on pages T28–T31.

*Frame numbers at point of use
(TR) Teaching Resources, Unit 3
(IT) Instructional Transparencies
(LI) Laboratory Investigations
(SD) *Science Discovery* Videodisc
    Correlations and Barcodes
(SRG) Study and Review Guide

CHAPTER MATERIALS

Title	Page	Materials
Discover by Doing	263	(per individual) camera light meter or separate light meter
Discover by Doing	264	(per class) projector, screen
Discover by Doing	267	(per individual) pencil, piece of clear glass, spoon, glass of water
Teacher Demonstration	267	small, rectangular fishtank; water; milk; flashlight or laser
Discover by Observing	268	(per group of 3 or 4) convex lenses, sunlight or distant light source, sheets of paper, metric ruler
Skill: Communicating by Using a Diagram	270	(per pair) mirrors (2), table, small objects
Activity: How can you find a virtual image?	273	(per individual) colored glass, 20 cm × 10 cm corrugated cardboard, straight pins (6), metric ruler, protractor
Discover by Doing	276	(per individual) wall mirror, sheets of plain and transparent paper, marking pens
Investigation: Finding the Properties of Images Formed by a Convex Lens	278	(per individual) white cardboard, convex lens, meter stick, lamp

ADVANCE PREPARATION

You might solicit students' help in order to have enough small mirrors, lenses, cameras, eyeglasses, and so forth for the activities in this chapter. Ask particularly for students to bring in eyeglasses and bifocals no longer in use.

TEACHING SUGGESTIONS

Field Trip
The school's photography or camera club would be a good place to visit or have come to the classroom. Advanced students can demonstrate how cameras work, different attachments, and so forth. If your high school has no such club, perhaps a local college or junior college would be willing to oblige.

Outside Speaker
Ask a local optician or optometrist to speak to your class. The lecture can focus on how eyes work, how they are tested, or modern day environmental threats to the eyes, depending upon your speaker's interests.

CHAPTER 10 LIGHT AND LENSES

CHAPTER THEME—SYSTEMS AND STRUCTURES

In Chapter 10 the students will learn about the similarity between the human eye and devices that use lenses—systems that capture and focus light to form images. The students also will investigate the way in which light interacts with other material through reflection and refraction. A supporting theme in this chapter is **Environmental Interactions.**

MULTICULTURAL CONNECTION

Before the invention of the microscope, lenses had to be perfected. The first convex lenses (similar to those used in the first microscope) were invented in what is now Tunisia—a country in northern Africa—about 300 B.C. After the invention of the lens, the knowledge was passed through trade routes to the Middle East. Improvement of these lenses and a better understanding of how they worked were made by many people during this time, notably Al-Hazen (965–1038 A.D.) Eventually, the information was translated into Latin, and the European community became aware of it.

The Dutch had a great interest in lenses, and, in fact, they were the first to use a lens to make a microscope (around 1590). This first microscope was made by combining two double-convex lenses in a tube. In 1680 Leeuwenhoek, a Dutch biologist, built a simple microscope that magnified objects up to 200 times.

MEETING SPECIAL NEEDS

Mainstreamed

Have the students observe their shadow at three different times on a sunny day. The students should note the time of day that they make their observations and the position of the sun in the sky. This will help the students observe how the shadows are affected by the changing angles from which the light is coming.

▶ 260 CHAPTER 10

CHAPTER 10 LIGHT AND LENSES

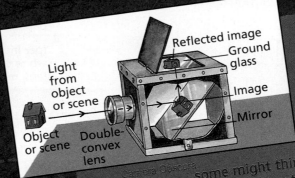

Camera Obscura

Telescopes fitted with the first glass lenses brought new worlds closer to Europe in the 1600s. With the telescope, scientists discovered moons and stars never before seen by human eyes. Yet the glass lens also brought the familiar world closer. In the hands of a Dutch master artist, the lens revealed an inner world of marvelous light.

Jan Vermeer is admired as one of the greatest painters of the 1600s. However, some might think Vermeer was something of a cheat. He probably composed his vivid interior scenes using an early camera-like device called the *camera obscura.*

Camera obscura is Latin for "darkened room." For centuries scientists had known that a tiny shaft of light entering a darkened chamber projected an image on the back wall. By Vermeer's time, inventors had added glass lenses to focus the image and made the camera obscura into a portable box.

How did the images created by the camera obscura affect Vermeer's paintings? Historians find several clues that Vermeer used the camera obscura to trace scenes such as the geographer gazing out a window. Notice that the geographer's table and woven tablecloth seem overly large. The camera devices of Vermeer's day enlarged objects near

CHAPTER MOTIVATING ACTIVITY

A spectacular demonstration is possible if you can effectively darken the classroom or another room. Keep the room completely dark except for a 5-cm hole in a sheet of poster board that is taped against a window. This serves as a pin-hole; the entire room becomes a camera obscura. The whole scene outside the room will appear, upside down, on the wall opposite the window.

For Your Journal

The students' answers to the journal questions will help you assess their knowledge about light and lenses. The students will have a variety of responses to the journal questions. (Accept all answers, and tell the students they will have a chance to refer back to the questions.)

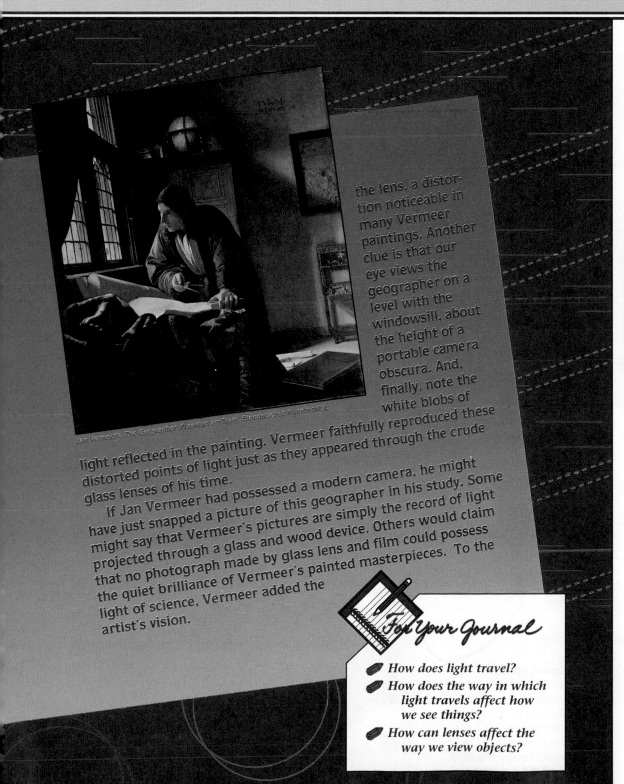

Jan Vermeer's The Geographer, Frankfurt am Main, Staedelsches Kunstinstitut.

the lens, a distortion noticeable in many Vermeer paintings. Another clue is that our eye views the geographer on a level with the windowsill, about the height of a portable camera obscura. And, finally, note the white blobs of light reflected in the painting. Vermeer faithfully reproduced these distorted points of light just as they appeared through the crude glass lenses of his time.

If Jan Vermeer had possessed a modern camera, he might have just snapped a picture of this geographer in his study. Some might say that Vermeer's pictures are simply the record of light projected through a glass and wood device. Others would claim that no photograph made by glass lens and film could possess the quiet brilliance of Vermeer's painted masterpieces. To the light of science, Vermeer added the artist's vision.

ABOUT THE ARTIST

Jan Vermeer (1632–1675) painted very few canvases. His subject was often the beauty and simplicity of ordinary Dutch people engaged in everyday activities. Vermeer is considered a master of the interior scene and the pictorial representation of light, particularly that of the sun pouring through a window.

Johannes Vermeer, "The Kitchenmaid," Oil on canvas, 45.5 x 41 cm. Rijksmuseum, Holland.

You can further utilize the illustration by having the students look at the art and relate it to the Chapter Motivating Activity. Ask the students why the image in a camera obscura is upside-down. (Light rays travel in a straight line. Those from the top enter the pin-hole and strike the wall lower than those from the bottom, which do just the opposite.)

For Your Journal

✎ How does light travel?
✎ How does the way in which light travels affect how we see things?
✎ How can lenses affect the way we view objects?

FOCUS

This section focuses on how light travels and the measurement of its speed and brightness. Theories of light are used to explain the properties of light.

MOTIVATING ACTIVITY

Cooperative Learning Have the students work in pairs or small groups to draw and collect pictures of everyday things they encounter that produce or process light. (Examples include various kinds of indoor lights, flashlights, fire, sunlight, cameras, eyes of humans and animals.) Have the students arrange the pictures in two groups—objects that produce light and objects that process light. Encourage the students to share their work with classmates.

SECTION 1

What Is Light?

Objectives

Recognize *that the brightness of light depends upon the distance from the light source.*

Diagram *how light travels using the ray model.*

Summarize *the theories that have been proposed to explain the nature of light.*

Imagine that you've joined the school photography club. For your first meeting, the club members go downtown to a sporting goods store where an athletic apparel company is having a photography session for some new advertisements. Two teenage models, wearing T-shirts and caps with the company logo, are taking directions from the photographer. The photographer sets up her camera and lights. She places framed white sheets behind the models.

Even though it is a cloudy and cool day, the models look warm under the hot lights. The scene itself looks bright and sunny because light reflects off the white material. The models pose over and over again, shot after shot, until the photographer captures the exact image she wants.

The Brightness of Light

Before each shot, the photographer checks to make sure that the camera setting matches the reading from a light meter. She allows a photography club member to look at an extra light meter on the set. You can try the same thing.

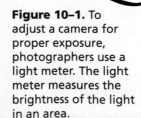

Figure 10–1. To adjust a camera for proper exposure, photographers use a light meter. The light meter measures the brightness of the light in an area.

● **Process Skill:** *Comparing*

Ask the students to explain how a lamp and a flashlight differ in the light they send out. (A lamp sends a shower of light in all directions; a flashlight sends light as a beam, with little spread. Lamps are used to light entire rooms or areas, while flashlights are used to aim light beams at specific, limited areas.)

● **Process Skill:** *Observing*

Ask the students for everyday examples of light traveling in straight lines. For instance, what do they often see coming through a break in the clouds? (sunlight) Why do drivers have to slow down as they go around curves at night? (because the light from the headlights does not curve as they round the bend)

DISCOVER BY *Doing*

Using the light meter of a camera or a separate light meter, measure the brightness at 10 different locations. Make a table of your measurements, and compare it with the tables of your classmates. Which areas are the brightest? ✎

The photography club's advisor, Mr. Tallchief, explains that photographers are not the only people who measure the brightness of light. Scientists measure brightness also. They use a unit of measure called the **lux.** Sunlight may be up to 50 000 lux. A lamp by which you might read a book is about 1000 lux. The dim light in a hall or theater lobby is about 1 lux. The closer you are to a light source, the brighter it is, so the higher the lux.

▼ **ASK YOURSELF**

How does the lux measurement change as the brightness of light increases?

The Ray Model of Light

You know that the closer you are to a light source, the brighter it is. Why? The reason is that light moves in straight lines outward from a source. What does that have to do with brightness? As light leaves the source, the rays spread apart. The farther the light travels, the more the rays spread apart.

Consider the models at the photo shoot. The closer they stand to the lights, the more rays actually strike them. Therefore, the closer they stand, the brighter the light. As they move away from the lights, fewer rays strike them. So the light is less bright.

A diagram of how light travels, such as the one shown here, can help you understand brightness. Light rays, just like the other waves you have studied, move in straight lines until something happens to make them bend. You can experiment with light rays by doing the next activity.

Figure 10–2. A ray model can help you understand how light behaves.

DISCOVER BY *Doing*

Brightly lit interiors and outdoor areas on a sunny day are among the brightest.

INTEGRATION– Fine Arts

The representation of light is a challenge to painters' techniques. Remind the students that when they look at a painting that depicts a certain kind of light, they are really seeing the artist's use of color.

You might extend this point by having the students bring from home or the library reproductions of artwork that depict light of various kinds—outdoor, dusk, sunset, heavy sun, indoors. Encourage them to find information on their favorite artists and put it in their journal. If possible, set aside a section of the classroom to display the artwork.

SCIENCE BACKGROUND

Besides the lux, another measurement for the brightness of light is the candela. Until 1940 it was called a candle and was based on the luminous intensity put out by a candle made from the wax of sperm whales. (Whale oil was generally used for lamps before the invention of the electric light bulb.)

ONGOING ASSESSMENT
▼ **ASK YOURSELF**

The brighter the light, the higher the lux measurement is.

SECTION 1   **263** ◀

GUIDED PRACTICE

Help the students calculate how long it takes light to travel from the moon to Earth, a distance of 3.8×10^8 m. Remind them that the speed of light is 3.0×10^8 m/s. (Since time is equal to distance/speed, time in this example would be

$$\frac{3.8 \times 10^8 \text{ m}}{3.0 \times 10^8 \text{ m/s}} = 1.3 \text{ s})$$

INDEPENDENT PRACTICE

 Have the students provide written answers to the Section Review. In their journals, have them write a brief description of the lighted areas in their homes, comparing the brightness levels in several places.

EVALUATION

 PORTFOLIO ASSESSMENT Have the students work independently or in pairs to write a brief debate between Isaac Newton and Christiaan Huygens about the nature of light. Have the students put their written dialogue between Huygens and Newton in their science portfolios.

DISCOVER BY *Doing*

No light shines on the areas your hands and body cover; the light is blocked. Light outlines the shape of your pictures. Accept all reasonable drawings. Encourage the students to share their drawings with classmates.

THE NATURE OF SCIENCE

The astronomer Galileo Galilei (1564–1642) also attempted to measure the speed of light. In 1630 he and an assistant, each carrying a candle lantern, climbed two neighboring hills. Galileo uncovered his lantern and flashed its light. When his assistant saw the light, he flashed his lantern in return. Galileo measured the time between the moment he flashed his lantern and the moment he saw the assistant's light. He then tried the same experiment at two hills farther apart.

Galileo found light travels so fast that the extra time taken to travel the longer distance was too small for him to measure. Although Galileo approached the problem as a scientist, he was at a disadvantage because sophisticated clocks had not yet been invented.

ONGOING ASSESSMENT
▼ ASK YOURSELF

The students' diagrams should show the light rays traveling from the lamp, to the page of the book, and then reflecting from the page. Accept all reasonable diagrams.

DISCOVER BY *Doing*

Use a projector to project light onto a screen or clear wall. Create shadow pictures with your hands. On the screen or wall, what happens to the area your hands and body cover? What outlines the shape of your pictures? Draw a diagram to show what happens.

You ask the photographer's assistant setting up the photo area if he knows how fast the light travels. He looks at you and chuckles. "It gets there as soon as I turn the lights on," he says.

Mr. Tallchief explains that the speed of light is not easy to calculate. In fact, it has been a long-time problem for scientists. Early scientists tried to measure the speed of light, but it couldn't be done by ordinary means. It wasn't until 1676 that the Danish astronomer Olaus Roemer (ROH muhr) successfully calculated the speed of light. By studying the eclipses of the moons of Jupiter, he was able to figure out how long it takes for light to travel to Earth. Mr. Tallchief shows you a diagram that explains how Roemer made his measurements.

Mr. Tallchief amazes you when he says, "In a vacuum, light travels at a speed of about 300 million m/s. That means that a light beam could get from Los Angeles to Atlanta in less than a hundredth of a second!"

Figure 10–3. Roemer was able to calculate the speed of light by comparing the distance between Jupiter's moon, Io, and Earth in two different positions in Earth's orbit.

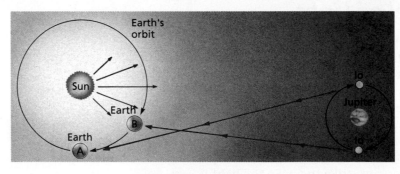

▼ ASK YOURSELF

Draw a diagram that shows how light rays travel from a reading lamp to the page of a book a person is reading.

Other Models of Light

At school the next day, you look up the subject *light* in an encyclopedia and several other library books. You discover that, through history, some scientists have believed that light behaves like waves while other scientists have believed that light behaves like streams of particles. You make a poster showing what you learned.

RETEACHING

Write this section's subheads on the chalkboard. Then have the students draft summary statements about the main idea for each subhead. Encourage the students to share their statements with classmates.

EXTENSION

Have the students research the two Greek schools that studied light. The Pythagorean school said that every visible object emits a stream of particles. Aristotelians concluded that light travels in waves.

CLOSURE

Cooperative Learning Have the students work in small groups to write a summary of this section. Ask a spokesperson from each group to read its summary aloud.

Summary of the Theories of Light

In 1672 the British scientist Sir Isaac Newton (1642–1727) wrote a paper about light. In this paper, Newton hypothesized that light consists of streams of tiny particles. Around the same time, the Dutch physicist Christiaan Huygens (HOY guhnz) (1629–1695) hypothesized that light behaves like waves, since light beams can pass through each other undisturbed—something waves can do, but particles cannot.

In 1804 the British scientist Thomas Young (1773–1829) performed an experiment that showed how light, like waves, has the property of interference. He passed a beam of light through two narrow slits onto a screen. The light produced a pattern of bands that proved light underwent interference.

A century later, Albert Einstein (1879–1955) explained how the energy of light comes out in tiny packages called photons. Photons are given off by hot objects. Einstein's photons are similar to Newton's light particles.

The particle theory of light

The wave theory of light

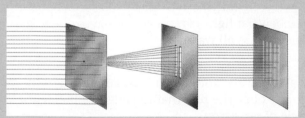

Thomas Young's experiment: On the screen, bright bands result from constructive interference and dark bands result from destructive interference.

▼ **ASK YOURSELF**

In what ways does light behave like waves?

SECTION 1 REVIEW AND APPLICATION

Reading Critically
1. Why does moving closer to a light source increase its brightness?
2. In what ways does light behave like a ray?
3. What are two models of the nature of light?

Thinking Critically
4. Astronomers measure distances in light-years. One light-year is the distance light travels in a year. How many kilometers are there in a light-year?
5. Make a sketch showing light from a lamp striking an object. Using the ray model, explain why the light source looks dimmer when viewed from a greater distance.

✧ **Did You Know?**
Thomas Young was a British physicist, physician, and Egyptologist who helped to decipher the Rosetta stone. He not only showed that light rays have the properties of waves, but also calculated their spectrum wavelengths.

ONGOING ASSESSMENT
▼ **ASK YOURSELF**

Light behaves like waves when light beams pass through one another undisturbed, something particles cannot do.

SECTION 1 REVIEW AND APPLICATION

Reading Critically
1. The closer you are to a light source, the more rays hit you; therefore, the light is brighter.

2. Light acts like a ray because it moves outward in straight lines from a source.

3. the particle theory and the wave theory

Thinking Critically
4. 9 460 000 000 000 km or 9.46×10^{12} km

5. Accept all reasonable sketches. Explanations should note that the light seems dimmer farther away because fewer rays enter your eyes.

FOCUS

This section describes how light rays are changed or bent by reflection and refraction. The characteristics of convex and concave lenses and how they affect light rays are also discussed.

MOTIVATING ACTIVITY

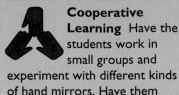

Cooperative Learning Have the students work in small groups and experiment with different kinds of hand mirrors. Have them answer questions such as these:

How can mirrors be used to signal someone far away? How can they throw light on the classroom walls? (Mirrors can be used to reflect light toward another person or thing.)

PROCESS SKILLS
• Interpreting Data
• Communicating

POSITIVE ATTITUDES
• Creativity • Questioning

TERMS
• convex lens • concave lens
• focal length

PRINT MEDIA
"Digital Photography" from *Popular Science* magazine by Daniel and Sally Grotta (see p. 211b)

ELECTRONIC MEDIA
The Simple Lens BFA Educational Media (see p. 211c)

SCIENCE DISCOVERY
• Concave lens
• Convex lens

BLACKLINE MASTERS
Laboratory Investigation 10.1
Study and Review Guide

REINFORCING THEMES
Environmental Interactions

Point out to the students that refraction in the air varies with temperature. When a layer of air near the ground is heated and then covered with a layer of cooler air, light will curve in such a way as to make distant objects visible. The air temperature conditions might cause objects on the ground to appear upside-down in the air. This can cause mirages in deserts, where temperatures are extreme, or long refractive curves from headlights, which might be interpreted as UFOs.

SECTION 2

Objectives

Describe *the processes by which the direction of a light ray can be changed.*

Compare *the light rays that emerge from a lens when the light source is at various distances from the lens.*

Explain *what is meant by the focal length of a lens.*

Figure 10–4. When a light beam reflects from a mirror, the angle of incidence equals the angle of reflection.

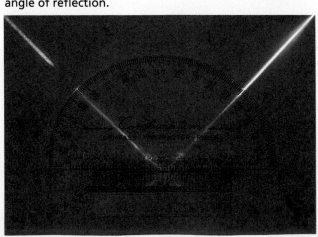

Using the Ray Model

Y ou arrive at the next photography club meeting. On a table you find an exhibit of optical devices, including mirrors, lenses of different shapes, and a 35 mm camera. You pick up the camera and look through it at a tree outside the window. The tree resembles something from a bad dream. The branches look as wide as the trunk, and the whole image looks fuzzy. Then you focus the lens and continue to look at the tree. Now it seems as though the tree is moving toward you. As it does, the image becomes clearer and clearer—no longer a dream, but a reality.

Bending Light Rays

Mr. Tallchief says, "Before you begin to use a camera, you should know what happens when waves of light energy reach optical devices like the mirrors and lenses you see here on display. Remember that light rays are really waves of energy, and they behave just as all waves do."

Reflection When light hits the smooth, flat surface of a mirror it acts as any wave does when it strikes a barrier—it reflects. Light rays obey the rule that the angle of incidence equals the angle of reflection.

Of course, most objects in the world do not have smooth, flat surfaces; they have rough and irregular surfaces. What happens when light from different directions strikes these irregular surfaces? It bounces off in many directions. This scattering is called *diffuse reflection*.

Refraction There's another way in which light changes direction. Mr. Tallchief sets up a simple demonstration to show what happens when light moves from one medium to another. You can do the same thing.

● **Process Skill:** *Synthesizing*

Have the students study the lenses in the illustration on this page. Then ask volunteers to formulate statements that are true for all lenses. (They are made of transparent materials such as glass or plastic; they bend or refract light rays; they have one or two curved surfaces.)

● **Process Skills:** *Experimenting, Predicting*

 Cover a small plane mirror, and place it in the sunlight. Ask the students to predict where the beam will strike when you uncover the mirror. Use various positions and angles. Have the students diagram the results in their journals.

DISCOVER BY Doing

Take a pencil, and view it through a piece of glass held at an angle to your line of sight. What happens to the image of the pencil? Stand a spoon in a glass half filled with water. What happens to the image of the spoon? Explain why you think the spoon looks as it does. 🖋

When light travels from air to glass, it moves from one medium to another medium. Light waves, like other waves, travel at different speeds in different media. When these waves of light energy change speeds, they also change direction, or refract. The part of the pencil behind the glass appears to move to one side, and the spoon seems to bend when it hits the water because the light waves refract.

▼ ASK YOURSELF

Choose an object close to you, and explain how light reacts to its surface.

Lenses and Light Rays

You are looking at the lenses on the table at the front of the room. Mr. Tallchief notices your interest and asks if you would like to know how the lenses work. "Of course," you respond, so he begins to tell you about the two basic types of lenses.

Mr. Tallchief explains that the shape of a lens determines how it focuses the light rays that enter it. He asks, "What can a prism do that a plain piece of glass can't do?"

"Oh, I learned in science class that a prism causes refraction as the light comes into it and as the light goes out of it too," you respond.

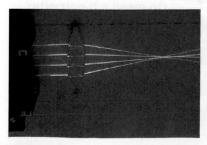
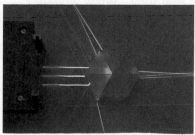

Figure 10–5. A convex lens causes parallel light rays to converge. Six prisms combined in a circle act just as a convex lens does.

DISCOVER BY *Doing*

The images of the pencil and the spoon appear split. The images appear as they do because the light waves break when they strike the glass or the water.

Demonstration

Use the following procedure to demonstrate refraction. Fill a small, rectangular fish tank with water, and add a small amount of milk to make the water slightly cloudy. Then use a flashlight or laser to direct a beam of light at the fish tank. Make the beam visible in the air by blowing chalk dust into the air.

The students will easily see the beam as it enters the water at an angle. If the beam is bright and sharp, the students will be able to see it bend away from the normal as it leaves the tank.

ONGOING ASSESSMENT
▼ ASK YOURSELF

The students' choices will vary. For example, if light hits your hand, your hand blocks the light, causing a shadow behind your hand. Accept all reasonable responses.

SKILL

Communicating by Using a Diagram

Process Skills: Interpreting data, Communicating

Grouping: Pairs

Objective
● **Describe** light reflections by using diagrams.

Procedure
5. Light from the object hit the first mirror and reflected off it at an angle to hit the second mirror. The second mirror reflected the light in the direction of the viewer's eyes. At that point, the viewer could see and identify the object.

▶ Application
1. You could view the final image through a magnifying lens.

2. The students' answers will vary. Examples include a jeweler's instrument for viewing minute jewelry parts, surgical instruments, and a microscope.

※ Using What You Have Learned
The students' periscopes should be based on their diagrams. Encourage the students to demonstrate their periscopes for classmates.

★ PERFORMANCE ASSESSMENT

Cooperative Learning
Encourage pairs of students to point out each part of the periscope they constructed. Check that the students' periscopes work.

You may wish to have the students place their diagrams in their science portfolios.

SKILL Communicating Using a Diagram

▶ MATERIALS
● object ● small mirrors (2)

▼ PROCEDURE

1. Get two small mirrors, and sit under a table as shown in the picture.

2. Ask a classmate to put an object you have not seen onto the table.

3. Hold one mirror above the table so light from the object can hit the mirror. That mirror will reflect the light downward to the second mirror, below the table. Then the second mirror will reflect the light to your eyes. You should be able to see the object on the table.

4. Draw and label a diagram showing the path of reflected light from the object to your eyes.

5. Use what you know about the ray model of light to explain why you can see the object on the table.

▶ APPLICATION
How could you use lenses to magnify the object you saw with the mirrors? What use would you have for an optical instrument that would enlarge an image? Name as many instruments of this type as you can.

※ Using What You Have Learned
A periscope is a device consisting of a tube that holds a system of lenses, mirrors, or prisms. A periscope is used to see objects not in your normal range of vision. One use for a periscope is on a submarine—when the submarine is submerged the periscope allows the sailors to see above the water. Use your diagram to construct a periscope.

Images

At the beginning of the next meeting of the photography club, you leaf through books of photographs. You admire the nature photographs, in particular. One photograph of a ship on a stormy sea is so real looking that you can feel your knees buckle as you stand on the tilted deck along with the other sailors. In another photograph you feel as though you are actually breathing fresh pine-scented air and touching craggy, mineral-encrusted rocks on the steep slope of a stately mountain.

You are so impressed with these photographs that you decide you'd like to take photographs of nature, from simple wildflowers to the vast grassy plains that stretch out in every direction from your town. Before you embark on your first photographic adventure, you should know more about images and the way the human eye sees.

Real Images from Lenses

A camera lens can focus on objects that are very close and those that are far away. "Remember what you just learned about the f-setting for a camera lens?" Mr. Tallchief asks. "Well, imagine that you are visiting Yosemite National Park, taking pictures of the mountain El Capitan. Tiny amounts of the reflected sunlight from the mountain enter the lens of the camera. The lens uses the light to form a picture on the film. That picture is an image. Now think of El Capitan as composed of tiny points. Light reflects off each one of those points and scatters in all directions. The camera lens focuses the light from each point onto the film so that each point is represented by a point of light on the film. The image created when the collection of points that make up El Capitan is focused by the lens is a **real image**."

Figure 10–8. El Capitan

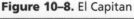

● **Process Skill:** *Inferring*

Remind the students that the farther they are from a light source, the dimmer it becomes, because fewer light rays fall on their eyes. If this is so, how can a camera, which has a small light opening, take a photograph of a distant object and produce a picture that is bright enough to see? What part does the lens

play? (A convex lens captures light from a large area and focuses it into a small point. This keeps the object's brightness high.)

● **Process Skill:** *Solving Problems/Making Decisions*

Ask the students to suggest what images they could produce that would prove that the image in a flat mirror is virtual. (Any image produced shows that the image is right side up; the virtual image appears to be as far behind the mirror as the object is in front of the mirror.)

● **Process Skills:** *Observing, Experimenting*

 Gather some hand lenses and have the students look at objects by moving the lenses back and forth. When does the image appear upside down? Right side up? When does the image change? Have them write their observations in their journals.

Remember that the distance from the center of the lens to the film is the focal length of the lens. This distance equals the focal length when the camera is set at "infinity." When you take a close-up photograph, the lens can still focus, but it has to be farther away from the film.

Figure 10–9. A lens is positioned further from the film for a close-up. Light passes through the lens and converges on the film to form an enlarged real image.

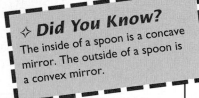

◇ Did You Know?
The inside of a spoon is a concave mirror. The outside of a spoon is a convex mirror.

▼ **ASK YOURSELF**

How can you focus your camera to take a photo of a distant mountain and to take a photo of a flower at the base of the mountain?

Not So Real Images

Have you ever used a hand lens to look closely at an object? A hand lens is a convex lens. It creates an image larger than the real object. When you use a hand lens, the object is placed close to the lens—at a distance less than the focal length. The rays from a point on the object do not come to a focus. The rays coming out of the lens are divergent. When some of these rays enter your eyes, they seem to be diverging from some point beyond the object. What you see is an enlarged image of the object. The image that forms does so only because the light seems to diverge from it. This type of image is called a **virtual image**.

A concave lens, also called a diverging lens, causes light to spread. Therefore, it can never focus light to form a real image.

Figure 10–10. With a convex lens, the light rays you see seem to diverge from points beyond the real image.

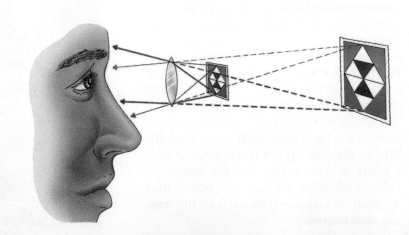

ACTIVITY

How can you find a virtual image?

Process Skills: Measuring, Comparing, Interpreting Data

Grouping: Individual

Hint

You can see through a piece of colored glass at the same time you see the reflection from its surface. If the students think that the image is at the surface of the glass, point out that the image appears in the same place, regardless of the angle from which it is viewed. This shows that it is possible to specify a definite location for the image.

▶ **Application**

1. Regardless of the angle from which it is viewed, the second pin and the image of the first pin are in the same place.

2. an angle of 90°; the angle of incidence is always equal to the angle of reflection, therefore, the light must be striking the glass straight on to make a 90° angle with the normal.

It can, however, form a virtual image. This type of image is always smaller than the actual object. See if you can find a virtual image by doing the next activity.

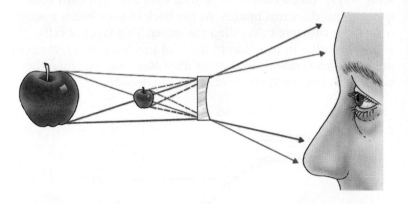

Figure 10–11. With a diverging lens, every point on the object is represented by a point on the virtual image. A concave lens always forms images reduced in size.

ACTIVITY

How can you find a virtual image?

MATERIALS
colored glass; corrugated cardboard, 10 cm × 10 cm; straight pins (6); metric ruler; protractor

PROCEDURE
1. Mount the glass on the cardboard with straight pins, as shown.

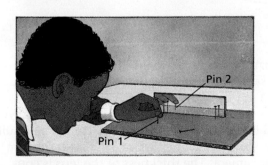

2. Place *Pin 1* upright in the cardboard, 10 cm in front of the glass.

3. Look into the glass at the image of *Pin 1*.
4. Place *Pin 2* behind the glass in the same place as *Pin 1*. Move your head to different positions to make sure *Pin 2* stays in the same place as the image of *Pin 1*.
5. Measure and record the distance from *Pin 1* to the glass and from *Pin 2* to the glass.

APPLICATION
1. What happens to the image if you change the point from which you look at the pin?
2. If you draw a line from *Pin 1* to the image, what angle does it make with the glass? Relate this to what you know about the angle of incidence and the angle of reflection.

▼ **ASK YOURSELF**
What is the difference between a real image and a virtual image?

ONGOING ASSESSMENT
▼ **ASK YOURSELF**

A real image is a focused collection of points of light, each point corresponding to a point on the object. A virtual image is a collection of light rays that seem to be coming from a set of points corresponding to points on the object.

SECTION 3   **273** ◀

● **Process Skill:**
Communicating

PORTFOLIO ASSESSMENT

Have the students copy and label the diagram of the human eye. Then have the students write a definition for each part (cornea, pupil, iris, lens, retina, ciliary muscles, optic nerve).

After the students have completed their eye diagrams, have them place their graphics in their science portfolios.

SCIENCE BACKGROUND

Cameras and eyes have some analogous parts. For example, both have lenses, diaphragms for regulating the intensity of entering light, and light-sensitive "film."

INTEGRATION– *Health*

Glaucoma and cataracts are eye diseases. In glaucoma, fluid builds up in the eye and causes pressure that can damage the optic nerve and cause blindness. Cataracts cause a clouding of the lens. Surgery can remove or replace the lens. If the lens is removed, special glasses must be worn to maintain sight. If the lens is replaced, an artificial lens is substituted for the original lens.

⟡ **Did You Know?**
Most eyeglasses also correct for astigmatism, which results from the imperfectly spherical shape of the eye's lens. Correction is made by making one surface of the eyeglass lens cylindrical instead of spherical.

The Human Eye

Mr. Tallchief explains that a camera is like the human eye in some ways. "Like a camera lens, your eyes use their own convex lenses to produce real images. At the back of your eye is a sensitive layer of nerve cells called the *retina*. This layer of cells, which is similar in function to the film in your camera, receives the image and codes it into nerve impulses. You see when the nerve impulses reach your brain and are decoded."

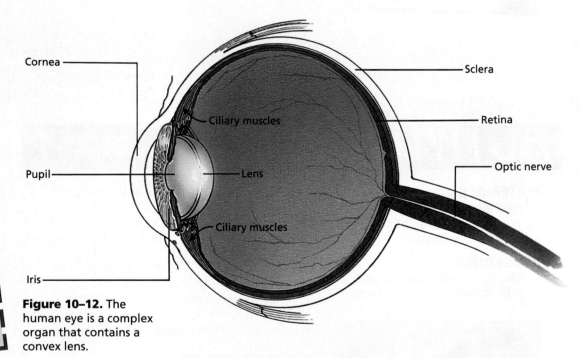

Figure 10–12. The human eye is a complex organ that contains a convex lens.

Adjusting Vision "Obviously," says Mr. Tallchief, "your eye can't focus the same way a camera does. You can't move your eyeballs in and out the way you do a camera lens."

"What happens is this," he explains. "The lens in your eye can change its focal length by changing its thickness. The result is the same as when you change the distance between the film and the lens in your camera. The thickness of the lens is controlled by the ciliary muscles attached to the edges of the lens. When you look at something far away, the ciliary muscles relax, thus thinning each lens. To see something close up, the ciliary muscles contract, allowing each lens to become a little thicker. This increases the converging ability of the lens so that the light rays focus on the retina."

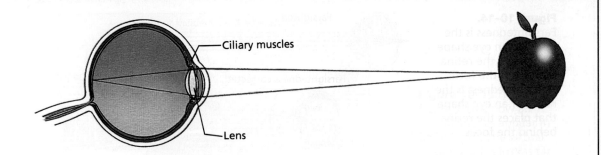

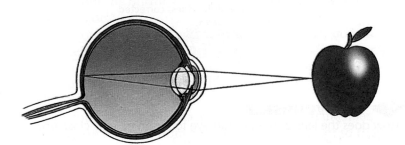

Figure 10–13. Action of the ciliary muscles change the shape of the lens. A thinner, flatter lens focuses on objects far away. The lens becomes thicker and more rounded to focus on objects close by.

Correcting Vision When one club member wants to know how close up or far away an object has to be for the eye to focus, Mr. Tallchief answers, "On average, a young person can focus objects at 7 cm, while 60-year-olds can't focus closer than 200 cm. Why? The lenses, like the rest of the human body, age and become less flexible. That is why many people, like myself, wear eyeglasses when they get to be around 40 years old. The glasses provide an additional convex lens for their eyes."

Another club member wants to know why, then, some young people wear eyeglasses. Mr. Tallchief explains, "If a person's eyeballs are a little too short from front to back, images of near objects are formed by their lenses behind the retina. This condition is called *farsightedness*. Eyeglasses containing convex lenses that allow extra convergence enable the image to focus on the retina."

"If a person's eyeballs are too long from front to back, images of distant objects are formed in front of the retina. This condition is called *nearsightedness,* and eyeglasses containing concave lenses that reduce the amount of convergence enable the image to focus on the retina."

Objective

● **Experiment** with the properties of virtual and real images.

Pre-Lab

Ask a volunteer to describe a convex lens. (a lens whose middle part is thicker than its edges so that the lens bends light rays together) Then explain to the students that in this *Investigation* they will observe the properties of images formed by a convex lens.

Hint

The students will discover that as the lens is brought closer to the lamp the image recedes. As the object distance approaches the focal length of the lens, the image rapidly moves away.

▶ Analyses and Conclusions

1. When the object is very far away, the image is at the focal length.

★ PERFORMANCE ASSESSMENT

After the completion of this Investigation, ask volunteers to demonstrate the procedure. Check the students' responses to the questions for accuracy.

2. With a smaller focal length, the image distances would always be less, and the images would be smaller.

3. A virtual image formed by a single element is always right side up, while a real image is inverted. A real image can be cast onto a screen; a virtual image cannot.

▶ Application

Students' diagrams should show that convex mirrors reflect divergent rays, creating a small image of a wide field of view.

4. Rays from the object enter the lens and are inverted as they come to a focus.

✳ Discover More

The images are virtual and right side up.

Post-Lab

Ask the students to summarize what they have learned in this *Investigation*.

INVESTIGATION

▶ Finding the Properties of Images Formed by a Convex Lens

▶ MATERIALS

● white cardboard ● convex lens ● meter stick ● lamp

▼ PROCEDURE

1. Place your cardboard screen facing the window of the room as shown. Move the lens until a sharp picture appears on the screen.

2. Measure and record the distance from the lens to the screen. Is the image real or virtual? right side up or upside-down? smaller or larger than the object? Explain the reasons for your results.

3. Turn on your lamp, and use it as the object. Place the lamp several meters away from the lens, and focus the image of the lamp on the screen. With an object distance of several meters, what is the image distance?

4. Move the lamp closer to the lens. As the object distance decreases, what happens to the image distance? As the image distance increases, what happens to the size of the image? Explain why.

5. Keep moving the lamp closer until you cannot see an image. What is the object distance at that point?

6. Now use the lens as a magnifying glass to read the letters on this page. Start with the lens about 1 cm from the page. Is the image real or virtual? right side up or upside-down? larger or smaller than the object?

7. Increase the object distance. What happens to the image? Why?

8. Keep increasing the object distance until you can no longer find an image. What is the object distance at that point?

▶ ANALYSES AND CONCLUSIONS

1. Describe the situations in this experiment that allowed you to find the focal length of the lens.

2. In what ways would you expect the results to be different if you used a lens with a smaller focal length?

3. Describe two ways in which you can tell whether an image is real or virtual.

4. Why is the image on the screen always upside-down?

▶ APPLICATION

A concave lens works like a convex mirror. This type of mirror is often used on the passenger side of cars so that the driver can see cars on the right and behind. Why does a convex mirror allow the driver to do this? Draw a diagram to support your answer.

✳ Discover More

Repeat this procedure using a concave lens. How do the results differ?

The Big Idea— Systems and Structures

Lead the students to an understanding of the eye and the camera as two basic systems—one natural, one artificial—for capturing and focusing light. Be sure the students understand the constant processes of reflection and refraction that go on around them as light interacts with other matter.

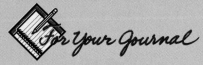

For Your Journal

The students' ideas should reflect an understanding of how light is affected by travel through different media. You might have the students share their new ideas with the rest of the class.

CONNECTING IDEAS

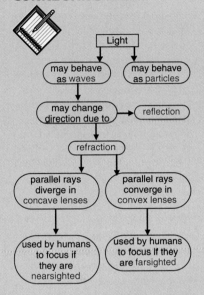

CHAPTER 10 HIGHLIGHTS

The Big Idea

Light may act and react as waves do or as particles do. Lenses and mirrors change the direction in which light moves. One way to understand the behavior of light is to study it using the ray model. The brightness of light rays depends upon an object's distance from the light source.

Light energy in the form of waves may reflect or refract. When light rays enter a lens, they refract. Through a convex lens, parallel light rays converge at the focal point and diverge after passing through it. Through a concave lens, parallel light rays diverge. When light is focused by a lens, a real image is produced. An image formed by a divergent lens is a virtual image.

Jan Vermeer, The Geographer. Frankfurt am Main, Staedelisches Kunstinstitut.

For Your Journal

Look back at the ideas you wrote in your journal at the beginning of the chapter. How have your ideas about light changed? Revise your journal entry to show what you have learned. Be sure to include information on how lenses can affect light.

Connecting Ideas

This concept map shows how the big ideas of this chapter are related. Copy the concept map in your journal. Extend the concept map to show how light and lenses interact to form images.

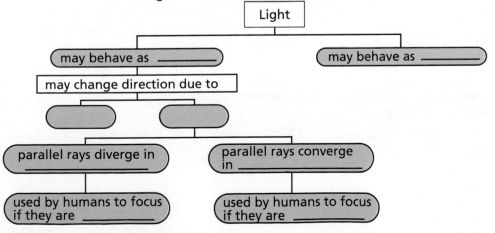

Understanding Vocabulary

1. (a) A convex lens causes the light rays to converge. It is thicker in the middle than at the edges. A concave lens causes light rays to diverge. It is thinner in the middle than at the edges.

(b) A real image is formed by the collection of points of light from an image, focused by a lens. A virtual image is formed because light only appears to diverge from an image.

(c) Lux is the SI unit of brightness. Light rays are models of how light travels; they are always straight unless something happens to bend them.

(d) The focus is the point at which light entering a lens converges. When incoming rays are all parallel, the distance from a lens to a focus is called the focal length.

Understanding Concepts

Multiple Choice

2. b

3. d

4. a

5. c

6. d

7. c

Short Answer

8. The ciliary muscles relax and contract to control the thickness of the eye's lens, which, in turn, focuses light rays on the retina.

9. The lens becomes less flexible as people age. Eventually, the lens loses its ability to thicken enough to focus on close objects.

Interpreting Graphics

10. Lens; a

11. ciliary muscles; c

12. retina; b

13. optic nerve; d

CHAPTER 10 REVIEW

Understanding Vocabulary

1. For each set of terms, explain the similarities and differences in their meanings.
   a) convex lens (268), concave lens (268)
   b) real image (271), virtual image (272)
   c) lux (263), light rays (263)
   d) focus (266), focal length (268)

Understanding Concepts

MULTIPLE CHOICE

2. When you see an image, it's because your brain decodes nerve impulses that have been registered on your
   a) sclera.　　**b)** retina.
   c) lens.　　　**d)** iris.

3. An important assumption of the ray model of light is that light
   a) always travels at a certain speed.
   b) acts like waves.
   c) acts like particles.
   d) travels in straight lines.

4. In seeing things, your eyes make use of light beams that are
   a) divergent.
   b) convergent.
   c) virtual.
   d) focused.

5. The Danish astronomer Olaus Roemer successfully determined
   a) the colors found in light.
   b) the angle of incidence.
   c) the speed of light.
   d) the degree of refraction.

6. Who was responsible for proving the theory that light acts like a wave?
   a) Isaac Newton
   b) Christiaan Huygens
   c) Albert Einstein
   d) Thomas Young

7. Convex lenses make objects look
   a) smaller than they really are.
   b) brighter than they really are.
   c) larger than they really are.
   d) darker than they really are.

SHORT ANSWER

8. Describe the function of the ciliary muscles.

9. What causes farsightedness in people as they grow older?

Interpreting Graphics

Answer each question below. Then write the letter of the part of the diagram that identifies your answer.

10. Which part enables your eye to focus light rays?

11. Which part changes the thickness of the lens to allow it to adjust the focus?

12. Which part captures the image that you eventually see?

13. Which part carries nerve impulses to the brain?

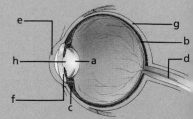

Reviewing Themes

14. Only a real image can be reproduced on film. A virtual image is not actually there—it only appears to be there—so it could not be developed into a photograph.

15. Since some light is always absorbed as the light passes through a material, the light's intensity decreases as it passes through frosted glass.

Thinking Critically

16. Apparent magnitude would be partially determined by the star's closeness or distance from the viewer's eyes.

17. The speed of light is the same in the two media. Any time light changes speed, it changes direction.

18. It would not be possible to start a fire with a glass concave lens because such lenses never focus (and concentrate) the sun's rays on a specific area. However, a convex lens, such as a magnifying glass, can do so.

19. The water differs from the air in that light travels through the water at a slower speed than the air. The water refracts the light, causing each water droplet to act as a lens. Since each "lens" is in a different position, each image is formed in a different place.

Reviewing Themes

14. Systems and Structures
Explain why a camera lens must produce a real image rather than a virtual image.

15. Environmental Interactions
Many light bulbs used in reading lamps are made with frosted glass. So are the glass doors in shower stalls and some inside walls in buildings. How does frosted glass affect the intensity of light passing through it?

Thinking Critically

16. Stars have different absolute magnitudes. In other words, each star has a degree of brightness determined by how much energy it produces. Astronomers assign each star a degree of *apparent* magnitude or brightness—how bright the star *seems* to viewers on Earth. What factor would affect this measurement?

17. A light beam passing from benzene into an alcohol solution at a high angle of incidence does not bend. What can you conclude? Explain your answer.

18. In the novel *Lord of the Flies,* a very nearsighted character starts a fire with his glasses. Explain why this could not happen.

19. One ballerina, but many images. Each drop of water is a tiny lens, producing an image of the dancing ballerina. Explain what it is about the water that makes it different from the surrounding air. Why does each drop of water act just like a tiny lens? Why does each droplet form an image in a different place?

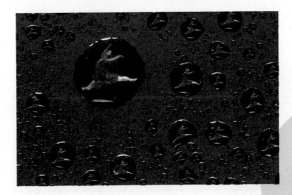

Discovery Through Reading

Gloeckner, Carolyn. "The Human Eye: A Powerful Little Package." *Current Health* 18 (October 1991): 4–9. This article includes fascinating facts about the human eye, including how we see in three dimensions and differentiate the colors of light.

THE SPECTRUM

PLANNING THE CHAPTER

B = Basic   **A** = Average   **H** = Honors

The coding Basic, Average, and Honors indicates subsections, features, and resources that might be appropriate for different levels of learners. For additional suggestions regarding choice of topic and depth of coverage, see the Pacing Chart on pages T28–T31.

*Frame numbers at point of use
(TR) Teaching Resources, Unit 3
(IT) Instructional Transparencies
(LI) Laboratory Investigations
(SD) *Science Discovery* Videodisc
    Correlations and Barcodes
(SRG) Study and Review Guide

CHAPTER MATERIALS

Title	Page	Materials
Discover by Problem Solving	286	(per individual) paper, pencil
Activity: How does a prism form colors from white light?	286	(per group of 3 or 4) tape, white poster board, light source, prism
Skill: Communicating Using Scientific Notation	288	(per individual) paper and pencil
Discover by Observing	289	(per group of 2 or 3) old dental X-rays
Discover by Observing	292	(per individual) paper, pencil
Discover by Observing	293	(per class) AM-FM radio
Discover by Problem Solving	300	(per individual) paper, pencil
Teacher Demonstration	302	slide projectors or powerful flashlights with red, green, and blue filters (3)
Discover by Doing	302	(per individual) paper, pencil
Discover by Observing	302	(per individual) paper, pencil
Discover by Problem Solving	303	(per individual) paper, pencil
Discover by Doing	305	(per individual) filter
Investigation: Studying the Three-Primary Model	306	(per group of 3 or 4) 15-watt bulb lamp, color filters: red, green, blue, cyan, yellow, magenta

ADVANCE PREPARATION

Determine in advance if any students are colorblind. Such students will likely have difficulty with some of the color-based activities. Have these students work with a student who is not colorblind for such activities.

TEACHING SUGGESTIONS

Career Day
You might organize a "Career Day" for the class by inviting several people with careers related to this chapter's concepts. Encourage the representatives to bring displays and other materials used in their line of work. Examples include X-rays, cameras, film, portable radar screens, and eye-examining instruments.

Ask the visitors to discuss why they chose the career they did, what specific training was required, and what a typical day at work involves.

Do-It-Yourself Career Day
If inviting several speakers proves impractical, have the students serve as their own career consultants. Encourage them to research career areas in which they are interested and prepare reports (including visuals) to present to the class.

CHAPTER 11 THE SPECTRUM

CHAPTER THEME—ENERGY

Chapter 11 will help the students understand the energy forms represented by the electromagnetic spectrum and its use in our lives. The students will learn how various kinds of electromagnetic waves, particu- larly those of visible light, inter- act with Earth and its life. A sup- porting theme in this chapter is **Technology.**

MULTICULTURAL CONNECTION

During the early eleventh cen- tury, an Arab physicist known as Alhazen recognized that vision is caused by the reflection of light from objects into the eyes. Alhazen stated that this reflected light forms optical images in the eyes. He believed that the colors seen in objects depended on the light striking the objects and on some property of the objects themselves.

MEETING SPECIAL NEEDS

Second Language Support

Have limited-English- proficient students identify, in their jour- nals, the colors they see in the van Gogh painting. Instruct them to tell how each color makes them feel.

CHAPTER 11 THE SPECTRUM

> Dear Theo:
> In my picture of the "Night Café" I have tried to express the terrible passions of humanity by means of red and green. The room is blood-red and dark yellow, with a green billiard table in the middle; there are four lemon-yellow lamps with a glow of orange and green. Everywhere there is a clash and contrast of the most alien reds and greens in the figures of little sleeping hooligans in the empty dreary room, in violet and blue. The white coat of the patron [landlord], on vigil in a corner, turns lemon-yellow, or pale luminous green.
>
> Vincent

From *Dear Theo: The Autobiography of Vincent van Gogh* edited by Irving Stone

***S**cientifically speaking, colors are created by light of different wave- lengths striking the eye. Traditionally, the spectrum of colors are also associated with a spectrum of human emotions. The color yellow suggests gaiety to most people, while blue is associated with sadness. That's why those smiling happy faces are colored yellow, but mournful singers croon the "blues."*

The artist Vincent van Gogh was fascinated by colors. He was also gripped by powerful emotions, often suffering bouts of depression. Influenced by Japanese art, van Gogh freed himself from the tra- ditional color spectrum in order to use colors more imaginatively. In one self-portrait he painted the shadows on his face in green. In another, painted after he cut off part of his ear, the background is red and orange.

In letters to his brother Theo, van Gogh explained the colors of his picture "The Night- Café." To van Gogh, the unusual colors expressed a wide spectrum of emotions, from darkness to "Japanese gaiety." The cafe was "a place where

Display photographs, illustrations, or actual examples of an X-ray, a microwave oven, and a radio. Ask the students to list ways in which the items are similar. Point out that the three items make use of various kinds of electromagnetic waves. Then work with the students to list the similarities and differences of the parts of the electromag-

netic spectrum. You might have to coach the students in order to elicit the information. (Similarities: all the waves carry electromagnetic energy; all have the basic characteristics of waves discussed in Chapter 8. Differences: frequencies, wavelengths, amplitudes, uses, and the effects on humans.)

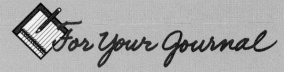
For Your Journal

The journal questions will help you identify the students' level of knowledge about the electromagnetic spectrum. (The students will have a variety of responses to the journal questions. Accept all answers, and tell the students that they will have a chance to refer to the questions and amend their answers at the end of the chapter.)

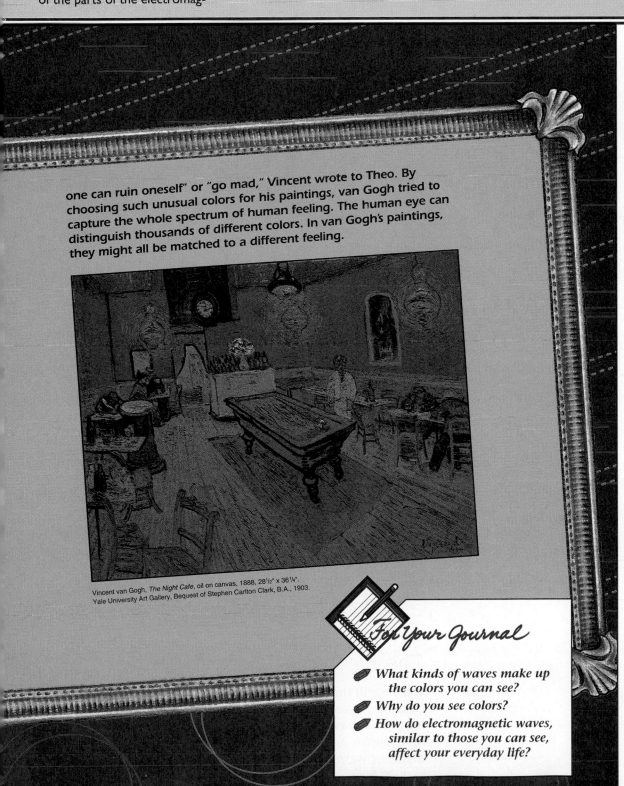

one can ruin oneself" or "go mad," Vincent wrote to Theo. By choosing such unusual colors for his paintings, van Gogh tried to capture the whole spectrum of human feeling. The human eye can distinguish thousands of different colors. In van Gogh's paintings, they might all be matched to a different feeling.

Vincent van Gogh, *The Night Cafe*, oil on canvas, 1888, 28½" x 36¼". Yale University Art Gallery, Bequest of Stephen Carlton Clark, B.A., 1903.

ABOUT THE ARTIST

Vincent van Gogh trained as a missionary in Belgium and worked with coal miners and their families. His religious society took issue with his complete identification with and care for the poor with whom he worked. The society subsequently retired van Gogh. In 1880 van Gogh decided to become a painter. Though his first works were very dark, he turned later to light colors, especially yellow. As a post-impressionist, van Gogh revolutionized the use of color as an expression of emotion in painting.

Subject to undiagnosed epileptic seizures throughout his life, van Gogh often felt he was going insane. His brother, Theo, was the only person in whom he found friendship and comfort, and Vincent's letters to Theo reveal much of what is known about his life. Ironically, the man whose paintings today are as famous and recognizable as any in the history of Western art sold only one canvas during his life. Sadly, van Gogh considered himself a failure.

For Your Journal

✏ **What kinds of waves make up the colors you can see?**

✏ **Why do you see colors?**

✏ **How do electromagnetic waves, similar to those you can see, affect your everyday life?**

SECTION 1

Introducing the Spectrum

Objectives

Compare the characteristics of electromagnetic waves to their position in the electromagnetic spectrum.

Relate frequency to wavelength.

Demonstrate the separation of white light into colors of the visible spectrum.

It's career day. Everywhere you turn, you find posters advertising careers for this afternoon's festival. There's no particular career that especially interests you. You do know, though, that you want excitement on the job and you like to be involved with people. Sometimes you imagine yourself as a famous surgeon, performing complicated operations and saving lives. While watching the evening news, you've often daydreamed about becoming the anchorperson on a national network, informing the nation of late-breaking, global events. Then again, you also find adventure in quiet contemplation, like when you draw pastels of magnificent sunsets over the river or abstract compositions that include mysterious images from your dreams. There are so many interesting career possibilities you want to investigate.

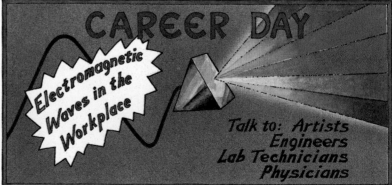

The Electromagnetic Spectrum

You are still thinking about careers as you walk into your first class. Ms. Oyama, your science teacher, is talking about **electromagnetic waves**—the invisible transverse waves that carry both electric and magnetic energy. She asks, "How do electromagnetic waves relate to the career choices you might someday make?" Everyone looks puzzled, but then she explains.

TEACHING STRATEGIES

"Say you want to become a radio disc jockey. What kinds of waves would you use?" One student confidently answers, "Sound waves." Ms. Oyama agrees that we hear music from a radio because of sound waves. "But," she goes on, "before radio signals are turned into sound waves, they're transmitted as radio waves to the receiver on your radio. Radio waves are just one type of electromagnetic wave. All the types of electromagnetic waves together are the **electromagnetic spectrum.**"

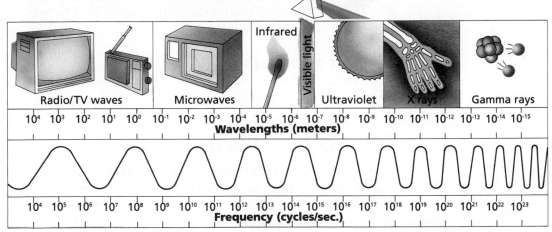

You wonder what electromagnetic waves have in common and what other kinds there are. You raise your hand and ask Ms. Oyama. She tells you that the frequencies of these waves are high, but each type differs in frequency and wavelength from the rest. The entire class looks confused. "OK," she says. "Electromagnetic waves have a certain range of frequency, from less than 100 cycles per second to more than 10^{23} cycles per second. Most technologies, however, use electromagnetic waves in the range of 10 000 Hz to 10^{23} Hz. Compare that to the highest sound you can hear, which has a frequency of 20 000 cycles per second. That's what I mean when I say that electromagnetic waves have a high frequency."

Look at the diagram of the electromagnetic spectrum. You can see that radio and TV waves have lower frequencies than X-rays and gamma rays.

Someone asks how fast these different waves travel. "Good question," she answers. "The speed of electromagnetic waves is another common property. They all travel at the speed of light—300 000 000 m/s. What's more, the speed of a wave, its frequency, and its wavelength are all related mathematically. The wavelength of an electromagnetic wave multiplied by its frequency equals the speed of light."

Figure 11–1. The electromagnetic waves in this diagram of the electromagnetic spectrum are presented in order, according to their wavelength and frequency. The visible spectrum takes up only a small portion of the spectrum.

● **Process Skills:** *Observing, Evaluating*

Introduce the visible spectrum by discussing rainbows. Challenge the students to name the colors of a rainbow in the order in which the colors occur. (red, orange, yellow, green, blue, and violet) Then have the students tell whether or not the colors of a rainbow are always in the same order. (yes)

GUIDED PRACTICE

Have the students provide definitions in their own words for each of the following terms: *frequency, speed,* and *wavelength.* Encourage the students to draw simple diagrams on the board to illustrate their definitions.

INDEPENDENT PRACTICE

Have the students provide written answers to the Section Review. In their journals, have them describe the visible light spectrum and tell how it can be produced from white light.

DISCOVER BY *Problem Solving*

A high-frequency wavelength is very short; the frequency of a long wave is very low.

ONGOING ASSESSMENT
ASK YOURSELF

Electromagnetic waves all carry electromagnetic energy and travel at the speed of light.

ACTIVITY:

How does a prism form colors from white light?

Process Skills: Analyzing, Observing

Grouping: Groups of 3 or 4

Hint
A slide projector is a good source of light for this activity.

▶ **Application**

1. All colors of the visible spectrum, in this order: red, orange, yellow, green, blue, violet

2. Some students will be able to see shades, others will not. Shades occur because of overlapping waves.

3. The shape of water drops separates the different wavelengths just as a prism does.

Ms. Oyama asks you to use this fact to clarify some relationships among these quantities. Try the next activity.

DISCOVER BY *Problem Solving*

How would you describe the wavelength of an electromagnetic wave with very high frequency? the frequency of a wave with a very long wavelength? Use the diagram in Figure 11–1 to help you. Explain your answer.

ASK YOURSELF
What characteristics do different kinds of electromagnetic waves have in common?

The Visible Spectrum

Ms. Oyama tells you that visible light—the light you can see—takes up only a small part of the electromagnetic spectrum. She explains that nearly 300 years ago, the British physicist Sir Isaac Newton passed a ray of sunlight through a prism. The ray refracted and produced the colors of the rainbow. You can do the same experiment.

ACTIVITY

How does a prism form colors from white light?

MATERIALS
tape, white poster board, light source, prism

PROCEDURE
1. Tape the white poster board on a wall or against a stand.
2. Darken the classroom, and turn on the light source. Hold the prism between the light source and the poster board until the colors of the rainbow appear on the poster board.
3. Move and turn the prism to see whether the colors change. Record your results.

APPLICATION
1. What colors could you see clearly? In what order did the colors appear?
2. Were there shades of colors in between the most noticeable colors? Why do you think this is so?
3. Rainbows in the sky are formed by refraction from water drops in the atmosphere. How do water drops act like prisms?

EVALUATION

Ask the students why we know about other kinds of electromagnetic waves besides visible light if we cannot see them. (There are other ways to detect electromagnetic waves. For example, receivers pick up radio waves.)

RETEACHING

Have the students list each area of the electromagnetic spectrum and describe its position, using wavelength and frequency information.

EXTENSION

Have the students use library resources to find information about Newton's other experiments with the spectrum. Encourage the students to summarize their findings in a written or oral report, or as a poster to be displayed in the classroom.

CLOSURE

Cooperative Learning Have the students work in small groups to frame three main ideas and three supporting details to summarize the section.

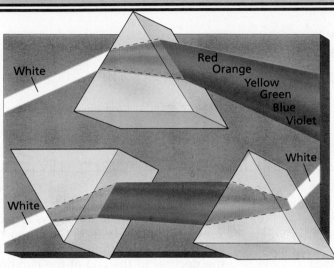

Figure 11–2. A diamond sparkles with color because it acts as a prism. A prism can separate white light into the colors of the visible spectrum.

The order of the colors in Sir Isaac Newton's experiment appeared from the color with the longest wavelength (red) to the color with the shortest wavelength (violet). This is the same order of colors in the rainbow—longest wavelength to shortest, or red to violet. The colors produced—red, orange, yellow, green, blue, violet—are the colors of the *visible spectrum*.

You ask Ms. Oyama, "Then how do you make white light?" Another student correctly deduces, "White light is the combination of the colors in the rainbow."

At the end of class Ms. Oyama asks everyone to stay alert during the afternoon career festival to find out the ways in which electromagnetic waves relate to many different careers.

 ASK YOURSELF

In what order do the colors of the visible spectrum appear? How does this order relate to wavelength?

SECTION 1   *REVIEW AND APPLICATION*

Reading Critically

1. What is the electromagnetic spectrum?

2. What is the relationship between wavelength and frequency?

3. What is the relationship between wavelength and color in the visible spectrum?

Thinking Critically

4. Why is it possible to separate the colors of the visible spectrum with a prism and then recombine the light waves to form white light?

5. Longer wavelengths of light bend less than the shorter ones. Why do you think this is true?

ONGOING ASSESSMENT
ASK YOURSELF

Red, orange, yellow, green, blue, violet; they appear in the order of longest wavelength (red) to shortest wavelength (violet).

SECTION 1 REVIEW AND APPLICATION

Reading Critically

1. all types of electromagnetic waves

2. Together, they always equal the speed of light. So, as the wavelengths become shorter, their frequencies increase; and as the wavelengths become longer, their frequencies decrease.

3. The wavelengths get smaller as they move from the red end of the spectrum to the violet end.

Thinking Critically

4. A prism slows down the waves and bends them. Another prism can bend the light again so that the waves recombine. The second prism must be upside-down with respect to the first prism.

5. Light bends as it enters the glass medium of the prism because it slows down. Light of different wavelengths travel at different speeds through the glass and are thus refracted at different angles.

SECTION 1   **287** ◀

Process Skills:
Communicating, Interpreting Data

Grouping: Individual

Objective
● **Communicate** using scientific notation.

▶ **Application**

1. 3.0×10^8 m/s

2. 3.2×10^{-4} m

3. 7.53×10^{-10} kg

4. 1.0×10^{12} m

✳ **Using What You Have Learned**

1. ease of reading and computation

2. Scientific notation is used with very large or very small numbers. Accept reasonable responses.

★ **PERFORMANCE ASSESSMENT**

Have the students show how they arrived at their answers in the "Application." Check the students' procedures and answers for accuracy.

SKILL *Communicating Using Scientific Notation*

▶ **MATERIALS**
● paper ● pencil

▼ **PROCEDURE**

When numbers are very large or very small, scientists and engineers write them using a method called scientific notation.

1. In scientific notation, a number is stated by writing the number 10 raised to some exponential power. For example:

a hundred = 100 = 10^2
a hundred thousand = 100 000 = 10^5

In each of these cases, the number of zeros after the 1 is the same as the exponent of the 10 in scientific notation.

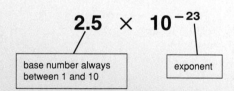

2. You can also write numbers that are not simply powers of ten when you use scientific notation. For example:

four hundred and fifty thousand =
450 000 = 4.5×10^5

The trick is to place the decimal point after the first digit of the number. This makes the base number always less than 10 and greater than or equal to one. Then count the number of places from the decimal to the end of the number to get the exponent number.

17 100 000 000 000 000 cycles per second is a frequency of ultraviolet light. To put this number into scientific notation, put the decimal after the 1. This gives you a base of 1.71. From the decimal point there are 16 digits to the end of the number. Therefore, this frequency of ultraviolet light is about 1.71×10^{16} cycles/second.

3. For numbers less than one, the exponent is negative. Place the decimal point after the first nonzero digit. Then count the number of places to the right of the decimal to get the negative exponent of 10.

0.000 000 0176 m is the wavelength of the ultraviolet light above. To change this number to scientific notation, put the decimal after the 1. This gives you a base of 1.76. From the decimal point there are 8 digits to the front of the number. Therefore, the wavelength of this ultraviolet light is about 1.76×10^{-8} m.

▶ **APPLICATION**
Write the following in scientific notation.

1. The speed of light in a vacuum is 300 000 000 m/s.

2. The thickness of a piece of gold foil is 0.000 32 m.

3. The mass of a dust particle is 0.000 000 000 753 kg.

4. A terameter is 1 000 000 000 000 m.

✳ ***Using What You Have Learned***

1. What purpose does using scientific notation serve?

2. Name three quantities, not discussed here, that would be easier to express in scientific notation.

Section 2:

THE NON-VISIBLE SPECTRUM

FOCUS

Section 2 discusses those waves on either end of the spectrum that are not visible to the human eye. The characteristics and uses of X-rays, ultraviolet waves, infrared waves, and radio waves are emphasized.

MOTIVATING ACTIVITY

Cooperative Learning Have the students work in small groups to list situations in which they or someone they know has been X-rayed. (Examples include X-rays of teeth, possible bone fractures, tumors, CAT scans, and so on.) Discuss how some procedures, such as a CAT scan or MRI, differ from others.

POSITIVE ATTITUDES
• Creativity • Curiosity

TERM
• telecommunications

PRINT MEDIA
Radar: The Silent Detector by Deborah Hitzeroth. (see p. 211b)

ELECTRONIC MEDIA
All About Matter and Energy Focus Media (see p. 211c)

SCIENCE DISCOVERY
• Television set
• Microwave oven

BLACKLINE MASTERS
Extending Science Concepts
Thinking Critically
Reading Skills
Study and Review Guide

The Nonvisible Spectrum

SECTION 2

At the career festival, the first booth you visit is called Medicine and Technology. Physicians and medical technicians from a nearby hospital are there to help you learn about their profession. You're amazed by what you see: X-rays that show a person's skull; infrared photographs that show the outline of the human body filled in with different colors; photographs of an operating room stocked with more equipment, you think, than a spaceship. Now you realize that this profession would satisfy the mechanic and engineer in you as well as your desire to help people.

You imagine a day in the future when over a hospital loudspeaker you hear: "Dr. Rey, your expertise is requested in the OR (as a famous brain surgeon, you know *OR* means "operating room," of course!). We have the virtual reality scanners ready and waiting for you."

Objectives

Distinguish between ultraviolet radiation and infrared radiation and explain the uses of each.

Identify the major uses of X-rays, radio waves, and microwaves.

REINFORCING THEMES
Technology

The information X-rays supply concerning the nature and properties of matter and the effect of X-rays on living tissue have made the discovery of X-rays one of the most important events in modern physics. X-rays cannot be seen, but they can penetrate nearly everything. X-rays are used not only in medicine but also in industry, security, design, and quality control.

DISCOVER BY *Observing*

Dark spots might be cavities. They are dark because they are softer than the surrounding area; X-rays pass through them.

X-Rays

At the Medicine and Technology booth, a medical technician tells students that X-rays were an important, early innovation in modern medical technology. The German physicist Wilhelm Roentgen (1845–1923) discovered X-rays in 1895 and in 1901 received the Nobel Prize for his work. Then the technician hands out old dental X-rays for students to study.

 BY *Observing*

Ask your dentist for an old dental X-ray. Light spots may be fillings or other metals. Are there any dark spots on the X-ray? What do you think the spots are? Why are the spots dark? How can a dentist find cavities on an X-ray?

X-rays are very high-frequency electromagnetic waves. They are usually produced when electrons traveling at high speeds strike a solid object. The technician explains that when

● **Process Skills:** *Evaluating, Inferring*

Discuss the parts of the X-ray machine shown on this page. Ask a volunteer to explain the function of the tungsten target. (The electrons produced in the tube strike the tungsten target, which emits X-rays.)

● **Process Skills:** *Analyzing, Predicting*

 Ask the students why they are given lead aprons to wear when they receive dental X-rays. Have them write their ideas in their journals. (Lead absorbs X-rays, so they cannot enter the body.)

✦ **Did You Know?**
ROSAT, an X-ray satellite that was launched in June 1991, stands for Roentgen satellite. Named for the discoverer of X-rays, the satellite is seeking new sources of X-rays in the universe.

 **SCIENCE TECHNOLOGY SOCIETY** The increasing depletion of the ozone layer is a threat to all life on Earth. Have the students work independently or in pairs to research the causes of ozone depletion and what steps countries and international groups are taking to slow down the process.

ONGOING ASSESSMENT
▼ ASK YOURSELF

Only the bone structure of the foot would show because X-rays pass through low-density matter, such as tissue and muscle, but are absorbed by high-density matter, such as bone.

INTEGRATION– Health

Even though the human body can manufacture its own vitamin D after exposure to sunlight, it is still necessary for people to consume foods such as dairy products, fish oil, and eggs to get the minimum daily requirement of vitamin D.

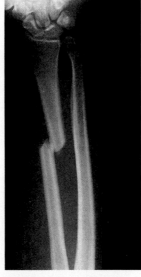

Figure 11–3. X-rays that pass through an object darken film to create an image.

part of your body gets X-rayed, electricity is passed through a tube and the electrons that are released hit a target. The X-rays produced shoot out of the tube through a small hole, pass through the object being X-rayed, and strike the film. The X-rays pass through matter that has low density, such as skin and muscle, but are absorbed by high-density matter, such as bones and teeth. Where X-rays are absorbed, a light area appears on the film.

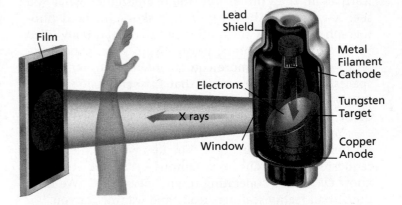

The technician explains that X-rays can be harmful because they can kill healthy body cells. X-rays can also kill cancer cells and are, therefore, used to treat certain types of cancer. Because X-rays are potentially harmful, the number of them you receive should be carefully monitored. Always ask for a protective lead apron to wear during an X-ray if you are not offered one.

▼ ASK YOURSELF
If an X-ray was taken of your foot, what parts of your foot would show? Why?

Ultraviolet Waves

You ask the medical technician about the photograph in the booth that interests you most—one that shows the interior of a hospital operating room. The technician points out the special lights and tells you they are UV lights. UV stands for ultraviolet. These lights produce *ultraviolet radiation*—that is, waves with a frequency just higher than visible light. UV waves are a part of the electromagnetic spectrum.

UV can kill bacteria; therefore, UV lights help keep an operating room sterile. The technician adds that UV radiation is invisible, but it is present in sunlight. In fact, it is the UV rays in sunlight that cause skin to tan or burn. She then tells everyone that while overexposure to UV can cause skin cancer, a certain amount of UV helps skin cells produce vitamin D—a substance necessary for healthy bones and teeth. She passes out copies of the following chart to students who visit her booth.

Table 11-1	Sunscreen Protection
No SPF	You may get a fast tan with these products, but you can also get a bad burn. They offer no protection.
SPF 4 to 6	These are good if you tan easily and rarely burn. It's the least you should start with on your face.
SPF 7 to 8	These provide extra protection that still permits tanning.
SPF 9 to 10	Excellent protection is provided by these products if you burn easily but would still like a minimal tan.
SPF 15+	These products provide the best protection against burns. They also prevent tanning.

The students who stop at the booth study the chart and discover that the higher the SPF (sun protection factor) number, the more completely the chemical blocks UV rays from your skin. You learn that you should cover your body with clothing, wear a hat, and use a sunscreen when you go outdoors in the sun.

 ASK YOURSELF

What are some ways in which ultraviolet waves help people?

Infrared Waves

The technician next shows photographs developed from film sensitive to infrared waves. *Infrared radiation,* or IR, is part of the electromagnetic spectrum. It has wavelengths slightly longer than those of visible light.

First the technician shows a photo that looks like an outline of a person, but it has many different colors. "This is a thermogram," the technician explains. "When an object receives IR waves, it absorbs the waves and becomes warmer. Then the object radiates the IR back into the environment. The different

INTEGRATION–
Life Science

Cold-blooded animals have no internal control over their body temperature. Their temperature depends upon that of their surroundings. Many cold-blooded animals use infrared waves to help them keep warm. Snakes, for example, seek cool caves and underground areas during the daytime heat of the desert. As the sun goes down, the rocks continue to radiate the infrared rays they have been absorbing all day. The snakes crawl out and lie on the rocks, where they soak up infrared rays and warm their bodies.

✧ Did You Know?
As the continent nearest Antarctica (where the hole in the ozone layer continues to grow) Australia has initiated a humorous ad campaign to get its inhabitants to protect themselves from the sun. The ads consist of three words: *Slip, Slap, Slop!*—slip on a shirt, slap on a hat, and slop on some sunscreen.

ONGOING ASSESSMENT
ASK YOURSELF

Ultraviolet rays kill harmful bacteria in operating rooms. They also help the body produce vitamin D.

● **Process Skill:** *Applying*

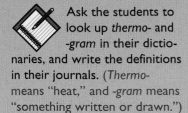

Ask the students to look up *thermo-* and *-gram* in their dictionaries, and write the definitions in their journals. (*Thermo-* means "heat," and *-gram* means "something written or drawn.") Using this information, have the students deduce the purpose of a thermogram. (a type of drawing that is based on heat emission)

● **Process Skills:** *Evaluating, Inferring*

Explain to the students that radio waves are the longest electromagnetic waves used in modern technology. They are emitted by natural objects in the universe and can be produced artificially on Earth. Radio waves do not interact with living organisms as other forms of electromagnetic energy do, and they are not absorbed by Earth's atmosphere. Ask the students whether radio waves are harmful to living things, or whether they can be felt or heard. (No, they do not interact with the body and do not affect it.)

MULTICULTURAL CONNECTION

African-American physicist Dr. George Carruthers is one of two naval research laboratory personnel responsible for the Apollo 16 lunar surface ultraviolet camera/spectrograph. Dr. Carruthers designed the instrument; Dr. William Conway adapted the camera for the lunar mission. The spectrographs were obtained from 11 targets and included the first photos of the ultraviolet bands of atomic oxygen that circle Earth. Dr. Carruthers received the NASA Exceptional Scientific Achievement Medal for his work on this project.

DISCOVER BY *Observing*

Have the students write their explanations in their journals. One hand is warm, the other is cool. Red and yellow indicate warm areas; blue and black indicate cool areas.

ONGOING ASSESSMENT
ASK YOURSELF

Infrared waves keep food in cafeterias hot, help locate people in dark and foggy areas, and help people with night blindness to see.

Figure 11–4. Infrared sensitive film can be used to photograph the IR radiation given off by objects.

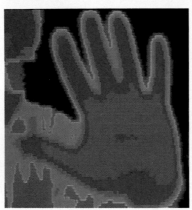
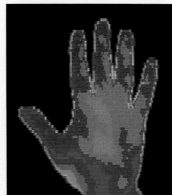

colors of a thermogram represent different amounts of IR waves, and therefore different temperatures, given off by the object."

The technician explains that infrared rays are often used in cafeterias to keep food hot. When special viewers are used that can "see" IR, the wearer can locate objects or people in foggy or dark areas because of the IR waves they give off. IR even helps people with night blindness see at night through special glasses. Then she shows you two photographs that show IR waves as they are given off by two hands.

DISCOVER BY *Observing*

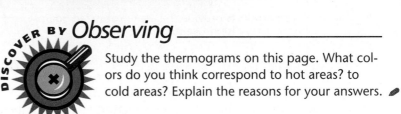

Study the thermograms on this page. What colors do you think correspond to hot areas? to cold areas? Explain the reasons for your answers.

ASK YOURSELF

How are infrared waves used to help people?

Radio Waves

The idea of seeing yourself nightly on television as a television journalist or hearing your voice make announcements on the radio lures you to the Communications booth at the career festival. As you arrive, a communications engineer begins a presentation on *radio waves*—electromagnetic waves with frequencies at the lower range of the electromagnetic spectrum. You're surprised to find out that the first voice radio broadcast took place in 1910. The program featured the famous opera singer Enrico Caruso, and it was broadcast live from the Metropolitan Opera House in New York City.

You have been listening to radio broadcasts for years. Sometimes you tune in to FM 105.9 to listen to rock music, and sometimes you like to listen to AM 680 for the new country sound. What do those numbers on a radio dial really mean?

DISCOVER BY *Observing*

Turn on a radio to AM, and move the tuning dial from station to station. How do the numbers of the radio stations relate to frequency? Now do the same thing with FM. How do these numbers relate to frequency? Which have longer wavelengths, radio stations that broadcast on AM or those that broadcast on FM? How do you know? ✐

You find out from the communications engineer that AM radio stations broadcast on frequencies from 535 kilohertz (kHz) to 1605 kHz. FM stations use much higher frequencies than AM stations, from 88.1 megahertz (MHz) to 107.9 MHz. When you tune in a radio station, you are really adjusting it to receive a particular frequency of radio waves. Then the engineer shows you an illustration.

Amplitude modulation (AM)

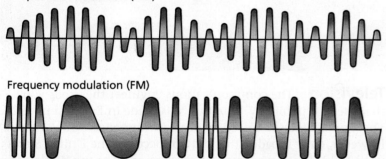

Frequency modulation (FM)

Figure 11–5. Radio waves are either amplitude modulated for AM broadcasts or frequency modulated for FM broadcasts. The frequency remains constant in an AM wave; the amplitude remains constant in an FM wave.

In the AM wave, the sound information is encoded by varying the amplitude of the wave. When the wave is picked up by an AM radio receiver, the sound, or audio, information is converted into an electrical audio signal. This signal is then amplified and sent to a loudspeaker, where it is converted into a sound wave.

In FM, the audio information is encoded by varying the frequency of the radio wave. The amplitude of the wave is kept constant. When the radio wave is picked up by an FM radio receiver, its frequency variations are converted into an electrical audio signal. This signal is then converted into a sound wave in the same manner as in an AM radio.

You ask the communications engineer why it is possible to hear an AM station farther away from its source than an FM station. He explains that AM waves reflect from the ionosphere, an

TEACHING STRATEGIES, continued

● **Process Skill:** *Evaluating*

Point out that radio transmitters send out two kinds of radio waves, those that go toward the sky and those that travel along the ground. What happens to each kind of AM wave? How does this affect AM reception? (Waves going toward the sky reflect off the ionosphere to Earth and can be heard beyond the Earth's curvature. Waves traveling horizontally can be heard until Earth curves, because they are long waves.)

Ask what happens to FM waves sent in the same manner. (FM waves heading toward the sky pass through the ionosphere and cannot be beamed back to Earth. Those traveling horizontally often do not even reach Earth's curvature because they are such short waves.)

SCIENCE BACKGROUND

Italian inventor Guglielmo Marconi developed the concept of using radio waves for communications. In 1895 Marconi demonstrated the wireless telegraph, using a transmitter and a radio receiver. Long-range radio reception became practical in 1913 when Edwin Armstrong patented the circuit for a regenerative receiver. Initially, radios were used mainly for ship-to-shore communications.

INTEGRATION–
Earth Science

The ionosphere is the highest layer of Earth's atmosphere. It is also called the thermosphere because it is a warm layer. Radiation from the sun produces electronically charged atoms, or ions, in the ionosphere. This phenomenon reflects radio waves that reach the ionosphere to Earth, permitting long-range radio communications.

✧ Did You Know?
Why are FM transmissions used at all if they are so limited in range? FM sound reproduction is superior to AM sound reproduction, and FM broadcasts are much less affected by static than are AM broadcasts.

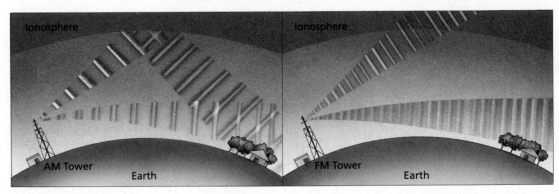

Figure 11–6. AM radio waves have a longer wavelength and are reflected by the ionosphere back to Earth. FM waves have a shorter wavelength, which allows them to pass through the ionosphere.

electrically charged layer of atmosphere, just as ultraviolet light is reflected from a windowpane. Due to this reflection, many AM broadcasts can be received over distances of thousands of kilometers or more.

FM waves, with their shorter wavelengths, pass through the ionosphere into space, just as visible light passes through a windowpane. They do not reflect back to Earth and, therefore, have a limited range. Shortwave radio broadcasts, with frequencies just above the AM broadcast band, reflect from the ionosphere even better than AM waves. This is why shortwave transmissions are used for international broadcasts.

Television The communications engineer then tells you that the first public TV broadcast was made in England in 1927. The United States followed with its own broadcast in 1930. However, daily broadcasting did not occur in the United States until 1939. That broadcast was of ceremonies at the World's Fair in New York City.

The engineer explains that television broadcasts are similar to radio broadcasts. Your TV antenna receives the waves and uses them to make electric currents, which are sent to your TV set. There the information is taken from the modulated waves, amplified, and sent to the picture tube.

In a color TV set, the colors are produced on the screen by little glowing dots. These dots are called *phosphor dots*. If you look at a color TV screen through a magnifying glass, you can see these dots. These dots occur in three colors—red, blue, and green—and are arranged in clusters, with one of each color per cluster.

The wave that carries the picture information is coded separately for each of the three colors. The separate information for each color becomes part of the electric signal that the antenna sends to the TV receiver. In the receiver, the information for

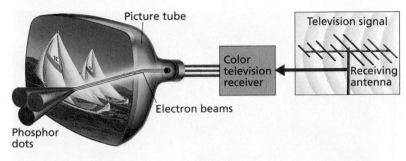

each color is separated and sent to the picture tube. There, the proper dots on the screen are stimulated, resulting in a color TV picture.

Figure 11–7. The dots on a color TV picture tube light up at different times to produce the moving television picture.

Radar

The communications engineer tells you that radar waves occur between radio wave frequencies and IR frequencies. They are often used to track planes in flight, to detect ships at sea, to locate storms, to track space vehicles and satellites in orbit around the earth, and even to detect speeding cars. He explains that the word *radar* is an acronym for **ra**dio **d**irection **a**nd **r**anging. A radar device sends out short pulses of radio waves. Any plane within a certain distance will reflect these

Figure 11–8. Video games would not be possible without the technology of phosphor dots.

● **Process Skill:** *Comparing*

Cooperative Learning Have the students work in groups to determine how radar is used to forecast weather. (Radar signals can also be reflected from raindrops or snow, thus providing weather forecasters a way to determine the size, location, and path of storms.)

● **Process Skill:** *Comparing*

Point out to the students that radar-absorbent paint on the leading edge of the high-altitude American B-2 Stealth bomber helps make the aircraft virtually invisible to enemy radar. The aircraft's fan-shaped wings show a low radar echo.

Similarly, modern submarines are coated in a thick layer of a top-secret resin. The resin is highly absorbent acoustically and reflects only a minute amount of the energy transmitted by sonar detectors.

✧ Did You Know?

During World War II, both Allied and Axis forces tried to jam the other's radar. A common practice was to have planes on bombing raids drop thousands of strips of foil, which reflected radio signals just as planes would. As a result, radar operators had much difficulty distinguishing the planes from the foil strips.

Figure 11–9. Among its many other useful applications, radar is used to track ships and airplanes.

waves. The radar receives the reflected waves and calculates the distance between the plane and the receiver. The radar does this by measuring the time it took for the echo of the wave to return to the receiver. The positions of detected objects can be seen on a screen similar to a TV screen.

Recently, engineers have developed "stealth" technology for military applications. This technology combines coatings that absorb commonly used radar frequencies with shapes that scatter the waves. Objects that employ this technology do not reflect commonly used radar frequencies and are therefore not detected by radar. In order to "see" the new stealth aircraft, radar installations would have to have new radar equipment that would detect different frequencies.

ONGOING ASSESSMENT
▼ ASK YOURSELF

In radio broadcasts, only sound information is encoded by the amplitude of the wave. In television broadcasts, both sound and light (video image) information is encoded.

▼ ASK YOURSELF
Compare radio and television broadcasts.

Figure 11–10. This stealth aircraft can "sneak" by normal radar undetected.

Microwaves

The communications engineer asks what you know about the use of *microwaves*—radio waves with wavelengths of about a centimeter. Most students know that these electromagnetic waves are used in microwave ovens; one student describes a microwave transmission tower she once saw while traveling in the Rocky Mountains.

The engineer tells you that in addition to being used to cook food, microwaves are used for **telecommunications**—the science and technology of sending messages over long distances, such as by telephone. Where telephone lines are difficult or impossible to install, microwave dishes are set up to send up to 100 telephone calls at one time. Then he adds that microwaves are used to transmit live broadcasts from foreign countries. International sporting events are often transmitted this way.

Next he explains how the familiar microwave oven works. Microwaves pass through some substances, such as glass, but are absorbed by other substances, such as water. Microwave ovens send microwave radiation through the food placed inside. The water in the food is heated by the microwaves, and the food is cooked. Although the container in which the food is cooked is not usually heated by the microwaves, it may still get hot as it absorbs heat from the food inside.

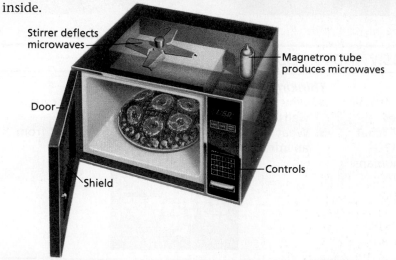

Stirrer deflects microwaves

Magnetron tube produces microwaves

Door

Controls

Shield

Figure 11–11.
Microwave transmission towers allow telephone messages to be sent to areas where the installation of telephone lines is difficult or impossible.

Figure 11–12.
Microwave ovens can quickly cook most types of food.

THE NATURE OF SCIENCE

The development of a vacuum tube called a *magnetron* made modern radar possible. The tube generates pulses of microwave energy used in radar systems. Also used in microwave ovens, the magnetron sends microwaves to the stirrer, a fanlike instrument. The blades of the stirrer scatter the microwaves in the oven, where they bounce from wall to wall until they enter the food being prepared.

 LASER DISC
1799

Microwave oven

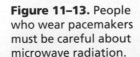
The engineer cautions that overexposure to microwave radiation may be harmful to humans since microwaves can kill healthy cells. They can even cause people who use pacemakers to have erratic heartbeats. Because of the danger, microwave ovens are constructed with shields and tight seals to keep radiation from escaping during use.

▼ ASK YOURSELF

What are two important uses of microwaves?

Figure 11–13. People who wear pacemakers must be careful about microwave radiation.

SECTION 2 REVIEW AND APPLICATION

Reading Critically

1. In what ways do the different electromagnetic waves resemble each other? What makes them different?

2. What problems can humans have if overexposed to X-rays, UV waves, and microwaves?

3. How can the problems in question 2 be prevented?

Thinking Critically

4. Why do you think radio waves can pass through walls, while radar waves bounce off?

5. What important information could you get from an infrared photograph of a house?

FOCUS

This section discusses color, including the relationship between color and wavelength, how the eye sees color, color-blindness, primary and secondary colors, and polarized light.

MOTIVATING ACTIVITY

Cooperative Learning Invite the students to be art critics. Have them work in small groups to discuss the artwork shown here in terms of the feelings each suggests. Discuss with the students whether these feelings are caused by the colors used by the artist, or by other factors in the paintings.

PROCESS SKILLS
• Analyzing • Expressing Ideas Effectively

POSITIVE ATTITUDES
• Enthusiasm • Precision

TERMS
• Primary colors
• Secondary colors

PRINT MEDIA
"There's a Lot More to Color Than Meets the Eye" in *3–2–1 Contact* magazine by Deborah Heiligman (see p. 211b)

ELECTRONIC MEDIA
Light, Color, and the Visible Spectrum Coronet Film and Video (see p. 211c)

SCIENCE DISCOVERY
• Effect of polarizing filters

BLACKLINE MASTERS
Laboratory Investigations 11.1, 11.2
Study and Review Guide

Color

As you pass by the Art booth at the festival, you remember Ms. Oyama's demonstration about the visual spectrum. You wonder how an artist uses this knowledge to create colorful paintings. You have used pastels to create images of a sunset, blazing in oranges and pinks. At a museum once, you marveled at early paintings by the artist Pablo Picasso (1881–1973), executed almost exclusively in various shades of blue. Near the Picasso blue-period paintings, you were startled by a Vincent van Gogh (1853–1890) self-portrait in striking greens. You imagine a self-portrait you would like to create, with deep purples and dazzling yellows in the background. As much as you savor drawings and compositions in black and white, art would be far less exciting for you without the use of luscious colors, thousands of them!

SECTION 3

Objectives

Describe the relationship between color and wavelength.

Relate rods and cones to the way people see color, including primary and secondary colors.

Identify the way in which certain filters polarize light.

INTEGRATION– Art

Picasso did have a "blue period" (1901–1905) early in his painting career. His works from this period show thin, wasted figures in postures of defeat and loneliness.

Encourage interested students to find books that illustrate some of Picasso's works, such as *Two Saltimbanques with Dog* (1905), *Young Girl on a Ball* (1905), or any others from those years. The students should compare the works' color schemes and the feelings these colors generate in the viewer with those of works from later in Picasso's career. (Note: *Saltimbanques* are circus workers.)

Picasso, "The Tragedy" (1903)

Courtesy of the Fogg Art Museum, Harvard University Art Museums, Bequest of collection of Maurice Wertheim, class of 1906.

Vincent van Gogh, "Self Portrait Dedicated to Paul Gauguin, 1888," oil on canvas, 60.5 × 49.4 cm.

Color and Wavelength

You learned in science class that the colors of the visible spectrum differ from one another in terms of wavelength. The commercial artist begins her presentation by saying, "Color is the

● **Process Skills:** *Observing, Solving Problems/Making Decisions*

Refer the students to the illustration on this page and explain that when white light strikes an object, some of the light is absorbed and some is reflected.

Then have the students name other objects that are red, yellow, or purple, and ask them to explain which wavelengths are absorbed by each object and which wavelengths are reflected. For example, does a red apple absorb wavelengths

with a higher or lower frequency than a banana does? (lower) Which wavelengths are absorbed by a white object? (None; white objects reflect all wavelengths of white light.)

way your eyes interpret the wavelengths of light. Ordinarily, each combination of wavelengths gives you its own special sense of color. When white light strikes most objects, certain colors are reflected while others are absorbed. The reflected light is what you see as color. So a blue object appears blue because it reflects mostly wavelengths of blue light. Most of the other colors are absorbed by the object. The human eye is capable of distinguishing among 17,000 colors!" Then the artist poses a problem for the students to solve.

Figure 11–14. White light consists of the colors of the spectrum. Objects absorb certain colors of light and reflect others when white light strikes them.

DISCOVER BY *Problem Solving*

Study Figure 11–14. Why does the banana look yellow, the apple look red, and the plum look purple? How do wavelengths of various colors determine the color of each object?

After each student responds, the commercial artist explains, "Your ability to distinguish between wavelengths of light is not perfect. The banana is yellow because it reflects almost the full spectrum of light to your eye. Only the shortest wavelengths are missing. The light from a sodium vapor lamp, like those that are now widely used on some streets and highways, is also yellow. That light contains only a single wavelength, yet its color is very similar to the color of the banana, which reflects many different wavelengths of light."

▼ ASK YOURSELF
Explain why a cherry looks red and a lemon looks yellow.

How You See Color

You ask the commercial artist how the eye works in order to see color. She explains, "In the retina of the eye, there are two types

● **Process Skills:** *Interpreting Data, Inferring*

Direct the students' attention to the graph, and discuss the three types of cone cells in the human eye. Point out the overlap of the lines in the graph. Remind the students that cone cells do not work well in dim light. Have the students write a paragraph in their journals that explains what the graph shows.

of cells—rods and cones—that respond to light. The *rods* are light sensitive and allow you to see in dim light. The *cones* detect color and control the sharpness of the image you see. The cones do not work in dim light. This is why you cannot see sharp images or colors when the light is very low."

She displays a graph and continues, "The cone cells of a human eye can be compared to the phosphor dots on a color TV set. Like the dots on the TV screen, there are three kinds of cone cells in the retina. Each kind of cell is sensitive to a different set of wavelengths. One kind of cone cell—the red receptors—receives only the longer wavelengths. The blue receptors respond only to the short wavelengths. The green receptors react to light in the middle of the visible spectrum."

SCIENCE BACKGROUND

People with dichromatic vision can see yellows and blues but cannot distinguish reds and greens. People with achromatic vision can see blacks, whites, and grays, much like the shades of a black-and-white photograph. More men than women have red-green colorblindness because it results from a sex-linked recessive gene.

THE NATURE OF SCIENCE

The cause of colorblindness is not known. There are several theories, including defective cones, too much or too little pigmentation in the eyes, or a malfunctioning optic nerve. So far, there is no cure for colorblindness.

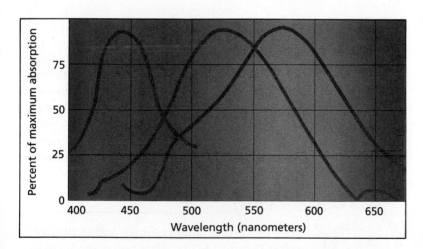

Figure 11–15. This graph shows how the cones in the retina of the human eye absorb colored light.

"If you study the graph (Figure 11–15), you can see that the wavelengths the cones receive overlap. This overlap means that the ability of the cones to receive colors overlaps. For instance, the green receptors are sensitive to some short wavelengths and some long wavelengths. However, they are most sensitive to wavelengths in the middle of the spectrum."

Then she explains that some people are unable to distinguish certain colors; these people are *colorblind*. About seven percent of men and less than one percent of women cannot distinguish between red and green. There is a rare kind of colorblindness in which a person cannot distinguish any colors at all. John Dalton (1766–1844), a British chemist, was the first person to detect colorblindness. In fact he was colorblind himself! Colorblindness, for a time, was also called *Daltonism*. The commercial artist then presents a test for red-green colorblindness.

● **Process Skill:** *Inferring*

Be sure the students understand that when the three primary colors of light are combined in equal parts, they produce white light. Other colors are produced by specific combinations of the three primary colors.

● **Process Skills:** *Observing, Comparing, Analyzing*

 Provide some magazines with colorful pictures and colored text. Have the students experiment with looking at the pages through different colored filters or pieces of cellophane. Ask why the objects on the magazine pages appear to be different colors and why some seem to disappear. Have the students record their observations in their journals. (Pictures or words in the same color as the filter being used seem to disappear because the light passing through the filter is of the same wavelength. Pictures or words of a different color than the filter change in color because the filter absorbs some of the light.)

DISCOVER BY *Doing*

The majority of the students should be able to identify the numbers. Students who cannot do so should be referred to an eye doctor for further evaluation.

ONGOING ASSESSMENT

▼ **ASK YOURSELF**

The rods respond to light and allow you to see in dim light. The cones respond to color and control the sharpness of the image.

DISCOVER BY *Observing*

Red + blue = magenta;
red + green = yellow;
green + blue = cyan.
White is a combination of all three colors.

Demonstration

Introduce this demonstration by asking the students to think of the kinds of filters with which they are familiar. Examples include filters used in coffee makers, over drains, and in science laboratories. Ask what all filters have in common. (They allow some materials to pass through and prevent others from doing so.) Then cover the lenses of three slide projectors or powerful flashlights with red, green, and blue filters. Focus the projector with the red filter onto a projection screen or wall. Add the other colors to make a color scheme similar to that in Figure 11-17.

Figure 11–16. These patterns are used to test for colorblindness.

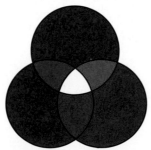

Figure 11–17. The primary colors of light are red, green, and blue. When two of these colors overlap, secondary colors are produced.

**DISCOVER BY** *Doing*

 Follow these instructions to find out whether or not your eyes can see green and red. Your color vision is considered normal if you can see a *5* in the top circle and an *8* in the bottom circle. If you do not see a number in the top circle and you see a *3* in the bottom circle, you have red-green colorblindness. What is the result of your test? 🖊

▼ **ASK YOURSELF**

What function do the rods and cones of the human eye perform?

A Model for Color

Next, the commercial artist shows a diagram of the colors people see. She explains that since the eye has three kinds of cone cells, the colors people see are made from these three colors—red, green, and blue—known as **primary colors.** By studying the three-primary model diagram shown to the students, you can understand primary colors and the colors formed by a combination of two primary colors—known as **secondary colors.**

DISCOVER BY *Observing*

Look at the diagram of primary and secondary colors (Figure 11–17). What two primary colors produce each of the secondary colors? How is white produced? 🖊

Filters   Then the commercial artist hands out different cellophane filters. When you look at white light through a filter of a primary color, the other two primary colors are absorbed and everything takes on the primary color of the filter. The world becomes totally red with a red filter, blue with a blue filter, and green with a green filter.

One student asks, "What happens if you look through a filter of a secondary color?" She answers, "If a filter of a secondary color is used, one primary color is absorbed—the one that is not necessary to make that color. For instance, a yellow filter, which is composed of green and red, absorbs blue light."

The artist then has everyone look again at the illustration of the banana, apple, and plum from earlier in her presentation.

Figure 11–18. The three primary pigments are yellow, magenta, and cyan. When these colors are combined with black for contrast, as they are when a book is printed, they produce a full-color photograph.

DISCOVER BY *Problem Solving*

Look again at Figure 11–17. Imagine that this image is in the form of a slide. When you send white light from the projector through the slide, each filter absorbs some colors and allows the other wavelengths through. Why is the image of the apple red? ✏

Pigments A student asks the artist how the color of paint is produced. She explains that pigments—colors used to tint other materials—give paints their colors. She tells you that in color film, color printing, and color TV screens, the three-primary model works well because the processes are carefully controlled. The simple three-primary model is not especially useful in mixing paints because the paints reflect many wavelengths of light to produce the color you see. Every artist and house painter has to learn by experience what kinds of pigmented paints to mix in order to get the desired color.

► **ASK YOURSELF**

What does the three-primary model tell you about the colors people see?

Have available several pairs of sunglasses, some with polarizing lenses and some with tinted lenses. Invite the students to compare the effect on light of both types of lenses. Ask the students how sunglasses with nonpolarizing lenses help block sunlight. (The tint in the lenses absorbs some of the light so it is not as bright.)

GUIDED PRACTICE

Have the students describe in their own words or with pictures the relationship between color and wavelengths of white light.

INDEPENDENT PRACTICE

 Have the students provide written answers to the Section Review. In their journals, have them record the results of their colorblindness tests.

THE NATURE OF SCIENCE

Polarized light is used in many ways. Microscopes using polarizers change colorless crystals into colorful ones. Scientists use a polariscope to find weak spots in eyeglasses and to identify and measure the sugar in solutions.

✧ Did You Know?

Photographers use thin plastic filters called polarizers to reduce glare reflected from water and other shiny surfaces. Edwin Land invented the polarized light filter in 1928. He also invented the Polaroid Land camera, first demonstrated in 1947.

 **LASER DISC**
41684–42271

Effect of polarizing filters

Figure 11–19.
Polarizing filters allow only certain light waves to pass. The first polarizer shown here screens all waves except those that vibrate up and down. No light passes through the second polarizer. Why?

Polarized Light

Outside the school, after the festival, students talk about future career opportunities. Some of Ms. Oyama's science students tell her what they found out about the ways in which electromagnetic waves are related to many careers and technologies. Ms. Oyama smiles, hiding the pleased look in her eye behind dark sunglasses. A student asks, "Ms. Oyama, do sunglasses really keep the glare of the sun out of a person's eyes?" "Yes," she answers.

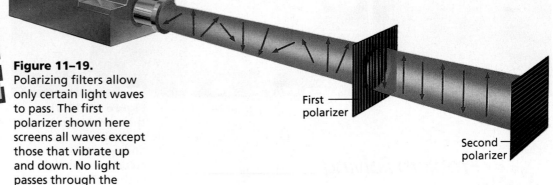

First polarizer

Second polarizer

In class the next day, Ms. Oyama uses a diagram to explain how sunglasses work. She shows her students that *polarized light* is light passed through a polarizing filter. The filter reflects all the waves except those that are vibrating in a particular direction. The light waves that come into a polarizing filter are transverse, like all electromagnetic waves. The vibrations of some of the waves are up-and-down, some are side to side, and others are at different angles. The first filter, however, blocks all waves but those that vibrate in a vertical, or up-and-down, direction. The light passing through the filter is polarized light because its waves only vibrate in one direction.

After being polarized by the first filter, the light then passes to the second filter. The second filter is horizontal, so it blocks all waves but those with horizontal vibrations. Since only light waves with vertical vibrations could pass through the first filter, the second filter allows no light to pass through.

Ms. Oyama shows her sunglasses and explains that they have a vertical polarizer, like the first filter in the diagram. The vertical polarizer blocks the glare from the sun. To help you understand more about polarized light, she has you do the following experiment.

EVALUATION

Discuss with the students why painting, photography, and other arts have contributed so much to our understanding of light and color.

RETEACHING

Have the students make their own diagrams of the primary colors and secondary colors. Encourage them to use their creativity to show the relationship between these two sets of colors.

EXTENSION

Have the students pick a favorite scene from a film and discuss how they would light it. Which colors would they use for the actors or a crowd scene? How would they create each of these colors? Have the students place their lighting ideas in their science portfolios.

CLOSURE

Cooperative Learning Have the students work in small groups to write three questions related to this section's content. Then have the groups exchange questions and answer them.

DISCOVER BY *Doing*

Use a polarizing filter, such as a lens from polarizing sunglasses, to identify polarized light around you. Look through the filter as you rotate it. If rotating the filter dims the light, the light is polarized. If the light does not dim, it is not polarized.

Figure 11–20.
Shown here are two photographs of the same store window. The photograph on the left, however, does not show the glare because it was taken through a polarizing filter.

It has been an exciting day. You had no idea that so many careers involved a working knowledge of electromagnetic waves. In fact, until today, you didn't know what an electromagnetic wave was! It seems strange that these invisible things have such an effect on your everyday life. Cooking food with microwaves, seeing, checking for broken bones or cavities in your teeth, rays from the sun—electromagnetic waves are pretty important.

ASK YOURSELF

How do polarizing sunglasses reduce glare?

SECTION 3 REVIEW AND APPLICATION

Reading Critically
1. What is the relationship between wavelengths and the way you see different colors?
2. What is colorblindness?

Thinking Critically
3. What color would pass through a combination cyan and magenta filter?
4. Why are polarizing sunglasses especially useful for drivers and pilots?
5. Make a list of 10 ways electromagnetic waves affect your daily life.

DISCOVER BY *Doing*

You can extend this activity by having the students demonstrate how the lenses of polarizing sunglasses polarize light. The students will need two pairs of polarizing sunglasses. Have them place one lens on top of the other, so that the axes of polarization coincide and light passes through normally. Then have them rotate one lens 90 degrees. No light can pass through.

ONGOING ASSESSMENT
ASK YOURSELF

Glare is horizontally polarized light, so it will not pass through a vertical polarizing filter.

SECTION 3 REVIEW AND APPLICATION

Reading Critically

1. When wavelengths of white light strike an object, all the colors in the white light are absorbed except those of the color of the object. Those colors reflect to create an image in the viewer's eyes.

2. Colorblindness is the inability either to distinguish certain colors, particularly reds and greens, or to see color at all.

Thinking Critically

3. blue

4. Polarized sunglasses reduce glare by absorbing the horizontally polarized light that causes glare. This helps drivers and pilots by increasing visibility.

5. Each student's list should reflect his or her own experiences.

Studying the Three-Primary Model

Process Skills: Analyzing, Expressing Ideas Effectively

Grouping: Groups of 3 or 4

Objective

● **Analyze** a three-primary color model.

Pre-Lab

Discuss with the students what they already know about filters. (The students might point out that filters generally hold back some substance(s) while allowing other substances to pass through.)

Hint

If your school does not have sets of filters, they can be purchased from most scientific supply houses or you can make some with cellophane. Homemade filters will probably not work as well as commercially produced ones, but they will give the students an idea of the effects of filters on color. You can do this demonstration for the entire class by using one set of filters and a slide projector.

	Red Filter	Green Filter	Blue Filter	Cyan Filter	Yellow Filter	Magenta Filter
Red Filter	red	brown	purple	purple	orange	red
Green Filter	brown	green	blue-green	blue-green	yellow-green	brown
Blue Filter	purple	blue-green	blue	blue	green	purple
Cyan Filter	purple	blue-green	blue	cyan	green	purple
Yellow Filter	orange	yellow-green	green	green	yellow	orange
Magenta Filter	red	brown	purple	purple	orange	magenta

Analyses and Conclusions

1. Filters absorb certain colors so that what you see is different from the image seen without the filter.

2. A combination of filters absorbs a combination of colors, letting fewer colors through.

3. Answers should include the information that filters support the three-primary color model.

Application

a. dark blue, green, brown
b. orange, yellow
c. polarized dark green, blue, brown

✳Discover More

The results would be similar to the results when colored filters are used. The water colors are not "true" primary colors and provide varying shades of the primary colors.

Post-Lab

Based on what they have learned in this investigation, have the students discuss how color filters can be used by photographers to produce better photos. (A photographer can use filters to adjust the amount of light in a photo. Color filters can also be used to change the intensity of certain colors.)

INVESTIGATION

*S*tudying the Three-Primary Model

▶ MATERIALS

● lamp with a 15-watt bulb ● color filters: red, green, blue, cyan, yellow, and magenta

▼ PROCEDURE

1. Make a table like the one shown.

2. Look at the lamp through the red filter. What color do you see?

3. Look at the lamp through each of the other filters. Record the colors you see.

4. Take the red filter, and put the green filter behind it so that the light goes first through the red filter and then through the green filter. Look at the lamp through the two filters. What color do you see?

5. Reverse the two filters so that the light goes through the green filter first. What color do you see?

6. Repeat steps 4 and 5, using all the possible combinations of filters. If little or no light comes through the filters, write *black* in the table.

TABLE 1: EFFECTS OF FILTERS ON LIGHT						
		Top Filter				
		Red Filter	Green Filter	Blue Filter	Cyan Filter	Yellow Filter	Magenta Filter
Bottom Filter	Red Filter						
	Green Filter						
	Blue Filter						
	Cyan Filter						
	Yellow Filter						
	Magenta Filter						

▶ ANALYSES AND CONCLUSIONS

1. How do the filters affect the colors that you see?

2. How does combining the filters affect the colors that you see?

3. Draw a conclusion about how these filters fit the three-primary color model.

▶ APPLICATION

Sunglasses come in many colors. Using your knowledge of filters, explain what color sunglasses would be best for (a) very sunny days, (b) overcast days, and (c) water activities such as fishing. How would you produce each of these tints?

✳ *Discover More*

Create watercolor paint mixtures with the same combination of colors. Compare the results to the results when colored filters are combined. Explain any differences.

The Big Idea—Energy

Lead the students to an understanding that the waves of the electromagnetic spectrum are a system of energy that interacts in different ways with other things in the universe and with human beings and their technology.

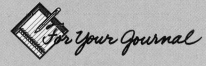

For Your Journal

The students' ideas should reflect their understanding of central concepts—the waves of the electromagnetic spectrum and their similarities and differences, the nature of color as a component of visible light, and so forth.

CONNECTING IDEAS

The students' illustrations should express an understanding of the nature of each kind of wave and its effects and uses.

CHAPTER
11

*H*IGHLIGHTS

Vincent van Gogh, *The Night Cafe*, oil on canvas, 1888, 28-1/2" x 36-1/4". Yale University Art Gallery, Bequest of Stephen Carlton Clark, B.A., 1903.

The Big Idea

The electromagnetic spectrum consists of electromagnetic waves—transverse waves that carry both electric and magnetic energy. The specific characteristics of these different waves determine their effects on humans and their usefulness in technology.

Electromagnetic waves have many characteristics in common, but they have some unique characteristics as well. Their specific qualities determine how they interact with other parts of the universe, from atoms to humans to planets. Humans can manipulate certain electromagnetic waves to perform useful functions.

For Your Journal

Review the questions you answered in your journal before you read the chapter. How would you answer the questions now? Add an entry to your journal that shows what you have learned from the chapter. Include new ideas about how electromagnetic waves affect your everyday life.

Connecting Ideas

By looking at the diagram of the electromagnetic spectrum and creating illustrations for each type of wave, you can show how the big ideas of this chapter are related. In your journal, illustrate each type of electromagnetic wave, except gamma rays. Write a paragraph that briefly describes each kind of wave you have illustrated.

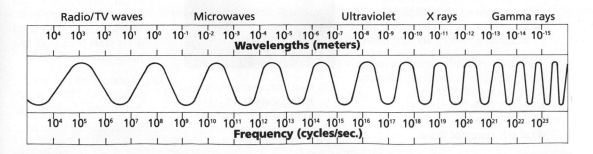

Understanding Vocabulary

1. (a) Electromagnetic waves are invisible, transverse waves that carry both electric and magnetic energy. All the types of electromagnetic waves together comprise the electromagnetic spectrum.

(b) Both terms describe types of color. The primary colors are red, blue, and green. The human eye has three kinds of cone cells that pick up these colors. The secondary colors— magenta, yellow, and cyan—are formed by a combination of two primary colors.

(c) Each term is related to sending messages. Telecommunications is the science and technology of sending messages over long distances. Radio waves—one of the first types of waves used in telecommunications—are electromagnetic waves with frequencies at the lower range of the electromagnetic spectrum. Microwaves—in addition to being used in microwave ovens—are radio waves with wavelengths of about 1 cm. Microwaves are also used in telecommunications, particularly in cellular-telephone communications.

Understanding Concepts

Multiple Choice

2. c

3. b

4. b

5. d

6. a

7. d

Short Answer

8. Answers can include objects such as microwave ovens (microwaves), televisions and radios (radio waves), heat lamps (infrared waves), and sun lamps (ultraviolet waves).

9. Radar uses electromagnetic waves at the upper range of radio waves. These waves are reflected by many objects and can be used to detect the presence of ships, planes, satellites, and storms. Radar has made air and sea navigation safer because it can "see" (detect) objects in the dark and in adverse conditions.

Interpreting Graphics

10. Yes, because stars emit visible light, as our sun does.

11. Around the edges of the disk to the left and in the center of the galaxy; hot areas show as red and yellow on infrared film.

12. Infrared waves are invisible to the human eye.

CHAPTER 11
REVIEW

Understanding Vocabulary

1. For each set of terms, explain the similarities and differences in their meanings.
   a) electromagnetic waves (284), electromagnetic spectrum (285)
   b) primary colors (302), secondary colors (302)
   c) telecommunications (297), radio waves (292), microwaves (297)

Understanding Concepts

MULTIPLE CHOICE

2. X-rays pass most easily through
   a) bones.　　**b)** metal.
   c) skin.　　**d)** teeth.

3. The electromagnetic waves given off by all warm objects are
   a) microwaves.　　**b)** infrared waves.
   c) ultraviolet waves.　　**d)** radio waves.

4. People with defective cone cells might be unable to
   a) protect themselves from uv rays.
   b) distinguish certain colors.
   c) adapt to seeing in bright sunlight.
   d) see in dim light.

5. Radio waves are used in all of the following except
   a) AM broadcasts.
   b) radar.
   c) FM broadcasts.
   d) thermograms.

6. The electromagnetic waves that cause people to tan and burn are
   a) ultraviolet waves. **b)** visible light.
   c) infrared waves.　　**d)** microwaves.

7. The different kinds of waves in the electromagnetic spectrum all travel
   a) at the speed of sound.
   b) at the speed of satellites.
   c) at different speeds.
   d) at the speed of light.

SHORT ANSWER

8. Name three objects found in the home that produce or use electromagnetic waves. Which type of waves does each object produce or use?

9. What is radar and how is it used? Describe how it has made air and sea navigation safer.

Interpreting Graphics

Photo A is a visible-light view of the Andromeda galaxy. Photo B shows the same galaxy on infrared film similar to a thermogram.

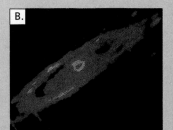

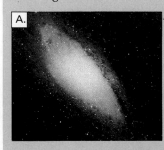

10. Can you see already existing stars in Photo A? Why or why not?

11. Dust and gases that eventually become stars must first get very hot. In which areas in Photo B are stars forming? Explain how you know.

12. Explain why the stars forming in Photo B cannot be seen in Photo A.

Reviewing Themes

13. Some forms of this energy are necessary to sustain life forms and can be usefully applied, while others are dangerous to life forms and must be avoided or controlled.

14. Doctors can take special pictures (thermograms), using special film to detect cancer cells. This film measures the amount of infrared radiation an object gives off. Since cancer cells are warmer than healthy cells, they would appear more prominently on the film.

Thinking Critically

15. Total colorblindness would make it hard to coordinate clothing, such as matching socks, and would make it difficult or impossible to work in certain fields, such as art, photography, and printing.

16. Since gamma rays have an even higher frequency than ultraviolet rays and X-rays, they would likely be very dangerous to humans.

17. Visible light and infrared radiation (heat); light is necessary for photosynthesis, which makes all life possible, and heat is necessary to sustain life.

18. The iridescence of a soap bubble is a result of the two reflections of the light rays entering the bubble. Some light rays reflect from outside of the soap bubble, and others reflect from the inside surface of the bubble. It is a combination of these reflections that produce the wide variety of colors on the surface of the bubble.

Reviewing Themes

13. *Energy*
Why is the interaction between the electromagnetic spectrum and life forms on Earth such a delicate balance?

14. *Technology*
Cancer cells are warmer than healthy cells. Explain how a doctor could use a type of electromagnetic wave to detect cancer cells.

Thinking Critically

15. Suppose you were totally colorblind—you could see only black, white, and shades of gray. How might your colorblindness affect your life?

16. Recall how ultraviolet and X-ray waves affect human beings. What effect do you think gamma rays would have on the human body? Reexamine the electromagnetic spectrum on page 285, if necessary.

17. The sun emits all forms of electromagnetic radiation. Which two forms are the most important for life on Earth? Why?

18. The shiny color, or iridescence, of this soap bubble is caused by reflected light. Light is reflected from the inside and outside surfaces of the bubble, causing two sets of reflections. These reflections can combine to form stronger colors or they might cancel each other out. Using your knowledge of reflection, refraction, interference, and color, explain how iridescence occurs.

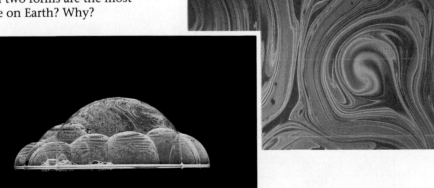

Discovery Through Reading

Kesten, Lou. "Future Visions: Tomorrow's Video Technology Looks Out of Sight!" *3-2-1 Contact.* 109 (September 1990) 24–27. Three-dimensional movies? A television that is smaller than a matchbook? Read about these technological advancements that could be closer to reality than you think!

USES FOR LIGHT

PLANNING THE CHAPTER

B = Basic   **A** = Average   **H** = Honors
The coding Basic, Average, and Honors indicates subsections, features, and resources that might be appropriate for different levels of learners. For additional suggestions regarding choice of topic and depth of coverage, see the Pacing Chart on pages T28–T31.

*Frame numbers at point of use
(TR) Teaching Resources, Unit 3
(IT) Instructional Transparencies
(LI) Laboratory Investigations
(SD) *Science Discovery* Videodisc Correlations and Barcodes
(SRG) Study and Review Guide

CHAPTER MATERIALS

Title	Page	Materials
Teacher Demonstration	314	any compounds of metallic salts (potassium chloride, barium chloride), nichrome loop, Bunsen burner, spectroscopes
Discover by Writing	315	(per individual) paper, pencil
Discover by Researching	318	(per individual) paper, pencil
Discover by Doing	322	(per individual) safety goggles, small piece of optical cable
Activity: How can you use an image conduit?	323	(per group of 3 or 4) safety goggles (4), laboratory aprons (4), Bunsen burner, image conduit, triangular file
Skill: Measuring Angles	325	(per group of 3 or 4) protractor, paper, ruler
Investigation: Determining Total Internal Reflection	326	(per group of 3 or 4) metric ruler, white paper, hemicylindrical plastic dish, flashlight, tape, protractor

ADVANCE PREPARATION

For the *Teacher Demonstration* on page 314, you will need to begin collecting cardboard tubes from paper towels, unless spectroscopes are available. You will need enough for each class member to make a spectroscope. For the *Discover by Doing* on page 322, you will need to obtain enough pieces of optical cable for each student to have a piece. You may wish to contact your local telephone service to acquire enough.

TEACHING SUGGESTIONS

Outside Speaker
Contact your local telephone company and arrange for a speaker to explain how optical fiber communication systems are set up. Transmission Engineers would be able to provide interesting information to your class. Also, a surveyor might be interested in coming to the classroom to explain how lasers are used in this field. Contact local construction companies to arrange for a speaker.

Field Trip
Arrange for a visit to the local planetarium to see their laser show. If this is not possible, you might investigate other forms of entertainment in your area that use lasers.

Chapter 12 presents the characteristics of laser light. It discusses how several different kinds of lasers work, the uses of lasers in medicine, how compact-disc players utilize lasers, and how holograms are produced. Fiber optics is also discussed.

Lasers are involved in many different systems. Communication systems as well as many medical systems rely upon lasers. Because the energy of light can be harnessed, society has been able to make great advances in medicine, space measurements, industry, and construction. A supporting theme in this chapter is **Energy**.

CHAPTER

12

USES FOR LASERS

COURTESY OF LUCASFILM LTD.™ & © Lucasfilm Ltd. (LFL) 1980. All Rights Reserved.

Lasers have long appealed to the makers of science fiction. In the old television series Star Trek, Captain Kirk cried "Set phasers on stun!" and zapped his enemies with laser-like beams. The Jedi knights of the Star Wars movies dueled with sabers of laser light.

Imagine a movie starring a young doctor whose sweetheart is blinded in an accident. The doctor daringly aims a laser beam at her eye! The eye is repaired, and the sweethearts live happily ever after. Medical science fiction? No, today's physicians use lasers to repair damage to the human retina.

Instead of "Ghostbusters," a future movie might star a team of "Scumbusters." These heroes could travel the

Display several objects that are produced by lasers or that use lasers. Examples include UPC symbols, objects with holograms on them (credit cards, paperback books, baseball

cards), compact discs, and a diamond or other gem cut by a laser. Discuss with the students how lasers are related to each object.

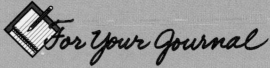

For Your Journal

The journal questions will help you identify any misconceptions students might have about the uses of light. In addition, these questions can help the students realize that they do possess knowledge about uses of light. (Students will have a variety of responses to the journal questions. Accept all answers, and tell the students that at the end of the chapter they will have an opportunity to refer back to these questions.)

ABOUT THE PHOTO

According to *Skywalking: The Life and Films of George Lucas,* by Dale Pollock, the film *Star Wars* had made almost $3 million in its premiere week. Five weeks later, the film had grossed nearly $30 million, a figure that grew to $68 million just two weeks later. Within three months, *Star Wars* totaled $100 million in revenue, reaching that amount in less time than any other film at that time.

MEETING SPECIAL NEEDS

Gifted

 Have the students read *The War of the Worlds* by H.G. Wells (1898) and write a follow-up report in their journals, naming those ideas in the book that are now a reality, and describing any ideas that are still science fiction. Conclude by encouraging the students to speculate whether or not these ideas might ever become a reality.

world restoring ancient buildings and statues nearly destroyed by air pollution. A laser gun could zap the coating of scum caused by decades of air pollution. The stone beneath the scum would reappear in all its beauty. In fact, such lasers are already being used to help restore damaged art works. In one restoration, a tiny laser beam cleaned silver threads in a precious drapery. The beam did not touch the fragile cotton threads woven into the same cloth.

Imaginative scientists are only beginning to tap the potential of laser tools. Already, their real progress surpasses some of the wildest inventions of science fiction!

For Your Journal

- What makes laser light different from ordinary light?
- How are lasers used in medicine?
- How can lasers be used to carry sound?

In this section, the characteristics of laser light are compared to those of ordinary light. Laser light is coherent light—light composed of waves of one frequency all in step.

Remind the students of the nursery rhyme "Twinkle, twinkle little star." As Earth cools, air currents are produced as cold air sinks and warm air rises. The index of refraction of air is related to its temperature. Light bends as it passes through air masses of different temperature. The result is a twinkling effect.

CAUTION: Tell the students not to look directly at the laser.

Shine the beam of a low-power laser (1 milliwatt) onto a screen in a darkened room. Place a plugged-in hot plate below the beam of the laser. Have the students observe the twinkle of the laser on the screen.

PROCESS SKILLS
• Comparing • Analyzing • Generating Ideas

POSITIVE ATTITUDES
• Curiosity • Openness to new ideas

TERMS
• coherent light • excited state • pumping

PRINT MEDIA
"The New Worlds Lasers are Conquering" from *Business Week* magazine by Neil Gross (see p. 211b)

ELECTRONIC MEDIA
The Laser: The Light Fantastic BFA Educational Media (see p. 211c)

SCIENCE DISCOVERY
• Lasers; explanation

BLACKLINE MASTERS
Connecting Other Disciplines
Reading Skills
Study and Review Guide

SECTION 1

The Laser

Objectives

Distinguish *between ordinary light and coherent light.*

Describe *how lasers are produced.*

Richard loves science fiction stories, especially those with wild, futuristic inventions. One rainy day he went to choose one of the many science fiction books from the family library. When he reached up to pull a book off the shelf, down fell a dusty old folder that had been buried behind the row of books. He blew off the dust, opened it up, and saw the name *Marsha Carroll* on the top sheet of paper. "My grandmother," he thought. "I had no idea she wrote science fiction."

Richard began to read with delight. He realized that his grandmother had written this story many years before the invention of technologies that are common today.

Figure 12–1. Lasers are used to produce light shows like this one at Stone Mountain, Georgia.

● **Process Skills:** *Expressing Ideas Effectively, Synthesizing*

Initiate a discussion on how lasers are used in entertainment. Encourage the students to describe examples they might have seen, such as laser light shows at concerts, in museums, at planetariums, at special events, and at fireworks shows. You might then lead the discussion into a comparison of ordinary light and coherent light.

● **Process Skill:** *Communicating*

There are many types of lasers: semiconductor lasers, metal-vapor lasers, gas-dynamic lasers, and chemical lasers. Each relies on a different element or compound to function.

Metal-vapor lasers rely on various metals to produce many different colors of laser light. Ask the students to look up the atomic number and element symbol of each of these elements used in metal-vapor lasers: cadmium, zinc, tin, lead, manganese, copper, indium, gold, calcium, and germanium.

Coherent Light

One passage in the story read, "This was not an ordinary flashlight. The beam from this light pierced the air like an arrow, thin and straight. It shone for miles and miles."

"Wow!" thought Richard. "Grandma Marsha's futuristic light sounds like a laser." This was a technology he had been learning about in science class.

The beam of a flashlight may be about 3 cm across when it leaves the flashlight, but it enlarges to about 40 cm by the time it reaches the other end of a room. In contrast, the beam produced by a laser is only 2 mm across, and it stays 2 mm wide even when it reaches the other end of a room.

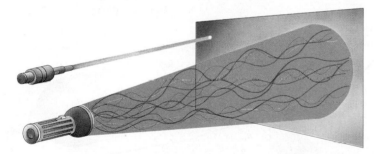

Figure 12–2. Laser light consists of light waves that are usually close in wavelength with a uniform pattern. Ordinary light consists of many different wavelengths that have a random pattern.

Unlike ordinary light from a flashlight, a laser beam's light contains only a few wavelengths. These wavelengths are usually very close together. For example, a helium-neon laser contains wavelengths in the long-wave part of the visible spectrum, that is, in the red part of the spectrum. Therefore, the light looks red. Other materials in lasers produce light of different colors.

Laser light is unique because all the light waves are in step. That is, crests travel with crests and troughs travel with troughs. In a beam of ordinary light many waves overlap, and there is a jumble of crests and troughs. When all the light waves in a beam are in step, the light is called **coherent light.** Coherent light may contain more than one wavelength, but for each wavelength the crests and troughs are in step.

Coherent light produces a very straight, concentrated beam. During research for his science class, Richard found out that when the subway tunnel was built under San Francisco Bay, its course was lined up by a laser beam that went from one side of the bay to the other. He also found out that the most accurate measurement of the distance to the moon was made by shining a laser at a reflector that was placed on the moon by an astronaut. Grandma's light beam was amazing, but not as amazing as the real thing!

Figure 12–3. This low-power helium-neon laser uses neon gas to produce coherent light.

NATURE OF SCIENCE

Although Dr. Theodore Maiman is credited with creating the first laser light in 1960, it was actually the culmination of efforts of teams of scientists over almost 100 years. Some of the first attempts to understand light were made by Isaac Newton in 1666. Then in 1917, Albert Einstein recognized the phenomenon of stimulated emission, but did not know it would be the key to making a laser. In 1951, Charles H. Townes and his brother-in-law, Dr. Arthur Schawlow, produced the first maser (Microwave Amplification by Stimulated Emission of Radiation). Townes and Schawlow published their ideas in 1958. They attracted considerable interest. Scientists worldwide tried to make a device that would give people control of light. Then in May 1960, Dr. Maiman put an end to all skepticism by producing lasing action in a ruby crystal rod, thus creating the long-sought new light.

INTEGRATION– *Mathematics*

Have the students calculate, in their journals, the distance between Earth and the moon if it takes a laser beam 2.5 s to travel to and from the moon. Have them use the formula $d = vt$. The speed of light is approximately 300 000 km/s. Remind them that 2.5 s is the round-trip time. [(1.25 s) × (300 000 km/s) or 375 000 km]

SECTION 1 **313** ◀

● **Process Skills:** *Formulating Hypotheses, Predicting*

After the students have reread the first paragraph, have them write, in their journals, an explanation of how lasers could be used in defense of our country in the next 20 years.

● **Process Skill:** *Expressing Ideas Effectively*

Cooperative Learning Explain to the students that since laser beams can readily travel through space, they make ideal weapons to knock out planes,

missiles, and satellites. However, such lasers would need to be extremely large and powerful. As a result, lasers—and the satellites that carry them—will be difficult and expensive to construct. In 1984 the United States committed about $200 million to developing a defense system nicknamed Star Wars.

Have the students work in small groups to research this system. Encourage the students to share their information with classmates as part of a follow-up discussion. Have the students place their reports on the Star Wars defense plan in their portfolios.

ONGOING ASSESSMENT
▼ ASK YOURSELF

Ordinary light is composed of many different wavelengths of light. Coherent light is composed of light waves that are all in step.

Demonstration

Demonstrate the raising of atoms to excited states by heating table salt and/or any other metallic salt. Use a Nichrome loop and a Bunsen burner. Have the students observe the display through spectroscopes.

If spectroscopes are unavailable, you might have the students make their own. For each spectroscope, obtain a cardboard tube (from a roll of paper towels), a square of black paper, and a piece of plastic diffraction grating. Use a razor blade to make a slit about 2.5 cm long in the square of paper. Tape the paper over one end of the cardboard tube and the grating over the other end. The students should look through the diffraction grating on one end and aim the slit at the colored flame. Conclude by having the students draw a picture of the spectrum they see for each metallic salt you burn.

LASER DISC
2242, 2243

Lasers; explanation

▼ ASK YOURSELF
What is the difference between ordinary light and coherent light?

How Lasers Work

In Grandma Marsha's story, a weapon shot out beams of straight light. The weapon looked like an old-fashioned dart gun. Richard learned in science class that lasers are often created in tubes filled with neon. He also learned that the way coherent light is produced from neon in tubes is similar to the way in which a dart gun is loaded for firing.

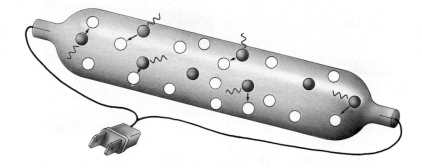

Figure 12–4. In a neon light, electric current raises many atoms to an excited state. The atoms quickly return to their ground state by emitting photons. This makes the bulb glow.

To load a dart gun, you compress the spring. This adds the potential energy of the compressed spring to the system. When you shoot the dart, the spring is released. The energy that was stored in the spring is transferred to the dart, which shoots toward the target.

An atom of a gas acts in a way similar to the spring of the dart gun. The atom can absorb and store energy. When an atom contains stored energy, it is in an **excited state**. In a neon light, the electric current going through the neon gas provides the energy to raise the atoms to an excited state. The atoms in the gas are pumped by the current. **Pumping** is the process of raising atoms to an excited state.

Now the atoms in the neon tube are like a loaded dart gun. Each of the atoms possesses stored energy. But there is no mechanism to hold the neon atoms in their excited state, so they quickly lose their energy. When atoms give off the energy they gained in an excited state, they return to their **ground state**.

When atoms return to their ground state, they release the extra energy by emitting a photon of light. A photon is the small packet of energy in light waves. The photon shoots out of

● **Process Skills:** *Analyzing, Constructing/Interpreting Models*

After the students have studied Figure 12-5, point out that only a photon with the correct amount of energy will cause the emission of an identical photon from the excited atom. These two photons will collide with two other atoms in excited states, producing a total of four photons.

Ask the students to predict the effect of a photon with the incorrect amount of energy striking an excited atom. (A photon with the incorrect amount of energy will not release two photons.)

GUIDED PRACTICE

Cooperative Learning Have the students work in groups of four and take turns explaining and describing what each of the following names or terms means: *coherent light, excited state, laser, pumping, ground state, photon,* and *stimulated emission.*

INDEPENDENT PRACTICE

Have the students provide written answers to the Section Review. In their journals, have the students write the names of the kind of lasers mentioned in this section and the meaning of the acronym, laser.

the neon atom just as a dart shoots out of a dart gun. ("Grandma was pretty smart," thought Richard.) In the neon tube, the photons are released in all directions, at all times. This means that the emitted photons do not form a coherent beam of light.

Stimulated Emissions of Photons A neon tube can be converted into a laser because of a unique property of neon atoms. When an excited neon atom is struck by a photon, the atom immediately releases two photons.

The two photons are identical. They leave the atom together, in the same direction and in step. The process of forcing identical photons into step is called **stimulated emission**. While Grandma probably didn't know about excited or ground states of atoms or stimulated emissions of photons, she was on the right track.

Stimulated emissions from one photon produce two photons. If each photon strikes an excited neon atom, four photons are produced. This process continues and takes place quickly. An excited atom must be stimulated to give off a photon before it has a chance to drop back to its ground state.

Richard decided that, with his parents' permission, he would try to modernize his grandmother's manuscript. How could Richard make a diagram that would help his readers understand the coherent light of lasers? Try the next activity to find out.

DISCOVER BY Writing

Illustrate the difference between ordinary light and coherent light. Include labels and captions to describe each kind of light. ✎

Making a Laser Years and years ago, when Grandma Marsha wrote her science fiction story, she had never heard of the word laser, which is an acronym for **l**ight **a**mplification by **s**timulated **e**mission of **r**adiation. If you know what the word means, you can understand how a laser works. To create a laser, you need more than photons that are in step. The ends of the tube of neon gas must be coated with a reflecting surface to cause any photons traveling up and down the tube to bounce back and forth between the ends. The bouncing photons stimulate emission from other atoms as they pass by. Soon, there is an enormous flow of photons back and forth in the tube,

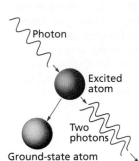

Figure 12–5. When a photon strikes an excited atom, the excited atom will emit two photons in step with one another. This process is called stimulated emission.

DISCOVER BY Writing

Students' diagrams for coherent light should show that the wavelengths are the same and the crests and troughs are in step. Diagrams for ordinary light should show a variety of wavelengths crossing each other.

REINFORCING THEMES
Technology

Reinforce the concept that the production of a laser requires the complex interaction of many systems. Have the students count how many steps are involved in producing a laser beam. (1. pumping the atoms; 2. stimulated emission of photons; 3. light directed into a beam)

INTEGRATION– *Health*

Misused laser beams pose a serious threat to vision. The Federal Drug Administration (FDA) regulates radiation-emitting devices. Depending on the type of laser, various radiation safety features are required. The FDA advises those thinking about attending laser light shows to follow these guidelines.

1. Call the local health department to find out whether safety checks have been made.
2. Never look directly or through binoculars or cameras into any intense light source, laser or otherwise.
3. If there is any doubt about a laser show's safety, it is better to not attend.

 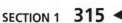

EVALUATION

Ask the students to write the definitions of *coherent* and *incoherent* in their journals. Then ask a volunteer to explain why *coherent* is an appropriate term to apply to harnessed light energy. Answers should demonstrate an understanding of coherent light and of light as energy.

RETEACHING

Use gas-vapor tubes to demonstrate stimulated emission. Have the students draw a diagram of the apparatus in their journals and label each part.

EXTENSION

Point out to the students that the first working laser was built in 1960 by T.H. Maiman. Then have the students work independently or in pairs and use library resources to write a report on Maiman and his invention. Encourage the students to accompany their reports with illustrations.

CLOSURE

Cooperative Learning Have the students work in small groups to write a summary of this section. Then ask each group to read aloud its summary.

ONGOING ASSESSMENT

▼ **ASK YOURSELF**

A bouncing photon is produced by energizing or exciting an atom.

✧ **Did You Know?**

Mars was nicknamed the Red Planet by many ancient cultures. Mars was the Roman god of war. The planet was given the name because its color was similar to the color of blood.

SECTION 1 REVIEW AND APPLICATION

Reading Critically

1. Coherent light occurs when all of the light waves in a beam have the same wavelength and are in step—crests travel with crests, and troughs travel with troughs.

2. One mirror at the end of a laser tube is only partially mirrored so that the laser beam can escape and form a beam.

Thinking Critically

3. A laser can be used to produce an image of the American flag through its reflective properties. The laser beam can be bounced off an actual American flag and then projected onto a screen. Holography could also be used.

4. A biologist would have to focus a laser beam to a tiny spot by sending it through a lens or lens system.

Figure 12–6. Use the key to follow the bouncing photons created in this neon tube from A to D. The photons bounce off the reflective surface at the ends of the tube. One end is only partially mirrored. This is the end from which the laser beam passes out of the tube (D).

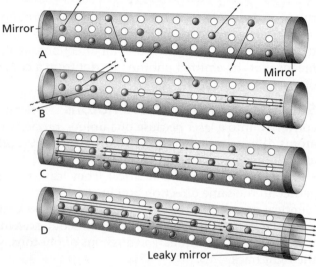

KEY

○ Ground-State atom

◗ Excited atom

~ Photons

Figure 12–7. The laser is being used to cut engine bores.

all in step with one another. If one of the mirrored ends has a thinner coating, some of the light can pass through. From this end, five to ten percent of the coherent light escapes to form a laser beam.

Richard learned that any material that can produce photons in step can produce a laser beam. He also found out that energy other than electric energy, like chemical reactions or flashes of light, can stimulate the production of photons in step. Lasers, Richard discovered, are produced when reflective ends cause a standing wave of light inside a tube. His science teacher explained in class that coherent light has been observed coming from Mars. Scientists hypothesize that it is produced naturally by the action of sunlight on the carbon dioxide in the Martian atmosphere.

Richard writes down this new information. It is giving him great ideas that may help him to modernize Grandma's story.

▼ **ASK YOURSELF**

How are bouncing photons produced?

SECTION 1 *REVIEW AND APPLICATION*

Thinking Critically

1. Explain what coherent light is.

2. Why is the mirror at one end of a laser tube only partly mirrored?

Reading Critically

3. Explain how lasers could create an image of the American flag on a screen.

4. How could a biologist use a laser to destroy the nucleus of a single cell?

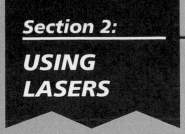
FOCUS

This section focuses on the uses of lasers in medicine. The use of lasers to play compact discs and to record on compact discs is also discussed.

MOTIVATING ACTIVITY

Ask the students to share any examples they know that involve the use of lasers in medicine. Examples include people who have had cataract surgery or have had birthmarks removed. Other experiences might include the sealing of blood vessels in ear surgery. Briefly discuss each example, and explain to the students that they will learn more about medical uses of lasers.

PROCESS SKILLS
• Synthesizing
• Constructing/ Interpreting Models

POSITIVE ATTITUDES
• Enthusiasm for science and scientific endeavor • Curiosity

TERMS
none

PRINT MEDIA
How Did We Find Out About Lasers? by Isaac Asimov (see p. 211b)

ELECTRONIC MEDIA
Introduction to Holography Britannica Video (see p. 211c)

SCIENCE DISCOVERY
• Laser applications

BLACKLINE MASTERS
Extending Concepts Laboratory Investigations 12.1, 12.2 Thinking Critically Study and Review Guide

Using Lasers

SECTION 2

In addition to fantastic futuristic inventions, Richard's grandmother had a knack for creating colorful characters. Her main character, Jack Ratty, had enjoyed adventures around the galaxy before he settled in at Gulch Creek, Earth. One day as he leaned over a simple household gadget, he seriously injured his eye. A visiting venusian friend shone a thin blue beam of light from her forehead into the injured eye. Immediately, Jack regained perfect vision. Later in the manuscript, a small yellow beam of light was struck against a mirror, and galactic music was amplified throughout space.

At the end of the manuscript are the words *The End* and the date: 1940. Grandma's imagination was well ahead of her time.

Objectives

Construct a chart explaining medical procedures that use lasers.

Explain how music is stored and played on a compact disc.

Lasers and Medicine

If the character Jack Ratty had suffered a detached retina in his eye in the 1990s, an earthling rather than a venusian could have helped him. To treat the condition, laser pulses are fired onto many points around the edge of the detached region. Each pulse produces a tiny bit of scar tissue that holds the retina in place. If he had suffered broken blood vessels in his eyes due to diabetes, a laser could have been focused on a micrometer-sized spot on the retina. A short, intense pulse of light would close the bleeding vessels. No bleeding occurs because the laser heats the tiny spots of tissue so much that it fuses the tissues together.

 SCIENCE TECHNOLOGY SOCIETY New findings in laser research are reported on an almost daily basis. Researchers at the University of California Medical Imaging Science Group are testing a new technique with lasers that might be able to detect lung cancers before they are 1 mm long. Doctors elsewhere are attempting to correct nearsightedness and astigmatism by reshaping the cornea of the eye with precisely placed burns on the front of the eye. When the tiny burn area is healed, it shrinks, tightening the focus of the eye. Dental researchers are experimenting with laser light to attempt to make tooth surfaces stronger.

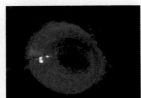

Figure 12–8. An argon gas laser (left) is often used in performing delicate eye surgery (right).

 LASER DISC
2245, 2246

Laser applications

TEACHING STRATEGY

● **Process Skill:** *Formulating Hypotheses*

Discuss how compact discs compare with vinyl records. Have the students closely examine a vinyl record. Point out that vinyl records have grooves. Music is recorded as different patterns of grooves. A needle is placed in the groove while the record turns.

Have the students speculate on the advantages and disadvantages of each technology. (Vinyl records are larger and can warp and scratch easily. Compact discs are smaller, will not wear out, and produce higher-quality sound.)

GUIDED PRACTICE

Explain that in the production of CDs, music comes to the microphone as a sound wave and is converted into wavelike electric energy. This signal is then coded into pulses or bits. Have the students think of other examples when this process occurs. (radio broadcasts, radar detectors, television, and photosynthesis.)

INDEPENDENT PRACTICE

Have the students provide written answers to the Section Review. In their journals, have the students develop a definition of the binary system as used in computer programming.

ONGOING ASSESSMENT
ASK YOURSELF

Lasers are used in surgery because they can quickly alleviate bleeding and are as precise as a scalpel. Lasers can also reach hard-to-reach areas in the human body.

MEETING SPECIAL NEEDS

Gifted

Ask interested students to use library materials and other resources to find out how scientists get the special crystals they use for the various types of lasers. The types of crystals used are YAG, ruby, neodymium, and gallium arsenide.

NATURE OF SCIENCE

The YAG laser can be operated in the visible light region to produce a green beam. These lasers produce a beam that is ideal for eye surgery.

A laser was first used in eye surgery on a nine-year-old girl suffering from a membrane growing across her eye. The girl had undergone three previous operations, each of which had failed. In February 1968, Dr. Francis A. L'Esperance of Columbia Presbyterian Hospital in New York used a laser scalpel to cut through most of the membrane. The laser's heat sealed each blood vessel as he worked, thus preventing bleeding. The next day, surgeons completed the successful operation and restored the girl's vision.

Figure 12–9. A CD player uses a laser, as illustrated here, to "read" the code on the CD.

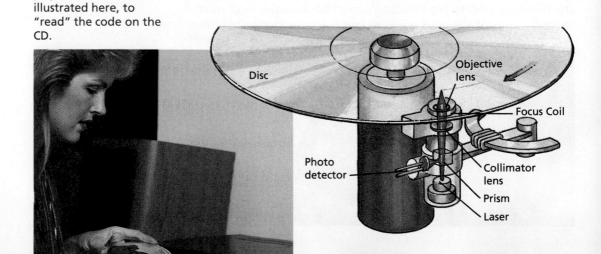

In some cases, lasers can be used instead of scalpels during surgery. An infrared laser made from carbon dioxide can focus continuous light to kill a tumor while sealing blood vessels around the tumor at the same time. A YAG laser, made from a crystal of **y**ttrium **a**luminum **g**arnet, makes an intense infrared beam, focused to a fine point, that can reach areas within the middle ear where a scalpel cannot be used. Since Richard wants to find out about other medical procedures in which lasers are used, he decides to take a trip to the library.

 ### DISCOVER BY *Researching*

Go to the library or talk to a laser surgeon, and collect as much information as possible on the use of lasers in medicine. Make a chart about the procedures in which lasers are used and the specific type of laser that is needed for each procedure.

ASK YOURSELF
Why are lasers used in surgery?

The Compact Disc

Richard's grandmother may have thought that a music machine played by a light beam was a wild idea, considering it was 1940. But CDs, or compact discs, played with laser light, are common today.

The surface of a CD is touched only by a microscopic spot of infrared light. The light produced by a tiny laser is focused by a lens system to a pinpoint spot. This spot of light reflects from the tiny pits on the mirrored surface of the disc. The reflected light is converted into an electric signal, which is then amplified and sent to the speakers.

Music is recorded on a CD in a digital code. The music comes to the recording microphone as a sound wave. The microphone converts the sound wave into a wavelike electric signal. This signal is coded into a series of tiny pulses. Each pulse stands for the number one. A missing pulse stands for the number zero. Each number is called a *bit*. About 16 billion bits will put one hour of music on the disc.

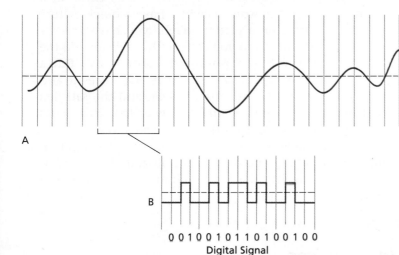
Digital Signal

Figure 12–10. In recording sound on a CD, the sound wave is converted by a microphone into an electric signal (A). This signal is then coded into a series of pulses (B). The digital signal turns the laser beam on and off to record sound on a CD.

▼ **ASK YOURSELF**
How is music coded onto a CD?

SECTION 2   REVIEW AND APPLICATION

Reading Critically
1. How does a laser beam allow us to hear music on a CD?
2. What is a bit?

Thinking Critically
3. A laser can be used to remove a red birthmark called a portwine stain. This type of birthmark contains many blood vessels. Why do you think a laser is used for this procedure?
4. Since a surgical laser uses infrared light, the surgeon cannot see the cutting beam. Suggest a way that another laser might be added to the system to solve the problem. Justify your answer.

PROCESS SKILLS
- Experimenting
- Formulating Hypotheses
- Observing • Measuring
- Analyzing • Interpreting a Graph

POSITIVE ATTITUDES
- Curiosity • Enthusiasm for scientific endeavor
- Openness to new ideas

TERMS
- Hologram • Total internal reflection • Light pipe
- Optical fiber

PRINT MEDIA
The Holographic World by Michael Talbot (see p. 211b)

ELECTRONIC MEDIA
Fiber Optics Britannica Video (see p. 211c)

SCIENCE DISCOVERY
- Hologram
- Fiber optics; telephone communication

BLACKLINE MASTERS
Study and Review Guide

MEETING SPECIAL NEEDS

Second Language Support

Assign the short story "Dreaming is a Private Thing," by Isaac Asimov. Instruct limited-English-proficient students to make a list of words they do not understand in the story. Then have a classmate help the student determine the meaning of the words. Encourage the students to use context clues before they refer to a dictionary.

FOCUS

Section 3 discusses the use of light in holograms and fiber optics. Holograms are three-dimensional images produced by the interference of light. The science of fiber optics is based on the principle of total internal reflection, which allows light to be carried inside thin glass wires.

MOTIVATING ACTIVITY

Obtain a bottle of children's play bubbles that includes a wand. Blow a series of bubbles and ask the students to write in their journals a description of the bubbles. Then ask the students to write a hypothesis explaining why the bubbles appear as they do. (The rainbow-like patterns are due to the interference of light. If the light waves are in step, constructive interference occurs and the entire spectrum is visible on the bubble. If the light waves are out of step, destructive interference occurs and no spectrum is visible.)

SECTION 3

Using Light Waves

Objectives

Describe the image produced by a hologram.

Evaluate the necessary conditions for practical communication by light waves.

In Grandma Marsha's manuscript, Jack Ratty enters a vivid dream machine and sees each dream image in three dimensions. This reminds you of a favorite Isaac Asimov story, "Dreaming Is a Private Thing," in which people learn to dream in more than three dimensions. You look up the story in your family's library and reread a favorite passage: "I don't mean think as in reason. I mean think as in dream. . . . Now when I think of a steak, I think of the word. Maybe I have a quick picture of a brown steak on a platter. Maybe you have a better pictorialization of it and you can see the crisp fat and the onions and the baked potato. I don't know. But a dreamer . . . He sees it and smells it and tastes it and everything about it, with the charcoal and the satisfied feeling in the stomach and the way the knife cuts through it and a hundred other things all at once."

Richard laughs as he thinks about what his grandmother's reaction would be to a holographic image, or even virtual reality machines. Grandma surely would think she was dreaming!

Figure 12–11. The two photos above are the same hologram, viewed from different angles.

Holograms

Richard saw his first hologram in science class a few weeks ago. The picture showed an astronaut floating in space. By turning the hologram, he could see the figure in three dimensions.

The hologram looked like a sheet of dirty transparent plastic, but it was actually covered with tiny, closely spaced squiggly lines that can only be seen under a microscope. If a light is placed behind the hologram and a person looks through the plastic sheet, a tiny astronaut floating in space appears. The image has a three-dimensional quality, a feeling of reality, that does not exist in other optical images.

Richard learned that a holographic image changes when you look at it from a different angle. From one position, the astronaut's left arm hid the instruments on his abdomen (top photograph). Then, you moved to the left to look behind that arm and gradually saw something different as you moved (bottom photograph). When you moved even farther to the left, the

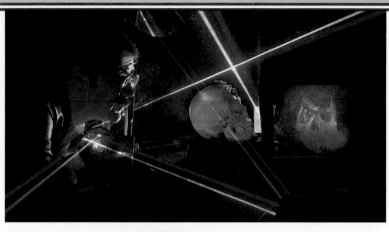

picture changed even more. A hologram can do this because it produces a set of light waves exactly like those coming from a real object.

A **hologram** is an interference pattern created by a laser on a piece of film. The laser beam is split into two parts. One part goes directly to the film. The other part goes to the object being photographed and is reflected onto the film. When the film is developed, a series of dark and light lines appears. The dark lines show where interference of the two beams is constructive. At the light lines, the film was not exposed. There was no light at those points because crests of one beam arrived at the same time as troughs of the other. That is, there was destructive interference.

When light comes through the hologram, the interference pattern blocks some of it. What comes through is exactly like the beam that made the hologram. This transmitted light has all the properties of the light reflected from the object.

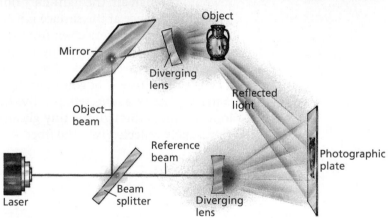

Figure 12–12. A hologram is made by splitting a laser beam into two beams and then reuniting the beams on the film. The interference pattern formed when the two beams meet produces the holographic image.

ASK YOURSELF

How is a hologram made?

Fiber Optics

There is another futuristic invention Grandma Marsha never even dreamed of—glass fibers that carry telephone conversations. You can find out more about these glass fibers by doing the next activity.

ONGOING ASSESSMENT
ASK YOURSELF

A hologram made by a laser beam is split into two parts. One beam goes directly to the film. The other goes to the object being photographed and is reflected onto the film. The interference pattern that results produces a hologram.

NATURE OF SCIENCE

In 1947 Dr. Dennis Gabor, a Hungarian scientist, was trying to improve the electron microscope by producing three-dimensional images. His idea was to divide the light from a single source into two beams. Dr. Gabor bounced one beam off the object and onto a photographic plate. The second beam was aimed at a mirror, reflecting the beam onto the same photographic plate. The two waves bumped and created an interference pattern on the photographic plate. Dr. Gabor called this new approach holography, from the Greek words *holos* (whole) and *graphos* (picture). However, Dr. Gabor could not actually make a hologram because he had no source of coherent light. After the invention of the laser scientists had a way to create holograms.

LASER DISC
44449–44965

Hologram

● **Process Skill:** *Observing*

Display a light pipe, and encourage the students to make observations about its weight, length, and function. Direct the students to see the light pipe's ability to curve, thus bringing light to otherwise unreachable places. Then ask the students to speculate if all the light inside a light pipe undergoes total internal reflection. (No; those light rays that strike the surface of the light pipe at a steep angle are refracted and escape.)

● **Process Skill:** *Constructing/Interpreting Models*

Remind the students that although light travels in a straight line, it can be directed with the use of mirrors. Display a diagram of a periscope, and ask a volunteer to explain the path the light travels in the periscope. Be sure the volunteer points out the mirrors. Explain that light behaves the same way in a periscope as in a light pipe, zigzagging its way through the pipe.

DISCOVER BY *Doing*

Be sure to call the telephone service to get some samples of optical cables. The company might not give it to the students. The tiny fibers will be coated with a colored insulation, which can be gently scraped off. Have the students use a hand lens to observe the clear filaments. Put the final diagrams in the students' science portfolios.

LASER DISC
42272–42512

Fiber optics; telephone communication

THE NATURE OF SCIENCE

Light signals carrying a telephone conversation through an optical fiber must be converted into electrical signals before they can be amplified. Research is being conducted on systems called integrated optical circuits (IOCs), which can amplify light signals without first converting them to electrical signals.

MULTICULTURAL CONNECTION

Earl D. Shaw, an African-American scientist, and Kumar Patel, an Asian-American scientist, worked together to invent a laser that can tune in to different wavelengths. This type of laser is especially useful for analyzing air pollution.

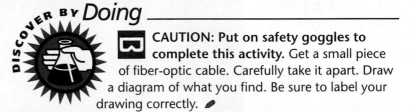

DISCOVER BY *Doing*

CAUTION: Put on safety goggles to complete this activity. Get a small piece of fiber-optic cable. Carefully take it apart. Draw a diagram of what you find. Be sure to label your drawing correctly. ✎

Light flows through these optical cables. Today, long distance lines transmit conversations in the form of tiny pulses of light. A new kind of laser makes this system work. When light travels through a medium in which it travels more slowly than it would through air, the light may not be able to get out. If the angle of incidence at the surface is too great, the light is reflected inward. This reflection from the inside surface is called **total internal reflection.** When total internal reflection occurs, the light waves act like a rock does when you skip it across a quiet pond. Instead of going through the surface and into the water, the stone is reflected into the air. The light waves travel long distances through the tiny glass fibers because they are continually reflected into the fiber.

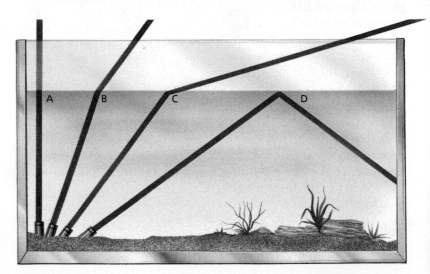

Figure 12–13. This illustration of lights in an aquarium shows how a beam of light reflects from the internal surface of water. When the angle of incidence is great enough there is total internal reflection (D).

Figure 12–14 shows a surgeon viewing the throat of a patient with a light pipe. A **light pipe** is a clear plastic rod that can be bent into a curved shape to carry light into otherwise unreachable places. Total internal reflection is used in the light pipe. As long as the curvature is not too great, the light strikes the surface at a high angle of incidence and is reflected inward.

● **Process Skill:** *Generating Ideas*

Before beginning the activity, have the students name characteristics of the conduit that make it appropriate for its purpose. (Characteristic properties include sturdiness, resistance to high temperatures, resistance to freezing temperatures, durability, and malleability.)

Cooperative Learning Prompt the students to recall times when they heard crosstalk during a telephone conversation. Have them work with a partner to develop a hypothesis that might explain this phenomenon. (Possible reasons are: dampness in the lines; damaged insulation allowing a connection between fibers through which conversations jump as a result of animals chewing on the cables or toxic wastes seeping into the ground and causing corrosion.)

How can you use an image conduit?

Process Skills:
Experimenting, Observing, Formulating Hypotheses

Grouping: Groups of 3 or 4

Hints

Remind the students that an image conduit consists of two types of fibers: flexible and nonflexible.

Procedure

4. the letter e

5. the same image

Application

1. The image comes through undistorted because the conduit accepts the image as separate bits of information, each of which is carried in a separate fiber to the other end of the conduit. The fibers maintain their relative positions through the change in shape of the conduit as a whole.

2. Image conduit has important medical uses. It enables doctors to see inside the stomach, trachea, lungs, or intestines. In general, it can give visual access to many places too difficult to examine.

Richard learned from his science teacher that a long, thin light pipe is called an **optical fiber.** Then he looked up a picture in his science book that shows how many optical fibers make up an image conduit. Each fiber, made of glass and encased in plastic, is considerably thinner than a human hair. Light travels faster in the plastic than it does in the glass, making total internal reflection possible. Each fiber picks up one small portion of the light from the object and carries it to the other end of the conduit. The conduit can be bent and twisted into all sorts of shapes without losing the image. You can see this for yourself in the next activity.

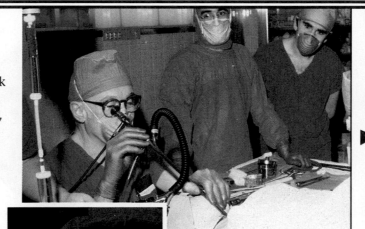

Figure 12–14. This physician is using a light pipe to check a patient's throat. The light pipe provides the view shown on the bottom.

ACTIVITY

How can you use an image conduit?

MATERIALS
safety goggles, laboratory aprons, Bunsen burner, image conduit, triangular file

PROCEDURE
1. CAUTION: Put on safety goggles and a laboratory apron, and leave them on throughout this activity. Turn on and adjust the Bunsen burner. Hold the image conduit by the ends, and heat the center of it in the burner flame.
2. When the image conduit softens, bend it, twist it, or stretch it to change its shape. Remove it from the flame, and let it cool.
3. With an edge of the file, cut a nick in the conduit at some point beyond where you have bent it, as shown. Then break the glass at that point.

4. Put one end of the conduit directly on top of the letter *e* somewhere on this page. Look into the other end of the conduit. What do you see?
5. Reverse the conduit so that the other end is on the letter *e*. Now what do you see?

APPLICATION
1. Why is the image of the letter *e* able to come through the conduit undistorted?
2. Suggest some uses for an image conduit.

⭐ PERFORMANCE ASSESSMENT

After the completion of the *Activity,* have individual students demonstrate the procedure. You can check the student's work by viewing the letter e through the image conduit.

ONGOING ASSESSMENT

▼ ASK YOURSELF

Total internal reflection is the reflection from the inside surface. When total internal reflection occurs, the light waves are reflected back into the fiber. As long as the light does not strike the surface at a high angle of incidence, the light is reflected inward.

SECTION 3 REVIEW AND APPLICATION

Reading Critically

1. Signals in optical fibers travel faster, do not have to be amplified as often, carry more information, and require smaller conduits than electric signals.

2. Total internal reflection occurs when all the light that enters an object stays inside. The telephone signal is digitized and sent as light waves through optical fibers.

Thinking Critically

3. A three-dimensional image could be formed without damaging the organism. This would allow scientists to study both the structure and behavior of the organism.

4. A bit is either an "on" or "off" signal, so it is not subject to noise distortion.

▶ **324** CHAPTER 12

Figure 12–15. The laser used to send light in an optical fiber is very small. One such laser is shown here—lying on George Washington's eye on this quarter.

Figure 12–16. Each of the optical cables (right) consists of 144 individual fibers shown on the left.

Telephone cables made of optical fibers also consist of fine glass fibers encased in plastic. When you talk on the telephone, your voice is converted into an electric signal. At a local telephone station, this signal is changed into a digital code, like the code used on a CD. The electric bits of the code are used to trigger a tiny laser, no bigger than a grain of salt. The laser codes your voice as a series of flashes of infrared light, which travels through the glass fiber.

When you speak, you leave spaces between words and syllables. In the fiber, these intervals are filled in with someone else's conversation! A cable of 12 fibers can carry 50 000 conversations at once.

Grandma Marsha would marvel at the new technologies available to earthlings in the twentieth century. Richard is looking forward to updating her wonderful story.

▼ ASK YOURSELF
What is total internal reflection?

SECTION 3 REVIEW AND APPLICATION

Reading Critically
1. What are the advantages of optical fibers for transmitting telephone conversations?
2. How is total internal reflection used to transmit telephone conversations?

Thinking Critically
3. Biologists have developed a technique for making holographic pictures through a microscope. How would such pictures be useful in the study of microscopic organisms?
4. Why do CDs, optical fibers, and digital audio tapes all represent sound in the form of bits rather than waves?

EXTENSION

Have the students use library materials to investigate careers that use laser technology. They should include in their reports information regarding the amount of training or education required, average salaries, and location of job opportunities. The book, *Laser Careers,* by Dennis Eskow is a good reference.

CLOSURE

Cooperative Learning Have the students work in pairs to write a summary of this section. Then ask them to share their summaries with classmates and revise their summaries based on classmates' feedback.

SKILL

Measuring Angles

Process Skills: Measuring, Analyzing, Observing

Grouping: Groups of 3 or 4

Objectives
● **Construct** angles of various degrees.
● **Demonstrate** the use of a protractor and ruler.

► **Application**

1. The vertex point is the point of origin for the angle.

2. The result should be an angle of 40°.

3. 78°

✳ **Using What You Have Learned**

By knowing how to measure the angles in the science lab, you can more effectively use all of the space that is available.

★ **PERFORMANCE ASSESSMENT**

After the completion of this *Skill* activity, have individual students demonstrate the use of a protractor on other angles. Check each student's technique for accuracy.

PORTFOLIO ASSESSMENT Have the students put the completed *Skill* in their science portfolios.

SKILL *Measuring Angles*

▶ **MATERIALS**
● protractor ● paper ● ruler

▼ **PROCEDURE**

1. Look at your protractor carefully. Note that it is shaped like a half circle, with a scale along its rounded edge. The scale indicates that the number of degrees in a half circle is 180. The center of the flat side of your protractor is marked with a short line. (In some protractors there is a hole at the center point.)

2. To measure an angle, place the protractor flat on the paper, with its center at the vertex, or point, of the angle. Move the protractor until the 0° line is on one ray of the angle. Be sure to keep the protractor center on the vertex. The measure of the angle is the point where the other ray intersects the protractor's scale. The number of degrees in the angle can be read directly from the scale.

3. To construct an angle of a specific measurement, use a ruler to draw a line on your paper. On the line, mark the point where you want the vertex of the angle to be. Place the protractor on the paper with its center point at the vertex of

the angle. Adjust the protractor until the 0° line is on the line you have drawn. Then mark a dot on the paper at the desired degree position on the scale. Remove the protractor, and draw a straight line through the vertex and the mark you just made.

4. Use a ruler to draw an angle on a sheet of paper. Label the vertex and rays. Measure the angle.

5. Construct five other angles, and measure them. Mark each of your angles with its correct measure in degrees.

6. Construct angles of 25°, 30°, and 60°.

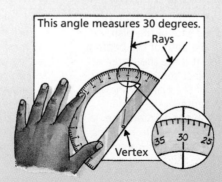

This angle measures 30 degrees.
Rays
Vertex

35 30 25

▶ **APPLICATION**

1. Why is it important to locate the vertex of the angle?

2. What is the result if you add an angle of 25° to an angle of 15°? Use a protractor to see whether your answer is correct.

3. How many degrees are there in the angle between the vertical and a road that slopes at 12° from the horizontal? Diagram the road.

✳ ***Using What You Have Learned***

You have decided to enter a contest to design a new science lab. How can knowing how to measure angles help you with your design? Design a new lab to test your ideas.

Determining Total Internal Reflection

Process Skills: Observing, Interpreting a Graph, Measuring

Grouping: Groups of 3 or 4

Objectives
● **Observe** total internal reflection.
● **Determine** the critical angle of water.

Pre-Lab
Review with the students what they know about reflection, refraction, and total internal reflection.

Hint
Emphasize to the students that they should focus the impact of the *Investigation* around the angle of incidence. The students should realize that if the angle of incidence is larger than the critical angle, the light will be reflected back into the water.

Analyses and Conclusions

1. As the beam leaves the water, the angle of refraction is larger than the angle of incidence.

2. The ray bends away from the normal because it speeds up as it enters the air.

3. When the angle of incidence is larger than the

critical angle, the beam is reflected back into the water.

4. The critical angle of incidence of water is 48°.

Application
The transmission of telephone conversations through optical fibers is also dependent on total internal reflection. In order for no distortion to occur, the signals must not hit the sides at an angle greater than the critical angle of incidence for optical fibers.

✳ Discover More
When the angles are larger than the critical angle of incidence, the beams are reflected back into the water. When total internal reflection occurs, the angle of incidence is always greater than the critical angle of incidence. Then all reflections occur within the medium.

Post-Lab
Have the students discuss how the results of this *Investigation* might differ if a liquid other than water were used. (Results would vary depending upon the viscosity of the liquid.)

After the completion of this *Investigation,* have individual students demonstrate the procedure. You can check the students' work by reviewing their data on angle of incidence and angle of refraction.

INVESTIGATION

Determining Total Internal Reflection

▶ MATERIALS
● metric ruler ● white paper ● hemicylindrical plastic dish ● flashlight
● tape ● protractor

▼ PROCEDURE

1. Using a ruler, draw a line down the center of a sheet of white paper. Place the plastic dish on the paper in such a way that the dish is centered on the line. The line on the paper is now the normal to the flat side of the dish.
2. Add water to the dish until it is about three-fourths full.
3. With your ray box, send a beam of light through the curved side of the dish. Adjust the path of the beam to make it leave the water at the exact center of the flat side of the dish, as shown in the figure.
4. On the paper, mark the path of the beam entering and leaving the water. Then remove the dish, and con-

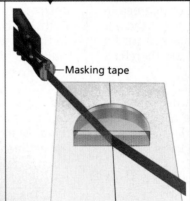

—Masking tape

nect the lines. Use the protractor to measure the angles of incidence and refraction at the flat side of the dish. Remember that the angles are measured from the normal.
5. Prepare a chart like the one

shown. Enter the size of the angle of incidence and the size of the angle of refraction in the chart.

TABLE 1: TOTAL INTERNAL REFLECTION	
Angle of Incidence	Angle of Refraction

6. Increase the angle of incidence, and measure the two angles again. Continue taking readings until you find the largest angle of incidence that allows the beam to come out of the water. Record the angle at which total internal reflection begins to occur.

▶ ANALYSES AND CONCLUSIONS
1. How does the angle of refraction compare with the angle of incidence?
2. Why does the beam bend away from the normal as it leaves the water?
3. The *critical angle of incidence* is the largest angle of incidence that will allow the beam to leave the water. What happens to the beam when the angle of incidence is larger than this?
4. From your data, what is the critical angle of incidence of water?

▶ APPLICATION
Relate what you learned in this investigation to the transmission of telephone conversations through optical fibers.

✳ *Discover More*
Continue the procedure with angles larger than the critical angle of incidence. What are your results? How do the angles of incidence and reflection when total internal reflection occurs compare to the angles of incidence and reflection when total internal reflection does not occur? Why is this so?

The Big Idea—Technology

Lead the students to an understanding of the production of a laser beam, focusing on the interactions that must occur for the beam to be made. Point out that a laser beam depends upon stimulated emisson to produce coherent light.

Remind the students that holograms are also a product of numerous systems and their interactions. Finally, be sure the students understand that many technological advances are the result of the interaction of complex systems—each harnessing a different energy source.

CHAPTER

12 HIGHLIGHTS

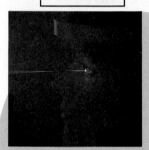

The Big Idea

Light energy has many uses, some of which are made possible by modern technology. The development of new technologies has produced the laser and the optical fiber. Both of these technologies are based on the behavior of light under certain circumstances: coherent light must be produced to create a laser beam; total internal reflection must be achieved to allow optical fibers to carry light. These systems and their interactions have made possible technological advances in medicine, manufacturing, telecommunications, the arts, and many other fields.

For Your Journal

Lasers probably have more effect on your everyday life than you thought. What have you learned about lasers in this chapter? Look back at the ideas you wrote in your journal at the beginning of the chapter. How have your ideas changed? Revise your journal entries to show your new ideas.

Connecting Ideas

The following situations are made possible by either lasers or fiber optics. Decide which. Then write a description in your journal that tells how lasers or fiber optics made the situation possible.

For Your Journal

The students' ideas should reflect an understanding of how lasers affect their everyday life and that lasers are not simply futuristic dreams but are present reality. You might encourage the students to share their new ideas with classmates.

CONNECTING IDEAS

1. Talking on a telephone is made possible through fiber optics.

2. A laser made the music possible by reading a compact disc.

3. Using a credit card with a hologram is made possible by lasers, which read the code on the card. The hologram makes the card more difficult to forge.

4. Laser surgery is an innovation made possible by the combined research of many scientists.

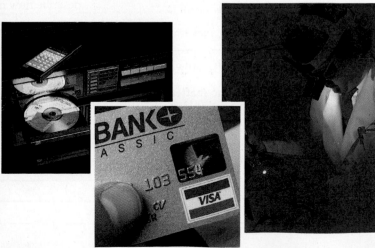

Understanding Vocabulary

1. (a)Coherent light is light in which all the waves in a beam are in step—that is, crests travel with crests, and troughs travel with troughs. An optical fiber is a very thin fiber made of glass and encased in plastic. It is through optical fibers that digital information in the form of flashes of infrared coherent light travels.

(b)Pumping is the process of raising atoms to an excited state. The excited state is a state in which an atom contains stored energy.

(c)The ground state is the state to which atoms return after they give off the energy gained in an excited state. Stimulated emission is the process of forcing identical photons to leave an atom together in the same direction and in step.

(d)A laser produces light in a thin beam that can be focused on a very tiny spot. A hologram is an interference pattern created by a laser on a piece of film.

(e)Total internal reflection is reflection from the inside surface of a fiber. Light waves are continually being reflected back into the fiber. The light pipe is a clear plastic rod that can be bent into a curved shape to carry light into otherwise unreachable places.

Understanding Concepts

Multiple Choice

2. a
3. b
4. c
5. a
6. b

Short Answer

7. Answers may include the following examples: in medicine, in construction for measuring and aligning, in industry for cutting and drilling, in weapons manufacturing, and in communications.

8. A hologram is an interference pattern that begins with a single beam of coherent light. The beam must be split into an object beam and a reference beam. These two beams then will interfere with each other, causing a holographic image.

9. A neon sign consists of long glass tubes formed into letters or other shapes. Each tube is filled with an unreactive gas. (Neon is not the only gas used.) An electric current passes through the gas when the sign is turned on. The electric current provides the energy necessary to raise the atoms to the excited state. The atoms are then considered pumped. They then begin to release energy in the form of light and photons, return to their ground state, and the process repeats itself. In a neon sign, the photons are being released in all directions, which causes the sign to glow or light up.

Interpreting Graphics

10. A

11. 1, 4

CHAPTER

12

REVIEW

▶ Understanding Vocabulary

1. For each set of terms, tell how they are related.
   a) coherent light (313), optical fiber (323)
   b) pumping (314), excited state (314)
   c) ground state (314), stimulated emission (315)
   d) laser (315), hologram (321)
   e) total internal reflection (322), light pipe (322)

▶ Understanding Concepts

MULTIPLE CHOICE

2. Light does not escape from an optical cable because
   a) it travels more slowly in that medium than it does through air.
   b) it travels faster in that medium than it does through air.
   c) the cable is curved.
   d) the cable is too thin.

3. Coherent light can be produced only by
   a) electric lights.
   b) lasers.
   c) neon tubes.
   d) stars.

4. One characteristic of a laser beam is that
   a) it spreads out the further it travels.
   b) it has more wavelengths than an ordinary beam of light.
   c) it maintains its width no matter how far it travels.
   d) its waves are continuously overlapping.

5. Which of the following is not a type of laser?
   a) quartz crystal
   b) YAG
   c) helium-neon
   d) argon gas

6. Constructive interference and destructive interference is used to a unique advantage in
   a) the taking of X-rays.
   b) the making of a hologram.
   c) laser surgery.
   d) the making of a compact disc (CD).

SHORT ANSWER

7. List and briefly describe as many uses as you can think of for lasers.

8. Explain why the beam of a laser must be split in order to make a hologram.

9. Using what you know about atoms, photons, ground state, and excited state, describe how neon signs work.

Interpreting Graphics

10. Which of these two figures represents coherent light?

11. How many frequencies are represented in Figure A? Figure B?

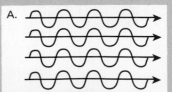

Reviewing Themes

12. *Energy*
Light is a form of energy. Discuss some aspects of light that demonstrate its ability to do work.

13. *Technology*
Most people use a telephone on a daily basis. Without realizing it, we rely on immense communications systems to provide telephone service. Describe the ways in which this system has changed our lives and the ways the telephone system can be used.

Thinking Critically

14. Explain why many credit card companies have included holograms on their credit cards.

15. In addition to the example given in this chapter, what other surgical techniques could be aided by the use of the light pipe? Describe at least three examples.

16. The Northern Lights are rainbow-colored lights appearing at night. They are caused by huge explosions on the sun. These explosions emit millions of particles which travel through space and reach Earth's atmosphere, where they collide with other particles and produce light. Are the Northern Lights an example of coherent light or ordinary light? Explain your answer.

17. Shown here are two examples of laser-cut greeting cards. Many new types of cards and stationery are made using industrial lasers. Using your knowledge of lasers, explain why this type of cut-out card can be made using lasers. How else could cards of this type be made? What is the advantage of a laser over the other methods you can think of?

Discovery Through Reading

Pennisi, E. "Etching Technique Lights Up Porous Silicon Three Ways." *Science News* 141 (June 27, 1992): 423. Read how new techniques are being developed to improve holographic data storage.

Discussion

● **Process Skills:** *Generating Ideas, Evaluating*

This *Science Applications* is about energy from the sun. Many students might have experienced warming themselves with sunlight. Ask the students to describe instances when they have used the heat from infrared rays of the sun to warm themselves. (Answers might include standing in a sunny area instead of a shady area.)

● **Process Skills:** *Formulating Hypotheses, Expressing Ideas Effectively*

Most students are also aware that exposure to sunlight can cause sunburn. Have the students discuss what they can do to reduce their chance of

---

Science PARADE

Science PARADE

SCIENCE APPLICATIONS

ENERGY FROM THE SUN

Waves of energy flow from the sun. All life on Earth depends on the energy of these waves—they allow plants to make their own food through the process of photosynthesis. Without plants, animals would have no food and would die.

Visible and Invisible

The largest part of the energy from the sun is in the form of visible light. However, the energy comes in other forms also. Waves of invisible infrared rays and ultraviolet rays bombard you every minute the sun is shining. While your eyes cannot detect these waves, they affect your body in many ways.

Have you ever seen a cat seek out a sunny spot for sleeping? The warmth the cat feels when it lies in the sunshine comes from infrared

Some animals, such as this cat, like to lie in the sunshine and absorb infrared waves.

waves. These waves are also known as radiant heat. When was the last time you felt the warmth from infrared waves?

Ouch! The ultraviolet waves coming from the sun can cause sunburn. On the other hand, ultraviolet waves do have an important beneficial effect. They act on a chemical in the skin to produce vitamin D. Every child needs vitamin D to form strong bones and teeth. Adults need it to help keep their bones healthy. About 15 minutes a day of bright summer sunshine provides the human body with all the vitamin D it needs.

getting a sunburn. (Use sun-screen and reapply it as necessary; stay out of the sun during midday, when the sun's rays are most intense; wear a hat and protective cloth-ing; use common sense.)

Journal Activity
After completing the Unit and reading and discussing the *Science Applications*, ask the students to consult the list of questions they recorded in their journals from the discussion at the beginning of the Unit. (You might ask the students to turn back to the Unit Opener on pages 212–213.) Lead a class discus-sion during which the students can verbalize what they have learned in order to answer their original questions. You might have the students record their answers and new ideas in their journals.

Good (and Bad) Things Come in Small Packages

All the sun's wave energy comes in tiny packages called *photons*. Ultraviolet rays come in two forms: ultraviolet A (UVA) and ultraviolet B (UVB). In the short wavelength ultraviolet, UVB, each photon has a lot of energy. In the longer wavelength ultraviolet, UVA, the energy is divided into many small-er packets. UVB can be much more damaging than UVA. UVB is like a shower of bricks; UVA is more like a shower of sand.

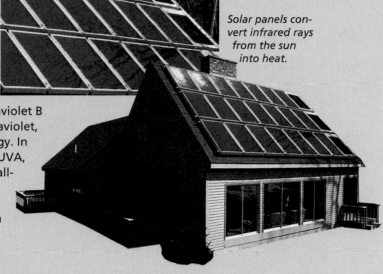

Solar panels con-vert infrared rays from the sun into heat.

To UVB or Not to UVB

Sunburn is caused by UVB waves. These waves are most intense in the middle of the day. As you probably know, people differ greatly in their susceptibility to sunburn. The dark pig-ment in the outer layer of the skin offers some protection against sunburn. Fair-skinned, freckled redheads have practically no dark pigment; therefore, any exposure to UVB can be dangerous to them. On the other hand, very dark-skinned people are much less likely to get sunburned. Between these two skin variations are peo-ple with some pigment in their skin. Whether the skin tans or burns depends on the amount of pigment in the skin and on the length of exposure to the sun. Exposure to the sun in the morning or late afternoon, when there are few UVB waves

in the light, produces a UVA tan. Sunscreens and tanning lotions promote a UVA tan. Most of these lotions contain a chemical that blocks the UVB and lets the UVA get through to the lower living layers of the skin. Since UVA waves are less damaging than UVB waves, some people believe that a UVA tan is healthy.

Is a "healthy tan" really healthy? NO! Even mild UVA waves can damage the connective tissue fibers in the lower layers of the skin. UVB waves can produce even worse damage: wrinkling, coarse-ness, mottling, and roughness. Every suntan you see is a step to-ward skin growing old too soon. Worse yet, repeated exposure to sunshine may lead to cancer of the skin.

The next time you see a sunburned per-son, what will you think about? Think about using common sense and sunscreen. Both of these things will help you enjoy the sunshine without endanger-ing your life. ◆

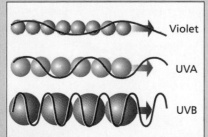

Violet

UVA

UVB

UVA consists of more photons than UVB, but each photon carries less energy.

Background Information

Many medications that are used to treat pets are actually drugs for humans. Veterinary equivalents of these drugs have not been developed for several reasons, primarily because drug manufacturers have not felt it is economically feasible to do so. Some medications are labeled "For use in dogs only" because manufacturers are not financially able to test all medications on all species of animals. Even though these medications are not labeled for use in specific species, when properly used by a veterinarian, they are very effective, and sometimes essential, for pets. These medications include insulin, antibiotics, chemotherapy medications, and hormones.

▼ READ ABOUT IT!

CAT Scans

High-Tech Medicine Can Save Animals, Too

by Lisa Feder-Feitel

from *3-2-1 CONTACT*

Marjorie Shaw was upset. Her normally active, bright-eyed 13-year-old was wandering around the house, bumping into furniture and walls. When it was clear that her teenager couldn't see, Marjorie rushed her to the hospital.

There, a team of medical experts used a CAT scan—a machine that makes computerized images of the inside of the body—to check out the patient's brain. As the teen was wheeled into the CAT scan room, a doctor placed a hand on her forehead and comforted her. Within moments—and without pain—the scan was complete. It showed doctors that the teenager had been suddenly blinded by a brain tumor.

Several times over the next few weeks, the patient lay on a table as invisible cobalt rays destroyed her tumor. By the end of the treatments, the patient had gotten her sight back.

This kind of modern medical "miracle" is not so unusual. What is unusual is where it took place: The patient's sight was saved at the Animal Medical Center in New York City. Marjorie Shaw's 13-year-old was her dog, Echo.

For many people—young and old alike—a pet is a member of the family. Because of this, many owners spend a great deal of money making sure that their pets stay healthy.

First-Class Care for Fido

Echo's story is not unusual. The same treatments, high-tech machines and medicines that were developed to help save people's lives are now being used to help save the lives of animals. So, animals, as well as humans, are living longer, healthier lives.

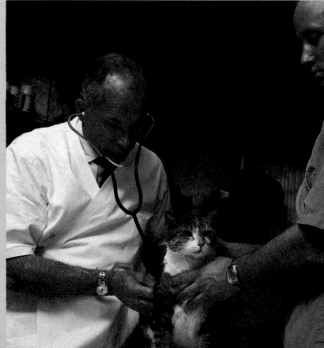

Discussion

● **Process Skill:** *Expressing Ideas Effectively*

Ask the students to share experiences they have had with high-tech machines and medicines used in treating their own pets. (Responses will vary depending upon the animal and its ailment.)

● **Process Skills:** *Applying, Synthesizing*

Remind the students that in Chapter 11 they learned about the nonvisible spectrum. Then ask the students what parts of the spectrum can be used by veterinarians to treat animals. (X-rays can be used to check for broken bones or dislocated joints, to locate tumors, or to examine the digestive tract for foreign objects.)

"Veterinary medicine can do anything on cats and dogs that we can do on human patients. I like to say we work out the 'bugs' on humans. Then when it's safe enough, we find ways to use the same procedures on animals," jokes Dr. Michael Garvey. He is a veterinarian at the Animal Medical Center in New York City.

Many of the latest medical techniques used on pets go beyond the emergency care that Echo received. Your dog may have her cataracts (a cloudiness that fills the lens of the eye, causing blindness) removed with laser beams, just as humans do. Or your cat (like people) may have an artificial hip placed in his body if arthritis (a painful disease of the joints) sets in.

For pups whose hearts don't beat to a regular rhythm, there are human pacemakers. Pacemakers are tiny, battery-operated machines placed in a person's body to keep the heart beating in a strong and regular rhythm. The exact same machine—and procedure—is now used to keep a dog's heart in tip-top shape. In St. Petersburg, FL, the Pinellas Foundation gives donated human pacemakers to vets, who use them on needy dogs.

Not all pets are physically ill, however. Some have emotional problems or fears. If a pet's problem is psychological, there are therapists to handle that, too. Does your dog hide and shake under the bed during thunderstorms? Suzanne Johnson of Beaverdam, VA, might be able to calm the pooch down.

Ms. Johnson is one of about 40 animal psychologists in the U.S. who treat fearful, aggressive or unhappy pets. To treat a fear of thunder, for instance, she may put the dog in a room and play a tape recording of a storm. She'd start at low volume, and increase it until the dog hardly notices the booming noise. It may frighten the owner, though!

Good Health at Great Cost

Now that veterinarians are using high-tech equipment, advanced techniques and new medicines, pets are living longer and healthier lives. That's great news for animals and their owners. But there's also a down side to this story. A pet's good health can carry a very high price tag.

Marjorie Shaw's bill for Echo came to $3,000. Cataract surgery can cost between $600 and $1,200 an eye. A pacemaker can be fitted for $1,500. Americans spend nearly $6 billion a year on veterinary services.

Laser surgery on cats (and dogs) with eye problems is common. An infrared light is used as part of the operation to remove cataracts.

Troubling Questions

These higher costs and new treatments may also force veterinarians and pet owners to make difficult choices: When should care for the pet continue, and when is it more kind to the pet to stop it?

Dr. Garvey of New York City's Animal Medical Center uses the example of a 15-year-old cat whose kidneys stopped working. "The animal may respond to treatment and do very well for months to years," he says. "On the other hand, the cat may live only for a few days or weeks. It will cost the owner several hundred dollars to find out. What should be done?"

Dr. Mike Shires of the University of Tennessee at Knoxville has dealt with another troubling problem: an owner's inability to see what is best for his or her pet.

"Many people seek all sorts of extreme medical treatments for their pets that weren't available a few years ago," he says. In cases where continued treatment is cruel, he recommends that the pet be put to sleep. If an owner still insists, he says, "we tell them to take their pet some place else."

Some doctors believe in treating pets and owners with equal care. Dr. Jane Mason, from Chantilly, VA, says, "I don't push the medical advances on owners with older pets or ones whose chances of a pain-free life are not good."

Deciding how far a person should go in saving his or her pet is becoming a common question. Humans who are seriously ill can usually say how much they want done to help them. But animals can't talk. The choice is ultimately up to the animal's owner.

A veterinarian checks instruments as she prepares her animal patient for an operation.

An operating room in an animal hospital looks very much like one in a hospital for humans. High-tech medical equipment that was developed for people is used to treat animals.

Health Food Heals Pets

Kiss the canned dog food goodbye. Instead, treat your pup to a dish of yogurt or steamed broccoli. Or if your pet has fleas, try soaking its collar in oil made from eucalyptus (*say: you-kah-LIP-tus*) leaves or sprinkling garlic on its supper. Yes, garlic!

Healthful foods make healthy pets, says Dr. Monique Maniet, a veterinarian in Takoma Park, MD. She practices a kind of care for pets that some doctors use on humans. It's about as far from high technology as Maryland is from Madagascar. Called holistic medicine, it encourages the use of natural foods and soothing herbs.

Vaccines and antibiotics do have a place in Dr. Maniet's medicine cabinet, however. But she prefers to use more natural cures. If your pet suffers from swollen joints, for example, she might inject it with some honeybee venom. You might say that the poor pet gets stung twice, but Dr. Maniet says it helps a lot more than it hurts!

Should people use holistic or high-tech treatments with their sick pets? Always discuss it with your veterinarian. And never try any treatments, or give your pet any medicines, yourself. Always put your pet under a doctor's care. ◆

THEN AND NOW

Thomas Young (1773-1829)

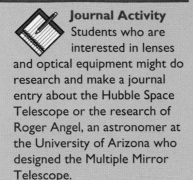

Today's scientist is expected to be a specialist, but scientists in the 1800s were mostly amateurs. That means that they did science for fun! Thomas Young was one example—he was a practicing physician, but he was interested in many other academic fields.

Young's skill as a linguist appeared early. By the age of 14 he was fluent in Latin, Greek, French, Italian, Hebrew, Persian, and Arabic. His language skills were later put to use when the famous Rosetta stone was found in Egypt. This stone, which contained an inscription in Egyptian hieroglyphics and a Greek translation, was the key to deciphering the Egyptian written language. Young was one of the first to make progress understanding it.

Young is best known, however, for reviving the wave theory of light. He conducted an experiment in which he showed that separate bands of light appeared where there should have been nothing but shadow. These bands of light diffracted around corners and could not be explained by particle theory. This experiment allowed Young to calculate the wavelength of visible light and determine what kinds of waves they were.

Thomas Young was also interested in other forms of energy. In fact, he initiated the use of the word *energy* in the modern sense we use today. During his lifetime, Young established his reputation as a distinguished linguist, physician, physicist, and writer. ◆

Jean M. Bennett (1930-)

Jean M. Bennett is a member of the Research Department of the Naval Air Warfare Center at China Lake, California. In 1979 she was elected Woman Scientist of the Year at the Center. She was chosen to be a Senior Fellow for her scientific contributions to the Center in 1989.

Bennett was the first woman to receive a doctorate in physics from Pennsylvania State University in 1955, and has done important research in several fields, including the physics of the atmosphere and infrared spectroscopy. She has written two books about her current specialty, studies of the surface roughness and scattering of optical light.

Her main work in recent years has been in the field of optical surfaces and coatings. Most lenses in optical equipment are polished to a very smooth finish and are coated with thin films that prevent ghost reflections from interfering with the sharpness of the image. The effectiveness of the coatings is limited because they are not perfectly smooth. Irregularities in the surfaces tend to scatter some of the light.

Bennett has made many contributions to the understanding of light scattering and has worked with manufacturers to make smoother surfaces, and coatings that have lower scatter levels. ◆

Discussion

● **Process Skills:** *Formulating Hypotheses, Generating Ideas*

After the students have read the feature on Shrinivas Kulkarni, ask them why Kulkarni compares pulsars to lighthouses.

(A lighthouse sends out a beam of light as its lamp rotates. You see the light each time the lamp points in your direction. Pulsars send out radio waves as they turn. Each time they rotate a radio telescope can pick up the radio waves.)

● **Process Skills:** *Synthesizing, Applying*

Many students may have heard of *serendipity*—a fortunate discovery made by accident. Ask the students where in the feature there is an example of serendipity. (When Kulkarni forgot to slow down the data, he found the pulsar he was looking for.)

SCIENCE AT WORK

SHRINIVAS KULKARNI
Astronomer

"I don't remember even once looking at the sky and saying, 'Wow, look at the stars...,'" Shrinivas Kulkarni says. While growing up in southern India, Kulkarni never thought that one day he would be an astronomer. In college he decided to study physics and engineering.

Kulkarni discovered astronomy by accident. He had to choose a place for summer study. India can be very hot in the summer, so he chose a school in a cool place. Astronomy was taught at that school. "I went there, and I instantly liked it," explains Kulkarni. After college in India, he came to the United States to study astronomy and has lived here ever since.

Kulkarni says that astronomy may have begun with people looking at the sky. Perhaps you think that this is all modern-day astronomers do, but it isn't. "People think that I go out on a cold night and look through a telescope," Kulkarni says. In fact, astronomers do not always look through the large telescopes that they use. They can program a computer to tell the telescope what to look for. Then astronomers look at the pictures and information that the telescope collected.

Shrinivas Kulkarni's special interest is how stars die. A star dies with a big explosion and becomes a neutron star. Neutron stars give off powerful radio waves as they rotate. These stars are called *pulsars* because the telescopes on Earth record something like pulses of light as the neutron stars rotate.

"A pulsar is like a lighthouse," Kulkarni explains. A lighthouse sends out a beam of light as the lamp rotates. You see the light each time the lamp turns and points toward you. Each time a pulsar turns, telescopes can record the radio waves from the pulsar. The study of pulsars is called *radio astronomy*.

Most known pulsars rotate 30 times a second or less. At least, that was the speed of those that had been discovered. Kulkarni decided to look for a pulsar that rotated 1000 times a second. He thought that he could find more slowly rotating pulsars by slowing down his computer data. Kulkarni meant to slow down the data when he looked at it the next day—but he forgot. Because he forgot, Kulkarni found the pulsar he was looking for; it was rotating at a rate of 642 times a second!

Today, Kulkarni is a professor of astronomy at the California Institute of Technology in Pasadena, California, where he continues his research on pulsars. ◆

Discover More

For more information about careers in astronomy, write to the

Astronomical Society of the Pacific
390 Ashton Avenue
San Francisco, CA 94112

Background Information

Have the students read the article on laser fusion and speculate how the development of fusion energy would affect human life on Earth. Then point out to the students that one of the problems with fusion on Earth is that, thus far, no container can hold matter at the required temperatures for very long. Some fusion projects use a doughnut-shaped reactor that creates a circular magnetic field to confine the superheated deuterium and tritium.

Journal Activity
Encourage interested students to read the following magazine articles and to make a journal entry about the information.

Cook, W. J. "Harnessing the Physics of the Sun." *U.S. News and World Report* 111 (November 25, 1991): 62. This article describes the making of the largest amount of fusion ever by the Joint European Torus.

Stover, Dawn, ed. "A First for Hot Fusion." *Popular Science* 63 (March 1992): 22. This is a brief, easy-to-read description of the Joint European Torus fusion reactor.

SCIENCE/TECHNOLOGY/SOCIETY

Laser Fusion

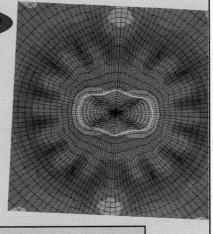

This computer simulation shows the core of a fusion fuel pellet just before fusion. Fusion occurs at the core, where the fuel is hottest and most dense.

Someday, scientists hypothesize, we will have a power source as powerful as the sun and as inexhaustible as the sea. In the next century, scientists expect that nuclear fusion will produce a large part of the world's electric energy. Today's most promising experiments involve the fusion of molecules by laser beams.

Explosive Energy!

Nuclear fusion happens when small atoms, such as hydrogen, suddenly combine to form larger atoms, such as helium. The fusion releases enormous amounts of energy, as we see in the explosion of a hydrogen bomb. Of course, that kind of explosive energy is not useful for generating electricity!

The major problem with fusion energy is the very high temperatures required to fuse atoms. Fusion of certain small nuclei takes place at almost 100 million degrees Celsius. In a hydrogen bomb, this temperature is reached by exploding a fission bomb.

The Promise of Fusion

The first usable fusion energy will probably come from a combination of laser technology and fusion science. In laser fusion, scientists begin with small pellets of deuterium and tritium. Deuterium and tritium are "heavy" hydrogen atoms found in ordinary sea water. To heat the deuterium fuel, scientists use a powerful laser beam. The laser beam is split into several separate beams, and each beam is amplified. When the amplified beams converge on the fusion pellet, they heat its nuclei enough to cause fusion and release heat energy.

Today's laser fusion devices all consume more energy than they create. However, using computer simulations, scientists hypothesize that a laser fusion reactor could produce 100 times more energy than it uses. A reactor would ignite ten deuterium fuel pellets every second and generate a billion watts of electricity. Rocket scientists also hope that brief bursts of fusion energy can be harnessed to power rockets. If the promise of fusion energy is someday realized, the world may never again suffer an energy shortage. ◆

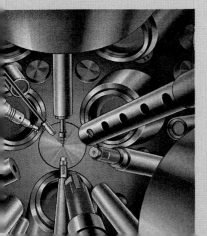

In this painting you can see the laser beams being focused on a fuel pellet. The fuel pellet is the small, bright sphere at the center of the painting.

UNIT 4

FIELDS—STRUCTURED ENERGY

UNIT OVERVIEW

Unit 4 presents information on electricity and magnetism, including static electricity and electric current. The relationship between magnetism and electricity is investigated and how both are used in doing work is presented. The operation of motors, video and audio equipment, televisions, and telephones are also discussed.

Chapter 13: Electricity, page 340

This chapter presents the characteristics of static and current electricity. The methods used to create and control electric current are discussed, as are the units and methods used to measure electricity.

Chapter 14: Magnetism, page 366

This chapter presents the characteristics of the different kinds of magnets. The magnetic fields of magnets are described according to their arrangements and direction in which they move. The relationship between magnetism and electricity is also discussed.

Chapter 15: Electricity in Use, page 388

This chapter presents the many uses of electricity, including home and audio-visual devices and radio and television systems. Also discussed are the processes of converting electric energy to other forms of energy, as well as methods for converting mechanical energy into electrical energy.

Science Parade, pages 416-423

The articles in this unit's magazine provide an overview of the increasing role that electricity will play in the development of the automobile. The use of superconducting magnets is also discussed. Biographies of a physicist and a computer systems developer are presented. The training and skills needed by a meter technician are also presented.

UNIT RESOURCES

PRINT MEDIA FOR TEACHERS

Chapter 13

Johnson, J.T. "The Hot Path to Solar Electricity." *Popular Science* 236 (May 1990): 82-85. Solar power is used to generate electricity in the Mojave Desert. Clear diagrams and extensive discussion explain the process.

Chapter 14

Crabb, Charlene. "Magnetic Minds." *Discover* 14 (January 1992): 70. Researchers discover that brain tissue has magnetite crystals that may provide a link between electromagnetic fields and disease.

Chapter 15

Sussman, Vic. "Home Tech Fever." *U.S. News and World Report* 111 (November 25, 1991): 68–72. Innovations in electric high-tech gadgets are discussed.

PRINT MEDIA FOR STUDENTS

Chapter 13

Amdahl, Kenn. *There are No Electrons: Electronics for Earthlings.* Clearwater Publishers, 1991. Here is a book that explains electricity in terms simple enough for every level of learner.

Sugarman, Ellen. *Warning: The Electricity Around You May be Hazardous to Your Health.* Simon & Schuster, 1992. Electromagnetic fields (EMFs) are generated by all electrical devices, such as lights, appliances, powerlines, and office equipment. This book is an invaluable guide to the risks of EMFs and the steps you can take to protect yourself.

Watson, Bruce. "Let There Be Light." *Cobblestone* 11 (October 1990): 14–17. The birth of electric power stations is outlined from the Edison Illuminating Company to the building of dams for hydroelectric power.

Chapter 14

Fishman, Joshua. "Giant Moments." *Discover* 12 (June 1991): 26–27. New studies of the magnetism of iron are presented.

Uerschuur, Gerrit L. *Hidden Attraction: The Mystery and History of Magnetism.* Oxford University Press, 1993. Magnetism was once the subject of many superstitions. This book traces our fascination with magnetism from the discovery of magnets in Greece to the state-of-the-art theories that see magnetism as a basic force in nature.

Zubrowski, Bernie. *Blinkers and Buzzers: Building and Experimenting with Electricity and Magnetism.* William Morrow & Co., 1991. A wealth of experiments and projects in electricity and magnetism are carefully explained and illustrated. Most of the materials used are found around the home or can be obtained easily from hardware stores or electronic supply stores.

Chapter 15

Lampton, Christopher. *Telecommunications: From Telegraphs to Modems.* Watts, 1991. The history of telecommunications is presented in this illustrated book.

Mattis, Richard. "The War of the Currents." *Cobblestone* 11 (October 1990): 18–19. George Westinghouse joined the power struggle with Thomas Edison to provide electricity to homes and businesses. Westinghouse preferred AC over Edison's DC power lines. One of them won the current war.

Weiner, Leonard. "Going Wireless." *U.S. News and World Report* 111 (November 25, 1991): 72–73. Wireless innovations are highlighted. A discussion of headphones, speakers, and VCRs is included.

ELECTRONIC MEDIA

Chapter 13

Electronics: An Introduction. Film or videocassette. BFA Educational Media. 9 min. Electronics and electronic devices are presented.

Making Circuits. Computer software. Cambridge Development Laboratory, Inc. These computer simulations and diagrams explain the basic concepts in the operation of simple, series, and parallel electrical circuits.

Static and Current Electricity. Film or videocassette. Coronet Film and Video. 16 min. Live-action experiments and animation present the behavior of static and current electricity. Circuits also are discussed.

Chapter 14

Electricity and Magnetism. Film or videocassette. Coronet Film and Video. 17 min. The relationship among motion, electricity, and magnetism is illustrated through animation.

Magnetism and Fields of Force. Film or videocassette. Barr Films. 13 min. Substances are studied to find whether or not they can be made into magnets or have magnetic properties.

What Is Magnetism? Videocassette. 8 min. Encyclopaedia Britannica Educational Corporation. This video demonstrates magnetic attraction and repulsion and discusses the earth's magnetic field and poles.

Chapter 15

Electrical Circuits. Videocassette. 12 min. Britannica. Using examples from everyday life, this program explains electricity and circuits.

Hot Line: All About Electricity. Videocassette. Focus Media, Inc. 15 min. Simple experiments that can be done in the home are used to illustrate the properties of electricity.

Investigating Matter and Energy Series—MMV Electricity. Computer Software. Cambridge Development Laboratory, Inc. This software exposes students to the process of science and methods of inferring, while they learn science concepts about electricity.

Discussion Lightning experiments produce strong and visible electric fields. Ask the students to describe natural lightning that they have observed during thunderstorms. (They might mention bright flashes or streaks occurring between clouds or between a cloud and the ground.) Then ask the students whether they can explain how the spark that sometimes occurs when they touch a doorknob and lightning are alike. (The students might mention that both are electricity.)

UNIT 4

FIELDS — Structured Energy

CHAPTERS

13 Electricity 340
Today's artists are making artistic statements in neon light. High-voltage lighting, sometimes called "electrical advertising," decorates places throughout the world.

14 Magnetism 366
Science frequently helps in law enforcement, from computerized fingerprint banks to DNA profiles. But few of today's methods cause as much excitement as the time the science of magnetism helped catch a criminal!

15 Electricity in Use 388
What would you do if all the lights went out and stayed out for hours? The power-line grid that supplied the Northeast electric utility companies failed and caused the biggest power failure in history.

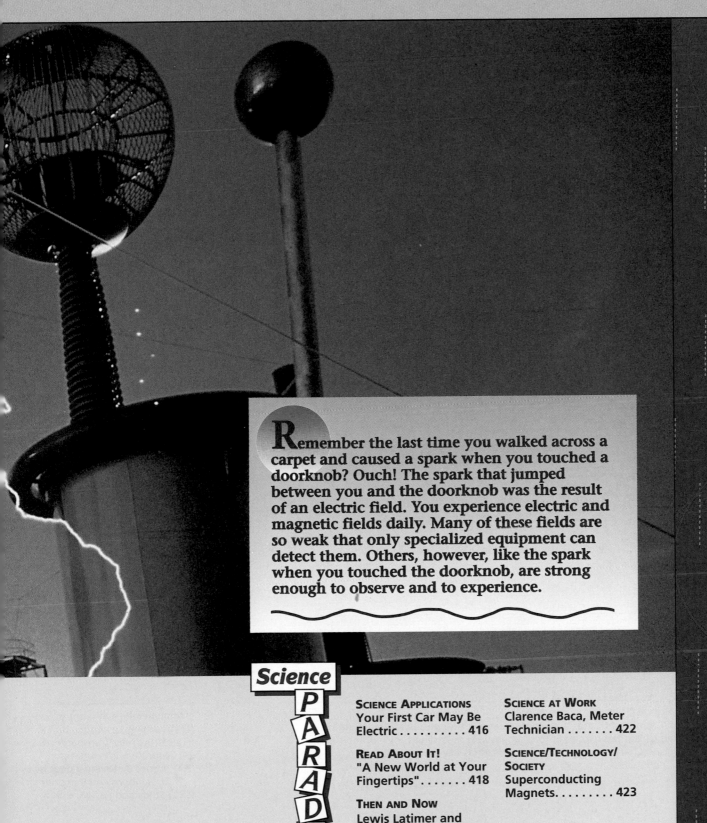

Journal Activity You can extend the discussion by asking the students to describe their experiences with magnets. Students might mention toys or games that use magnets or magnets that are used to attach notes or pictures to refrigerators or other metal objects. You might also bring in the following Unit topics: What is electricity? What is magnetism? How can electricity be used to create magnetism? How can magnetism be used to create electricity? How are electricity and magnetism used to do work? Ask the students to keep a record in their journals of their questions. After the class has completed the Unit, have the students look again at their questions. A repeat discussion might help the students realize how much they have learned.

Remember the last time you walked across a carpet and caused a spark when you touched a doorknob? Ouch! The spark that jumped between you and the doorknob was the result of an electric field. You experience electric and magnetic fields daily. Many of these fields are so weak that only specialized equipment can detect them. Others, however, like the spark when you touched the doorknob, are strong enough to observe and to experience.

Science PARADE

CHAPTER 13

ELECTRICITY

PLANNING THE CHAPTER

Chapter Sections	Page	Chapter Features	Page	Program Resources	Source
CHAPTER OPENER	340	*For Your Journal*	341		
Section 1: STATIC ELECTRICITY	342	Discover by Observing **(B)**	342	*Science Discovery**	SD
• Electric Charge **(A)**	342	Discover by Doing **(A)**	343	Connecting Other	
• Conservation of Charge **(H)**	344	Activity: How can you produce and study electrical charges? **(A)**	344	Disciplines: Science and Art, Displaying Electricity in Art **(B)**	TR
		Section 1 Review and Application	347	Investigation 13.1: Making an Electric Maze **(A)**	TR, LI
				Thinking Critically **(B)**	TR
				Study and Review Guide, Section 1 **(B)**	TR, SRG
Section 2: ELECTRIC CURRENT	348	Discover by Doing **(B)**	349	*Science Discovery**	SD
• Producing Electric Current **(B)**	348	Discover by Doing **(A)**	351	Investigation 13.2:	
• Potential Difference **(A)**	350	Discover by Calculating **(A)**	353	Classifying Electrical Insulators and Conductors **(A)**	TR, LI
• Keeping Up With Current Events **(B)**	353	Section 2 Review and Application	353	Batteries **(B)**	IT
				Study and Review Guide, Section 2 **(B)**	TR, SRG
Section 3: ELECTRIC CIRCUITS	354	Discover by Problem Solving **(B)**	354	*Science Discovery**	SD
• Making an Electric Circuit **(B)**	354	Discover by Doing **(B)**	355	Reading Skills: Using Flow Charts to Show Sequence **(A)**	TR
• Current and Resistance **(A)**	356	Discover by Doing **(A)**	360		
• Series Circuits **(A)**	357	Section 3 Review and Application	360	Extending Science Concepts: Electronic Circuit Diagrams **(A)**	TR
• Parallel Circuits **(A)**	359	Skill: Interpreting a Circuit Diagram **(A)**	361	Record Sheets for Textbook Investigations **(A)**	TR
		Investigation: Making Series and Parallel Circuits **(A)**	362	Study and Review Guide, Section 3 **(B)**	TR, SRG
Chapter 13 HIGHLIGHTS	363	The Big Idea	363	Study and Review Guide, Chapter 13 Review **(B)**	TR, SRG
Chapter 13 Review	364	For Your Journal	363		
		Connecting Ideas	363	Chapter 13 Test	TR
				Test Generator	

B = Basic **A** = Average **H** = Honors
The coding Basic, Average, and Honors indicates subsections, features, and resources that might be appropriate for different levels of learners. For additional suggestions regarding choice of topic and depth of coverage, see the Pacing Chart on pages T28–T31.

*Frame numbers at point of use
(TR) Teaching Resources, Unit 4
(IT) Instructional Transparencies
(LI) Laboratory Investigations
(SD) *Science Discovery* Videodisc Correlations and Barcodes
(SRG) Study and Review Guide

CHAPTER MATERIALS

Title	Page	Materials
Discover by Observing	342	(per individual) comb, paper
Discover by Doing	343	(per individual) balloon, wheat puffs (2),
Activity: How can you produce and study electric charges?	344	(per group of 3 or 4) buret clamp, ring stand, wooden rod, thread (2 pieces), vinyl strip, wool, graphite-coated plastic-foam balls (2), acetate strip, plastic food wrap
Discover by Doing	349	(per individual) dry cells (2), flashlight
Teacher Demonstration	351	pennies (10), nickels (10), pieces of wire (2), flashlight bulb, tape, paper, salt water
Discover by Doing	351	(per individual) bar magnets (2)
Discover by Calculating	353	(per individual) journal
Discover by Problem Solving	354	(per individual) 1.5-volt battery and holder, switch, wires with alligater clips (3), 1.5-volt light bulb
Discover by Doing	355	(per individual) journal
Discover by Doing	360	(per individual) flashlight bulbs (2), battery, switch, wire
Skill: Interpreting a Circuit Diagram	361	(per individual) paper, pencil
Investigation: Making Series and Parallel Circuits	362	(per group of 3 or 4) 6-volt lantern battery, 6-volt bulbs (2), sockets for flashlight bulbs (2), key switches or knife switches (2), wire leads with alligator clips (8)

ADVANCE PREPARATION

For the *Teacher Demonstration* on page 351, you will need 10 pennies and 10 nickels.

TEACHING SUGGESTIONS

Outside Speaker
Invite a meteorologist to class to discuss lightning. Ask the speaker to discuss how lightning produces light, heat, and noise. Encourage the speaker to discuss the hazards associated with lightning and the safety precautions students should use.

Field Trip
Take a trip to a local utility company or to an electrician's office where students can observe an electrician at work. You may want to have the electrician describe his or her job, including education and training. Have the students prepare questions before the field trip. Ask the electrician to discuss hazards associated with electricity and electrical devices.

CHAPTER THEME—ENERGY

In Chapter 13 the students will learn that electricity is a form of energy and that it can be changed to other forms of energy, such as heat and light.

The energy of electrons is discussed in the context of electric currents and circuits. A supporting theme in this chapter is **Systems and Structures**.

MULTICULTURAL CONNECTION

African-American inventor Granville T. Woods surpassed most others in the field of electricity in the number and variety of his inventions of important and significant electric systems and devices. In 1881 Woods invented a telephone transmitter that could carry the voice over much longer distances than the transmitter in use, with louder and more distinct sound. One year later, Woods patented a device that combined the telephone with the telegraph, coining the word telegraphone for it.

In 1887, Woods introduced his railway telegraphy. The invention could send messages between moving trains, and from moving trains to a railway station.

MEETING SPECIAL NEEDS

Mainstreamed

Have mainstreamed students work individually. Provide each student with two or three magazines, and have them cut out pictures that illustrate uses of electricity. Challenge the students to find as many pictures as possible. Have them use the pictures to make a collage titled "How We Use Electricity." You might display the collages in the room and refer to them at appropriate points in the chapter. After the students have completed this chapter, have them place their collages in their science portfolios.

▶ **340** CHAPTER 13

CHAPTER 13

ELECTRICITY

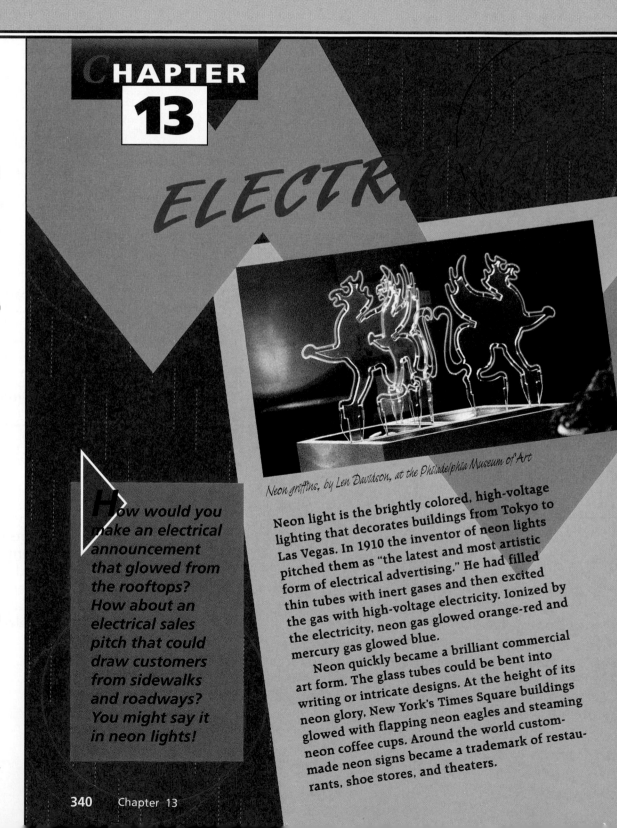

Neon griffins, by Len Davidson, at the Philadelphia Museum of Art

How would you make an electrical announcement that glowed from the rooftops? How about an electrical sales pitch that could draw customers from sidewalks and roadways? You might say it in neon lights!

Neon light is the brightly colored, high-voltage lighting that decorates buildings from Tokyo to Las Vegas. In 1910 the inventor of neon lights pitched them as "the latest and most artistic form of electrical advertising." He had filled thin tubes with inert gases and then excited the gas with high-voltage electricity. Ionized by the electricity, neon gas glowed orange-red and mercury gas glowed blue.

Neon quickly became a brilliant commercial art form. The glass tubes could be bent into writing or intricate designs. At the height of its neon glory, New York's Times Square buildings glowed with flapping neon eagles and steaming neon coffee cups. Around the world custom-made neon signs became a trademark of restaurants, shoe stores, and theaters.

CHAPTER MOTIVATING ACTIVITY

Discuss with the students neon signs they have seen that are particularly interesting or visually appealing. If possible, have the students bring in pictures of the signs. You might also display a picture of a neon-lined street to lead them into a discussion about our dependence on electricity.

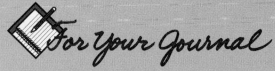

For Your Journal

The journal questions will help you identify any misconceptions the students might have about electricity, their knowledge of electricity, and their appreciation of its usefulness. The students will probably be familiar with everyday uses of electricity. However, they might not be aware of its characteristics. (The students will have a variety of responses to the journal questions. Accept all answers, and tell the students that at the end of the chapter they will have a chance to refer back to the questions.)

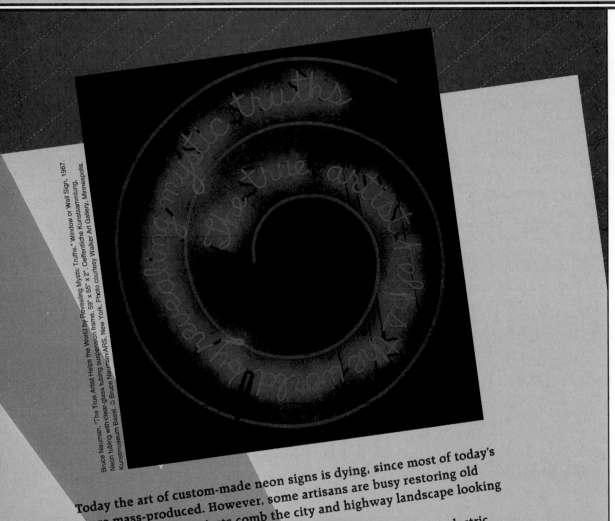

Bruce Nauman, "The True Artist Helps the World by Revealing Mystic Truths." Window or Wall Sign, 1967. Neon tubing with clear-glass tubing suspension frame, 59" x 55" x 2". Oeffentliche Kunstsammlung, Kunstmuseum Basel. © Bruce Nauman/ARS, New York. Photo courtesy Walker Art Gallery, Minneapolis.

ABOUT THE PHOTOGRAPHS

Neon is a colorless, odorless gas found in tiny amounts in Earth's atmosphere (about 0.001 percent). It was discovered in 1898 by two British chemists, William Ramsay and Morris W. Travers, while they were studying liquid air. They named the gas for the Greek word *neos*, which means "new."

Neon signs are made by removing air from glass tubing and filling the tubing with neon gas at low pressure. Electrodes are sealed in the tube, through which about 15 000 volts of electricity pass. The charge causes the neon atoms to become ionized. Orange-red light is produced as the neon ions return to their normal state. Other gases produce different colors of light when they are ionized. Helium produces a pink light; argon, a blue-white light; mercury, a green-blue light. The gases can be mixed to produce additional colors.

Today the art of custom-made neon signs is dying, since most of today's signs are mass-produced. However, some artisans are busy restoring old neon signs. Neon enthusiasts comb the city and highway landscape looking for rusting neon classics.

At the same time, artists are making new artistic statements in electric light. Artists like neon light because it combines the bright colors of painting and the shapeability of sculpture. One neon artist includes the lights' high-voltage electric transformers in his neon creation. He treats the transformer's hum as part of the art work. Neon lights enable artists to make vivid statements about their work. One neon work reads, "The true artist helps the world by revealing mystic truths." It's not the same as "Eat at Joe's," but it's written in the same glowing electric script.

For Your Journal

- List as many characteristics of electricity as you can think of.
- What can electricity do?
- How do you use electricity every day?

FOCUS

This section provides demonstrations and descriptions of how the negative and positive charges of static electricity behave with different kinds of materials. The law of conservation of charge is also discussed.

MOTIVATING ACTIVITY

Cooperative Learning Provide small groups of students with two balloons, two pieces of string, and a wool cloth. Have the students inflate the balloons, tie a string to each, and gently rub one balloon with the wool cloth. Direct them to suspend both balloons from the strings, keeping the rubbed balloon close to the other balloon for about a minute. Have them record their observations in their journals and provide a possible explanation of their observations. Conclude by telling the students they will refer to their journal entry later in the section.

PROCESS SKILLS
• Observing • Inferring

POSITIVE ATTITUDES
• Curiosity • Enthusiasm for science and scientific endeavor • Skepticism

TERM
• static electricity

PRINT MEDIA
There Are No Electrons: Electronics for Earthlings by Kenn Amdahl (see p. 337b)

ELECTRONIC MEDIA
Static and Current Electricity Encyclopaedia Britannica Educational Corp. (see p. 337b)

SCIENCE DISCOVERY
• Electrostatic energy; electroscope
• Electrostatic energy; Van de Graaff
• Lightning

BLACKLINE MASTERS
Connecting Other Disciplines
Laboratory Investigation 13.1
Thinking Critically
Study and Review Guide

DISCOVER BY *Observing*

The bits of paper are attracted to the comb because the comb has an electric charge.

MEETING SPECIAL NEEDS

Second Langauge Support

After limited-English-proficient students read the information under each subhead, help them relate the main points in their own words. Encourage the students to note key words in their journals for future reference.

SECTION 1

Static Electricity

Objectives

Describe the effects produced by the accumulation of a static electric charge.

Apply the law of conservation of charge to specific cases.

Y ou've probably heard someone say that clothes dryers eat socks. You may have even thought so yourself as you sorted clean clothes. However, the socks usually turn up, stubbornly clinging to a shirt or towel. The crackling sound that you hear when peeling the clothes apart is entertaining though. Okay, so the dryer hasn't eaten the socks, but something got into them!

Electric Charge

You may have heard that crackling sound before. But where? You've probably heard that noise while combing your hair. Your hair wants to stand up instead of lie down or cling to the comb. You can investigate a comb's mysterious attractive force for yourself.

DISCOVER BY *Observing*

Tear some paper into small pieces. Then run a comb through your hair a few times, and place the comb close to, but not touching, the paper. What do you observe? How can you explain your observations?

Figure 13–1. The cat's fur is attracted to the charged balloon.

In the activity, you discovered that the comb picks up the bits of paper. It can do this because it has an electric charge. Many objects can become electrically charged. For example, have you noticed how a nylon shirt wants to cling to your body? Notice the effect of the charged balloon on the cat's fur in the photograph. All of these effects are due to the accumulation of an electric charge by one object, which causes another object to be attracted to it.

Charge It!   An electric charge can be transferred from one object to another. For example, a vinyl strip stroked with a piece of wool will gain an electric charge. The charge has been rubbed off the wool. When you touch a graphite-covered plastic-foam ball with the vinyl strip, the ball jumps away. If you repeat the exercise using an acetate strip and a piece of plastic food wrap, the same reaction occurs.

Now forget about the ball, and try touching the vinyl and the acetate together. What happens? Instead of jumping away, the two strips are attracted to each other. All charges are *not* the same; some charges repel objects while others attract them. There are two kinds of electric charge. The charge on the vinyl is a negative charge (-), and the charge on the acetate is a positive charge (+). The general rule, as shown in the illustration, is this: Like charges repel one another, and unlike charges attract one another.

A Mutual Attraction   Most things don't have any charge in their normal state. But objects can become charged by static electricity. *Static* means "not moving." **Static electricity** is a buildup of an electric charge on an object; it is one form of electric energy. You can discover one way that objects acquire static electricity.

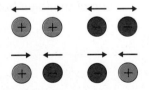

Figure 13–2. Only opposites attract.

Discover by *Doing*

Put five wheat puffs in a balloon, and blow it up. Rub the balloon on your shirt; then hold it close to your hair. What do you observe? Try rubbing the balloon with different materials, and test its "attraction" each time. In your journal, explain what is happening in terms of positive and negative charges. ✐

You just found that a balloon can do the same thing to hair that a comb sometimes does! So when you have trouble getting your hair just right, don't blame the comb. Blame static electricity. To investigate further, do the following activity.

ACTIVITY

How can you produce and study electric charges?

Process Skills: Observing, Inferring

Grouping: Groups of 3 or 4

Hints

Prepare a number of foam balls in advance. Use 1.25-cm polystyrene spheres coated with graphite. You can obtain them prepared or coat them yourself. Use the finest (thinnest) thread available. Thread a needle, and string a number of balls onto the thread. Remove each ball with about 20 cm of thread, and tie the ball in place.

Procedure

2. It attracts the ball.

3. Yes. It repels the ball.

4. The ball loses charge.

6. The balls attract because opposite charges attract.

▶ Application

1. The foam balls were charged by stroking them with different materials.

2. The objects were attracted to each other.

3. The charge is transferred from the strip to the ball, causing repulsion.

ONGOING ASSESSMENT
▼ ASK YOURSELF

The two objects would repel each other.

① no

② Rubbing moves charges from one object to another.

● **Process Skills:** *Inferring, Analyzing*

You might extend the Section 1 Motivating Activity on page 342 by having a volunteer hold a charged balloon near his or her other hand. The balloon should be repelled by one side of the hand and pushed to the other side of the hand. Ask the students to explain why the balloon is attracted to the hand. (The negatively charged balloon repels the electrons in the hand, pushing them as far as possible to the other side of the hand, thus creating a positive charge on the side of the hand nearest the balloon. Encourage the students to add new information to their journal entries.)

● **Process Skill:** *Analyzing*

Explain to the students how humidifiers help decrease static electricity in buildings. The humidifier adds moisture to the air. Static charges tend to leak off through films of surface moisture, thus reducing the incidence of static electricity.

ACTIVITY

How can you produce and study electric charges?

MATERIALS
buret clamp, ring stand, wooden rod, thread (2 pieces), vinyl strip, wool, graphite-coated plastic-foam balls (2), acetate strip, plastic food wrap

PROCEDURE
1. Set up the apparatus as shown.

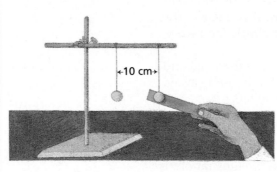

←10 cm→

2. Stroke the vinyl strip with the wool a few times. Then bring the strip near one of the foam balls. What happens?

3. Stroke a foam ball with the charged vinyl strip. Is the ball now charged? Explain.

4. Touch the same ball with your hand. What happens?

5. Repeat steps 2 through 4 using the acetate strip after it has been stroked with plastic food wrap to redistribute the electric charges.

6. Charge one ball from the vinyl and the other from the acetate. Slide the threads closer together without touching the balls. What happens? Explain the results.

7. After the balls touch, test them with the vinyl and acetate strips to see whether they are charged.

APPLICATION
1. How did you charge the foam balls?
2. How could you tell when two of the objects had opposite charges?
3. Explain why each ball has the same charge as the strip from which it was charged.

▼ ASK YOURSELF
How would two negatively charged objects act if brought near each other?

Conservation of Charge

Now you know that static electricity can make your hair do strange things. Does the static electricity stay in your hair forever? What is there about rubbing one kind of material against another that produces an electric charge? When you stroked the vinyl strip with the wool, the strip became negatively charged. Something else also happened—the wool became positively charged.

It would appear that stroking a vinyl strip with wool creates two kinds of charges. Actually, rubbing does not produce any charges—it just moves them around. The object that loses negative particles becomes positively charged because its charges are no longer completely balanced. The object that gains negative particles becomes negatively charged because it now has more negative particles than positive particles.

Figure 13–3. The placement of particles, or electrons, has been redistributed.

As shown in the illustration, wool and vinyl become charged when they are rubbed together because particles move from the wool to the vinyl. The particles that move are called *electrons*. The vinyl becomes negative when it gains electrons because the electrons have a negative charge. The wool becomes positive when it loses some electrons.

Keeping It Equal The *law of conservation of charge* states that an electric charge cannot be created or destroyed. The atoms that make up all materials have positive and negative charges. When something is uncharged, its positive charge is equal to its negative charge. The positive and negative charges cancel each other, so that the total charge is zero. An object with zero charge is said to be neutral. This is when your hair behaves and your socks don't stray.

When an object *is* charged, it may have either a negative or a positive charge. An object has a positive charge if it has given some of its negative electrons to another object. The object that receives the extra electrons has a negative charge. Remember, only the negatively charged electrons move.

You'll Get a Charge Out of This Just when you've pulled yourself together and put your socks on and combed your hair down, the car attacks you! You reached to open the

INTEGRATION–
Mathematics

The use of algebraic signs to designate the two kinds of charges originated with the discovery by an Arab named al-Rhazes that the addition of opposite, but equal, charges produces zero charge.

✧ **Did You Know?**
Gasoline trucks are equipped with tires made to conduct an electric charge or are connected to an electric "ground" while delivering fuel to a gas station. These steps prevent the buildup of charge on the truck, because the charge is conducted to the ground. As a result, not enough charge builds up to spark a fire or an explosion.

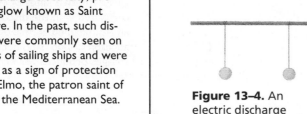

Figure 13–4. An electric discharge neutralizes both objects.

Figure 13–5. This student is experiencing static electricity firsthand!

car door and suddenly received a shock. If it had been dark enough, you might have seen the spark that caused the "Ouch."

When a negatively charged object and a positively charged object are brought together, the effects of both charges are cancelled out. Electrons move from one object to another until both objects are neutral. If the charge is strong enough, the objects do not even have to be touching for this exchange to take place. Electrons can jump a gap between two oppositely charged objects. When this happens, the charge heats the air enough to make a spark. After the spark, both objects are electrically neutral. This process of transferring electric charges is called *electric discharge.*

Static electricity may be discharged if a charged object comes in contact with an object that will accept the charge. When you shuffle across a carpet and touch a doorknob, the static electricity is discharged and in the process you receive a small electric shock. You may even see a small spark as the static electricity is discharged.

The Biggest Shock of All While lightning is certainly fascinating and dangerous, it's just a spark. Okay, so it's a really big spark. How is this enormous spark, or electric discharge, produced? Obviously no one is rubbing wool or plastic wrap over the clouds.

Electric charges in a thunderhead become redistributed as rapidly rising air currents carry water drops and hailstones upward. The upper part of the cloud in the illustration has a positive charge, and the lower part has a negative charge. The negative charge of the lower portion of the cloud becomes strong enough to make charges in the earth redistribute themselves. What type of charge must the earth's surface have?

RETEACHING

Refer the students to Discover by Doing on page 343. On the chalkboard, work with the students to outline the steps of the activity. Ask the students to describe the movement of electrons between the objects in each step and to explain how the total charges were conserved.

EXTENSION

 Have the students use library references and other sources to learn about Benjamin Franklin's experiments with static electricity, which led to his invention of the lightning rod. Suggest that the students prepare a written report and share it with classmates.

CLOSURE

Cooperative Learning Have small groups of students write outlines that summarize the concepts of this section. You might extend this activity by having the groups share the content of their outlines and giving them the opportunity to revise their outlines with additional information.

A lightning bolt occurs when the charge is neutralized. The negatively charged bottom of the cloud discharges either to the positively charged upper part of the cloud or to the positively charged earth. How do these types of discharges affect where the lightning appears?

While the illustration shows how lightning strikes the ground, it is important to remember that lightning may also go from cloud to cloud or even from the ground to the cloud!

▼ **ASK YOURSELF**

How does a lightning bolt demonstrate the law of conservation of charge?

Figure 13–6. Lightning is a spark that forms when opposite charges are neutralized during a storm.

SECTION 1 REVIEW AND APPLICATION

Reading Critically

1. When you stroke an inflated balloon on your hair, the balloon acquires a negative charge. How is your hair affected?
2. When wool is stroked with vinyl, the wool acquires a positive charge. How does the law of conservation of charge apply to the vinyl and the wool?

Thinking Critically

3. How can you determine whether or not a balloon has an electric charge?
4. You discover your pen can pick up scraps of paper. Tell how you can determine whether the charge on the pen is positive or negative, using only a rubber rod, a woolen glove, and a suspended plastic-foam ball.
5. Explain how electric charges act with respect to one another.

SCIENCE BACKGROUND

Electric charges in clouds become separated by collisions between light particles (water droplets) and heavy particles (hailstones) in the clouds. The collisions cause the heavy particles to acquire a negative charge and fall to the bottom of the cloud. The lighter particles become positively charged and rise to the top of the cloud.

ONGOING ASSESSMENT
▼ **ASK YOURSELF**

A lightning bolt neutralizes the charges between the bottom and top of a cloud or between the bottom of a cloud and Earth's surface.

SECTION 1 REVIEW AND APPLICATION

Reading Critically

1. Your hair gets a positive charge.

2. While the wool acquires a positive charge, the vinyl acquires an equal negative charge. The vinyl acquires the electrons that the wool loses.

Thinking Critically

3. Charge another balloon to see whether the two balloons repel each other.

4. Stroke the rubber rod with the glove. The charge on rubber, by definition, is negative. Charge the ball by contact with the rod to see whether the pen attracts or repels the ball.

5. Like charges repel, and unlike charges attract.

FOCUS

In this section the students learn the characteristics of electric current and how batteries produce current. The measuring units—amperes, volts, and watts—are also discussed.

MOTIVATING ACTIVITY

Use a low-voltage dry cell, wires of different gauges, a light bulb, a switch, and a voltmeter to set up a simple electric circuit. Have the students observe as you open and close the switch. Remind the students to also note the voltage. Then ask a volunteer to describe what happened. (When the circuit was closed, electric current could flow. When the circuit was open, current could not flow.) Finally, close the switch, but replace the wires with wires of a different gauge. The students should observe that the voltage changes with the different gauge of wire.

POSITIVE ATTITUDES
• Curiosity • Skepticism

TERMS
• electric current • direct current • alternating current • potential difference

PRINT MEDIA
"Let There Be Light" from *Cobblestone* magazine by Bruce Watson (see p. 337b)

ELECTRONIC MEDIA
Electronics: An Introduction BFA Educational Media (see p. 337b)

SCIENCE DISCOVERY
• Battery, car
• Battery, rechargeable
• Battery, mercury cells

BLACKLINE MASTERS
Laboratory Investigation 13.2
Study and Review Guide

 **SCIENCE TECHNOLOGY SOCIETY** The phenomenon in which light can stimulate the flow of electricity in certain materials was first discovered in the 1800s, but it remained a laboratory curiosity for many years. The first commercial use was in a battery developed by the Bell Telephone Laboratories in 1954. The battery was built to supply power to amplifier stations on rural telephone lines. The efficiency of this device was about 10 percent. The theoretical conversion efficiencies of modern cells are near 28 percent.

SECTION 2

Electric Current

Objectives

Explain *what electric current is and what the function of a battery is.*

Measure *the units in which electric current and potential difference are measured.*

Utilize *equations to determine the values of electric power and energy.*

Although lightning may seem like a giant Terminator, you now know that a more accurate nickname would be the Equalizer. Lightning can knock down power lines, leaving you without electricity. Are you really without electricity?

What about a flashlight? It doesn't have a cord and a plug, but still uses electricity to produce light. How? Inside the flashlight are batteries that generate electric charges. It is the flow of these electric charges that causes the flashlight to light.

Producing Electric Current

As you found out when you touched the car, too many electrons in one place produce an electric charge. The clouds in a thunderstorm develop electric charges. When the charges neutralize, electrons move, or flow, from one place to another. The flow of electrons is called **electric current.** Electric current is a form of electric energy. A bolt of lightning is a large electric current flowing through the air.

Figure 13–7. Lightning is one type of electric discharge.

● **Process Skills:** *Observing, Inferring*

As the students complete the *Discover by Doing* on p. 349, encourage them to change the orientation of the negative and positive terminals of the batteries in the flashlight. Then ask them to explain in their journals why the flashlight does not light.

● **Process Skill:** *Analyzing, Comparing*

Display a cutaway diagram of a wet cell, such as an automobile battery. Point out that each cell of the battery contains a series of plates. Some plates contain lead only; some contain lead dioxide. The plates are immersed in a solution of sulfuric acid. As the battery functions, the plates and sulfuric acid undergo a sequence of reactions. As the reactions occur, a continuous flow of electrons move from one type of plate to the other, producing electric current.

The energy from electric currents can be used for many purposes. As you switch on your flashlight, you are using the electric current produced by an *electric cell*. An electric cell converts chemical energy into electric energy and contains two terminals (positive and negative) and an electrolyte. An *electrolyte* is a substance that allows an electric current to be conducted.

The cells in a battery may be either dry cells or wet cells. A dry cell, like those used in flashlights, contains a pastelike electrolyte. A wet cell, like those used in some car batteries, contains a liquid electrolyte. Examine a dry cell for yourself.

DISCOVER BY *Doing*

Caution: Do not break open the dry cell, since the electrolyte can injure skin tissue. Obtain a dry cell. Compare the dry cell to the drawing, and identify the positive and negative terminals. Get a flashlight, and place two dry cells in it. Examine the way the dry cells contact the parts of the flashlight. Make a sketch in your journal to show how electricity flows from the cells through the bulb and back to the cells. ✏

This diagram shows the inside of a dry cell. The positive and negative terminals of the cell are made of carbon and zinc. An electrolyte separates the two terminals. When the battery is attached to a device such as a light bulb, the electrolyte reacts with the negative terminal of the battery to release electrons. The electrons flow from the negative terminal through the bulb and back to the positive terminal. Remember only the electrons move, so the flow has to be from where there are more electrons to where there are fewer electrons. The current will flow until the chemicals in the cell are used up. When this happens, you will be in the dark.

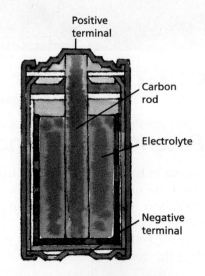

Positive terminal

Carbon rod

Electrolyte

Negative terminal

Figure 13–8. A cross section of a dry cell

DISCOVER BY *Doing*

PORTFOLIO ASSESSMENT

The students' sketches should accurately depict the two dry cells and the bulb setup in the flashlight. After the students have completed the *Discover by Doing*, have them place their sketches in their science portfolios.

THE NATURE OF SCIENCE

In 1791 Luigi Galvani, an Italian surgeon, observed that a dead frog's legs twitched when a spark from an electric machine accidentally struck them. After further experimentation, Galvani decided that the electricity originated in the nerves in the frog's legs. However, another Italian scientist, Alessandro Volta, disagreed with Galvani. Volta believed that a metal object near the frog, such as a scalpel, produced the current. Volta used his ideas about the frog to develop the first crude electric battery. The battery consisted of copper and tin disks and a salt solution.

● **Process Skill:** *Comparing*

Relate AC current waves to sound waves and the other longitudinal waves studied in Chapter 8. Note that, as in any longitudinal waves, the particles merely vibrate back and forth, but the wave has direction.

● **Process Skill:** *Analyzing*

Have the students review what they learned from *Discover by Doing* on page 349. Then challenge the students to determine why you should remove a dead battery from a device as soon as the battery stops working. (A dead battery still has the electrolyte in it. The electrolyte can eat through the battery container and leak into the device, causing permanent damage to the device.)

SCIENCE BACKGROUND

Light bulbs do not flicker when AC changes direction because the change of direction occurs very rapidly (50 to 60 cycles per second; 50–60 Hz). For a 50-Hz frequency, the current builds up to a maximum in one direction and drops to zero in the first 0.01 second. It builds up to a peak in the opposite direction and drops to zero again in the next 0.01 second, thus completing the entire cycle in 0.02 seconds.

ONGOING ASSESSMENT
⟁ ASK YOURSELF

The major difference between DC and AC is the direction of the flow of electrons. DC flows steadily in one direction; AC changes direction over and over again.

⟡ **Did You Know?**

The film *The Day the Earth Stood Still* (1951) tells about a visitor from another planet who comes to Earth to warn people to stop fighting among themselves. Fighting, the visitor tells them, will lead to their destruction. To demonstrate his power, the visitor turns off all the power in the world for 30 minutes.

① The students will probably point out that their lives would be dramatically changed. For example, electric appliances and lights would no longer work. Accept all reasonable responses.

How much more energy is in a lightning bolt than in a flashlight battery? Lightning is hard to measure, but electric current can be measured with an instrument called an *ammeter*. The ammeter determines how much charge passes through it every second. This movement, or current, is measured in units called *amperes* (A). A device using 2 amperes is using electricity twice as fast as a device using 1 ampere. Twice as many electrons flow through the 2-ampere device every second.

The current produced by a battery flows steadily in one direction from the negative terminal to the positive terminal. This kind of current is called **direct current (DC).** Does this mean that a power station is nothing more than a gigantic battery? No. There are other sources of current besides batteries. Your electric company produces electricity with generators. This kind of current is not DC. Instead of flowing in one direction, the current repeatedly changes direction. The current is alternating its direction, so it is called **alternating current (AC).**

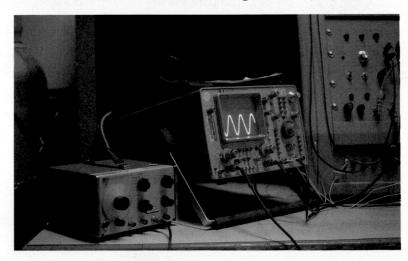

Figure 13–9. This oscilloscope is being used to monitor AC current. The crests of the wave indicate current traveling in one direction. The troughs indicate current traveling in the opposite direction.

⟁ ASK YOURSELF
What is the major difference between DC and AC?

Potential Difference

Sitting in the dark on a stormy night may seem fun for a while, but it can quickly become a bore. If you never got the electricity back, how would this affect your life? Electricity is used everywhere. What is there about electricity that enables it to carry ① energy? Perhaps an experiment with magnets can help to answer the question.

● **Process Skill:**
Constructing/Interpreting
Models

Explain to the students that the potential difference of electricity can be illustrated by water pressure. Water tends to flow from a higher level to a lower level. Have the students think of a pipe connecting two containers

at different levels. Water will flow from the higher container through the pipe to the lower container. In other words, the water flows from a position of higher potential energy to a position of lower potential energy. This potential difference is responsible for the force that moves water from one container to another.

● **Process Skill:** *Comparing*

Point out to the students that the lamp pictured in Figure 13–10 is an incandescent lamp. Then explain that a fluorescent lamp uses only about 20 percent as much electricity as an incandescent one uses to produce the

same amount of light. It also produces only about 20 percent as much heat for the same amount of light. You might encourage interested students to do additional research on fluorescent lamps and share their findings with classmates. Encourage the students to place their research in their science portfolios.

DISCOVER BY *Doing*

Obtain two bar magnets. Bring the south pole of one magnet close to the north pole of the second magnet. As with opposite electric charges, the opposite poles attract. Slowly separate the magnets. Do you feel the force of attraction between the poles? How do you know that energy is required to separate the opposite poles?

You supplied the energy to separate the magnets. Since negative and positive electric charges attract each other, it takes energy to keep them apart. This energy is stored any time negative and positive charges are separated from each other. In ordinary matter, electrons carry the negative charge and atomic nuclei carry the positive charge. (You will learn more about this when you study Chapter 16.) Negatively charged electrons can move easily, but atomic nuclei cannot. In a battery, a generator, or a thunderhead, electrons are moved away from positively charged nuclei. When they are separated from the nuclei, the electrons have stored energy. This situation is analogous to a rock being raised above the earth. The rock has stored energy because of earth's gravitational attraction for the rock. The electrons have stored energy because of the electrical attraction between them and the positive nuclei. The rock's stored energy is called gravitational potential energy; the electrons' stored energy is called *electrical potential energy.*

In a thunderhead, the lightning flash reunites the separated charges and produces a spectacular display of light, heat, and noise—enough to shake you up. When negative electrons move toward and reunite with positive nuclei, electrical potential energy is converted into other forms of energy. In other words, moving electrons can be used to do work, just like falling water can.

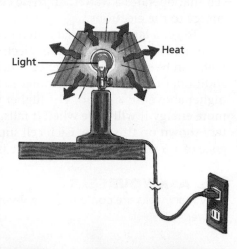

Figure 13–10. When energy passes through a lamp or an appliance in the form of moving electrons, some energy is converted into heat and light.

DISCOVER BY *Doing*

The students should feel the attraction between the poles and should also observe that they need to pull on the magnets to separate them.

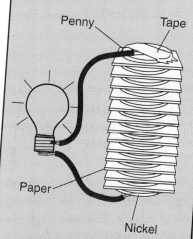

Demonstration

Construct a battery using ten pennies, ten nickels, two pieces of wire, a flashlight bulb (or galvanometer), tape, and heavy paper soaked in a strong saltwater solution.

Assemble the battery as shown, using tape to attach the wires. Light the bulb by holding the free ends of the wires to the bulb. Ask the students to point out the battery's two terminals (a penny and a nickel) and the electrolyte (salt water absorbed by the paper)

SECTION 2 351 ◀

GUIDED PRACTICE

Have the students use the formula $P = I \times V$ (power = current × potential difference) to calculate the power of a toaster and a hair dryer. The toaster has 4.0 A of current flowing through it at 110 V. The hair dryer has 2.5 A of current flowing through it at 120 V. Conclude by asking the students which device uses more power. (The toaster uses 440 W. The hair dryer uses 300 W.)

INDEPENDENT PRACTICE

 Have the students provide written answers to the Section Review. In their journals, have the students draw a diagram of a dry cell and explain how it produces electric current. The students might refer to the diagram in their text, if necessary.

EVALUATION

 Cooperative Learning Have small groups of students explain the difference between current, potential difference, and power. (Current is the amount of charge that passes per second. Potential difference is related to electrical potential energy. Power is the rate at which energy does work.)

INTEGRATION– Health

The human body is quite a good conductor of electricity—once the electricity gets inside it. This is because the human body is largely water, with many types of salts dissolved in it.

The most dangerous condition to be in is to have a wet skin and bare feet. This makes bathrooms potentially dangerous, electrically; special care is needed. All bathroom switches and outlets should be a "power-breaker" type that instantly cut off current when water comes in contact with a switch or outlet.

◇ Did You Know?

Lithium batteries last much longer than carbon-zinc batteries because lithium can give more electrons per unit volume.

LASER DISC

2122

Battery; mercury cells

2123

Battery, rechargeable

2124

Battery, car

ONGOING ASSESSMENT
▼ ASK YOURSELF

A 9-V battery has 6 cells, each with a voltage of 1.5 V.

The strength of batteries, generators, or thunderclouds is measured by a value related to the electrical potential energy of the electrons called **potential difference.** Although you can calculate potential difference if you know what the electrical potential energy is, it is a lot easier to measure it directly. Potential difference is expressed in *volts* (V) and can be measured by a voltmeter.

Potential difference is created by having extra electrons in one place and missing electrons in another. The negative terminal of a battery has extra electrons, for example, and the positive terminal is missing electrons. This can be compared to having a higher level of water on one side of a dam than the other. A battery is like an "electron dam" that keeps more electrons at the negative terminal than at the positive terminal. Just as the flow of water over a dam can be used to drive a water wheel, a flow of electrons can be used to light a flashlight bulb.

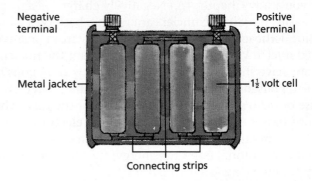

Figure 13–11. This is a cutaway view of a dry cell battery containing four 1.5-volt cells.

A dry cell has a potential difference of about 1.5 volts. A voltmeter will read 1.5 volts whether it is connected to the terminals of a flashlight battery, a large laboratory cell, or a tiny cell that operates a watch. All these cells give about the same energy to the electrons.

To get a larger potential difference, several cells may be combined to form a battery. An electron that is boosted to 1.5 volts can be sent into another cell that will raise its energy an additional 1.5 volts. This is similar to raising a rock higher and higher above the ground. The higher you raise the rock, the more energy it will have when it falls. Look at the dry cell battery shown on this page. Each cell supplies a potential difference of 1.5 volts. What total voltage is supplied by the battery?

▼ ASK YOURSELF

How many cells are contained in a 9-volt battery? Explain your answer.

RETEACHING

 Review with the students how the terms *volt, watt, ampere, alternating current,* and *direct current* are related. Then have the students write a brief paragraph about each in their journals. You might encourage the stu- dents to share their paragraph with classmates and revise their paragraph based on classmates' feedback.

EXTENSION

Thomas Edison started the first electric company. He transmitted direct current to homes close to his generators. Have the students use library references and other resources to find out why direct current (DC) is no longer used by electric companies and why Edison opposed changing to alternating current (AC).

CLOSURE

 Have the students work independently to write a paragraph that summarizes the major concepts in this section.

Keeping Up with Current Events

You know that light is a form of energy. The rate at which energy is delivered is called *power.* The rate at which electrical energy is delivered to an appliance is the power the appliance uses. Electric power depends on two things: the potential difference (in volts) and the amount of charge that goes through the appliance each second, or current (in amperes). Power can be calculated by using the following formula.

$$\text{power} = \text{current} \times \text{potential difference} \quad \text{OR} \quad P = I \times V$$

If current is measured in amperes and potential difference is measured in volts, the power is expressed in *watts* (W). You may recall that mechanical power is also measured in watts. Try the next activity to find out more about power.

 DISCOVER BY Calculating _____

Different appliances and light bulbs in your home use different amounts of power. Check several bulbs and appliances, including a toaster, TV, radio, and a room air conditioner, if you use one, for information on the power (watts) that they use. All of these appliances are likely to be connected to 120-volt lines. Calculate the current that each appliance and bulb uses. In your journal, make a table showing the power and current usage for each. ✎

▶ **ASK YOURSELF**

How can you calculate the amount of electric energy you use in your home?

SECTION 2 REVIEW AND APPLICATION

Reading Critically
1. Compare the amount of charge per second passing a point when the current is 2 amperes and when it is 10 amperes.
2. What is the function of a battery?

Thinking Critically
3. The electric system of a small boat operates from a 12-volt battery. What do you think would happen if you used a household light bulb on this boat?
4. A toaster on a 120-volt line uses 7.5 amperes of current. How much power is the toaster using?

SCIENCE BACKGROUND

The watt was introduced in an earlier chapter as one joule of energy expended or work done each second. Since charge Is measured in coulombs (a meter-kilogram-second unit of electric charge equal to the quantity of charge transferred in one second by a steady current of one ampere) and current is rate of flow of charge, 1 ampere is 1 coulomb per second.

 DISCOVER BY Calculating

Have the students place their tables in their science portfolios.

ONGOING ASSESSMENT
ASK YOURSELF

The amount of energy can be calculated by multiplying the total power you use by the length of time you use it.

SECTION 2 REVIEW AND APPLICATION

1. At 10 A, five times as much charge passes a point every second as at 2 A.

2. A battery converts chemical energy into electric energy.

Thinking Critically
3. The household bulb is designed to light on 120 V. On a 12-V circuit, the current would be only one-tenth as much as the bulb needs. The bulb would not light.

4. 900 W

PROCESS SKILLS
- Communicating
- Constructing/
Interpreting Models
- Experimenting

POSITIVE ATTITUDES
- Experimenting • Observing
- Expressing ideas effectively

TERMS
- electric circuit
- conductor • insulator
- resistance • series circuit
- parallel circuit

PRINT MEDIA

Warning: The Electricity Around You May be Dangerous to Your Health by Ellen Sugarman (see p. 337b)

ELECTRONIC MEDIA
Making Circuits Cambridge Development Laboratory, Inc. (see p. 337b)

SCIENCE DISCOVERY
- Electrical switches
- Stove burner
- Switch construction; lab setup

BLACKLINE MASTERS
Reading Skills
Extending Science Concepts
Study and Review Guide

① The train stops because the current does not have a complete path along which it can flow.

DISCOVER BY
Problem Solving

 The bulb doesn't light because current is not flowing through it. Ask the students to place their sketches in their science portfolios.

FOCUS

This section discusses the parts that make up an electric circuit and distinguishes between conductors and insulators. Emphasis is given to the conditions that make a current flow in a circuit and methods for diagramming and measuring it. The differences between a parallel circuit and a series circuit are also described.

MOTIVATING ACTIVITY

 Cooperative Learning Ask the students to define the word *circuit*. (Examples include a circular path, a path with no beginning or end, and a route.) Then ask a volunteer to explain why the train track in the illustration is a circuit. (The track is continuous.)

In small groups, have the students brainstorm other examples of circuits. (Examples include a circulatory system, a bus route, a race track, and a roller coaster.) Finally, initiate a discussion of how these circuits would function if the circuits were not complete. (They would not function.)

SECTION 3

Electric Circuits

Objectives

Describe the parts that make up a simple electric circuit.

Compare and **contrast** series and parallel circuits in terms of the way currents and potentials are distributed in them.

P erhaps you have seen an electric train set. The engine races around a track, pulling the cars behind it. If one section of track becomes disconnected, even if the train is on a section where the track is complete, the train comes to a stop! Why does the train stop just because one piece of track is broken? The train set is still plugged in. Try the next activity to find some answers.

Figure 13–12. Plugging in the train set provides current. So why isn't this train moving? ①

Problem Solving

Obtain a 1.5-volt battery and holder, a switch, three wires with alligator clips, and a 1.5-volt light bulb. Connect the items in such a way as to make the bulb turn on when you throw the switch. Make a sketch of your setup. Why doesn't the bulb light if the switch is open?

Making an Electric Circuit

If you were able to light the bulb in the last activity, then you made a complete path along which electricity could flow. To

● **Process Skill:** *Evaluating*

Have the students bring to class different materials that can be examined in terms of being conductors or insulators.

● **Process Skill:** *Generating Ideas*

Have the students carefully examine the circuit diagram in Figure 13–14. Discuss the conditions that cause the current to flow in the circuit. Ask the students to suggest in their journals how the circuit could be changed and still remain a complete circuit. (You can vary the locations of the different parts.)

make current flow through the bulb, two things are needed: a potential difference and a complete circuit. The potential difference can be supplied by a battery. An **electric circuit** is the path electricity follows. A complete circuit from one terminal of the battery to the other is necessary for electricity to flow.

In the circuit shown in illustration 13–13, the electrons flow out of the negative terminal of the battery and through a wire to the switch. They pass through the closed switch into another wire, through the bulb, and back through another wire into the battery. Compare this illustration to the sketch you made when you did the last activity. Explain whether or not you made a complete circuit.

Electricity has an easier time moving through some materials than others. A material through which electric current can flow is called a **conductor.** Metals, especially silver, copper, aluminum, and gold, are good conductors of electricity.

Just as some materials allow the energy flow of electricity, others prevent it. An **insulator** is a material that will not carry current. Rubber and plastic are good insulators. An electric cord contains metal wire, usually aluminum or copper, surrounded by an insulator, usually rubber. What could happen if part of the metal wire were exposed because the insulator wore away? ①

A switch contains a conductor and an insulator. A switch is an electric device that is used to start and stop the flow of current through a conductor. When the switch is open, no current can flow. What part of the switch is an insulator? ②

A good way to describe what is happening in a circuit is to use a circuit diagram. A circuit diagram is an illustration that shows how the parts of a circuit are connected. The circuit diagram shown here uses standard symbols for a battery, a switch, and a light bulb. Wire connections are indicated by straight lines. The diagram shows clearly that there is a complete conducting path around the circuit as long as the switch is closed. The next activity gives you some practice using circuit diagrams.

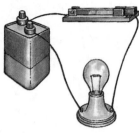

Figure 13–13. A battery, a switch, a light bulb, and some wire are all you need for an electric circuit.

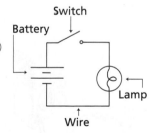

Figure 13–14. A circuit diagram consists of standard symbols that represent the actual parts of a circuit.

LASER DISC
2117

Electrical switches

371, 372, 373, 374

Switch construction; lab setup

▲ **MEETING SPECIAL NEEDS**

Gifted

Have interested students set up a simple circuit, using a low-voltage dry cell, a small light bulb, and an ammeter. Have them determine the approximate resistance of the bulb in the circuit.

① If part of the metal were exposed, an electric fire might result.

② The handle of the switch is an insulator.

DISCOVER BY *Doing*

After the students complete *Discover by Doing,* have them place their circuit diagrams in their science portfolios.

DISC VER BY *Doing* _____

In your journal, make a circuit diagram of the circuit you made in the Discover by Problem Solving on page 354. Refer to the original sketch you made of your circuit. Be sure to use the standard symbols shown in Figure 13–14 as you complete your circuit diagram. ✐

TEACHING STRATEGIES, continued

● **Process Skill:**
Constructing/Interpreting Models

Have the students stand in a circle. Give each student a paper clip. Ask one student to be the energy source or starter. The other students should imagine they are conductors and the paper clips are electrons. Have

the starter pass his or her clip to the student on the right. Explain to the students that they cannot hold onto more than one "electron" at a time, so when they have two "electrons," they must pass one to the next student.

Continue having the students pass the extra "electron." Then

have a volunteer become an "insulator" by clenching both fists. Point out to the students that since the student now cannot accept a clip, the flow of "electrons" stops.

REINFORCING THEMES
Energy

 Reinforce the theme of **Energy** by having the students list as many things as they can see in the classroom that use electricity. You might have the students repeat the procedure at a different location in the school or at home. Encourage the students to place their lists in their science portfolios.

> ❖ **Did You Know?**
> Only 2–3 percent of the energy supplied to an incandescent lamp is turned into visible radiation (light). Much of the remaining energy is infrared radiation (heat), which is invisible.

Figure 13–17. In a series circuit, current is the same throughout, as you can see by reading the ammeters shown here.

the energy of the electrons. Remember, the energy supplied by an electron is equal to the potential difference across the bulb. This is measured by connecting a voltmeter from one side of the bulb to the other. If the outlet supplies a potential difference of 120 volts, the voltmeters across each of the six bulbs will read 20 volts. You just divide the total 120 volts by the 6 bulbs that must share it to find each bulb getting 20 volts.

The potential differences in a series circuit have to add up to the potential difference of the whole circuit. Figure 13–18 shows the same circuit as Figure 13–17, but with voltmeters added to measure potential differences. Add up the potential differences across the resistances. Their sum must equal the potential difference supplied by the battery.

Figure 13–18. The potential difference within a series circuit varies with the amount of resistance. The sum of the potential differences equals the total potential supplied to the circuit.

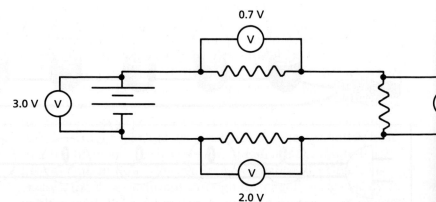

ONGOING ASSESSMENT
▼ ASK YOURSELF

1.5 amps; the ammeters are in series with the circuit.

▼ ASK YOURSELF
If you inserted two ammeters in a 120-volt, 1.5-amp series circuit, what would the reading on each ammeter be? Explain.

GUIDED PRACTICE

PORTFOLIO ASSESSMENT Have the students draw and label a diagram of a series circuit and a diagram of a parallel circuit. Encourage the students to place their diagrams in their science portfolios.

INDEPENDENT PRACTICE

Have the students provide written answers to the Section Review. In their journals, have the students write a few paragraphs about the safe use of electricity in their homes.

EVALUATION

Set up a simple parallel circuit that includes two or more lamps or other electric devices. Have the students determine whether the circuit is a series or a parallel circuit and explain their choice.

Parallel Circuits

As you've learned, a string of 100 twinkle lights in a series circuit with one bad bulb is a problem. The circuit diagram in Figure 13–19 shows how a string of ornamental lights is usually arranged today—in a parallel circuit. A **parallel circuit** has two or more separate paths through which electricity can flow. The current enters the circuit and travels to the first branching point. Some of the current goes through the first bulb. The rest continues past the branching point to the rest of the bulbs. At the second bulb, the circuit branches again. Eventually, all the current comes back together again before it goes back into the wall outlet. As you can see, if one bulb burns out, there is a path for the electricity to take around the open part of the circuit and the remaining bulbs stay lit. The electric outlets around your house and school are in a parallel circuit with each other. This is why you can use any outlet you want and it will provide the flow of electricity to power your appliance. In the next activity, you'll have a chance to make an electric circuit and to determine whether it is a series or parallel circuit.

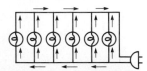

Figure 13–19. If one bulb in this parallel circuit burns out, the others will continue operating.

INTEGRATION– Engineering

A variety of symbols are used to diagram electricity. Share the symbols below with the students. You might draw each symbol on the board and have the students predict what it represents.

Circuit Symbols

ammeter	—Ⓐ—
battery	—⊦⊦------⊦⊦
bell	🔔
buzzer	⊓
cell	—⊣⊦—
lamps	—⊗— —◯—
loudspeaker	▷☰
microphone	▷☰
motor	—⊏Ⓜ⊐—
resistor	—Ɱ—
switches	—o⁄o— —o‾o—
voltage divider	—▭—
voltmeter	—Ⓥ—

RETEACHING

Tell the students to draw both series circuit and parallel circuit diagrams that include wires, batteries, two lamps, and a switch. Direct the students to label the parts of the circuits. Conclude by having them trace the flow of electricity through each circuit.

EXTENSION

Cooperative Learning Have groups of students research safety precautions people should follow when handling or using electric equipment or devices. Encourage the students to summarize their findings in the form of a pamphlet.

CLOSURE

PORTFOLIO ASSESSMENT Have the students make a chart that defines and gives examples of conductors, insulators, series circuits, and parallel circuits. Ask the students to place their charts in their science portfolios.

DISCOVER BY *Doing*

You might have the students exchange circuit diagrams and identify them as series or parallel circuits. Discuss any disagreements the students might have.

ONGOING ASSESSMENT
ASK YOURSELF

If the appliances were wired together in a series circuit, they could not be used independently of each other.

SECTION 3 REVIEW AND APPLICATION

Reading Critically

1. Remove one light bulb from the circuit, and plug in the lights. If the other bulbs do not light, it is a series circuit. If the other bulbs remain lit, it is a parallel circuit.

2. 4 V

Thinking Critically

3. very low current at high potential

4. 120 W

DISCOVER BY *Doing*

Gather two flashlight bulbs, a battery, a switch, and some wire. Put them together in such a way that both bulbs light up. Make a diagram of your circuit. Is it a series or parallel circuit? Now try to make the other type of circuit.

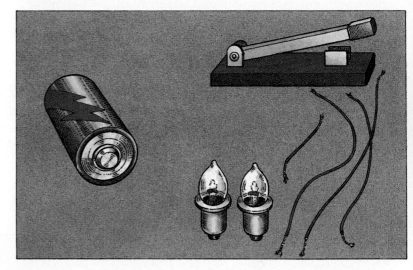

ASK YOURSELF

As you read, the outlets around your house are wired in parallel circuits. Why is that important? How would they work if they were all wired together in a series circuit?

SECTION 3   REVIEW AND APPLICATION

Reading Critically

1. How could you determine whether a string of lights is in a series circuit or a parallel circuit?

2. Three devices are connected in series to a 12-volt battery. The potential differences across the first two devices are 3 volts and 5 volts, respectively. What is the potential difference across the third device?

Thinking Critically

3. Under what conditions would you want to use a wire that has a very thin copper core and an extremely thick insulating cover?

4. Three electric devices are connected in series to a 120-volt circuit. In the first device, the current is 2 amps and the potential difference is 40 volts. The potential difference across the second device is 20 volts. How much power is used by the third device?

SKILL

Interpreting a Circuit Diagram

Process Skills:
Communicating,
Constructing/Interpreting
Models

Grouping: Individuals

Objectives
● **Observe** and **identify** relationships among devices and other parts of an electric circuit.
● **Interpret** ideas from a diagram of an electric circuit.
● **Communicate** ideas based on a drawing of an electric circuit.

Discussion
The circuit is the simplest possible series-parallel combination. While there is no reference to series-parallel combinations in the text, the students should be able to answer the questions by applying the most basic principles to the circuit. The students only need to know that current does not change until there is a branching point, currents add at a branch, a switch breaks the circuit, and parallel branches have the same potential difference.

Interpreting a Circuit Diagram

▶ MATERIALS
● pencil ● paper

▼ PROCEDURE

This circuit diagram shows a battery, three electric devices, four ammeters, and two switches in a circuit. The circuit is a combination of series and parallel arrangements. Study and analyze the circuit carefully.

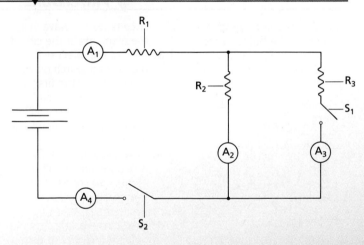

▶ Application
1. A_4
2. R_1 and R_2
3. All read zero.
4. R_3
5. A_2
6. R_2

✳ Using What You Have Learned

Students' diagrams should show two bulbs in parallel, both controlled by one switch in series.

▶ APPLICATION
Answer the following questions about the diagram.

1. If both switches are closed, which ammeter must have the same reading as ammeter A_1?

2. If switch S_1 is opened and switch S_2 is closed, which two devices will be in a series?

3. What will the ammeters read if switch S_2 is opened?

4. Which device is rated for the same potential difference as R_2?

5. Which other ammeter reading must be added to the reading on A_3 to give the reading on A_1?

6. Which device is in a parallel circuit with R_3?

✳ *Using What You Have Learned*
Two bulbs in a lamp are to be operated at their rated value. Both must be controlled by the same switch so that both of them are always either on or off. Draw a circuit diagram showing how the bulbs and the switch are connected to the wall outlet.

★ PERFORMANCE ASSESSMENT
At the completion of this *Skill,* encourage volunteers to share their diagrams with classmates. Check the students' diagrams for accuracy.

Have the students place their circuit diagrams and answers from the *Skill* in their science portfolios.

Making Series and Parallel Circuits

Process Skill:
Experimenting

Grouping: Groups of 3 or 4

Objectives
● **Construct** series and parallel circuits, and **study** their properties.

Pre-Lab
Have the students recall what they have learned so far about series and parallel circuits. Then explain to the students that in this Investigation they will construct series and parallel circuits and study the properties of each.

Hints
CAUTION: Warn the students that investigations of this type would be extremely dangerous if done using a household electric supply.

Procedure
2. One light shines; one circuit; battery, switch, bulb

3. Both bulbs shine at the original brightness; two complete circuits

5. Neither bulb shines; no complete circuit

6. Both bulbs shine dimly; one complete circuit

Analyses and Conclusions
1. the parallel circuit only

2. 3 V

3. The lights are dim in the series circuit.

4. Either switch, if open, breaks the circuit.

▶ Application
She should connect the instruments in a series circuit. Since there is only one path for current flow in a series circuit, the current will be the same at each instrument.

✳ Discover More
Adding a third bulb will make the bulbs in the series circuit

dimmer but will not affect the brightness of the bulbs in the parallel circuit.

Post-Lab
Discuss with the students what they learned from this Investigation.

INVESTIGATION

Making Series and Parallel Circuits

▶ MATERIALS
● 6-volt lantern battery ● 6-volt bulbs (2) ● sockets for flashlight bulbs (2)
● key switches or knife switches (2) ● wire leads with alligator clips (8)

▼ PROCEDURE

1. Follow circuit diagram A to connect the bulbs to the battery in a parallel circuit. Leave the switches open.

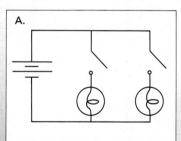

A.

2. Close one switch. What happens? How many complete

circuits do you have? List the elements in the circuit.
3. Predict the effect of closing the second switch on the brightness of the first bulb. Now close the second switch. What happens? How many complete circuits do you have?
4. Now disconnect all the wires. Reconnect all the elements in a series circuit, following circuit diagram B. Leave the switches open.
5. Predict the brightness of the bulbs in this circuit compared with the brightness of

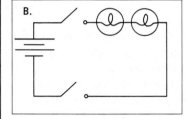

B.

the bulbs in the parallel circuit. Close one switch. What happens? How many complete circuits do you have?
6. Close the other switch. What happens? How many complete circuits do you have?

▶ ANALYSES AND CONCLUSIONS
1. Both the battery and the bulbs are rated at 6 volts. In which of the two circuits do the bulbs receive the potential difference for which they are designed?
2. When the bulbs are connected in series, what is the potential difference across each bulb?
3. What is the evidence that the voltage across each bulb in the series circuit is less than the voltage across each bulb in the parallel circuit?
4. In the series circuit, why do you have to close both switches to make the bulbs light?

▶ APPLICATION
In designing an experiment, an engineer finds that she must have exactly the same current in four different devices. How should she connect them? Explain.

✳ Discover More
Obtain a third bulb and socket. Predict the effect on the brightness of the bulbs if you include this bulb in a series circuit and then in a parallel circuit. Test your prediction.

PERFORMANCE ASSESSMENT

Cooperative Learning Ask each group how they might prove to someone not familiar with electricity and circuits that one circuit is in series and the other is in parallel. Check that each group's circuits are functional.

CHAPTER 13 HIGHLIGHTS

The Big Idea— Energy

The students should understand that both static electricity and current electricity are forms of energy. They can be converted to other forms of energy, such as heat and light. The transformability of electricity has led to the development of many useful devices and appliances.

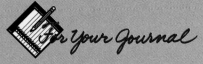

For Your Journal

The students' responses might include a description of charged particles, the movement of electrons in circuits, and the conversion of electricity to other forms of energy.

CONNECTING IDEAS

The concept map should be completed with the following words.

1. electrical charges
2. static electricity
3. current electricity
4. parallel circuit

CHAPTER 13 *H*IGHLIGHTS

The Big Idea

Energy takes many forms and has many roles. Both static electricity and electric current are forms of electric energy, one more usable than the other. Electric energy involves the buildup (static) or flow (current) of electric charges. Both types are the same in that the total charge is always conserved. The electrons themselves can be neither created nor destroyed. Electricity is a transformable energy that can be put to almost limitless use.

For Your Journal

Look back at the ideas you wrote in your journal at the beginning of the chapter. How have your ideas changed? Revise your journal entry to include more characteristics of electricity. What can electricity do, and how is it used?

Connecting Ideas

This concept map shows how the ideas in this chapter are related. Copy the concept map into your journal and complete it.

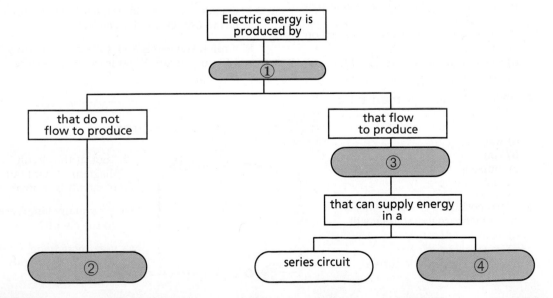

Understanding Vocabulary

1. (a) Static electricity is a buildup of an electric charge on an object; electric current is a flow of electrons.

(b) Direct current is the flow of current in one direction. Alternating current is the flow of current in alternating directions. Potential difference is a measure of the electrical potential energy in a direct current or alternating current system.

(c) A conductor is a material through which electric current can flow; an insulator is a material through which electric current does not flow. Resistance is a measure of how much a material opposes the flow of electricity.

(d) An electric circuit is the path electricity follows. In a series circuit, there is only one path for current to flow through. In a parallel circuit, electricity can flow through two or more separate paths.

Understanding Concepts

Multiple Choice

2. c

3. d

4. c

5. d

6. a

Short Answer

7. Since a single dry cell only produces about 1.5 V, eight cells are needed to produce 12 V.

8. 4.167 amperes

Interpreting Graphics

9. a series circuit

10. two lamps

11. Both switches are open. In order for the switches to be closed, the lines would have been unbroken.

CHAPTER 13
REVIEW

Understanding Vocabulary

1. Compare the meanings of the terms in each set.

   a) static electricity (343), electric current (348)

   b) direct current (350), alternating current (350), potential difference (352)

   c) resistance (356), conductor (355), insulator (355)

   d) electric circuit (355), series circuit (357), parallel circuit (359)

Understanding Concepts

MULTIPLE CHOICE

2. An electric iron is rated at 800 watts. This tells you the

   a) potential difference to which it should be connected.

   b) amount of current it takes.

   c) rate at which it uses electric energy.

   d) total amount of energy it uses.

3. When a vinyl strip is stroked with wool

   a) new negative charges are produced on the wool.

   b) new negative charges are produced on the vinyl.

   c) new positive charges are produced on the wool.

   d) the total amount of charge does not change.

4. Which unit measures the number of electrons passing through an electrical appliance per second?

   a) watt

   b) volt

   c) ampere

   d) ohm

5. If two positively charged objects are put close together, they will

   a) attract each other.

   b) have no charge.

   c) produce an electric current.

   d) repel each other.

6. The reason a switch must be placed in series with the device it controls is that

   a) when a series circuit is opened, the current stops.

   b) the switch must use some of the energy of the circuit.

   c) there must be a high potential difference across a closed switch.

   d) there must be no current passing through the switch.

SHORT ANSWER

7. Explain why a radio that works on 12 volts has to use several dry-cell batteries.

8. A microwave uses 500 watts of power on a 120-volt circuit. What is the microwave's amperage?

Interpreting Graphics

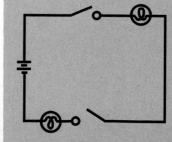

9. Look at the circuit diagram. What kind of circuit is shown?

10. How many lamps are in the circuit?

11. Are the switches open or closed? How can you tell?

Reviewing Themes

12. Each light bulb uses one-third of the total energy.

13. The opposition of a substance to the flow of electrons causes the energy of the moving electrons to be changed into heat energy. When enough heat is produced, the substance radiates light energy.

Thinking Critically

14. The charge travels through the person's body and deposits on the hair. The hair stands on end because all the hairs have the same charge and repel each other.

15. Since the two lamps carry different amounts of current, they must be connected in a parallel circuit.

16. Electric current flows through the wires inside a toaster. The resistance of the wires causes them to becomes heated, thereby toasting the bread.

17. Stroke the vinyl strip with the wool, which will make the vinyl negatively charged. Then, negatively charge the foam ball by contact with the vinyl. Hold the pen near the ball to see whether it is repelled or attracted. If attracted, the pen is positively charged. If repelled, it is negatively charged.

18. The wire is not covered with insulation because the air around the wire acts as insulation. This method of insulating wire is possible because the wire are elevated high enough to prevent accidental contact with animals or objects that provide an electrical path to the ground. Birds can safely sit on the wires because they are not grounded and thus do not provide an alternative path for current flow. However, very large birds (such as eagles) have been known to touch two wires at the same time and have been electrocuted. They were electrocuted because their bodies provided a path for current to flow between the wires.

Reviewing Themes

12. *Energy*
Three identical light bulbs are connected in a series circuit. What fraction of the total energy of the electrons in the circuit does each light bulb use?

13. *Systems and Structures*
Explain how resistance in conductors produces heat energy and light energy.

Thinking Critically

14. Explain how it is possible to make someone's hair stand on end by giving the person an electric charge.

15. Two lamps in a circuit carry currents of 1 ampere and 2 amperes, respectively. In what kind of circuit must the lamps be connected?

16. Explain how a toaster toasts bread.

17. How can you determine whether the charge on a pen is positive or negative, using only a vinyl strip, a piece of wool, and a suspended plastic-foam ball?

18. This transmission line carries current at a potential difference of 240 000 volts. The line is made of bare wire, kept isolated from the steel tower by ceramic insulators. Why isn't this wire covered with insulation? Birds can sit on this line and nothing will happen to them. Why is this so?

Discovery Through Reading

Elmer-Dewitt, Philip. "Mystery—and Maybe Danger—in the Air." *Time* 136 (December 24, 1990): 67–69. This article discusses evidence which suggests that exposure to electric devices can cause cancer in humans.

MAGNETISM

PLANNING THE CHAPTER

Chapter Sections	Page	Chapter Features	Page	Program Resources	Source
CHAPTER OPENER	366	*For Your Journal*	367		
Section 1: PERMANENT MAGNETS	368	Discover by Observing **(B)**	369	*Science Discovery**	SD
• A Magnetic Attraction **(B)**	368	Discover by Doing **(A)**	370	Extending Science Concepts: Mapping Magnetic Fields **(A)**	TR
• Making Magnets **(B)**	369	Activity: How can you make a permanent magnet? **(A)**	371	Investigation 14.1: Mapping Magnetic Fields with Iron Filings **(A)**	TR, LI
• Inside Magnets **(H)**	371	Discover by Writing **(B)**	373		
• Magnets in Use **(A)**	372	Section 1 Review and Application	373	Study and Review Guide, Section 1	TR, SRG
Section 2: MAGNETIC FIELDS	374	Discover by Doing **(A)**	375	*Science Discovery**	SD
• The Magnetic Earth **(B)**	374	Discover by Doing **(B)**	377	Reading Skills: Taking Notes **(B)**	TR
• The Fields Around Magnets **(A)**	376	Discover by Doing **(A)**	378	Thinking Critically **(A)**	TR
		Section 2 Review and Application	378	Magnets and Magnetic Fields **(A)**	IT
		Skill: Constructing Models **(A)**	379	Earth as a Magnet **(B)**	IT
				Study and Review Guide, Section 2	TR, SRG
Section 3: MAGNETISM AND ELECTRICITY	380	Discover by Doing **(A)**	382	*Science Discovery**	SD
• Electricity Makes Magnetism **(A)**	380	Section 3 Review and Application	383	Connecting Other Disciplines: Science and Language, Writing About Electromagnets **(A)**	TR
• Magnetism Makes Electricity **(A)**	382	Investigation: Studying Electricity and Magnetism **(A)**	384	Investigation 14.2: Analyzing the Magnetic Field Around a Current-Carrying Wire **(A)**	TR, LI
				Record Sheets for Textbook Investigations **(A)**	TR
				Study and Review Guide, Section 3	TR, SRG
Chapter 14 HIGHLIGHTS	385	The Big Idea	385	Study and Review Guide, Chapter 14 Review **(B)**	TR, SRG
Chapter 14 Review	386	For Your Journal	385	Chapter 14 Test	TR
		Connecting Ideas	385	Test Generator	

B = Basic   **A** = Average   **H** = Honors
The coding Basic, Average, and Honors indicates subsections, features, and resources that might be appropriate for different levels of learners. For additional suggestions regarding choice of topic and depth of coverage, see the Pacing Chart on pages T28–T31.

*Frame numbers at point of use
(TR) Teaching Resources, Unit 4
(IT) Instructional Transparencies
(LI) Laboratory Investigations
(SD) *Science Discovery* Videodisc Correlations and Barcodes
(SRG) Study and Review Guide

CHAPTER MATERIALS

Title	Page	Materials
Discover by Observing	369	(per individual) bar magnets (2), iron nails
Discover by Doing	370	(per individual) bar magnet, iron nails, paper clips
Activity: How can you make a permanent magnet?	371	(per group of 3 or 4) hacksaw blade, mounted compass needle, permanent magnet, paint
Discover by Writing	373	(per individual) journal
Discover by Doing	375	(per individual) astronomer's night light (or flashlight covered with a red filter), compass, long sticks (2), protractor
Teacher Demonstration	375	steel curtain rod, compass
Discover by Doing	377	(per individual) bar magnet, sheet of paper, iron filings
Discover by Doing	378	(per individual) bar magnets (2), sheet of paper, iron filings
Skill: Constructing Models	379	(per individual) paper, pencil
Discover by Doing	382	(per group of 3 or 4) 6-volt dry cell, 15 cm-long iron nail, 75 cm-insulated bell wire, paper clips
Investigation: Studying Electricity and Magnetism	384	(per pair) nail, 1.5 m wire, batteries (2), wire connectors (2), flashlight bulb and socket, paper clip, compass, tape

ADVANCE PREPARATION

Remember that all nails and paper clips will retain their temporary magnetism for a period of time following the magnet experiments in this chapter. This makes them unsuitable for reuse in new experiments. Use fresh nails and clips for each experiment, and then put those used during activities into a separate box. When you run out of fresh nails and clips, those in the boxes should have lost their magnetism.

TEACHING SUGGESTIONS

Field Trip
A trip to a local wilderness area or park would make a good compass activity. Try to schedule the trip for a cloudy day, as the students will not be able to rely on the sun for direction. Have students test their compasses from a variety of areas and note what things attract the compass needle besides magnetic north.

Outside Speaker
Most major businesses have some involvement with electromagnetism. Speaker suggestions include automotive workers and manufacturers of computers, stereos, and appliances. Ask the speaker to discuss how electromagnetism plays a role in his or her business.

CHAPTER THEME—*SYSTEMS AND STRUCTURES*

This chapter will help the students understand magnetism as a force found in and on Earth and used in a variety of ways. The students will also learn how the interaction between magnetism and electricity leads to our modern technology, particularly in the areas of transportation and electronics. A supporting theme in this chapter is **Environmental Interactions.**

CHAPTER 14

MAGNETISM

MULTICULTURAL CONNECTION

Although many early cultures discovered the magnetic properties of the lodestone (the mineral magnetite), it was the Chinese who put it to practical use. In 376 B.C., Haung Ti, a Chinese general, first used a lodestone as a compass. Compasses were used by the Chinese military for hundreds of years, but only on land. In the thirteenth century, the Chinese went to sea with compasses. Soon Arab nations acquired this useful device through trade.

MEETING SPECIAL NEEDS

Second Language Support

Have limited-English-proficient students illustrate some of the situations they will observe during the Chapter Motivating Activity. Encourage the students to share and explain their illustrations to classmates. Encourage the students to place their illustrations in their science portfolios.

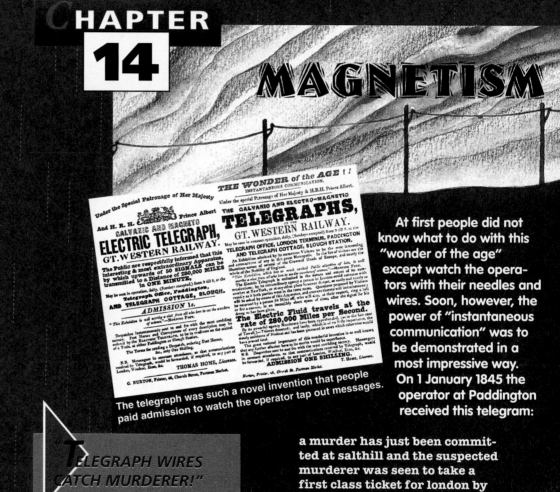

The telegraph was such a novel invention that people paid admission to watch the operator tap out messages.

At first people did not know what to do with this "wonder of the age" except watch the operators with their needles and wires. Soon, however, the power of "instantaneous communication" was to be demonstrated in a most impressive way. On 1 January 1845 the operator at Paddington received this telegram:

"TELEGRAPH WIRES CATCH MURDERER!"
Those might have been the London headlines in 1845 when a telegraph and its electromagnetic needles helped catch a dangerous criminal. The following is a historian's account of how the telegraph helped nab a murderer.

a murder has just been committed at salthill and the suspected murderer was seen to take a first class ticket for london by the trains which left slough at 7:42 a.m. he is in the garb of a kwaker with a brown greatcoat on which reaches nearly down to his feet. he is in the last compartment of the second first class carriage.

Display several kinds and sizes of magnets (including several lodestones) and objects that are both attracted and not attracted by the magnets. Have the students spend a short period examining these objects and then develop a statement about the objects' behavior. (Example: Magnets can hold nails and paper clips but not erasers or pencils.) Have the students keep track of their observations about the properties of magnets during the chapter and write their findings in their journals.

For Your Journal

The journal questions will help you identify the students' level of knowledge about magnetism. (The students will have a variety of responses to the journal questions. Accept all answers, and tell the students that they will have a chance to refer to the questions again at the end of the chapter.)

The operator could not make out what was meant by "kwaker," and sent a query back to Slough. He was told that "kw" stood for the letter "between p and r;" there was no "q" on the telegraph panel. He hurried to the police station and delivered the message. When the train arrived at Paddington, two plain-clothes policemen were on the look-out for the "kwaker." They shadowed him across London on a horse-bus, and finally arrested him.

The murder trial of John Tawell was the sensation of 1845. The policemen gave evidence how the telegraph delivered the suspect into their hands. Tawell confessed and was executed, and Londoners said: "Them cords [the telegraph wires] have hung John Tawell!"

from *A History of Invention*
by Egon Larsen

The first transatlantic cable successfully carried a message from Queen Victoria of England to U.S. President James Buchanan.

For Your Journal

- List as many uses for magnets as you can think of.
- What can these magnets do?
- How do you use magnets every day?

ABOUT THE INVENTION

The early telegraph consisted of several small electromagnets that controlled six magnetic needles and a disk on which the letters of the alphabet were printed. A transmitter sent an electric current through the wires of the electromagnets, setting up a magnetic field. Each letter sent caused a needle to point to that letter on the dial. This was very likely the device, with its "needles and wires," used to catch the murderer.

The person who perfected the telegraph was not a scientist but an American artist. Samuel Morse (1791–1872) gained fame as a portrait painter. Some of his paintings of historical figures are still displayed in public buildings. After his wife died, he stopped painting and traveled to Europe. On his return to America, he saw something that inspired him to change careers.

A physician on board the ship entertained the passengers with an electromagnet. He showed that a piece of soft iron would become a temporary magnet when current was sent through wire around it. However, it would become nonmagnetic again when the current was stopped. Morse watched closely and suddenly had an idea: If electricity could be made visible in any part of an electric circuit closed by a switch, then couldn't signals also be transmitted by electricity? The artist became an inventor.

FOCUS

This section concentrates on the force of magnetism, the properties of temporary and permanent magnets, and the materials used to make them. It also presents some of the uses of magnets.

MOTIVATING ACTIVITY

Cooperative Learning Have the students work in groups with lodestones to figure out how early sailors might have used the stone as a compass. Encourage each group to invent a way in which the stone can be used. Have the students write their results in their journal.

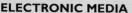

PROCESS SKILLS
• Observing
• Communicating

POSITIVE ATTITUDES
• Skepticism • Curiosity

TERM
• magnetism

PRINT MEDIA
Hidden Attraction: The Mystery and History of Magnetism by Gerrit L. Uerschuur (see p. 337b)

ELECTRONIC MEDIA
Magnetism and Fields of Force Barr Films (see p. 337b)

SCIENCE DISCOVERY
• Magnet lifts spoon
• Magnetic declination

BLACKLINE MASTERS
Extending Science Concepts
Laboratory Investigation 14.1
Study and Review Guide

MULTICULTURAL CONNECTION

Remind the students that when Columbus arrived, America was not unpopulated. In all the Americas, there were more than 70 million people. In addition, the Native Americans at Columbus' time were not primitive. Many tribes had a simple way of life; others had a complex society. You might have the students do research about the advanced civilizations of the Mayas, the Aztecs, or the Incas and discuss that civilization's contribution to science.

SECTION 1

Permanent Magnets

Objectives

Describe the properties of a magnet.

List some uses of permanent magnets.

Use a model to explain the properties of magnets.

When Christopher Columbus first set out on his voyages, he was looking for a shortcut to Asia. He didn't even know he was lost when he found the Americas. How did any of these early explorers find anything?

The explorers did have a few things going for them. One of these was the lodestone, a natural magnet that was used on sailing ships as part of a compass. By looking at the compass, Columbus knew immediately which direction was north, even if he didn't know precisely where he was.

Figure 14–1. Columbus used a compass as he sailed west in search of Asia.

A Magnetic Attraction

A compass is just a small magnet. A magnet is a material that attracts certain materials, such as iron, nickel, and cobalt. Any bar magnet that is free to rotate will act like a compass. The magnet will swing back and forth until it comes to rest lined up approximately north and south. For example, a bar magnet demonstrates magnetic force, or **magnetism,** a force of repulsion or attraction between like or unlike poles. The end that points north is called the north-seeking pole, or north pole, and is marked "N." The south-seeking pole, or south pole, is marked "S." You can investigate the properties of magnetism by trying the next activity.

● **Process Skills:**
Communicating, Inferring

Thirty-two points or directions were marked on old-style compasses. Sixteen points are more common. North, south, east, and west are called the cardinal points. Northeast, southeast, southwest, and northwest are the intercardinal points.

Between north and northeast is a direction called north-north-east, and between northeast and east is east-northeast. Ask students to name the direction on the 16-point compass that would fall between south and southwest (south-southwest) and between west and north-west (west-northwest). The stu-

dents might make a complete list or diagram of the 16 directions for reference: north, north-northeast, northeast, east-northeast, east, east-south-east, southeast, south-south-east, south, south-southwest, southwest, west-southwest, west, west-northwest, north-west, north-northwest.

DISCOVER BY *Observing*

Obtain two bar magnets and some iron nails. Observe how the two north poles of the bar magnets behave when brought near each other. Do the same with the south poles. Now see what happens when a north pole and a south pole are brought close together. Pick up a single nail with one pole of a magnet so that only one end of the nail is attached to the magnet. Then try picking up a second nail by touching it with the hanging end of the first nail. What can you infer about a nail that is touching a magnet? Record your observations in your journal.

In the activity, you should have discovered that an N pole and an S pole attract each other. On the other hand, the N pole of one magnet repels the N pole of another. Likewise, an S pole repels another S pole. Magnetic force both attracts and repels.

▼ ASK YOURSELF
Why are the poles of a magnet marked "N" and "S"?

Making Magnets

Where would an explorer be without a compass? He or she would probably be lost! If Columbus had lost his compass, could he have made a new one? Yes, if he knew what materials he needed to make a magnet. Try making a magnet yourself.

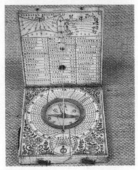

Figure 14–2. A variety of historical compasses

DISCOVER BY *Observing*

The students should infer that the nail also becomes a magnet. You might extend the activity by asking the students to predict what happens to the nail's mag-netism when it is no longer attached to a magnet. (The nail will lose its magnetism.) Then have the students test their predictions.

ONGOING ASSESSMENT
▼ **ASK YOURSELF**

Magnets are marked with "N" and "S" to show which pole the magnets are seeking.

INTEGRATION– *History*

Suggest that the students choose a particular place in the world, research what was happening there in 1492, and write about how that place has changed as a result of Columbus' exploration. Have the students place their research in their science portfo-lios. Students might consult the following sources.

Fritz, Jean. *The World of 1492.* Holt, 1992.

Hawke, Sharryl, and James E. Davis. *Seeds of Change: The Story of Cultural Exchange After 1492.* Addison-Wesley, 1992. (A teacher's guide is also available.)

Ventura, Piero. *1492: The Year of the New World.* Putnam, 1992.

SECTION 1 **369** ◀

TEACHING STRATEGIES, continued

● **Process Skills:** *Predicting, Observing*

Ask the students what they think will happen if a permanent magnet picks up one paper clip and that clip touches another. (The permanent magnet will temporarily magnetize the first paper clip so that it attracts the second.) Have them experiment to find out. How many clips can be picked up in this way? On what does that depend? (The number will vary. It depends on the strength of the magnet.)

● **Process Skill:** *Inferring*

Ask the students to explain why ceramic magnets are used at times even though they break easily. (They are less costly to manufacture, they work well when protected, as on a refrigerator door, and more powerful magnets are not needed with certain products.)

● **Process Skill:** *Communicating*

You might share the following book with the students. It provides detailed drawings and explanations that show how Columbus was able to make his voyages.

Yue, Charlotte and David Yue. *Christopher Columbus: How He Did It.* Houghton Mifflin, 1992.

✧ Did You Know?

Columbus navigated by a method called dead reckoning. To determine how far, fast, and in what direction his ships had traveled each day, he used a compass, an hourglass, and a long rope tied with knots at equal spaces. The compass showed the direction he was sailing. To measure speed, a log was attached to the end of the rope. The log was thrown overboard as the ship sailed, and the hourglass was turned over. By counting how many knots slipped over the railing in a half minute, sailors determined the ship's speed in "knots"—a term still used by sailors today. Knowing his speed and relying on his instinct, Columbus had an idea of the distance he had traveled each day.

THE NATURE OF SCIENCE

Have the students look at Figure 14–3. Tell the students that according to legend, a shepherd boy named Magnus once rested the iron-tipped end of his staff on a stone. When he tried to move the staff, it seemed to be stuck. This stone was the natural magnet now called lodestone or magnetite.

One of the earliest scientists to experiment with lodestone was a Greek named Thales (640–546 B.C.), who lived near Magnesia in Asia Minor and called the lodestone a magnetic stone.

Figure 14–3. This map shows the location of Magnesia, where lodestone, also known as magnetite, was discovered. Lodestone is a naturally occurring magnetic material made of mixed oxides of iron.

DISCOVER BY Doing

Obtain a bar magnet, an iron nail, and some paper clips. Stroke the iron nail with one end of the bar magnet. Stroke the nail only in one direction. Do this at least 50 times. Then see if the nail will attract the paper clips. How many paper clips can you lift with the nail?

As you discovered, some metals, such as soft iron, can easily be changed into magnets. However, because they also lose their magnetism easily, they are known as *temporary magnets*. Harder metals, such as steel, are more difficult to magnetize but tend to keep their magnetism better. A magnet made of materials that keep their magnetism is called a *permanent magnet*. Permanent magnets are made of a mixture of iron, aluminum, nickel, and cobalt.

Ceramic magnets are made of mixed oxides of iron and other metals. They are light in weight, and they break easily. Soft, flexible magnets, such as the ones you may have on your refrigerator, are made of powdered magnetic ceramics embedded in rubber.

How might Columbus have made a magnet that would stay magnetized long enough to get him home? In the following activity you can find out.

Have the students answer the following in their journals:
Which part of a compass is a permanent magnet? (the needle) Why must the rest of the compass be made of nonmagnetic material? (If another part of the compass were made of magnetic material, the needle would always point to it, making the compass useless as a directional finder.)

ACTIVITY

How can you make a permanent magnet?

Process Skills: Observing, Communicating

Grouping: Groups of 3 or 4

Discussion

Since a compass needle will be attracted to an unmagnetized piece of steel, only repulsion will provide proof that the hacksaw blade is magnetized.

Procedure

5. N and S; the end of the strip opposite the paint spot will be an S pole.

▶ **Application**

1. Every piece always has an S pole at one end and an N pole at the other.

2. Iron and copper cannot be magnetized permanently.

ACTIVITY

How can you make a permanent magnet?

MATERIALS
hacksaw blade, mounted compass needle, permanent magnet, paint

PROCEDURE
1. Test the hacksaw blade to see whether it is magnetized. You can do this by bringing one end of it near the compass needle. If it attracts, try the other end. Magnetization is indicated when you find repulsion.
2. Stroke the hacksaw blade in one direction only against one of the poles of the permanent magnet. Test it again to see whether the blade is now magnetized.
3. Using the compass needle as a reference, determine which end of the blade is the N pole. Mark the N pole with a spot of paint. (Remember, the N of the compass is the north-seeking end and will be attracted to the south pole of your hacksaw-blade magnet.)
4. Break the hacksaw blade in half. Mark the end that is closest to the blade's N pole.
5. Test the piece of blade with the marked pole against the compass needle. What kinds of poles does it now have?
6. Test the other half of the hacksaw blade.
7. Repeat steps 4, 5, and 6.

APPLICATION
1. In your journal, describe the poles that are always present after you have broken the magnetized blade. Label all the ends of the broken pieces "N" or "S."
2. Why is it necessary to use something made of steel for this demonstration rather than iron or copper?

★ PERFORMANCE ASSESSMENT

After the completion of this *Activity*, have individual students explain to you what would happen if they were to cut a permanent magnet in two along the lines of magnetization rather than across them. (There would be two magnets, each the same length and half the width of the previous one.) Evaluate the student's response for an understanding that cutting does not change the orientation of the atoms in the magnet.

ONGOING ASSESSMENT
▼ **ASK YOURSELF**

Magnetism is a force that attracts and repels objects with like and unlike poles.

▼ **ASK YOURSELF**
What is magnetism?

Inside Magnets

If you dip a bar magnet into a box of small iron nails, it will come out with nails clinging to it. Notice in the photograph that many nails are clinging to one end of the bar magnet.

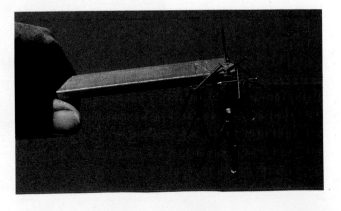

● **Process Skills:**
Communicating, Inferring

As the students look at Figure 14–6, share the following information with them. A tape recorder records sound on a tape coated with powder that contains a magnetic substance. Sound is changed into varying electric currents by a microphone. The current goes into the recording head and is changed into a varying magnetic force, which then is recorded on the tape. When the tape is played, magnetic variations on the tape yield an electric current that speakers then convert into sound.

Review the characteristics of magnetic poles with the class, and have the students use a model to explain why the strength of a magnet is at the poles.

 Have the students provide written answers to the Section Review. In their journals, have them note two areas of their homes in which the use of magnets is necessary.

MULTICULTURAL CONNECTION

Tseng Kung-Liang's *Compendium of Important Military Techniques* in 1084 A.D. described magnetized iron "fish" that floated in water and could be used for finding south. About this time the Chinese began to use the compass for navigation, most likely using the iron "fish." In 1086 A.D., Shen Kua's *Dream Pool Essays* refered to the use of a magnetic compass for navigation.

ONGOING ASSESSMENT
▼ **ASK YOURSELF**

At one end of a magnet, the N poles of many of the atoms do not face the S poles of other atoms. At the other end of the magnet, the S poles of many of the atoms do not face the N poles of other atoms.

LASER DISC
32632, 32633–32744

Magnet lifts spoon

543

Magnetic declination

MULTICULTURAL CONNECTION

In 1050 B.C., the Duke of Chou in China built an early magnetic compass called a *south-pointing carriage.*

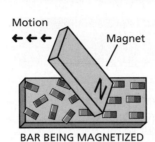

Motion
←←←
Magnet

N

BAR BEING MAGNETIZED

BAR MAGNETIZED

Figure 14–4. Iron bars contain atoms that have magnetic poles. When a magnet is brought close to an iron bar, the poles of the atoms all point in the same way.

The model shown here explains why a magnet's force is strongest at the poles. Within a magnetized rod, there are many N and S poles facing each other. They attract each other but have little effect on any outside object. At one end of the magnetized rod, there are many N poles that are not facing any S poles; this produces a strong N pole at the end of the rod. At the other end, lots of S poles face nothing but air. This end becomes the S pole of the rod. If the rod is cut, its atoms, still lined up in the same way, will continue to produce N and S poles at the ends of the shorter pieces.

 ASK YOURSELF
Why does every magnet have two poles?

Magnets in Use

Columbus needed a magnet for his compass, and of course you need magnets to hold notes and coupons on the refrigerator. Are magnets really that important or useful today?

You may be surprised to find out that a magnet might have provided you with a quick, hot meal recently. Every microwave oven contains a magnet. The poles of the magnet face each other, and a vacuum tube that produces electrons is between the poles. The magnetic field causes the electrons to follow a circular path. As they travel, they lose kinetic energy. This energy is used to produce the microwaves that cook food.

Your home is probably full of magnets, and you don't even know it. Electric clocks, motors, stereos, loudspeakers, and televisions all contain magnets. One magnet that is easy to find is the magnetic catch that holds your refrigerator door closed. In

Figure 14–5. Microwave ovens contain strong magnets that guide the flowing electrons that produce the microwaves.

EVALUATION

Ask the students what would happen if many additional natural materials besides magnetite were magnetic. Would this make the use of magnetic compasses easier or harder? Why? (Students' responses should demonstrate an understanding of the concepts. Accept all reasonable responses.)

RETEACHING

Have the students draw models in their journal, showing how magnets attract and repel each other. Ask the students to place their drawings in their science portfolios.

EXTENSION

Have interested students do research to find out how a gyrocompass differs from a magnetic compass. The students should also find out why and where such compasses are used.

CLOSURE

Cooperative Learning Have the students work in small groups to find three main ideas and three supporting details to summarize this section. Encourage the groups to share their ideas and details with other groups.

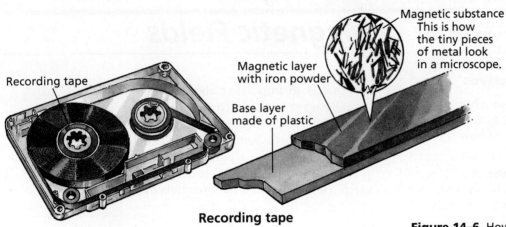

Recording tape

Magnetic substance This is how the tiny pieces of metal look in a microscope.

Magnetic layer with iron powder

Base layer made of plastic

Recording tape

Figure 14–6. How magnets are used in recording tape

fact, magnets are much more important today than they were in Columbus's time. Try the following activity to get an idea of how you would be affected if all magnets suddenly disappeared.

DISCOVER BY *Writing*

Think about the uses that you just read about for magnets around the home. Find other ways that you use magnets in your home. Then, in your journal, write about an imaginary day when none of the magnets in your home worked. How were you affected? What discoveries about magnetism did you make?

ASK YOURSELF

What are some uses for magnets in the home?

SECTION 1   REVIEW AND APPLICATION

Reading Critically

1. If you take a bar magnet and cut it into three pieces, what is the total number of magnetic poles you will have? Explain.
2. What kinds of magnetic poles repel each other? What kinds of poles are attracted to each other? Why?

Thinking Critically

3. Suppose you want to test an iron rod with a compass to find out whether it is magnetized. Why would you have to show repulsion before you can be sure the rod is a magnet?
4. Suppose you had a bar magnet and you drilled a hole in it. Into the hole you put a compass. Toward which pole of the magnet would the N pole of the compass point? Explain your answer.
5. Why should you never expose a magnet to high temperatures?

DISCOVER BY *Doing*

You might have the students place their *Discover by Writing* themes about an imaginary "magnetless day" in their science portfolios.

ONGOING ASSESSMENT
ASK YOURSELF

Magnets hold cupboard and refrigerator doors closed; they store information on tapes and disks; and they are parts in televisions, telephones, and radios.

SECTION 1 REVIEW AND APPLICATION

Reading Critically

1. Six; after each cut, the poles of the atoms at the ends will not face opposite poles of other atoms.

2. Similar poles (such as two north poles) repel, and opposite poles (such as one north and one south pole) attract.

Thinking Critically

3. An unmagnetized rod will attract either pole of a compass, so attraction proves nothing.

4. The N pole of the compass points to the N pole of the magnet. Every field line forms a closed loop, entering at or near the S pole of the magnet and continuing through the metal to emerge near the N pole.

5. High temperatures cause the atoms to vibrate rapidly. This will bring them out of alignment and destroy the magnetization.

SECTION 1   **373** ◀

FOCUS

Section 2 describes a magnetic field and magnetic lines of force and tells how they can be identified using different kinds of magnets. This section also distinguishes geographic poles from magnetic poles. In addition, Earth's magnetic lines of force are illustrated.

MOTIVATING ACTIVITY

Cooperative Learning Have the students work in small groups to list some of the things they encounter every day that are invisible. (Examples include heat, sound, electricity, and X-rays.) Point out that we have different ways to measure and record these forces, such as thermometers, tape recorders, voltmeters, and film, and that magnetism and magnetic fields are two such phenomena. Conclude by telling the students that they will learn about magnetic fields in this section.

PROCESS SKILLS
• Constructing/Interpreting Models • Expressing Ideas Effectively

POSITIVE ATTITUDES
• Curiosity • Creativity

TERM
• magnetic field

PRINT MEDIA
"Giant Moments" from *Discover* magazine by Joshua Fishman (see p. 337b)

ELECTRONIC MEDIA
What Is Magnetism? Encyclopaedia Britannica Educational Corp. (see p. 337b)

SCIENCE DISCOVERY
• Magnetic field

BLACKLINE MASTERS
Reading Skills
Thinking Critically
Study and Review Guide

SECTION 2

Magnetic Fields

Objectives

Explain *what is meant by a magnetic field and how it is detected.*

Describe *the magnetic field of the earth.*

Diagram *the magnetic fields around different kinds of permanent magnets.*

What is so important about north? Columbus didn't actually want to go north, but that is where the compass always pointed. Why do compasses point north? What is attracting the magnet in the compass? It must be something big that could make every compass everywhere point north. There really is only one answer. Only the earth itself could affect all magnets everywhere.

The Magnetic Earth

When Columbus returned from his voyages, he noted that in the lands he had visited, his compasses did not point directly north. The "north" that Columbus meant was north according to the direction a compass in Europe pointed. It seemed that north was a somewhat different direction in the lands Columbus visited.

Today it is known that compasses do not point to the *geographic* North Pole. Earth's axis, the imaginary line it spins around, goes through the geographic North Pole. Compasses point to Earth's north *magnetic* pole. This point is in northern Canada more than 2000 km away from the geographic pole. And the direction to the north magnetic pole depends on your location on Earth. You can see what difference, if any, exists between true north and magnetic north for your location by doing the following activity.

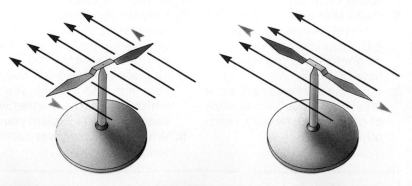

Figure 14–7. The needle of a compass will stop turning only when it is aligned with the earth's magnetic field.

● **Process Skills:** *Interpreting Data, Inferring*

Explain to the students that since the north magnetic pole is located a distance from the north geographic pole, there are few places on Earth where the poles line up in one direction. Ask the students to determine where, on most of Earth, the compass needle points in relation to the north geographic pole. Have them use a globe for assistance. (A compass needle will point either east or west of true north. For example, in California the north geographic pole lines up west of the north magnetic pole.)

● **Process Skill:** *Communicating*

You may wish to share the following information with students. Birds have been found to contain a small grain of magnetite—a magnetic crystal—between their brain and their skull. This might explain many birds' startling ability to migrate the same long distances each year. The magnetite might allow the birds to "read" Earth's magnetic field.

Scientists have recently discovered magnetite in human brain tissue. Why our brains make the mineral and its biological function remain a mystery.

DISCOVER BY *Doing*

Obtain a compass. Use it outdoors at night to find magnetic north. Use a stick to mark the direction of magnetic north. Look at the diagram shown here, and then find Polaris, the North Star. Polaris is directly over the geographic North Pole, or true north. Over the stick that indicates magnetic north, place a second stick to mark the direction of geographic north. Compare the two sticks. Use a protractor to measure the number of degrees difference between magnetic north and geographic north. Write your measurements in your journal. What do you know now about the difference between magnetic north and geographic north? ✐

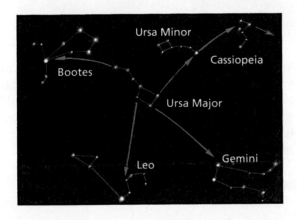

The Two Norths

In many places the difference between the two norths is so slight that it can be ignored when using a compass. But in some places the difference is great enough to cause problems in navigation. Imagine how annoyed explorers Admiral Peary and Matthew Henson would have been if they had not reached the geographic North Pole but the magnetic north pole—a difference of 2000 km! Fortunately, maps used for navigation and in hiking usually indicate this difference so that adjustments can be made when choosing a direction.

Feeling the Pull

No matter where you are on Earth, something causes the N pole of a compass to point one way and the S pole to point the other way. Any region in which magnetic forces are present is a **magnetic field.** The earth itself is surrounded by a magnetic field. When a compass needle comes to rest, its poles are lined up with Earth's magnetic field.

Magnetic fields are represented in diagrams by magnetic lines of force. These lines identify the position and strength of

Figure 14–8. A compass rose shows magnetic north.

DISCOVER BY *Doing*

Using a regular flashlight will destroy the students' night vision and complicate the activity. Acquire some sky observer lights used by astronomers, or have the students cover their flashlight with a red filter. Students' responses should indicate that magnetic north and geographic north are not located in exactly the same place.

Demonstration

Hold a steel curtain rod, pointed north-south, at a 70° angle downward to the north. (Note: In most of the United States, Earth's magnetic lines of force are positioned at about this angle with the horizontal.) Test the rod with a compass needle to make sure it is not magnetized. Now tap the rod about 12 times. Test again to see whether it is magnetized. Try reversing the rod and tapping again. Ask the students why tapping a magnetized steel rod in a east-west direction will demagnetize it. (The tapping shakes the magnetized atoms in the rod out of line. The east-west position keeps Earth's lines of force from entering and causing the atoms to line up in a north-south direction.)

● **Process Skills:** *Interpreting Data, Inferring*

Tell the students that Earth's magnetic poles have not always been as they are: the magnetic north and south poles have reversed many times. Scientists know this from studying igneous rocks that contain magnetic minerals. The direction of the

magnetic poles at the time the rock was formed is recorded in the minerals. Within the last four million years, there have been four main periods of normal and reversed polarity. Scientists use this knowledge to determine the relative ages of the rocks on the spreading ocean floors.

● **Process Skills:** *Interpreting Diagrams, Evaluating*

Have the students describe what the magnetic lines of force look like around a magnet. They can refer to Figure 14–12, if necessary. Then ask the students what happens to the lines of force once they have pushed out into the space around a

magnet. (They stay there without moving until disturbed by the presence of other iron objects or magnetic fields.)

THE NATURE OF SCIENCE

As scientists attempt to discover the source of Earth's magnetism, they have proposed this theory: Earth's core is made up of iron and nickel. The liquid outer part of the core contains matter that flows and causes an electric current. This electric current is responsible for a magnetic field. Since the material that contains the electric current is everchanging, the magnetic field likewise shifts.

Figure 14–9.
Geographic north and magnetic north are in different locations. Geographic north is at the earth's point of rotation.

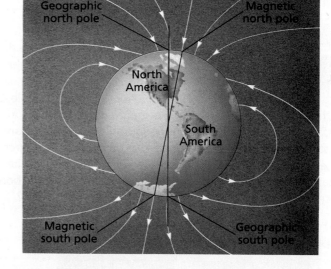

the magnetic field around an object. Figure 14–9 shows magnetic lines of force drawn to show the magnetic field around Earth. They tell which way the N pole of a compass would point from any given location. They also tell how strong the field is. The field is strongest where the lines are closest together. Where ① is Earth's magnetic field strongest?

▼ **ASK YOURSELF**
How would a compass behave south of the equator? Explain your answer.

The Fields Around Magnets

If all magnets are affected by the earth's magnetic field, then what happens to each magnet's own magnetic field? If you move a compass around a magnet, the compass needle turns. This happens because the compass is affected by the magnetic field produced by the magnet. At any location in the field, the compass needle comes to rest in line with the magnetic lines of force. Look at the photograph that shows several compasses around a bar magnet. Notice that each needle points toward a pole of the bar magnet.

① Earth's magnetic field is strongest at the magnetic poles.

ONGOING ASSESSMENT
▼ **ASK YOURSELF**

A compass needle would still point to the north magnetic pole, because this pole remains the same below the equator as it does above the equator.

INTEGRATION–
Earth Science

Earth's magnetic field causes beautiful lights in the sky known as the *aurora borealis*. These lights are seen close to Earth's north magnetic pole. Similar displays are seen close to Earth's south magnetic pole.

Figure 14–10. The needle of each compass shows the direction of the magnetic field around the magnet.

GUIDED PRACTICE

Have the students look at Figure 14–12 showing the lines of force, and ask them to predict the lines of force on magnets with other shapes.

INDEPENDENT PRACTICE

 Have the students provide written answers to the Section Review. In their journals, have the students write a short paragraph that distinguishes the geographic and magnetic poles of Earth.

EVALUATION

Ask the students whether an object's magnetic field and its lines of force are the same. Why or why not? (A magnetic field really exists, even though it is invisible. Magnetic lines of force are a way to represent a magnetic field in a diagram.)

Like the earth's field, the magnetic field around a bar magnet can be represented by magnetic lines of force. These lines are not visible, yet they can be used to predict which way a compass needle will point. You can investigate the magnetic lines of force around a bar magnet in the following activity.

Obtain a bar magnet, a sheet of paper, and some iron filings. Place the bar magnet under the sheet of paper, and sprinkle iron filings on the paper. Gently tap the paper so that the iron filings line up with the magnetic lines of force of the magnet. In your journal, make a sketch showing the pattern made by the iron filings. ✒

Is the Force Field Really There?

The magnetic force of the bar magnet made little magnets of the iron filings. These magnetic filings lined up, forming long, thin chains in the magnetic field of the bar magnet.

The illustration below shows a diagram of the field around a bar magnet and a horseshoe magnet. The arrowheads show the direction of the magnetic lines of force, which come out of the N poles of the magnets and enter the S poles. The concentration of magnetic lines of force at the poles shows that the field is strongest there.

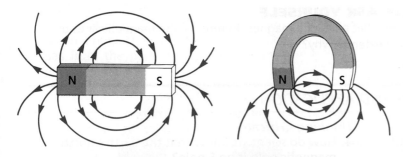

Do Not Cross!

Magnetic lines of force never cross each other. When two or more magnets produce fields that overlap, the result is a combined field. Figure 14–13 shows the fields produced when the poles of two bar magnets are brought near each other. If opposite poles are aligned, most of the lines run from the N pole of one to the S pole of the other. Investigate the overlap of magnetic fields for yourself in the next activity.

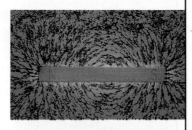

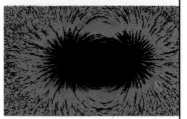

Figure 14–11. These iron filings show the magnetic field that surrounds the magnet.

Figure 14–12. These diagrams show the magnetic lines of force around a bar magnet and around a horseshoe magnet.

DISCOVER BY *Doing*

 If there is a surplus of iron filings available, the students might make permanent prints of their magnetic field lines by spraying the paper with an art fixative. The students might also obtain clear plastic bags that seal tightly. By filling the bags with iron filings, the students can manipulate their experiment more easily.

REINFORCING THEMES
Systems and Structures

Use Figure 14–9 to show how different systems of measurement interact with each other to tell us about Earth and our place on it. Point out that all these systems consist of imaginary lines put in place by human beings, although magnetic north and Earth's magnetic field really exist in and around Earth.

 LASER DISC
34144

34145–34290

Magnetic field

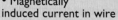
SECTION 3

Magnetism and Electricity

Thanks to a compass and Earth's magnetic field, Columbus was able to make several voyages to the Americas. Not only was he able to get back home, but he also was able to find the Americas more than once. Compasses are still useful today, but there are more powerful magnets that are even more useful. It was 300 years after Columbus sailed the Atlantic that these powerful magnets were discovered.

Objectives

Recognize that magnetic fields and electric currents are related.

Describe the structure of an electromagnet, and list some of its uses.

Analyze the relationship between a magnetic field and an electric current.

Electricity Makes Magnetism

In 1820 Professor Hans Christian Oersted of Denmark was giving a lecture on electricity to his students. He happened to have a compass nearby. When he closed a switch to demonstrate the flow of current, he noticed that the compass needle turned. That seemed odd to Oersted. After all, weren't magnets affected only by other magnetic fields? Oersted realized that the electric current must be generating a magnetic field.

Look at the setup in the illustration to see how this magnetic field can be detected. Before the wires are connected to the battery, all of the compasses point in the same direction. To what magnetic field are the compasses responding? When the wires are connected to the battery, current flows in the wires, moving upward through the tube in the platform. The compasses on the horizontal platform show the direction of the magnetic field at various points. The magnetic lines of force are shown in red. They form circles around the wire. Notice what happens when the connections to the terminals of the battery are reversed. How would you explain this in terms of magnetic poles? ①

Figure 14–14. Compasses are used here to detect the magnetic field around an electric current. Notice how the compasses behave when the battery is disconnected or connected as shown.

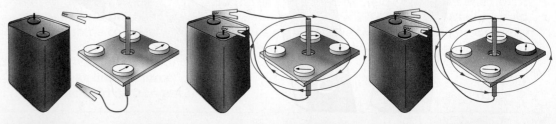

TEACHING STRATEGIES

● **Process Skills:** *Comparing, Interpreting Data*

Direct the students to examine Figure 14–15 and compare the compass readings when the electric current passes in different directions through the wire. Then ask volunteers to describe what actually happens to the compass needle in each case.

(The compass needle aligns itself with the lines of force at any particular point.)

● **Process Skills:** *Observing, Inferring*

Point out to the students the detail of the wrapped wire in the photograph of the solenoid. Why are the wires wrapped so closely together? (The closer the wires are to each other, the greater the number of coils that can be made. More coils make a stronger magnet.)

● **Process Skills:** *Evaluation, Solving Problems*

Ask the students whether an electromagnet is a temporary or permanent magnet, and have them explain why. (An electromagnet is a temporary magnet. It only acts as a magnet when an electric current is operating through it.)

Cranking It Up

To produce an even stronger magnetic field, the wire can be wound into a spiral. A wire spiral through which a current can flow is called a **solenoid.** When a current flows through a solenoid, the magnetic fields produced by the coils of wire combine. The result is a strong magnetic field in the center of the solenoid. Increasing the number of coils in the solenoid increases the strength of the magnetic field. Increasing the current flowing through the coils also strengthens the magnetic field.

Look at the photograph of a solenoid and the diagram of its magnetic field. If you compare the solenoid's magnetic field with the field around a bar magnet, you will see that the fields are exactly alike. The lines of force come out of the N pole of a solenoid and go in at the S pole.

Electromagnets

If an iron core is placed inside a solenoid and a current is passed through the coil, the field of the solenoid acts on the atoms in the core. They line up with the solenoid's field, making the field hundreds of times stronger. A solenoid with a core is called an **electromagnet.** If the core is made of a material that can be made into a permanent magnet, the solenoid's field will turn it into one.

Walking on the ceiling, like the person in this photograph, might be fun, although it is not particularly useful. Turning a magnetic field on and off at will is useful, though. That's really one of the things you're doing when you turn on an appliance. Electromagnets operate motors that run appliances. Electromagnets are also used in television sets, circuit breakers, and meters that measure the use of electric energy.

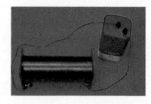

Figure 14–15. When current passes through a solenoid (top), the current produces a strong magnetic field (bottom).

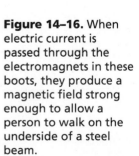

Figure 14–16. When electric current is passed through the electromagnets in these boots, they produce a magnetic field strong enough to allow a person to walk on the underside of a steel beam.

SCIENCE TECHNOLOGY SOCIETY

Electromagnetism is the basis for much of today's technology. It makes possible our systems of transportation, appliances, home entertainment, computers, and many other industries. When Professor Oersted made his serendipitous discovery, electricity and magnetism were still thought to be separate and distinct forces. The importance of Oersted's work is shown by the fact that the short pamphlet describing his experiment and discovery was translated into five different languages within weeks of its publication.

MEETING SPECIAL NEEDS

Gifted

Have advanced students do research on the use of electromagnets in automotive ignition systems. Encourage the students to draw illustrations to accompany their work and present their findings to classmates.

 LASER DISC

33813, 33814–34143

Electromagnet

2120

Electromagnet; wrecking yard

32745, 32746–33021

Magnetically induced current in wire

● **Process Skills:**
Communicating, Solving Problems

Explain to the students that in an electric generating plant, water or steam is used to push the blades of a device called a turbine, which in turn is connected to a generator. The coils of wire on the generator then turn within a magnetic field, producing an electric current. Follow up by asking volunteers to suggest another way to produce a current without turning the coils. (Introduce a moving magnet within the coils.)

GUIDED PRACTICE

Have the students write as much as they can recall about how to create electricity using magnetism. Then have them check their work against the text and revise their notes. Ask the students to place their notes in their science portfolios.

INDEPENDENT PRACTICE

Have the students provide written answers to the Section Review. In their journals, have them write a few paragraphs explaining why the electromagnet is so important in modern industry.

MEETING SPECIAL NEEDS

Mainstreamed

You might have hearing-impaired students work independently or together to prepare a report on the role of electromagnets in devices such as the TTY, doorbell-light system, and other devices they use regularly.

DISCOVER BY *Doing*

Cooperative Learning You might have the students work in groups to vary the following elements of the experiment: core (nail) size, number of coils, and amount of current. Then have the groups compare their results.

⬦ **Did You Know?**
André Ampère (1775–1836), a French physicist, discovered that both a loop and a coil of wire acted like magnets while current was passed through them.

ONGOING ASSESSMENT
▼ ASK YOURSELF

A solenoid is a wire spiral through which a current can flow. This produces a stronger magnetic field.

Figure 14–17.
The ability to turn magnetism off and on has many uses.

To see a powerful electromagnet in action, visit a junkyard. The crane that moves junked automobiles from one place to another has a large, strong electromagnet. You can make a small electromagnet in the following activity.

DISCOVER BY *Doing*

Obtain a 6-volt dry cell, an iron nail at least 15 cm long, and about 75 cm of insulated bell wire. Remove some of the insulation from each end of the wire. Leaving about 20 cm of wire free at one end, begin wrapping the wire around the nail, starting just under the head. When the wire is about 2 cm from the nail's tip, begin wrapping upward, toward the head. Wrap wire about halfway to the top of the nail. Connect the ends of the wire to the dry cell. Get some paper clips and other metallic objects. See how strong your electromagnetic crane is. ✐

So, how successful were you in building an electromagnet? Ready to go into the junk business? Or are you ready to junk your electromagnet?

▼ **ASK YOURSELF**
What is a solenoid?

Magnetism Makes Electricity

If electricity can create a magnetic field, can a magnetic field create an electric current? Your local energy plant can demonstrate that relationship.

In electric generating plants, huge generators are operated by engines. These generators make the electric energy network function and can be used to produce an electric current.

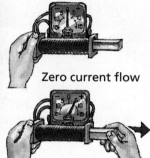

Zero current flow

Current flow

Figure 14–18. Moving a bar magnet within a coil of wire produces an electric current by electromagnetic induction. When the bar is stationary, no current is induced.

The process by which a magnetic field produces an electric current is called **electromagnetic induction.** You can see this process at work by moving a bar magnet through the coils of a solenoid. As shown in the photograph and the illustration, if the ends of the solenoid are connected by wires to a sensitive ammeter, the ammeter will indicate the current in the solenoid.

Current will be induced in the solenoid whenever there is a *changing* magnetic field in the solenoid. If you move a bar magnet into the solenoid, the ammeter will show a current as long as the magnet is moving. As soon as you stop moving the bar magnet, the current will stop. Thus it is not just a magnetic field but a changing magnetic field that induces the flow of electricity. The greater the number of coils in the solenoid, the stronger the current that will be induced. The more powerful the magnet, the greater the current that will be induced.

 ASK YOURSELF

How can a magnetic field be used to produce electric current?

SECTION 3 REVIEW AND APPLICATION

Reading Critically

1. How might electromagnets be useful in industry?
2. How does an electromagnet demonstrate that an electric current produces a magnetic field?
3. Why does an iron core make the field of a solenoid so much stronger?

Thinking Critically

4. You find that an electromagnet you are using is not strong enough. Suggest two ways in which you might make a stronger electromagnet.
5. When an electromagnet is at rest inside a solenoid, no current flows in the solenoid. If the core is pulled out of the electromagnet, current will flow in the solenoid while the core is being removed. Explain why this happens.

SECTION 3 **383**

INVESTIGATION

Studying Electricity and Magnetism

Process Skills: Inferring, Constructing/Interpreting Models

Grouping: Pairs

Objectives
● **Construct** a model
● **Observe** the cause and effect between electricity and magnetism

Hints

Use insulated wire for this *Investigation* (either bell-wire or enamelled magnet wire). The light bulb in the circuit will make the batteries last longer, but the magnetic effects may be reduced.

The more turns in a coil, the greater the effect. Remind the students to tap the compass to make sure the needle is free to turn.

Procedure

1. No

2. The bulb lights up.

3. It is attracted to the nail.

4. The nail will either attract or repel the south pole of the compass, depending on the circuit.

5. The other end of the needle will be attracted to the nail.

7. A galvanometer

8. It turns 90° from its original orientation.

9. It turns 90° the other direction

Analyses and Conclusions

1. The nail becomes an electromagnet.

2. The coil of wire itself, even without the nail, is magnetic. It confirms the fact that electric currents produce magnetic fields.

▶ **384** CHAPTER 14

Application

Coil the wire around the nail and connect the ends of the wire to the battery. Tape the nail to the battery and suspend the entire unit with the piece of thread. Once the thread unwinds, the electromagnet will point north and south.

✳**Discover More**

When current passes through the coil, it produces a magnetic field that is attracted or repelled by the permanent magnet inside the galvanometer. This causes the needle to move to the right or the left depending on the direction of the current.

Have the students place their observations, results, and answers from this *Investigation* in their science portfolios.

INVESTIGATION

Studying Electricity and Magnetism

▶ **MATERIALS**
● nail ● wire, 1.5 m ● batteries (2) ● wire connectors (2)
● flashlight bulb and socket ● paper clip ● compass ● tape

▼ **PROCEDURE**

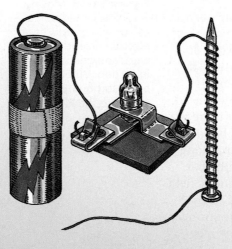

1. Wrap 30 turns of wire around a nail and set up an electric circuit as shown in the diagram. Before making the final connection, bring the nail head near a paper clip. Does anything happen to the clip?

2. Now make the final connection to complete the circuit. How can you tell that a current is flowing?

3. While the current is still flowing, bring the nail near the paper clip again. What happens to the clip this time?

4. Next, with the circuit still complete, bring the head of the nail near the compass. Does it attract or repel the south pole of the compass needle?

5. Repeat step 4 using the point of the nail. What happens this time?

6. Remove the wrapped nail from the circuit and set it aside.

7. Now wrap some fine insulated wire around the compass about 30 times along its north-south axis. What would you call the device you just made?

8. Line up the compass needle under the wire and connect the ends of the wire to the battery. What happens to the needle when the circuit is completed?

9. Reverse the connections on the battery. What happens to the needle this time?

▶ **ANALYSES AND CONCLUSIONS**

1. What does the wire-wrapped nail become once it is connected to an electric circuit?

2. What was demonstrated by wrapping a coil of wire around the compass and connecting it to the battery? What does this confirm about electricity?

▶ **APPLICATION**

Using the materials in this Investigation and a piece of thread, explain how you could make a compass.

✳ *Discover More*

A galvanometer contains a coil of wire and a permanent magnet. When it is connected to a source of current, its needle is deflected to one or the other side of zero. Explain how this works.

The Big Idea—Systems and Structures

Lead the students to understand that the relationship between magnetism and electricity powers much of the world's technology.

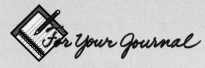

For Your Journal

The students' ideas should reflect an understanding of how magnetism works as a force and how it can be used with electricity in industry.

CHAPTER
14
*H*IGHLIGHTS

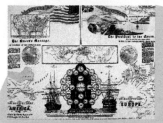

The Big Idea

The discovery of magnetic forces led to many useful applications. Magnets are used in many technologies including robotics, manufacturing, engineering, and the television and recording industries. Electric currents produce magnetic fields, and changing magnetic fields produce electric currents. This interaction between electricity and magnetism plays a key role in many systems. The magnetic field produced by an electromagnet operates doorbells, motors, and many other electrical devices. The electric current produced by changing magnetic fields is the primary source of electricity for homes and industry.

For Your Journal

Look back at the list of uses for magnets you wrote in your journal. What new uses can you add to your list as you complete this chapter? You may wish to do research on electromagnets and continue to add to your list.

CONNECTING IDEAS

Students concepts maps should be completed with words similar to the ones shown below.

1. electromagnets

2. electromagnetic induction

3. microwave ovens

4. televisions

Connecting Ideas

This concept map shows how ideas about magnetism are related. Copy the concept map into your journal, and extend the concept map to include ideas that link magnetism and electricity.

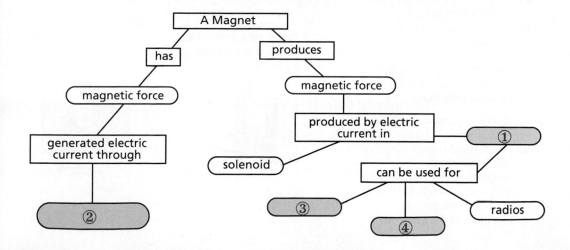

CHAPTER
14
REVIEW

ANSWERS

Understanding Vocabulary

1. (a) Magnetism is the force of repulsion or attraction between like and unlike poles. Any region in which magnetic forces are present (either repulsion or attraction) is a magnetic field.

(b) A solenoid is a wire coil through which a current flows to produce a magnetic field. An electromagnet is a solenoid with a core made of material that can be made into a permanent magnet. Electromagnetic induction is the process by which a magnetic field produces an electric current.

Understanding Concepts

Multiple Choice

2. b

3. d

4. a

5. d

6. c

Short Answer

7. An unmagnetized piece of iron or steel will be attracted to either pole of a magnet. To find out whether the steel is magnetized, bring it near the other pole of the compass. If the steel is a magnet, it will repel the S pole of the compass.

8. The field makes the atoms of the core turn to line up with the solenoid's field. This adds to the strength of the solenoid's field.

Interpreting Graphics

9. Sample conclusion: A horseshoe magnet can attract objects through cardboard. Accept all reasonable conclusions the students can justify.

10. Sample conclusions: The magnet is not located close enough to the objects to attract them. The objects are not made of materials that are attracted to a magnet. Accept all reasonable conclusions the students can justify.

CHAPTER
14
REVIEW

▶ Understanding Vocabulary

1. Compare and contrast the meanings of the terms in each set.
a) magnetism (368), magnetic field (375)
b) solenoid (381), electromagnet (381), electromagnetic induction (383)

▶ Understanding Concepts

MULTIPLE CHOICE

2. The N pole of a compass needle points to the north magnetic pole of Earth because Earth's magnetic pole is
a) a positive pole.
b) an S pole.
c) an N pole.
d) a negative pole.

3. If the poles of two magnets repel each other,
a) both poles must be S poles.
b) both poles must be N poles.
c) one is an S pole and the other is an N pole.
d) both poles are either S or N poles.

4. Magnetizing a piece of iron is a process by which
a) the poles of existing atoms are brought into line.
b) magnetic atoms are added to the iron.
c) each atom in the iron is converted into a magnet.
d) magnetic lines of force are brought into line.

5. An electric motor uses an electromagnet to change
a) mechanical energy into electrical energy.
b) magnetic fields in the motor.
c) magnetic poles in the motor.
d) electrical energy into mechanical energy.

6. The ability of any object to generate a magnetic field depends upon its
a) size.
b) location.
c) composition.
d) direction.

SHORT ANSWER

7. You bring a piece of steel near a compass and find that the steel attracts the N pole of the compass. Explain why this does not prove that the steel is magnetized.

8. Why does an iron core make the magnetic field of a solenoid stronger?

▶ Interpreting Graphics

9. Write a statement that might be a conclusion you can draw from Figure A.

10. Write a statement that might be a conclusion you can draw from Figure B.

A.

B.

Reviewing Themes

11. It must be constantly changing.

12. The crane operator positions the magnet on top of the car. The current is turned on, and the electromagnet lifts the heavy car. The crane operator turns the crane toward the railroad car and places the old car in it. Then the crane operator turns off the current to the electromagnet, thus breaking the magnetic attraction between the car and the electromagnet.

Thinking Critically

13. When you use a compass, you are determining direction based on a horizontal field, stretching from the north to the south magnetic poles. At the poles themselves, the field becomes vertical and is impossible to register on a compass.

14. Both items are encoded using magnetic tape. Then the tapes are reproduced either as sound or visual content. If the tapes were accidentally demagnetized by the book check-out devices, the tapes would be erased.

15. Energy made from wind is inexpensive because the energy of wind is free. Also, wind energy does not produce any air pollution or water pollution. The main disadvantage of wind-generated energy is that it is unpredictable. The generators only produce electricity when the wind is blowing. In addition, wind-operated generators are often noisy and considered by many people to be unattractive.

Reviewing Themes

11. *Environmental Interactions*
In order to produce an electric current, what must be true of the magnetic field in a solenoid?

12. *Systems and Structures*
Suppose a salvage yard has an electromagnet attached to the end of a crane. An automobile that has been crushed for scrap metal must be placed in a railroad car. Explain how the crane operator can move the crushed automobile from the ground and place it in the railroad car for transport.

Thinking Critically

13. Why do you think a compass does not work at the magnetic poles of Earth?

14. Many libraries use a security system based on magnetism for checking out books. Magnetic strips are put into each book. When books are checked out, they are demagnetized by being run across or through a device that sits on the check-out desk. Then the books are again magnetized when they are returned. (A sensor system in the door buzzes when books are taken from the library without this processing.)

Why do you think library audio cassettes and videotapes are generally not run through this magnetizing/demagnetizing system?

15. On this wind farm, every windmill is attached to a generator. In each generator, a coil spinning in a magnetic field produces electricity. Other methods of producing electricity are also being tried. What do you think are the advantages and disadvantages of using the wind to make electricity?

Discovery Through Reading

Normile, Dennis. "Superconductivity Goes to Sea." *Popular Science* 241 (November 1992): 80–85. The *Yamato 1*, an ocean-going ship with a magnet-driven propulsion system, took its first voyage in the summer of 1992. Read this article to find out how it works.

CHAPTER 15

ELECTRICITY IN USE

PLANNING THE CHAPTER

Chapter Sections	Page	Chapter Features	Page	Program Resources	Source
CHAPTER OPENER	388	*For Your Journal*	389		
Section 1: THE POWER GRID	390	Discover by Doing **(A)**	393	*Science Discovery**	SD
• Fleeing from a Hurricane **(B)**	391	Discover by Doing **(A)**	395	Investigation 15.2:	
• Generators **(H)**	392	Section 1 Review and Application	399	Studying the Relationship	
• Motors **(H)**	394	Investigation: Testing		Between Heat and Electrical	
• Transformers **(A)**	397	Electromagnetic Induction **(A)**	400	Energy **(A)**	TR, LI
				Motors and Generators **(H)**	IT
				Study and Review Guide, Section 1 **(A)**	TR, SRG
Section 2: HOUSEHOLD CIRCUITS	401	Discover by Observing **(B)**	403	*Science Discovery**	SD
• Circuits **(A)**	402	Activity: How does a fuse work? **(A)**	404	Reading Skills: Making an Outline **(B)**	TR
• Overloads **(A)**	403	Discover by Doing **(B)**	405	Extending Science Concepts:	
		Section 2 Review and Application	405	Heat Loss During Energy Transmission **(A)**	TR
		Skill: Calculating Your Electric Bill **(A)**	406	Study and Review Guide, Section 2 **(A)**	TR, SRG
Section 3: ELECTRONICS	407	Discover by Writing **(A)**	410	*Science Discovery**	SD
• Hello, Is Anyone There? **(B)**	408	Section 3 Review and Application	412	Investigation 15.1: Using Electrical Communicating Devices **(A)**	TR, LI
• The Number Please **(A)**	410			Thinking Critically **(B)**	TR
• Radio and Television **(A)**	410			Connecting Other Disciplines: Science and Computer Science, Creating Flow Charts **(A)**	TR
				Audio Devices **(A)**	IT
				Record Sheets for Textbook Investigations **(A)**	TR
				Study and Review Guide, Section 3 **(A)**	TR, SRG
Chapter 15 HIGHLIGHTS	413	The Big Idea	413	Study and Review Guide, Chapter 15 Review **(A)**	TR, SRG
Chapter 15 Review	414	For Your Journal	413		
		Connecting Ideas	413	Chapter 15 Test	TR
				Test Generator	

B = Basic   **A** = Average   **H** = Honors
The coding Basic, Average, and Honors indicates subsections, features, and resources that might be appropriate for different levels of learners. For additional suggestions regarding choice of topic and depth of coverage, see the Pacing Chart on pages T28–T31.

*Frame numbers at point of use
(TR) Teaching Resources, Unit 4
(IT) Instructional Transparencies
(LI) Laboratory Investigations
(SD) *Science Discovery* Videodisc Correlations and Barcodes
(SRG) Study and Review Guide

CHAPTER MATERIALS

Title	Page	Materials
Teacher Demonstration	393	generator, galvanometer, automobile storage battery
Discover by Doing	393	(per group of 3 or 4) small toy motor, wires (2), galvanometer
Discover by Doing	395	(per group of 3 or 4) compass, wire 60 cm-long, alligator clips (2), 6-V battery
Investigation: Testing Electromagnetic Induction	400	(per group of 3 or 4) galvanometer, solenoid, insulated wire leads (2), bar magnets (3)
Discover by Observing	403	(per individual) journal
Activity: How does a fuse work?	404	(per group of 3 or 4) safety goggles (each), 15-Amp plug fuse, table lamp
Discover by Doing	405	(per individual) journal
Skill: Calculating Your Electric Bill	406	(per individual) paper, pencil
Discover by Writing	410	(per individual) journal

ADVANCE PREPARATION

For the *Discover by Writing* activity on page 410, make a photocopy of the alphabet in binary form for each student.

TEACHING SUGGESTIONS

Field Trip
Plan a field trip to a local power plant or to a transformer station. Encourage students to prepare their questions ahead of time. Students may wish to ask questions about blackouts such as the one that they will read about in the Chapter Opener.

Outside Speaker
Electrical engineers design, test, and help produce electrical and electronic devices, such as generators, motors, televisions, communication satellites, and computers. Invite an electrical engineer to speak to the class about the kind of work he or she does. You might contact an electrical engineer by calling businesses that are involved in the production of electrical or electronic devices. Ask the speaker to describe a typical day on the job, educational requirements, and future career opportunities.

CHAPTER 15 ELECTRICITY IN USE

CHAPTER THEME—*TECHNOLOGY*

This chapter presents the many uses of electric energy and potential hazards associated with its use. Also discussed are the processes of converting electric energy into other forms of energy, as well as methods for converting mechanical energy into electric energy. A supporting theme in this chapter is **Energy.**

CHAPTER 15 ELECTRICITY IN USE

If the supply of electricity to your home fails for some reason, all the lights go out. You probably fumble to find a flashlight.

Now imagine such a blackout darkening the entire northeastern United States during evening rush hour. And imagine all of New York City waiting for the lights to come on. In 1965, it really happened!

CHAPTER MOTIVATING ACTIVITY

 Cooperative Learning Have the students work together in small groups to write a short story about an imaginary power failure that lasted three days. The students should include in their stories how the characters coped with the power outage during this time. Encourage each group to share its story with classmates.

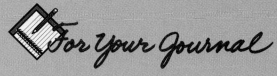

For Your Journal

The journal questions will help you identify any misconceptions the students might have about the uses of electricity. In addition, these questions will help the students realize that they do possess knowledge about electricity. (The students will have a variety of responses to the journal questions. Accept all answers, and tell the students that at the end of the chapter they will have a chance to refer back to the questions.)

On November 9, 1965—in Buffalo, in Albany, in Boston, in New York City—the clocks in the Megalopolis sputtered to a standstill. Lights blinked and dimmed and went out. Skyscrapers towered black against a cold November sky, mere artifacts lit only by the moon. Elevators hung immobile in their shafts. Subways ground dead in their tunnels. Streetcars froze in their tracks. Street lights and traffic signals went out—and with them the best-laid plans of the traffic engineers. Airports shut down. Mail stacked up in blacked-out post offices. Computers lost their memories. TV pictures darkened and died. Business stopped. Food started souring in refrigerators.

For the very young and the very old, it remained a fearful night—but numberless others seemed to be enjoying themselves. Strangers helped one another in the dark and sent thank-you notes and flowers the next morning. An off-Broadway show went on, by candlelight, to a gallery of seven people and two dogs.

But the experience was also sobering to the Megalopolitans—a rather unsettling lesson in how totally their lives are wired to electricity. Black Tuesday had been an epiphany of the push-button age, a demonstration of the final vulnerability of the mightiest society on earth. Could it happen again? Maybe so, maybe not—but it was a memory that all the lights in all the cities could not dim.

from *Newsweek* magazine
November 22, 1965

The black sections of the map represent areas that were affected by the blackout.

For Your Journal

- Why is it important to be able to transform electricity for long-distance transmission?
- How does an electric generator work?
- List as many electrical safety features as you can think of.

ABOUT THE PHOTOGRAPH

The power outage, or blackout, on November 9, 1965, affected a large urban area. This large urban area, referred to as a megalopolis, spanned 207 200 square kilometers across parts of eight states and southern Canada. Over 30 million people were affected. The outage lasted for varying lengths of time: 13 hours in New York City; more than 5 hours in Boston; more than 4 hours in Rochester, Albany, and Toronto; and more than 2 hours in Providence.

Why did the power outage occur? The power-line grid that joined the electric utility companies of the Northeast has been compared to a safety rope linking mountain climbers. One company fell so suddenly that it took all the other companies down with it and caused the biggest power failure in history.

Investigators inspected transmission lines, generators, and circuit breakers to try to explain the blackout. Their explanation mentioned that a series of flickers in the grid had combined in an improbable sequence that brought the entire network down. It shouldn't have happened, but it did.

FOCUS

This section describes how generators produce electric energy and how motors convert electric energy into mechanical energy. The use of transformers for transmitting electric current over long distances is also discussed.

MOTIVATING ACTIVITY

 Cooperative Learning Review with the students the concept of electromagnetic induction, discussed in Chapter 14. In small groups, have them design and sketch a device that uses electromagnetic induction to produce electricity. Have the students save their sketches to refer to later in the section.

PROCESS SKILLS
• Experimenting • Inferring
• Analyzing

POSITIVE ATTITUDES
• Creativity • Precision

TERMS
• power grid • electric motor • transformer

PRINT MEDIA
"The War of the Currents" from *Cobblestone* magazine by Richard Mattis (see p. 337b)

ELECTRONIC MEDIA
Hot Line: All About Electricity Focus Media, Inc. (see p. 337b)

SCIENCE DISCOVERY
• Generator, gas powered
• Motor construction; lab setup • Motor demonstration
• Motor generator; lab setup
• Motor, electric

BLACKLINE MASTERS
Laboratory Investigation 15.1
Study and Review Guide

INTEGRATION–
Earth Science

As the air soaks up warm moisture from the sea, thunderclouds form. Atmospheric winds cause the clouds to swirl. As the storm continues to move over warm water, it grows bigger and more violent. A hurricane is said to be at full force when a calm center, or eye, forms inside the swirling winds. The strongest winds are found just outside the eye, but it can also be very windy and rainy hundreds of kilometers from the eye. More than 40 million Americans live in coastal areas that are vulnerable to hurricanes. In many coastal areas, the safest thing to do when a hurricane is approaching is to get into a car and drive inland!

SECTION 1

The Power Grid

Objectives

Apply *the principle of electromagnetic induction to the production of electric energy in generators.*

Compare *and* **contrast** *a generator with a motor.*

Explain *why it is important to be able to transform electric potentials for long-distance transmission.*

Roberto dropped his books on the counter as he got home from school. "Ugh! I hope we move fast through this electricity stuff we are studying. It's so boring."

His father was home early from work. That was unusual. "I hope you're learning as much as you can about electricity," said his dad. "Over the next few days it might be important. By the way, pack a bag with clothes to last you several days."

"Wow!" Roberto jumped, "Are we going on vacation?"

Roberto's dad looked at him very seriously. "No, son," he replied. "We've been asked to evacuate this area because there is a hurricane headed our way."

Figure 15–1. Though these electronic signals clearly showed the immensity of the storm, meteorologists could not predict how many lives and how much property would be affected once Hurricane Andrew hit the coast of Florida.

TEACHING STRATEGIES

● **Process Skill:**
Communicating

Ask the students to identify components of their local power grid that they have seen. (Answers might include overhead power lines, transformer stations, long-distance transmission lines, power stations, local transformers, home power sup-

ply lines, and electric meters.) Discuss with the students how all these components work together to deliver electricity to homes in an efficient and safe manner.

● **Process Skills:**
Communicating, Comparing

Refer the students to Figure 15–1. Ask them to compare the size of the storm to the total area of the state. Point out that electromagnetic waves are used to make radar pictures such as the one shown.

● **Process Skill:** *Generating Ideas*

Ask the students to research the usefulness of radar pictures in preparing people for hurricanes and other natural disasters. Encourage the students to share their research with other classmates.

Fleeing from a Hurricane

Roberto and his family live in Florida. In August 1992, south Dade County, the area around Miami, was targeted by a fast-approaching hurricane—Hurricane Andrew.

"Roberto, bring along your science book," his dad continued. "I can help you study about electricity while we drive to your sister's in Orlando." A groan was heard from the hallway as Roberto headed for his room.

Finally, all was packed in the car, and they were ready to go. Roberto knew that hurricanes were dangerous, but he had never seen one, and he wished they could stay. His parents were very worried, but he thought all this was exciting. Just before they left the house, Roberto's dad switched off the main circuit breaker to the house. "Why did you do that, Dad?" Roberto asked.

"Just as a safety precaution," his dad replied. "I'll explain later. Let's hit the road."

Once they were underway, his dad relaxed a little and returned to the subject. "Roberto, you probably don't think about it when you flick a light switch, but do you realize how much our family depends on electricity? Think of how different your life would be if we had to live even one week without it. Why don't you read aloud from your lesson on generating electricity. We can discuss any part you don't understand. It'll be bumper-to-bumper traffic all the way to Orlando, so we'll have lots of time to talk." Roberto opened his science book and began to read.

Thousands of powerful generators stretch kilometer after kilometer across the country. They produce the electricity that is needed for use in homes, schools, and businesses. This energy is fed through a network of transmission lines called a **power grid**. You have probably seen these transmission lines in your neighborhood.

ASK YOURSELF
How different would your life be if you had to live without electricity for one week?

Figure 15–2. These electric power lines make up a small fraction of the power grid that feeds electric current to all parts of the United States.

MEETING SPECIAL NEEDS

Mainstreamed

Refer to the illustrations and diagrams in this chapter as much as possible to help mainstreamed students understand the concepts discussed. Have the students point to appropriate parts of the diagrams as they explain processes or functions.

SECTION 1   **391** ◀

● **Process Skill:** *Analyzing*

Explain to the students that the word *generator* comes from the Latin word *generate*, meaning "to produce." Then ask a volunteer to describe how the function of a generator is related to its word origin. (A generator produces electric current.)

● **Process Skills:** *Communicating, Inferring*

Ask the students to locate on Figure 15–3 the magnetic field in the generator. (between the north and south poles of the *magnet*) Point out the movement of the coil within the magnetic field.

● **Process Skills:** *Constructing/Interpreting Models, Evaluating*

 Have the students look at the sketches they made as part of the Section 1 Motivating Activity on p. 390. Ask a spokesperson from each group to describe any similarities between the device his or her group designed and the generator shown in Figure 15–3. Discuss whether the students' "generators" would work.

THE NATURE OF SCIENCE

In 1879 Thomas Edison invented the incandescent lamp, which produces light by burning a filament inside a glass bulb. Prior to Edison's invention, it was known that a burning filament gives off light for a few minutes. Edison and up to 3000 workers spent a year testing hundreds of different kinds of filaments to find one that would burn for a longer period of time. During their investigations, Edison discovered that removing air from the bulb produced a brighter light that lasted longer. Edison's first perfected incandescent lamp burned continuously for 40 hours.

MULTICULTURAL CONNECTION

Lewis Latimer, an African-American inventor who worked with Edison, refined Thomas Edison's incandescent light bulb. Tell the students that they will have an opportunity to learn more about Lewis Latimer in the Unit 4 Science Parade, p. 421.

Generators

Generators use electromagnetic induction to produce electricity. Recall that this process was explained in the previous chapter. Faraday's law of induction summarizes the principle. This law states that a current may be induced in a loop of wire by changing the magnetic field that is passing through the loop. The illustration shows how an electric generator works. In this case, the loops of the wire coil are moved through the magnetic field between the magnet's north and south poles. Whether the loop or the magnet does the moving, the effect is the same: a current flows through the wire.

Figure 15–3. A simple generator

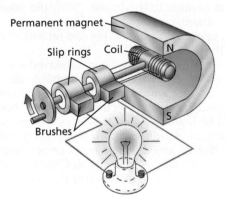

A simple generator consists of a loop of wire or a coil of several loops placed in a magnetic field. The coil shown in Figure 15–3 is between the north and south poles of a magnet. As the coil rotates, it cuts through the magnetic lines of force. This results in a current flowing through the wire. If the loop spins continuously, the orientation of the coils in the magnetic field changes constantly. Current flows in the loop, changing direction every half cycle. The flowing current passes through the slip rings and brushes and is then transmitted to the power user. In the diagram, the power user is the light bulb. A much larger generator would transmit the electric current through the power grid of an entire city or region.

Figure 15–4. Huge generators are used in electric energy plants to produce the electric current used in homes and in industry.

Figure 15–5. Small generators are available for home use in case of power failure or in rural areas without access to a power grid.

Roberto interrupted his reading to ask, "Dad, you bought a generator for the house this week, didn't you?"

"Yes," his father replied. "I thought we might need one if the hurricane came close to Miami and the power went out. I thought we could run some lights with it."

Roberto looked at the diagram of the electric generator in his science book again. "Dad, this generator reminds me of Grandma's hand-cranked ice-cream maker. Would we have to take turns cranking a handle to keep the coil in the generator rotating?"

Roberto's dad chuckled, "No, it has a small gasoline engine that will do that for us. It will run for about eight hours on one tank of fuel."

DISCOVER BY *Doing*

Obtain a small toy motor, two wires, and a galvanometer. A galvanometer is used to measure current. Attach one end of each wire to the terminals of the galvanometer. Then attach one of these wires to one lead of the motor and the other to the other lead. Spin the motor shaft between your fingers, first in one direction and then in the other. What do you observe about the galvanometer's readings? What have you turned the motor into? What energy changes are taking place? Record your answers in your journal. 🖋

SECTION 1

● **Process Skill:** *Evaluating*

Discuss the usefulness of portable generators such as the one Roberto's father purchased. Ask the students to identify in their journals businesses or organizations that might have backup generators to use in the event of a power failure. (Answers include hospitals, restaurants, police and fire stations, and stores.)

● **Process Skills:** *Measuring, Comparing*

Have the students use the information in the text to calculate the difference in power production between home generators and power plants. (Power plants produce more than six thousand times more power than home generators. 34 000 000 watts/5000 watts = 6800.) Discuss with the students the size that generators in power plants must be in order to produce this much power.

SCIENCE BACKGROUND

The voltage produced by a generator can be increased by increasing the strength of the magnetic field, increasing the speed of the turning coil or magnet, or increasing the number of loops in the coil.

Figure 15–6. The generators of a hydroelectric-energy plant are operated by large turbines. The force of the water draining from the dam causes the turbines to spin.

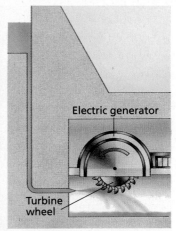

Electric generator

Turbine wheel

Most home generators produce from 200 to 5000 watts (W) of power. The large generators in power plants may produce 34 000 kilowatts (kW) or more. In these large generators, the coil is stationary and the magnetic field rotates. In most large power plants, a burning fuel, such as coal, or a nuclear reaction is used to turn water into steam. The moving steam causes blades in a turbine to spin. This in turn moves the magnets. In a hydroelectric plant, falling water causes the turbine blades to spin.

ASK YOURSELF
In a generator, how is current in a wire produced?

ONGOING ASSESSMENT
ASK YOURSELF

A current is generated in a wire by either moving a loop of wire through a magnetic field or a magnet around a coil of wire.

 **LASER DISC**

2125

Generator, gas powered

383, 384

Motor construction; lab setup

30698, 30699–30987

30988–31652

Motor demonstration

385, 386

Motor generator; lab setup

1852

Motor, electric

Motors

 "Dad," Roberto asked, "How can a motor run our generator to make electricity when we need electricity to run the motors in things like the refrigerator, the vacuum cleaner, and the food processor?"

"What you call the motor on the generator is really a small engine," said his dad. "People usually use the term *motor* when the power source is electricity. Actually, an electric motor works a lot like a generator—but in reverse. There's probably something about it in your book. Keep reading."

A diagram of an electric motor would appear very similar to that of a generator. An **electric motor** is a device that converts electric energy into mechanical energy, or movement. A generator does the opposite: it turns mechanical energy into electric energy. That is, a generator uses the circular movement of a wire to produce a current, while a motor uses a current to produce a circular movement.

● **Process Skill:** *Comparing*

Refer the students to Figures 15–3 and 15–7. Ask them to identify structural similarities between the generator and the electric motor. (Both have a magnet and a wire loop.) How

are they different? (The electric motor has a power source and a commutator. The generator does not.)

● **Process Skill:** *Communicating*

Use Figure 15–7 to help the students understand how the reversal of the coil's magnetic poles causes it to spin. Point out that the coil in the motor

becomes an electromagnet when current flows through it. The north pole of the coil is attracted to and lines up with the south pole of the stationary magnet, and vice versa.

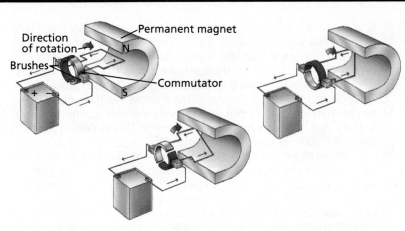

Figure 15–7. A simple electric motor

DISCOVER BY *Doing*

The compass needle will be deflected perpendicular to the alignment of the wire when current flows through the wire. The needle reverses direction when the alligator clips are switched, because the polarity of the electromagnet (the wire) is reversed. Have the students place their observations and answers in their science portfolios.

Electric motors operate as a result of the interaction of two magnetic fields. In the simple motor shown in the illustration, there is a permanent magnet with a wire loop between its poles. As a current moves through the wire, it creates an electromagnet that generates its own magnetic field. The magnet exerts an upward, attractive force on one side of the loop and pushes down on the other side of the loop. This makes the loop rotate. The loop acts in the same way as a compass needle. It aligns itself with the magnetic field of the magnet.

In order to be a motor, the loop must turn all the time. The loop in a motor keeps turning because the current flowing into the loop keeps changing direction. Circuits in homes and businesses are supplied with alternating current (AC), which reverses continuously. Many appliances contain AC motors, in which the direction of the current, and therefore the magnetic poles, constantly change. This switching of magnetic poles causes the coil of wire to spin inside the magnet. This produces the motion of the shaft that protrudes from the motor.

DISCOVER BY *Doing*

Place a compass on top of a wire approximately 60 cm long. The wire should have alligator clips at each end. Align the wire so that it is parallel to the compass needle. Connect the ends of the wire to a 6-V battery, and observe what happens to the compass needle. Without moving the compass, switch the alligator clips so that the one attached to the positive terminal is now on the negative terminal and vice versa. What happens to the compass needle? How is this similar to what happens in a motor? ✐

● **Process Skill:** *Inferring*

Ask the students why the current flowing through the wire loop inside a motor must continuously change direction. (If the current did not change direction, the magnetic poles of the coil would not reverse and the coil would remain station-ary. No circular movement would be produced.)

● **Process Skill:** *Classifying*

Cooperative Learning To help the students understand the importance of motors in their lives, have small groups of students list devices that run on motors. Ask the students whether the motor is AC or DC. (The motor is AC if the device is plugged into an outlet; the motor is DC if it is powered by a battery.)

SCIENCE BACKGROUND

The AC motor was invented in 1888 by Nikola Tesla, a Serbian engineer. Tesla was a prolific inventor who was particularly interested in high-voltage, high-frequency electricity. He initially worked for Thomas Edison in Europe, but later he set up his own laboratory in New York City. Unfortunately, Tesla did not keep detailed records of his work, which other scientists might have used to duplicate his ideas and discoveries. Consequently, some of Tesla's inventions were lost forever.

MEETING SPECIAL NEEDS

Mainstreamed

Obtain an old electric motor, such as the ones found in hair dryers or fans. Take the motor apart, and let the students examine the internal structures. Then help the students draw and label a diagram in their journals of an electric motor.

ONGOING ASSESSMENT
▼ ASK YOURSELF

The loop moves back and forth until it aligns with the magnetic field.

It was late at night now, and the traffic was still heavy. Even stopping for gas was a problem. So many people were fleeing north from the hurricane that the family had to wait in a long line of cars at the gas station on the turnpike. Roberto decided to pass the time by figuring out the motor diagram in his science book.

"Dad," he asked, "some of the motors we have at home run on batteries. This book even shows a motor connected to a battery. But batteries make direct current. Direct current doesn't alternate, so how can it run a motor?"

Roberto's dad examined the diagram. "Look right here," he said, pointing to Figure 15–7. "The ends of the loop are connected to a split ring called a *commutator*. Two brushes touch the commutator. The battery's current feeds in through one brush and leaves through the other. From the 'in' brush, current flows through one-half of the ring and on through the loop. The magnetic field this produces causes the loop to turn. This brings the other half of the ring into contact with the 'in' brush, sending current into the loop from the opposite direction and reversing its magnetic field. In a DC motor, the commutator changes the direction of the current. As a result, the loop spins just as it does in the AC motor."

Roberto's father continued, "A single loop of wire like the one in the diagram would carry only a small amount of current. Real motors have many loops of wire that can carry a large amount of current. Also, most motors use electromagnets to generate the magnetic field."

"I get the picture," said Roberto. "Then the spinning loop can be connected to a shaft, causing it to spin. The spinning shaft can be connected to other parts, like the blades in a mixer, causing them to turn to do work. Pretty cool."

Figure 15–8. This cutaway view shows the inside of an electric motor. Notice that many loops of wire are used instead of the single loop shown in Figure 15–7.

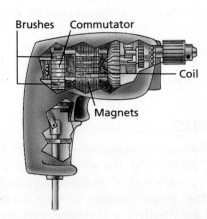

▼ ASK YOURSELF

In what way does the loop of a motor behave like a compass needle?

Transformers

Roberto's family reached Orlando after a tiring seven-hour drive that would normally take only four hours. Even though it was late, Roberto and his parents could not fall asleep. They listened to the news on television and watched the weather radar that showed Hurricane Andrew moving closer to Miami. The newscaster warned people not to go out after the storm because there might be dangerous high-voltage lines lying on the ground.

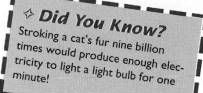

Figure 15–9 Some of these lines might have an electric current of 30 000 volts (V) or more.

✧ **Did You Know?**
Stroking a cat's fur nine billion times would produce enough electricity to light a light bulb for one minute!

"Dad," Roberto said, "I don't understand something. If the electricity that gets to our house has 30 000 volts, why don't all our appliances get burned to a crisp? They all say they run on either 220 or 110 volts."

"Good question, Roberto," his Dad replied. "Have you ever heard of transformers?"

"Sure," Roberto grinned. "When I was younger, I had a car that transformed into an awesome robot!"

"No, not those transformers," his father smiled. "Let's look in your science book and read about them."

A **transformer** is a device that uses electromagnetic induction to change the electric potential (voltage) of alternating current. In practical terms this means that wire coils of different sizes can be used to either increase or decrease voltage. You may have seen transformers attached to electric poles. They look like large, gray cylinders. It would be harmful to send thousands of volts into your household circuits. Appliances would burn out, and fires would start. A *step-down* transformer decreases voltage so that electricity reaches your home in a manageable form.

Figure 15–10. This particular transformer lowers the voltage of electric current from about 30 000 V to the 120 V you receive in your home.

SECTION 1   **397** ◀

GUIDED PRACTICE

Help the students calculate the change in voltage by a step-up transformer receiving 120 V if the proportion of loops in the primary to secondary coil is 1 to 4. (The voltage would increase fourfold, from 120 V to 480 V.) Have them practice calculating other changes in voltage by step-up and step-down transformers. Ask the students to place their calculations in their science portfolios.

INDEPENDENT PRACTICE

Have the students provide written answers to the Section Review. In their journals, have the students write an outline of how a generator or an electric motor works. They might include a diagram with their outline.

EVALUATION

Ask the students how the function of a generator and an electric motor are related. (In a generator, a current is induced by turning a wire coil or by moving a magnet. In a motor, a magnetic field is produced by having alternating current flow through the coil.)

SCIENCE TECHNOLOGY SOCIETY Scientists have turned to electricity to improve meals for U.S. soldiers. Bacteria and other undesirable organisms in liquid food can be destroyed by pulses of high-voltage electricity rather than traditional heating methods, such as frying or baking. The new technique destroys bacteria without subjecting food to high temperatures, which can produce a dull flavor and decrease the vitamin content of the food.

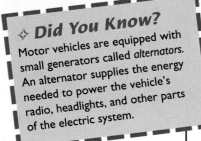

Figure 15–11. The glowing of this heating element is due to the resistance of the burner coil.

☆ Did You Know?

Motor vehicles are equipped with small generators called *alternators*. An alternator supplies the energy needed to power the vehicle's radio, headlights, and other parts of the electric system.

There are also *step-up* transformers. A step-up transformer increases voltage so it can be used for long-distance transmission. A power plant generates electricity at a relatively low voltage. A step-up transformer changes it to many thousands of volts before sending it to a city many kilometers away. This is done to reduce the amount of energy that is wasted during transmission through the wires.

Power, Voltage, and Current

In a previous chapter you learned about power, voltage, and current. Recall that $P = V \times I$. That means that in order to increase the power, either voltage or current must be increased For example, if you wanted to generate 60 W of power, here are two possibilities:

$$P = V \times I$$

$$60 \text{ watts} = 6 \text{ V} \times 10 \text{ A OR } 60 \text{ watts} = 10 \text{ V} \times 6 \text{ A}$$

The equation is like a seesaw. When the voltage of a current is increased by a transformer, the current is reduced. Likewise, when the voltage is reduced, the current increases.

There is one problem, and that is resistance. A wire offers resistance to the flow of electrons through it. This resistance makes the wire heat up. You make use of this fact when you cook on an electric stove. The burner glows red due to the resistance of the burner coil. When electricity is sent long distances, some of the electrical energy is converted to heat energy. This heat energy is wasted since it is not put to any practical use. The amount of energy lost in transmission lines depends on the amount of current in the lines. More current in the wires means more energy is wasted due to resistance. Power companies need to limit the amount of energy wasted when electricity is transmitted over long distances. So the power is sent using very high voltage with very small currents. When the energy reaches its destination, the voltage is dropped and the current is raised so it can be used in homes and businesses.

Figure 15–12. These are examples of step-up and step-down transformers.

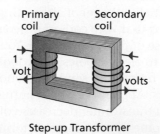

Step-up Transformer   Step-down Transformer

RETEACHING

On the chalkboard write the headings "Generator" and "Motor." Under the headings, have volunteers list similarities and differences between the two types of devices. (Examples should parallel each other. For example, a generator changes mechanical energy into electric energy; a motor changes electric energy into mechanical

energy. A generator uses circular motion to produce a current; a motor uses current to produce circular motion.)

EXTENSION

Have the students use reference resources to learn about three main types of motors: series-wound, shunt-wound, and compound-type.

Suggest that interested students make a poster that illustrates and explains the differences among the motors.

CLOSURE

Cooperative Learning Have the students work in small groups to write a travelogue of Roberto's journey from Miami to Orlando. The travelogue should include what Roberto learned about electricity at specific points along his journey.

What's Inside a Transformer?

As you can see in 15–12, a transformer is an iron core with coils of wire wrapped around it. Alternating current is sent into the primary coil. Recall that a constantly changing current will produce a changing magnetic field. The iron core magnifies this change because it has magnetic properties. This constantly changing magnetic field induces current in the secondary coil. If the secondary coil has more loops of wire than the primary coil, the induced voltage in the secondary coil will be correspondingly greater than the voltage in the primary coil. For example, if the secondary coil has twice as many loops as the primary coil, the voltage on the secondary coil will be twice the voltage on the primary coil. This is a step-up transformer. As you might infer, reversing this system produces a step-down transformer.

Suppose that a power plant produces electricity at 22 000 V. A step-up transformer changes it to 250 000 V for long-distance transmission. Your local station uses a step-down transformer to reduce the potential to 30 000 V for distribution to homes. Finally, a transformer near each home steps down the voltage again to 110 V. As you can see, electricity can be manipulated to provide the best possible efficiency for the job to be done.

Figure 15–13. Thomas Edison's first generator had no transformers. The current generated was the current delivered to the homes.

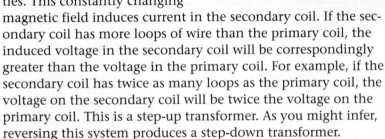

ASK YOURSELF

Why is it not practical for a power plant to send electricity to your house with a high current and a low voltage?

SECTION 1 REVIEW AND APPLICATION

Reading Critically

1. How can turning a loop of wire generate electricity?
2. How are the primary and secondary coils of a step-up transformer different from those of a step-down transformer?
3. Why is the current in a typical generator coil always AC?

Thinking Critically

4. Explain why a transformer will not operate on DC.
5. Why do you think high potentials are used to transmit electric power over long distances?
6. Look again at Figure 15–13. How do you think the scene of Edison's generator would compare with a modern generating plant?
7. How is an electric motor similar to a generator? How is it different?

ONGOING ASSESSMENT

ASK YOURSELF

Sending high current would waste a lot of energy due to resistance.

SECTION 1 REVIEW AND APPLICATION

Reading Critically

1. Turning a loop of wire in a magnetic field induces a flow of current in the wire.

2. In a step-up transformer, the primary coil has fewer loops of wire than the secondary coil. In a step-down transformer, the primary coil has more loops of wire than the secondary coil.

3. The current reverses direction every half cycle.

Thinking Critically

4. A direct current will not produce a changing magnetic field in the transformer.

5. High potentials are used to reduce the amount of current carried by the transmission lines, and thus reduce energy loss.

6. Answers might include the use of computers instead of meters on the wall; nonflammable floors, turbine generators instead of Edison's DC generator; and that today's generators are more complex.

7. Both electric motors and generators include a magnet and a wire loop. However, a generator converts mechanical energy into electric energy. A motor converts electric energy into mechanical energy.

Testing Electromagnetic Induction

Process Skills:
Experimenting, Inferring, Analyzing

Grouping: Groups of 3 or 4

Objective
● **Induce** a current into a coil of wire.

Pre-Lab
Discuss with the students what types of measuring they have done during this course. (Examples include measure distances, volume, and temperature. Accept all reasonable responses.) Explain to the students that in this Investigation they will use a device called a galvanometer to measure electromagnetic induction.

Procedure
3. Current is induced only when the magnet is in motion.

4. As the magnet is removed, the current is in the opposite direction from that induced while the magnet was being moved into the solenoid.

5. Reversing the poles reverses the direction of the current.

6. If the magnet is moved more rapidly, there is more current, but it lasts for a shorter period of time. (Note: This effect can be seen only with fairly slow motion. If the magnet is moved too fast, the galvanometer will not be able to respond rapidly enough to show a short pulse of high current.)

7. More magnets produce a stronger field and will induce larger currents.

Analyses and Conclusions
1. There is no current unless the magnet is in motion, changing the magnetic field within the solenoid.

2. Using a stronger magnetic field or moving it faster produces more rapid changes in the field and more current.

▶ Application
Rotating a coil within a magnetic field in a generator results in a current that reverses direction every half cycle.

✳ Discover More
Moving the solenoid induces a current just as moving the magnet did.

Post-Lab
Have the students hypothesize what would happen if they replaced the bar magnets with plastic or wooden bars. (These items would not create electromagnetic induction because they do not have magnetic fields.) If possible, allow the students to test their hypotheses.

INVESTIGATION

Testing Electromagnetic Induction

▼ MATERIALS
● galvanometer ● solenoid ● insulated wire leads (2) ● bar magnets (3)

▼ PROCEDURE

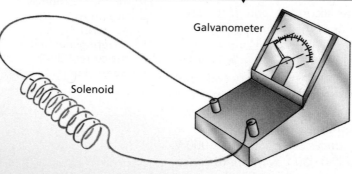

Galvanometer

Solenoid

1. Examine the galvanometer. If the galvanometer has more than one setting, be sure it is set to its most sensitive range.
2. Set up the apparatus as shown. With this arrangement, any current induced in the solenoid will pass through the galvanometer. The galvanometer measures the amount of current.

3. Take one of the bar magnets, and insert its N pole into the solenoid, while observing the galvanometer needle. What happens?
4. Pull the magnet out while observing the galvanometer. What happens?
5. Turn the magnet around and insert its S pole into the solenoid. Compare the result of moving the S pole

in and out with that of moving the N pole in and out. Record both results.
6. Vary the speed with which you move the magnet. Does the speed of the magnet affect the amount of current induced? What happens if you do not move the magnet at all?
7. Create a stronger magnetic field by putting two bar magnets alongside each other, with N poles and S poles together. Test this stronger magnetic field. Record your observations. Then increase the strength of the magnetic field again by using three magnets. How does the amount of current induced depend on the strength of the magnetic field?

▶ ANALYSES AND CONCLUSIONS
1. What evidence did you find that current is induced by a changing magnetic field?
2. What two kinds of evidence did you find to show that more current is induced if the field changes rapidly?

▶ APPLICATION
How is the principle you have just investigated used in making a generator that produces alternating current?

✳ Discover More

Hold the bar magnet steady within the solenoid, and, while observing the galvanometer, move the solenoid back and forth. What happens? How is this result similar to that of the Investigation you carried out?

FOCUS

This section discusses household circuits, including methods for protecting against overloads and short circuits.

MOTIVATING ACTIVITY

Cooperative Learning Write the terms *ground, hot wire, short circuit, overload, fuse,* and *circuit breaker* on the board. In small groups, have the students discuss the meaning of each term, decide on the best definition, and write it on a sheet of paper. Have the students keep their definitions to refer to at the end of this section.

PROCESS SKILLS
• Observing • Inferring
• Interpreting Data
• Measuring

POSITIVE ATTITUDES
• Curiosity
• Cooperativeness

TERMS
• short circuit • fuse
• circuit breaker

PRINT MEDIA
"Going Wireless" from *U.S. News and World Report* by Leonard Weiner (see p. 337b)

ELECTRONIC MEDIA
Electrical Circuits Encyclopaedia Britannica Educational Corp. (see p. 337b)

SCIENCE DISCOVERY
• Electric meter
• Appliance wattage; table

BLACKLINE MASTERS
Reading Skills
Extending Science Concepts
Study and Review Guide

Household Circuits

SECTION 2

After the hurricane had crossed the state of Florida, Roberto and his family headed for home. Everyone in the car was in a serious mood. They had seen the television reports of the hurricane damage. It showed areas of southern Dade county totally destroyed, with block after block of flattened houses looking as if they had been bombed. Roberto's family wondered whether they still had a home to go to.

The newscaster had shown pictures of power lines down, sparking on the wet ground. The preliminary reports said that some areas would be without electricity for weeks. Roberto realized now why his dad had asked him to think about how he would live without electricity. Imagine having no air conditioning or refrigeration with daytime temperatures in the 90s!

Roberto remembered that his dad had switched off the main circuit breaker before they left. Now he wanted to know more about how that device worked and how electricity moved around his house. Maybe he could help his family make repairs. He decided to look at his science book again.

Objectives

Describe the components of a typical household electric circuit.

Compare and **contrast** the roles of fuses, circuit breakers, ground wires, and grounded appliances as safety features.

✧ **Did You Know?**
Hurricane Andrew, which reached a maximum speed of 262 kilometers per hour, killed 33 people and left 300 000 people homeless.

INTEGRATION– Social Studies

The following magazine article shows the widespread devastation from Hurricane Andrew and tells how scientists have learned more about hurricanes.

Gore, Rick. "Andrew Aftermath." *National Geographic* 183 (April 1993): 2–37.

Figure 15–14.
Hurricane damage in Homestead, Florida

● **Process Skill:**
Communicating

Have the students describe in their journals some of the experiences they have had with electric wiring in their homes. (Examples include setting up entertainment units and computer equipment, and rewiring appliances or electric outlets.) Emphasize that electric wiring is not always correctly installed in buildings and can pose dangers.

● **Process Skill:** *Analyzing*

Display an assortment of electric cables, each with some of its insulation stripped off. Have the students compare the wires within each cable and try to identify the hot wire in each.

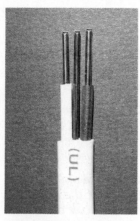

Figure 15–15. Electric cable such as that shown here is used to carry electric current in homes and other buildings.

Circuits

You are familiar with electric outlets, ceiling fixtures, and wall switches. You might also have seen a box with fuses or circuit breakers somewhere in your home. All of these devices are connected to wires inside the walls.

A typical cable used in wiring houses is composed of a group of three wires enclosed in a plastic casing. Two of the wires, one covered with gray insulation and one covered with blue insulation, carry alternating current to the outlets. Touching either of these wires could be dangerous. The third wire is not insulated and acts as a safety feature. If there is any leakage of current, the electricity will flow to the earth, or ground, through the ground wire rather than through you.

Roberto wondered whether his house had been damaged by any of the nearby trees falling on it. If one of them had broken some of the electric wiring, maybe his video games had been ruined. He expressed his fears to his dad.

"Roberto, that is the least of your worries," his father said. "Before we left, I turned off the main circuit breaker so that no electricity could enter the house, even if lightning struck a nearby electrical pole. Your mom and I also unplugged all of the appliances. The house is protected against short circuits."

Roberto wondered what a short circuit was, but his dad was now talking to his mom about insurance, so he decided to look in his science book.

Accidental Connections Worn insulation or a poor connection can create a short circuit. A **short circuit** is any acci-

● **Process Skill:** *Applying*

Ask the students why current would pass through the ground wire rather than through their body if they touched the metal shell of an appliance. (Dry human skin has a fairly high resistance, whereas the ground wire has a very low resistance. Current will flow through the

path of least resistance.) Point out to the students, however, that while dry skin has a resistance of 500 000 ohms, wet skin has a resistance of only about 1000 ohms.

● **Process Skill:** *Problem Solving*

Have the students describe power failures they might have experienced. Discuss with the students how such problems might be avoided or eliminated. Have the students place their

descriptions in their science portfolios. (Possible answers include frequent checking of the conditions of electric cords, replacing defective or frayed cords, repairing malfunctioning appliances, adding more outlets, not plugging too many appliances into one outlet, and distributing the wires evenly among outlets.)

dental connection that allows current to take an unintended path instead of passing through an appliance. It gets its name because the circuit the electricity takes is shorter than its intended path. For example, a short circuit in a radio would keep the radio from working even when it's turned on. The current would flow through the short circuit instead of the radio's electronic components and speakers.

A 110-V potential difference can produce an enormous current if there is no appliance in the circuit to provide resistance. If a short circuit occurs, the wires carry more current than they were designed to carry. The wires can overheat and cause fires.

Safety Features In modern homes, the ground wire in the electric cable is designed to protect people against short circuits. The metal shell of an appliance is connected to the ground wire through the round, third terminal of the plug. If the "hot" wire touches the shell, the current goes directly to ground through this low-resistance path. If you touch the shell of a grounded appliance, very little of the current will go through your body because the appliance is already grounded. An appliance with a plastic shell insulates the user from the current. Such an appliance does not need a grounding plug. You can find out whether the appliances in your home have this safety feature by trying the next activity.

 Observing

Examine the plugs on the appliances you have at home. Which of them have a third, round prong? Which of your appliances do not need a third prong? Why? Do you own some without a third prong that should have one? What could you do about this? 🖊

▼ **ASK YOURSELF**

Why are short circuits considered a safety concern?

Overloads

As Roberto's family approached their house, Roberto couldn't believe the destruction. Most trees were uprooted or simply gone. Those that remained were stripped of leaves. All the road signs had blown away. It was difficult to recognize the once-familiar routes. Along the way Roberto

DISCOVER BY *Observing*

The students will probably find that large appliances, such as washers, dryers, microwaves, and refrigerators, have three-pronged plugs. Any appliances with high-wattage requirements should be grounded.

SCIENCE TECHNOLOGY SOCIETY To the untrained eye, satellite images of South Florida after Hurricane Andrew resemble images of two pictures of the same place—but in two different colors. Remote sensing experts at the Florida Marine Research Institute in St. Petersburg know that the images mean acres and acres of dead mangroves and other ecological and environmental damage.

The satellite images will assist biologists who surveyed South Florida by foot, vehicle, boat, and airplane since Hurricane Andrew struck on August 24, 1992.

Using the satellite images, agencies from the state Department of Transportation to the state Department of Environmental Regulation charted everything from highways to patterns of wetlands destruction. The satellite images have documented this damage to the environment:

• About 10 000 acres of mangroves have been damaged

• Roughly $4.5 milion is damages occured to South Florida beaches

• State parks suffered about $7.5 million in damage

GUIDED PRACTICE

Have the students prepare drawings that show potential electric hazards in a home. (Examples include overloaded circuits or outlets, use of electric appliances near water, and frayed electric cords.) Then tell the students to exchange drawings and circle the hazards shown. Have the students explain the hazards they identified.

INDEPENDENT PRACTICE

 Have the students provide written answers to the Section Review. In their journals, have the students describe specific steps they can take in their own home to help prevent electric fires.

EVALUATION

Have the students explain how a short circuit leads to an overload. Encourage the students to accompany their explanation with an illustration. (When current flows through a shorter path, the wire carries more current than intended, resulting in an overload.)

SCIENCE BACKGROUND

Two basic types of circuit breakers are thermal circuit breakers and electromagnetic circuit breakers. A thermal breaker has a bimetallic strip. When the current flowing through the strip exceeds the maximum level allowed, enough heat will be generated to bend the strip, snap open the contacts, and break the circuit. Thermal circuit breakers react rather slowly to overload conditions, but they break circuits quickly in case of a short circuit.

ACTIVITY

How does a fuse work?

Process Skills: Observing, Inferring

Grouping: Groups of 3 or 4

▶ **Application**

1. The fuse wire melted because too much current was flowing through it.

2. No electricity is flowing through the fuse when the lamp is turned off.

★ **PERFORMANCE ASSESSMENT**

Have the students relate their experiences with other safety features. Examples include circuit breakers, thermostats, and automatic shut-off valves. Evaluate their examples for appropriateness.

▶ **404** CHAPTER 15

Figure 15–16. These are fuses found in some buildings.

saw transformers that had been destroyed. His dad told him that the damage to the transformers could have been caused by lightning or by flying debris. After their conversation, Roberto began to read the section about overloads.

Have you ever plugged in too many appliances at once? If you have, you may have overloaded the circuit. Your home is probably wired with four, five, or more parallel circuits. Several outlets are connected to each circuit. If too many appliances on one circuit are turned on at the same time, too much electric current moves through the wire at once. This can cause an overload, and that circuit will shut down, cutting off the electricity.

Too much current can cause wires to heat up and melt their insulation. Then current can contact the material of the walls and cause a fire. Building codes require that homes be equipped with safety devices to prevent this from happening. These devices are designed to turn off the current if an overload occurs, before an electric fire can start.

Protecting the Circuits Protection against overloads may be provided by a fuse. A **fuse** is a safety device containing a short strip of metal with a low melting point. If too much current passes through the metal, it melts, or "blows." This causes a break in the circuit, and the current can no longer flow. This is your signal to find and correct the overload and to replace the fuse. Why should you correct the overload before you replace the fuse? Try the next activity to find out.

ACTIVITY   How does a fuse work?

MATERIALS
safety goggles,
15-A plug fuse, table lamp
CAUTION: Wear safety goggles throughout this activity.

PROCEDURE
1. Examine the fuse. Where are the terminals? Look in the little window to see the fuse wire.
2. Turn off the lamp, and remove its plug from the wall outlet.

3. Screw the fuse into the lamp socket, and plug the lamp into the wall outlet.
4. While wearing safety goggles, look carefully at the fuse wire, and, at the same time, turn on the lamp.

APPLICATION
1. Explain what happened to the fuse wire in step 4.
2. Why is it that nothing happens to the fuse until you turn the lamp on?

RETEACHING

Display a new fuse, a blown fuse, and a cir-cuit breaker. Review with the students how each device

PORTFOLIO ASSESSMENT

works. Then have the stu-dents make a chart in their jour-nals that summarizes the simi-larities and differences between fuses and circuit breakers.

Encourage the students to include a drawing of each item with their charts. Ask the students to place their charts and drawings in their science portfolios.

EXTENSION

Interested students might find out about the building codes for electric wiring in their commu-nity by calling the city or village hall for the department to con-tact for information. Ask the students to report their findings to the class. Alternatively, you might invite a building inspector to speak to the class about local codes for electrical wiring.

CLOSURE

Cooperative Learning Refer the students to their list of defini-tions from this section's Motivating Activity. Based on what they learned in this sec-tion, have the students modify their definitions as needed. Then review the definitions with the class.

Break It Up Another device that protects circuits from over-loads is a **circuit breaker**. One type of circuit breaker is a switch attached to a bimetallic strip. When the metal gets hot it bends, which opens, or "trips," the circuit. This action does not harm the circuit breaker. After the overload has been corrected, the circuit breaker can be reset.

Roberto's dad had switched off the main circuit breaker before they left the house. This may have protected Roberto's video games from an overload. He remembered that the electrical box in the garage contained many circuit breakers. They were labeled according to the appliances or areas of the house they protected. Once when his mom had connected too many appli-ances to the same outlet, one of the circuit breakers had tripped. When she unplugged two of the appliances and switched the tripped circuit breaker back on, everything was fine.

Figure 15–17. Circuit breakers interrupt the flow of current.

DISCOVER BY *Doing*

PORTFOLIO ASSESSMENT

After the students have completed the *Discover by Doing* activity, have them place their tables in their science portfolios.

ONGOING ASSESSMENT
ASK YOURSELF

When a circuit is overloaded, a circuit breaker need only be reset, not replaced. A fuse must be replaced.

DISCOVER BY *Doing*

Have a parent or other adult help you locate and examine the fuse box or circuit-breaker box in your home. How many separate circuits supply your home? How many of them are 15 A? How many are 20 A? Each time a fuse blows or a circuit breaker trips, note the outlets that are affected. In your journal, make a table showing which outlets in your home are on the same circuit. ✐

ASK YOURSELF
What is the advantage of a circuit breaker over a fuse?

SECTION 2   *REVIEW AND APPLICATION*

Reading Critically
1. Why are houses wired with a ground wire?
2. What steps must you take if a circuit breaker in your home trips?
3. Why is a grounding wire unnecessary in an electric mixer that has a plastic casing?

Thinking Critically
4. Explain why the third prong of a grounded plug should not be removed to make the plug fit a two-pronged outlet.
5. What is the largest number of 100-W bulbs that can be turned on in a 115-V circuit without blowing a 15-A fuse?
6. If you were an insurance agent examining a building someone wanted you to insure, what would you look for in its electric system?

SECTION 2 REVIEW AND APPLICATION

Reading Critically

1. The ground wire protects people against short circuits.

2. Turn off some appliances on that circuit, and reset the circuit breaker.

3. A plastic casing is a noncon-ductor.

Thinking Critically

4. In case of a short circuit, the shell of the appliance could become electrically hot. If you were to touch this ungrounded appliance, your body could pro-vide a path to ground and you could be shocked.

5. 17 (fuse protects up to 115 × 15 = 1725 watts: 17 bulbs at 100 watts = 1700 watts

6. Answers include making sure the house is wired with the right kind of cables, checking for circuit breakers or fuses, and verifying that appliances are properly grounded.

SKILL

Calculating Your Electric Bill

Process Skills: Interpreting Data, Measuring

Grouping: Individuals

Objectives
- **Interpret** data from a meter.
- **Calculate** the amount of energy used from the data.

Discussion
The students should have little difficulty in reading the meter twice, subtracting, and multiplying by the rate.

Procedure
2. You should read the lowest number that is closest to the pointer.

▶ Application
1. 3988

2. $2.72

3. $3.47

4. In a 30-day month, $0.86 could be saved by using 60 W bulbs.

Using What You Have Learned

1. Answers will depend upon the electric bill of the students' families. If possible, check students' calculations.

2. The students should calculate the average daily cost and multiply it by the number of days in a month.

★ PERFORMANCE ASSESSMENT

Have the students record the number of hours they watch TV in one week and calculate the cost of watching TV. Check the students' calculations, which should be based on the wattage of the TV, the number of hours watched, and the rate of charge.

PORTFOLIO ASSESSMENT Have the students place their observations and answers from the *Skill* in their science portfolios.

SKILL *Calculating Your Electric Bill*

An electric utility company charges customers according to the amount of electric energy they use. Electric power, measured in watts, is the amount of energy used per second.

The unit of measure for electric energy is the *kilowatt-hour* (kWh). A kilowatt hour represents 1000 watts of power used for an hour.

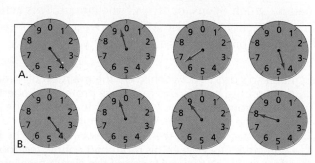

The energy used in a household is measured by a kilowatt-hour meter. It is usually located either in the basement or outside the house. A meter reader comes once a month or every other month to read the meter so that the company can calculate the bill.

▼ PROCEDURE

1. To find the amount of electric energy used, multiply the power (in kW) used by the length of time (in hours) it was used.
2. The illustration shows the dials on a typical kWh meter. Each dial stands for a digit of a 4-digit number. Note that all dials are read clockwise. The dials in the top meter read 3964 kWh. Why is the third dial read at six instead of seven?

▶ APPLICATION
1. Read the meter at the bottom, and record the reading.
2. Suppose the rate the electric company charges for energy is $0.08 per kWh. If a meter reads as in Figure A at one time and as in Figure B later, how much will the electric bill be for the time between the readings?
3. How much does it cost to run a 350-W television for an average of four hours every day for the month of March?
4. The three bulbs in a kitchen light fixture burn for an average of three hours a day. How much could be saved every month by reducing the power of the bulbs from 100 W each to 60 W each?

※ Using What You Have Learned
1. Look at the electric bill for your home. What rate does your family pay for electric energy per kWh?
2. Read your home's electric meter every day for five days. Using your electric power company's table of rates, calculate your bill for each day. What do you think your bill will be for the entire month?

FOCUS

This section explains how some electronic devices work and discusses differences between analog signals and digital signals.

MOTIVATING ACTIVITY

Ask the students what the word *electronics* means to them. Point out that electronic devices use electricity to process or send information. Have the students list electric devices and electronic devices on the board in two separate columns. (Electric devices include motors, toasters, and light bulbs. Electronic devices include radios, CD players, computers, and televisions.) Conclude by telling the students that they will learn about four kinds of electronic devices in this section: telephones, speakers, radios, and televisions.

Electronics

SECTION 3

§ When Roberto's family reached their own neighborhood, he noticed that most of the electric poles around the neighborhood were down. He knew it would be a long time before electricity was restored. At least they still had a roof over their house. The house next door was missing half of its roof. The yard was filled with debris—pieces of other houses, trees, and scraps of metal. They couldn't get the car into the garage because debris was in the way. Roberto got out of the car and began to clear the driveway.

Later, when Roberto walked into the house, he was amazed to hear the telephone ringing. His sister was calling from Orlando to see how things were. "But Mom," said Roberto, "how can the phone be working if there is no electricity?"

Objectives

Distinguish between an analog signal and a digital signal.

Describe the operation of a telephone and a television.

Compare electronic circuits with other types of electric circuits.

● **Process Skill:** *Generating Ideas*

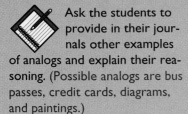

Ask the students to provide in their journals other examples of analogs and explain their reasoning. (Possible analogs are bus passes, credit cards, diagrams, and paintings.)

● **Process Skills:** *Constructing/Interpreting Models, Analyzing*

Display a telephone handset, and remove the cap from the mouthpiece and earpiece so that the students can examine the handset's internal compo-

nents, such as those shown in Figure 15–18. Point out how the disk in the mouthpiece can move. Encourage the students to identify other components, if they are readily visible.

SCIENCE BACKGROUND

During related experiments with the telegraph, Alexander Graham Bell (1847–1922) found that a device containing vibrating reeds could generate sound. From that idea, Bell made a steel disk vibrate by speaking against it. The disk was placed near an electromagnet and made a pattern of vibrations on the current passing through the electromagnet. The pattern could be duplicated so a listener at the other end could hear the vibrations.

In order to make it work, Bell hit upon the idea of modulating current to parallel the wave forms of the voice. He modulated the signal by varying some characteristic of one wave in accordance with another.

The invention of electronics led to the development of audio surveillance, which includes telephone wiretapping and eavesdropping with hidden microphones. The legality of electronic surveillance has been a long-standing issue debated over many years. Various laws that attempt to limit the use of eavesdropping have been passed since the 1930s. For example, a federal law was passed in 1968 allowing government agencies to use surveillance devices in certain criminal investigations if a court order had been previously obtained.

Hello, Is Anyone There?

Roberto's mom told him that the telephone network is separate from the electric power grid. "The telephone lines in this neighborhood are underground," she explained, "so they did not suffer any damage. I'm sure that's not the case in many places. We were lucky." Roberto remembered what his book said about how a telephone works.

Converting Sound Waves Once electricity reaches your home, it is changed by various appliances into heat, light, and motion. Electricity is also used by circuits in electronic devices such as telephones, VCRs, computers, and televisions.

The telephone converts the sound waves produced by your voice to an audio signal. The audio signal is a series of electric impulses. This signal is an analog of the sound wave. An **analog** is something that is a substitute for something else. Let's say your mom writes a check to pay a bill. The check is not cold cash, but it does represent money. It can be converted into real dollars. In the same way, an analog stands for and can be converted into something else, such as sound.

When you speak into a phone, the sound waves you produce cause a metal disk in the telephone to vibrate. These vibrations produce an audio signal that is the analog of your voice. This audio signal is then amplified and directed through a switching system to your friend's telephone. There it causes another disk in the earpiece to vibrate, producing the same pattern of sound waves that was received by your mouthpiece.

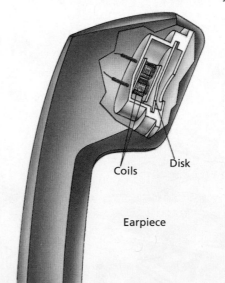
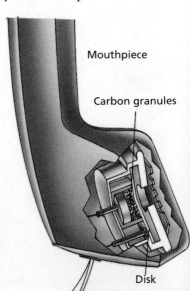

Mouthpiece

Carbon granules

Coils

Disk

Earpiece

Disk

Figure 15–18. These cutaway views of a telephone show the internal parts of the mouthpiece and earpiece.

● **Process Skill:** *Comparing*

Ask a volunteer to draw an analogy between digital signals and a light switch. (Digital signals are a series of pulses that stand for two digits, 0 and 1. Each pulse stands for the number 1. A missing pulse stands for the number 0. When a light switch is off, no electric current can flow; it is similar to the number 0. When a light switch is on, current can flow; it is similar to the number 1.)

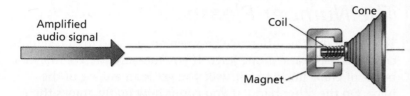

Figure 15–19. The basic makeup of each of these speakers is shown in the diagram. The reaction between the magnetic field of the coil and that of the magnet causes the cone to vibrate to produce sound.

You can see a similar procedure in a large speaker. Changes in electric current cause a coil in back of the speaker cone to be attracted to and repelled from a permanent magnet. You can actually see the speaker cone in a large speaker vibrate back and forth. Go to your nearest electronics store, and check it out.

As telephones have become more complex and their use has increased, more efficient methods of transmission have become necessary. Analog signals carry some unavoidable noise. Every amplifier and electronic circuit adds more noise to the signal. Perhaps you have occasionally noticed the static on the line during a telephone call.

Digital Signals   Technology has made it possible to replace analog signals with digital signals. A **digital signal** is a series of electric pulses that stand for two digits, 0 and 1. These digits make up binary numbers. Each of the digits in a binary number is called a *bit*. An electrical signal can carry bits in the form of a string of pulses. Each pulse stands for the number 1; a missing pulse stands for the number 0. Information is coded by the sequence of 1s and 0s.

Sailors use a flashing light to communicate from ship to ship. They use a lantern that can click on and off at different speeds. One quick flash followed by a longer signal could mean hello. They can carry on long-distance conversations as long as they are able to decode the sequence of light bits. This is similar to a digital signal. A digital telephone is connected directly to an electronic circuit that produces a string of bits.

Digital signals offer many advantages over analog signals. There is no noise in a digital signal, because digital information is a count, not a measurement. Many instruments use digital signals in their electronic circuits. Videodisc players, computers, and fax machines are examples. Many telephone companies have already changed their systems to digital signals. This change has permitted information to be faxed through phone lines. It also allows for computers to transmit data via modem through phone lines.

▼ **ASK YOURSELF**

What is a digital signal?

REINFORCING THEMES
Energy

Point out to the students that a telephone works by converting one form of energy into another. In the mouthpiece, sound energy (voice) is converted into mechanical energy (vibrating disk), which is converted into electrical energy (audio signals). In the earpiece, electric energy (audio signals) is converted into mechanical energy (vibrating disk), which is converted into sound energy (voice).

INTEGRATION–
Mathematics

Our system of numbers has a base of ten. The binary system has a base of two. For example, the number 11 in base 10 is equal to 1011 in base two.

| Base Ten | | | |
Thousands	Hundreds	Tens	Ones
0	0	1	1

| Base Two | | | |
Eights	Fours	Twos	Ones
1	0	1	1

ONGOING ASSESSMENT
▼ **ASK YOURSELF**

A digital signal is a series of electric pulses that represent the digits of binary numbers.

● **Process Skills:**
Communicating, Inferring

Discuss the process of radio transmission with the students. Then have them draw a simple diagram of the transmission from a radio station to a radio. Ask the students why radio

waves travel best through space. (There is nothing to interfere with the electromagnetic waves.)

● **Process Skill:**
Communicating

 Refer the students to the electromagnetic spectrum shown in Chapter 11. Have them identify the wavelengths of radio waves. (10^{-1} to 10^4 meters) Interested students might research the different kinds of radio waves and

write their research results in their journals.

DISCOVER BY *Writing*

 After the students complete the *Discover by Writing* activity, have them place their letters and its digital codes in their science portfolios.

SCIENCE TECHNOLOGY SOCIETY Telecommunication companies are field-testing digital car phones to replace current models. A digital car phone system has several advantages—improved sound quality, nearly no static or interference, and security since phone conversations cannot be picked up by radio scanners.

ONGOING ASSESSMENT
▼ **ASK YOURSELF**

Digital signals have less noise than analog signals.

 LASER DISC

30399, 30400–30697

Animation; electric meter

1863, 1864

Energy use, appliance; table

Figure 15–20. An analog signal can be compared to a pitcher of grape juice. A digital signal can be compared to a bunch of grapes. A quantity of juice must be measured. However, a quantity of grapes can be counted.

The Number Please

To understand the difference between analog and digital signals, consider grapes and grape juice. If you measure the volume of a bottle of grape juice, the answer you get is an analog of the juice. On the other hand, if you count how many grapes there are in a bag, the answer is a whole number, a digital statement. Now suppose you pour the juice from one bottle into another. You will introduce some inaccuracy, some "noise." You might leave the original bottle wet; there might have been some dirt in the second bottle. If you move some grapes from one bag to another, you might introduce some noise. Maybe the grapes were wet, or the bag was dirty. However, in the digital signal the noise does not matter. The number of grapes, the digital information, has not changed.

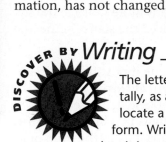 DISCOVER BY *Writing*

The letters of the alphabet can be expressed digitally, as a series of 1s and 0s. Use the library to locate a source that shows the alphabet in binary form. Write a brief letter in your journal, and translate it into digital code. See whether a classmate can decode the letter. ✍

▼ **ASK YOURSELF**

Why are digital signals being used more and more in preference to analog signals?

Radio and Television

ʃ It had been two weeks since Hurricane Andrew struck southern Florida. In that time Roberto and his family had cleared all the debris from around the house and made some minor repairs on the roof. They were able to run a few things from their generator, but the area still had no electricity. The latest estimate was that it would take at least another two weeks to restore power. People tried to avoid going out to buy food or fuel. Without traffic lights, the whole area was in a constant traffic jam. It took two hours to get to the nearest supermarket and back. The heat and the mosquitos made the evenings hard to bear.

Being without electricity showed Roberto how important it is. Anything electronic now interested him. He had begun to enjoy finding out how things work. "Maybe," he thought, "I'd like to become an electrician or an electrical engineer."

GUIDED PRACTICE

Ask the students to explain how electronic devices send information. (They send information by electric signals in analog or digital form.)

INDEPENDENT PRACTICE

 Have the students provide written answers to the Section Review. In their journals, have the students write one or two paragraphs about the impact of television on their lives.

EVALUATION

Cooperative Learning The students might need practice in reading diagrams. In small groups, have them use the diagram of the telephone on p. 408 to make a poster that explains the operation of it. Ask them to reproduce the diagram on their poster with labeled arrows detailing how the device works.

Electromagnetic Waves

While Roberto's school was closed, his only source of amusement or information was a battery-powered radio. "Dad," he said one hot afternoon, "I bet the radio station we're listening to uses audio signals the same way a phone does. But how do these signals get to us since the radio isn't connected to any wires?"

"Audio signals are converted to electromagnetic waves," his dad explained. "Your radio antenna picks up these waves and changes them back into electric signals. The audio signal is then amplified and sent to the speakers."

"What about a television, Dad?" Roberto asked. "Besides the air conditioning, I miss television most of all. How does it work?"

"Roberto," his dad replied, "let's look together at your science book to find some answers."

You probably don't realize that the pictures you see on television are painted by electrons. The moving pictures are produced on the inside of picture tubes by beams of electrons. In a black-and-white television, a single electron beam is produced inside the picture tube and aimed at the screen. The screen is coated on the inside with chemicals called **phosphors.** When the electron beam strikes the phosphors, they glow. The electron beam very quickly scans the inside surface of the screen.

MULTICULTURAL CONNECTION

The word television comes from the Greek word *tele*, meaning "far," and the Latin word *videre*, meaning "to see." Thus, television means "to see far."

Today, there are more than 200 million television sets in the United States. Japan has about 31 million television sets—about five times more than any other Asian nation. While television is still a small industry in some countries, the industry is growing rapidly in other countries, including Israel, Kuwait, and South Korea.

RETEACHING

Display a telephone that has a mouthpiece and earpiece with removable covers. Allow time for the students to examine the telephone's components. Ask them to find the metal disks in the mouthpiece and earpiece, and compare their functions.

(In the mouthpiece, the vibrating disk produces an audio signal. In the earpiece, it produces sound waves when it vibrates.)

EXTENSION

Invite a TV repairperson to speak to the class and to show the students the internal components of a color television and a black-and-white television. Ask the speaker to describe the kind of repair problems he or she most frequently encounters.

CLOSURE

Have the students write an outline that summarizes the concepts in this section. You might prepare a master outline on the chalkboard that combines the students' individual outlines. Encourage the students to place their outlines in their science portfolios.

MEETING SPECIAL NEEDS

Gifted

Have the students use reference resources to find out how video and audio signals are transmitted from a color TV camera to a television set. Encourage the students to draw a detailed diagram of the process.

ONGOING ASSESSMENT
▼ ASK YOURSELF

A telephone uses electricity to transmit information. Other kinds of electric systems use electricity to operate motors.

SECTION 3 REVIEW AND APPLICATION

Reading Critically

1. An analog signal is a measurement; a digital signal is a count.

2. A black-and-white television has one electron gun and colorless phosphors. A color television has three electron guns and three colors of phosphors.

Thinking Critically

3. A photograph is an analog because it represents the scene of which it was taken.

4. Analog: Answers should include items that are weighed to determine the price. Digital: Answers should include items that are premarked with the price.

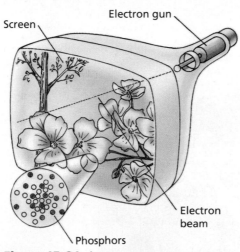

Figure 15–21. An electron beam causes phosphors on the screen to glow.

Figure 15–22. Televisions that use digital signals can show more than one program on the screen.

Phosphor Dots Television signals travel to houses from the station by electromagnetic waves. But this wave carries both the audio signal and the video signal. The video signal causes fluctuations in the strength of the electron beam. How much the phosphors glow depend on the strength of the electron beam at that particular instant. The changing brightness of the phosphors is what creates an image on the television screen.

In a color television, the screen has phosphors of three different colors. These *phosphor dots* are arranged in groups of three. Each group, called a *pixel*, contains a red dot, blue dot, and green dot. There are three electron guns in a color television. One scans only the red phosphor dots in each pixel, another the blue, and the third the green as their beams move across the screen. The signal coming into the television controls the fluctuations of each of the electron beams. You can actually see the small phosphor dots by looking at a television screen with a magnifying glass. Because the dots are so small, your eyes see blends of colors rather than individual colored dots.

> While Roberto's parents went to help their neighbor repair his roof, he sat thinking about how devastating this hurricane had been. "I wouldn't want to go through this again," Roberto said to himself, "but I certainly learned a lot the hard way—about electricity and its uses!"

▼ **ASK YOURSELF**
In what way are electronic systems, such as that of a telephone, different from other kinds of electric systems?

SECTION 3   REVIEW AND APPLICATION

Reading Critically
1. Explain the difference between an analog and a digital signal.
2. How are black-and-white televisions different from color sets?

Thinking Critically
3. Why can a photograph be considered an analog?
4. When you shop, some items are charged by analog measurements and some are charged digitally. Name three examples of each kind of item.

CHAPTER 15 HIGHLIGHTS

The Big Idea—Technology

Use an example from the chapter to review the concept of energy conversion. You might refer to energy conversions that occur in a generator, motor, telephone, radio, or television.

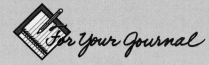
For Your Journal

The students' entries might include references to how electricity can be generated, how motors work, or how electronic devices operate.

CHAPTER 15 HIGHLIGHTS

The Big Idea

 In today's world, we depend on many forms of energy: motion, heat, light, and sound. Our many modern conveniences and connections to the world depend on energy conversions. It all begins with coils of wire moving through magnetic fields. After that it's just a matter of harnessing the energy so that we can run motors, heat and light our homes, and enjoy many forms of communication.

For Your Journal

Think about what you have learned from this chapter about electricity in use. Review what you have written in your journal, and revise or add to the entries. Include information about the importance of electricity to your everyday life.

CONNECTING IDEAS

The concept map should be completed with words similar to those shown here.

1. light energy
2. sound energy
3. heat energy
4. heat energy
5. mechanical energy

Connecting Ideas

Copy the concept map into your journal. Fill in the missing types of energy to show how electricity is transformed.

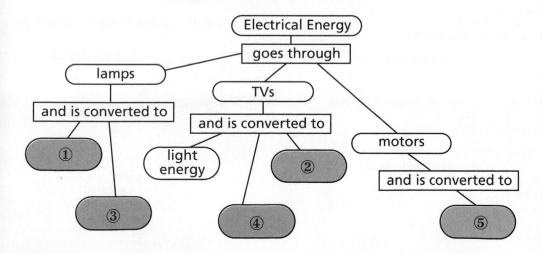

Understanding Vocabulary

1. (a) Each term is related to the path of electric current. A short circuit is any accidental connection that allows current to take an unintended path. A circuit breaker is a device that protects circuits from current overloads. A fuse is another kind of device that protects against current overloads.

(b) Each term is related to the transmission or conversion of electric energy. A power grid is a network of energy transmission lines. A transformer is a device that uses electromagnetic induction to change the electric potential of alternating current from the power grid. A transformer can either "step up" or "step down" the electric potential. An electric motor is a device that converts electric energy into mechanical energy.

(c) Each term is related to electronics. An analog is something that is a substitute for something else. In the case of electronics, an analog signal stands for and can be converted into something real, such as sound. In contrast, a digital signal is a series of electric pulses that stand for two digits, 0 and 1, that make up binary numbers, or bits. An electric signal can carry bits as a string of electric pulses. An analog signal is a measurement, while digital information is a count. Phosphors are chemicals that glow when electron beams strike them. They are used on the inside of television screens. The strength of the electron beam regulates the degree to which the phosphors glow.

Understanding Concepts

Multiple Choice

2. b

3. d

4. b

5. b

Short Answer

6. The force of moving water turns turbines that move the magnets in a generator, thereby inducing a current in the generator's coil.

7. The gray insulated wire carries current with a potential of 110 volts. The blue insulated wire has a potential of 0 volts and allows current to flow to and from an appliance. The uninsulated wire is the ground wire and carries any current leakage to the ground.

Interpreting Graphics

8. When direct current flows through the wire loop, the loop becomes magnetized. Its N and S poles are then attracted to the opposite poles of the permanent magnet. But when the commutator turns to the other side, the current direction is reversed, also reversing the magnetic poles of the loop. This continues constantly, spinning the loop.

CHAPTER

15

*R*EVIEW

▶ Understanding Vocabulary

1. For each set of terms, explain the similarities and differences in their meanings.
  a) short circuit (403), circuit breaker (405), fuse (404)
  b) power grid (391), transformer (397), electric motor (394)
  c) analog (408), digital signal (409), phosphors (411)

▶ Understanding Concepts

MULTIPLE CHOICE

2. The wire loop of an electric motor spins because
  a) the current flowing through the wire is DC.
  b) its magnetic poles are attracted to the opposite poles of the magnet.
  c) a turbine rotates the loop.
  d) its magnetic poles do not reverse.

3. What device changes the voltage of current entering your home?
  a) step-up transformer
  b) power grid
  c) circuit breaker
  d) step-down transformer

4. A device that opens a circuit in case of overheating is a
  a) conductor.
  b) fuse.
  c) transformer.
  d) short circuit.

5. Which of the following converts electrical energy into mechanical energy?
  a) generator
  b) electric motor
  c) transformer
  d) circuit breaker

SHORT ANSWER

6. Explain how hydroelectric power plants produce electricity.

7. Electrical cables used to wire buildings have three wires. Describe the function of each wire.

Interpreting Graphics

8. Explain the movement of the wire loop in the illustration of the electric motor.

9. How might you make the motor turn in the opposite direction?

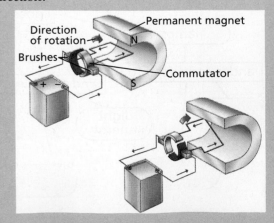

9. The motor would turn in the opposite direction If students reversed the connections to the battery or reversed the position of the wires connected to the loop of wire.

Reviewing Themes

10. The generator does not create energy; rather, it converts one form of energy (mechanical) into another (electric).

11. AC allows easier changes of voltage.

Thinking Critically

12. Since the ground wire is connected to the electric ground of the building, there is no need to insulate it to prevent accidental contact. Also, since the hot wires of the system are insulated, there is enough protection against a short circuit with the bare ground wire.

13. A windmill is a turbine that can be connected to a generator. The movement of the windmill turns the magnet located in the generator.

14. This connection is necessary to provide a short-circuit path to ground in the event that any part of the shell of the system touches a hot wire.

15. A coil in the earpiece becomes an electromagnet when current flows through it. Its changing magnetic field causes a metal disk in the earpiece to vibrate and produce sound waves.

16. To produce current, the coil and the magnet in a generator must be moving relative to each other. Thus, no current would be produced if they were moving simultaneously in the same direction.

17. From the three colors of dots in a color TV, all colors can be made. The dots must be red, blue, and green because these are the primary colors of light. Combinations of these colors will produce any other color.

18. The fuse must be connected in series because the total current flowing through the circuit must pass through the fuse in order to protect the entire circuit.

19. Since the body uses electric signals to stimulate muscles, it is possible for a computer to be used to duplicate these electric signals to stimulate paralyzed muscles artificially.

Reviewing Themes

10. *Energy*
Does a generator create energy? Explain your answer.

11. *Technology*
Explain why AC is sometimes more useful than DC.

Thinking Critically

12. Why is it unnecessary for the grounding wire of a plastic-insulated cable to have its own separate insulation?

13. How can wind power be used to generate electricity?

14. Explain why a home's electrical supply system must be connected to the ground at some point.

15. How is an electromagnet used in the earpiece of a telephone to produce sound waves?

16. Would a generator produce electricity if both the coil and the magnet moved simultaneously in the same direction? Explain.

17. Why does a color-TV screen need only three different colors of dots—red, blue, and green?

18. Should a fuse be connected in series or in parallel with the appliances in a circuit? Explain your answer.

19. The nerves in your body send messages to your muscles by using electrical signals. When the nerves are damaged, they cannot send the proper signals. The woman in the photograph is paralyzed from the waist down, yet she is able to ride a bicycle with the aid of computer output devices attached to her legs. Explain how the computer can "talk" to the muscles so that this woman can ride a bike and walk.

Discovery Through Reading

Vizard, Frank. "Previewing Widescreen TV." *Popular Mechanics* 169 (May 1992): 126–131. This article discusses television technology that produces large, detailed images and digital sound.

Background Information

One noted computer designer predicts that, by the year 2000, a person will be able to sit in front of the terminal of his or her personal computer, put on 3-D glasses and a data glove, and be able to ride on an electron or in a car.

Scientists know that about half of the human brain is dedicated to visual processing. Therefore, virtual reality is the most natural way for people to understand and manipulate computerized data—and be able to do so in three dimensions.

Discussion

● **Process Skills:** *Recognizing Time/Space Relations, Predicting*

Ask the students to speculate on what it might be like learning to drive a car using virtual reality. How might initially learning to drive this way before driving an actual car be safer? (Students might mention that they could practice driving on wet, icy, or snowy roads and if they made a mistake and had a virtual accident, they wouldn't be injured or damage the car.)

A New World at Your Fingertips

from *National Geographic World*

It has many faces and almost as many names. Scientists and computer hackers call it "artificial reality," "cyberspace," "telepresence," and "virtual reality." "Weird but fun" is what Janaea Commodore, of Novato, California, calls it.

Janaea uses wired goggles and gloves to explore a world that exists only in computers. When she moves her gloved hand, the glove instantly sends information about her movements to a computer. The robot hand, projected here on a giant screen, shadows her movements.

On the page the robot hand looks flat, or two-dimensional. But Janaea sees the action on two small television screens inside goggles that block her view of the real world. Each screen shows a slightly different view of the action. As a result, Janaea sees her computer world in three dimensions. It has height, width, and depth, so it looks like the real world. Well, *somewhat* like the real world. . . .

Janaea describes it like this: "I was in a room and could go through doors and windows. I tried to catch balls and pick up objects. At first it was hard. You don't know where your hand is going. But it got easier."

To grab something, Janaea made a fist. To let go, she opened her hand. "When you point your finger, you fly," she says. "That's really cool!"

Janaea spent only about 20 minutes in the wacky world of computer-based reality. She got the chance to take this fantasy trip because her mother works for a company that makes virtual reality software for computers.

The software can do a lot more than provide spacey and spectacular fun. Doctors can use it to travel inside the human body. Architects can walk through buildings that haven't been built. A weather forecaster can fly above the earth. Space scientists can skim the surface of distant planets. Pilots can safely train for emergency situations.

Virtual reality has a lot in common with computer games. But it goes much further. A person actually seems to enter another real world. "In a virtual reality experience you are totally or at least partially in a 3-D world," explains Howard Rheingold, the author of a book on the subject. "You have the ability to navigate within that world and to interact with the objects in it."

To interact—touch or move objects, change your position, or create sounds—you might put on goggles and a wired glove, just as Janaea did. You might climb into a suit that senses every move you make or walk on a wired treadmill. You might stand in front of specialized video cameras, pull objects out of a framed screen, use a joystick, or simply speak.

▶ *Different Drums.* At a computer show, two men drum up sounds in space. You see the drummers as they really are. You see them as they appear on a video screen. A camera feeds their images into a computer. It compares those images with its stored images of drums. A "hit" produces a computer sound.

One of the best things about artificial space is that two people who are really far apart can share it. For example, with their computers linked, they can practice a surgical procedure or play racquetball together.

Artificial reality is still so new that no one can be sure what forms it will take. And no one can predict all the things it may help us do in the future.

Right now the equipment is bulky and expensive. But people who work with computers think that will change within a few years. Someday you may spend part of your time in the real world and part of it in a virtual world that you and computers create together.

▶ *Power Play.* Using a special glove, Jack Menzel, of Napa, California, plays a ball game on his television screen. "To catch the ball, you make a fist," Jack says. "A hand on the screen makes a fist when you do." When you use such a glove, a flat screen seems more like virtual reality. Although you don't see the action in three dimensions, you can move objects simply by moving your hand. You don't need a computer mouse. Jack has also tried the "goggles-and-gloves" virtual reality. He describes it as "a little confusing, but neat."

▶ *Busy Bodies* can be created by computers. This "virtual body" will help medical students learn more about how parts of a real body work. The leg image projected on the screen shows up in three dimensions inside the goggles. Students can look at the leg from all angles and move parts of it around. Surgeons could use a virtual body to practice difficult operations before performing them on real patients.

David Zeltzer of MIT's Media Lab predicts that virtual reality will someday change movies forever because the viewer will be able to become part of a movie through virtual reality.

If you or your students are interested in learning more about virtual reality, you might read the following.

Rheingold, Howard. *Virtual Reality*. Touchstone Books, 1992. This book gives a thorough overview of the history and status of virtual reality. Rheinghold explores virtual reality's potential applications in fields ranging from architecture to medicine and the social implications of the ability to generate artificial realities.

Making A Hit with A Computer

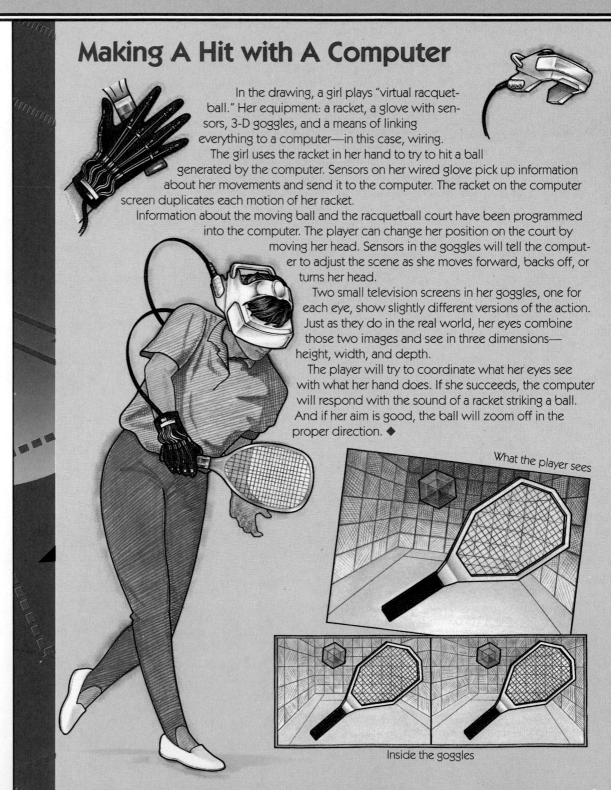

In the drawing, a girl plays "virtual racquetball." Her equipment: a racket, a glove with sensors, 3-D goggles, and a means of linking everything to a computer—in this case, wiring.

The girl uses the racket in her hand to try to hit a ball generated by the computer. Sensors on her wired glove pick up information about her movements and send it to the computer. The racket on the computer screen duplicates each motion of her racket.

Information about the moving ball and the racquetball court have been programmed into the computer. The player can change her position on the court by moving her head. Sensors in the goggles will tell the computer to adjust the scene as she moves forward, backs off, or turns her head.

Two small television screens in her goggles, one for each eye, show slightly different versions of the action. Just as they do in the real world, her eyes combine those two images and see in three dimensions—height, width, and depth.

The player will try to coordinate what her eyes see with what her hand does. If she succeeds, the computer will respond with the sound of a racket striking a ball. And if her aim is good, the ball will zoom off in the proper direction. ◆

What the player sees

Inside the goggles

THEN AND NOW

Lewis Latimer (1848-1928)

The son of a fugitive slave, Lewis Latimer was a pioneer in the great age of American invention during the late 1800s. Latimer is best known for his refinement of Thomas Edison's incandescent light bulb. He also assisted in Alexander Graham Bell's invention of the telephone.

Lewis Latimer was born in Massachusetts in 1848. He served in the Union navy during the Civil War. In Boston after the war, Latimer taught himself drawing. Despite discrimination against African Americans, he found work as a draftsperson, often drawing other people's inventions.

Latimer was befriended by a local teacher who was designing a device to help his hearing-impaired wife communicate. The teacher, Alexander Graham Bell, employed Latimer to make drawings and descriptions of his new "telephone." Latimer's sketches helped gain the patent for Bell's first telephone, issued in 1876.

Latimer's work attracted the attention of Thomas Edison, who in 1879 had invented the incandescent light bulb. Latimer went to work for Edison and soon made his most important contribution. Latimer devised a method for manufacturing carbon filaments that lasted much longer than Edison's first efforts. Soon, Latimer-style electric lamps were in wide use. In 1890 he published a textbook on electric lighting and continued to patent his own inventions. ◆

An Wang (1920-1990)

An Wang was an electronics wizard whose company pioneered the computer and communications systems now common in offices. More important to sports fans, Wang's firm designed the first computerized scoreboard. The electronic messages and images now commonplace in arenas and stadiums can be traced to this inventor.

Wang was born in Shanghai, China, and came to the United States in 1945 to study applied physics. At that time, computers were enormous clattering machines that filled entire rooms. Wang revolutionized computer memory by inventing a magnetic device for storing data. His doughnut-shaped magnetic core was the forerunner of today's semiconductor chips.

With $600 in savings, Wang founded his own company, Wang Laboratories, in 1951.

His tiny firm concentrated on special applications of computer technology. One famous example of this technology was the digital scoreboard at New York's Shea Stadium, which opened in 1964. Instead of mechanically placing the numbers and letters to show the score, the electronic scoreboard flashed scores and customized messages.

Wang's company became a pioneer in office automation, bringing electronics to the ordinary workplace. He credited his success to a combination of ancient Chinese wisdom and the American spirit of invention.◆

Background Information

Perhaps your students have seen the electric meter that measures the electric energy used in their homes. Point out to students that most appliances are marked with power ratings.

The amount of energy an appliance uses depends both on the power rating and how long the appliance is used. The higher the power rating, the greater the amount of energy used per second. Explain that appliances that heat or cool things generally have higher power ratings than appliances that only move things.

Discussion

● **Process Skills:** *Evaluating, Expressing Ideas Effectively*

After the students have read the feature on Clarence Baca, ask them why his job is important. (Baca makes sure that electric meters are correctly installed and are working properly. He also checks to be sure the correct amount of electricity is going into a building.)

SCIENCE AT WORK

CLARENCE BACA
Meter Technician

When you turn on the light switch in your classroom, you expect the lights to come on. At home you expect the refrigerator to be cold, the television to work, and the house to be warm on a winter day.

Clarence Baca is one of the people who makes all that possible. Baca is a meter technician. He makes it possible for the lights to come on and for television stations to broadcast your favorite programs.

Baca does many important jobs. The one he does most often is installing electric meters that measure the electric power that comes into a building. The utility company that supplies the electric power uses the meter to tell how much electricity has been used.

The meters that Baca installs in houses and apartments are called *watt-hour meters*. They measure electrical energy in kilowatt-hours. When electric power flows through the meter, a disk turns. Each time the disk goes around, a gear makes the pointer on a dial move to the next number. When electricity is not being used in a building, the disk does not turn. The utility company can tell from the numbers on the dials how much electric power has been used.

Baca also checks electric meters to be sure they are accurate. Sometimes people may think they have been charged too much money for the electricity that they used. When that happens, Baca checks the meter to see whether it is accurate. If necessary, he makes adjustments or repairs.

A meter technician cannot tell by looking at a wire whether it has power in it. Because it is not possible to see or hear electricity, people who work with it must be very careful. If you should touch a live wire with your bare hand, you could be badly shocked, burned, or killed. Meter technicians use tools that have nonconducting materials on the handles. They also wear gloves made of rubber and leather so that they will not be shocked or killed if they should touch a live wire.

Meter technicians spend about four years learning their trade. First, Baca became an apprentice. An *apprentice* learns a trade while he or she works at it. Then, Baca worked with experts who had been meter technicians for a long time. Baca also went to classes for two hours a day, twice a week, for four years. He then had to pass several tests to become licensed. It takes about as long to become a meter technician as it does to earn a college degree. ◆

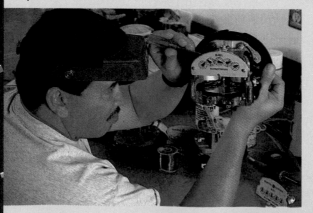

Discover More

For more information about careers for electricians and meter technicians, write to the

National Joint Apprenticeship and Training Committee for the Electrical Industry (NJATC) 16201 Tradezone Avenue, Suite 105 Upper Marlboro, MD 20772

Background Information

Magnetic propulsion may some-day make your vacation smoother! An experimental Japanese ship has a unique magnetic propulsion system. Its superconducting magnets shoot out jets of electrified sea water.

Journal Activity

Encourage interested students to research and make a journal entry about the magnetic-levitation trains that are under development in Germany; Japan; and Orlando, Florida. In addition, students might be interested in reading the following magazine article.

"Superconductivity Goes to Sea." *Popular Science* 64 (November 1992): 80–85. This article describes the *Yamato 1*, the first full-scale ship to use superconducting electromagnets for propulsion.

SUPERCONDUCTING MAGNETS

Something remarkable happens when certain metals are cooled to a point near absolute zero (zero kelvin). At such low temperatures, these metals conduct electricity without resistance. They become *superconductors!*

Power to Spare

Superconductors provide a huge advantage over ordinary electric wire. The wire in household circuits carries 15 amps safely, while wasting about 1.8 watts per meter of wire. A superconducting wire can carry 2000 amps for an infinite distance with no wasted power! Widespread use of superconductors could save billions of dollars in electric costs.

The drawback is that today's working superconductors must be kept at temperatures around 4 K (-269°C). The superconductors are bathed in liquid helium, requiring complicated and expensive refrigeration. Scientists are working on materials that superconduct at around 90 K. Liquid nitrogen is colder than that and is less expensive than liquid helium. So, liquid nitrogen can by used to cool the new superconductors.

Supermagnets

Superconductors are at the core of a new generation of supermagnets. With electric coils made of superconducting material, these magnets can produce fields 200 000 times the earth's magnetic field at the poles. Superconducting magnets provide the power for magnetic resonance imaging (MRI), the medical device that takes pictures of body tissues.

Superconducting magnets are also powering supercold refrigerators. Conventional refrigerators cool by compressing and expanding gas. Instead of gas, the magnetic refrigerator uses a crystal that heats and cools when passed through a magnetic field. Superconducting magnets help this superfridge to reach temperatures near absolute zero.

Magnet levitating over superconducting disk

This combination of a superconducting magnet and a conventional magnet is the strongest magnet ever made.

Take a Supermagnetic Cruise

Engineers in Japan have already tested trains and ships powered by superconducting magnets. The *maglev* (magnetic levitation) train rises by the force of mutually repelling superconducting magnets. The train floats on a magnetic field and is pulled forward by magnets along its sides. Without the friction of wheels or rails, the maglev train reaches speeds of 640 km/h, twice as fast as Japan's bullet trains.

Most applications of superconducting magnets remain experimental, since the magnets still require bulky and expensive helium refrigeration. Scientists are working to overcome these obstacles. ◆

Instead of propellers, this ship uses jets of electrified sea water, pushed out by superconducting magnets.

UNIT 5

THE NATURE OF MATTER

UNIT OVERVIEW

This unit describes the atomic structure of matter and uses the periodic table to identify the known elements. Metals and nonmetals are discussed in detail. Radioactive elements, their characteristics, and uses are also discussed.

Chapter 16: The Structure of Matter, page 426

This chapter discusses atomic theory and models of the structure of atoms, from those proposed by ancient Greeks to the current model. Comparisons of the relative masses and sizes of electrons, protons, and neutrons are presented. This chapter also discusses the arrangement of elements in the periodic table and the information that can be obtained from the table.

Chapter 17: Metals and Nonmetals, page 454

This chapter presents the characteristics of metals and nonmetals as they are classified in the periodic table. The elements in each group are examined and their occurrence and usefulness is carefully considered.

Chapter 18: Radioactivity, page 480

Radioactive elements and their characteristics are introduced in this chapter. Various types of radiation, as well as the use of radiation in medicine, industry, food storage, and dating artifacts are described. Natural radiation and its impact on living things is also discussed.

Science Parade, pages 506–513

This unit's magazine includes articles on how science joins hands with the arts to preserve priceless works of art and to produce the flashy—and hazardous—science of fireworks. Biographies of a Nobel Prize-winning physicist and a medical researcher are featured. The responsibilities and qualifications of a chemist are discussed and the technology of NMR spectroscopy is described.

UNIT RESOURCES

PRINT MEDIA FOR TEACHERS

Chapter 16

Freedman, David H. "The Top Bottoms Out—Again." *Discover* 14 (January 1993): 97–98. This article discusses one member of the quark family—the "top" quark.

Chapter 17

Nadel, Brian. "Ozone-Friendly Cooling." *Popular Science* 237 (July 1990): 57–63. CFCs damage the ozone layer. These chemicals are widely used as coolants in refrigerators and air conditioners. Now replacements for these chemicals are being produced.

Chapter 18

Geller, Nisa. "Old Masters." *Scientific American* 264 (April 1991): 23. Methods of dating cave paintings using carbon-14 are presented.

PRINT MEDIA FOR STUDENTS

Chapter 16

Asimov, Isaac. *Atom: Journey Across the Subatomic Cosmos.* Dutton, 1991. Asimov presents the atom in his own entertaining style.

Davies, Paul and John Gribbin. *The Matter Myth: Dramatic Discoveries That Challenge Our Understanding of Physical Reality.* Simon & Schuster, 1992. This book presents an up-to-date account of present-day physics and astrophysics.

Gleick, James. *Genius: The Life and Science of Richard Feynman.* Pantheon, 1992. This book discusses the Nobel Prize physicist who developed theories of superfluidity, diffusion, radioactive decay, and quarks.

Chapter 17

McGowen, Tom. *Chemistry: The Birth Of A Science.* Franklin Watts, 1989. Follows the early history of chemistry, from the time of the ancient world to the creation of modern chemistry in the late eighteenth century.

Dane, Abe. "Ozone Hole: What It Means." *Popular Mechanics* 167 (February 1990): 35. The interaction of the greenhouse effect and the ozone hole is presented. NASA graphics are included to illustrate the ozone hole.

Chapter 18

Codye, Corinn. *Luis W. Alvarez.* Raintree Publications, 1990. This book tells about the life of Luis Alvarez, the physicist who won the Nobel Prize in 1968 and contributed to the development of a radar system and the atomic bomb program during World War II.

Liptak, Karen. *Dating Dinosaurs and Other Old Things.* Millbrook Press, 1992. The explanations in this book help make difficult concepts easier to understand.

Zimmer, Carl. "Tritium Infusion." *Discover* 13 (January 1992): 75. Fusion physicists inject tritium into a machine along with deuterium and achieve a power output of almost 2 megawatts, 40 times the previous record.

ELECTRONIC MEDIA

Chapter 16

Mass of An Atom. Film or videocassette. BFA Educational Media. 12 min. Students visit a nuclear power plant to study atoms.

The Chemistry Help Series: Atomic Structure and Bonding. Computer Software. Focus Media, Inc. Colorful graphics and easy-to-follow text present parts of the atom, energy levels, and properties within periodic groups.

The Periodic Table. Film or videocassette. Coronet Film and Video. 23 min. The organization of the periodic table is illustrated.

Chapter 17

The Chemistry Help Series: The Periodic Table. Computer Software. Focus Media, Inc. Through graphics and text, this tutorial program presents the Periodic Law, periodic strands within groups and periods, transition elements, and metallic and nonmetallic properties.

Silicon. Videocassette. 13 min. Britannica. This video tells how silicon is processed for use in the microelectronics industry.

Chapter 18

Radioactivity. Videocassette. BFA Educational Media. 12 min. Radioactivity is demonstrated at a hospital.

Using Radioactivity. Videocassette. 22 min. Britannica. This video shows the way radioactivity is both helpful and harmful to humankind.

I Work in Atomic Energy. Videocassette. 25 min. Britannica. Looking at the human side of the nuclear industry, this program shows the workers—their concerns and interests.

Discussion The changes that occur in parts of a shipwreck depend on the matter involved. Ask the students to describe objects or photographs they have seen of objects salvaged from a shipwreck. (Answers might include descriptions of coins, jewelry, cannons, or weapons.) Ask the students how they think sea water affects different metal objects. (Some students might know that gold is not affected by contact with sea water, but that silver and iron are.) Have the students discuss the importance of what can be learned from shipwrecks.

UNIT
5

THE NATURE OF MATTER

CHAPTERS

Journal Activity You can extend the discussion by having the students list what metal objects might be found on a shipwreck. You might also bring in the following Unit topics: How do scientists describe the structure of atoms? What are the characteristics of metals and nonmetals? What are the characteristics of radioactive elements? How is radiation used?

Ask the students to keep a record in their journals of any questions they might have about these topics. After the students have completed the Unit, encourage them to look again at their questions. A follow-up discussion might help the students realize how much they have learned.

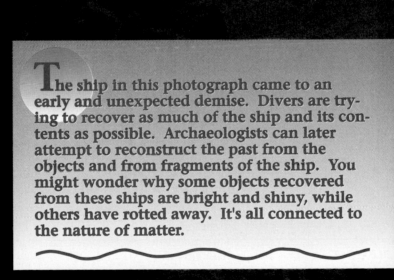

The ship in this photograph came to an early and unexpected demise. Divers are trying to recover as much of the ship and its contents as possible. Archaeologists can later attempt to reconstruct the past from the objects and from fragments of the ship. You might wonder why some objects recovered from these ships are bright and shiny, while others have rotted away. It's all connected to the nature of matter.

Science PARADE

CHAPTER 16

THE STRUCTURE OF MATTER

PLANNING THE CHAPTER

Chapter Sections	Page	Chapter Features	Page	Program Resources	Source
CHAPTER OPENER	426	**For Your Journal**	427		
Section 1: ATOMIC THEORY	428	Activity: How can you determine the shape of an unknown object? (A)	430	*Science Discovery**	SD
• Start at the Beginning (A)	428	Discover by Writing (A)	436	Reading Skills: Remembering What You Read (B)	TR
• Science Hits the Atomic Theory (A)	430	Section 1 Review and Application	436	Extending Science Concepts: Filling Electron Shells (H)	TR
• It's What's Inside that Counts (A)	431			Investigation 16.1: Comparing the Density of Different Materials (A)	TR, LI
• A Positive Step (A)	433			Atomic Theories	IT
• There'll Be Some Changes Made (A)	435			Study and Review Guide, Section 1 (B)	TR, SRG
Section 2: THE MODERN ATOM	437	Discover by Calculating (A)	439	*Science Discovery**	SD
• It All Works Together (B)	437	Section 2 Review and Application	442	Thinking Critically (A)	TR
• It's All Relative (B)	439	Investigation: Determining Relative Mass (A)	443	Connecting Other Disciplines: Science and Music, Making a Musical Periodic Table (A)	TR
• ID Please (B)	439			Investigation 16.2: Testing Element Reactivity (A)	TR, LI
• Do the Neutron Dance (B)	440			Isotopes	IT
				Record Sheets for Textbook Investigations (A)	TR
				Study and Review Guide, Section 2 (B)	TR, SRG
Section 3: THE PERIODIC TABLE	444	Discover by Observing (A)	444	*Science Discovery**	SD
• The Key is Organization (A)	444	Section 3 Review and Application	449	Periodic Table	IT
• Reading the Periodic Table (A)	448	Skill: Making Tables (B)	450	Study and Review Guide, Section 3 (B)	TR, SRG
Chapter 16 Highlights	451	The Big Idea	451	Study and Review Guide, Chapter 16 Review (B)	TR, SRG
Chapter 16 Review	452	For Your Journal	451	Chapter 16 Test	TR
		Connecting Ideas	451	Test Generator	

B = Basic   **A** = Average   **H** = Honors
The coding Basic, Average, and Honors indicates subsections, features, and resources that might be appropriate for different levels of learners. For additional suggestions regarding choice of topic and depth of coverage, see the Pacing Chart on pages T28–T31.

*Frame numbers at point of use
(TR) Teaching Resources, Unit 5
(IT) Instructional Transparencies
(LI) Laboratory Investigations
(SD) *Science Discovery* Videodisc Correlations and Barcodes
(SRG) Study and Review Guide

CHAPTER MATERIALS

Title	Page	Materials
Activity: How can you determine the shape of an unknown object?	430	(per group of 3 or 4) 40 cm x 60 cm cardboard, marble
Teacher Demonstration	433	books (2), small pane of glass, small pieces of paper, silk cloth
Discover by Writing	436	(per individual) journal
Discover by Calculating	439	(per individual) calculator
Investigation: Determining Relative Mass	443	(per group of 3 or 4) balance, beads (5), medium washers (5), small washers (5)
Discover by Observing	444	(per class) sodium, fluorine, potassium, chlorine, magnesium, copper, silver
Skill: Making Tables	450	(per individual) objects of assorted sizes and materials (15–20)

ADVANCE PREPARATION

Washers and beads for the *Investigation* on page 443 are available at home-improvement stores or craft stores. For the *Skill* on page 450, ask the students to bring in a variety of small objects.

TEACHING SUGGESTIONS

Field Trip
A field trip to a science museum is an excellent supplement to your instruction on the atom. Nearly all science museums have displays—some interactive—on the atom and the evolution of atomic theories. If possible, arrange a visit to a nearby government research facility, such as Fermilab. Encourage the students to use the opportunity to learn about the current research on other particles in the atom.

Outside Speaker
If a field trip is not feasible, invite a chemistry professor from a nearby college or university to speak about how chemists use the periodic table.

CHAPTER 16
THE STRUCTURE OF MATTER

CHAPTER THEME—SYSTEMS AND STRUCTURES

In Chapter 16 the students will learn the atom is an arrangement of smaller particles that form a unified whole. A supporting theme in this chapter is **Environmental Interactions.**

MULTICULTURAL CONNECTION

People from many cultural backgrounds have become atomic scientists. These scientists include Luis Alvarez, Shirley Jackson, and Robert Hofstadter, United States; Rudolf Mössbauer, Germany; Lev Landau, Russia; Eugene Wigner, Hungary; and J. Hans Jensen, Germany. Interested students might research how these scientists' cultural upbringings affected their choices to become atomic scientists.

MEETING SPECIAL NEEDS

Gifted

Help gifted students prepare atomic diagrams that show the chemical reaction that takes place in the Chapter Motivating Activity. Instruct them to label the diagrams with the names of the substances. Provide assistance as necessary.

▶ **426**   CHAPTER 16

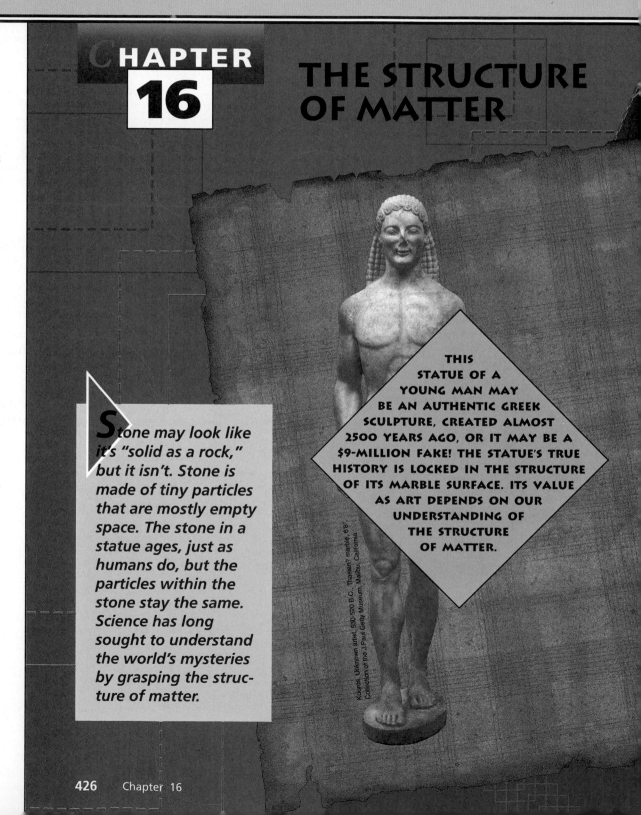

CHAPTER 16
THE STRUCTURE OF MATTER

Stone may look like it's "solid as a rock," but it isn't. Stone is made of tiny particles that are mostly empty space. The stone in a statue ages, just as humans do, but the particles within the stone stay the same. Science has long sought to understand the world's mysteries by grasping the structure of matter.

THIS STATUE OF A YOUNG MAN MAY BE AN AUTHENTIC GREEK SCULPTURE, CREATED ALMOST 2500 YEARS AGO, OR IT MAY BE A $9-MILLION FAKE! THE STATUE'S TRUE HISTORY IS LOCKED IN THE STRUCTURE OF ITS MARBLE SURFACE. ITS VALUE AS ART DEPENDS ON OUR UNDERSTANDING OF THE STRUCTURE OF MATTER.

Kouros. Unknown artist. 530-520 B.C. Thasian? marble, 6'8". Collection of the J.Paul Getty Museum, Malibu, California

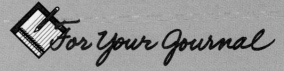
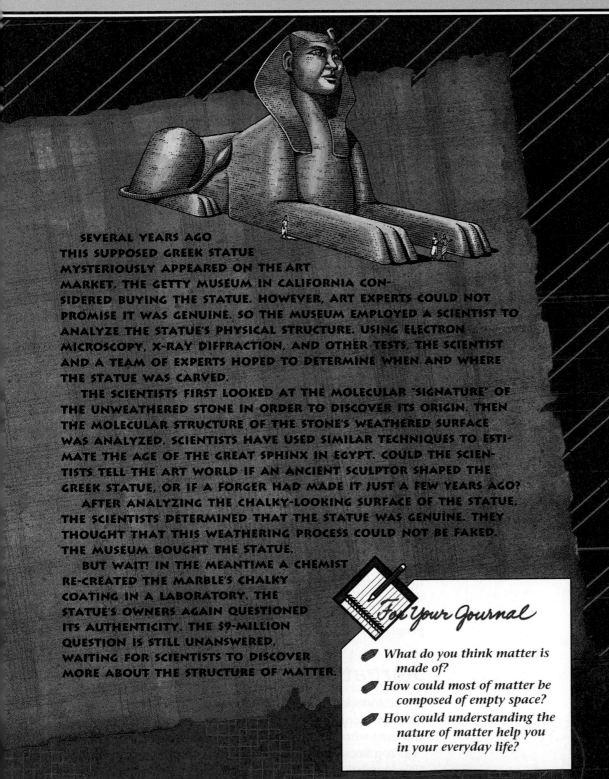

SEVERAL YEARS AGO THIS SUPPOSED GREEK STATUE MYSTERIOUSLY APPEARED ON THE ART MARKET. THE GETTY MUSEUM IN CALIFORNIA CONSIDERED BUYING THE STATUE. HOWEVER, ART EXPERTS COULD NOT PROMISE IT WAS GENUINE. SO THE MUSEUM EMPLOYED A SCIENTIST TO ANALYZE THE STATUE'S PHYSICAL STRUCTURE. USING ELECTRON MICROSCOPY, X-RAY DIFFRACTION, AND OTHER TESTS, THE SCIENTIST AND A TEAM OF EXPERTS HOPED TO DETERMINE WHEN AND WHERE THE STATUE WAS CARVED.

THE SCIENTISTS FIRST LOOKED AT THE MOLECULAR "SIGNATURE" OF THE UNWEATHERED STONE IN ORDER TO DISCOVER ITS ORIGIN. THEN THE MOLECULAR STRUCTURE OF THE STONE'S WEATHERED SURFACE WAS ANALYZED. SCIENTISTS HAVE USED SIMILAR TECHNIQUES TO ESTIMATE THE AGE OF THE GREAT SPHINX IN EGYPT. COULD THE SCIENTISTS TELL THE ART WORLD IF AN ANCIENT SCULPTOR SHAPED THE GREEK STATUE, OR IF A FORGER HAD MADE IT JUST A FEW YEARS AGO?

AFTER ANALYZING THE CHALKY-LOOKING SURFACE OF THE STATUE, THE SCIENTISTS DETERMINED THAT THE STATUE WAS GENUINE. THEY THOUGHT THAT THIS WEATHERING PROCESS COULD NOT BE FAKED. THE MUSEUM BOUGHT THE STATUE.

BUT WAIT! IN THE MEANTIME A CHEMIST RE-CREATED THE MARBLE'S CHALKY COATING IN A LABORATORY. THE STATUE'S OWNERS AGAIN QUESTIONED ITS AUTHENTICITY. THE $9-MILLION QUESTION IS STILL UNANSWERED, WAITING FOR SCIENTISTS TO DISCOVER MORE ABOUT THE STRUCTURE OF MATTER.

For Your Journal

- What do you think matter is made of?
- How could most of matter be composed of empty space?
- How could understanding the nature of matter help you in your everyday life?

ABOUT THE PHOTO

This Greek kouros was thought to have been carved between 540 and 520 B.C. It was reassembled from seven fragments and Is 206 cm tall. The statue is carved from dolomite, a highly stable type of marble, which is formed when limestone is subjected to high temperature and pressure; this may explain why it has a very thin patina. The proportions of stable carbon and oxygen isotopes vary depending on the specimen's geologic and geographic origin. Analysis of the isotopes provides clues to the structural and compositional changes caused by weathering and burial.

● **Process Skill:** *Comparing*

Ask the students to compare what they each hypothesized and discovered in the *Activity*. How does this *Activity* compare with modern scientific methods?

Process Skills:
Experimenting, Measuring, Observing

Grouping: Groups of 3 or 4

Hints

The unknown object can be cut from 2-cm plywood in the shape of a triangle, square, rectangle, or circle. The rebound angle should tell the students something about the shape of the object.

▶ **Application**

1. Answers should include information on rebound angles.

2. It is similar because you use what you know and you cannot see the object. It is different because you can collect data from measurements, which the Greeks could not do.

PERFORMANCE ASSESSMENT

Cooperative Learning

After the completion of the *Activity*, have each group of students brainstorm how the shape and size of an unknown object might be determined if the object could not be seen but could be handled. Each group should describe its proposed method and possible tools of measurement. Evaluate the choice of procedure and measurement tools for appropriateness.

ONGOING ASSESSMENT
▼ **ASK YOURSELF**

It would be possible to formulate hypotheses, but the Greeks did not have the ability to construct and conduct experiments to test their hypotheses.

ACTIVITY

How can you determine the shape of an unknown object?

MATERIALS
unknown object; rectangular cardboard, 40 cm × 60 cm; a marble

PROCEDURE
1. One member of the group will get the unknown object from the teacher. Without showing the object to the other members of the group, the student should place the object on a table and cover it with the cardboard. The other students should not look under the cardboard.

2. While the first student holds the cardboard, a second student should *gently* roll a marble under the cardboard to hit the object.

3. Use a sheet of paper to keep a record of the bounce-off angle. By rolling the marble from different directions, try to determine the shape of the object.

APPLICATION
1. Form a group conclusion about the object's shape. What led you to this conclusion?

2. How is this activity similar to the philosophizing of the ancient Greeks? How is it different?

▼ **ASK YOURSELF**

Why would it have been impossible for the ancient Greeks to create a scientific atomic theory?

Science Hits the Atomic Theory

You decide to go back to the holodeck and ask the computer to simulate a later period. The setting is now an English laboratory in the 1800s. A man is seated at a desk writing something entitled "Dalton's Atomic Theory." Ah . . . this must be John Dalton; you remember reading about him. You peek over his shoulder to read what is on the paper. Dalton's theory starts out with the same idea the ancient Greeks had: "All matter is made up of atoms."

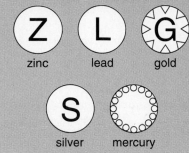

Unlike the Greek philosophers, however, Dalton believed that atoms may be distinguished by their mass. Furthermore, he stated that all atoms of the same element, such as oxygen, have the same mass. Atoms of different elements have different masses. In the process of developing his theory, Dalton had developed several experiments to determine the masses of atoms.

"How could he weigh an atom?" you ask. "Maybe I should try it." The holodeck computer breaks in and suggests that there is an experiment that would tell you what you want to know. That experiment is the Investigation on page 443.

Hydrogen   Oxygen
Carbon   Copper   Nitrogen
Sulfur   Phosphorus

Figure 16–1. John Dalton proposed some of the first symbols for the elements. Some of his original symbols are shown here.

◢ **ASK YOURSELF**

Why would weighing an atom be important to the development of an atomic theory?

It's What's Inside That Counts

Dalton's experiments for finding the masses of atoms gave scientists a way to study the elements. He was not concerned with the composition of the atoms.

You and everyone else in the twenty-third century know that atoms are composed of a nucleus and electrons. When did scientists find out about the parts of the atom? You ask the holodeck computer to advance you to the time when the first subatomic particles were discovered. Space seems to whirl about you. Suddenly, you are in a British lecture hall filled with people. It's 1897. A man at the front of the hall is using a cathode-ray tube to perform some experiments. You think to yourself, "What an antique!" The last time you saw a cathode-ray tube was in a museum of the twentieth century. There you learned that the screens of antique television sets were basically cathode-ray tubes.

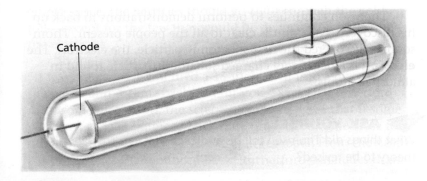

Cathode

Figure 16–2. A cathode-ray tube similar to this one was used by Sir Joseph Thomson in his experiments.

◆ **Did You Know?**

The Greeks and alchemists used symbols associated with the sun and planets of the solar system for some of the more well-known elements: sun/gold; moon/silver; Saturn/lead; Venus/copper; Mars/iron; Earth/antimony.

ONGOING ASSESSMENT
◢ **ASK YOURSELF**

Knowing the weight of an atom would allow you to identify which element it was. Also, the weights of the subatomic particles should add up to the weight of the whole atom.

THE NATURE OF SCIENCE

In 1814 Jöns Jakob Berzelius adopted Dalton's system for describing the elements, but he replaced the symbols with the first letter or letters of the Latin name of the element. A knowledge of Latin and its roots helps scientists of different languages to communicate.

Much modern scientific literature is published in English in journals such as the *Journal of the American Chemical Society*, the *Journal of Medicinal Chemistry*, and *Nature*, but French, German, and Russian journals also publish literature critical to a scientist doing research.

● **Process Skill:** *Comparing*

Compare an electron cloud model with a bicycle wheel. When the wheel is not moving, the separate spokes can be seen. But when the wheel is spinning, the spokes look fuzzy, like a cloud. Challenge the students to draw similar comparisons. Examples include the blades of an electric fan and the beaters of an electric mixer.

✧ **Did You Know?**

Joseph Thomson won a Nobel Prize in physics in 1906, and James Chadwick in 1935. Ernest Rutherford won a Nobel Prize in chemistry in 1908. All of these men were awarded this honor for their work on atomic structure.

ONGOING ASSESSMENT

▼ **ASK YOURSELF**

The density of the nucleus was too great to be made up only of the proton.

atoms have a very small but dense nucleus that is positively charged. Today, we call the positive subatomic particles within the dense nucleus **protons.**

OK. This entire process is beginning to make sense. You have seen scientists do experiments that confirmed the presence of electrons and protons. They could do these experiments because the particles were charged. You wonder how they are going to find out about the neutral subatomic particles that you know are also in the nucleus. You ask the computer for a summary of the experiments leading to the discovery of these particles.

According to the computer, the scientists of the early twentieth century knew that there was still something missing: protons alone could not account for the density of the nucleus. Hydrogen was known to be made of one electron and one proton. Helium contained two electrons and two protons. Therefore, the ratio of the mass of a helium atom to the mass of a hydrogen atom should, in theory, be 2:1. But the experimental ratio came out to 4:1.

In 1932 Sir James Chadwick discovered the reason. He performed an experiment similar to Rutherford's. He bombarded a thin sheet of beryllium with alpha particles. Unidentified, high-energy radiation was given off. Chadwick continued to experiment, eventually proving that the radiation was made up of electrically neutral particles that weighed about as much as the proton. Chadwick had discovered the neutron. **Neutrons** are subatomic particles found in the nucleus and have no charge— they are neutral. Now the 4:1 ratio of the mass of helium to the mass of hydrogen could be explained. The nucleus of hydrogen has one proton and no neutrons, but the nucleus of helium has two protons and two neutrons.

▼ **ASK YOURSELF**

How did scientists know there was another particle in the nucleus other than the proton?

GUIDED PRACTICE

Have the students use their knowledge of geometry to estimate the angle of deflection in Rutherford's gold-foil experiment.

INDEPENDENT PRACTICE

 Have the students provide written answers to the Section Review. Then have the students draw and label a time line in their journals. Instruct them to place the history of atomic discoveries along the line.

EVALUATION

Have the students describe and illustrate the subatomic particles with their charges and position in the atom. Answers should reflect an understanding of the latest model.

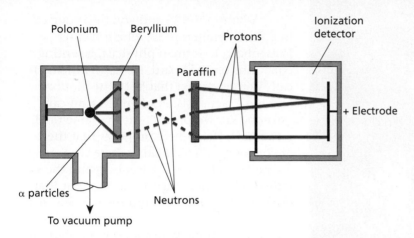

Figure 16–5.
Chadwick's experimental apparatus

There'll Be Some Changes Made

You ask the computer to take you back into the early twentieth century. You remember from your student days that the atomic theory continued to be revised throughout this century, and you want to see some of the highlights. The various atomic models developed during the early twentieth century are shown in Figure 16–6.

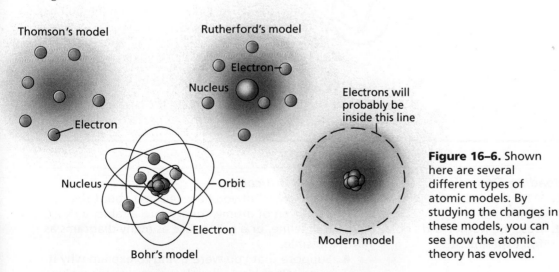

Figure 16–6. Shown here are several different types of atomic models. By studying the changes in these models, you can see how the atomic theory has evolved.

You find yourself back in Dr. Rutherford's lab. Niels Bohr, a Danish scientist working with Rutherford, is discussing his idea that the electrons circle the nucleus in orbits called **energy levels**. (You recall that he won the Nobel Prize for his work on atomic structure in 1922.)

SCIENCE BACKGROUND

In 1963 Dr. Maria Goeppert Mayer became only the second woman to win the Nobel Prize in physics, sharing it with German physicist J. Hans Jensen and American physicist Eugene Paul Wigner. (The first woman was Marie Curie in 1903.)

Mayer and Jensen, working independently, prepared nearly identical papers on the shell structure of atomic nuclei. They discovered that atomic nuclei possess shells similar to the electron shells of atoms. These shells contain varying numbers of protons and neutrons, which permit systematic arrangement of nuclei according to their properties.

✧ Did You Know?

Even though the nucleus accounts for 99.9 percent of an atom's mass, the diameter of the atom is about 100 000 times larger than the nucleus. If each atom could be pressed into a sphere as small as its nucleus, the Washington Monument could be crammed into a space the size of a pencil eraser.

RETEACHING

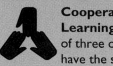

Cooperative Learning In groups of three or four, have the students draw pictures of each atomic model throughout history, without labels. Then have the groups exchange diagrams, and have them identify and label each other's diagrams.

EXTENSION

Have the students do research and write a report about Niels Bohr. Bohr's major contribution was the concept of energy levels. He noted that electrons move from one energy state to another only at definite energy levels. The energy absorbed or released as electrons change energy levels causes electro-magnetic radiation.

CLOSURE

Cooperative Learning Have the students work in small groups to write a summary of the section. Ask each group to read its summary aloud.

MEETING SPECIAL NEEDS

Second Language Support

Have limited-English-proficient students work with classmates to write down the names and phonetic pronunciations of the scientists studied in this section. Provide additional assistance as needed. Encourage the students to keep the information in their journals.

ONGOING ASSESSMENT
ASK YOURSELF

Because neutrons have no charge, they are not affected by electric fields as are electrons and pro-tons, and thus were not observed at first.

SECTION 1 REVIEW AND APPLICATION

Reading Critically

1. the ancient Greeks

2. He aimed alpha particles at a very thin piece of gold foil. A screen surrounding the apparatus glowed when hit by the alpha par-ticles. Some particles bounced back from the foil.

Thinking Critically

3. Diagrams should include the ancient Greeks, Dalton, Thomson, Rutherford, Chadwick, Bohr, and Heisenberg.

4. Answers will depend on the detail students choose to include. All answers should include the Greek and modern view of the atom. Accept reasonable responses.

A jump to 1926 and you find yourself in a small seminar room where Werner Heisenberg, a German physicist, is writing equations on the board. He has mathemati-cally demonstrated that both the motion and the exact position of an electron can never be known precisely at the same time. As a result, Heisenberg proposes that there are regions, called *orbitals,* where electrons are most likely to be. Heisenberg and the sci-entists who worked with him thought that electrons are found around the nucleus in a cloud.

You continue your journey through the "primitive" twentieth century. Eventually, you write what you learned about the atom in the ship's log. "Ship's log: star date 2254.7. I understand much better how hard it is to formulate an atomic theory and why it has been modified so often. I will proceed to summarize my findings . . . ."

DISCOVER BY *Writing*

Summarize your ideas about atomic theories in your journal. Include how experimentation has helped advance the ideas of science. ✎

 ASK YOURSELF

Why do you think the neutron was among the last subatomic parti-cles to be discovered?

SECTION 1 *REVIEW AND APPLICATION*

Reading Critically

1. Who first used the word *atom?*

2. Describe Rutherford's gold-foil experiment.

Thinking Critically

3. Organize what you have learned about the evolution of atomic theory into a table, a chart, a timeline, or a graph. Use as many diagrams as possible.

4. Suppose that you were trying to explain why it rains. What kind of explanation would you have used if you were an ancient Greek atomist? What kind of explanation would you give if you lived in the early part of the twentieth century? Compare and contrast both theories.

FOCUS

In this section, protons, electrons, and neutrons are discussed as part of the modern atomic theory. Atomic mass units, mass number, and atomic number are explained. Isotopes are defined.

MOTIVATING ACTIVITY

Ask the students to think of examples from their everyday experience in which an item is so small it is measured in larger numbers. For example, we don't buy one grape at the grocery store, but a bunch, which is usually sold by the pound.

The Modern Atom

SECTION 2

It is very hard to visualize particles that are so small you cannot see them. Visual models such as those in Figure 16-6 can help. How can you improve on the modern model in this figure?

It All Works Together

Back at the starship, you decide to get a bite to eat in your quarters. You are still trying to fit modern atomic theory and its historical development together. How can you visualize an electron cloud? Ah! You have it. You ask the food replicator to make you some cotton candy. You realize that the electron cloud is like a ball of cotton candy. It is not very dense on the outside, and if you touch it, it gives a bit. As you go toward the center, it gets more dense. It has a solid, dense center (the paper cone it is attached to). This center would be similar to the nucleus. "That's it!" you think to yourself. I shouldn't be thinking of electrons as solid balls orbiting a nucleus. That was Bohr's theory, which was outdated even in the twentieth century."

After finishing the cotton candy, you feel you have placed all the concepts in their proper place. You dictate the following summary into the ship's log.

Objectives

List the three major types of atomic particles and their properties.

Calculate the number of electrons, protons, and neutrons in a given atom, using the atomic number and atomic mass.

TEACHING STRATEGIES

● **Process Skill:** *Comparing*

Have the students think how the electron-cloud model fits the results of Rutherford's experiment. (The positive particles that Rutherford used would pass through the electron cloud with little chance of striking or coming near an electron, since the area is large compared with the size of the electrons.)

● **Process Skill:** *Interpreting Data*

Cooperative Learning Have the students work in small groups to make a three-dimensional model of an atom with any materials they want. Encourage the students to compare and contrast the different models. Ask them which model best represents the modern atomic theory. Ask the students to justify their answers.

An atom is the smallest unit of matter that retains the properties of matter. The modern model of the atom has electrons swirling around the nucleus in a large region, rather than orbiting in a fixed pattern. The region in which the electrons are found is called the *electron cloud.* As you move toward the center of the atom, chances of encountering a particle increase.

According to modern atomic theory, an atom is composed of three types of particles: protons, neutrons, and electrons. Protons are particles with a positive charge. Neutrons have no charge. The protons and neutrons make up the nucleus. Electrons have a negative charge. The negative charge of an electron is equal in strength to the positive charge of a proton. When the number of electrons equals the number of protons in an atom, the atom has no charge: it is a neutral atom.

It is very difficult to experiment with atoms since they are so small. It would take about 1×10^{18} atoms to make a period at the end of this sentence. Moreover, the nucleus of an atom is even smaller. The computer says that if the diameter of an atom were enlarged to the length of a football field, the nucleus would be slightly smaller than a pea!

ONGOING ASSESSMENT

▼ **ASK YOURSELF**

Heisenberg; Greeks; Bohr. Accept reasonable explanations.

▼ **ASK YOURSELF**

Match each of the following models with the scientist whose atomic theory it best represents: a swarm of bees; a tennis ball; a ball tied with a string and whirled above your head. Explain your choices.

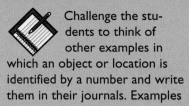
It's All Relative

DISCOVER BY Calculating

Suppose you knew the mass of three of your friends—25 kg, 50 kg, and 75 kg. You can make a set of relative masses by assigning your 50-kg friend a relative mass of 1 fmu (friendly mass unit). The friend that has a mass of 75 kg would have a relative mass of $1\frac{1}{2}$, or 1.5, fmu. Every 50 kg would be 1 fmu. What would your 25-kg friend's mass be in fmu?

The mass of an atom is also measured using relative units called *atomic mass units* (amu). Protons and neutrons each have about the same mass—nearly 1 amu. The **atomic mass** of an atom is equal to the sum of the relative masses of all its protons and neutrons. An atom's **mass number** is its atomic mass rounded to the nearest whole number. It indicates the total number of protons and neutrons in the atom.

Although the electrons take up most of the space in an atom, they contribute very little to the mass of the atom. In fact, it would take nearly 2000 electrons to equal the mass of one proton. Because electrons have so little mass, they can usually be ignored when calculating the mass of an atom.

▼ ASK YOURSELF

How do scientists determine the atomic mass of an element?

ID Please

Whether you live in an apartment, a house, or a boat at the marina, your home is identified by a number. This guarantees that you get your mail and that friends can find where you live. Your address is unique.

Figure 16–7. A street address is similar to an element's atomic number.

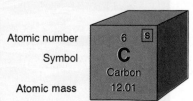

		State:
Atomic number	6 · S	S Solid
Symbol	**C**	L Liquid
	Carbon	G Gas
Atomic mass	12.01	X Not found in nature

RETEACHING

Allow the students who need assistance to look up the answers to the Section Review or to work together in groups.

EXTENSION

Have the students use library materials to read about deuterium and tritium, radioactive isotopes of hydrogen. These are found in nature and are used as fuel in fusion reactions where they combine to produce helium and a great deal of energy. Have the students make written reports of their findings. Keep the reports in the students' portfolios until you study Chapter 18. Return the reports at that time, and ask the students to revise them, using what they've learned about fusion.

CLOSURE

Cooperative Learning Have the students work in small groups to write a summary of the section. Ask each group to read its summary aloud.

MEETING SPECIAL NEEDS

Mainstreamed

Have mainstreamed students work with classmates to prepare a table listing the number of protons, electrons, and neutrons in the atoms of isotopes listed in Table 16–1.

ONGOING ASSESSMENT

ASK YOURSELF

Isotopes are atoms of an element that have different numbers of neutrons.

SECTION 2 REVIEW AND APPLICATION

Reading Critically

1. The atom has proton(s) with a positive charge, neutron(s) with no charge, and electron(s) with a negative charge. The protons and neutrons are in the nucleus, and the electrons form a cloud around the nucleus.

2. 14 protons, 14 neutrons, and 14 electrons; accept reasonable graphics

Thinking Critically

3. In ordinary matter under normal circumstances, there are as many protons as electrons, so the objects around you have no net charge.

4. Chlorine-35 has 17 protons, 18 neutrons, and 17 electrons. Chlorine-37 has 17 protons, 20 neutrons, and 17 electrons.

Table 16-1	Common Isotopes and Their Natural Abundance	
Isotope	**Abundance (%)**	**Mass (amu)**
Hydrogen-1	99.98	1.008
Hydrogen-2	0.02	2.014
Carbon-12	98.89	12.000
Carbon-13	1.11	13.003
Oxygen-16	99.76	15.995
Oxygen-17	0.04	16.999
Oxygen-18	0.20	17.999

As Dalton worked on his atomic theory, he assumed that all the atoms of a particular element had the same mass. We now know that atoms of the same element may have a different number of neutrons. This means that atoms of the same element could have slightly different masses. When scientists talk about atomic mass, they are referring to the *average atomic mass*. As the term implies, this is an average mass of all the isotopes for that element.

After examining the results of the test, Dr. Shawan realizes that Ensign Roark has a blockage in the artery to one leg. Dr. Shawan now knows how to treat the problem. Fortunately, the solution to this problem was fairly simple. Thank goodness it was not the dreaded Malsupian virus!

 ASK YOURSELF
What is an isotope?

SECTION 2 REVIEW AND APPLICATION

Reading Critically
1. Describe and define the subatomic parts of the atom.
2. A neutral atom has a mass number of 28 and an atomic number of 14. How many protons, neutrons, and electrons does it have? Construct a graphic similar to the one on page 440 that shows how you arrived at your answer.

Thinking Critically
3. Electricity is a flow of electrons. Electrons are part of all substances, yet you are not electrically shocked by all the objects around you. Explain.
4. Two isotopes of chlorine are chlorine-35 and chlorine-37. What can you specify about the atomic particles of these two isotopes?

Determining Relative Mass

Process Skill: Comparing

Grouping: Groups of 3 or 4

Objective

● **Understand** the concept of relative masses and why the masses are used in the study of atoms and molecules.

Pre-Lab

Have the students review the *Discover by Calculating* on page 439. Ask them how relative mass would be helpful in trying to determine the mass of atoms.

▶ Analyses and Conclusions

1. Students should note whether the number of objects they predicted is higher or lower than the actual number of objects. Encourage the students to justify their answers.

2. Atomic mass unit is a relative number. Relative to other atomic particles, the mass of the proton is 1. There are no units.

3. Atoms are extremely small. Their actual masses would be expressed in extremely small numbers that would be difficult to work with.

▶ Application

While students might find it helpful to know the relative mass of the opposing team, it would probably be more useful to know individual masses.

✳ Discover More

The Commission on Atomic Weights of the International Union of Pure and Applied Chemistry and the corresponding association for physicists adopted carbon-12 as a reference mass, because its mass can be accurately related to the atomic masses of other elements, using a mass spectrometer.

Post-Lab

Have the students use what they have learned to evaluate the usefulness of relative mass. Encourage them to write their ideas in their journals.

INVESTIGATION

Determining Relative Mass

▶ MATERIALS

● balance ● beads (5) ● medium washers (5) ● small washers (5)

▼ PROCEDURE

1. Make a table like the one shown.

2. Determine the mass of one bead and each type of washer. Record the data in your table.

3. Compile 2.00 g of beads, remove them from the balance, and count them. Record the number of beads, and compare your results with those of three other groups.

4. Repeat step 3 for each set of washers. Record the results in your table.

5. Determine the mass of all five beads. Then calculate the average mass of one bead by dividing the mass by 5. Record your results in the table.

6. Using the method in step 5, calculate the average mass of each type of washer.

7. To determine the standard relative mass, choose the smallest of the average masses you calculated. Then divide the average mass of each group by the standard. For example, if your average masses were:

Washer 1	Washer 2	Bead
0.50 g	0.11 g	0.01 g

the smallest mass would be 0.01 g, so the bead would be your standard. You would divide the mass of each object by the mass of the bead. For Washer 1 you would divide 0.50 g by 0.01 g, which equals 50.

This is the relative mass of Washer 1. It is called the *relative mass* because its measurement is relative to a standard. Record the relative masses of the objects in the table.

8. Without counting, use the balance to determine one-half of the relative mass for each object. For example, if you use the standard, its relative mass is 1.0 g. One-half of the relative mass would be 0.5 g. Now count the number of each object. Record your results.

TABLE 1: CALCULATION OF RELATIVE MASS					
Object	Mass of One	Number of Objects in 2 g	Average Mass	Relative Mass	Number in ½ Relative Mass
Bead					
Washer #1					
Washer #2					

▶ ANALYSES AND CONCLUSIONS

1. When you counted the objects in step 8, were there as many as you thought there would be? Explain.

2. How is the atomic mass unit similar to the relative mass units you worked with in this Investigation?

3. What are the benefits, in the case of atoms, to using relative masses rather than actual masses?

▶ APPLICATION

If you were a football player, would it help you to know the relative mass of an opposing team? Explain.

✳ Discover More

The atomic mass unit is defined as $\frac{1}{12}$ the mass of the carbon-12 atom. The relative mass of other elements is found by comparing their average atomic mass to that of carbon-12. Why do you think carbon-12 is used as the standard?

PERFORMANCE ASSESSMENT
Cooperative Learning

After the completion of this *Investigation*, have the students demonstrate the procedure, this time using large washers in place of the small washers. Check each group's results for accuracy.

● **Process Skill:**
Classifying/Ordering

Cooperative Learning Prepare sets of index cards (25 cards per set; one set for each group) that have the following information on them: element name, chemical symbol, atomic number,

atomic mass, physical properties (phase at room temperature, color, hard/soft, boiling point, melting point, conductivity, brittle, luster), and chemical properties (active/inactive/inert).

Divide the class into small groups, and give each group a set of index cards. Tell the groups to study the cards and

arrange them in some logical order. In their journals, have the students write the criteria they use for ordering the elements. When they have completed ordering all their elements, have the groups compare the criteria they used.

Relate the ordering of elements by each group to the method

Mendeleev must have used to organize the first periodic table.

THE NATURE OF SCIENCE

Mendeleev was not the only chemist who attempted to arrange the known elements in some kind of order. Others had noticed the periodic similarities in the properties of elements. A German chemist, Julius Meyer, proposed a periodic table similar to Mendeleev's. Mendeleev published his table in 1869; Meyer published his one year later. Mendeleev was the first person to leave gaps in the table where there were no known elements having the predicted properties. When three of these elements were later discovered and their properties fit those predicted by Mendeleev's table, his work was finally accepted.

LASER DISC

2353

Periodic table

SCIENCE BACKGROUND

The chief source of beryllium, the first of the Group 2 metals, is a complex beryllium aluminosilicate, $Be_3Al_2(SiO_3)_6$, called beryl. Compounds of beryllium do not occur in large quantities in nature, but small amounts are present in various minerals and in silicate rocks. Beryllium has a clean surface of silvery white (a tenacious oxide film gives this metal a dull surface), and it is a good conductor of electric current. The compounds of beryllium are extremely poisonous and should be handled very carefully.

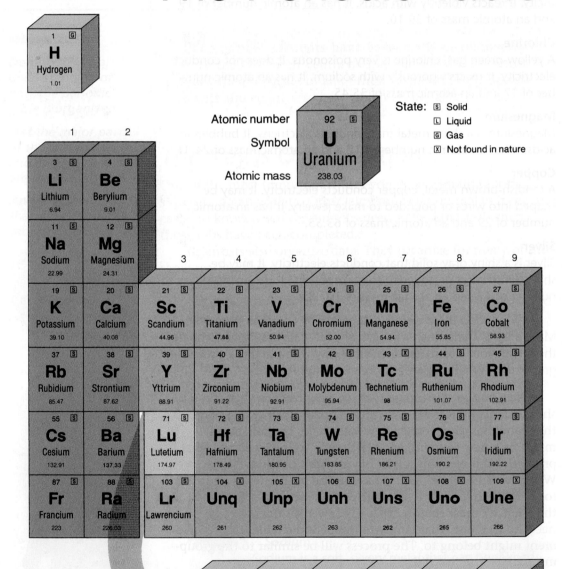

The Periodic Table

Legend

- Metals
- Transition Metals
- Nonmetals
- Noble gases
- Lanthanide series
- Actinide series

18

13	14	15	16	17	2 [G] **He** Helium 4.00
5 [S] **B** Boron 10.81	6 [S] **C** Carbon 12.01	7 [G] **N** Nitrogen 14.01	8 [G] **O** Oxygen 16.00	9 [G] **F** Fluorine 19.00	10 [G] **Ne** Neon 20.18
13 [S] **Al** Aluminum 26.98	14 [S] **Si** Silicon 28.09	15 [S] **P** Phosphorus 30.97	16 [S] **S** Sulfur 32.07	17 [G] **Cl** Chlorine 35.45	18 [G] **Ar** Argon 39.95

10	11	12						
28 [S] **Ni** Nickel 58.69	29 [S] **Cu** Copper 63.55	30 [S] **Zn** Zinc 65.39	31 [S] **Ga** Gallium 69.72	32 [S] **Ge** Germanium 72.61	33 [S] **As** Arsenic 74.92	34 [S] **Se** Selenium 78.96	35 [L] **Br** Bromine 79.90	36 [G] **Kr** Krypton 83.80
46 [S] **Pd** Palladium 106.42	47 [S] **Ag** Silver 107.87	48 [S] **Cd** Cadmium 112.41	49 [S] **In** Indium 114.82	50 [S] **Sn** Tin 118.71	51 [S] **Sb** Antimony 121.75	52 [S] **Te** Tellurium 127.60	53 [S] **I** Iodine 126.90	54 [G] **Xe** Xenon 131.29
78 [S] **Pt** Platinum 195.08	79 [S] **Au** Gold 196.97	80 [L] **Hg** Mercury 200.59	81 [S] **Tl** Thallium 204.38	82 [S] **Pb** Lead 207.2	83 [S] **Bi** Bismuth 208.98	84 [S] **Po** Polonium 209	85 [S] **At** Astatine 210	86 [G] **Rn** Radon 222

63 [S] **Eu** Europium 151.96	64 [S] **Gd** Gadolinium 157.25	65 [S] **Tb** Terbium 158.93	66 [S] **Dy** Dysprosium 162.50	67 [S] **Ho** Holmium 164.93	68 [S] **Er** Erbium 167.26	69 [S] **Tm** Thulium 168.93	70 [S] **Yb** Ytterbium 173.04
95 [X] **Am** Americium 243	96 [X] **Cm** Curium 247	97 [X] **Bk** Berkelium 247	98 [X] **Cf** Californium 251	99 [X] **Es** Einsteinium 252	100 [X] **Fm** Fermium 257	101 [X] **Md** Mendelevium 258	102 [X] **No** Nobelium 259

THE NATURE OF SCIENCE

As new technologies are invented and employed by chemists, the atomic mass of various elements are corrected. Depending on the publication used as a reference, the numbers may vary by as much as an entire unit, although discrepancies are usually much smaller. For that reason, if a student looks up the atomic mass of an element in another source, it may not agree with the atomic mass presented here.

MULTICULTURAL CONNECTION

Dr. Hunter Haveline Adams III, an African-American chemist, is one of many scientists who continue to investigate atoms. He is currently working at the Argonne National Laboratory atomic accelerator. He is investigating proton beam detection and the relationships between magnetic fields and the rise of various civilizations.

THE NATURE OF SCIENCE

Spectral analysis, or spectroscopy, was first developed about 1860, and was responsible for the discovery of many new elements. During a solar eclipse in 1868, a new set of spectral lines was found in the spectrum of the corona. These lines had never been seen before, and were thought to be from a new element, named "helium" after the Greek word for sun.

GUIDED PRACTICE

Cooperative Learning Have the students draw one of the squares in the periodic table and label each piece of information within the square. You might extend this activity by having the students work in pairs or small groups to construct a table showing how properties vary down groups and across periods in the periodic table. Students' tables should show that the atomic number and number of electrons in the outer energy level increase across a period; the atomic number and number of energy levels increase down a group; and the number of electrons in the outermost energy level remains the same down a group.

INDEPENDENT PRACTICE

Have the students provide written answers to the Section Review. In their journals, have the students draw as much of the periodic table as they can, labeling the groups and periods.

EVALUATION

Cooperative Learning Have the students work in pairs. Provide each pair with properties of certain elements, but not the names of the elements. Have the students try to predict which group an element would be in, based on its properties.

① They are alike because both group elements using periodic properties; they are different because Mendeleev put different groups with similar properties in rows in his table. In the modern table, groups with similar properties are in columns.

ONGOING ASSESSMENT

  ASK YOURSELF

Answers should indicate that the similarities of the properties elements are repeating, or periodic.

A periodic property is one that repeats every so often (periodically). The coming of winter, for example, is a periodic event. Some of the properties of the elements are also periodic. Compare Mendeleev's periodic table with your grouping of similar elements. How are they alike? How are they different? Why do you think the tables are different?①

▶ **ASK YOURSELF**

Why is the periodic table called periodic?

Reading the Periodic Table

Since Mendeleev developed the periodic table, many more elements have been discovered and added. What is surprising is that the modern periodic table is very similar to Mendeleev's original. The major difference, other than the addition of elements, is that the vertical columns on the modern table are like Mendeleev's horizontal rows.

Figure 16–11. Mendeleev's original manuscript is shown at the left. His findings were first published in the journal shown at the right.

The modern periodic table is a useful summary of many atomic properties. Knowing how to use the periodic table will help you understand many of the properties of elements without having to memorize the information.

Look at any square of the periodic table. In the square, you can see the following information: the symbol for the element, the atomic number, and the average atomic mass. In addition, the state of the element as it is found in nature is shown in the upper right-hand corner.

The current periodic table is arranged in order of atomic number. Scientists have found that this method best illustrates periodic properties.

If you look at the top of the table, between Group 13 and Group 14, you will see the beginning of a zigzag line. To the left of this line are the metals, and to the right are the nonmetals. We will focus on metals and nonmetals in the next chapter.

RETEACHING

Encourage the students who need assistance to look up the answers to the Section Review or to work together in pairs.

EXTENSION

Have the students do research on British scientist Henry Moseley, who used X-rays to determine the atomic numbers of the elements. Moseley found that when the elements were arranged according to atomic number instead of mass, the pattern produced when they were X-rayed was more regular. Have the students record their information about him in their journals.

CLOSURE

Cooperative Learning Have the students work in small groups to write a summary of this section. Tell each group to appoint a secretary to write down the summary. Have them appoint one person to read their summary aloud.

Elements arranged in vertical columns are known as **groups,** or families. These are similar to the "families" you used when you organized data about the elements. Some of these families have specific names. In Figure 16–12 you see some important groups. All the elements in any of these major groups have similar properties.

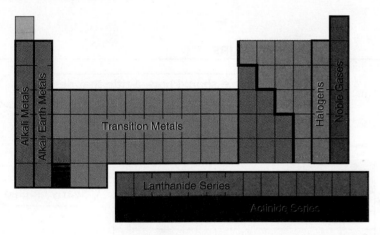

Figure 16–12. Some of the groups and periods are marked on this shell of the periodic table.

Thanks to the holodeck's computer, you have a better understanding of atomic theory and the elements. However, you still have some questions. What do subatomic particles have to do with how matter reacts? Why do some elements combine with others in a particular way? Why are some elements unreactive? There is still much to learn, but you are interrupted by the chief engineer—something is wrong with the engines, and you'd better take a look. Your research will have to wait.

ASK YOURSELF

What information would you find in one square of the periodic table?

SECTION 3 REVIEW AND APPLICATION

Reading Critically

1. Use the periodic table to find the elements whose atomic numbers are 12, 34, and 17. Write the period and the group of each element.
2. Why is the periodic table important to scientists?

Thinking Critically

3. Describe the usefulness of the periodic table.
4. You have discovered a new element and are trying to decide what group it belongs to in the periodic table. What information do you need to accurately place the element?

MEETING SPECIAL NEEDS

Second Language Support

Have limited-English-proficient students prepare three squares of the periodic table with the names of the element in the student's native language. Have the students use common elements, such as gold and iron, in order to obtain a wide variety of foreign names for the elements.

ONGOING ASSESSMENT

ASK YOURSELF

Each square of the periodic table includes the chemical symbol, name of the element, atomic number, average atomic mass, and physical state.

SECTION 3 REVIEW AND APPLICATION

Reading Critically

1. 12, magnesium, Group 2; 34, selenium, Group 16; 17, chlorine, Group 17

2. Scientists have to deal with a huge number of experimental facts. In order to use them effectively, they must be classified and organized in a way that brings out their similarities and differences.

Thinking Critically

3. It is useful because it correlates the physical and chemical properties of the elements and their compounds, and correlates atomic structure to periodic arrangement.

4. You need to know the atomic number and average atomic mass. It is also helpful to know whether it is a metal.

Making Tables

Process Skill:
Classifying/Ordering

Grouping: Individual

Objectives
● **Organize** data in a table to display relationships.
● **Classify** data to show similar characteristics.

Discussion
Scientists use tables to organize and classify recorded data. Remind the students they can use only characteristics they observe and not those they might attribute to the item from previous experience.

▶ Application
The students might include such characteristics as size, shape, color, mass, and type of material (metal, wood, plastic, glass, and so on).

Ask the students whether they can think of more objects that would be members of a particular group in their table.

✳ Using What You Have Learned
Similarities to periodic table: organize objects in rows; organize objects in columns; use symbols for objects or elements. Differences from periodic table: not elements, but objects; not classified by atomic number; not classified by atomic mass.

★ PERFORMANCE ASSESSMENT
After completion of this *Skill,* have individual students explain how they developed their table. Check each student's table for clarity in organizing their information.

SKILL Making Tables

▶ MATERIALS
● 15 to 20 objects

▼ PROCEDURE

1. All tables should be organized in a way similar to the diagram shown here.

A table should have a title that describes the information shown.

Each row should be clearly labeled.

Each column should be clearly labeled.

TABLE TITLE		
Column Heading	Column Heading	Column Heading
Row Heading		
Row Heading		
Row Heading		
Row Heading		
Row Heading		

Legend: This explains any symbols that are used in the table.

If you use symbols in your table, you must have a legend that explains them.

2. Choose 15 to 20 objects, and write down their names. Decide on ways to group the objects by characteristics they may have in common. For example, objects made out of wood may be one group.

Remember to use only those characteristics that you can observe. Do not take any of the objects apart. You might want to design symbols for the objects you choose for your table.

▶ APPLICATION
What characteristics did you use to separate the objects in the table into rows and columns? How did your choice of characteristics help you to classify the objects you were given?

✳ *Using What You Have Learned*
List at least three ways your table is like the periodic table and three ways in which your table is different from the periodic table.

CHAPTER 16 HIGHLIGHTS

Guide the students to an understanding that the particles in the atom are extremely small and invisible, but combined into elements, molecules, and compounds, they are the structure of matter in the world around us.

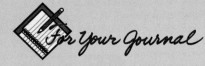

For Your Journal

The students' ideas should reflect an understanding of how the nature of matter could help them in their everyday lives. You might have the students share their new ideas with the rest of the class.

CHAPTER 16 *H*IGHLIGHTS

Kouros, Unknown artist, 530–520 B.C., Thasian? marble, 6'9". Collection of the J. Paul Getty Museum, Malibu California.

The Big Idea

All matter is made up of small particles called atoms. Scientists use models to show their ideas about the structure of an atom and the interaction of its particles. As scientists gain new information from experiments, they revise their models. Scientists today use the electron cloud model. This model, too, may change as new information is discovered.

Scientists organize what they know about matter in the periodic table. Identifying patterns and organizing them into tables helps scientists predict the properties of different kinds of matter.

For Your Journal

Look back at the ideas you wrote in your journal at the beginning of the chapter. Have your ideas changed? Revise your journal entry to show what you have learned. Be sure to include information on how understanding the nature of matter could help you in your everyday life.

CONNECTING IDEAS

The time line given on the page should be extended by the students to include modern atomic theory. The students might complete the time line in their journals.

Spaces on the concept map should be completed with ideas similar to those shown here.

1. Rutherford's gold-foil experiment
2. Helsenberg proposes orbital model of the atom.
3. Chadwick discovers neutrons.
4. Electron cloud model

Connecting Ideas

This timeline shows the basic development of the atomic theory. Copy the timeline into your journal, and use what you have learned in the chapter to complete it.

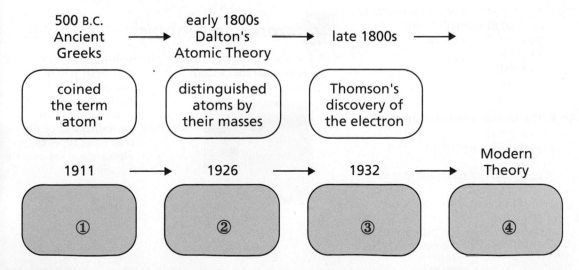

500 B.C.
Ancient Greeks → early 1800s Dalton's Atomic Theory → late 1800s →

- coined the term "atom"
- distinguished atoms by their masses
- Thomson's discovery of the electron

1911 → 1926 → 1932 → Modern Theory

① ② ③ ④

Understanding Vocabulary

1. (a) All are particles in the atom. Protons are in the nucleus and have a positive charge. Neutrons are in the nucleus and are neutral. Electrons are outside the nucleus and have a negative charge.

(b) Electrons circle the nucleus in orbits called energy levels.

(c) Atomic mass is equal to the sum of the relative masses of the protons and neutrons in an atom. The atomic number of an element is the number of protons in its nucleus.

(d) Atoms of the same elements that have a different number of neutrons are called isotopes.

Understanding Concepts

Multiple Choice

2. b

3. b

4. a

5. a

6. c

Short Answer

7. The modern periodic table is arranged in order of atomic number. Elements arranged in vertical columns are known as groups and have similar properties.

8. Bohr used the solar system as a basis for his model, with particles moving in definite orbital paths. According to modern theory, electrons form negatively-charged clouds around the nucleus. This region is subdivided into energy levels.

9. boron

10. Gases are along the right-hand side. Metals cover three-fourths of the table (center and left). Nonmetals cover one-fourth of the table (upper and right). Synthetic elements are at the bottom (number 93 and on). Most radioactive elements are along the bottom.

CHAPTER 16

Understanding Vocabulary

1. Explain the meanings of each term or set of terms below.
   a) proton (434), neutron (434), electron (432)
   b) energy levels (435)
   c) atomic mass (439), atomic number (440)
   d) isotopes (441)

Understanding Concepts

MULTIPLE CHOICE

2. According to modern atomic theory, which of the following is not a particle in an atom?
   a) electron
   b) nucleus
   c) neutron
   d) proton

3. The model of the atom that consisted of a positive center with electrons fastened to its surface was proposed by
   a) Rutherford.
   b) Thomson.
   c) Aristotle.
   d) Dalton.

4. The negative charge of an electron is equal in strength to the positive charge of the
   a) proton.
   b) electron.
   c) neutron.
   d) mass.

5. The atomic number of an atom is equal to the
   a) number of protons in the nucleus.
   b) number of electrons in the nucleus.
   c) number of neutrons in the nucleus.
   d) total number of protons and electrons in the nucleus.

6. The mass number equals the
   a) number of protons.
   b) number of electrons.
   c) total number of protons and neutrons.
   d) total number of protons and electrons.

SHORT ANSWER

7. Briefly explain the structure of the modern periodic table.

8. Compare and contrast the modern model of the atom and the Bohr model.

Interpreting Graphics

9. Look at this square of an element in the periodic table. What element is it?

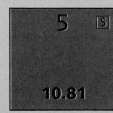

10. Using the periodic table and what you have learned about the elements, find the placement on the periodic table of the following types of elements: gases, metals, nonmetals, synthetic elements, and radioactive elements.

Reviewing Themes

11. *Systems and Structures*
Describe the three types of subatomic particles and explain how they interact.

12. *Systems and Structures*
Using the periodic table, explain how one element differs from the one before it and the one after it.

Thinking Critically

13. Explain why most of the positive particles in Rutherford's experiment passed through the solid gold foil. Why were some particles deflected at an angle? Why did some particles bounce back at the particle source?

14. One periodic property is atomic size, which is compared in terms of the radii (plural of *radius*) of the atoms. The trend is that atomic size decreases as you move along the period, as shown in the diagram below. Explain how this could occur, given that each successive atom in a series has one more proton and one more electron, as well as more neutrons, than the atom before it.

15. Why have all the explanations on the nature of the atom been theories instead of laws or principles?

16. In what way does the nucleus of an atom differ from the electron cloud?

17. If protons and neutrons are present in the nucleus of an atom and are surrounded by a much larger electron cloud, what parts of atoms are most likely to interact when atoms come together?

Li   Be   B   C   N   O   F

Discovery Through Reading

von Baeyer, Hans C. "Atom Chasing: Thanks to Sam Hurst, the Chemical Composition of Matter Is Now an Open Book." *Discover* 42(April, 1992). This article focuses on the founder of Atom Sciences, Inc., who has developed a resonance technique for detecting the smallest particles of matter.

METALS AND NONMETALS

PLANNING THE CHAPTER

B = Basic **A** = Average **H** = Honors
The coding Basic, Average, and Honors indicates subsections, features, and resources that might be appropriate for different levels of learners. For additional suggestions regarding choice of topic and depth of coverage, see the Pacing Chart on pages T28–T31.

*Frame numbers at point of use
(TR) Teaching Resources, Unit 5
(IT) Instructional Transparencies
(LI) Laboratory Investigations
(SD) *Science Discovery* Videodisc Correlations and Barcodes
(SRG) Study and Review Guide

CHAPTER MATERIALS

Title	Page	Materials
Activity: Can you change copper to gold?	457	(per group of 3 or 4) spatula, zinc dust, 400-mL beaker (2), 6 M sodium hydroxide solution, hot place, forceps, copper tongs, laboratory burner, paper towels
Discover by Researching	458	(per individual) journal
Discover by Observing	461	(per class) sodium, calcium, zinc, magnesium, copper, water, weak acid solution
Discover by Doing	462	(per group of 3 or 4) iron nail, zinc strip, small dish
Investigation: Testing Ions for Flame Colors	464	(per group of 3 or 4) wax pencils, test tubes (7), test tube rack, test solutions (7), laboratory burner, wire loops, HCl, 1 M
Teacher Demonstration	469	radon-detection kit
Discover by Writing	474	(per individual) journal
Discover by Writing	475	(per individual) journal
Skill: Drawing Conclusions	476	(per group of 3 or 4) carbon, lead, copper, sulfur, aluminum, hammer, anvil, light bulb and socket, battery and holder, wire, wire clips

ADVANCE PREPARATION

For the *Discover by Observing* on page 461, have sodium, calcium, magnesium, copper, and zinc ready. For the "Discover More" on page 464, ask the students the day before to bring in some household compounds to test. Be careful there are no explosive substances or very reactive substances in any of the products before heating them.

TEACHING SUGGESTIONS

Outside Speaker

Invite someone who works with metals to come and talk with the class. A welder, a jeweler, and a plumber are some possibilities. You might also have a local health official talk about the dangers of radon in the home and how testing for radon is done. Finally, you might invite an environmentalist to talk on the dangers of destroying the ozone layer and what the students and their families can do to prevent further destruction.

CHAPTER 17 METALS AND NONMETALS

This chapter will help the students understand that metals and nonmetals are grouped by their characteristics. Elements in each group have specific properties and electron configurations. Discussions of metals, nonmetals, alloys, metalloids, noble gases, flame tests, rusting, and halogens develop the greater concepts of structure and differences in substances. A supporting theme in this chapter is **Environmental Interactions**.

MULTICULTURAL CONNECTION

After working as a chemist in the Chicago Department of Health Laboratories, Dr. Lloyd A. Hall, an African American, became president and chemical director of a consulting laboratory—the Chemical Products Corporation in Chicago. During this time, he developed an improved method for preserving meat. The method involved "flash-drying" a solution of sodium chloride and nitrogen-containing salts (nitrate and nitrite). Dr. Hall's "flash-dried" crystals were far superior to any meat-curing salts ever produced and were widely used in the meat industry.

Dr. Hall was involved in other aspects of the food-chemistry industry throughout his career. His work included the sterilization of foods and substances associated with foods, such as spices. Dr. Hall spent considerable time in consulting work and had a very active interest in detergents, vitamins, and asphalt. He was a driving force behind the formation of the Institute of Food Technologies in 1939.

MEETING SPECIAL NEEDS

Second Language Support

Have limited-English-proficient students work with English-fluent students to identify the objects in the Chapter Motivating Activity and the properties those objects have.

CHAPTER 17

METALS AND NONMETALS

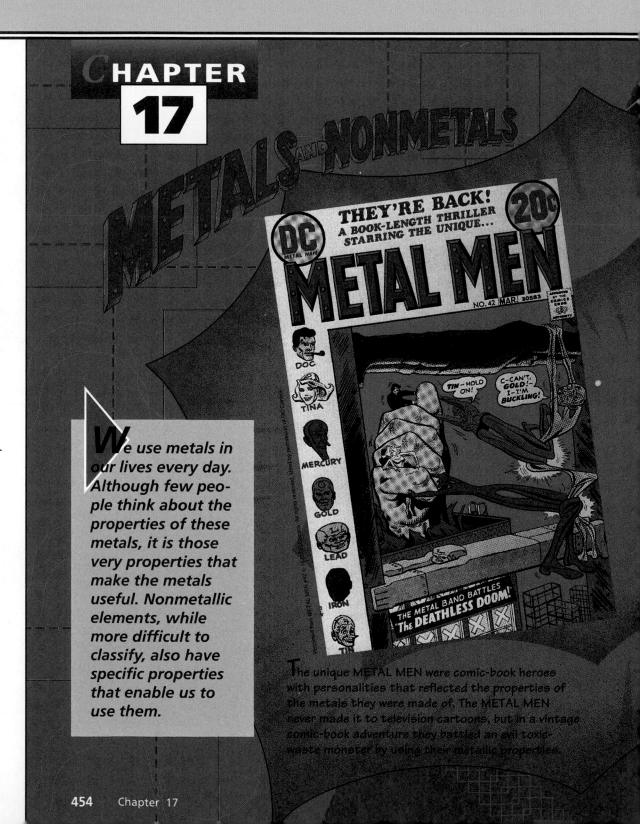

We use metals in our lives every day. Although few people think about the properties of these metals, it is those very properties that make the metals useful. Nonmetallic elements, while more difficult to classify, also have specific properties that enable us to use them.

The unique METAL MEN were comic-book heroes with personalities that reflected the properties of the metals they were made of. The METAL MEN never made it to television cartoons, but in a vintage comic-book adventure they battled an evil toxic-waste monster by using their metallic properties.

CHAPTER MOTIVATING ACTIVITY

Obtain and display some of the following objects and substances: mercury thermometer, sodium, glass cooking bowl, sheets of aluminum foil, copper or iron pot, iodine, crystals of sulfur, gold jewelry, brass bracelet, copper wire. Then ask the students to separate them into these two categories—metals and nonmetals. Ask the students what criteria they used for the categorizing and what properties make the substances useful. For example, copper and iron make good cooking utensils because these metals readily conduct heat. Conclude by telling the students they will learn more about the properties and uses of metals and nonmetals in this chapter.

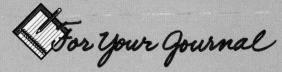

For Your Journal

The journal questions can help you identify misconceptions the students might have about metals and nonmetals. (Accept all answers, and tell the students that at the end of the chapter they will have a chance to refer back to the questions.)

There was IRON, the "muscular" metal, who could shape himself into girders and cranes. PLATINUM was an elegant metal who could stretch herself thinner than a human hair. The pushover in this crew of heroes was TIN, a soft metal who was always "buckling" under pressure. TIN just knew he would be a failure. The temperamental metal was MERCURY, who could feel himself expanding as he got angry.

The METAL MEN could not match today's high-tech cartoon heroes, but their comic-book story shows how science fiction anticipates the future. The METAL MEN battled "Chemo," a monster made of discarded toxic waste. Chemo's harmful chemicals corroded buildings and melted railroad tracks as if they were wax. The METAL MEN may be out of date, but they fought scientific problems that scientists and citizens still face today.

ABOUT THE LITERATURE

The first comic book—a collection of reprinted "Yellow Kid" strips—appeared in 1897. In the 1920s, the first comic books made up of original material were published in Japan. Comic books began to appear in their present magazine form in the 1930s. In the 1930s, adventure stories became popular. These adventure comics included "Dick Tracy," "Prince Valiant," and "Superman."

Comics lost some popularity after television became widespread in the late 1940s. Comic books especially suffered because a few publishers printed ones that contained obscenities and violence. In 1954 almost all U.S. comic book publishers adopted the Comics Code Authority, which prohibited such material. By the 1960s, comics had regained their popularity. While comics once appealed mostly to children, today's strips—such as "Peanuts" and "Calvin and Hobbes"—are enjoyed by all audiences.

For Your Journal

- How can you tell the difference between a metal and a nonmetal?
- How do you use metals in your daily life?
- How are nonmetals useful to you?

FOCUS

Section 1 discusses the general properties and uses of metals, the advantages of alloys, the properties of metallic families, the concept of rust, and the flame test of metals. The differences between the electron configurations of metals and nonmetals and the lack of specific properties of nonmetals are also examined.

MOTIVATING ACTIVITY

Demonstrate the electric conductivity of various substances by connecting one or two batteries to a light bulb and socket in series. (Note: Do not complete the circuit. Make certain you have enough power to light the bulb.) Place a piece of sulfur, copper, iron, or sodium into the circuit. Have the students observe how brightly the light shines, if it shines at all. Help the students compare the conductivity of the elements tested. Discuss ways to prove that the elements really did conduct electricity. Conclude by helping the students understand that all metals conduct electricity to some degree, due to the nature of their metallic bonds.

PROCESS SKILLS
• Analyzing • Experimenting
• Observing

POSITIVE ATTITUDES
• Curiosity • Openness
to new ideas

TERMS
• ion • alloy • metalloids

PRINT MEDIA
Chemistry: The Birth of a Science by Tom McGowen
(see p. 423b)

ELECTRONIC MEDIA
The Chemistry Help Series: The Periodic Table Focus Media, Inc. (see p. 423b)

SCIENCE DISCOVERY
• Rust; iron

BLACKLINE MASTERS
Laboratory Investigation 17.1
Reading Skills
Connecting Other Disciplines
Study and Review Guide

MULTICULTURAL CONNECTION

Ancient cultures knew and used many native metals. Gold was used for ornaments, plates, jewelry, and utensils as early as 3500 B.C. Gold objects showing a high degree of artistry have been excavated at the ruins of the ancient city of Ur in Mesopotamia. Silver was used as early as 2400 B.C. and was considered by many to be more valuable than gold because it was rarer in the native state.

SECTION 1

Characteristics of Metals and Nonmetals

Objectives

List the general properties of metals.

Name five metallic elements, and **describe** their uses.

Compare and contrast the properties of metals and nonmetals.

Some elements are found in nature as beautiful crystals; others occur as gases you breathe every day. Still others exist only under laboratory conditions. They range from very active to totally nonreactive, and they have a wide variety of colors, shapes, and smells. Some elements even produce characteristic colors when they ignite, as they do when fireworks explode.

You know that the periodic table is divided into metals and nonmetals. How do chemists decide whether an element is a metal or a nonmetal?

Figure 17–1. Properties of elements are as varied as the colors of a fireworks display.

From Lead to Gold

In ancient times metals were very important for making tools and weapons. So important, in fact, that historians have given certain periods of history the names of the metals that were used during the period, such as the Iron Age and the Bronze Age.

Knowledge of metals increased throughout the ages, from prehistoric times up to the Middle Ages, where we will take a

Figure 17–2. Alchemists worked in laboratories just as modern scientists do. They even used some of the same kinds of equipment.

TEACHING STRATEGIES

● **Process Skills:** *Observing, Analyzing*

Be sure the students observe the changes in the token. Help the students determine why the copper did or did not turn to gold.

● **Process Skills:** *Analyzing, Predicting*

Have the students speculate whether they would get the same results in the Activity if they replaced the copper token with a new penny. (Results would differ since the penny is not solid copper, but rather has a zinc center.)

---

short stop to look at some research. The science of chemistry was being born. Early experimenters were called *alchemists*; the subject they studied was called *alchemy*—a word that comes from an old name for Egypt. The term *chemist* is derived from the original *alchemist*.

Alchemists had two main objectives: to find the elixir of perpetual youth and to convert ordinary metals to gold. They dreamed of "striking it rich" by finding a way to convert copper, tin, or iron into beautiful, shiny, valuable gold. You, too, can experiment in alchemy. Try changing copper into gold.

ACTIVITY
Can you change copper to gold?

MATERIALS
safety goggles; laboratory apron; spatula; zinc dust; beaker, 400 mL (2); sodium hydroxide solution, 6 M; hot plate; forceps; copper penny, pre-1982; tongs; paper towels; laboratory burner

CAUTION: Zinc dust can ignite. Sodium hydroxide solution can cause burns. Be sure to wear safety goggles and a laboratory apron for this activity.

PROCEDURE
1. Use the spatula to put about 5 g of zinc dust into a beaker. Add enough sodium hydroxide solution to cover

the zinc, and gently warm it on the hot plate until the mixture is steaming.
2. Using forceps, place the copper penny in the mixture. Use the tongs to remove the beaker from the hot plate.
3. Observe the penny. Record any changes in your journal.
4. Use the forceps to remove the penny from the solution. Swish the penny in a beaker of clean water, and dry it with a paper towel.
5. Hold the penny with the forceps, and insert it into the outer part of the burner flame for several seconds until you observe a change. Immediately place the penny back into the beaker of water. What has happened?

APPLICATION
1. Cut the penny, and describe what the inside looks like. Explain what you observe.
2. Did you turn copper to gold? Explain your answer.

Did you make gold? How do you know? What metal coats your token? To find out, you can do what scientists do every day—research.

ACTIVITY
Can you change copper to gold?

Process Skills: Observing, Experimenting, Analyzing

Grouping: Groups of 3 or 4

Hint
Use pre-1982 pennies since post-1982 pennies are copper-plated zinc. Be sure the copper pennies do not overheat since the coated copper will over oxidize and turn black.

Procedure
5. The copper token has changed color.

▶ Application
1. Since the reaction occurs only on the surface of the token, the unaffected copper should be visible when the token is cut.

2. No, the copper did not turn to gold. Brass was produced when the copper reacted with the zinc.

REINFORCING THEMES
Systems and Structures

Use the results from the *Activity* to discuss the interaction of elements, such as the interaction between zinc and copper that produces brass.

● INTEGRATION– *History*

For nearly a thousand years, alchemists were the closest thing to scientists the Western world knew. Alchemists believed that all kinds of matter were very much the same. Everything was made from a combination of three or four basic elements. Substances differed only in the way these basic elements were combined. Therefore, changing a common metal, such as iron, into a gold or silver seemed feasible.

SECTION 1 457 ◀

● **Process Skill: *Analyzing***

The names and symbols of many elements are derived from Latin. For example, Cu, or copper, comes from the word *cuprum.* (The Romans obtained copper from the island of Cyprus.) Silver (Ag) comes from the word *argentum.* Lead (Pb) is derived from the Latin *plumbum.* Point out to the students that the Romans used lead to line their water pipes. Then challenge the students to name other water-related words that are derived from *plumbum.* (Answers include plumber and plumbing.)

● **Process Skill: *Communicating***

Point out to the students that platinum—one of the heaviest substances known—is even more precious than gold. Then challenge the students to name examples of how platinum is used. (Examples include use in jewelry, in automotive catalytic converters, and in surgical instruments.)

---

① Aluminum could be combined with iron to make a durable, but light, alloy.

② Aluminum would be useful in the manufacture of aircraft bodies because it is a very strong, though very lightweight, metal.

③ Since tin turns to powder at a relatively low temperature, the pan would be impractical for cooking.

INTEGRATION– *Environmental Sciences*

Mercury in the environment is hazardous chiefly because its poisonous compounds have been found in plants and animals that people use for food. Mercury acts as a cumulative poison—that is, the body has difficulty eliminating it. Thus, it can collect over a period of time, eventually reaching dangerous levels.

Government and industry are working to keep mercury out of the environment. In the early 1970s, the U.S. and Canadian governments began to prohibit dumping of industrial wastes that contain mercury.

ONGOING ASSESSMENT
▼ ASK YOURSELF

Alloys have the properties of both metals or of the metal and nonmetal they are made of. Therefore, scientists and engineers can choose the properties they need for an object and produce the perfect substance to make it out of.

Figure 17–6. Aluminum is a lightweight metal. Why would it be useful in the manufacture of aircraft bodies? ②

"My name is Tina, and my platinum properties allow me to stretch myself into a strand over 85 km long."

Alloys have dozens of uses. Suppose you wanted to make a very light form of steel for airplane wings. It would have to be strong enough to resist heavy winds but light enough for the plane to fly. Steel is an alloy of iron and other substances. What light, shiny metal that you have in your kitchen could you combine with iron to make a durable but light alloy? ①

Sometimes people need metals that have very specific properties. They may have a use for a very strong metal or a metal that is a better conductor of heat than any of the metallic elements alone. Mixing metals and other elements can sometimes produce these specific properties. For example, brass, the mixture of copper and zinc that you made, is harder and far more durable than either copper or zinc alone.

Think about the pots and pans in your kitchen. You may have some pans made of iron, some of aluminum, some of steel. You may even have some that are made of glass! The properties of the elements in these materials enable you to cook food efficiently. What would happen if you had a pan made of tin? ③

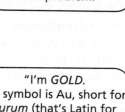

"I'm *TIN.* I'm softer than zinc but harder than lead. I turn to powder at 200°C. Cooking with me could be a problem."

"I'm *GOLD.* My symbol is Au, short for *aurum* (that's Latin for "gold"). An ounce of me can be stretched into a wire about 85 km long. Not bad, huh?"

Gold is a very soft metal. For this reason it cannot be used alone to make durable items, such as jewelry. Therefore, gold jewelry is made using an alloy. For instance, copper is combined with gold to make yellow gold. Platinum or silver is added to gold to make white gold.

▼ ASK YOURSELF
Why are alloys important?

● **Process Skill:**
Communicating

Point out that alkali metals are in the first column of the periodic table. Explain to the students that alkali metals are quite soft compared with other metals. Then ask the students why

alkali metals are found naturally only in compounds. (Alkali metals are so reactive that they cannot be found in the purest form or free state in nature.)

● **Process Skill:** *Inferring*

Tell the students that the diameter of the atoms increases going down the column of the periodic table from Li to Cs. Ask the students to infer what happens as the metallic atom becomes larger. (Softness

increases; lithium is a little harder than lead, sodium and potassium can be cut with a knife, rubidium is as soft as butter, and cesium is a liquid at 28°C.)

Metallic Families

Metals tend to lose electrons to form positive ions. Group 1 metals tend to form +1 ions, while Group 2 metals tend to form +2 ions. The metals in the other groups may form more than one positive ion. That is, they can form +1, +2, or even +3 ions.

The properties of the metals depend on how easily they lose electrons. The easier it is for them to lose electrons, the more reactive they are. In order to analyze the reactivity of metal families, your teacher will demonstrate how they react in water or acid solutions. Make careful observations, and write them in your journal.

DISCOVER BY *Observing*

Your teacher will place the following metals in water: sodium (Na), calcium (Ca), magnesium (Mg), copper (Cu), and zinc (Zn). What happens? Your teacher will place the metals that didn't react in water into a weak acid solution. Acids are more reactive with metals than water is. Record your observations.

Was the reaction in acid the same as the reaction with water? Describe the similarities and differences.

Now, place the metals in order of reactivity, with the most reactive metal at the top of the list. Then, look up the position of the metals in the periodic table. How does their position in the table relate to their reactivity? ✎

Potassium (K) wasn't used in the activity because it reacts so violently with water that it is safer to show you a photograph (Figure 17–7). Where would you place potassium in your scheme? ④

Now that you have organized the metals in order of reactivity, can you see any particular trend? In general, the reactivity of metal families increases as you move from the metals on the top right on the periodic table to the bottom left-hand corner. The alkali metals are the most reactive of the metals, with francium—the last member of the family—being the most reactive.

The transition metals do not follow this trend exactly; there are some exceptions. Which element(s) from the activity did not follow the trend? ⑤

Rust Protection How can the concept of reactivity be used? You may be using it right now to protect your bike. Let's discover how.

Figure 17–7. Alkali metals are very reactive in water. If potassium is placed in water, it bursts into flame.

INTEGRATION–
Vocational Education
Cooperative Learning

Have the students work in pairs to interview people whose occupations involve working with metals. Occupations include chemist, jeweler, welder, materials scientist, sculptor, mining engineer, and dentist. Have the students write summaries of their interviews in their journals.

DISCOVER BY *Observing*

The students' answers should reflect that the metals are more reactive in acids than they are in water. The metals at the left of the chart are the most reactive.

④ Potassium would be placed at the top of the chart.

⑤ copper

SCIENCE TECHNOLOGY SOCIETY Have the students do research on the transition elements cadmium, nickel, mercury, and lead. They should realize that these elements are hazardous to health and that society must prevent poisoning or pollution by these elements.

INVESTIGATION

Testing Ions for Flame Colors

Process Skill: Observing

Grouping: Groups of 3 or 4

Objective

● **Observe** the flame colors created by different compounds

Pre-Lab

Have the students predict the colors metal ions will produce when flame tested.

Hints

Dip the loop into each salt moistened with a little HCl. Hold it in the outer margin of the flame. Clean the loop between trials by dipping it in HCl and then burning it off.

▶ Analyses and Conclusions

1. All of the compounds have different flame colors, even though they all contain chlorine. The metal ion must be giving the color, since it is different in each compound.

2.

Metal	Ion	Color
lithium	Li+	red
sodium	Na+	yellow
potassium	K+	blue
copper	Cu++	green
strontium	Sr++	reddish-purple
barium	Ba++	blue-green

PERFORMANCE ASSESSMENT

Cooperative Learning

Have each group of students think of an example in which the color a compound produces when burned would be useful. (Examples include fireworks and flares. Accept reasonable examples the students can justify.)

---

3. If you mix two compounds, you will see a color that is a combination of the separate flame colors. Note: If one of the compounds is in greater concentration, its color might trail off at the end.

▶ Application

Answers will depend on the compounds in the salt substitute.

✳ Discover More

Answers will depend on the household compounds tested.

Post-Lab

Have the students evaluate the predictions they made at the beginning of the lab.

Were their criteria valid? Have them write a new criteria for judging flame color.

INVESTIGATION

Testing Ions for Flame Colors

▶ MATERIALS

● laboratory apron ● safety goggles ● wax pencils ● test tubes (7) ● test tube rack ● test solutions (7) ● laboratory burner ● wire loops ● HCl, 1 M

CAUTION: Put on an apron and goggles before starting this experiment. Slight spattering may occur.

▼ PROCEDURE

1. You will be observing the flame colors created by different compounds dissolved in water. Make a table like the one shown to record your observations.

TABLE 1: FLAME TESTS	
Compound	**Color Observed**

2. Your teacher will list the compounds on the chalkboard. Copy the names of the compounds into your table.

3. Use a wax pencil to label the test tubes A through G. Place them in the test tube rack in order. Place a small sample of each compound in the corresponding test tube and moisten with a few drops of HCl.

4. CAUTION: Flames are hot and can burn you. Do not get your hands too close. Light your burner. You may wish to refer to page 632 for instructions on adjusting the burner flame.

5. Insert the wire loop into compound A. Hold the loop in the flame. Record the color you see.

6. Repeat step 5 for each of the compounds in the test tubes.

▶ ANALYSES AND CONCLUSIONS

1. You tested different compounds that contained the same element. For instance, LiCl, NaCl, and KCl all contain chlorine. Do all of these compounds have the same flame color? Explain.

2. The elements in each compound are present as ions. In your table, add the names of the ions in each compound and identify their specific colors.

3. What do you think will happen if you mix two of the compounds? Explain. After you have answered the question, try it. Were the results what you expected? Why? Why not?

▶ APPLICATION

Salt substitute usually contains some compound that can be used to replace sodium chloride (NaCl). Do a flame test on some of this substitute and, without looking at the label, determine what metal is present. Look at the label when you are done to see if you were correct.

✳ Discover More

Choose some household compounds and test them. Determine from their flame colors which metals they contain.

FOCUS

Section 2 discusses nonmetals. The reasons for the lack of specific properties of nonmetals, the noble family, the halogens, the electron configurations of nonmetals, and environmental concerns over ozone and radon are also discussed.

MOTIVATING ACTIVITY

Cooperative Learning Have the students work in pairs or small groups to brainstorm what could be done to make people aware of the diminishing ozone layer.

Some students might research what can be done to protect the ozone layer, while other students design posters or develop slogans that focus on the issue.

PROCESS SKILLS
• Analyzing • Communicating
• Interpreting data

POSITIVE ATTITUDES
• Creativity • Environmental awareness

TERMS
• diatomic molecules
• allotropes

PRINT MEDIA
"Ozone Hole: What it Means" from *Popular Mechanics* magazine by Abe Dana (see p. 423b)

ELECTRONIC MEDIA
Silicon Britannica Video (see p. 423b)

SCIENCE DISCOVERY
• Neon light • Neon sign • Iodine

BLACKLINE MASTERS
Laboratory Investigation 17.2
Extending Science Concepts
Thinking Critically
Study and Review Guide

Nonmetals

SECTION 2

The comic book characters in the beginning of the chapter had only metallic properties. Could we have nonmetal people? *IODINE* would be purple since that is the color of iodine crystals. Iodine easily sublimes (converts from a solid directly into a gas) into a purple vapor. *IODINE* could turn into a purple gas and move through keyholes and under doors. Imagine the possibilities! What other nonmetal can you personify? What properties would be interesting? In order to find out, you will need to study the properties of nonmetals further.

Objectives

Compare the noble gases with other elements in the periodic table.

Diagram the electron configuration for elements in Periods 1 and 2 in the periodic table.

Explain what is meant by diatomic, and give two examples of diatomic elements.

✧ **Did You Know?**
Fewer than two dozen elements are nonmetals.

● **Process Skill:** *Identifying*

Have the students identify the noble gases on the periodic table. Explain to the students that the noble gases were considered inert because they would not combine with other elements. Discuss with the students why the possibility of noble gases forming chemical bonds is unlikely.

● **Process Skill:** *Formulating Hypotheses*

Ask the students to hypothesize why argon is unreactive, even though its outermost energy level (the third level) does not have 18 electrons, which is its capacity. (The outer energy level has eight electrons, or a stable octet. Even though more electrons might be added to the third level, the presence of eight electrons makes the element stable and unreactive.)

SCIENCE TECHNOLOGY SOCIETY

Tell the students that a controversy still exists over whether to add fluoride to the drinking water. Encourage interested students to do research on the advantages and disadvantages of adding fluoride to drinking water. Have the students summarize their findings in their journals.

MEETING SPECIAL NEEDS

Mainstreamed

Help mainstreamed students use a map of Kansas to find the location of Dexter, Kansas (where helium was first discovered in the United States).

Figure 17–10. Helium, which is lighter than air, is often used to fill balloons.

Figure 17–11. The hydrogen that was used to fill the *Hindenburg* was very reactive. Today, dirigibles are filled with helium.

A Stable Octet Is Not Eight Singing Horses!

The stability of the noble gases is due to the number of electrons in their outer energy levels. All of these elements except helium have eight electrons in their outer energy level. This seems to be a stable configuration and, in fact, is called a *stable octet*. Helium has only two electrons, but it has a filled first energy level, which also gives it great stability.

When elements react, they lose, gain, or share electrons to achieve an electron arrangement like one of the noble gases. Here are some examples. Sodium forms Na^+, a positive ion, by losing one electron. This gives the sodium ion the same electron arrangement as argon, a noble gas. Chlorine, a nonmetal, forms a negative ion, Cl^-, by gaining one electron. This gives the chlorine ion the same electron arrangement as argon. Both elements form stable octets as their electron arrangements change.

Stability in Your Life

The noble gases have many practical uses. You know about helium balloons, but did you know that helium is also used in blimps? Since helium is less dense than air, it makes sense to use it in blimps and other airships. But this has not always been the case. Once, hydrogen was used in giant airships called zeppelins. Hydrogen is also less dense

● **Process Skill:** *Comparing*

 Discuss with the students the uses of hydrogen and helium in blimps. Ask volunteers to do research on the *Hindenberg*, write their findings in their journals, and share their findings with classmates.

● **Process Skill:** *Communicating*

Reinforce that a diatomic molecule is a molecule composed of two atoms, and that free halogens are diatomic. Then have the students locate the five halogens on the periodic table. (fluorine, chlorine, bromine, iodine, astatine)

● **Process Skill:** *Communicating*

Some students might be concerned that the fluorine and iodine in products might be dangerous for human consumption. Assure the students that elements, such as fluoride in toothpastes and most drinking water and iodine in table salt, are present as compounds with other elements and therefore they are not poisonous.

than air, so the zeppelins floated. Hydrogen, however, has one significant characteristic that helium doesn't have. It explodes! The passenger ship *Hindenburg* exploded in 1937 when its hydrogen ignited. Since that time, nonreactive helium has been used to fill airships.

There are other noble gases that you use or see used every day. Neon, as you have probably guessed, is used in neon signs. It produces a red color when an electric current passes through it. Argon is used to fill regular light bulbs. It provides an inert atmosphere for longer-lived bulbs. Xenon and radon are rarer and, therefore, have fewer applications. Xenon is used in photographic flashbulbs for high-speed photography, while radon is used to treat cancer.

Radon, however, has a downside. Radon is produced when radium undergoes radioactive decay. (You will learn more about this type of decay in the next chapter.) This radon is released from soil and rocks and can leak into houses through the cracks in the foundation or basement. Usually any radon is diluted by natural circulation and open windows. However, in new, energy-efficient buildings radon can build up to dangerous levels. If inhaled in large enough quantities, radon can cause lung cancer.

▼ **ASK YOURSELF**

Why do metals tend to form positive ions?

Demonstration

Obtain a radon-detection kit. Display and discuss the kit's components with the students. You might expand this demonstration by setting up the kit in the basement of your home and analyzing the test's results with the students.

SCIENCE BACKGROUND

The helium atmosphere used in underwater habitats distorts the sound of divers' voices, making them sound higher in pitch. Helium is lighter than the nitrogen it replaces. Therefore, it causes the vocal cords to vibrate much faster.

ONGOING ASSESSMENT
▼ **ASK YOURSELF**

The electrons in the outer levels of a metallic element are not held closely by the nucleus. They can give up an electron easily, thus forming a positive ion.

SECTION 2 **469** ◀

● **Process Skill:** *Inferring*

Ask the students how phosphorus-containing detergents can be beneficial and harmful at the same time. (While phosphorus increases the cleaning power of detergents, phosphorus in waste waters causes algae to multiply excessively. As the algae dies and decomposes, they use up all the oxygen, eventually "suffocating" a body of water.)

✧ **Did You Know?**
Under usual conditions, fluorine and chlorine are gases. Iodine is a solid, while bromine is a liquid. Astatine is not a naturally occurring substance.

LASER DISC

2350

Iodine

INTEGRATION–
Language Arts

Many elements are named after one of their properties. For example, bromine comes from the Greek word *bromos*, which means "stink." Challenge the students to describe the nature of argon, knowing that *argos* means "idle."

Nonmetal Families

Besides the noble gases, only one other nonmetal family has a commonly used specific name. The halogens, Group 17, are among the most reactive of the nonmetals. Recall that the most reactive metals, the alkali metals, are on the opposite side of the periodic table. In their uncombined form, the halogens are poisonous and highly corrosive. All the halogens exist naturally as molecules composed of two atoms, or **diatomic molecules**. For example, chlorine gas is Cl_2, and fluorine gas is F_2.

No doubt you would like to test the reactivity of nonmetals in the same way as you did for the metals. Unfortunately, many of them are toxic and very dangerous, so we will tell you about their reactivity.

I'm Always Hungry!   The halogen fluorine is the most reactive of all the nonmetals. In fact, it is so reactive that it is often called the *Tyrannosaurus rex* of elements. That means that

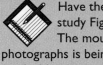
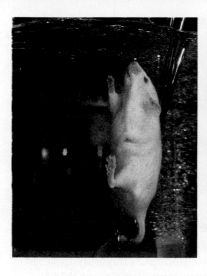

Figure 17–12. One fluorine compound can be "breathed" and is being used experimentally for deep-sea dives. The mouse shown here is being submerged in the compound. It "breathes" in the liquid and gets plenty of oxygen. After the compound is drained from its lungs, the mouse is fine!

✧ **Did You Know?**
Only 12 elements were known in ancient times. Between the 1500s and 1870, 43 more elements were discovered. By 1925, 88 elements were known. Since then, 21 more elements have been discovered or created artificially.

INTEGRATION– Film

An example of a human breathing a highly oxygenated fluorine compound, similar to the one the mouse is breathing, can be seen in *The Abyss*, a movie about an underwater drilling platform.

it will "eat" (combine with) almost anything. Fluorine compounds have many uses. One fluorine compound, Teflon, is used to make nonstick surfaces on items ranging from cookware to adhesive bandages. Another fluorine compound is used as a blood substitute in emergencies. Yet another fluorine compound can be breathed, as shown in Figure 17–12.

The reactivity of the nonmetals increases as you move toward fluorine. That is, as you move from the lower left to the upper right of the periodic table, the reactivity of the nonmetals increases. Don't forget, though, that the noble gases don't fit the trend.

What Phase Are You?

Have you noticed that the nonmetals described have all been gases? This is not true of all nonmetals. Bromine, another halogen, is a red liquid. Some bromine compounds are used in agriculture for insect control. The element iodine is a purple crystalline solid. The liquid antiseptic with which you might be familiar is a tincture of iodine, that is, iodine crystals dissolved in alcohol.

In the nitrogen family, Group 15, only nitrogen is a gas. In fact, it is a diatomic gas, N_2. All the other elements in this family are solids.

Figure 17–13. Iodine crystals change into a gas, or sublime, at room temperature.

● **Process Skill:** *Inferring*

Tell the students that the name for the chemical phosphorus came from the Greek *phosphoros* meaning "light-bearing." The Greek *phosphoros* was the ancient name for the planet Venus when it appeared at the morning star. Ask the students why they think the chemical was named as it was. (Several forms of phosphorus ignite spontaneously in air, producing a very bright light.)

INTEGRATION–
Earth Science

Phosphorus is never found free in nature, but is common in combination with minerals. Phosphate rock, which contains the mineral apatite (a form of calcium phosphate), is an important source for phosphorus. Phosphate rock is mined from large deposits found in such places as Morocco, Florida, Tennessee, Utah, and Idaho.

Figure 17–14. White phosphorus, when exposed to air, bursts into flame.

What's the Same but Different?
Twins. Have you ever known a set of twins? Even though they have the same birthday and often look just alike, they are different. They may have different like and dislikes, or they may excel at different things. Some nonmetals are like twins. They are the same but different. These nonmetals exist in two or more molecular forms called **allotropes**.

Phosphorus, a member of the nitrogen family, exists in two major forms—red and white. Red phosphorus is fairly stable. White phosphorus, however, reacts violently with oxygen in the air, so it must be stored under water.

Figure 17–15.
Which twin is white phosphorus? How do you know?

● Process Skill: *Inferring*

Display a world map and ask a volunteer to point out the Arctic and Antarctic. Explain to the students that holes in Earth's ozone layer have been found in these areas. Then ask the students to speculate how the holes in the ozone layer could affect these regions. (Some scientists think that the regions' ecological systems could be adversely affected by increases in ultraviolet light reaching the earth's surface.)

● Process Skill: *Inferring*

Remind the students that oxygen is the most abundant element in nature. Ask them why oxygen is also the most important element on Earth. (Since most organisms require oxygen to live, without oxygen there would be only a limited number of simple life forms that could live without oxygen.)

The differences in the properties of red and white phosphorus are due to the arrangement of atoms in their molecules. White phosphorus exists as molecules of four atoms, whereas red phosphorus molecules are large disorganized clumps of atoms.

Other families also have allotropes. Oxygen, for example, exists in two forms. The oxygen you breathe is diatomic oxygen (O_2). Oxygen can also be found in a triatomic state (O_3) known as *ozone*.

In the Ozone Diatomic oxygen is converted to ozone by the sun's ultraviolet rays. The ozone forms a layer high in the atmosphere that screens out or absorbs most of the ultraviolet rays from the sun. However, ozone is a pungent and toxic gas and, when found close to the earth, is considered a pollutant.

Ultraviolet rays cause sunburn, and long-term exposure to them may cause skin cancer. The ozone layer is therefore a very important defense against the sun's harmful rays. Recently, scientists have found holes in the ozone layer. These holes, which seem to open and close at random over different parts of the earth, might be caused by certain gaseous molecules that reach the upper atmosphere and break down O_3 into O_2. One of the culprits might be a class of compounds known as chlorofluorocarbons, or CFCs. (You may have heard them called fluorocarbons.) Read the following article to find out more about CFCs.

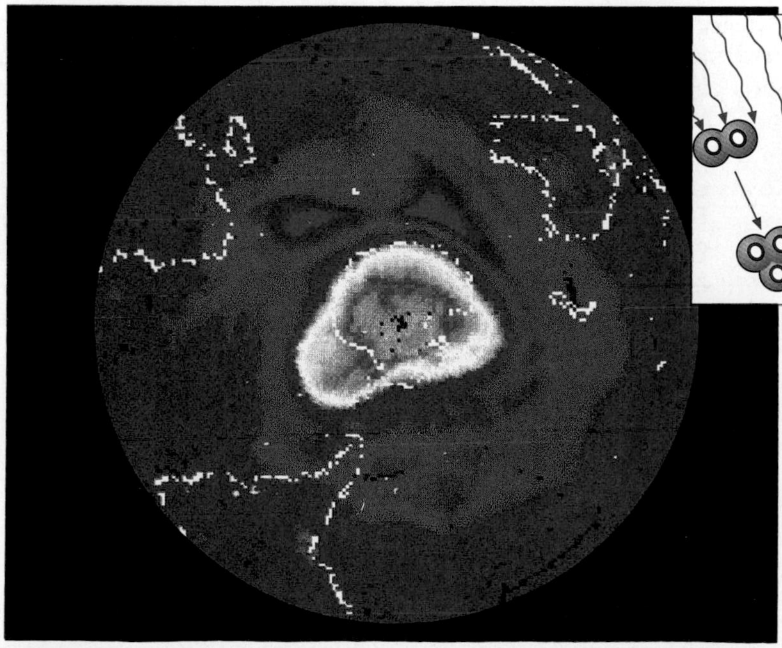

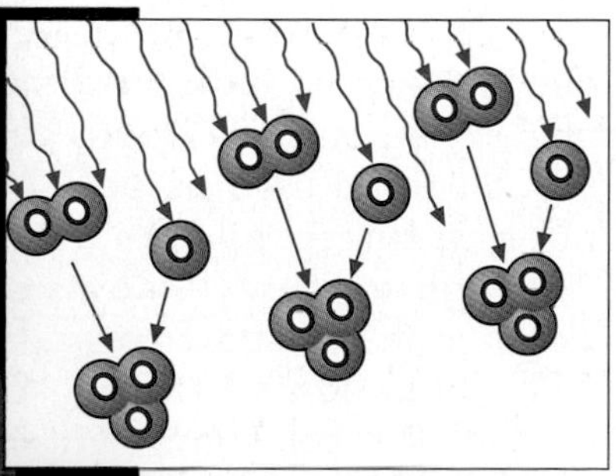

Figure 17–16. Oxygen (O_2) is converted to ozone (O_3) by ultraviolet radiation from the sun. CFCs may be creating a hole in the ozone covering of the earth.

INTEGRATION–*Language Arts*

Have the students look up the prefixes *di–* and *tri–* in the dictionary and write the definitions in their journals. Ask them how these prefixes will help them distinguish between oxygen and ozone. (*Di–* means "two, twice, or double." Oxygen is a diatomic molecule composed of two oxygen atoms. *Tri–* means "three." Ozone is a triatomic molecule made up of three oxygen atoms.)

✧ *Did You Know?*

As a result of the widespread use of freons and other related chemicals, the protective ozone layer around the earth is being destroyed. In an effort to protect the earth's ozone layer, 46 countries around the world have agreed to reduce the use of ozone-destroying chemicals. The agreement was signed in 1987 and will remain in effect through 1999.

Have the students think of and list aerosol products that once contained or now contain CFCs. (Examples might include household cleaners, personal-hygiene products, and spray paints.) Then have the students think of products that could be used in place of aerosol-based products. (Examples include using a stick or roll-on antiper- spirant instead of an aerosol antiperspirant and using furni- ture oil or cream instead of an aerosol furniture wax.)

Have the students provide written answers to the Section Review. In their jour- nals, have the students choose a common element and write as much as they know about the element.

Have the students describe and summarize the characteristics of metals and nonmetals. En- courage the students to name several specific metals and non- metals in their summaries.

THE NATURE OF SCIENCE

The use of CFC-containing pack- aging increased significantly during the past decade. Today the trend is to find and use alternative means of packaging. Have the stu- dents think of and list items that are currently packaged in CFC (plastic foam) containers. Examples include fast-food items and electronic equipment. Then have the students brainstorm ways the items could be pack- aged, using alternate means of packaging. You might extend this activity by having the students do research on what is being done to recycle CFC packaging.

THE TROUBLE WITH CFCs

from Odyssey

They're at the heart of refrigerators and air condi- tioners. They're useful as solvents to clean electronic parts. They help make insu- lation for your house, con- tainers for your hamburger, and nice soft cushions for your family-room couch. They're the chlorofluorocar- bons (CFCs for short), a type of chemical that has become almost indispens- able in the modern world. (CFCs can be used as pro- pellants in spray cans. Although this use is now illegal in the United States, CFC propellants are still used in other countries.)

Prized for years be- cause they seemed to be nontoxic and inert—that is, they don't react easily with other substances—these ap- parently harmless chemicals have turned out to be wolves in sheep's clothing.

When a car's air con- ditioner leaks, or when a foam hamburger container is crushed, chlorofluorocar- bons are released into the atmosphere. Some of the CFCs eventually make their way up to the stratosphere.

In the stratosphere, sun- light breaks up the CFC, re- leasing chlorine from it. The chlorine destroys ozone (O_3) by turning it into regu- lar oxygen (O_2).

When there is less ozone in the stratosphere, more damaging ultraviolet light from the Sun can reach Earth's surface. An increase in ultraviolet light could have serious consequences. It is bad for plants, including food crops. It can hurt some sea creatures. It can cause skin cancer in humans.

To avoid these serious health and environmental problems, scientists agree, we must act to keep ozone from being depleted any more. We must cut back on our use of CFCs, either by recycling some CFCs or by finding new chemicals to re- place them. We don't have a lot of time to waste. The CFCs already in the stratos- phere will affect the ozone layer for more than a century.

In September 1987 an international panel worked out an agreement to limit future use of chlorofluoro- carbons. Production of these chemicals was halted at the 1986 level; by 1999 it must be cut in half. The U.S. Envi- ronmental Protection Agency is calling for even stricter cut- backs. We're taking our first steps in the right direction.

DISCOVER BY Writing

Look up another recent article that talks about the holes in the ozone layer. Write a summary of the article in your journal. Be sure to include the possible causes of and solutions to the problem that the article suggests.

Carbon Has Many Disguises Look at the photograph shown here. What do you think these "things" have in common? ①

All of the substances in the figure are allotropes of carbon. Carbon black, one common allotrope, is very similar to charcoal. It exists as disorganized atoms of carbon. A second allotrope of carbon is something you use every day. It is called graphite, and it is used in pencil "leads." Its atoms are arranged in sheets, which makes pure graphite very slick. The third allotrope has carbon atoms in a very rigid crystal structure; it is a diamond. What useful things could *GRAPHITE MAN* do? (He could give the word *graffiti* new meaning!) What useful properties would *DIAMOND WOMAN* have? ②

Carbon black

Graphite

Diamond

 ## Writing

Create your own nonmetal person. Describe his or her properties and special abilities in detail. Make a sketch of what your character would look like. ✏

What Do All These Things Have in Common? Did you have a hard time at the beginning of the section listing the properties of nonmetals? Now you can understand why. Nonmetals are as varied in properties as they are in color, appearance, and physical state. They generally do not conduct electricity or heat, and they tend to form negative ions by gaining electrons. Aside from that, these elements have exciting and varied properties that scientists continue to explore and utilize.

▼ **ASK YOURSELF**

Explain why allotropes can have different properties.

SECTION 2 REVIEW AND APPLICATION

Reading Critically

1. How is chemical activity of the nonmetals organized in the periodic table?

2. Distinguish between metals and nonmetals based on their properties.

Thinking Critically

3. Element *X* is a gas that has a mass number of 16. It has 6 electrons in its outer energy level. Which element is it?

4. Why are the noble gases particularly useful?

5. Metals tend to form positive ions while nonmetals tend to form negative ions. In what ways are these tendencies useful?

Process Skills:
Communicating, Interpreting Data, Analyzing

Grouping: Groups of 3 or 4

Objectives:
● **Observe** properties of various elements.
● **Interpret** data and draw conclusions from observations.

Discussion

Various tests are used in science to observe properties. Point out to the students that, in this case, all of the characteristics listed must be present if the element is a metal.

▶ **Application**

Aluminum, copper, and lead are metals. (Copper is a transition metal.) Each can be shaped by hammering, can conduct electricity, and has some degree of luster.

✳ **Using What You Have Learned**

The students' experiments should be designed to collect data and to draw conclusions to solve the problem.

PERFORMANCE ASSESSMENT

Cooperative Learning

Have each group of students summarize its results. Evaluate the summaries by questioning the group, as appropriate.

PORTFOLIO ASSESSMENT You may wish to have the students place their experimental designs from Using What You Have Learned in their science portfolios.

SKILL *Drawing Conclusions*

▶ **MATERIALS**
● safety goggles
● aluminum ● carbon ● copper ● lead ● sulfur ● hammer ● anvil
● light bulb and socket ● battery and battery holder ● wire ● wire clips

▼ **PROCEDURE**

1. Copy the table shown.

TABLE 1: TESTS FOR METALS AND NONMETALS			
Element	**Shaped by Hammering**	**Electric Conductor**	**Luster**
Aluminum			
Carbon			
Copper			
Lead			
Sulfur			

2. CAUTION: Put on safety goggles and leave them on throughout this entire activity. Test each of the elements listed to determine whether it is a metal or a nonmetal. Remember that all of the characteristics listed must be present if the element is a metal.

3. Test to see if the shape of the element can be changed by hammering.

4. To test for electric conduction, set up an electric circuit like the one shown. A metal will conduct electricity, and the bulb will light.

5. Observe the element to determine whether it is shiny. If it is, it has luster.

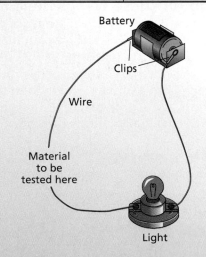

Battery
Clips
Wire
Material to be tested here
Light

▶ **APPLICATION**
Use the results of your experiment to draw conclusions about the elements you tested. Which elements are metals? Which are nonmetals? In what ways does your data support your conclusions?

✳ **Using What You Have Learned**
Think of a problem, and write a description of it. Exchange problems with a classmate. Design an experiment that will help you collect data to solve the problem, complete the experiment, and draw conclusions to solve the problem.

The Big Idea— Systems and Structures

Lead the students to an understanding that a substance's atomic framework—or electron configuration—affects the properties of a substance. Also help the students to understand that by dividing elements into metal and nonmetal categories, their reactivities are more clearly understood.

CHAPTER 17

HIGHLIGHTS

The Big Idea

Have you ever watched a large building under construction? It is hard to guess the final appearance of that building when only the steel girders are up. Yet this framework is essential and allows the building to take the shape the architect designed. It will determine whether the building is massive and solid or airy and covered in glass.

In this chapter, you have looked into a different sort of framework, a different structure. You have learned how the atomic framework, or the electron configuration, affects the properties of a substance. Science seeks to discover the links between the structure of matter and its properties. By dividing the elements into metal and nonmetal categories and looking at their atomic structure, you have gained some understanding of their reactivities.

For Your Journal

Look back at the ideas you wrote in your journal at the beginning of the chapter. How have your ideas changed? Revise your journal entry to show what you have learned. Be sure to include information on how different elements are useful in your daily life.

Connecting Ideas

Below are some photographs of elements. Can you tell by looking at them whether they are metals or nonmetals? Do the names of the elements help you figure it out? Look at the photographs, and make a table that shows what you know about each element. Then write a summary statement that tells how you can differentiate between metals and nonmetals.

For Your Journal

The students' ideas should reflect an understanding of the characteristics of metals and nonmetals and how various elements are used in daily life. Encourage the students to share their new ideas with classmates.

CONNECTING IDEAS

Calcium and mercury are metals; sulfur and bromine are nonmetals. While the students cannot determine by sight alone whether the substances are metals or nonmetals, they might note that the information is available on the periodic table. The students' tables should accurately reflect information about each element. The students' summaries should note that metals have similar properties, such as the ability to conduct heat and electricity. Nonmetals do not have many properties in common. In addition, they do not conduct heat or electricity.

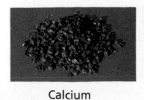

Calcium

Mercury

Sulfur

Bromine

Understanding Vocabulary

1. (a) Early experimenters in the Middle Ages were alchemists. The term *chemist* is derived from the word *alchemist*.

(b) Mixtures of two or more metals are alloys.

(c) Some metalloids, which have properties of metals and nonmetals, can be used as semiconductors.

(d) Noble gases, which used to be called inert, have a stable electron configuration and do not combine easily with other elements.

(e) All halogens exist as molecules composed of two atoms, or diatomic molecules.

(f) Oxygen can exist either as O_2 or as ozone, O_3.

Understanding Concepts

Multiple Choice

2. a

3. b

4. a

5. c

Short Answer

6. Nonmetals do not conduct heat or electricity.

7. The stability of noble gases is due to the number of electrons in their outer energy levels. Most noble gases have eight electrons in their outer energy levels.

8. Metals are good conductors of electricity and heat. They can be shaped into sheets and formed into wires. Metals have luster; some are shiny while others tarnish.

9. Alkali metals are very reactive and may explode when mixed with water.

10. Copper is a good conductor of heat. However, tin is a soft metal that turns to powder at high temperatures.

Interpreting Graphics

11. 109

12. on the left-hand side of the periodic table

13. fluorine, chlorine, bromine, iodine, and astatine

14. The reactivity of the nonmetals increases as you move toward fluorine, except for the noble gases. The most active metals are on the left-hand side of the periodic table.

CHAPTER 17

REVIEW

▶ Understanding Vocabulary

1. For each set of terms, explain the relationship in their meanings.
   a) alchemist (457), chemist (457)
   b) metals (458), alloys (459)
   c) metalloids (462), semiconductors (462)
   d) noble gases (466), inert gases (466)
   e) halogens (470), diatomic molecules (470)
   f) oxygen (473), ozone (473)

▶ Understanding Concepts

MULTIPLE CHOICE

2. Elements that have some properties of metals and some properties of nonmetals are
   a) metalloids.
   b) alloys.
   c) semiconductors.
   d) ions.

3. Because they are naturally found as two-atom molecules, nitrogen, oxygen, and fluorine are examples of
   a) more stable molecules.
   b) diatomic molecules.
   c) ionic molecules.
   d) less stable molecules.

4. Substances that are good conductors of electricity and heat are
   a) metals.
   b) nonmetals.
   c) noble gases.
   d) inert gases.

5. Which element is *not* an example of a halogen?
   a) fluorine
   b) chlorine
   c) iron
   d) iodine

SHORT ANSWER

6. What are some characteristics of nonmetals?

7. Explain why noble gases are so stable.

8. What are some properties of metals?

9. Explain why it is unwise to mix water with alkali metals.

10. Why are some cooking pots made of copper but not tin?

Interpreting Graphics

Refer to the periodic table in this chapter to answer the following questions.

11. How many different types of elements are found on the periodic table?

12. Where are the metals located on the periodic table?

13. Which elements are part of the halogen family?

14. What pattern of chemical activity is reflected in the periodic table?

Reviewing Themes

15. *Systems and Structures*
How do chemists decide whether an element is a metal or a nonmetal?

16. *Systems and Structures*
Discuss how you might use metals and nonmetals in your daily life.

Thinking Critically

17. What are some advantages of using alloys?

18. Why is helium rather than hydrogen used to fill blimps?

19. You are an engineer designing a new airplane. You know you want to use an aluminum alloy. What properties would you want in this alloy?

20. How are metals held together? How does this arrangement affect the properties of metals?

21. Explain why carbon is considered a metalloid. Give specific examples to support your explanation.

22. How are the Group 1 metals similar to the halogens? How are they different?

Discovery Through Reading

Berger, Melvin. *Our Atomic World.* New York: Franklin Watts, 1990. This book provides a clear, interesting discussion of atoms in our environment.

RADIOACTIVITY

PLANNING THE CHAPTER

Chapter Sections	Page	Chapter Features	Page	Program Resources		Source
CHAPTER OPENER	480	**For Your Journal**	481			
Section 1: RADIOACTIVE ELEMENTS	482	Discover by Problem Solving **(A)**	484	*Science Discovery*		SD
• What's in a Name? **(A)**	482	Discover by Doing **(A)**	490	Investigation 18.1: Using Isotope Notation **(H)**		TR, LI
• Jittery Isotopes **(A)**	485	Section 1 Review and Application	490	Connecting Other Disciplines: Science and Mathematics, Predicting **(A)**		TR
• The Downside of Radiation **(A)**	486			Investigation 18.2: Balancing Equations for Nuclear Reactions **(H)**		TR
• Detecting the Invisible, Hearing the Silent **(A)**	489			Extending Science Concepts: Balancing Nuclear Equations **(H)**		TR
				Radiation **(A)**		IT
				Study and Review Guide, Section 1 **(A)**		TR, SRG
Section 2: USING RADIATION	491	Discover by Writing **(A)**	492	*Science Discovery**		SD
• Atomic Tailors **(A)**	491	Section 2 Review and Application	495	Thinking Critically **(A)**		TR
• Atoms that Keep on Ticking **(A)**	494	Skill: Communicating Using a Graph **(A)**	496	Reading Skills: Solving Problems **(A)**		TR
				Study and Review Guide, Section 2 **(A)**		TR, SRG
Section 3: NUCLEAR REACTIONS	497	Activity: What is a chain reaction? **(A)**	498	*Science Discovery**		SD
• To Split or Not to Split **(A)**	497	Discover by Researching **(A)**	499	Record Sheets for Textbook Investigations **(A)**		TR
• Fusion **(A)**	499	Section 3 Review and Application	501	Carbon-14 Dating **(A)**		IT
• How Much Is Too Much? **(A)**	500	Investigation: Making a Spinthariscope **(A)**	502	Fission and Fusion **(A)**		IT
				Radiation Exposure in the United States **(A)**		IT
				Study and Review Guide, Section 3 **(A)**		TR, SRG
Chapter 18 HIGHLIGHTS	503	The Big Idea	503	Study and Review Guide, Chapter 18 Review **(A)**		TR, SRG
Chapter 18 Review	504	For Your Journal	503	Chapter 18 Test		TR
		Connecting Ideas	503	Test Generator		

B = Basic   **A** = Average   **H** = Honors
The coding Basic, Average, and Honors indicates subsections, features, and resources that might be appropriate for different levels of learners. For additional suggestions regarding choice of topic and depth of coverage, see the Pacing Chart on pages T28–T31.

*Frame numbers at point of use
(TR) Teaching Resources, Unit 5
(IT) Instructional Transparencies
(LI) Laboratory Investigations
(SD) *Science Discovery* Videodisc Correlations and Barcodes
(SRG) Study and Review Guide

CHAPTER MATERIALS

Title	Page	Materials
Discover by Problem Solving	484	(per individual) journal
Teacher Demonstration	489	wool cloth, hard rubber rod, electroscope, radioactive samples of either radioisotopes or radioactive minerals
Discover by Doing	490	(per individual) iron filings, thin cardboard, bar magnet
Discover by Writing	492	(per individual) journal
Teacher Demonstration	494	fresh celery with leaves, 10 microcuries radioactive Na_3PO_4 (available from chemical supply house), lead foil, jars (2), Polaroid sheet film and holder/developer
Skill: Communicating Using a Graph	496	(per group of 3 or 4) pennies (100), shoe box with lid, graph paper
Teacher Demonstration	497	60 corks
Activity: What is a chain reaction?	498	(per group of 3 or 4) domino set, small wooden block
Discover by Researching	499	(per group of 3 or 4) journal
Investigation: Making a Spinthariscope	502	(per group of 3 or 4) 2.5 cm × 3.5 cm matchbox, clear nail polish, thick needle, radioactive source, shellac, hand lens

ADVANCE PREPARATION

For the *Skill* on page 496, you might ask some students to bring shoe boxes.

For the *Investigation* on page 502, you may wish to ask the students to bring matchboxes.

TEACHING SUGGESTIONS

Field Trip
A field trip to a nearby nuclear power plant is an excellent addition to your instruction on radioactivity. Most plants allow small groups to tour their facility. If such a trip is not feasible, a field trip to the nuclear medicine department of a hospital would also be a worthwhile supplement to this chapter's concepts.

Outside Speaker
Invite a nuclear medicine specialist to speak on the uses of radioisotopes in medicine. You might also contact a museum or university to arrange for an archaeologist or chemist to talk about radiochemical dating of artifacts.

CHAPTER 18

RADIO-ACTIVITY

CHAPTER THEME—CHANGES OVER TIME

Radioactive elements and their characteristics are introduced in Chapter 18. The students will find out why some atoms are radioactive and others are not and what happens to an atom after a nuclear reaction. A supporting theme in this chapter is **Systems and Structures.**

MULTICULTURAL CONNECTION

The Manhattan Project was a secret scientific program sponsored by the United States. Its purpose was to develop a nuclear-fission weapon. Enrico Fermi, an Italian fleeing fascism, proposed and then worked on the project. Other team members were Eugene Wigner and Leo Szilard, both Hungarian, Lloyd Quarterman, an African American who had worked with Fermi in Chicago, J. Robert Oppenheimer, an American and a graduate of Harvard, and Maria Goeppert Mayer, an American research physicist born in Poland. Although the project was directed by the Americans, it was carried out by an international panel of scientists.

MEETING SPECIAL NEEDS

Second Language Support

Cooperative Learning Pair a limited-English-proficient student with a partner. Have them review the description of the Trinity explosion and write down any words that the limited-English-proficient student does not understand. Encourage the partner to help the limited-English-proficient student fully understand the magnitude of the first atomic detonation by defining the unfamiliar terms in his or her own words.

CHAPTER 18

RADIOACTIVITY

A PINPRICK OF BRILLIANT LIGHT PUNCTURED THE DARKNESS, SPURTED UPWARD IN A FLAMING JET, THEN SPILLED INTO A DAZZLING CLOCHE OF FIRE THAT BLEACHED THE DESERT TO A GHASTLY WHITE....

The power of the atom has haunted scientists for years. Once this power was discovered and unleashed as a weapon, science has endeavored to find a way to harness the power for the good of humanity. The puzzle of nuclear reactions and the radiation they cause continues to be studied.

For a fraction of a second the light in that bell-shaped fire mass was greater than any ever produced before on earth. Its intensity was such that it could have been seen from another planet. The temperature at its center was four times that at the center of the sun and more than 10,000 times that at the sun's surface. The pressure, caving in the ground beneath, was over 100 billion atmospheres, the most ever to occur at the earth's surface. The radioactivity emitted was equal to one million times that of the world's total radium supply.

No living thing touched by that raging furnace survived. Within a millisecond the fireball had struck the ground, flattening out at its base and acquiring a skirt of molten black dust that boiled and billowed in all directions. Within twenty-five milliseconds the fireball had expanded to a point where the Washington Monument would have been enveloped. . . . At 2000 feet, still hurtling through the atmosphere, the seething ball turned reddish yellow, then a dull blood-red. It churned and belched forth

Present the students with information on the life of Andrei Sakharov, "the father of the Soviet hydrogen bomb." Discuss with the students why the knowledge of nuclear chemistry is important to a nation. Emphasize that in the future, knowledge of nuclear chemistry will influence the way in which the issues of nuclear weapons and nuclear energy are handled.

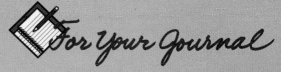

For Your Journal

The journal questions will help you identify misconceptions the students might have about radioactivity. These questions will also help the students realize they do possess some prior knowledge about radioactivity. (The students will have a variety of responses to the journal questions. Accept all answers, and point out that at the end of the chapter the students will have an opportunity to refer back to the questions.)

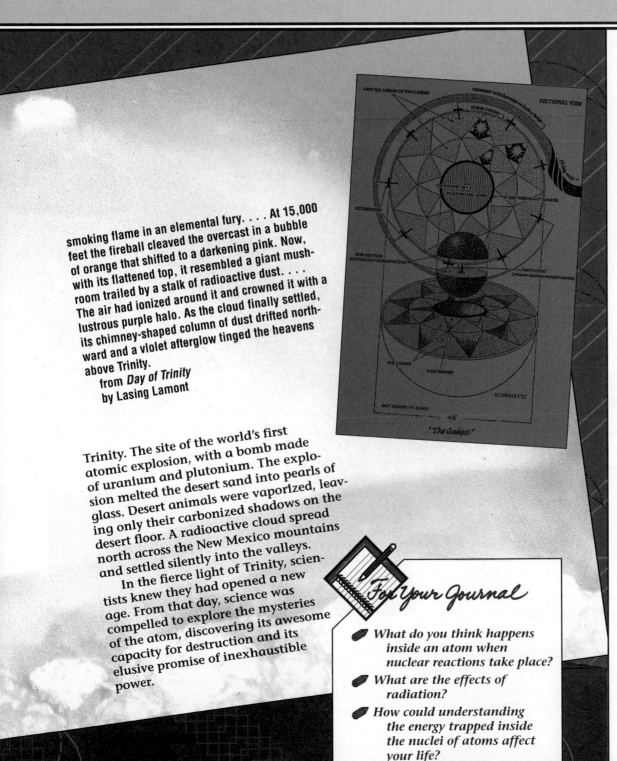

smoking flame in an elemental fury. . . . At 15,000 feet the fireball cleaved the overcast in a bubble of orange that shifted to a darkening pink. Now, with its flattened top, it resembled a giant mushroom trailed by a stalk of radioactive dust. . . . The air had ionized around it and crowned it with a lustrous purple halo. As the cloud finally settled, its chimney-shaped column of dust drifted northward and a violet afterglow tinged the heavens above Trinity.

from *Day of Trinity*
by Lasing Lamont

Trinity. The site of the world's first atomic explosion, with a bomb made of uranium and plutonium. The explosion melted the desert sand into pearls of glass. Desert animals were vaporized, leaving only their carbonized shadows on the desert floor. A radioactive cloud spread north across the New Mexico mountains and settled silently into the valleys.

In the fierce light of Trinity, scientists knew they had opened a new age. From that day, science was compelled to explore the mysteries of the atom, discovering its awesome capacity for destruction and its elusive promise of inexhaustible power.

For Your Journal

- What do you think happens inside an atom when nuclear reactions take place?
- What are the effects of radiation?
- How could understanding the energy trapped inside the nuclei of atoms affect your life?

ABOUT THE LITERATURE

Trinity, a 22-kiloton implosion-type fission device, was detonated in Los Alamos, New Mexico. It was the culmination of the Manhattan Project, which had been carried out in the United States' own backyard.

Upon witnessing the explosion, [Physicist Robert] Oppenheimer in that blinding instant thought of fragments from the sacred Hindu epic, *Bhagavad-Gita*:

If the radiance of a thousand suns
Were to burst at once into the sky,
That would be like the splendor
of the Mighty One . . .
I am become Death,
The shatterer of worlds.

This successful test convinced U.S. leaders that they could build fission weapons and use them to attack Japan. The first American nuclear weapon—dropped on Hiroshima, Japan, on August 6, 1945—was a gun-type fission bomb with a yield of about 13 kilotons. Three days later, the United States dropped a 22-kiloton implosion-type fission bomb on Nagasaki, Japan. The United States demonstrated to the world that it had cracked the riddle of the atom. As a result, Japan agreed to surrender five days later, thus ending World War II.

SECTION 1

Radioactive Elements

Objectives

Explain *what is meant by radioactivity.*

Differentiate *among the three types of radioactive decay.*

List *two ways in which radioactivity can be detected.*

Antoine-Henri Becquerel was a French physicist who was interested in the absorption of light. Becquerel's research in the 1890s centered around fluorescent compounds. He wanted to know if fluorescence and X-rays were related.

In 1896 Becquerel was studying the effect of sunlight on certain minerals. He stored a sample of uranium wrapped tightly in black paper in his desk drawer. In the same drawer there was an undeveloped photographic plate. A few days later, after he had developed several photographic plates, including the one he thought had not been exposed, he found an image on the unexposed plate. The image was the same shape as the uranium sample! Becquerel concluded that something in the uranium had come through the black paper and exposed the plate.

Does this account of such a small observation sound unimportant, even boring? If so, consider that the discovery of radioactivity has led to increased production of electricity, nuclear medicine, and the construction of the atomic bomb.

Figure 18–1. Becquerel discovered radioactivity while studying how certain minerals glow after being exposed to sunlight. The rock shown here contains willemite and calcite. When exposed to ultraviolet light, the willemite glows red and the calcite glows green.

 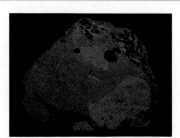

What's in a Name?

You probably already have some ideas about nuclear energy. Think about the terms you would use to describe radioactivity and nuclear energy. Chances are that you would use more negative terms than positive ones. You might remember hearing about the Chernobyl disaster or the accident at Three Mile Island. You may know something about the problem of the disposal of radioactive waste. The idea of radiation might send you running in the opposite direction!

● **Process Skill:** *Analyzing*

Point out to the students that the mineral pitchblende is a source of uranium, radium, and two rarer elements, cerium and polonium. A student of Henri Becquerel—Marie Curie—conducted research on pitchblende with her husband, Pierre. Together, they discovered two elements that gave off even more powerful rays than uranium. They named one of the new elements radium. The other element they named polonium, after Poland, Marie Curie's birthplace. For their work, Becquerel and the Curies won the Nobel Prize in physics in 1903.

● **Process Skills:** *Interpreting Data*

Share some of these facts with the students: serendipity in science is evidenced by Becquerel's discovery; X-rays were so called because scientists did not know what they were; Marie Curie died of leukemia, caused by the action of radiation. You might encourage interested students to check reference materials for additional facts related to radioactivity and share them with the class.

Studying the Unknown

Remember the marble shooting activity in Chapter 16? It wasn't easy to determine the unknown object's shape. It's hard to gather information about an object you can't see or touch. Scientists spend years working on such puzzles. What kind of person would spend a lifetime trying to find out about the unknown? About 100 years ago, there lived such a person; her name was Manya Sklodowska. The world knows her as Marie—Marie Curie.

Marie Curie was born in Warsaw, Poland, in 1867. By 1898, when she was 31, she was already a well-known scientist. Marie was a student of Becquerel and was immediately interested in his discovery of radioactivity. Marie and her husband Pierre were investigating a mineral called *pitchblende*. The Curies observed that pitchblende gave off a strange form of energy, similar to that given off by uranium.

After two years of work, the Curies announced that they had discovered two new elements in pitchblende. The Curies named one of the new elements *radium*, because it gave off rays. They named the other *polonium*, after Marie Curie's beloved homeland, Poland.

In order to learn more about radioactivity, elements were isolated from pitchblende and then purified. By 1902 Madame Curie was able to produce a nearly pure sample of radium. She had many questions: What made radium radioactive? Why are radioactive elements unstable?

Decaying Atoms   Madame Curie knew that certain elements were unstable and that they spontaneously broke apart and gave off particles or high-energy radiation. That is, they were **radioactive.** Madame Curie's hypothesis was that this instability came from the nucleus of these atoms. She believed that as particles or energy are released from the nucleus, the atom becomes more stable. This process is known as *radioactive decay.*

Why are some elements less stable than others? Let's see how well you can hypothesize about the answer.

Figure 18–2. Marie Curie (left) and her family

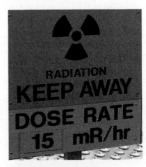

Figure 18–3. This symbol is used to label radioactive substances or areas where radiation is present.

● **Process Skill:** *Formulating Hypotheses*

Have pairs of students study the graph and discuss the interpretation. Direct the students to write their own hypotheses in their journals and compare it with their partners'.

● **Process Skill:** *Identifying/Controlling Variables*

Ask the students to explain why the ratio of protons to neutrons is important in the process of radioactive decay. (The most stable atoms are small atoms that have one neutron for every proton. A large nucleus is more likely to be unstable, because there are more repulsive forces than attractive forces. The nucleus of an unstable atom will break apart and give off radiation.)

MULTICULTURAL CONNECTION

Chien-shiung Wu was born in Liuhe, China, in 1912. After moving to the United States in 1936, Wu received her doctorate degree from the University of California at Berkeley.

In 1957 she conducted an experiment that disproved the law of the conservation of parity, a law most physicists had accepted as a universal principle for about 30 years. Using atoms of cobalt-60, Wu demonstrated that beta particles were more likely to be emitted in a particular direction that depended on the spin of the cobalt nuclei. Wu's experiment confirmed a theory proposed in 1956 by two Chinese-born American physicists, Tsung Dao Lee and Chen Ning Yang. Lee and Yang shared the 1957 Nobel Prize in physics for their theory.

ONGOING ASSESSMENT
▼ ASK YOURSELF

An atom is radioactive if the nucleus is unstable. Often in these cases, there are not enough neutrons to separate the positive charges of the protons, so particles of radiation are given off to stabilize the element.

DISCOVER BY *Problem Solving*

Look at the graph below. It shows the number of neutrons and protons for several stable (nonradioactive) isotopes. The shaded area represents a belt of stability. The solid line shows when the number of neutrons is equal to the number of protons. Notice that many of the smaller nuclei are stable if the number of protons equals the number of neutrons. Why do you think this is so? (Hint: What do you know about like charges?) Write your hypothesis in your journal. As the atomic number gets larger than 20, the belt of stability drifts away from the solid line and seems to favor more neutrons than protons. Why do you think the trend changes? Write your ideas in your journal. ✐

Why is carbon-12 stable and carbon-14 radioactive? The full explanation of what makes an isotope radioactive is too complicated to cover here, but you might have come close to the answer in your hypothesis. Protons, as you know, repel each other because they have like charges. These forces would tend to push the nucleus apart if they were all that was at work. But, besides these repulsions, there are forces of attraction between particles called *quarks* that make up protons and neutrons. These forces tend to hold the nucleus together and make it stable. If there are not enough neutrons, the repulsive forces between protons are greater than the attractive forces between the quarks that make up protons and neutrons. When forces of repulsion are greater than forces of attraction, the nucleus is unstable. As the nuclei become larger, more neutrons are needed in order to maintain a greater number of attractive forces than repulsive forces. That is why the graph shows the belt of stability dropping below the line.

Unstable radioactive atoms lie somewhere away from the belt of stability shown on the graph. In order to become stable, some of these atoms release particles from the nucleus in the form of radiation. You will find out more about radiation in the next section.

▼ ASK YOURSELF

Why are some atoms radioactive while others are not?

Jittery Isotopes

You know that radioactive substances give off radiation. You may even know that some kinds of radiation are more dangerous than others, but do you know why?

Three major types of radiation are given off by radioactive elements: alpha particles, beta particles, and gamma rays. Radiation occurs when a radioactive nucleus decays. When the nucleus decays, it shoots out or emits particles or energy waves or both. This flow of particles or waves is called *radiation*.

When an atom undergoes **alpha decay,** it emits an alpha particle, which is made up of two protons and two neutrons. An example of alpha decay is shown below.

Beta decay occurs when a beta particle shoots out of the nucleus at high speed. A beta particle is formed when a neutron is converted into a proton and an electron. The electron shoots out of the nucleus much like a cannonball shoots out of a cannon when the gunpowder inside it explodes.

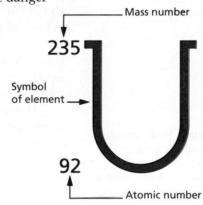

Mass number
Symbol of element
Atomic number

Figure 18–4.
Radioactive isotopes are written as shown here.

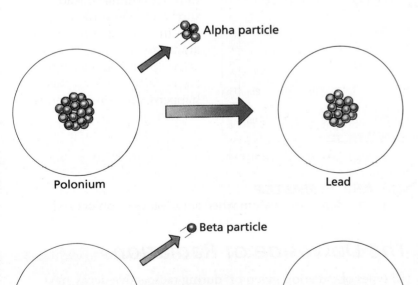

Alpha particle

Polonium

Lead

Beta particle

Phosphorus

Sulfur

Figure 18–5. Alpha decay is the result of the expulsion of a helium nucleus from the nucleus of an atom. Polonium becomes lead as a result of alpha decay.

Beta decay is the result of the expulsion of an electron from the nucleus of an atom. Radioactive phosphorus undergoes beta decay to form sulfur.

● **Process Skill:** *Analyzing*

Ask the students to think of ways that human beings know they are in danger. (Humans use their five senses to detect danger of all kinds. Sensory information coupled with learning often warns humans when danger is present.) Then challenge the students to think of ways humans can detect radiation danger, using their five physical senses. (There are no ways.) Conclude by helping the students understand that humans cannot detect radiation through their senses.

SCIENCE BACKGROUND

Gamma rays are a form of energy similar to X-rays, but with shorter wavelengths. Gamma rays can be emitted alone or with alpha and beta particles. The sun emits gamma rays naturally; however, most are filtered out in the atmosphere. If gamma rays do reach Earth, they can cause genetic mutation, since their shorter wavelengths give them immense penetrating power.

MEETING SPECIAL NEEDS

Second Language Support

Mainstreamed

 Have limited-English-proficient students and mainstreamed students work with classmates to write these terms and their definitions in their journals: *alpha decay, beta decay, gamma decay, radioactive.* Then have the students draw a picture illustrating each term next to its journal entry.

ONGOING ASSESSMENT
◤ ASK YOURSELF

The nucleus can split, releasing a radioactive particle—either alpha or beta—and energy in the form of heat or gamma radiation.

Gamma decay, which does not involve a particle, is the emission of high-energy radiation similar to X-rays. This radiation has no mass and no charge. How can radiation that has no particles make an atom more stable? This process is similar to the one in which an electron drops to a lower energy level. When an atom gives off energy, it moves to a lower energy level, and lower energy levels are more stable than higher ones. Gamma rays can penetrate tissue more deeply than alpha or beta particles, so this type of radioactive decay is considered the most dangerous.

When an atom undergoes radioactive decay, it is changed in atomic number, mass number, or both. Remember that the atomic number of an element is the number of protons in the nucleus. If the atomic number changes, the atom changes into another element. These changes are called *nuclear reactions.* Nuclear reactions occur in an element until a stable, nonradioactive atom is formed. Sometimes an element undergoes many types of decay before it becomes stable. An example is the decomposition of uranium to lead, shown in Figure 18–6.

In contrast, a polonium atom can become nonradioactive by emitting a single alpha particle. When this happens, an atom of lead is formed.

Figure 18–6. Uranium decays into many different elements before it forms lead as shown in this diagram.

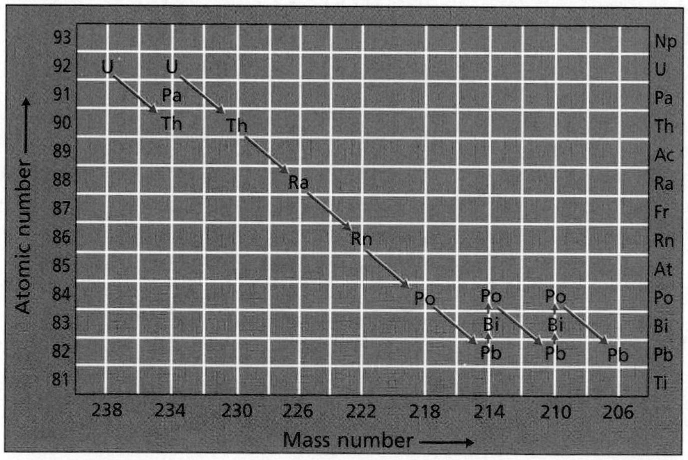

◤ ASK YOURSELF
What can happen to an atom when a nuclear reaction occurs?

The Downside of Radiation

All types of radiation given off during radioactive decay have the power to go through objects. On April 26, 1986, the world learned more about this power than it ever wanted to know. The Chernobyl disaster was the world's introduction to the dangers of nuclear power production. The following article, written one year after the accident, is a firsthand account of what happened.

● **Process Skills:** *Comparing, Generating Ideas*

Have the students study the Chernobyl photos. Then have them write in their journals as many visible effects of the accident as they can see and list as many unseen effects as they can

think of. Finally, have the students label each effect as either long- or short-term. Conclude by encouraging volunteers to share their ideas with classmates.

● **Process Skill:** *Expressing Ideas Effectively*

Cooperative Learning Divide the class into two teams. Have them debate the statement, "Fission reactors are safe enough to use." Encourage the teams to

research their side of the issue before debating.

"Chernobyl—One Year After"

from *National Geographic*

The air smelled of scorched metal, and to breathe without a mask was to cough. Helicopters swung low on quick bombing runs, dropping sacks of lead, boron carbide, sand, clay, dolomite. Their target was a tangle of machinery and pipe, visible through a gaping hole in a 70-meter-high building.

On the ground moved a veritable army, hastily and desperately assembled. In white garments, with masks and caps, many looked like physicians dressed for the operating room. Army personnel carriers, their armor augmented by slabs of lead, rumbled to and fro on deadly serious taxi duty.

Now and again buses passed, removing the people of whole villages. In the

nearby city of Pripyat, where 45,000 people had lived, laundry still hung on clotheslines, and a carnival carousel spun empty in the wind. An elderly woman departed her home carrying only her identity card, her spectacles, and her house key. Though she might never see her house again, she locked the door.

At times some of the people in this frantic scene paused to think of fellow workers—firemen, a doctor, two paramedics, a woman guard—who were rushed to hospitals in the early hours of April 26, 1986.

I describe the nightmare at the Chernobyl Nuclear Power Plant as it was a year ago, on, say, about the second of May. With four working reactors and two more being

built, Chernobyl was to be one of the most powerful nuclear power stations in the Soviet Union. At 1:24 on the fateful April morning, one or possibly two explosions blew apart reactor No. 4—the worst reported accident in the history of the harnessed atom.

The blast(s) knocked aside a thousand-ton lid atop the reactor core and ripped open the building's side and roof. Reactor innards were flung into the night. These included several tons of the uranium dioxide fuel and fission products such as cesium 137 and iodine 131, as well as tons of burning graphite. Explosion and heat sent up a five-kilometer plume laden with contaminants.

SCIENCE BACKGROUND

Two large steam explosions destroyed one of four nuclear reactors at Chernobyl in Ukraine. The explosions and resulting fire raged for 10 days and released 7000 kg of radioactive material into the environment. Thirty-one deaths, 1000 immedlate injuries, and $3 billion in losses were attributed to the accident.

Within days, much of Europe was experiencing the highest levels of radioactive fallout ever recorded there. Within two weeks, minor radioactivity was recorded in Tokyo, in Washington, D.C., and throughout the Northern Hemisphere.

In the aftermath of such a tragedy, and in the hope of preventing it from recurring, the Soviet government designated the area around Chernobyl as an ecological reserve. The reserve is open to scientists of all nations to carry out experiments on the impact of radiation on the natural environment.

✧ **Did You Know?**
Dolomite was dropped by helicopter on Chernobyl. It served to further block rays being emitted by the damaged reactor and seal the reactor.

● **Process Skill:** *Evaluating*

Ask the students to describe in their own words the penetrating power of alpha, beta, and gamma rays. Have a volunteer explain why alpha particles are the most easily stopped. (They have the lowest penetrating power.) Then ask the students to determine what kind of radi- ation is emitted from a nuclear power plant that is built with thick lead walls. (Gamma rays are being emitted.)

MEETING SPECIAL NEEDS

Gifted

Assign interested students *The Truth about Chernobyl* by Grigori Medvedev, a leading Russian nuclear physicist. It is a minute-by-minute account of the disaster and coverup. Have the students log their reactions in their journals after they read each chapter. Be sure the students date their entries. Then spend a class period allowing the students to share their reactions.

ONGOING ASSESSMENT

▼ ASK YOURSELF

Gamma decay does not release particles. Rather, it involves the emission of high-energy radiation (similar to X-rays) that has no mass and no charge. Gamma decay is the most dangerous and most penetrating type of radiation.

✧ Did You Know?

Scientists don't really know how long the land around Chernobyl will remain contaminated. They continue sampling to determine the amount of radiation present. Current estimates range from hundreds to tens of thousands of years.

Why were authorities dumping lead on Chernobyl? To understand, you need to know that all radiation has the ability to go through solid matter. This ability is called *penetrating power*. Alpha particles have the lowest penetrating power; they can be stopped by several sheets of paper. Beta particles have a greater penetrating power, but they can be stopped by a piece of aluminum 0.75 mm thick. That is about as thick as five pieces of aluminum foil. The penetrating power of gamma radiation is the greatest. Thick layers of lead and cement are needed to contain this radiation.

Figure 18–7. The penetrating powers of the three types of radioactive decay are shown here.

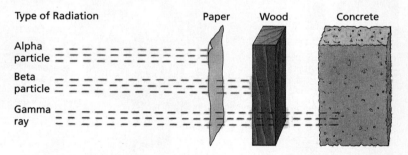

Type of Radiation		Paper	Wood	Concrete
Alpha particle				
Beta particle				
Gamma ray				

The lead that was dumped on the Chernobyl reactor would help contain the gamma rays—the most dangerous type of radiation. Boron carbide acts in the same way as a control rod does inside a reactor. It absorbs neutrons. The sand and clay dropped by helicopters would act as a shield for the alpha and beta particles. They would also put out some of the fires. Later on, the reactor was encased in concrete and steel to hold the radiation that would be there for thousands of years.

Clouds containing radioactive substances, which were the result of alpha and beta decay, traveled away from Chernobyl. These substances have been detected in the Scandinavian countries, in Greece to the southeast, and even as far away as the west coast of the United States. Milk and fresh produce were contaminated throughout much of Europe. None of these products could be eaten because alpha particles, even though they do not have much penetrating power, will cause radiation damage to internal tissues of living things if they are eaten.

As you can see, radioactivity is filled with potential hazards. Later in this chapter we will also look at how it can be helpful.

 ASK YOURSELF

How is gamma decay different from other types of radioactive decay?

GUIDED PRACTICE

Cooperative Learning In groups of four, have the students take turns explaining and describing each of the following names and terms: *Trinity, fluorescence, pitchblende, radioactivity, alpha decay, beta decay, gamma decay, Chernobyl, electroscope, cloud chamber, radioactive fallout.*

INDEPENDENT PRACTICE

Have the students provide written answers to the Section Review. Then direct the students to write in their journals one or two paragraphs defining radiation. They should include the three types of radiation and various methods of detection.

EVALUATION

Cooperative Learning Group the students into three teams. Provide each team with a molecular model, and ask them to demonstrate for the class one of the three types of radioactive decay. Have each group choose a spokesperson to narrate the demonstration, one or two stu- dents to handle the model, and one or two students to write the summary.

Detecting the Invisible, Hearing the Silent

Many of the initial researchers in radioactivity were not aware of its dangers. Both Marie Curie and her daughter Irene, who continued her mother's research, died of leukemia—a form of cancer believed to be produced by exposure to radiation.

Today, researchers may use any of several devices to detect even small amounts of radiation. The simplest device for detecting radioactivity is the *electroscope*.

Tracks in the Clouds Usually, however, more sophisticated devices are used to detect radioactivity. One instrument that is commonly used is the *cloud chamber*. A cloud chamber is a compartment filled with a gas kept at a temperature just below its condensation point. Water and alcohol are two substances often used in cloud chambers. When a radioactive particle passes through the chamber, it forms ions. The ions cause the gas to condense into drops of liquid. These drops attach to the ions the same way that dew condenses on the grass. The ions move very fast, causing the droplets to make lines in the cloud chamber. These lines are called *tracks* because they show where the particles have been, just as animal tracks in the snow show the movements of animals. Try the next activity to find out more about how a cloud chamber works.

Figure 18–8. When an electroscope is charged with static electricity, the leaves of foil inside the container repel each other. When placed near a natural radioactive source, the leaves of the electroscope collapse.

Figure 18–9. A cloud chamber shows the tracks of atomic and subatomic particles as the result of ionizing the gas inside the chamber.

Demonstration

Obtain wool cloth, a hard rubber rod, an electroscope, and radioactive samples of either radioisotopes or radioactive minerals.

CAUTION: Carry out all work with radioactive materials in a laboratory. Wear gloves whenever working with radioactive materials. Make a careful survey with a radioactivity-detecting device if a spill occurs. The area, clothing, hands, and equipment should be checked for radiation. Store all radioactive wastes and materials in appropriately labeled containers. Have any cut or wound sustained while working with radioactive materials checked by a doctor immediately. Before leaving the lab, check the work area thoroughly for any radioactive materials left unattended.

1. Rub the wool over the rod for about 30 seconds. Charge the electroscope by bringing the activated rubber rod close to the top of it. Ask the students why the foil leaves separate. (They become electrically charged with the same type of charge, so they repel each other.)

2. Bring a radioactive sample close to the knob on the electroscope. Have the students observe what happens and hypothesize about why it happens.

Have the students draw three columns in their journals and label the columns alpha decay, beta decay, and gamma decay. Then have them list as much as they know about each item. Encourage the students to compare their lists with classmates to see what information, if any, they are missing.

Have the students research the Three Mile Island nuclear accident of March 28, 1979. In this plant near Harrisburg, Pennsylvania, a highly improbable series of mechanical failures and human errors led to a partial meltdown of the reactor's core and heavily contaminated the plant. In their reports, the students should include facts regarding the anticipated long-term health hazards resulting from this accident.

Have the students write a summary of the section and read their summaries aloud. Encourage the students to revise their summaries after listening to classmates' summaries.

✧ Did You Know?

The Geiger counter, also called the Geiger-Müller counter, was invented by Hans Geiger and Walther Müller in 1912.

ONGOING ASSESSMENT

▼ ASK YOURSELF

Large doses of radiation can lead to health problems or even death.

SECTION 1 REVIEW AND APPLICATION

Reading Critically

1. Although alpha particles have low penetrating power, they can be ingested and go directly to inner tissues where they can do extensive damage.

2. by using radiation detection devices such as a cloud chamber

Thinking Critically

3. atomic number 90, atomic mass 234, $^{234}_{90}$Th

4. Gamma radiation has the ability to penetrate tissue, where it can cause genetic damage and cancer.

DISCOVER BY Doing

Sprinkle some iron filings on a thin piece of cardboard about the size of a sheet of paper. Bring a bar magnet close to the bottom of the cardboard piece but do not let it touch the cardboard. Move the magnet underneath the cardboard. What happens? How is this similar to the way ionizing radiation affects a cloud chamber?

Radiation Counts Detecting radiation is important, but measuring the amount of radiation present is also important. This measurement may be critical if a person has been exposed to radiation. One instrument that can measure the level of radiation is the **Geiger counter.** A Geiger counter contains a compartment filled with gas that forms ions when radiation passes through the compartment. The ions trigger a flow of electric current through a circuit, which is connected to a meter. The meter displays the amount of current produced, and the current is proportional to the level of radiation.

Some people work in places where there could be hazardous levels of radiation. They need to have a radiation detection device with them at all times. Devices, smaller than the Geiger counter, such as the dosimeter, have been developed for these people.

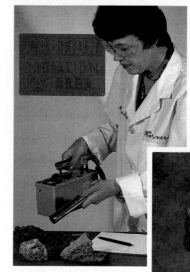

Figure 18–10. A Geiger counter (left) is a portable device used to detect radiation. A dosimeter (right) is a personal radiation device that can be worn or put into a pocket.

▼ ASK YOURSELF

Why is it important to know whether you have been exposed to radiation?

SECTION 1   REVIEW AND APPLICATION

Reading Critically

1. Under what circumstance can alpha radiation be harmful?
2. How can scientists study radioactive particles?

Thinking Critically

3. Element X is produced when uranium emits an alpha particle. What is the atomic number of the new element? What is its atomic mass? What is its symbol?
4. Explain why gamma radiation is the most dangerous type of radiation for humans.

FOCUS

This section discusses the use of radioisotopes in medicine, food storage, and industry. Half-life is defined, and a table showing the half-lives of some common radioisotopes is included. Radiochemical dating with the isotope carbon-14 is introduced as a technique used to date ancient artifacts.

MOTIVATING ACTIVITY

Discuss with the students anything they might know about the use of radiation in medicine. For example, ask them whether they have had or know of anyone who has had any treat-ments involving radiation. Examples include upper or lower GI and X-rays. Then discuss with the students the link between cancer and ionizing radiation.

PROCESS SKILLS
• Communicating •
Interpreting Data

POSITIVE ATTITUDES
• Curiosity • Openness to new ideas

TERMS
• none

PRINT MEDIA
Dating Dinosaurs and Other Old Things by Karen Liptak (see p. 423b)

ELECTRONIC MEDIA
Using Radioactivity Britannica Video (see p. 423b)

SCIENCE DISCOVERY
Food preservation; history

BLACKLINE MASTERS
Thinking Critically
Reading Skills
Study and Review Guide

Using Radiation

SECTION 2

Have you ever put tincture of iodine on a cut? Did you know that this chemical is poisonous? You know it prevents infection. Why? In small amounts, the chemical will kill the microorganisms that might infect your cut, but it won't harm you.

The effect of iodine is similar to the effects of radiation. Despite the dangers of radioactivity, there are many useful applications. Many common elements have radioactive isotopes that are used in medicine and scientific research.

Objectives

State *three uses for radioisotopes.*

Explain *what half-life is.*

Compare *and* **contrast** *methods of radiochemical dating.*

Atomic Tailors

What would happen if you shot a marble at a hollow glass ball? You guessed it, the ball would shatter. What would happen if the marbles were made of clay? With the right amount of energy, the clay might stick to the glass ball, forming a new shape. This is the principle used to create new elements in a particle accelerator.

Particle accelerators move particles at high speeds and smash them into atoms of other elements. If two nuclei collide and remain together, a new type of atom is formed. All the elements heavier than uranium have been created by smashing atoms together under conditions of intense heat and pressure. These elements are all radioactive.

Figure 18–11. A particle accelerator may be either linear or circular. Shown here is the main ring of the Fermi Accelerator in Batavia, Illinois.

MEETING SPECIAL NEEDS

Mainstreamed

In this section, you might pair a mainstreamed student with a partner. Encourage the partner to point out to the mainstreamed student any medical word or reference in this section. Then have the mainstreamed student write the word in his or her journal and look up the definition. Encourage the partner to give assistance as needed.

TEACHING STRATEGIES

● **Process Skill:** *Analyzing*

List the following radioisotopes on one side of the board. Then list their uses in random order on the other side. Have the students try to match the two lists. Cobalt-58 (traces intake of vitamin B-12); iodine-131 (used in diagnosis and treatment of thyroid malfunction); cobalt-60 (used in treatment of cancer); cesium-137 (used in treatment of shallow tumors).

DISCOVER BY *Writing*

PORTFOLIO ASSESSMENT

Have the students place their short stories from *Discover by Writing* in their science portfolios. Assess them on the basis of creativity and thoroughness of observation.

① Sodium-24 was first produced in 1934 by Irène Joliot-Curie and her husband, Frédéric. They were bombarding table salt with neutrons and produced sodium-24, as well as radioactive chlorine. Synthetic sodium-24 is now manufactured for use in medical treatments.

② Chemical tracers follow certain metabolic processes in the body. The observer can trace the flow of the process.

✦ Did You Know?

Availability and storage of medical radioisotopes pose special problems that are not common to most chemicals used in medicine. Most medical radioisotopes have short half-lives and cannot be stockpiled for any reasonable length of time. The isotopes must be generated on site as needed in cyclotrons or small-scale nuclear reactors, or shipped via rapid-delivery system.

LASER DISC

2824, 2825, 2826

Food preservation; history

Figure 18–12. The photograph on the left shows a physician administering a radioactive tracer. The photograph on the right is the resulting scan of the patient's thyroid gland.

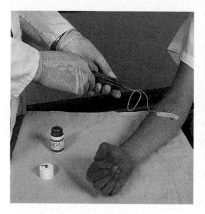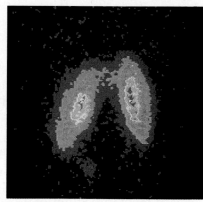

Figure 18–13. The mushrooms shown here were grown under the same conditions and were harvested at the same time. The mushrooms on the left were irradiated.

DISCOVER BY *Writing*

Suppose you are an alchemist of the Middle Ages. You are brought forward in time and are taken to visit a particle accelerator. Has science achieved what you have tried to do? Write a short story that examines how you, as an alchemist, view the creation of new elements in the particle accelerator. ✎

In Chapter 16, Ensign Roark—one of the ship's crew members—was sick. Do you remember the test that was done to check his leg? (See page 441 to refresh your memory.) The radioactive isotope that was used was sodium-24. This isotope is uncommon in nature. So where does it come from? ①

Scientists can make radioactive isotopes. They use particle accelerators to change a stable nucleus into an unstable, radioactive one. These manufactured radioisotopes have many practical uses, especially in medicine. For example, radiation therapy for the treatment of cancer uses X-rays and gamma rays produced by cobalt-60 or cesium-137. Both are manufactured radioisotopes. Natural and manufactured radioisotopes are also used as *tracers*. Tracers are radioactive chemicals that follow certain reactions inside living organisms. The major advantages of using radioisotopes as tracers are that they can be given in very small amounts and they can be directed to specific places in the body. The sodium-24 injected into Ensign Roark's leg is an example of a tracer. Why do you think these chemicals are called tracers? ②

Certain radioisotopes can be used to treat food. As a result of this treatment, called *irradiation*, the food can be stored for a long time without refrigeration. How safe is irradiated food? Read the following article to get some ideas.

Irradiation is used to treat food, but it has many other industrial uses. Encourage the students to research other uses for irradiation. (Possible ideas include sterilization of medical equipment, sealing plastic containers.)

Show the students a geologic time scale and ask a volunteer to point out where 4.5 billion years ago is on the scale. Then have them make a list of the major types of organisms that existed then. Point out that this is the half-life of uranium-238, the substance released at Chernobyl. Finally, have the students speculate in their journals what Earth will be like when all the uranium-238 has decayed.

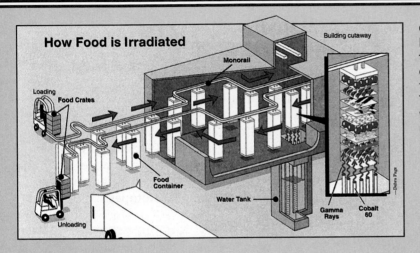

How Food is Irradiated

Building cutaway
Monorail
Loading
Food Crates
Food Container
Water Tank
Unloading
Gamma Rays
Cobalt 60

Crates of food are loaded into containers that pass over rods of cobalt-60 that bathe the food with gamma rays. The rods are stored in water when they are not in use.

SCIENCE TECHNOLOGY SOCIETY

Every being receives background radiation at low levels from natural and human sources. Natural sources include:

- cosmic rays (high-energy particles that bombard Earth from outer space);

- radioisotopes in rocks, soil, and ground water (^{235}U, ^{238}U);

- radioisotopes in the atmosphere (radon, ^{222}Rn, and its decay products, including polonium, ^{210}Po).

Human sources of background radiation include:

- fallout from nuclear weapons testing;

- increased exposure to cosmic radiation during air travel;

- medical X-rays;

- radioisotopes released into the environment during generation of nuclear power.

How Safe is Irradiated Food?

from *Current Science*

If you were told that the food sold in your local grocery store had been zapped with gamma radiation, would you be worried? Many Americans are, now that irradiated food is available for the first time in the United States.

But are the fears of these people realistic? Is irradiated food really dangerous to human health? Is it radioactive?

To answer these questions, it's helpful to know that the idea of food irradiation was hatched many years ago in an effort to improve food quality. Exposing fruits, vegetables, and meats to gamma radiation kills organisms (molds, insects, bacteria, parasites) that spoil food or cause food poisoning. About 6.5 million Americans contract food poisoning every year, and some elderly people and young children die from the poisoning.

Is Food Radioactive?

Contrary to popular belief, irradiated food is not *radioactive*—it doesn't give off radiation. When food is irradiated, invisible gamma rays pass through the food (see drawing). The gamma radiation does not make food radioactive, just as microwaves passing through food in a microwave oven do not make food radioactive.

However, exposure to gamma rays does rob food of some vitamins. But so do other forms of food preservation, including freezing and canning. The U.S. Food and Drug Administration (FDA) has concluded that irradiation does not destroy enough vitamins to jeopardize public health.

The biggest fear about irradiation is that it might make harmful chemical changes in food. Numerous studies have found that irradiated food contains tiny quantities of *radiolytic products (RPs)* These are created when ionizing energy splits food molecules (fats, carbohydrates, proteins).

Some scientists say that some RPs are *mutagenic*—they have the ability to change DNA. Also, excess exposure to mutagens may cause cancer. However, many years of testing have failed to find cancer in any laboratory animals fed large amounts of irradiated food.

Risky Business

Still, some people worry about RPs—they want to be absolutely certain that eating irradiated food is risk free. But not many foods are truly risk free. Many foods contain chemicals that are identical or highly similar to the RPs in irradiated foods, and eating too much fatty food may lead to heart disease and cancer.

Reviewing all the scientific evidence, the FDA has decided that the risk of eating irradiated food is acceptable. From here on, all irradiated foods will carry a special label so that your choice is clear—and now it is up to you.

ONGOING ASSESSMENT
▼ ASK YOURSELF

Radioisotopes are useful in medicine because they can be used to trace specific metabolic processes. Since they have very short half-lives, they do not linger in the body.

▼ ASK YOURSELF
How are radioisotopes useful in medicine?

GUIDED PRACTICE

Cooperative Learning Have the students work in pairs to solve this problem: You are given $1,000 and told that you can spend one-half of it during the first year, one-half of the balance during the second year, and so on. (One year is equivalent to the half-life of the $1,000.) If you spend the maximum allowed each year, at the end of what year would you have $31.25 left? (5 years) How much would be left after 10 half-lives? (98 cents)

INDEPENDENT PRACTICE

Have the students provide written answers to the Section Review. In their journals, have the students make a list of formerly living items that might be dated using the carbon-14 method.

EVALUATION

Have the students answer review question 3 in their journals. Encourage volunteers to read their answers aloud.

Demonstration

Obtain fresh celery with leaves, 10 microcuries of radioactive Na_3PO_4 (available from chemical supply house), lead foil, 2 jars, Polaroid sheet film and holder/developer.

See CAUTION on page 489.

1. Place two or three celery stalks in a jar containing a solution of 10 microcuries of radioactive sodium phosphate in 200 mL of water. Cover the container with lead foil.

2. In another jar, place some celery stalks in plain water as a control. Allow both jars to stand for 24 hours.

3. Prepare autoradiographs of the leaves and sections of the stem for both the control and the radioactive celery. Autoradiographs can be made by pressing cut sections of stalks or whole leaves on the Polaroid film. You will have to experiment with the amount of time required for exposure. It will depend on how much radiation has been absorbed. One day or more is probably necessary.

4. Develop the film as indicated in the film's instructions. The control should not show any image on the film. The outlines of the section of celery and the leaves should be visible in the developed print made with the radioactive celery.

Atoms that Keep on Ticking

When doctors use radioactive isotopes, they are limited to those that give off small amounts of radiation and decompose into nonradioactive elements in a short time. But how do scientists know how fast the radioisotopes will decompose?

Scientists have found that each radioactive element breaks down at a specific rate. The time required for one half of a sample to decay is known as the *half-life* of that substance. The half-life is constant regardless of what you do to the element. The table shows the half-lives of several radioisotopes.

Table 18-1 **Half-Lives of Some Common Radioisotopes**

Isotope	Type of Decay	Half-Life
$^{238}_{92}U$ (uranium)	alpha decay	4.5 billion years
$^{14}_{6}C$ (carbon)	beta decay	5730 years
$^{3}_{1}H$ (hydrogen)	beta decay	12.26 years
$^{32}_{15}P$ (phosphorus)	beta decay	14.3 days
$^{131}_{53}I$ (iodine)	beta decay	8.1 days

Some of the radioisotopes released during the Chernobyl accident were uranium-238, which has a half-life of 4.5 billion years, and cesium-137, with a half-life of 30 years. The land around Chernobyl cannot be used for years due to the long lifetime of the radioactive wastes that spilled out of the reactor.

Figure 18–14. All living organisms contain carbon-14. Knowing the half-life of this element allows scientists to determine the age of organic remains

Carbon-14, in the form of carbon dioxide, is used by plants that are eaten by animals.

One-half of the carbon-14 remains in the skeleton after 5 730 years.

One-fourth of the carbon-14 remains in the skeleton after 11 460 years

One-eighth of the carbon-14 remains in the skeleton after 17 190 years

RETEACHING

Cooperative Learning Have the students work in pairs to brainstorm a list of ancient materials that have probably been dated using carbon-14. (Answers could include such items as King Tut's tomb, articles from ancient civilizations such as the Aztecs, and Noah's ark.)

EXTENSION

Interested students might investigate the half-lives of these elements: uranium-234 (250 000 years), thorium-230 (80 000 years), polonium-218 (3 minutes), lead-214 (27 minutes), carbon-14 (5730 years). Have the students make a chart of these elements and arrange them in the order of their half-lives. The students might also write a paragraph describing which elements are the most dangerous and which ones might be useful in medical applications.

CLOSURE

Cooperative Learning Have the students work in pairs to write a summary of the section. Ask each group to read its summary aloud. Encourage classmates to make notes as they hear additional information and use them to revise their summaries.

Scientists have put the idea of half-life to work in dating artifacts. Various isotopes can be used in this process, but the most common is carbon-14. Carbon-14 is used because it is found in all living things. Since carbon-14 exists in the atmosphere, plants absorb it in the form of carbon dioxide. Animals get carbon-14 by eating plants or by eating animals that have eaten plants. So, the amount of carbon-14 in a living organism stays constant as long as the organism is alive. When the plant or animal dies, carbon-14 keeps on disintegrating, but it cannot be replaced. By knowing how much carbon-14 was in an organism while it was alive and how much remains in the object being tested, scientists can estimate the age of the material. This method is called *radiochemical dating.*

Other radioactive elements can be used for dating nonliving materials. (Remember that *nonliving* means that the material has *never* been alive.) Rubidium-87 is often used because it undergoes beta decay to form strontium-87. Since strontium-87 is a stable isotope, the amount of strontium-87 in a sample can tell scientists how old the sample is. Rubidium dating was used to determine the age of lunar rocks brought back as part of the Apollo 15 mission.

Figure 18–15. The rocks gathered from the moon were dated using rubidium-87. Why didn't scientists use carbon-14 to date the rocks?

▼ ASK YOURSELF
Why is carbon-14 used to date only once living material?

SECTION 2   REVIEW AND APPLICATION

Reading Critically
1. Why are radioactive tracers useful in medicine?
2. What safety measures must scientists take to make irradiated food safe to eat?

Thinking Critically
3. Carbon is a part of all living tissue, which should make it ideal for use as a medical tracer. Why would doctors not use a radioisotope of carbon for medical purposes?
4. Oxygen-15 has a half-life of 2.0 minutes. Using a line graph, illustrate the decay process of 20 g of oxygen-15. Label the vertical axis "Mass" and the horizontal axis "Time." Choose an appropriate scale for each axis. Calculate the amount of oxygen-15 remaining every 2.0 minutes.

ONGOING ASSESSMENT
▼ ASK YOURSELF

Only things that were once living contain carbon-14.

SECTION 2 REVIEW AND APPLICATION

Reading Critically
1. Tracers can be given in very small amounts and directed to specific places in the body. Also, tracers have very short half-lives, so they pose no threat from long-term exposure to radiation.

2. Irradiated food must be tested for radioactivity before being shipped. Scientists should continue to look into the production of radiolytic products to ensure they pose no threat to consumers.

Thinking Critically
3. Carbon-14 would not be used because radioactive tracers must have a short half-life. In addition, tracer carbon-14 would be confused with carbon-14 that is naturally present.

4. The students' diagrams should resemble the diagram shown here:

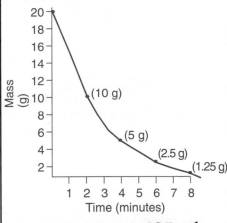

Process Skills:
Communicating, Interpreting Data

Grouping: Groups of 3 or 4

Objectives

● **Construct** a graph to show the relationship between variables.
● **Interpret** data from graphs.

Discussion

Point out to the students that straight-line graphs are used in science to analyze, interpret, and draw conclusions from recorded data. In this lab, as the number of shakes (or half-lives) increases to five, the number of "heads up" pennies decreases. The two are inversely proportional.

Procedure

1. 1600 years; 0.25 g

PERFORMANCE ASSESSMENT

Cooperative Learning Ask each group of students to determine why the line on the graph is not identical to the one on the text page. (The text graph shows an absolutely consistent decrease in the mass of radium remaining. The group's graph, however, relies on a 50/50 chance that a penny will land "tails up." Therefore, their graphs should resemble the text graph, but cannot be assured of being identical to the text graph.) Check each group's work by examining their graphs for accuracy in recording the data.

Application

The graphs for half-life of the "heads up" pennies and radium are similar, because they are both gently curving lines.

✳ Using What You Have Learned

You could organize the data by counting pennies that are "tails up" (no longer radioactive), rather than those that remain in the box. It might be easier to count pennies as you take them out of the box. The graph, however, will not indicate a decrease in the number of radioactive "atoms." It will show the increase in stable atoms.

REINFORCING THEMES— *Changes Over Time*

Discuss how the purpose of the *Skill* is related to this chapter's themes—stability and changes over time. Point out that through the process of radioactive decay, the radioactive isotopes become more stable. The decay process is a steady operation. How long it takes an isotope to totally stabilize depends upon the type of isotope being studied.

SKILL *Communicating Using a Graph*

▶ MATERIALS

● 100 pennies ● shoe box with lid ● graph paper

▼ PROCEDURE

1. Line graphs are a good way to organize some kinds of data, such as the half-life of a radioactive element. Look at the graph of the half-life of radium. How long is radium's half-life? If you started with one gram of radium, about how much would there be after 3200 years?

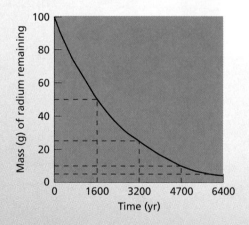

2. Suppose you wanted to make a graph of the half-life of a radioactive element you had just discovered. First you would need to gather data about the element. Then you could make a graph similar to this one. You can use simple materials to make a model that will provide the data you need for your graph.

3. Place the pennies in the box, all "heads up." Each penny represents an atom of a radioactive element.

4. Put the lid on the box, and shake it for 5 seconds. Take the lid off the box, and remove all the pennies that are "tails up." These pennies represent atoms that have broken down into some other element. Record the number of pennies left in the box.

5. Repeat step 4 four more times. Make a line graph of the results. Remember that the atoms of an element, unlike the pennies, take time to change. In a real half-life investigation, you would need to check on the element after certain time periods.

▶ APPLICATION

Compare the graph you made with the graph for radium. How are the two graphs similar?

✳ *Using What You Have Learned*

How else could you organize data about the half-life of a radioactive element? What would be the advantages and the disadvantages of organizing the data this way?

FOCUS

This section concentrates on the two nuclear reactions, fission and fusion. The process of fission involves chain reactions, which must be controlled to use nuclear power. Fusion is a possible energy source for the future.

MOTIVATING ACTIVITY

Show the students either photographs or a short film clip of an atomic blast and its aftermath. Ask them how they think something as small as an atom could release such an enormous amount of energy. Have them write their hypotheses in their journals.

Nuclear Reactions

SECTION 3

Remember the comparison between the marble-shooting activity and a particle accelerator? What would happen if you did it again, only this time you had a cluster of real marbles at the center of a ring and shot at it with a well-aimed marble? The marbles would fly apart.

Bombarding the nucleus of an atom with a neutron can cause the same type of reaction. Instead of a group of marbles separating, however, the nucleus would split into two nuclei. This process is called *nuclear fission*.

Objectives

Differentiate *between fission and fusion.*

Evaluate *the dangers and benefits of radiation.*

PROCESS SKILLS
- Experimenting • Observing
- Inferring • Measuring
- Analyzing

POSITIVE ATTITUDES
- Caring for the environment
- Curiosity • Openness to new ideas

TERMS
- chain reaction • plasma

PRINT MEDIA
"Tritium Infusion" in *Discover* magazine by Carl Zimmer (see p. 423b)

ELECTRONIC MEDIA
I Work in Atomic Energy Britannica Video (see p. 423b)

SCIENCE DISCOVERY
- Nuclear fission reactor • Cooling tower

BLACKLINE MASTERS
Study and Review Guide

To Split or Not to Split

When an atom of uranium-235 is struck by a neutron, energy and three neutrons are given off and krypton-92 and barium-141 are formed, as shown in Figure 18–16. The three neutrons released by this reaction can go on to hit three more atoms of uranium, each of which would release three more neutrons, for a total of nine neutrons! Imagine this occurring over and over again. The reaction would soon be out of control—limited only by the number of available uranium atoms. The uncontrolled nuclear fission of uranium-235 is the basis of the atom bomb.

When a single action causes the same reaction to occur over and over, the reaction is known as a **chain reaction**. A chain reaction is similar to a landslide. In a landslide, a rock falls and rolls down a mountain, striking other rocks in its path. These rocks move, hitting more rocks, and so on. Try the next activity to find out more about how a chain reaction works.

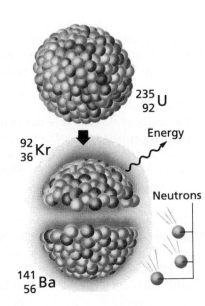

Figure 18–16. Uranium-235 undergoes fission to form krypton-92 and barium-141. Three neutrons and energy are given off.

$^{235}_{92}U$

$^{92}_{36}Kr$

Energy

$^{141}_{56}Ba$

Neutrons

Demonstration

Use this activity to demonstrate chain reaction. Give each student two corks (or other light objects), which represent neutrons, and tell the students to shut their eyes. Begin the chain reaction by tossing a cork in the direction of the students' desks. If hit, that student must toss a cork in a random direction. Repeat the process with the students huddled together on the floor. Under these conditions, the reaction is usually self-sustaining, and most corks get thrown.

TEACHING STRATEGIES

● **Process Skill:**
Communicating

Obtain a diagram or overhead projection of a nuclear power plant. Discuss the main structures and functions of the plant. Ask the students why it is necessary to have control rods. (They absorb excess neutrons and prevent the chain reaction from getting out of control.) Then have the students hypothesize what might happen should the coolant line to the reactor core break. (Overheating or meltdown would occur.) Finally, have the students develop criteria for site selection of a new nuclear power plant. (Criteria might include: not in an earthquake zone, not too near or too far from a big city, a large amount of cooling water available, solid soil foundation, and quick evacuation possible for those living within 15 km.)

ACTIVITY

What is a chain reaction?

Process Skills:
Experimenting, Observing, Inferring

Grouping: Groups of 3 or 4

▶ **Application**

1. Energy is required to start the reaction, and an average distance between atoms or dominoes must be maintained.

2. The first neutron is equivalent to your finger, and each domino represents an atom of uranium.

Cooperative Learning After the completion of this *Activity*, have each group of students describe one way it arranged the dominoes and tell what happened when it set off the "chain reaction." Then have the students repeat the set-up and see if they obtain the same results.

ONGOING ASSESSMENT
▼ ASK YOURSELF

An uncontrolled fission reaction could result in a nuclear accident or meltdown.

▶ **498** CHAPTER 18

ACTIVITY

What is a chain reaction?

MATERIALS
domino set; small wooden block

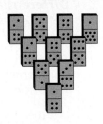

PROCEDURE
1. Set up the dominoes as shown. Push the first domino with your finger so that all the dominoes fall. If the chain reaction does not work, investigate the cause.
2. Once the triangular arrangement works, try the following variations: (a) place a small wooden block in the middle of the triangle of dominoes; (b) place some of the dominoes on their sides and some on end; (c) experiment with another design. Record your observations for all variations.
3. The fission reaction of uranium-235 releases three neutrons, which then hit three other atoms of uranium, and so on. Using your dominoes, create a model of this reaction.

APPLICATION
1. Based on the activity, what two things do you think are necessary for a chain reaction to occur?
2. What symbolizes the first neutron in your model of uranium-235? What does each domino represent?

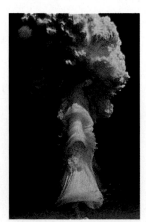

Figure 18–17. An uncontrolled fission reaction results in an explosion. Such a reaction is produced by an atom bomb, which creates the characteristic mushroom cloud shown here.

What can be done to stop a chain reaction? In the activity, the reaction was limited by the number of dominoes. Similarly, fission can be controlled by limiting the number of neutrons available for the reaction. This is what is done in a nuclear reactor, where cadmium rods are inserted between the rods of radioactive fuel. The cadmium rods absorb neutrons, so the number of available neutrons can be limited to one neutron per fission reaction. When there is more than one neutron, the reaction can accelerate at a dangerously rapid rate. If there are no neutrons, the reaction stops.

One problem with fission reactors, however, is that the waste products they produce are radioactive. Disposing of our daily garbage is a problem. Imagine how hard it is to find a dump site for radioactive wastes!

Another problem we must face is the possibility of a nuclear accident like the one at Chernobyl. Are we ready to cope with such a problem? As petroleum and natural gas become scarcer, should we turn to nuclear fission for electricity? The answers are not easy.

 ASK YOURSELF
Why must a fission reaction be controlled?

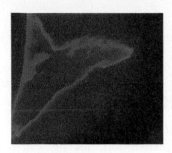

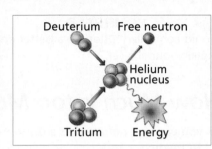

Figure 18–18. Fusion occurs naturally in the sun. Nuclear fusion results when two hydrogen atoms combine to form helium. A free neutron and a great deal of energy are released during this reaction.

Fusion

You may have heard someone say that the sun is our greatest source of energy. Have you ever wondered where the energy comes from? It comes from nuclear reactions, but not from fission. The sun's energy is the result of nuclei combining and giving off huge amounts of energy. This process is called *fusion*. Figure 18–18 shows you an example of a fusion reaction.

Fusion reactions produce much more energy per gram of fuel than fission reactions. Even though fusion reactors might depend on fission reactors for the creation of fuel, scientists predict that fusion reactors would generate much less radioactive waste and would be safer than fission reactors. So, what's the catch? Fusion can occur only in a special state of matter called a *plasma*. A **plasma** is a low-density, super-heated gas in which the electrons and nuclei are free to move at random. That means the electrons and nuclei are not found together as they are in atoms.

Once a fusion reaction starts, the reaction releases enough energy, in the form of heat, to keep itself going as long as there is fuel. One problem is the very high temperature required for such a reaction to occur—close to 10 million degrees Celsius. For this temperature to be achieved, a way to contain the hot material will have to be found. Researchers are experimenting with ways in which to contain the material and to produce the temperature necessary for fusion to take place.

DISCOVER BY *Researching*

A search is underway to understand what is known as "cold fusion." Go to the library and find out what cold fusion is and what has been done in this area. In your journal, record three sources you used to gather information. Share what you learn with your classmates. ✎

GUIDED PRACTICE

 Review the domino chain reaction with the students, and have them explain in their own words what the one-by-one arrangement represented and the difference between it and the triangular arrangement. The students should write their answer in their journals. Have them check their answer against a classmate's answer. Encourage the students to discuss any differences between their explanations.

INDEPENDENT PRACTICE

 Have the students provide written answers to the Section Review. Have the students draw and label a diagram of a nuclear fission reactor in their journals.

EVALUATION

Ask the students whether they feel they need to be concerned about getting sick from any kind of normal radiation exposure. (Answers should reflect an understanding of natural radiation, which is largely harmelss, and the greater doses of radiation near a nuclear plant or from medical treatment.)

ONGOING ASSESSMENT
▼ ASK YOURSELF

Controlled fusion is a self-sustaining reaction. It is also inexpensive and clean; it produces a much greater amount of energy than fission and produces few radioactive waste products.

SCIENCE BACKGROUND

One-time Exposure (rems)	Effects
0–25	no visible immediate effects on human body
25–50	small decrease in white blood cell count
50–100	marked decrease in white blood cell count; development of lesions
100–200	radiation sickness—nausea, vomiting, loss of hair; blood cells die
200–300	hemorrhaging, ulcers, death
300–500	acute radiation sickness, 50 percent of population dies in a few weeks
>700	100 percent of population eventually dies

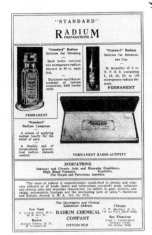

Figure 18–19. Radium was once sold as an over-the-counter medication for the treatment of general aches and pains.

▼ ASK YOURSELF

Would controlled fusion be a better energy source than fission? Explain your answer.

How Much Is too Much?

As you can see, radiation is a double-edged sword. It can be useful in medicine, industry, and energy production, but it is lethal if not managed correctly.

If radiation can kill you, it must be incredibly dangerous. Is any amount of radiation safe? If so, how much exposure is okay? Is a very small exposure a problem?

The fact is, you are exposed to radiation every day of your life. No, don't run for cover. You won't glow in the dark—that only happens in science fiction. The radiation you are exposed to is natural. It comes from the sun and other stars, from soil, and even from building materials. This natural amount of radiation is called *background radiation.*

To measure the exposure of living things to radiation, a unit called the *rem* is used. A rem is a unit of absorbed radiation. Exposure to radiation is usually measured in millirems, or one thousandth of a rem. Table 18–2 shows the average radiation exposure for the average person during a year.

Table 18-2	Average Annual Radiation Exposure in the United States	
Source		**Millirems**
Natural sources		
From outside the body		
Cosmic radiation		50.0
The earth		47.0
Building materials		3.0
From inside the body		
Inhalation of air		5.0
Elements found naturally in the body		21.0
Manufactured sources		
Medical		
Diagnostic X-rays		50.0
Radiotherapy, radioisotopes		10.0
Internal diagnostics		1.0
Atomic energy industry, laboratories		0.2
Luminous watch dials, television tubes, microwaves, radioactive industrial wastes, and so on		2.0
Radioactive fallout		4.0
Total exposure		193.2

RETEACHING

Write the properties of both fission and fusion on the chalk-board in random order. Have the class discuss each property and categorize it correctly.

EXTENSION

Plan a field trip to a local nuclear power plant. (Most plants allow small, prearranged groups to visit.) If this is not possible, you might invite a public relations speaker from the nearest plant.

CLOSURE

Cooperative Learning Have the students work in small groups to summarize the section. Encourage the groups to trade summaries and critique each other's summaries for thoroughness and clarity of expression.

How much radiation is too much? Experts disagree on the maximum number of rems to which a human can be safely exposed. A generally accepted figure is 5 rems (5000 millirems) per year. How much radiation were the people close to Chernobyl exposed to? One estimate suggests that the 24 000 people closest to the reactor received more than 45 rems of radiation after the reactor exploded. That is nine times the safe dose for an entire year!

All types of radioactive decay are potentially dangerous because they cause atoms in living things to form ions. These ions can interrupt the normal processes of living things. Gamma and beta radiation can penetrate deep into a living organism and cause extensive tissue damage. The heavy alpha particles cannot penetrate the skin. However, alpha particles may enter the system through ingestion of contaminated food, giving internal tissue a full dose of this radiation.

Did you know that smokers are exposing themselves to small amounts of radiation every day? The phosphate fertilizers used to grow tobacco are rich in uranium and some of its decay products. The smoke particles that are inhaled by the smoker are rich in lead-210. Lead-210 decays into bismuth-210 and polonium-210. These radioisotopes continue to build up in certain internal organs and expose them to alpha and beta particles. Lead-210 has a half-life of 20.4 years. So you see, it is not just the nicotine and tar that are bad for you.

Figure 18–20. Exposure to radiation may cause severe burns.

ONGOING ASSESSMENT
ASK YOURSELF

Although the penetrating power of alpha particles is small, they can be ingested and go directly to inner tissue, where they can do significant damage.

SECTION 3 REVIEW AND APPLICATION

Reading Critically

1. When a single action causes the same reaction to occur over and over, the reaction is a chain reaction. Example: In a landslide a rock falls and rolls down a mountain, striking other rocks in its path. These rocks move, hitting more rocks in a chain reaction.

2. Fusion is clean and inexpensive, produces four times as much energy as fission, and is a self-sustaining reaction.

Thinking Critically

3. to protect you from X-rays, which are ionizing radiation and could cause cell damage.

4. Both fusion and fission are chain reactions. Fusion is the combination of two atoms to form one atom with a heavier nucleus. It is a self-sustaining reaction because of the great quantity of energy released. Fission is the process by which a nucleus is bombarded with neutrons and split into two nuclei. Its self-sustaining reaction can be controlled by keeping the amount of neutrons constant with the use of cadmium rods.

▽ **ASK YOURSELF**

If alpha particles cannot go through your skin, why are they a dangerous source of radiation?

SECTION 3 *REVIEW AND APPLICATION*

Reading Critically

1. What is a chain reaction? Give an example of a reaction that simulates a chain reaction.

2. Why do some scientists believe that fusion is the answer to the energy needs of tomorrow?

Thinking Critically

3. Why does your dentist give you a lead apron when your teeth are X-rayed?

4. Are both fusion and fission chain reactions? Explain why or why not. You may wish to draw a diagram to support your explanation.

Making a Spinthariscope

Process Skills: Measuring, Observing, Analyzing

Grouping: Groups of 3 or 4

Objectives
● **Observe** the effect of radioactive particles on a spinthariscope.
● **Learn** how to handle radioactive substances safely.

Hints
Students might bring in old watches or alarm clocks with luminous dials from which the paint can be scraped off. Iodine-131 in the form of potassium iodide salt may be obtained from your local hospital or radiologist. Other radioisotopes are available from scientific supply houses.
CAUTION: Treat all radioactive material as hazardous. Review common laboratory practices in handling toxic materials with the students. The isotopes in this investigation are safe in these small amounts but the students should avoid touching the material or spreading it around on the table. Wash hands thoroughly when finished.

★ **PERFORMANCE ASSESSMENT**

Cooperative Learning After the completion of this investigation, have a student from each group summarize how the group assembled the spinthariscope. Then ask a second student to summarize the group's results.

► Analyses and Conclusions

1. A display of light should be observed if the room is quite dark.

2. Answers should note that radiation from the radioisotope hits the zinc sulfide particles, producing a burst of light.

3. Uranium has a half-life of 4.5×10^9 years, so the spinthariscope should shine for at least a billion years.

► Application
Uranium in watch dials would not be hazardous unless the watch were worn for a very long time—20 years or more.

✳ Discover More
The students should find out that a significant number of workers developed leukemia or other forms of cancer over an extended period of time. These conditions were attributed to the workers' exposure to the radium.

INVESTIGATION

Making a Spinthariscope

► MATERIALS
● matchbox, 2.5 cm × 3.5 cm ● clear nail polish ● zinc sulfide powder
● thick needle ● radioactive source ● shellac ● hand lens

▼ PROCEDURE

CAUTION: Treat all radioactive material as extremely hazardous. Avoid touching the material or spreading it on the table. Wash your hands thoroughly when you have finished.

1. Remove the inner box from the matchbox. Brush one of the outer ends of the inner box with clear nail polish, and sprinkle it with a thin layer of zinc sulfide powder. Allow the nail polish to dry.

2. Push the needle through the zinc sulfide coated end of the box, as shown in the figure. If the needle does not seem secure, you can fasten it in place by taping it to the inside of the box.

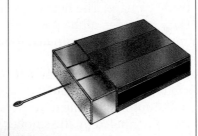

3. Carefully follow these instructions to attach a tiny piece of the radioactive material to the eye of the needle: Use the box as a holder to avoid touching the needle or the radioactive source. Place several drops of shellac on the radioisotope itself, and carefully

mix the two together with the eye of the needle. Most of the material should adhere to the eye. You have made a spinthariscope (spihn THAWR uh skohp).

4. Push the inside box partially back into the matchbox. Look at the zinc sulfide layer with the hand lens. The matchbox cover can be pushed forward or backward to get the best view with the hand lens. For best results, use the spinthariscope in a dark room.

5. Return all your materials to your teacher for proper disposal.

► ANALYSES AND CONCLUSIONS
1. Describe what you observe.
2. What do you suppose is causing it?
3. If the radioactive material you are using was from the luminous dial of an old watch, it probably contains tiny amounts of uranium. How long do you think your spinthariscope will work?

► APPLICATION
Do you think the amount of uranium found on watch dials would be hazardous to the wearer? Explain.

✳ *Discover More*
Many older watches had radium numbers on the dials so they would glow in the dark. The people who painted the numbers on would dampen the brushes in their mouths. Go to the library and find out what effect this had on the workers.

CHAPTER 18 *HIGHLIGHTS*

The Big Idea—Changes Over Time

Radioactive elements are radioisotopes. Their characteristics are due to instability caused by unequal forces of attraction and repulsion within the nucleus. Radioisotopes go through three basic types of decay to achieve stability: alpha decay, beta decay, and gamma decay. The result is a lighter and more stable nucleus. The decay processes occur in predictable patterns called half-lives.

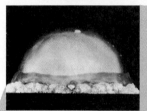

The Big Idea

Within every atom, interactions between matter and energy go on constantly. These interactions eventually form atoms with greater stability.

Some naturally occurring elements, such as uranium, are not stable. To achieve stability, they give off energy in the form of alpha, beta, or gamma radiation. As radiation is given off, the nucleus of the element changes, producing a more stable atom.

By altering the stability of naturally occurring atoms, scientists can create new radioactive elements. As with their natural counterparts, these elements emit alpha, beta, or gamma radiation until they become stable again. Returning to a nonradioactive state may require thousands of years.

For Your Journal

Look back at the ideas you wrote in your journal at the beginning of the chapter. How have your ideas about radiation changed? Revise your journal entry to show what you have learned. Don't forget to add what you have learned about how radioactivity is used in medicine.

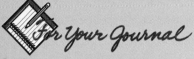

For Your Journal

Student entries should reflect an understanding of radioactivity and radioactive elements. You might encourage the students to share their entries with classmates.

CONNECTING IDEAS

The concept map should be completed with words such as *alpha decay, beta decay, gamma decay, half-life, fission,* and *fusion.* The students may need to refer to previous chapters to complete the sections on physical and chemical change.

Connecting Ideas

Substances may change in several ways. They may undergo nuclear change, chemical change, or physical change. Below is the beginning of a concept map. Copy it into your journal, and add what you know about how atoms and molecules change to complete the map. Be sure to show how the types of change are related.

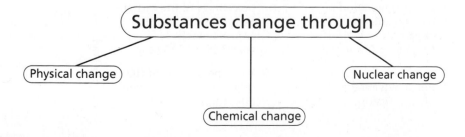

Substances change through

Physical change — Chemical change — Nuclear change

Short Answer

6. Pros: Exposing fruits, vegetables, and meats to gamma radiation kills such organisms as molds, insets, bacteria, and parasites. Irradiation helps prolong the shelf life of perishable foods and could prevent many cases of food poisoning.

Cons: Some nutritional value of the food is lost through irradiation. Some scientists believe that the RPs formed by irradiation are mutagins which cause changes in DNA. Some people fear that irradiated food is still radioactive.

7. New technology: medical treatments, growth in space exploration, nuclear power plants, advances in weaponry.

Problems: disposal of nuclear wastes, fear of new weaponry, hundreds of thousands of victims from nuclear explosions in Nagasaki and Hiroshima, risk of catastrophic accidents at nuclear power plants.

Interpreting Graphics

8. approximately 25%; approximately 50%

9. 50% of the previous 50%

Understanding Vocabulary

1. (a) Radioactive elements are unstable elements that spontaneously break apart and give off particles or high-energy radiation. Half-life is the time required for one half of a given sample of a radioactive substance to decay.

(b) Each term describes a type of radioactive decay. Alpha decay occurs when an atom emits an alpha particle made up of two protons and two neutrons. Beta decay occurs when an atom emits a beta particle from its nucleus at very high speed. The beta particle is formed when a neutron is converted into a proton and an electron. Gamma decay is the emission of high-energy radiation, similar to X-rays. Gamma radiation has no mass and no charge. Of the three types of radiation, gamma decay is the most dangerous.

(c) A chain reaction is related to nuclear fission. It is a self-sustaining reaction in which the fission of nuclei produces neutrons that cause the fission of other nuclei. A plasma is a low-density, super-heated gas in which nuclear fusion can be carried out safely.

(d) A Geiger counter is a portable device used to detect radiation. A dosimeter is a wearable instrument used by a person who works with radioactive materials. It allows the person to monitor the amount of radiation to which he or she is exposed.

Understanding Concepts

Multiple Choice

2. b   3. d

4. a   5. a

CHAPTER 18

REVIEW

Understanding Vocabulary

1. For each set of terms, explain the similarities and differences in their meanings.
   a) radioactive (485), half-life (496)
   b) alpha decay (487), beta decay (487), gamma decay (488)
   c) chain reaction (499), plasma (501)
   d) Geiger counter (492), dosimeter (492)

Understanding Concepts

MULTIPLE CHOICE

2. Which of the following is not a use of radioactive isotopes?
   a) used as tracers in the human body
   b) used in the production of paints and other commercial products to increase durability
   c) used in medicine to treat many forms of cancer
   d) used to treat food so that it can be stored for long periods of time without refrigeration

3. In a fusion reaction
   a) large atoms split into smaller atoms.
   b) dangerous waste products are produced.
   c) cadmium rods are used to absorb neutrons.
   d) the nuclei must be at very high temperatures.

4. What type of radiation is given off in the following transformation?
   $$^{226}_{88}\text{Ra} \rightarrow\ ^{222}_{86}\text{Rn}$$
   a) alpha radiation
   b) beta radiation
   c) gamma radiation
   d) none of the above

5. Which of the following is not an advantage of nuclear fission?

   a) Fuel is inexpensive and easily obtain
   b) Fission could replace natural gas and petroleum as an energy source.
   c) Large quantities of energy are produ
   d) Fission can be controlled by limiting number of neutrons available for the reaction.

SHORT ANSWER

6. Discuss the pros and cons of the irradia of food.

7. Scientists opened a new age when the f atom bomb was detonated. Make a list technological advances and problems t have occurred in the nearly 50 years sir their experiment.

Interpreting Graphics

8. Look at this line graph that contains data about carbon-14. What percentage of the carbon-14 breaks down after approximately 2865 years? after 5730 years?

9. What percentage of the original material remains after each half-life?

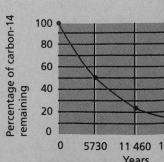

Reviewing Themes

10. Stability is reached in radioactive atoms by the release of energy in the form of alpha, beta, or gamma radiation. Alpha and beta radiation are both particulate, while gamma radiation is the emission of high-energy waves. When an atom undergoes one of these forms of decay, it then reaches a more stable state. Examples will vary.

11. Carbon-14 is used in radioactive dating because it is found in all living things. The amount of carbon-14 stays constant in an organism as long as it is alive because the organism ingests food that replenishes the carbon-14. When the organism dies, the level of carbon-14 begins to diminish because it is no longer being replaced. By comparing how much carbon was in an organism while it lived to how much remains at the time of examination, scientists can estimate the age of the organism.

Thinking Critically

12. Nuclear fission is a reaction in which the nucleus of a large atom breaks apart to form two smaller atoms. This reaction is used to generate power in nuclear power plants. Nuclear fusion is a reaction in which smaller nuclei join to form larger atoms. Both reactions release large amounts of energy. Both types of reactions have disadvantages. Nuclear fusion requires very high temperatures to occur and is hard to control. Nuclear fission is easier to control, but produces large quantities of radioactive waste products.

13. Individuals who work around radiation should wear overall protective clothing, wear a dosimeter at all times, wear a lead apron when directly exposed to radiation (such as X-rays), and follow all safety guidelines provided. Adherence to such guidelines is essential to avoiding radioactive contamination. While very brief and infrequent exposure to radiation carries minimal risks, the risks of developing conditions such as leukemia and radiation poisoning increase as exposure to radiation increases.

14. The top number, 238, represents the mass number of the atom, the total number of protons and neutrons. The bottom number, 92, represents the number of protons in the atom.

15. The beta particle would be attracted to the positively charged plate and would be detected on the left. The alpha particles would be on the right, since they are positively charged. The gamma rays are not charged particles; they would not be deflected at all and would be detected at the center.

16. Students should mention that the alpha particles cause a change in the atoms of the aluminum, transforming them into radioactive substances. The fact is that some aluminum atoms are converted to radioactive phosphorus-32. This experiment—performed by Frederic Joliot Curie—was the first example of artificial radioactivity. Accept reasonable hypotheses.

17. These two elements have equal numbers of protons and neutrons. The loss of one proton, however, diminishes the forces of repulsion between protons, making the nucleus more stable.

Reviewing Themes

10. *Changes Over Time*
In nature, radioactive atoms become stable over time by releasing particles or rays. Explain this phenomenon in terms of alpha, beta, and gamma radiation. Cite specific examples in your explanation.

11. *Systems and Structures*
Radioactive decay occurs in specific patterns and at specific rates called half-lives. The half-life of carbon-14 is 5730 years. Discuss how carbon-14 is used to date once-living material.

Thinking Critically

12. Compare and contrast nuclear fission and nuclear fusion.

13. List some ways in which people who work with radioactive materials can protect themselves. Explain why this is an important thing to do.

14. The notation for a radioactive isotope of Uranium is $^{238}_{92}$U. Explain what both the upper and lower numbers mean. Why do you think radioactive isotopes have a special kind of notation?

15. Ernest Rutherford used the apparatus on the right to identify radioactive particles. Beams emitted by radioactive materials would travel through the charged plates and hit the photographic plate. By developing the film, Rutherford could find information about the type of particle emitted and its charge. Using the figure shown, where do you think each of the three types of radiation should appear on the photographic plate?

16. If an aluminum sheet is bombarded by alpha particles, neutrons and other particles are given off. These particles continue to appear after the radioactive source is taken away. What can you conclude from this experiment?

17. Two atoms, $^{22}_{11}$Na and $^{26}_{13}$Al, undergo an unusual type of radioactive decay not covered in this chapter. They emit protons. What effect does this have on the stability of the products? Why does this type of decay occur?

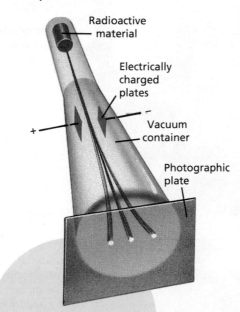

Radioactive material

Electrically charged plates

Vacuum container

Photographic plate

Discovery Through Reading

Miller, Peter. "A Comeback for Nuclear Power? Our Electric Future." *National Geographic* 180 (August 1991): 60–89. This article presents information on the usefulness and practicality of nuclear power.

Discussion

● **Process Skills:**
Communicating, Analyzing

This *Science Applications* discusses the preservation of objects found on shipwrecks and the shipwrecks themselves. Begin by asking the students to compare the photographs of the sword and the treasure recovered from the shipwreck. (The metal of the sword has dissolved, but the gold appears to be unchanged.)

● **Process Skills:** *Solving Problems/Making Decisions, Inferring*

Ask the students to explain why a salvage diver needs to know about the properties of silver and gold when looking for silver coins and gold goblets on a shipwreck. (The silver coins

Science
PARADE

The Art of Preservation

Gold bars, silver goblets, glittering jewels—who hasn't read or dreamed about the adventure of finding a sunken ship loaded with treasures? Sunken ships sometimes do contain treasures of gold and silver. They also contain valuable information that can tell scientists many things about the time in history when the ship sailed. For example, the materials that were used for trade and commerce can often be identified. The tools used to build the ship and to navigate it are often found. Sometimes cooking implements and eating utensils can provide information about the foods that were carried on long voyages.

How Do Elements Behave?

The greatest challenge to archaeologists and divers is to restore and preserve the ancient objects they uncover. To do this, they must have scientific knowledge of the properties of metals and nonmetals.

A famous treasure hunter once said, "Gold shines forever." Notice that the luster of the gold objects in the diver's hands is bright, even though they have been on the ocean floor for hundreds of years. Gold is one of the most stable metals. It does not tarnish, and it resists corrosion by most acids. Therefore, gold is not difficult to restore.

Silver, however, reacts very differently to sea water. The diver in the photograph is also holding silver coins stuck together by chemical reactions. Besides the outer crust, which can be chipped off, there is a blackened crust covering each coin. This crust is silver sulfide—formed by the reaction of the silver with the dissolved sulfur in sea water.

Silver sulfide can be cleaned off silver objects by a process known as *electrolytic reduction*. This process

X-ray of the sword (right) shows that all the metal has dissolved and that only the crust remains.

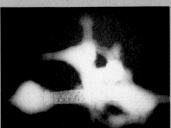

react with the sulfur in the sea water, forming a blackened crust over the coins. Thus, the coins could be overlooked. However, gold does not react with sea water, so goblets would remain as shiny as when the ship sank.)

 Journal Activity After completing the Unit and reading and discussing the *Science Applications*, ask the students to consult the list of questions they recorded in their journals from the discussion at the beginning of the Unit. (You might have the students turn back to the Unit Opener on page 424–425.) Lead a discussion during which the students can verbalize what they have learned. Encourage the students to record their new ideas in their journals.

As a follow-up to the class discussion, you might have the students prepare reports on the discoveries and salvaging of other shipwrecks. Encourage the students to share their reports with classmates.

involves dipping the pieces of silver into a solution of zinc and caustic soda (sodium hydroxide) through which an electric current is passed. The sulfur separates from the silver and reacts with the zinc in the solution.

Drawing of the Atocha, *which sank in 1622*

Gone, but Not Forgotten

Iron is probably one of the most difficult metals to restore. Iron rusts very easily, combining readily with oxygen in the water to form iron oxide. Moreover, rusted objects tend to disintegrate when exposed to air.

Most iron objects completely rust away on the ocean floor. Before this happens, though, sand and shells may form crusts on the objects' surfaces. Even as the objects continue to rust, a crust remains as a permanent record of the original shapes.

In some cases, the crust that formed around a long-dissolved iron object can be used as a mold. The crust can be cut in half, and any remaining iron oxide washed out. Then the crust can be filled with plaster or a rubber compound to form a cast of the original object. In other cases, an encrusted object can be X-rayed. The resulting X-ray shows the original shape of the metal object underneath.

Wood is another material that disintegrates easily once recovered from the ocean floor. If the wood is allowed to dry in the air, it will crack and warp. To avoid this damage, wood from ancient ships is submerged in several freshwater baths to remove all traces of salt and silt. If the wooden object is large, such as a ship, it is placed under sprinklers.

Restore with Care!

After the wooden object is washed, any material that is stuck to Its surface is carefully chiseled away. Then a fluid such as alcohol or polyethylene glycol is injected into or painted onto the wood. This fluid must be added to the wood to reinforce its structure and prevent warping and cracking.

The work of the archaeologist, diver, and restorer is always challenging. A knowledge of how elements, especially the metals, behave is essential. It is also necessary to work with the utmost care in order not to destroy valuable clues to the past. ◆

Archaeologist examining a timber from a sunken ship

Diver holding treasure from the Atocha

Background Information

The ancient Chinese are credited with inventing fireworks. They noted that some substances imparted colors to flames and used them to make colorful fireworks. In many countries, fireworks are used to celebrate festivals and national holidays. The most common types of fireworks are stars, shells, Roman candles, rockets, and fountains.

Chemical compounds are added to the fuel-oxidizer mixture to produce color in fireworks; the metallic ion in the compound is responsible for the color produced. Sodium compounds produce a bright yellow color, potassium compounds produce a violet color, copper compounds produce bright green, strontium compounds produce bright red, calcium compounds produce orange-red, and lithium compounds produce deep red.

The Flashy Science of

FIREWORKS

from *Current Science*

WHOOOMPH! Sssssssss—POP! "Ooooh, Aaaaah." Pause. Then, *BOOM! BADA-BOOOOOOOOOM!*

Don't you just love fireworks? The dazzling colors, the intricate patterns, all that really great noise? If so, you'll probably be one of the many millions of Americans watching fireworks displays this July 4.

For centuries, fireworks makers passed on their secret formulas only to family members. Many of the formulas remain secret today. But research over the last few years has revealed much about *pyrotechnics,* the science of fireworks. Here's some of what scientists now know about pyrotechnics.

Oxygen Fuels the Fireworks

Each firework shell contains a fuel that burns with explosive force and an *oxidizing agent.* The type of oxidizing agents used in fireworks are chemicals that readily give off oxygen. The burning of a substance—a process called *combustion*—requires the presence of oxygen or an oxidizing agent. In short, oxygen fuels the fuel.

Combustion in a firework shell occurs when the oxidizing agent (usually potassium perchlorate or ammonium perchlorate) ignites and releases oxygen. The sudden spurt of oxygen allows the fuel to burn. Fireworks fuels burn rapidly, producing sudden blasts of white light.

By adding other chemicals to the fuel-oxidizer mixture, a fireworks designer can add various colors to the fireworks flare. Barium chlorate, for instance, produces a green color. Strontium carbonate makes red.

Blue presents the toughest challenge for pyrotechnics experts. Getting a rich blue color requires a precise mixing of chemicals. Says fireworks expert Dr. John Conkling, "I pay close attention to flame colors. If a decent blue color appears, I am always impressed."

Discussion

● **Process Skills:** *Applying, Observing*

Ask the students to describe fireworks displays they have seen. (They might mention the colors and different types of fireworks.)

● **Process Skills:** *Evaluating, Formulating Hypotheses*

Ask the students why many states have prohibited the selling of most kinds of fireworks to the public. (Students will probably mention the danger involved with fireworks and the lack of knowledge most people have about them.)

Journal Activity
Have the students research the history of fireworks and record their findings in their journals. Then encourage the students to share the information they find with classmates.

How Fireworks Work

1. Technician or computer ignites fuse.
2. Short fuse ignites black powder, sending shell skyward.
3. Delay fuse ignites first chamber, creating colorful blast.
4. Explosion of first chamber ignites delay fuse to second chamber, which explodes, usually with different color.
5. Delay fuse ignites flash and sound mixture—BANG!

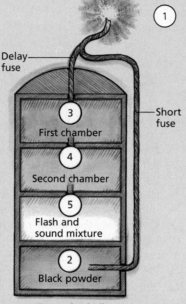

Delay fuse

Short fuse

First chamber

Second chamber

Flash and sound mixture

Black powder

Fireworks shell

Color + Timing = WOW!

Color alone, however, doesn't make a great fireworks show. More and more, fireworks designers launch fireworks in rhythm to particular pieces of music. Here's how.

Each shell contains two or more *fuses*, cords or wires that burn or melt at precise speeds. To launch a shell, a worker lights a rapidly burning fuse at the base of the shell. This fuse ignites a packet of explosive black powder, which sends the shell aloft at 390 feet (117 meters) per second.

The fuse also ignites a slower burning fuse. This fuse allows the shell to reach a safe height before exploding. By carefully timing the lighting of the fuses, workers can make, say, a red and blue shell explode at just the right moment during "The Star-Spangled Banner."

"We think of ourselves as artists," says Butch Grucci, president of Fireworks by Grucci, a prominent family-operated outfit. "The sky is our canvas." ◆

Discussion

● **Process Skills:** *Generating Ideas, Inferring*

Ask students to consider how the Curies' work has affected their lives. (Curies' work with radioactivity probably influenced the development of nuclear weapons and nuclear energy.)

Journal Activity
Students who are interested in finding out more about the life of Rosalyn Yalow and other Nobel laureates may want to read the following book and write a brief report in their journals.

Dash, Joan. *The Triumph of Discovery: Women Scientists Who Won the Nobel Prize.* Messner, 1991. This book is about determination, sacrifice, conflict, and the problems that exist for women today in science careers. (NSTA Outstanding Trade Book)

EXTENSION

Students who are interested in finding out more about the uses of radioactivity might want to read the following magazine article.

Shreeve, James. "The Dating Game" Discover 13 (September 1992): 76–83. This article describes how archaeologists are using some new methods involving radioactivity to date human artifacts.

◆ **THEN AND NOW**

Marie Sklodowska Curie (1867-1934)

Born in Warsaw, Poland, Marie Sklodowska left her native country to obtain an education. She went to Paris, where she studied mathematics, physics, and chemistry. While in Paris, she met a French physicist, Pierre Curie. They eventually married and worked together.

Shortly after her marriage, Marie Curie began working with radioactive elements. In fact, she was the first to use the word *radioactivity.* Together, she and Pierre discovered that the radiation coming from uranium ore was much more than the amount the uranium alone could account for. They hypothesized that there must be other radioactive material present in the ore. From tons of uranium ore they isolated two new radioactive elements, which they named *polonium* and *radium*. For their discoveries, they received the 1903 Nobel Prize in physics.

In 1911 Marie Curie was awarded a second Nobel Prize, this time in chemistry. Aided by her daughter Irène, Marie worked on the application of X-rays in medicine. After Marie's death, Irène and her husband, Frédéric Joliot-Curie, continued research on radioactive materials. ◆

Rosalyn Yalow (1921-)

What is a physicist doing in a Veterans Administration hospital working with physicians? Earning a Nobel Prize in medicine is the answer that Rosalyn Yalow might give you. Born in the Bronx, New York, Yalow graduated with honors from Hunter College in Manhattan. Following graduation she was accepted as a graduate student in the College of Engineering at the University of Illinois. While studying for her degree, Yalow became interested in the measurement of radioactive substances, a skill she later put to use as a medical physicist.

In 1947 Yalow converted a closet at the veterans' hospital where she worked into a

radioisotope laboratory. Soon after this, she met Solomon Berson, a physician, with whom she would work for more than 22 years. In 1977 Yalow and Berson won a Nobel Prize in medicine for a technique that they developed. The technique, called *radioimmunoassay,* is a test that uses radioactivity to identify and measure small traces of substances in the blood.

Today Yalow travels throughout the world to lecture. She has received many awards and prizes from universities, medical societies, and associations and has also been awarded the National Medal of Science. ◆

Background Information

NMR, also known as magnetic resonance imaging (MRI), is used to create detailed pictures that reveal the inner workings of the human body. NMR imaging machines use computers to create detailed pictures of body structures and chemical compositions. The process can be used to uncover such problems as brain damage and nerve degeneration. Scientists at EEG Systems Laboratory in San Francisco have used NMR imaging to learn how different parts of the human brain work.

Discussion

You may wish to share this information about Theophilus Leapheart with the students: Leapheart develops new compounds on a small scale. His reactions use very small amounts of chemicals. If his reactions are successful, he turns the idea over to chemical engineers. They conduct the reaction on a much larger scale and use thousands of liters of materials. The chemical engineeers make sure that the reaction can be done safely in large quantities.

SCIENCE AT WORK

Theophilus Leapheart, Chemist

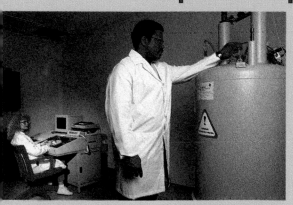

Each year chemists make thousands of new chemical compounds. These compounds can be very complex molecules. How do chemists know the structures of the compounds they make? Theophilus Leapheart, a chemist from Midland, Michigan, uses special instruments to find out the structures of new chemical compounds and medicines he makes.

Leapheart works as part of a team of scientists. "Chemistry is very team oriented," he explains. "No one can put the whole puzzle together alone. Chemists, engineers, mathematicians, and physicists all need to work together to solve problems and to develop new products. To be successful, you have to know how to work with others."

One of the tools that Leapheart uses to find out the structure of a compound is called a *nuclear magnetic resonance* (NMR) instrument. When Leapheart makes a new compound, he takes a small sample of it and dissolves it in a solvent. He then puts the sample into the NMR's strong magnetic field. The magnetic field causes the molecules to line up, just like the earth's magnetic field causes a compass needle to line up.

Next, Leapheart passes radio waves through the sample. This causes the molecules to wobble. When the radio waves are shut off, the molecules try to line up again with the magnetic field. As they do, they give off energy. The frequency of the energy depends on the kind of molecule. These frequencies show on a graph called a *spectrum*.

To a trained scientist, the NMR spectrum of a compound is like its signature. No two compounds have exactly the same NMR spectrum. A spectrum can tell what atoms are present in the compound and how many of each are present. Leapheart uses computers to run the NMR machine. The computers tune the instrument, much like you would tune a television to get a clear picture. The computer also collects the data and stores it for later use.

To prepare for his career, Leapheart took many science and mathematics classes. However, he also thinks that English classes are important. "Writing is very important," Leapheart adds. "In any setting, you must be able to communicate what you have done, how you did it, and how someone else can follow through on your work after you are gone." ◆

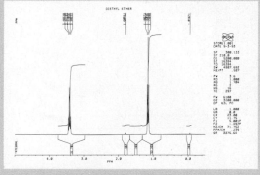

NMR printout

Discover More

For more information about NMR and careers in chemistry, write to the

Office of Pre-High School Chemistry
American Chemical Society
1155 16th Street, N.W.
Washington, DC 20036

Discussion

● **Process Skills:** *Comparing, Evaluating*

Have the students read the article on NMR spectroscopy and speculate on why some scientists compare its value to modern medicine to that of the X-ray, discovered by Roentgen in 1895. (Students might mention that X-rays first allowed physicians to see bones inside the body and that NMR spectroscopy allows physicians to view living tissue and determine the chemicals that make up those tissues.)

SCIENCE/TECHNOLOGY/SOCIETY

NMR Spectroscopy

Nuclear magnetic resonance (NMR) spectroscopy has become a powerful tool in chemistry and medicine. Chemists use it to determine the structure of complex substances. NMR is also used by physicians to make pictures of the soft tissues in the body and to help diagnose some diseases.

NMR spectroscope

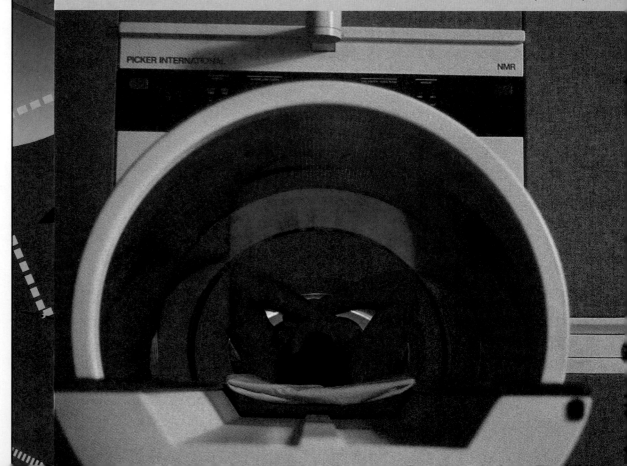

Nuclei— Line Up!

NMR uses isotopes of elements that have magnetic nuclei. Hydrogen-1, carbon-13, and phosphorus-31 are often used. If a chemist is trying to find the exact structure of a substance, a sample of the substance is placed in a strong magnetic field. This field is 30 000 to 240 000 times greater than the magnetic field of the earth. The field causes the nuclei of the unknown sample to line up. This happens for the same reason that a compass needle points to the magnetic north pole.

Sending Out a Signal

When the nuclei are lined up, radio waves are sent through the sample. These waves make the nuclei wobble in much the same way that a rotating top does. After the radio waves are discontinued, the nuclei try to line up again. As they line up, they send out radio waves like the ones that they received. These waves can be detected by a computer. The computer can take information from the waves and make a graph that shows chemists such as Leapheart where the atoms are located in a molecule.

Many important molecules in living things contain phosphorus-31. For this reason, NMR can be used to study how cells and organs function. The use of NMR makes it possible to learn about the structure of molecules in living organisms without killing the subject or even performing surgery!

An Image— a Diagnosis

The application of NMR in medicine is known as Magnetic Resonance Imaging (MRI). MRI gives physicians a two-dimensional picture of the human body, but it does not use or produce any harmful radiation. Therefore, it provides physicians with a safe tool for diagnosing illnesses.

MRI can also be used to study what is happening in living tissue. The image on this page shows the inside of a human brain. The colors indicate different areas of the brain and give important information on the chemicals that make up the tissue of the human brain. You can see how using MRI enables physicians to study what is happening in living systems. Additionally, physicians can study what happens to tissue or body systems when they are under stress, under the influence of drugs, or diseased. ◆

MRI of the human brain

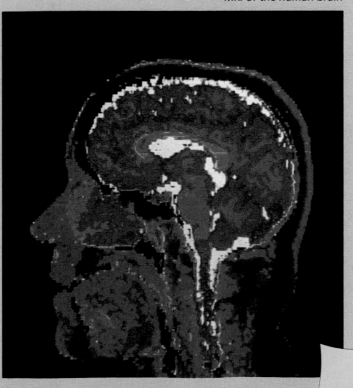

CHANGES IN MATTER

UNIT OVERVIEW

This unit presents and describes the many changes that matter can undergo, from the formation of a simple mixture to biochemical reactions in living organisms. Different types of mixtures are presented; types of chemical reactions are explained; writing and balancing chemical equations is taught.

Chapter 19: Interactions of Matter, page 516

This chapter reviews mixtures and compounds and presents three kinds of mixtures: solutions, suspensions, and colloids.

Chapter 20: Bonding and Chemical Reactions, page 540

In this chapter bonding and chemical reactions are introduced. The use of chemical formulas and balanced equations is discussed as ways of representing bonding and chemical reactions.

Chapter 21: Acids, Bases, and pH, page 564

This chapter defines acids and bases in terms of hydronium ions and hydroxide ions and provides examples of each class of compound. The pH scale is introduced and the use of indicators to show the strength of acids and bases is discussed. The formation and uses of salts are also presented.

Chapter 22: Carbon Compounds, page 584

This chapter presents the characteristics of carbon and other members of the carbon family, and organic compounds. Included within this chapter is a discussion of the many uses of organic compounds.

Science Parade, pages 614–623

The articles in this unit's magazine focus on chemical communication in nature, the use of anabolic steroids, and advancements made in the development of artificial body parts. The biography feature focuses on a biochemist and a biomedical engineer. The career feature discusses the qualifications needed to be a dietitian.

UNIT RESOURCES

PRINT MEDIA FOR TEACHERS

CHAPTER 19

Cox, Beverly and Martin Jacobs. "Spirit of the Harvest." *Native Peoples* (Spring 1992): 54–57. The author presents two Native American recipes and explanatory text.

CHAPTER 20

"Go With the Grain." *Scientific American* 262 (May 1990): 26–27. Tiny crystals have been discovered that can help to make metals very strong, ceramics ductile, and composites have strong structural and electrical properties. The uses in electronics or space are innumerable.

CHAPTER 21

Horgan, John. "Liquid Sky." *Scientific American* 263 (November 1990): 32. Liquid crystals are being studied to propose an analogy between the phase transitions of the crystals and the phase transitions of the cosmos.

CHAPTER 22

Crabb. Charlene. "More Fun With BuckyBalls." *Discover* 14 (January 1993): 72–73. Ever since chemists created 60-carbon-atom "buckyballs," the search has been on to find these spheres in nature. In 1992 mineralogist Semeon Tsipursky of Arizona State University did it.

PRINT MEDIA FOR STUDENTS

CHAPTER 19

"Glass" *Kids Discover* 3 (April 1993): 2–19. It looks like a solid, but it really is a liquid. This article discusses glass and its properties.

Peterson, Ivars. "Singling Out Molecules in Solution." *Science News* 139 (May 4, 1991): 287. This article discusses how molecules in solution have been identified through the use of a laser beam.

Schwartz, David. "Serving Up Science for Everyday Cooks and Gourmets Alike." *Smithsonian* 23 (December 1992): 110–119. Harold McGee, a "kitchen sleuth," looks at food and cooking from a scientific point of view. He has also published the books *On Food and Cooking: The Science and Lore of the Kitchen* and *The Curious Cook: More Kitchen Science and Lore.*

CHAPTER 20

Dane, Abe. "Endothermic Fueled Jet Could Break Mach 5." *Popular Mechanics* 168 (August 1991): 15. Compound that absorbs heat accelerates jet to five times the speed of sound.

Robson, David P. "Car Cooler." *Chem Matters* 11 (February 1993): 11. A car that's been sitting in the heat can reach a temperature of 130°F. What should you do? You could use a spray that's made of water and ethyl alcohol to cool down the interior.

Winter, Ruth. *A Consumer's Dictionary of Household, Yard and Office Supplies.* Crown, 1992. Toxic chemicals are a fact of modern life—we use them at home and at work. Minimizing the adverse effects of these substances is a concern. This book is a good introduction to this subject.

CHAPTER 21

Roberts, Leslie. "Learning from an Acid Rain Program." *Science* 251 (March 15, 1991): 1302–1305. The effects of acid rain are highlighted in this article.

Raloff, Janet. "Lime For Your Drink? Here's a New Twist." *Science News* 137 (February 24, 1990): 127. Methods of liming waters to neutralize acidic lakes and rivers are outlined.

CHAPTER 22

Hanson, Betsy. "Yews in Trouble." *Discover* 13 (January 1992): 55. Laboratories are coming to the rescue of the Pacific yew—a tree that is the source of taxol, a drug used to treat cancer.

Marsella, Gail. "Aspirin." *Chem Matters* 11 (February 1993): 4–7. This article discusses the history and chemistry involved with aspirin.

Regis, Ed. "Diamonds in the Rough." *Discover* 12 (March 1991): 66–71. Earth has been making diamonds since the beginning of time. However, John Angus has been making them since the 1960s. He might have found a better way to produce these precious gems.

ELECTRONIC MEDIA

CHAPTER 19

All About Matter—Mixtures: Separate, Yet Together. Videocassette. Focus Media, Inc. 14 min. Through real-life situations, this videocassette presents properties of mixtures, solutions, suspensions, and the processes of separation, filtration, and distillation.

Solutions. Film or videocassette. BFA Educational Media. 13 min. Sizes of particles in solutions and mixtures are illustrated and discussed.

The Chemistry Help Series: Elements, Compounds, and Mixtures. Computer Software. Focus Media, Inc. This tutorial program studies the differences among elements, compounds, and mixtures.

CHAPTER 20

Chemical Bonding and Atomic Structure. Film or videocassette. Coronet Film and Video. 23 min.

Metallic, ionic, and covalent bonds are presented. Also included are interatomic bonds, atoms, valence, and polarity.

Chemical Reactions. Videocassette; two-part series. Focus Media, Inc. 16 min. each. Students learn how to balance equations and observe chemical reactions in which there are energy changes.

Formulas and Equations. Film or videocassette. Coronet Film and Video. 11 min. The symbols used in formulas and equations and the concept of balanced equations are introduced.

CHAPTER 21

Acids, Bases, and Salts. Film or videocassette. Coronet Film and Video. 20 min. Reactions between acids and bases, producing energy changes, neutralization, and salts, are explored. The meaning of pH is defined.

Special Topics in Chemistry: Acids and Bases. Videocassette; four-part series. Focus Media, Inc. 63 min. total. Real-life applications are used to present acids and bases.

CHAPTER 22

How Atoms Combine. Film or videocassette. Coronet Film and Video. 11 min. Covalent bonding, polar covalent bonding, and ionic bonding are clearly and simply illustrated.

Special Topics in Chemistry: Biochemistry. Videocassette; four-part series. Focus Media, Inc. 63 min. total. Strong visual reinforcement is used to help students study everyday applications in biochemistry.

Special Topics in Chemistry: Organic Chemistry. Videocassette; four-part series. Focus Media, Inc. 63 min. total. Real-life situations help to illustrate applications of organic chemistry.

Discussion The bee in the photograph is one example of animals that use chemicals to communicate. Certain chemicals, such as those that create the odor of some flowers, attract bees. Other flowers give off chemical odors that attract moths or bats. Ask the students to name any chemicals that are used to repel insects. (The students might mention flea and tick repellents that are used on family pets or repellents used by people to prevent being bitten by mosquitoes.)

UNIT 6 CHANGES IN MATTER

CHAPTERS

19 **Interactions of Matter** 516

How many ways can substances interact? A chef could tell you more about this than you think.

20 **Bonding and Chemical Reactions** 540

Do people bond and react in the same way as chemicals? People once thought so.

21 **Acids, Bases, and pH** 564

Air pollution, dead fish, dying trees . . . what do these things have in common? Our knowledge of acids, bases, and pH can help us preserve the fragile environmental balance.

22 **Carbon Compounds** 584

Carbon is a very versatile element. It is a part of many of the products we use every day, from the bagel you had for breakfast to the plastic your school desk is made of. Even clothes and vehicles have carbon in their structures.

Journal Activity You might also bring in the following Unit topics: What are the types of matter? How do solutions differ from suspensions and colloids? How do atoms recombine during chemical reactions? How can you determine whether a substance is an acid or a base? What are organic compounds and where are they found? Ask the students to keep a record in their journals of any questions they might have about these topics. After the students have completed the Unit, encourage them to look again at their questions. A follow-up discussion might help them realize how much they have learned.

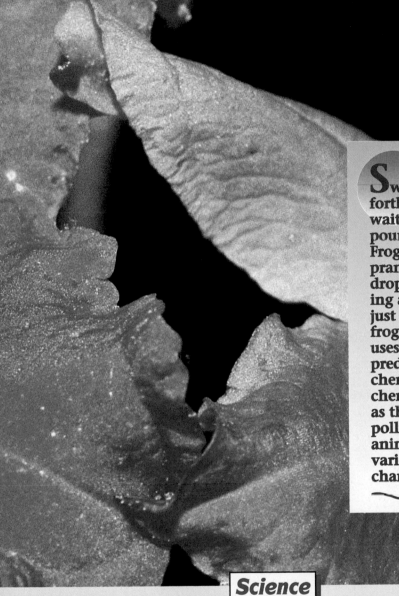

Swishing its tail back and forth, the house cat crouches, waiting for the right moment to pounce on its prey. Success! Frog in mouth, the cat begins to prance away. Suddenly, the cat drops the frog and begins foaming at the mouth. The cat has just discovered that nature, the frog in this example, sometimes uses chemistry to discourage predators. Nature also uses chemicals to attract. Plants use chemistry to invite insects, such as the bee in this photograph, to pollinate their flowers. Both animals and people have a variety of uses for chemistry, or changes in matter.

Science PARADE

CHAPTER 19

INTERACTIONS OF MATTER

PLANNING THE CHAPTER

Chapter Sections	Page	Chapter Features	Page	Program Resources	Source
CHAPTER OPENER	516	*For Your Journal*	517		
Section 1: TYPES OF MATTER: A REVIEW	518	Discover by Doing (B)	518	*Science Discovery**	SD
• Mixtures (B)	518	Discover by Doing (B)	518	Connecting Other Disciplines: Science and Home Economics, Changing Matter in the Kitchen (B)	TR
• Cooking the Chemist's Way (B)	519	Section 1 Review and Application	521	Investigation 19.1: Investigating Chemical Composition (A)	TR, LI
		Skill: Problem Solving (A)	522	Investigation 19.2: Separating Mixtures (A)	TR, LI
				Ratios of Elements in Covalent Compounds	IT
				Study and Review Guide, Section 1 (B)	TR, SRG
Section 2: SOLUTIONS	523	Discover by Doing (B)	524	*Science Discovery**	SD
• Making Solutions (B)	523	Discover by Problem Solving (B)	528	Solubility	IT
• What's the Solvent? (B)	524	Discover by Doing (B)	530	Thinking Critically (A)	TR
• A Little or a Lot (B)	526	Section 2 Review and Application	530	Study and Review Guide, Section 2 (B)	TR, SRG
• Now You See It, Now You Don't (B)	529				
Section 3: SUSPENSIONS AND COLLOIDS	531	Discover by Doing (A)	531	*Science Discovery**	SD
• Suspensions (B)	531	Section 3 Review and Application	535	Extending Science Concepts: Separation of Blood Serums by Paper Electrophoresis (H)	TR
• Colloids (B)	533	Investigation: Separating Mixtures (A)	536	Reading Skills: Determining Likenesses and Differences (B)	TR
				Record Sheets for Textbook Investigations (A)	TR
				Study and Review Guide, Section 3 (B)	TR, SRG
Chapter 19 HIGHLIGHTS	537	The Big Idea	537	Study and Review Guide, Chapter 19 Review (B)	TR, SRG
Chapter 19 Review	538	For Your Journal	537	Concept Map: Forms of Matter	IT
		Connecting Ideas	537	Chapter 19 Test	TR
				Test Generator	

B = Basic   **A** = Average   **H** = Honors
The coding Basic, Average, and Honors indicates subsections, features, and resources that might be appropriate for different levels of learners. For additional suggestions regarding choice of topic and depth of coverage, see the Pacing Chart on pages T28–T31.

*Frame numbers at point of use
(TR) Teaching Resources, Unit 6
(IT) Instructional Transparencies
(LI) Laboratory Investigations
(SD) *Science Discovery* Videodisc Correlations and Barcodes
(SRG) Study and Review Guide

▶ **515a**

CHAPTER MATERIALS

Title	Page	Materials
Discover by Doing	518	(per group of 3 or 4) mixtures of: iron filings and sulfur; fine sand and pebbles; sugar and salt crystals; salt water; copper sulfate and water
Discover by Doing	518	(per group of 3 or 4) shallow glass dish, hot plate, salt water, copper sulfate solution
Teacher Demonstration	520	plaster of Paris, water, pencil or marble
Skill: Problem Solving	522	(per group of 2 or 3) plastic tub, water, cooking oil, graduated cylinder, tissues, liquid soap, wooden sticks or long pencils, plastic wrap
Discover by Doing	524	(per group of 3 or 4) turmeric powder stain on counter top material or white tile, paper towels, soap, water, various detergents, journal
Activity: What are the effects of solvents on tincture of iodine?	525	(per individual) cotton swabs (3), tincture of iodine, water, ethanol
Discover by Problem Solving	528	(per individual) graph paper, pencil, journal
Teacher Demonstration	529	pans (2), warm water, ice water, bottles of carbonated beverage (2)
Discover by Doing	530	(per group of 3 or 4) regular sand, magic sand, 500-mL beaker, water
Discover by Doing	531	(per group of 3 or 4) oil, water, glass jar with lid, blender
Investigation: Separating Mixtures	536	(per individual) safety goggles, laboratory apron (per group of 3 or 4) cotton plug, glass tube (7 cm × 1 cm), activated charcoal, 150-mL beaker, clamp or clothespin, methylene blue solution, medicine droppers (2), copper sulfate solution, test tubes (4), spatula or scoop, "mineral ore," corks (2), 2-mL of mineral oil, wax pencil, unknown liquid, boiling chip, 2-hole stopper with glass tubing and thermometer inserted, ring stand, clamp, iron ring, wire gauze, laboratory burner, ice water, beaker

ADVANCE PREPARATION

For the *Discover by Doing* on page 524, you may be able to find broken or irregular white tiles or counter top material at a discounted price at a home improvement or hardware store. For the *Discover by Doing* on page 530, you can pre-order Magic Sand as Astro sand from Clifford W. Estes Co., Inc., P.O. Box 907, Lyndurst, NJ 07071. For the *Discover by Doing* on page 531, you may wish to ask students to bring in large glass jars with tight fitting lids from home.

TEACHING SUGGESTIONS

Field Trip

A field trip to a local photo processing plant is an excellent addition to your instruction on solutions and emulsions. Have a photo processor explain to the students how various solutions and emulsions are prepared and used in film developing. Ask the processor to explain the safety and precautions that are used when working with these mixtures.

CHAPTER THEME—*ENVIRONMENTAL INTERACTIONS*

This chapter will help the students understand that when there is continual activity on a molecular level in a system and when dynamic equilibrium at the molecular level exists, the system will exist in a balanced state. The balanced state represents an overall stability of the system as it exists without further additions or subtractions of matter or energy. A supporting theme in this chapter is **Systems and Structures.**

MULTICULTURAL CONNECTION

Throughout history, various cultures named different types of matter based on what they thought the matter looked like. This is how many foods were named. For example, the type of food that results from skillfully controlling the interaction of flour with butter, cheese, tomatoes, or spinach and forming this food into long string-like pieces is commonly called *pasta* or *spaghetti* in America.

A popular story tells that Marco Polo introduced pasta to Italy from China. However, at least 50 years before Marco Polo left Italy to travel East, Indians and Arabs were eating pasta. Indians called pasta *sevika*, meaning "thread." The Arabs called pasta *rishta*, which is the Persian word for thread.

MEETING SPECIAL NEEDS

Gifted

Some students might be interested in reading parts of J. Brillat-Savarin's book *The Physiology of Taste*. Translations of this book can be found in many libraries. Have the students prepare a brief report on a section of the book they find particularly interesting. Encourage the students to keep their reports in their science portfolios.

▶ **516** CHAPTER 19

CHAPTER 19 INTERACTIONS OF MATTER

In the early 1800s, a person needed only careful observation to know what he or she was eating. "Tell me what you eat, and I shall tell you what you are," wrote the Frenchman J. Brillat-Savarin in his classic book on eating, *The Physiology of Taste*. Today, however, you need to be a scientist to know what holds your food together.

If you have ever eaten Italian dressing, you probably know that water and oil do not stay mixed. You may not know, however, that at one time people had to knead margarine or else a watery liquid seeped out of the vegetable shortening. Today, chemical additives hold margarine's liquid and oil

Cooking is the art of combining ingredients to make foods look and taste good. In other words, good food results from skillfully controlling the interaction of matter.

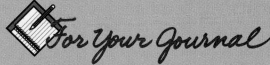
together. The same chemicals, called *emulsifiers,* also keep oil from floating to the top of peanut butter. But you are not likely to find a label that reads "Emulsified Peanut Butter!"

If you've ever drunk chocolate milk, you've probably drunk red algae with it. To keep cocoa suspended in milk, manufacturers must use an additive called a *stabilizer.* One of the most common stabilizers is derived from red algae that grow off the coast of Ireland. The same algal extract is added to those "lean" hamburgers sold at fast food restaurants. Algae burgers! Now you may think twice when you order lunch!

ABOUT THE LITERATURE

Jean Anthelme Brillat-Savarin was a French lawyer and politician who lived from 1755 to 1826. He published several works on politics, economics, and law before his book titled *Physiologie du goût, ou meditation de gastronomie transcendante, ouvrage théorique, historique et l'ordre du jour* was published in 1825. The book was translated into English in 1925 and was thus titled *The Physiology of Taste, or Meditation on Transcendent Gastronomy, a Work Theoretical, Historical, and Programmed.*

The Physiology of Taste is more of a series of witty anecdotes and observations about food than an actual cookbook. Only a few recipes are interspersed throughout the book. Today, *The Physiology of Taste* is mainly appreciated as a historical document that describes the lively appreciation of good food during the late eighteenth and early nineteenth centuries.

For Your Journal

- What are mixtures and compounds? How are they formed?
- What is a solution? What kinds of solutions do you use everyday?
- You use colloids all the time? What do you think colloids are and how might you use them?

FOCUS

In this section, mixtures and compounds are reviewed and homogeneous and heterogeneous mixtures are introduced. Ways of separating mixtures are compared.

MOTIVATING ACTIVITY

 Have the students work in groups of four to play "Matter Jeopardy." One member of each group will be the moderator; the other three will be contestants.

Make a list of 20 compounds and mixtures. Distribute a copy of the list to each moderator, who should read each item. The first contestant to raise his or her hand gets to answer in the form of a question. The master of ceremonies may correct wrong answers. Groups need not keep score since this activity should just get the students to recall prior knowledge of compounds and mixtures and motivate them to learn more.

PROCESS SKILLS
- Experimenting • Solving Problems • Predicting
- Evaluating

POSITIVE ATTITUDES
- Initiative and persistence
- Enthusiasm for science and scientific endeavor

TERM
- compound

PRINT MEDIA
"Serving up Science for Everyday Cooks and Gourmets Alike" from *Smithsonian* magazine by Harold McGee (see p. 513b)

ELECTRONIC MEDIA
The Chemistry Help Series: Elements, Compounds, and Mixtures. Focus Media, Inc. (see p. 513c)

SCIENCE DISCOVERY
- Desalination plant
- Compounds, common

BLACKLINE MASTERS
Connecting Other Disciplines
Laboratory Investigations 19.1, 19.2
Study and Review Guide

DISCOVER BY *Doing*

Display mixtures of: iron filings and sulphur, fine sand and pebbles, sugar and salt crystals, salt and water, and a copper sulfate and water.

① You could place the mixture over a large screen or strainer—the fine sand would fall through, but the pebbles would not.

② The students might suggest the liquids be heated or otherwise evaporated. Accept all reasonable responses.

SECTION 1

Objectives

Distinguish between homogeneous and heterogeneous mixtures, and give examples of each.

Propose a method by which you could separate the parts of a mixture.

Types of Matter: A Review

Have you ever tried to wash a stain from a shirt? Have you ever made salad dressing or fruit gelatin? Maybe you have helped paint the house or some furniture. What do detergent, salad dressing, and paint have in common? They are all matter, but are they alike in other ways?

 ### DISCOVER BY *Doing*

Examine the different substances your teacher gives you. Briefly describe each in your journal. Decide which ones could be made of two or more substances. Explain your choices. ✎

Mixtures

A mixture is a combination of two or more types of matter, each of which keeps its own characteristics. This means that if you separate a mixture, the parts are just the same as when they were put together. Because the parts of a mixture keep their properties, they can be separated by physical means, such as filtering or evaporation. Go back and look at your ideas from the activity. Which of the substances you observed was obviously a mixture? How would you separate the mixture into its components? ①

In the next activity, you will be given two liquids. One is colorless, and the other is blue. Do you think they are mixtures or pure substances? How could you find out? ②

 ### DISCOVER BY *Doing*

CAUTION: Put on safety goggles and a laboratory apron, and leave them on for the entire activity. Pour a small amount of each liquid into a shallow glass dish, and warm each dish gently on a hot plate. What do you observe? Are the liquids pure substances or mixtures? Explain what happened that led you to this conclusion. ✎

● **Process Skill:** *Classifying*

Ask the students to list familiar materials, identify them as mixtures or compounds, and classify each mixture as homogeneous or heterogeneous. For example, salt is a compound, and brass is a homogeneous mixture.

● **Process Skills:** *Generating Ideas, Communicating*

Ask the students to brainstorm examples of separating mixtures and write them in their journals. Guide them to think along the lines of environmental issues such as recycling, pollution control, and waste management.

● **Process Skill:** *Expressing Ideas Effectively*

Point out to the students that because substances that form a mixture can be combined in any amounts, mixtures do not follow the law of definite proportions. The properties of mixtures vary as the amounts of the component substances change. Then ask the students to extend this concept by providing examples. For example, some shirts are made of fabrics that are mixtures (or blends) of 80% cotton and 20% polyester. Other shirt fabrics might be made of 50% cotton and 50% polyester. The physical properties of these fabrics differ.

There are two types of mixtures: homogeneous and heterogeneous. A homogeneous mixture is uniform; it has the same composition through out. When you look at a homogeneous mixture, you might not be able to tell it is a mixture. Which of the mixtures in the activity were homogeneous? ③

A heterogeneous mixture is not uniform throughout. Instead it is made up of different substances that can be easily seen. Vegetable soup is an example of a heterogeneous mixture.

▼ **ASK YOURSELF**

What is the difference between a homogeneous and a heterogeneous mixture?

Cooking the Chemist's Way

Meet Chef Pierre. He boasts that he is the master of all edible matter. He can convert eggs, flour, sugar, chocolate, and a few other choice ingredients into a delicious heterogeneous

Figure 19–1. You encounter many types of heterogeneous mixtures every day, such as this tossed salad.

③ The salt water solution and copper sulfate solution were homogeneous mixtures.

ONGOING ASSESSMENT
▼ **ASK YOURSELF**

A homogeneous mixture is uniform; it has the same composition throughout. A heterogeneous mixture in not uniform throughout; it is obviously made up of different substances.

MEETING SPECIAL NEEDS

Second Language Support
Limited-English-proficient students may have difficulty distinguishing between the terms *homogeneous* and *heterogeneous*. Point out that the prefix *homo-* means "same" and *hetero-* means "different" and that *geneous* means "kind." Encourage the students to share words or expressions from their native languages that denote same and different.

INTEGRATION– Life Science

Many organisms separate mixtures in order to obtain the nutrients they need. For example, some whales use a fringelike structure called baleen to strain food organisms from water. Humans breathe air (a mixture), absorb some of the oxygen, and breathe out the rest.

● **Process Skill:** *Evaluating*

To avoid any misunderstandings, explain to the students that conventional table salt is mined from underground deposits or evaporated from sea water. Ask the students why the method demonstrated by Kimmie is not commonly used to make table salt. (It would be too impractical and too expensive to produce large quantities of table salt.)

Have the students practice separating the mixture of fine sand and pebbles using a fine screen or a strainer. Be sure the students understand they can separate these components by physical means because the original matter was a mixture.

Have the students provide written answers to the Section Review. Then have them list and describe in their journals the compounds and mixtures they have touched or seen today. The students should specify if mixtures were homogeneous or heterogeneous.

Demonstration

Make a wet paste by adding water to some plaster of Paris. Pour the paste over an item such as a pencil or a marble. When dry, the plaster will preserve the shape of the item. Point out to the students that plaster of Paris is calcium sulfate. Calcium sulfate reacts with water to form a hydrate, which hardens quickly:

$$CaSO_4 + 2H_2O \rightarrow CaSO_4 \cdot 2H_2O.$$

Then ask the students whether the physical properties of the plaster change as the plaster hardens. (Yes, the plaster becomes solid.) Finally, guide the students to understand that a chemical change (hydration) has occurred, thus the change in physical properties.

✧ **Did You Know?**

When sea water is evaporated, the salt that remains is called *solar salt*. Evaporating sea water is the oldest method of obtaining salt for use as a seasoning.

LASER DISC

2517, 2518, 2519, 2520

Desalination plant

2368, 2369

Compounds, common

▶ **520** CHAPTER 19

Figure 19–2. This is how chlorine and sodium look when they are pure elements.

Figure 19–3. This is how sodium and chlorine react with each other.

mixture known as a chocolate chip cookie. He can turn cream, broth, and seasonings into the most delicate of sauces. But Chef Pierre can make only mixtures. Even with all his skill, he cannot make a pure compound. His friend Kimmie, a chemist, has asked the chef to visit her lab and watch her make a compound that he uses every day.

Kimmie is very secretive about what she will cook up. She asks Chef Pierre to stand back and watch. "I will be making a compound," she says. A **compound** is a pure substance composed of elements that have been combined *chemically*. The compound has totally new properties that are different from those of the original materials.

Kimmie opens a gas cylinder that is connected to a flask and allows a yellow-green gas to flow into the flask. She covers the flask so the gas will not escape. "This is chlorine gas, and it is very poisonous. That is why we are doing this under the fume hood in the lab." Chef Pierre would rather be in his safe kitchen, but he is interested in this compound Kimmie is making.

Kimmie cuts a piece of a soft metal that has been stored in a liquid. The metal has a grayish sheen. Kimmie tells Chef Pierre that the metal is sodium. "Sodium catches on fire in water; therefore, I have to store it in oil or kerosene."

"Now, we are going to combine these two elements to form a compound that I am sure you have on your kitchen shelf, Pierre." "*Cordon bleu!* That is impossible," exclaims Pierre. "I would not have such dangerous substances in my kitchen."

A Flash and a Bang Kimmie places the piece of sodium on a spoon with a very long handle. She opens the flask containing chlorine and drops the sodium into it, capping it immediately. Chef Pierre is amazed. The piece of sodium ignites with a bright flame. It seems as though the whole flask is on fire. Even though he is not close to it, Chef Pierre knows the reaction is giving off a lot of heat. The green gas is also disappearing. Whatever was inside the flask is turning colorless!

Kimmie opens the flask under the hood to get rid of any leftover chlorine. She asks Pierre to come closer and look inside the flask. "What do you see? Look closely."

Pierre notices that there are small, colorless crystals around the inside of the flask. "Are these little crystals in my kitchen?" he asks.

Kimmie laughs. "The compound that you see is made of sodium and chlorine. I call it sodium chloride; you, Pierre, call it table salt."

"Fondue! Is it possible that substances can change so much when they combine?" Kimmie explains that chemists perform even more amazing feats every day.

Breaking Up Is Hard to Do
Chef Pierre looks worried. Will his table salt break down into its component elements and burst into flame? Kimmie assures him that while mixtures can be separated by physical means such as filtration, evaporation, or distillation, the elements in a compound cannot be separated in this way. In order to separate the elements in a compound, chemical changes involving energy are required.

There is another difference between a compound and a mixture. Substances that form a mixture can be combined in any amounts. For instance, you can add a pinch of sugar to a cup of tea, or you can add several spoonfuls. The result is still sweetened tea. Compounds, however, can be made only by combining elements in definite ratios. A water molecule, for example, is formed by combining two atoms of hydrogen with one atom of oxygen. Water cannot be formed with any other ratio of hydrogen and oxygen. Furthermore, once formed, water cannot be easily reseparated into hydrogen and oxygen.

Figure 19–4. Two atoms of hydrogen and one atom of oxygen form water. Two atoms of hydrogen and two atoms of oxygen form hydrogen peroxide, a common disinfectant.

  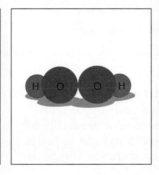 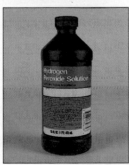

▼ **ASK YOURSELF**

How is making a compound different from making a mixture?

SECTION 1 *REVIEW AND APPLICATION*

Reading Critically
1. Give two examples of heterogeneous mixtures that might be found in your home.
2. Compare and contrast mixtures and compounds.

Thinking Critically
3. How would you separate a mixture of sand, sugar, oil, and water?
4. Why do you think the properties of sodium chloride are so different from the properties of the elements that are combined to make it?

Process Skills: Experimenting, Solving problems, Predicting, Evaluating

Grouping: Groups of 2 or 3

Objectives:
- **Design** a method to remove oil from an oil-water mixture.
- **Predict** the success of the chosen method.
- **Evaluate** the results of the chosen method, and revise the method, if necessary.

Procedure

3. There will probably be some oil left on the water. Accuracy of predictions of effectiveness will depend on the students' predictions.

▶ **Application**

1. The students' answers should indicate that the oil and water are a heterogeneous mixture and can be separated by physical means.

2. The students' revised plans should be well thought out. Accept all reasonable responses.

3. One practical application is to clean up oil spills from tankers in lakes or oceans.

※ **Using What You Have Learned**

This miniature oil spill is similar to an oil spill in the ocean. In both cases, the oil separates and floats on top of the water and can be removed by physical means. The experiment differs from a real oil spill because there are no currents in the pan to move the oil and cooking oil is much thinner than crude oil.

SKILL *Problem Solving*

▶ **MATERIALS**
- plastic tub ● water ● cooking oil, 200 mL ● graduate ● tissues
- liquid soap ● wooden sticks or long pencils ● plastic wrap

▼ **PROCEDURE**

1. Fill a plastic tub half-full of water. Put 200 mL of cooking oil into the water.

2. Using the equipment provided, design a plan to remove the oil from the water. Write down the steps you will use in your cleanup operation. Predict how effective your procedure will be.

3. Carry out your plan. Does any oil remain in the water? How does your result compare with your prediction?

4. Revise your plan. Predict how changes in your plan will affect how much oil is removed from the water.

▶ **APPLICATION**

1. How did you use what you know about mixtures to devise your plans?

2. How accurate were your predictions? What knowledge did you use to revise your plan and to increase the accuracy of your predictions?

3. What are some practical applications of this procedure in the real world? How could your ideas be used to solve environmental problems?

※ **Using What You Have Learned**

How is this miniature oil spill like an oil spill in the ocean? How is it different?

Solutions

C̲hef Pierre and Kimmie are sitting down to a nice glass of iced tea. Kimmie notices that Chef Pierre is putting three spoonfuls of sugar in his tea! She laughs. "Pierre," she says, "most of that sugar will end up on the bottom of the glass. Trust me! We chemists know these things."

Making Solutions

A soft drink is a mixture of water, flavorings, and carbon dioxide gas. A soft drink, however, looks like a single material; it is homogeneous. A homogeneous mixture is called a **solution**. Every solution has two parts, the solvent and the solute. The **solvent** is the material in which the solute is dissolved. Because there is always more solvent than any other component in a solution, the solvent is the major component. In a soft drink, water is the solvent.

The **solute** is the substance that is dissolved in the solvent. It is the minor component. The gas and flavorings in a soft drink are the solutes.

Although many common solutions are made from solutes added to liquids, solutions can also be combinations of liquids, solids, and gases. Table 19–1 lists examples of solutions.

Table 19–1 Types of Solutions		
Example	**Solvent**	**Solute**
Air	Gas	Gas
Soda water	Liquid	Gas
Silver amalgam	Solid	Liquid
Brass	Solid	Solid

SECTION 2

Objectives

Define and **give** examples of a solution, a solute, and a solvent.

Describe factors that affect the rate at which a solid dissolves in a liquid.

Interpret a solubility graph to determine whether a solution is saturated, unsaturated, or supersaturated.

◣ **ASK YOURSELF**
Give an example of a solid/liquid solution not mentioned in the section. Tell which component is the solute and which is the solvent.

Figure 19–5. This soda water is a solution of carbon dioxide gas in water.

● **Process Skill:** *Classifying*

Ask the students to list familiar solutions and classify them based on the phase of the solute and solvent, as in Table 19–1. Help the students understand that solutions are not just liquids with something solid mixed in. Solutions can be mixtures of a solid, liquid, or gas solute in a solid, liquid, or gas solvent.

● **Process Skill:** *Inferring*

Mix equal amounts of vegetable oil and water in a jar with a tight-fitting lid. After the students have observed the mixture for a few minutes, have one student add several drops of liquid soap to the mixture and shake it well. After the students have a chance to observe the effect of soap on the mixture, ask them why they think soap is used to wash greasy hands and clothes. (The soap helps the grease dissolve—like dissolves like— therefore, the soap molecules must be similar to the grease molecules.)

DISCOVER BY *Doing*

CAUTION: Do not use bleach because chlorine and ammonia react to form a poisonous gas.

Rub tumeric powder on some pieces of counter top material or white tiles with a wet paper towel. Be sure the yellow stain is quite visible. The students should try water and soap first. Ammonia-based detergents work to some degree, but do not take out the stain completely. They do, however, change the color of the stain to a pinkish red since turmeric is a natural acid-base indicator. Rubbing alcohol will remove the stain.

SCIENCE BACKGROUND

Whether one substance will dissolve in another depends on the distribution of electrons in the two substances. In some molecules, one part of the molecule has a slight positive charge and another part has a slight negative charge, although the molecule as a whole is neutral. Molecules like this are called polar molecules.

Water molecules are polar and therefore readily dissolve other polar molecules. For example, the positive ends of water molecules attract negative chloride ions in a crystal of sodium chloride and pull the chloride ions from the crystal—thus dissolving it. The negative ends of water molecules attract the positive sodium ions in a crystal of sodium chloride and pull the sodium ions from the crystal— thus dissolving it.

What's the Solvent?

Chef Pierre is back in his kitchen trying to solve a problem. One of his helpers has spilled an Indian spice called turmeric, and it has left a yellow stain on his white counter top. Your teacher has a similar counter top material with a turmeric stain. See if you can get it clean.

DISCOVER BY *Doing*

Try to get the stain out. In your journal, keep a careful record of what you do and the results you achieve. 🖋

Now you can understand Chef Pierre's problem; turmeric is a hard stain to get out. Chef Pierre calls Kimmie and explains his problem. According to Kimmie, scientists have a rule that might help: Like dissolves like. That is, chemically similar substances dissolve in each other. Nail polish, for example, does not dissolve in water. This is actually a good thing, because if

● **Process Skill:**
Communicating

Ask the students to list as many uses of the words *concentrated* and *diluted* as they can. Then have them use the words in sentences that explain their usage. (Answers include such uses as concentrated frozen fruit juice that is diluted when water is added.)

● **Process Skill:** *Generating Ideas*

Many materials used in arts and crafts, such as glue, paints, and wood finishes, use solvents (thinners) other than water. Many of these solvents are volatile and potentially toxic. These solvent containers have warning labels that alert the user to use the solvent only in a well ventilated area. Ask the students to generate ideas for preparing well ventilated areas. (Answers include opening windows and doors, using a fan to carry inside air outside, or even using these solvents outdoors.)

it did, you couldn't wash your hands without the nail polish coming off. To take off the nail polish, you use a chemical that is similar in structure to the nail polish—nail polish remover.

"The principal ingredient in turmeric," says Kimmie, "is an organic compound. Organic compounds contain carbon. You can probably get the stain out if you use another organic compound." The chef asks what organic compound he might have in his kitchen. "Do you have some cooking sherry? Try that."

You cannot use sherry in the classroom, but you can use rubbing alcohol. Alcohol is an organic compound, similar to the turmeric. How well did it work on the stain? Do you think that Pierre will get good results using his cooking sherry? ①

Water is sometimes called a universal solvent. This is because water can dissolve many different substances. Did water dissolve the turmeric stain? Finding solvents for substances that are not water soluble is important to the cleaning industry. Many dry cleaners have used fluorocarbons or chlorocarbons as stain removers. Since some of these compounds are either toxic or harmful to the ozone layer, they might have to be replaced.

Figure 19–6. Dry cleaners use solvents to remove stains.

① It should work since sherry contains ethanol—another alcohol.

ACTIVITY

What are the effects of solvents on tincture of iodine?

MATERIALS
cotton swabs (3), tincture of iodine, water, ethanol

PROCEDURE
1. Use a cotton swab dipped in tincture of iodine to make two small spots on the palm of your hand.
2. Dip a second cotton swab in water, and wash one iodine spot with it. Dip another swab in ethanol and wash the other spot.

APPLICATION
1. What happened when you put water on the iodine spot?
2. What happened when you put ethanol on the iodine spot?
3. Which solvent removed the iodine spot better? Why?

Removing stains is quite a science when you think of it. You have to find the proper solvent to remove a particular stain without bleaching away the color of the stained material or eating through it. Many industrial chemists dedicate long hours to the manufacture of good detergents and cleaners.

▼ **ASK YOURSELF**
What is meant by the "like dissolves like" rule?

ONGOING ASSESSMENT
▼ **ASK YOURSELF**
The "like dissolves like" rule tells that substances that are chemically similar dissolve in each other.

● **Process Skill:** *Constructing Models*

Cooperative Learning Have the students work in groups to construct a model of dynamic equilibrium in a saturated solution. The students may choose their own materials for the model,

but one possible model could use pencils to represent solute molecules and an empty beaker to represent the solvent. For example, for each pencil that is "dissolved" in a saturated solution, another pencil must "come out of solution." Ask the students to explain the process as they present their model to the

class. Encourage the students to be creative, but make sure the models depict the concept of dynamic equilibrium correctly.

REINFORCING THEMES
Systems and Structures

Lead the students to understand that a saturated solution is stable with respect to the number of molecules of solute that can be dissolved in the solvent. Even though more solute may be added to a saturated solution and dissolve, an equal amount of solute will come out of solution. This dynamic equilibrium is constant only as long as the temperature and pressure of the solution remain constant. Therefore, a saturated solution is stable, but only at a specific temperature and pressure.

① As the water evaporates, the sugar returns to its original state, crystals.

MEETING SPECIAL NEEDS

Second Language Support

The word *solution* has two common meanings in the English language—a liquid mixture and an answer or resolution to a problem. Help limited-English-proficient students understand both meanings.

Figure 19–7. Rock candy is formed when sugar crystalizes out of a saturated solution.

Figure 19–8. When a sugar cube is put into water, the water molecules surround the sugar molecules, causing them to go into solution, or dissolve.

Figure 19–9. The watch glass (left) contains crystals of copper sulfate that were formed after a solution of copper sulfate in water was evaporated. In the salt pond (right), the same principle is used to retrieve salt from sea water.

A Little or a Lot

Chef Pierre is making some rock candy to decorate one of his desserts. In order to make the candy, Pierre must dissolve as much sugar as he can in very hot water. He needs a saturated solution. Solutions can be *dilute* (not much solute) or *concentrated* (a lot of solute). A **saturated solution** has the maximum amount of solute that can be dissolved in a given amount of solvent at a given temperature.

When Pierre sees that no more sugar will dissolve after he stirs his mixture, he knows he has a saturated solution. The next step is to allow the solution to cool untouched for a few days.

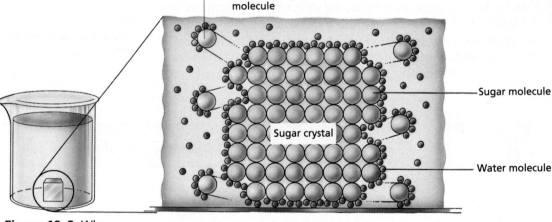

He places a weighted string in the jar holding the sugar syrup. Sugar crystals will form on the string. They will be large and beautiful and—of course—edible. They will look like diamonds on his award-winning chocolate torte. What makes the sugar come out of solution and form crystals? ①

A solution is saturated when no more solute seems to dissolve. In reality, however, some solute continues to dissolve. For example, salt added to saturated salt water still dissolves in the

water. What happens is that for every grain of salt that dissolves, another grain comes back out of solution. The solution has reached equilibrium. Equilibrium in a solution is a state of balance between what is going into solution and what is coming out of solution.

A saturated solution contains as much solute as the solvent can hold at a given temperature. In the case of sugar water, you may wonder how much sugar makes the solution saturated. Look at the graph in Figure 19–10. What usually happens when you increase the temperature of a solution? For most substances ② the amount of solute that can be dissolved increases. There are a few exceptions to the rule, as you can see with $Ce_2(SO_4)_3$.

The maximum amount of solute that can be dissolved at a specified temperature and pressure is known as **solubility**. Solubility is usually expressed as grams of solute in 100 g of solvent.

Which of the solutes in Figure 19–10 dissolves most easily in hot water? Which dissolves most easily in cold water? Which ③ ④ is the least affected by temperature change? Each curve on the ⑤

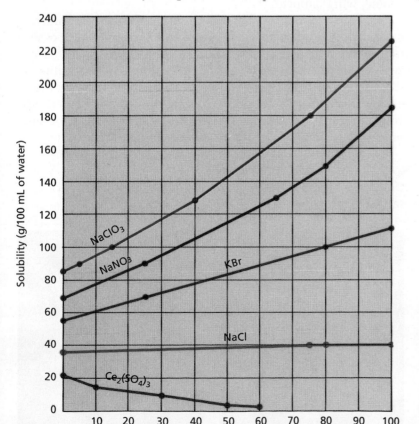

② The solubility of the solution increases.

③ NaClO₃

④ NaClO₃

⑤ NaCl

THE NATURE OF SCIENCE

Computer simulations have helped scientists answer many perplexing chemical questions. One such question is why nonpolar solvents are better than water in dissolving electrically neutral gas molecules, such as methane. Two professional chemists, Lawrence R. Pratt of the Los Alamos National Laboratory and Andrew Pohorilles of the NASA Ames Research Center, have used computer simulations of water and five nonpolar solvents to calculate the likelihood of finding an atomic-sized cavity into which a solute molecule would fit. They found that although water has more overall cavity space than nonpolar solvents, its cavity spaces are smaller and inflexible. The chemists concluded that water apparently has more difficulty than nonpolar solvents in rearranging its molecules to make room for small solute molecules.

Figure 19–10. This graph shows the solubility (in water) of several substances at different temperatures. Notice that most of the substances have higher solubilities at higher temperatures.

● **Process Skill:** *Comparing*

Some students might have difficulty identifying which substance is the solvent and which is the solute in a solution. Tell the students that if the two substances were originally in different phases, the substance that appeared to change phase is the solute. Then encourage the stu-

dents to think of examples that reflect this concept. (For example: Any solutions made with liquid water always results a liquid; the water is the solvent.) In a case in which two substances of the same phase dissolve in one another, the substance that is present in the larger amount is usually considered the solvent.

DISCOVER BY
Problem Solving

The students might have difficulty relating the concept shown in Figure 19–10 to Chef Pierre's rock candy. Help them use the graph to see what happens to a saturated solution if the temperature is lowered. Point out that the solution will become supersaturated. That is, the solute cannot stay dissolved in the solvent at that temperature. As Chef Pierre's sugar solution cooled, the sugar could not stay dissolved in the water (it crystallized out), and the rock candy formed.

ONGOING ASSESSMENT
▼ ASK YOURSELF

One possible answer is wanting to have very sweet iced tea, but not wanting to drink the gritty sugar that would not dissolve.

● INTEGRATION–
Life Science

Fish get the oxygen they need from air dissolved in water. As water moves past a fish's gills, oxygen diffuses into the gill capillaries. When the water is warm, less oxygen stays dissolved in it. Trout need lots of oxygen because of their active lifestyle. They can therefore only live in colder water. Sometimes fish become trapped in shallow water that has been heated by the sun and, therefore, does not contain much available dissolved oxygen. If fish cannot get enough oxygen to carry out cellular respiration, they suffocate and die.

Figure 19–11. A supersaturated solution (left) contains more solute than it can hold under normal circumstances. Scratching the side of the beaker will force the solute to come out of solution (right).

---

graph shows the maximum amount of solute that can be dissolved at a given temperature. The area below each curve gives the amount of solute in an unsaturated solution. Above each curve is the area indicating that no more solute would dissolve. Solubility is a characteristic property of each substance.

DISCOVER BY *Problem Solving*

Why do you think Chef Pierre has to let his saturated sugar solution cool in order to get crystals? Sketch a graph showing what happens, and write the explanation in your journal. ✐

It is sometimes possible to dissolve more solute in a solvent than you could normally. When this happens, the solution is *supersaturated*. Supersaturated solutions are not stable because they are no longer in equilibrium. As Chef Pierre's solution cooled, what happened?

A supersaturated solution is similar to a tightrope walker teetering on a rope before falling into the safety net. It doesn't take much of a push to cause the tightrope walker to fall. Similarly, it doesn't take much to make the solute in a supersaturated solution come out of solution. Often, you can bring the excess solute out of solution by cooling the solvent. Sometimes you can bring it out by stirring the solution, scratching the side of the container, or by adding just one crystal of the solute.

▼ ASK YOURSELF

Describe a condition under which you would want to have a saturated solution.

GUIDED PRACTICE

 Have the students write one or two paragraphs in their journals that compare and contrast the characteristics of dilute, concentrated, and saturated solutions. Encourage the students to read their paragraphs aloud and revise their paragraphs based on your and classmates' feedback.

INDEPENDENT PRACTICE

 Have the students provide written answers to the Section Review. Then have them list in their journals the times that they dissolved something recently. Help the students identify the solute, solvent, method of dissolving, relative temperature of the solutions, and the use for the resulting solution. Encourage the students to share and compare their lists with classmates.

EVALUATION

Have the students describe saturated, unsaturated, and supersaturated solutions in terms of solubility. (The maximum amount of solute that can be dissolved in a particular amount of solvent at a specific temperature and pressure—saturated; less than the maximum—unsaturated; and more than the maximum—supersaturated.)

Now You See It, Now You Don't

The dissolving of a solute can be affected by many factors. You are probably aware of at least three but have never really thought about them. Think about the answers to these questions; they may help you identify the three factors. Which is easier to dissolve in water—a spoonful of sugar or a sugar cube? When adding sugar to lemonade, what do you do to speed up its dissolving? Does sugar dissolve faster in iced tea or hot tea?

Let's see how many correct answers you have. The speed of dissolving can be increased by powdering the solid or breaking it into small pieces. When you make the pieces smaller, more of the solute is exposed to the solvent. So, a spoonful of sugar dissolves more quickly than a sugar cube.

Stirring also brings more of the solute into contact with the solvent and breaks up clumps of the solute. Therefore, stirring is another way to speed up the rate of solution. You stir your lemonade to dissolve the sugar.

Figure 19–12. The rate of solution can be increased by several methods, including stirring.

As you saw in Figure 19–10, an increase in temperature often increases solubility. It can also make the solute dissolve faster. An exception to the effects just described is the dissolving of a gas in a liquid. If you have ever opened a soda bottle that is warm or has been shaken, you know that the dissolved gas fizzes out of the bottle, spilling some of the liquid. When that happens, it shows that the gas is coming out of solution. Shaking or stirring the liquid causes the gas molecules to escape. Agitation allows more of the bubbles to reach the surface.

> ### Demonstration
>
> Most carbonated beverages contain dissolved carbon dioxide in the liquid portion of the beverages. Demonstrate how temperature can affect the amount of solute that a particular solvent can hold by setting up a pan of very warm water and a second pan of ice water. Place an unopened bottle of carbonated beverage in each pan, and allow the bottles to sit for about 10 minutes. Then have the students observe as you open both bottles. Ask then to explain what happened. (When the bottles were opened, the warm bottle overflowed as the carbon dioxide rushed out. Most of the carbon dioxide in the cold bottle remained in solution. The cold beverage was able to hold more dissolved carbon dioxide than the warm beverage.)

Cooperative Learning Have the students work in pairs. Provide each pair with a sugar cube, a beaker filled with water, and a stirring rod. Have the students prepare a sugar-water solution. As they prepare the solution, ask them the following questions: What does stirring do? (helps break up and dissolve the sugar cube) When is your solution unsaturated? (when more sugar can be dissolved in it) When is your solution saturated? (when no more sugar can be dissolved in it)

Solubility is a characteristic property of a substance. A chemist trying to identify an unknown substance would measure several characteristic properties. Encourage interested students to use library resources to research other identification methods, such as spectroscopy and chromatography.

Have the students prepare a graphic outline of the section. Outlines should include the following concepts: like dissolves like; the solubilities of most solids increase at higher temperatures, but the solubilities of gases decrease; stirring, increased temperature, or decreased particle size increases the rate of solution.

★ DISCOVER BY *Doing*

Magic sand may be purchased as Astro Sand from Clifford W. Estes Co., Inc., P.O. Box 907, Lyndhurst, NJ 07071. A simpler version can be made by treating ordinary sand with several applications of Scotchguard and allowing it to dry.

You might introduce the students to the word *hydrophobic*. Ask a volunteer to find the word's etymology in a dictionary. Then discuss with the students the meanings of the prefix *hydro-* and the suffix *-phobic*.

ONGOING ASSESSMENT
▼ ASK YOURSELF

The solubility of a gas in a liquid can be increased by decreasing the temperature.

SECTION 2 REVIEW AND APPLICATION

Reading Critically

1. a. solvent: water; solute: carbon dioxide
    b. solvent: nitrogen; solute: other gases
    c. solvent: water; solute: salt

2. Crushing and stirring bring more solute molecules into contact with solvent molecules.

Thinking Critically

3. Ice cubes made from cold water are cloudy because dissolved gas gets trapped inside. Hot water has less dissolved gas.

4. a. unsaturated
    b. supersaturated
    c. saturated

Figure 19–13. Factories and power plants sometimes return heated water to lakes and rivers.

Sometimes factories and power plants use water from rivers and lakes to cool their machinery. If this heated water is not cooled before being returned, the river's or lake's water temperature can be increased. This is called *thermal pollution*. Because less gas can stay dissolved in warmer water, the amount of oxygen dissolved in the water is reduced. This can have a drastic effect on the fish and other living things in the water that depend on the dissolved oxygen to stay alive.

★ DISCOVER BY *Doing*

Pour regular sand into a 500-mL beaker half-filled with water. Describe what happens. Do the same with some "magic sand." Describe what happens. Did you expect this? Can you explain what is so special about magic sand?

Try to separate the regular sand from the water by pouring the water into another beaker. Do the same thing with the magic sand. Describe what you observed. Relate what you observed to what you know about solutions and solubility. ✏

▼ ASK YOURSELF

How can you increase the solubility of a gas in a liquid?

SECTION 2 *REVIEW AND APPLICATION*

Reading Critically

1. Identify the solvent and solute in each of the following solutions:
    a. carbonated water
    b. air
    c. salt water

2. Why do crushing and stirring speed up dissolving?

Thinking Critically

3. Why do you think that ice cubes made with cold water are often cloudy, while ice cubes made with hot water are usually clear?

4. Using Figure 19–10, determine whether the solutions listed below are saturated, unsaturated, or supersaturated.
    a. 40 g of $NaNO_3$ in 100 mL of water at 40°C
    b. 70 g of KBr in 100 mL of water at 20°C
    c. 20 g of NaCl in 50 mL of water at 100°C

Section 3:

SUSPEN-SIONS AND COLLOIDS

FOCUS

In Section 3, the properties and uses of heterogeneous mixtures—suspensions and colloids—are presented. The differences between suspensions and colloids in terms of size and constant random motion of molecules are discussed. Three types of colloids are described: gels, aerosols, and emulsions.

MOTIVATING ACTIVITY

Darken the classroom. Have one student clap two chalkboard erasers together while another student shines a flashlight on the chalk dust. Ask volunteers to describe what they see and classify the chalk dust in the air as a homogeneous or heterogeneous mixture. (heterogeneous)

Next, mix a handful of fine sand in a jar of water. Place the jar on a desk, and have the students observe this mixture for about five minutes. Ask the students to describe what they see and then classify this mixture as homogeneous or heterogeneous. (heterogeneous) Finally, have the students describe the components of both mixtures in terms of settling particles.

Suspensions and Colloids

SECTION 3

Chef Pierre's assistant is making salad dressing. He mixes oil, vinegar, and spices for an Italian dressing. As soon as he stops shaking it, the dressing separates into two layers and the parsley and oregano drift down to the bottom. "Chef, I will have to shake vigorously each time I serve the dressing. I have 50 customers to please—I am a chef, not a weight lifter—can you help me?" Chef Pierre proceeds to show his assistant how to prepare a creamy salad dressing that will not separate as easily. He takes the oil and vinegar mixture and puts it in the blender for a few minutes. "*Voilà!*" says Pierre. "That should take care of the problem. This dressing should remain creamy for the duration of the banquet."

Objectives

Distinguish among a suspension, a solution, and a colloid.

Describe and **give** examples of different types of colloids.

Suspensions

What happened to the oil and vinegar mixture as Chef Pierre mixed it in the blender? You can try a similar experiment.

DISCOVER BY *Doing*

Combine a small amount of oil and water in a container and shake it. What happens? Now take the same oil and water, put it in a blender, and whip it for a few seconds. Describe any changes. What has happened? Will the oil and water separate? Observe for several hours or overnight. 🖊

What you have made is a suspension. A **suspension** is a heterogeneous mixture in which one of the parts is a liquid. Visible particles will settle out of any suspension. This is often because the two substances are insoluble in each other. That is, one will not dissolve in the other. Remember oil and water do not mix. Is the oil-water mixture that you placed in the blender a solution or a suspension? How do you know? ①

DISCOVER BY *Doing*

Use large, clear plastic jars with tight fitting lids for shakers.

① The oil-water mixture is a suspension because the oil is insoluble in water.

● **Process Skills:** *Inferring, Expressing Ideas Effectively*

Have the students offer other examples of suspensions. (Examples include muddy water, fruit juice with pulp, and vegetable soup.) Then ask the students whether colored water is a suspension, and help them devise a method to test their answer. (Colored water is not a suspension because after sitting for long periods of time, no visible particles settle out.)

● **Process Skill:** *Inferring*

Ask the students to name a disadvantage of medicines that are suspensions. (If these medicines are not shaken very thoroughly, the patient will not get the correct amount of each ingredient.)

MEETING SPECIAL NEEDS

Gifted

Have gifted and talented students research the use of medical suspensions and prepare a report on their findings. Have them include in their reports types of medicines that commonly are suspensions, such as eardrops.

◇ **Did You Know?**

Stabilizers are used to form stable emulsions in foods such as salad dressings, pickle relish, and citrus juices. One such stabilizer is a plant gum called *tragacanth*. This plant gum is derived from a small Astragalus bush that grows in dry, mountainous areas of Turkey and Iran.

MULTICULTURAL CONNECTION

Ancient Mesopotamia was a land of technology. By 3000 B.C., the Sumerians had made the first soap from vegetable oils. To make other substances used for magic rituals, cosmetics, or medicine, Sumerian chemists used filtering vessels, crucibles, and drip bottles. They also made perfume and incense, often guided by women who developed the formulas.

Figure 19–14. In order to make different colors of paint, pigments are suspended in a liquid.

Figure 19–15. The microcrystalline polymer suspension shown here can be made from old plastic bags.

You have probably used thousands of suspensions during your life. They are everywhere. Suspensions are used frequently as medicines. Liquid medicines that require shaking before they are taken are suspensions. Some paints contain solid pigments in a liquid. These pigments must be finely ground so that the paint does not separate readily. You may have noticed that paint stores have special machines that shake the cans of paint in order to mix the components.

The usefulness of suspensions is demonstrated by the many applications of MCP (microcrystalline polymer) suspensions. MCP suspensions were discovered by American chemist and inventor Dr. Orlando A. Battista. He discovered that he could break down synthetic materials such as nylon and polyester into tiny crystalline structures—MCPs—with the aid of acid.

MCP suspensions can cover hard-to-coat surfaces such as glass and aluminum. When sandwiched between two pieces of glass, MCP suspensions bind the sheets together, producing one of the cheapest forms of safety glass. Suspensions of MCPs in water result in a substance similar to petroleum jelly. These suspensions can be used as greaseless cosmetic bases. The best thing about MCPs is that they are made from waste products such as discarded nylon stockings and plastic bags!

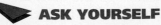

▼ **ASK YOURSELF**

What is a suspension?

Colloids

Eventually, particles in suspensions settle out. What if we could make the particles small enough so that they would stay in solution? There are mixtures like these; they are called **colloids**. In a colloid, the particles are very, very small. So, even though they do not dissolve, they do not settle out nor can they be filtered out. The reason for this is that colloidal particles are in constant random motion. Because of this motion, the particles do not settle out of the mixture.

Chef Pierre will share his recipe for a particular colloid. Look through Chef Pierre's recipe. Are there any ingredients that will not dissolve in each other? What can you do about it? What is Chef Pierre's colloid normally called? ①

Chef Pierre's Colloidal Sandwich Spread

1 egg
1 tablespoon lemon juice
1 teaspoon salt
1/4 teaspoon pepper
1/2 cup oil

Combine first four ingredients in a blender or food processor. Gradually add oil until mixture becomes colloidal.

Figure 19–16.
Mayonnaise is an example of an emulsion. The chef shown here is adding oil to the mayonnaise he is making.

SCIENCE BACKGROUND

Colloids were studied around 1860 by Thomas Graham. He found that substances such as glue, gelatin, starch, and albumin (a protein in egg white) diffuse much more slowly than do salt and sugar. When Graham put a mixture of sugar and glue in a cellophane bag in running water, the sugar soon passed through the bag, leaving the glue behind. This process is called *dialysis*.

Graham also found that he could not crystallize the "slow" substances, which he named *colloids*. He thought there was a fundamental difference between colloids and ordinary substances. Chemists now know that the difference is in the size of the molecules.

SCIENCE BACKGROUND

Colloidal particles are in Brownian motion, that is, they are being hit by solvent molecules in constant random motion. They are colloids because the collisions with the solvent molecules overcome the pull of gravity on very tiny particles.

Cooperative Learning Have the students work in pairs, and have them take turns naming a suspension and a colloid that are eaten and a suspension and a colloid that are used as cosmetics or cleansers. Examples include oil-and-vinegar salad dressing (suspension), gelatin dessert (colloid), liquid dishwasher detergent (suspension), and hair gel (colloid).

INDEPENDENT PRACTICE

Have the students provide written answers to the Section Review. In their journals, have the students describe a suspension and a colloid they have recently purchased and the uses for these mixtures.

EVALUATION

Have the students choose one suspension and one colloid, describe their uses and properties, and explain why each is classified as it is. (Answers should reflect an accurate understanding of suspensions and colloids.)

Tiny liquid and solid particles that enter the atmosphere from natural sources and from human-made sources, such as burning fossil fuels, are termed *aerosols.* These aerosols reflect sunlight back out to space, which causes a cooling of Earth's surface. The largest concentrations of aerosols are found over the industrialized regions in the Northern Hemisphere. Have the students predict the effects of the aerosol concentration and write their ideas in their journals.

Figure 19–17. Gelatin (above) and agar (right) are both examples of gels.

There are probably many colloids at your home right now. Gels, aerosols, and emulsions are all forms of colloids. *Gels* are liquid particles spread out in a solid. Gelatin, jelly, and stick deodorant are gels. *Emulsions* are colloids made of two liquids. You are familiar with many oil-water emulsions such as mayonnaise, hand cream, and milk.

Another type of colloid is an *aerosol.* Aerosols are formed when either solid or liquid particles are suspended in a gas. Fog and smoke are examples of aerosols. A spray can shoots out a colloid of particles finely dispersed in a gas. Chef Pierre uses an aerosol nonstick cooking spray when he cooks.

You might wonder how anyone could tell the difference between a colloid and a true solution. They both seem homogeneous. There is one easily observable way: colloids scatter light that passes through them and solutions do not. You can see this phenomenon in the photographs in Figure 19–18.

Figure 19–18. The photograph at the top shows the light of a slide projector shining through one glass containing a solution and one containing a colloid. You can see the beam of light in the glass containing the colloid. Fog (below) is a natural colloid; it scatters light as the sun's rays pass through it.

ASK YOURSELF

How are colloids different from solutions? How are they the same?

SECTION 3   REVIEW AND APPLICATION

Reading Critically

1. List five colloids that might be found in your home.
2. What are two differences between a colloid and a suspension?

Thinking Critically

3. How could you determine experimentally if a liquid is a solution, a suspension, or a colloid?
4. What advantage would a medicine in the form of an emulsion have over a medicine that is a nonemulsified suspension?
5. Natural milk contains cream that forms a layer when it rises to the top. The milk you buy at the store does not separate into layers because it has been homogenized. What do you think is involved in the process of homogenization?

SECTION 3   **535** ◀

Separating Mixtures

Process Skills: Analyzing, Comparing, Experimenting, Inferring, Observing

Grouping: Groups of 3 or 4

Objective:
- **Separate** mixtures by filtration and distillation.

 CAUTION: Copper sulfate may be toxic if ingested.

CAUTION: Be sure that the students' distillation apparatus is tightly capped to avoid fire hazards. The liquid collected in test tube B should not touch the end of the glass tube.

Hints

Part A: methylene blue solution should be light blue, 0.05 g or less per 500 mL of water.

Part B: the "mineral ore" is fine sand and ground Pb_3O_4.

Part C: suitable solutions include water with methanol and methanol with isopropyl alcohol.

To save time, prepare the stopper setup in advance. The tube should not extend far into test tube A. Adjust the length so that the tube goes into test tube B at least halfway.

Pre-Lab

Explain that this procedure resembles a process used in the mining industry. Ask students to hypothesize why this might be true.

Analyses and Conclusions

1. In part A, the methylene blue solution became clear, but the copper sulfate

solution stayed blue. In part B, the "ore" was held up in the oil foam, while the sand sank. In part C, the temperature remained steady as one substance in the mixture distilled out, then rose to the boiling point of the next substance and remained constant again.

2. to cool the vapors, causing them to condense

3. Answers will depend on the solutions used.

▶Application

Students have learned which processes are most effective in separating different mixtures. Accept all reasonable results.

✳ Discover More

Students' results will depend

upon the solutions and filtering methods chosen.

Post-Lab

To separate minerals from rock particles, air is blown through an oil-water mixture that contains finely ground ore. The "oil foam" absorbs the minerals. Have students revise their ideas about mining techniques using what they have learned.

INVESTIGATION

Separating Mixtures

▶ MATERIALS

PART A
- safety goggles ● laboratory apron
- cotton plug ● glass tube (7 cm × 1 cm)
- activated charcoal ● beaker, 150 mL ● clamp or clothespin ● methylene blue solution ● droppers (2) ● copper sulfate solution

PART B ● test tubes (2) ● spatula or scoop
- "mineral ore" ● corks (2) ● mineral oil, 2 mL

PART C ● test tubes (2) ● wax pencil ● unknown liquid ● boiling chip ● two-hole stopper with glass tubing and thermometer ● ring stand ● clamp
- iron ring ● wire gauze ● laboratory burner
- beaker ● ice water

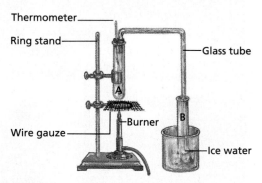

Thermometer — Ring stand — Glass tube — Burner — Wire gauze — Ice water

CAUTION: Put on safety goggles and a laboratory apron, and leave them on throughout the investigation.

▼ PROCEDURE

PART A
1. Place a cotton plug in one end of the glass tube. Add enough charcoal through the other end to reach a height of 2.5 cm. Suspend the tube over the beaker.
2. Using a dropper, fill the tube with the methylene blue solution. Collect the filtered liquid in the beaker.
3. Use the same dropper to collect the filtered liquid and put it through the tube again. Repeat until you observe a color change. Record your observations.

4. Rinse all the equipment, and repeat steps 1–3 using copper sulfate solution.

PART B
5. Fill two test tubes halfway with water. Use the spatula to add a pinch of the "mineral ore" to each tube. Stopper and shake.
6. To one of the test tubes, add 2 mL of mineral oil. Stopper and shake. Let the test tubes stand until you see two layers in the test tube with the mineral oil. Compare the test tubes.

PART C
7. Label one test tube "A" and one test tube "B."
8. Fill test tube A one-fourth full of the unknown liquid. Add a boiling chip.
9. Set up your equipment as shown. Be sure the glass tubing is not plugged up in any way.
10. Heat the liquid gently until it boils. Adjust the heat for even boiling.
11. Describe what you see in test tube B. Add a boiling chip, and heat until the test tube is almost dry. What do you observe?

▶ ANALYSES AND CONCLUSIONS
1. Summarize what happened in parts A, B, and C.
2. Why do you use ice water in part C?
3. In part C, did all groups get the same results? If not, explain why.

▶ APPLICATION
How could what you have learned here help you separate a mixture in your kitchen?

✳ Discover More
Choose three solutions, and filter them in the ways you learned in the investigation. How does each method affect the solutions?

CHAPTER 19

The Big Idea— Environmental Interactions

Lead the students to an understanding of equilibrium in solutions, suspensions, and colloids in terms of molecules and/or particles. Also, help the students recognize that the stability of systems can change and be reestablished.

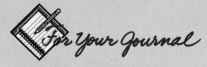

For Your Journal

The students' ideas should reflect an understanding of how matter interacts in mixtures and compounds. You might have the students share their new ideas with classmates.

CHAPTER 19

*H*IGHLIGHTS

The Big Idea

The process of making solutions, suspensions, and colloids could be seen as an exercise in equilibrium. A saturated solution is in a state of dynamic equilibrium. For every solute particle that dissolves, a dissolved particle comes out of solution. This is a balanced state.

Scientists study different examples of equilibrium in order to understand and to control them. Then the equilibrium can be shifted in one direction or another. Understanding how systems interact and how we can affect the stability of them enables us to manipulate reactions.

For Your Journal

How have your ideas about the interactions of matter changed? Revise your journal entry to show your new ideas about mixtures and compounds. Include information on how solutions are used in the everyday world.

Connecting Ideas

This concept map shows how the big ideas of this chapter are related. In your journal, complete the concept map to show how different types of mixtures fit into this arrangement.

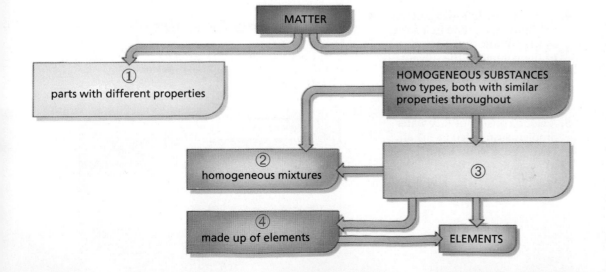

CONNECTING IDEAS

The concept map should be completed with information similar to that shown below:

1. HETEROGENEOUS MIXTURES
   parts with different properties

2. SOLUTIONS
   homogeneous mixtures

3. PURE SUBSTANCES
   have a characteristic composition

4. COMPOUNDS
   made up of elements

CHAPTER 19
REVIEW
ANSWERS

Understanding Vocabulary

1. (a) Both are types of matter. A mixture is a combination of two or more types of matter, each of which keeps its own characteristics. The parts of a mixture can be separated by physical means. A compound is a combination of elements that have been combined chemically. The properties of the compound are different from those of the elements. The elements in a compound cannot be separated by physical means but can be separated by chemical means.

(b) Both are parts of a solution. A solvent is the major component of the solution in which the solute, the minor component, is dissolved.

(c) A solution is a homogeneous mixture consisting of a solute and a solvent. A solution is saturated when no more solute seems to dissolve in the solvent at that temperature. Solubility is the maximum amount of solute that can be dissolved in 100 g of solvent at a specific temperature.

(d) Both are heterogeneous mixtures. The particles in a suspension are large enough that they eventually settle out. In a colloid, the particles are so small that they do not settle out.

Understanding Concepts

Multiple Choice

2. c

3. a

4. d

5. a

6. b

Short Answer

7. The major difference is the amount of solute. A saturated solution contains the maximum amount of solute possible. An unsaturated solution contains less than the maximum amount of solute possible.

8. Gels, aerosols, and emulsions are all forms of colloids. Gels are liquid particles spread out in a solid. An example of a gel is jelly. Aerosols are either liquid or solid particles suspended in a gas. An example of an aerosol is fog. Emulsions are liquid particles spread out in a liquid. An example of an emulsion is milk.

Interpreting Graphics

9. When sugar is placed into warm water, the water molecules surround the sugar molecules, causing the sugar molecules to go into solution, or dissolve.

CHAPTER 19
REVIEW

▶ Understanding Vocabulary

1. For each set of terms, explain the similarities and differences in their meanings.
a) mixture (518), compound (520)
b) solute (523), solvent (523)
c) solution (523), saturated solution (526), solubility (527)
d) suspension (531), colloid (533)

▶ Understanding Concepts

MULTIPLE CHOICE

2. Which of the following is a homogeneous mixture?
a) tossed salad
b) soil
c) lemonade
d) vegetable soup

3. Suppose you add a teaspoon of salt to a cool salt water solution and stir vigorously until the salt dissolves. The solution you started with was
a) unsaturated.
b) supersaturated.
c) saturated.
d) heterogeneous.

4. Which of the following affects the solubility of a solute in a solvent?
a) the volume of the solute
b) the volume of the solvent
c) the mass of the solute and solvent
d) the temperature of the solution

5. A cloud consists of tiny water droplets suspended in the air. Therefore, a cloud is a(n)
a) aerosol.　　b) gel.
c) emulsion.　　d) solution.

6. Suppose your doctor prescribes eardrops for an ear infection. The label on the eardrops says "Shake well before using." This medicine is probably a
a) solution.　　b) suspension.
c) colloid.　　d) gel.

SHORT ANSWER

7. What is the difference between a saturated solution and an unsaturated solution that share these characteristics: the same solvent and solute; the same volume of solvent; the same temperature?

8. Compare and contrast gels, aerosols, and emulsions. Give an example of each.

Interpreting Graphics

9. Study the drawing below. Refer to it to explain how sugar dissolves in warm water.

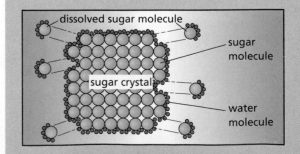

dissolved sugar molecule

sugar molecule

sugar crystal

water molecule

Reviewing Themes

10. A saturated sugar solution is stable because for every grain of sugar that dissolves, another grain comes back out of solution. A saturated solution is in dynamic equilibrium. The amount of solute capable of being dissolved remains constant. A supersaturated sugar solution is unstable: it is no longer in dynamic equilibrium. The amount of solute capa-

ble of being dissolved is not constant. It doesn't take too much to make the sugar come out of solution in a supersaturated solution.

11. Situations should show how dynamic equilibrium leads to overall stability. For example, one situation could be a candy machine. For every coin inserted, a piece of candy comes out—but the overall value (i.e., coin equals value of candy) remains stable.

Thinking Critically

12. Shaking the soda solution causes the gas molecules to escape from the liquid. You could cool the can by placing it in ice or in the refrigerator. By cooling the solution, you cause the gasses to dissolve back into the liquid. Or, you can wait for a longer time for the solution to return to equilibrium at room temperature.

13. One way is to stir a small amount of solid into the solution. If it dissolves, the solution was unsaturated. If the solid does not dissolve, the solution was saturated. If more solid comes out of solution than was added, the solution was supersaturated.

14. The salt could be dissolved in water, the solution could be filtered to remove suspended particles, and finally evaporated to leave the pure salt.

15. You should move the aquarium from the windowsill to a cooler, shadier spot or shield it somehow from the sun and heat. Because less gas can stay dissolved in warmer water, the amount of oxygen in warmer water is reduced. A reduced supply of oxygen could negatively affect the fish.

16. The membrane in contact with the skin allows diffusion of the active ingredients at a premeasured rate. Not all medicines can be absorbed through the skin, and not all will reach the correct location for treatment if administered in this manner.

Reviewing Themes

10. *Environmental Interactions*
Compare a saturated sugar solution at a specific temperature with a supersaturated sugar solution at the same temperature in terms of dynamic equilibrium and stability.

11. *Systems and Structures*
Describe a day-to-day situation that shows how dynamic equilibrium leads to stability.

Thinking Critically

12. You accidentally shake up a can of soda that you were planning to drink. What could you do so that the soda will not spray all over when you open the can?

13. How could you determine experimentally whether a solution of a solid in a liquid is saturated, unsaturated, or supersaturated?

14. The salt we use to flavor foods is usually mined from the ground. However, when mined, the salt is often combined with soil and minerals. How could this crude material be treated to yield pure salt? Be specific.

15. Suppose you have a fish aquarium sitting on a window sill in your bedroom. You know that the weather is going to be hot and sunny for the next two weeks. What should you do with the aquarium? Explain why.

16. One method of taking medicine is known as transdermal infusion. The medicine, enclosed in a small, soft package, is absorbed slowly through the skin. Medicines to prevent motion sickness or to treat some coronary conditions are sometimes administered this way. Explain how you think the transdermal infusion systems works. Could all medication be given this way? Explain your answer.

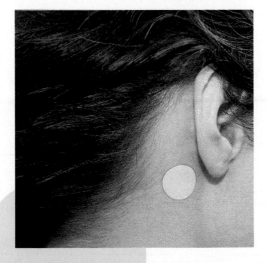

Discovery Through Reading

Soviero, Marcelle. "A Cure For Soggy Sandwiches." *Popular Science* 240 (March 1992): 23. Chitin—a natural substance found in nature—is resistant to solvents. Read this article to find out how this substance is used to prevent foods from becoming soggy.

BONDING AND CHEMICAL REACTIONS

PLANNING THE CHAPTER

B = Basic   A = Average   H = Honors
The coding Basic, Average, and Honors indicates subsections, features, and resources that might be appropriate for different levels of learners. For additional suggestions regarding choice of topic and depth of coverage, see the Pacing Chart on pages T28–T31.

*Frame numbers at point of use
(TR) Teaching Resources, Unit 6
(IT) Instructional Transparencies
(LI) Laboratory Investigations
(SD) *Science Discovery* Videodisc Correlations and Barcodes
(SRG) Study and Review Guide

CHAPTER MATERIALS

Title	Page	Materials
Discover by Doing	543	(per individual) 250-mL beaker, water, calcium chloride
Discover by Problem Solving	549	(per individual) journal
Teacher Demonstration	549	10-cm lengths of insulated copper wire (3), scissors, flashlight bulb socket and bulb, dry cell, small beaker each of distilled water, sugar solution, sodium-chloride solution
Discover by Doing	550	(per group of 3 or 4) index cards, pencil, periodic table
Activity: How do different crystals polarize light?	551	(per group of 3 or 4) wax pencil, microscope slides (8), dropper bottles with solutions (4), light microscope, polarizing film (3 cm²), tape, scissors
Discover by Doing	558	(per individual) measuring spoon, baking soda, alum, powered gelatin, large beaker or glass, stirring rod, vinegar
Skill: Drawing Conclusions	559	(per group of 3 or 4) paper and pencil
Investigation: Observing Chemical Reactions	560	(per student) safety goggles, laboratory apron (per group of 3 or 4) test tubes (8), test-tube rack, labels or wax pencil, dropper bottles of sodium fluoride, sodium chloride, potassium bromide, potassium iodide; graduated cylinder, droppers, calcium nitrate solution, silver nitrate solution

ADVANCE PREPARATION

For the Investigation, prepare the following solutions the day of the activity: 0.5 M solution of $Ca(NO_3)_2$ [161.4 g $Ca(NO_3)_2$ dissolved in water to make 1 L of solution]; 0.1 M solutions each of NaF, NaCl, and $AgNO_3$ [8.5 g NaF, 5.8 g NaCl, and 17.0 g $AgNO_3$, each dissolved in water to make 1 L of solution]; 0.2 M solutions each of KBr and KI [23.8 g KBr and 33.2 g KI, each dissolved in water to make 1 L of solution].

TEACHING SUGGESTIONS

Field Trip

Arrange to have the students visit a laboratory in a paint manufacturing plant, plastics manufacturing plant, or similar facility to observe the role chemistry plays in the manufacture of products. Encourage the students to bring their journals and take notes of the manufacturing processes so they might discuss their observations when they return to class.

Outside Speaker

Ask a sports physiologist or trainer to speak to the students on the importance of maintaining the body's electrolyte levels during exercise and the effects of drugs, such as steroids, on the body. You might also ask a representative of the Environmental Protection Agency (EPA) to discuss and demonstrate field tests at your school for contaminants in the soil, water, and air.

CHAPTER THEME—*ENVIRONMENTAL INTERACTIONS*

Chapter 20 will help the students realize that compounds are substances composed of atoms held together by chemical bonds. These bonds form as a result of atoms transferring valence electrons or sharing valence electrons. A supporting theme in this chapter is **Systems and Structures.**

CHAPTER

20 Bonding and Chemical Reactions

Often science's most intriguing ideas are interestingly wrong. In the early 1800s, scientists believed living things behaved according to the same laws as nonliving substances. The German author Johann W. von Goethe (GUHR tuh) based an entire novel on this fascinating —but scientifically inaccurate— principle. Goethe called the principle "elective affinity."

Listen as the novel's two couples discuss the laws of human interaction:

"Wait a moment and see if I have understood what you are driving at," Charlotte said. "Just as everything relates to itself, so it must have some relation to other things."

"And that relation will be different according to a difference in the elements involved," Eduard eagerly went on. "Some will meet quickly like friends or old acquaintances and combine without any change in either, just as wine mixes with water. But others will remain detached like strangers and refuse to combine in any way—even if they are mechanically mixed with, or rubbed against, each other. Oil and water may be shaken together, but they will immediately separate again."

[The Captain replies:] "Imagine an A so closely connected with a B that the two cannot be separated by any means, not even by force; and imagine a C in the same relation to a D. Now bring the two pairs into contact.

A will fling itself on D, and C on B, without our being able to say which left the other first, or which first combined with the other."

from *Elective Affinities* by Johann W. von Goethe

***M**atter combines in certain prescribed ways. It may form compounds or mixtures. Each of these combinations also react in predictable ways.*

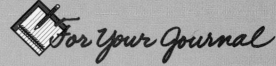
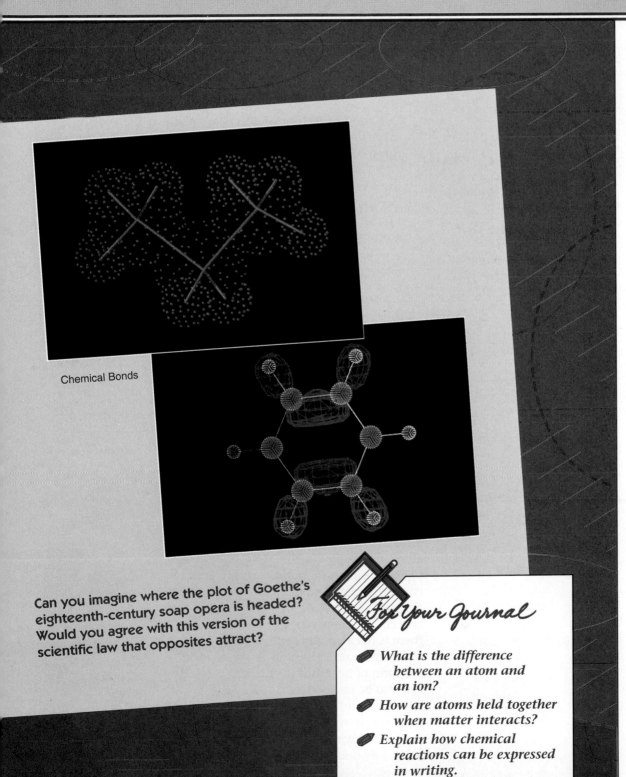

Chemical Bonds

Can you imagine where the plot of Goethe's eighteenth-century soap opera is headed? Would you agree with this version of the scientific law that opposites attract?

ABOUT THE AUTHOR

Johann Wolfgang von Goethe published *The Sufferings of Young Werther* (*Die Leiden des jungen Werthers*) in 1774. The great popularity of the novel made its main character, Werther, the model of the romantic hero and its author a leader of the German romantic movement. Goethe's greatest literary work is his dramatic poem *Faust.* Along with his literary pursuit of poetry, essays, and fiction, Goethe published works in philosophy and science.

His work in morphology led to his discovery of one of the small bones that make up the human jaw. He also proposed an explanation that refuted Newton's explanation of the refraction of white light by a prism. His friendships with chemists and interest in their work led him to incorporate theories of the physical sciences into his descriptions of human behavior. This viewpoint was articulated in 1809 in *Elective Affinities* (*Die Wahlverwandtschaften*).

For Your Journal

- What is the difference between an atom and an ion?
- How are atoms held together when matter interacts?
- Explain how chemical reactions can be expressed in writing.

This section relates chemical behavior to the electron arrangement in atoms. The role of valence electrons in chemical bonds is discussed. Ionic and covalent bonds are introduced and compared.

Tie off the end of an inflated rubber balloon, and hold it against a wall. Have the students predict what will happen if you release the balloon. Have them observe the balloon as it falls after being released. Then ask the students to hypothesize ways that would keep the balloon from falling. (Accept reasonable answers such as taping the balloon, propping the balloon with a pole.) If no one suggests charging the balloon, rub the balloon against your clothing and touch it to the wall. Ask the students to explain why the balloon remains stuck to the wall. (It is attracted by static electricity.)

PROCESS SKILLS
• Predicting • Observing • Formulating Hypotheses

POSITIVE ATTITUDES
• Creativity • Initiative and persistence

TERMS
• chemical bond • ionic bond
• covalent bond

PRINT MEDIA
"Car Cooler" in *Chem Matters* magazine by David P. Robson (see p. 513b)

ELECTRONIC MEDIA
Chemical Bonding and Atomic Structure. Coronet Film and Video. (see p. 513c)

SCIENCE DISCOVERY
• Fireworks • Rust; iron • Periodic table

BLACKLINE MASTERS
Extending Science Concepts
Reading Skills
Study and Review Guide

 LASER DISC
1807, 1808

Fireworks

2525

Rust; iron

How Atoms Combine

Objectives

List the types of chemical bonds.

Differentiate between ionic and covalent bonds.

Demonstrate your knowledge of chemical bonding by identifying the type of bond formed, given the elements involved.

"The San Antonio Rattlers are ready at the goal line," says the announcer. "Will they pass or let Number 18 run it in?"

The chances to score in a football game depend on the arrangement of the players and how close they are to the goal. Atoms often behave as football players do. You might say atoms pass or hand off electrons to other atoms to form compounds.

In Chapter 17 you learned about the reactivity of elements. There is an important connection between how elements behave and their electron arrangement.

Dance of the Electrons

Juan is using glue to put his kite together. The glue bonds the plastic, wood, and paper pieces together through some "sticky" interaction. The atoms in a compound are held together by similar "sticky" interactions known as chemical bonds. A **chemical bond** is the force that holds the atoms in a compound together. Chemical bonds contain a great deal of energy. When bonds are broken, the energy in them is released. This energy is most often seen in the form of heat, but it can also be released as light or sound. Find out for yourself what happens when bonds are broken by doing the next activity.

● **Process Skills:** *Comparing, Analyzing*

Remind the students that the cartoon shows that sodium has one electron in its outer shell, while neon has eight electrons in its outer shell. However, if sodium loses that one electron, it too would have eight electrons in its outer shell. Then display the periodic table, and have volunteers identify pairs of elements that reflect the concept illustrated in the cartoon. (Examples include argon and potassium, krypton and rubidium, and xenon and cesium.)

DISCOVER BY *Doing*

Get a 250-mL beaker and put 100 mL of water into it. Add 10 g to 15 g of CaCl₂ (calcium chloride) to the water. Touch the outside of the beaker. What do you observe? What do you think causes the change? ✐

Chemical bonding is a very complex process. So complex, in fact, that it would take years to understand all its different aspects. Since you don't have years to read this chapter, we are going to use a simplified explanation.

Suppose you have a baseball card collection. You exchange cards with friends to complete your collection. What if you have two identical cards? You might trade one of your cards with a friend for a card you don't have.

Atoms do something similar. They exchange electrons to achieve a stable noble gas configuration. That is, atoms trade electrons until they resemble the noble gas closest to them in the periodic table. The electrons involved are the electrons in the outer shell, or the *valence electrons*.

▼ ASK YOURSELF

How do atoms form compounds?

Give, Take, or Share

When atoms lose or gain electrons, they are no longer electrically neutral; they become ions. As you know, ions carry an electric charge. Metals, you might remember, lose electrons to form positive ions. Nonmetals may gain electrons to form

DISCOVER BY *Doing*

Have the students write their observations in their journals. The reaction of CaCl₂ in water is extremely exothermic. Using 10 g to 15 g of the solid will produce enough heat so the students can easily feel it. If you wish to use smaller amounts of CaCl₂, you can measure the temperature change using a thermometer.

ONGOING ASSESSMENT
▼ ASK YOURSELF

Atoms form compounds by losing or trading valence electrons until each atom resembles the noble gas closest to it on the periodic table.

REINFORCING THEMES
Systems and Structures

Stress that noble gases are almost inert, that is, they do not react because they are chemically stable. Their stability is due to the configuration of their valence electrons.

LASER DISC
2353

Periodic table

● **Process Skills:** *Comparing, Analyzing*

In their journals, have the students show that, even though the potassium and chloride ions are charged, the compound potassium chloride is electrically neutral. (K^+ has 19 protons, 18 electrons; Cl^- has 17 protons, 18 electrons; KCl has 36 protons, 36 electrons.) Then ask the students why neutrons are not counted. (Neutrons are electrically neutral.)

negatively charged ions. Keep in mind also that opposite charges attract. The bond resulting from the attraction between a positive ion and a negative ion is called an **ionic bond**. This bond is similar to two magnets stuck together. A compound formed by ionic bonds is called an *ionic compound.*

Suppose you wanted to make a compound with potassium and chlorine. Ask yourself these questions: What must potassium do to achieve a noble gas configuration? What must chlorine do to achieve a noble gas configuration? If you look at the periodic table, you'll notice that potassium is closest to argon. Potassium needs to lose one electron to have the same electron arrangement as neon. Chlorine, on the other hand, needs to gain an electron to have the same electron configuration as argon.

Madame, I would be pleased if you would accept this extra electron.

When potassium loses an electron, it becomes an ion with a +1 charge. This is because it now has one more proton, which has a positive charge, than it has electrons, which have a negative charge. Chlorine was neutral but then gained an extra electron. So now chlorine is an ion with a −1 charge. Both the potassium ion and the chloride ion have an electron arrangement like a noble gas. Once the ions are formed, their opposite charges attract so the potassium and chlorine are now held together like two tiny magnets. This attraction forms the ionic bond.

Of course, in order to make some compounds, more than one electron must be transferred. In the formation of magnesium bromide, for example, the charge for magnesium is +2. The charge of each bromide ion is −1. What must magnesium do to gain a stable configuration? What does bromine do to gain a noble gas configuration? To form the compound, a magnesium atom loses two electrons.

GUIDED PRACTICE

Cooperative Learning Have the students form small groups to list these elements on a sheet of paper: Li, F, Mg, S. In a second column, have the students write the symbol of the noble gas closest to each element (He, Ne, Ne, Ar). In a third column, have the students indicate

whether the element can achieve this noble gas configuration by gaining electrons (F, S); by losing electrons (Li, Mg); or by sharing electrons (F, S).

INDEPENDENT PRACTICE

Have the students provide written answers to the Section Review. In their journals, have them summarize how ionic and covalent bonds are formed.

EVALUATION

Have the students describe why KCl is ionic and Br₂ is a covalent.

When compounds are formed in this manner, the total number of electrons released must equal the electrons taken. The compound is, therefore, electrically neutral, even though the individual atoms are not. The +2 charge of the magnesium ion is balanced by the −1 charges of the two bromides.

 ASK YOURSELF

How are ionic compounds formed?

Atoms Apply for Loans?

Fluorine gas exists as F_2, or two atoms of fluorine bonded together. Which atom gives or takes an electron? How many electrons must each F atom give or take to achieve a noble gas configuration? None, of course. Puzzled? Both fluorines need one electron to become like the noble gas neon. If each fluorine gives one electron to the other, nothing is changed. How can both atoms gain an electron without losing any? They can share electrons with each other.

Okay, let me get this straight. I'll share one with you, and you'll share one with me. *Right?*

Consider this example. You have a chocolate bar, but your mother says you can't eat it because your brother doesn't have one. You can either save the chocolate, you can buy your brother his own chocolate bar, or you can offer to let your brother have some of your chocolate. The best solution is to share. Most atoms in a similar situation share electrons.

Each fluorine atom needs one electron to complete a stable octet. Transferring one electron from another fluorine atom would create one stable fluoride ion and one fluorine that needed two electrons. This wouldn't help much. Each fluorine atom can, however, share one electron with another fluorine

ONGOING ASSESSMENT

▼ **ASK YOURSELF**

Ionic compounds are formed when two or more oppositely charged ions form ionic bonds.

MEETING SPECIAL NEEDS

Gifted

Have students plan and carry out a presentation on metallic bonds. The presentation should include examples of substances that contain metallic bonds, a model of the bond, and what physical and chemical properties the metallic bond can be used to explain.

INTEGRATION– Chemistry

Beginning in 1912 Dr. Gilbert Lewis, one of the most distinguished chemists in the United States, built a powerful, creative chemistry department at the University of California at Berkeley. In 1916 Dr. Lewis published a paper titled "The Atom, and the Molecule." In it, he defined covalent bonding. Dr. Lewis' theories became the foundation for the electron theory of chemical structure. He summarized his work in his 1923 book *Valence and the Structure of Atoms and Molecules.*

RETEACHING

Prior to class, obtain four one-dozen egg cartons. Cut off four cups from the end of two cartons and five cups from the other cartons. Cut a single cup from one of the discarded ends and attach it to the bottom of one of the eight-cup cartons.

Then invert the cartons on a table. Explain to the students that the bumps on the cartons

represent valence electrons. Then show the electron configuration for F, Ne, and Na. Show that if F gained an electron and Na lost its outermost electron, each would have a configuration similar to Ne. Finally, tape two seven-cup cartons together to show how F forms covalent bonds.

EXTENSION

Have the students use reference resources to investigate how electron dot diagrams can be used to show the formation of covalent and ionic bonds.

CLOSURE

Cooperative Learning Have the students work in small groups to write a summary of the section. Ask a member of each group to read the group's summary aloud.

ONGOING ASSESSMENT
▼ ASK YOURSELF

Atoms form covalent bonds by sharing valence electrons.

SCIENCE BACKGROUND

The energy contained in the covalent bonds in the molecules in 1 L of water is 286 kJ. If converted to electric energy, it would be enough energy to operate a 100-W light bulb for almost 45 minutes.

SECTION 1 REVIEW AND APPLICATION

Reading Critically

1. A valence electron is any electron in the outer shell of the atom.

2. Ionic bonds form between oppositely charged ions, which were formed by atoms either gaining or losing valence electrons. Covalent bonds form between atoms sharing valence electrons.

Thinking Critically

3. A molecule of hydrogen peroxide contains four atoms—two hydrogen atoms and two oxygen atoms bonded. A molecule of water contains three atoms—two hydrogen atoms and one oxygen atom bonded.

4. a. covalent; You may need to remind students that hydrogen is a nonmetal.
   b. ionic
   c. ionic

Figure 20–1. Acrylic glues form covalent bonds between the substances glued together. The man shown here has had his shoes glued to a platform. The glue forms bonds so strong that they are able to support his weight.

Figure 20–2. These computer models show what covalent bonds look like.

atom. Then, each atom has a stable octet. This sharing of electrons is called a **covalent bond.** Compounds made in this way are called *covalent compounds*. The smallest particle of a covalent compound is called a *molecule*.

▼ ASK YOURSELF
How do atoms form covalent bonds?

SECTION 1   REVIEW AND APPLICATION

Reading Critically
1. What is a valence electron?
2. What are the differences between ionic and covalent bonds?

Thinking Critically
3. Hydrogen peroxide (H_2O_2) is a covalent compound made of hydrogen and oxygen.

Water (H_2O) is also a covalent compound made of hydrogen and oxygen. Explain how they are different.
4. Use the periodic table to answer this question. How will the following elements combine to form a compound? Indicate whether the compound formed is ionic or covalent.
   a. hydrogen and bromine
   b. magnesium and oxygen
   c. calcium and chlorine

FOCUS

Section 2 discusses the tendency of metals and nonmetals to form ionic bonds and the tendency of nonmetals to form covalent bonds. The students will write chemical formulas, given the elements or ions in a compound. Properties of ionic and covalent compounds are also discussed.

MOTIVATING ACTIVITY

Write the acronym *NASA* on the chalkboard. Ask what the term means. (National Aeronautics and Space Administration) Then ask the students how this word that stands for the name of the agency was formed. (The term contains the first letter of each major word in the name.)

Finally, have the students brainstorm the origins of other acronyms such as *AIDS* (Acquired Immune Deficiency Syndrome), *SCUBA* (self-contained underwater breathing apparatus). Conclude by telling the students that they will learn how to use shortened forms to identify substances.

PROCESS SKILLS
• Generating Ideas •
Observing

POSITIVE ATTITUDES
• Creativity • Skepticism

TERMS
None

PRINT MEDIA
"Endothermic-Fueled Jet Could Break Mach 5" in *Popular Mechanics* magazine by Abe Dane (see p. 513b)

ELECTRONIC MEDIA
Formulas and Equations. Coronet Film and Video. (see p. 513c)

SCIENCE DISCOVERY
• Calcium
• Chlorine

BLACKLINE MASTERS
Laboratory Investigation 20.1
Thinking Critically
Study and Review Guide

Chemical Formulas

SECTION 2

It was a dark and stormy night. The pair knew their only hope of survival was to find the lost message and get it to central command. They prepared for the treacherous journey to the drop point. Silently they glided through the wet streets, looking in every direction for their enemies. Finally . . . success! They reached the drop and took the message to their contact. The man opened the small container and gazed at the message. "It's in code!" he cried . . . .

Objectives

Write the formula and **name** the compound when you are given the elements or ions in the compound.

Summarize the crisscross method of formula writing.

Analyze the importance of covalent compounds to the human body.

LASER DISC
2392

Calcium

2389

Chlorine

Chemical Codes

Have you ever read about spies using secret codes to send messages? Maybe you have even made up your own code to use with a friend. It might seem that formulas are a chemist's secret code, yet anyone can learn to decipher them. A *chemical formula* contains the symbol for each atom in the compound. The formula for an ionic compound gives the simplest ratio of the different ions in the compound. For example, the formula for calcium chloride is $CaCl_2$. The 2 after the Cl, written lower than the symbol for the element, is called a *subscript*. If no subscript is written, as for Ca in this formula, a 1 is understood. By looking at the formula, you know that the crystal contains 2 chlorine

● **Process Skills:** *Inferring, Predicting*

Ask the students to classify hydrogen as a metal or non-metal. Most students will say hydrogen is a metal because of its position in the periodic table. Point out that hydrogen is unique because it can behave in several ways. Then ask the stu-dents what type of ion hydrogen becomes if it loses an electron (positive ion, H^+) or if it gains an electron (negative ion, H^-). Point out to the students that the electron configuration of H^- is similar to that of helium, a noble gas. Ask the students to explain how hydrogen can also form covalent bonds. (by shar-ing a pair of electrons)

● **Process Skill:** *Applying*

Give each stu-dent a piece of paper and have them label it "polyatomic ions." Using their knowledge of the electron structure of hydrogen, nitrogen, and oxygen atoms, they should show on paper that the charges of the polyatomic ions NH_4 and OH are +1 and −1, respectively. Have the students share their methods with classmates.

You might have the students place their "polyatomic ions" sheets in their science portfolios.

MEETING SPECIAL NEEDS

Mainstreamed

Have mainstreamed students use a hand lens to study crystals of table salt (NaCl). Then have the students observe crystals of a commercially available table-salt substitute. The students should draw what they observe and compare their drawings.

① NaCl, KF, and MgO are ionic compounds.

✧ Did You Know?

It is not really appropriate to refer to salt substitutes as "no salts" because these products contain potassium chloride, which is also a salt.

INTEGRATION– Language Arts

Explain to the students that the prefix *poly-* means "many." Then have volunteers look up the meaning of each the following terms: *polygon* (many-sided figure), *polynomial* (sum of two or more algebraic expressions), and *polysyllabic* (word with many sylla-bles). Point out that some metals, such as copper, are polyvalent because copper can exist as Cu (I) and Cu (II). Ask a volunteer to define the term. (An element is polyvalent when it has more than one valence.)

Figure 20–3. The crystal structure of the ionic mineral halite—or table salt—is distinctive and beautiful.

ions for every calcium ion. Compounds made from a metal and a nonmetal are nearly always ionic compounds, while compounds made of combinations of non-metals are covalent. By looking at the for-mula, therefore, you can tell an ionic compound from a covalent one. Use the periodic table to help you determine which of the following are ionic com-pounds: NaCl, KF, CO_2, MgO. ①

All for One Some ionic compounds contain ions made from several nonmetals that are bonded together covalently. These ions form a unit, and they have a single charge. They are called *polyatomic ions*. Table 20–1 lists some common poly-atomic ions.

Table 20–1	Common Polyatomic Ions	
Ion	**Formula**	**Charge**
Ammonium	NH_4	+1
Hydroxide	OH	-1
Sulfate	SO_4	-2
Phosphate	PO_4	-3
Nitrate	NO_3	-1

Charge It! In any compound, the net charge of the com-pound must equal zero. That is, the number of positive charges must equal the number of negative charges. If you know the charge of the ions in an ionic compound, it is easy to write the correct formula by using the crisscross method. Look at the examples below:

Example 1

$$Mg^{+2} + F^{-1}$$
$$Mg\quad F_2 \qquad MgF_2$$

Example 2

$$K^{+1} + S^{-2}$$
$$K_2\quad S \qquad K_2S$$

Example 3

$$Al^{+3} + O^{-2}$$
$$Al_2\quad O_3 \qquad Al_2O_3$$

Example 4

$$Mg^{+2} + OH^{-1}$$
$$Mg\quad (OH)_2 \qquad Mg(OH)_2$$

● **Process Skill:** *Comparing*

Ask the students what the terms *coordinate, cooperate,* and *copilot* have in common. (They all begin with *co-*; each pertains to some kind of sharing.) Then have a volunteer describe the relationship between these terms and *covalent.* (Covalent also pertains to a type of sharing, in this case, a sharing of valence electrons.)

● **Process Skill:** *Applying*

Ask the students to explain why the formula of calcium oxide is CaO and not Ca_2O_2.

DISCOVER BY *Problem Solving*

Look at the examples on page 548, and figure out what the crisscross method is. Write your explanation in your journal. Make sure your explanation works by writing the formulas for compounds made of potassium and bromine, aluminum and sulfur, and ammonium and hydroxide ions. ✎

After you have completed the activity, take a look at your formulas. When you are working with ionic compounds, you should check to be sure that the ratio of elements is the simplest possible. For example, Ca_2O_2 is not the simplest whole number ratio. CaO would be the correct formula.

In contrast to ionic formulas, formulas for covalent compounds tell the actual number of atoms of each kind that make up the molecule. They are never reduced to the simplest whole number ratio. For example, in the compound hydrogen peroxide, H_2O_2, every molecule has two hydrogen atoms and two oxygen atoms.

Consider some simple sugars. Glucose, the sugar made by plants during photosynthesis, has a formula of $C_6H_{12}O_6$. Sucrose, or table sugar, has a formula of $C_{12}H_{22}O_{11} \cdot H_2O$. (Sucrose loses a water molecule as it forms.) If you reduce glucose and sucrose to their simplest whole number ratio, you would get the same compound: CH_2O. This compound is the basic building block for all sugars—how many of them there are in a compound determines the type of sugar.

What's Your Name?

The general way to name a compound is to use the name of the first element followed by the root of the second's element name with the suffix *-ide* added. This is similar to calling someone from Florida a Floridian and someone from Boston a Bostonian. Consider two of the formulas you wrote in this section: potassium + bromine (potassium bromide) and aluminum + sulfur (aluminum sulfide).

Sometimes more than a suffix is added to the name. Prefixes can be used to tell you the number of atoms in an element. For example, carbon monoxide is CO (*mono-* means "one"). Carbon monoxide has one oxygen atom. The formula for carbon dioxide is CO_2 (*di-* means "two"), so carbon dioxide has two oxygen atoms.

To get some practice forming and naming compounds, try the next activity.

Floridian

Bostonian

What would you call him?

DISCOVER BY
Problem Solving

The charge of one ion in a compound becomes the subscript of the other: KBr, Al_2S_3, NH_4OH.

Demonstration

Construct a conductivity tester. Use scissors to strip about 1 cm of insulation from each end of three 10-cm lengths of copper wire. Use one wire to connect one terminal of a small flashlight bulb socket and bulb to one terminal of a dry cell. Attach one wire to the vacant terminal of the socket and one to the vacant terminal of the dry cell. Prepare a small beaker each of distilled water, sodium-chloride solution, and sugar solution. Dip the ends of the wires into each solution. The bulb should glow when the ends of the wires are placed in the sodium-chloride solution but not when placed in the other two solutions. Point out that solutions of ionic compounds—that is, those that contain electrolytes—can conduct electric current.

TEACHING STRATEGIES, continued

● **Process Skill:** *Comparing*

Point out to the students that ionic bonds are particularly strong because they extend throughout the entire crystal. Therefore, most ionic compounds are hard solids at room temperature.

● **Process Skills:** *Comparing, Analyzing*

Based upon the melting points below, have the students identify which compounds are ionic and which are not: carbon dioxide, –56°C (no); hydrogen oxide, 0°C (no); lithium fluoride, 864°C (yes); potassium fluoride, 858°C (yes)

MEETING SPECIAL NEEDS

Second Language Support

Cooperative Learning Have limited-English-proficient students work in groups to create posters illustrating the differences between the physical properties of ionic and covalent compounds. Provide assistance, as needed.

ONGOING ASSESSMENT
▼ ASK YOURSELF

The formulas for ionic compounds are balanced using the crisscross method. The number of the charge on one ion becomes the subscript for the other ion.

MULTICULTURAL CONNECTION

Emeralds were among the gemstones mined in ancient Egypt and Mesopotamia and were used to decorate ceremonial and religious objects. In the Western Hemisphere, emeralds were mined by the Incas prior to the arrival of the Spanish. Today, some of the finest emeralds are mined in Colombia.

▶ **550** CHAPTER 20

Figure 20–4. Ionic compounds have many practical uses.

DISCOVER BY *Doing*

Using index cards, make flash cards for each of the elements in the first two periods of the periodic table. Make four cards for each element. With a team of three classmates, shuffle the cards and place them upside down. Have a player draw two cards and tell whether a compound can be made from the elements chosen. If a compound can be named, the player must give the correct formula. ✏

I have two oxygens, and you only have one.

▼ ASK YOURSELF
Explain how you balance a chemical formula.

Ionic or Covalent . . . Who Cares?

Why would knowing whether a compound is ionic or covalent be of any interest? For one thing, their properties are different. Different properties make these compounds useful for different purposes. For example, ionic compounds tend to have much higher melting and boiling points than covalent compounds. A second difference is that pure ionic compounds are usually crystalline solids at room temperature, while covalent compounds may be solids, liquids, or gases. If you needed to fill a bicycle tire, would you use an ionic or a covalent compound?

Both ionic and covalent compounds have many uses in daily life. You use thousands of different compounds that you are not even aware of! Let's take a look at a few.

Ionic Compounds in Action Many ionic compounds dissolve easily in water. Your body must keep a precise amount of ions in solution in order for it to function properly. These

GUIDED PRACTICE

Write the following chart on the chalkboard:

aluminum sulfate	Al	SO_4
aluminum sulfide	Al	S
sulfur monoxide	S	O
sulfur dioxide	S	O

Ask how each symbol in the pair is related to the name of the compound on the left. (name of first element, name of polyatomic ion; name of first element, root of second element's name with *-ide*; name of first element, root of second element's name with *mono-* and *-ide* added; name of first element, root of second element's name with *di-* and *-ide* added) Then have volunteers write the formulas for the compounds ($Al_2(SO_4)_3$, Al_2S_3, SO, SO_2)

INDEPENDENT PRACTICE

Have the students provide written answers to the Section Review. In their journals, have them write a brief paragraph explaining the difference between a chemical symbol and a chemical formula.

EVALUATION

Have the students identify three electrolytes and name the compounds that might be sources of these electrolytes.

ions are known as electrolytes. The most important are Na^+, K^+, Cl^-, and Ca^{+2}.

Electrolytes? Doesn't that have something to do with electricity? There is no electricity in your body, you might say. Pinch your arm. The reason you can feel the pinch is because nerve impulses deliver small electric messages to your brain. Without the proper concentration of electrolytes, your nerve impulses cannot travel to your brain. When you sweat due to heavy exercise, you lose some of these ions. Often, athletes use special drinks to keep their electrolytes in balance.

"Oh, what a pretty ring! What kind of stone is that?" Ionic, that's what kind. Ionic compounds are found in some of the minerals in the earth. Many ionic compounds are also very beautiful. You can find out more about crystals by doing the next activity.

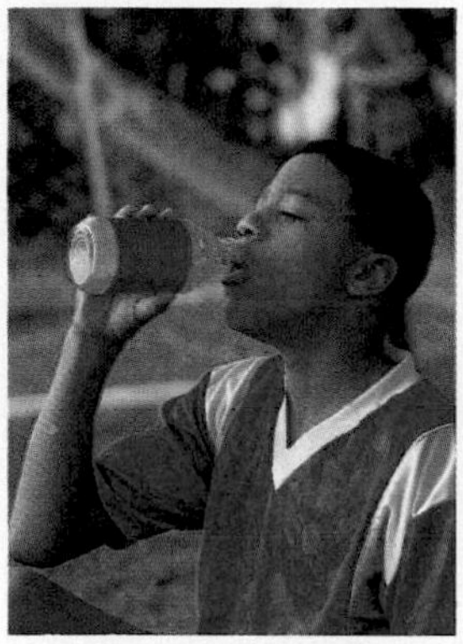

ACTIVITY

How do different crystals polarize light?

MATERIALS

wax pencil, microscope slides (8), dropper bottles with solutions (4), light microscope, polarizing film (3 cm^2), tape, scissors

PROCEDURE

1. Using a wax pencil, label each microscope slide with the name of the solution to be used. Add two drops of one solution to the center of a slide. Use another slide to smear the solution across the first slide. Set it aside to dry. Prepare a slide for each solution provided. Allow each slide to dry.
2. Observe each chemical, using the microscope. On your paper, sketch the crystal shapes.
3. After you have looked at all the crystals, place a piece of polarizing film on the microscope stage and fasten it with tape. Cut a circular piece of film to fit over the eyepiece. Place a slide over the film on the stage, and look at the slide. Rotate the eyepiece film to get the best color. Repeat for each slide, and record your results.

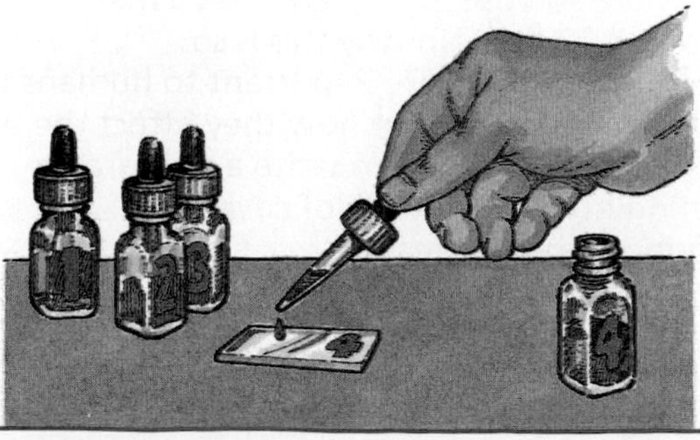

APPLICATION

1. Summarize in a short paragraph what you have observed.
2. The compounds were given to you in aqueous (water) solutions. What type of compounds are these? How do they exist in solution? How do they exist as crystals?

ACTIVITY

How do different crystals polarize light?

Process Skill: Observing

Grouping: Groups of 3 or 4

Hints

Thirty mL of solution should be adequate for each class. Each solution is 10 percent by mass. Compounds that might be used are sodium chloride, sugar, magnesium sulfate, boric acid, and potassium ferrocyanide. Polarizing film is available from Edmund Scientific Company, 101 East Gloucester Pike, Barrington, NJ 08007–1380.

Application

1. Light that passes through polarizing film consists of waves that vibrate in only one direction. This light is refracted by the crystals, which act like prisms, producing an array of colors. Rotating the eyepiece changes the direction of the light waves and therefore the pattern of light.

2. Except for sugar, the compounds are ionic. In solution, ionic compounds exist as ions, but sugar exists as molecules. When ionic compounds crystallize, they form a lattice according to their charges.

RETEACHING

Cooperative Learning Have small groups of students copy the following formulas on a sheet of paper: K(OH), Na₂(NO₃)₂, C(O)₂, and Mg₂(OH)₂. Have the students correct the error in each compound and name the correct compound. Have a member of each group explain a correction.

EXTENSION

Have the students use reference resources to research crystals and their structures. Have interested students build models of crystals to show and to discuss with classmates.

CLOSURE

Have the students each choose a compound that they think best represents covalent compounds and one that best represents ionic compounds. Have them write short explanations of their choices in their journals. You might have the students share their explanations with the class.

ONGOING ASSESSMENT
▼ ASK YOURSELF

Ionic compounds exchange electrons as they form ionic bonds; covalent compounds share electrons in covalent bonds. Ionic compounds have higher melting points and boiling points than do covalent compounds. Ionic compounds exist as crystalline solids at room temperature, while covalent compounds might be solids, liquids, or gases.

SECTION 2 REVIEW AND APPLICATION

Reading Critically

1. The formula of an ionic compound indicates the simplest ratio of the different ions in the compound. The formula of a covalent compound indicates the exact number of each type of atom in the compound.

2. CaF₂, calcium fluoride; Na₂S, sodium (mono)sulfide; AlCl₃, aluminum chloride

Thinking Critically

3. Ionic compounds, such as NaCl, have high melting points. Covalent compounds, such as sucrose, tend to have lower melting points.

4. The students might note that covalent compounds make up most of the atmosphere, make up living cells, and provide the source of energy in burning fossil fuels.

▶ **552** CHAPTER 20

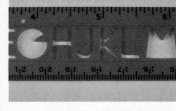

Figure 20–5. Covalent compounds have many practical uses.

Covalent Compounds to the Rescue One, two, three, four. One two, three, four. Exercise is good for you. "Breathe," shouts your instructor. "I can't," you cry. "I have a shortage of covalent compounds!"

Covalent compounds make up most of the atmosphere and, therefore, the air we breathe. They also are major components in living cells. Your response to your instructor could mean you couldn't get your breath, or it could mean you didn't have the muscle power to get the job done!

Covalent compounds made of carbon are the basis for fossil fuels. These fuels provide us with a great percentage of the energy we use daily. Without them, at least until new technology is implemented, you might not have lights in your classroom!

So, who cares if a compound is ionic or covalent? You do.

▼ ASK YOURSELF

Name three differences between ionic and covalent compounds.

SECTION 2 REVIEW AND APPLICATION

Reading Critically

1. What is the difference between the formula of an ionic compound and that of a covalent compound?

2. Write the formulas and names for the compounds made from the ions of the following elements:
 a. calcium and fluorine
 b. sodium and sulfur
 c. aluminum and chlorine

Thinking Critically

3. Sodium chloride, NaCl, has a melting point of 801°C, while sucrose, C₁₂H₂₂O₁₁, has a melting point of 185°C. Explain why this is so.

4. Covalent compounds are important to humans in many ways. Make a list of how they affect the way your body works, and then write a paragraph explaining how an absence of covalent compounds would affect you.

Section 3:

CHEMICAL REACTIONS

FOCUS

In Section 3, the students learn to write and balance chemical equations. The three basic types of chemical reactions are introduced.

MOTIVATING ACTIVITY

 CAUTION: Silver nitrate solution stains skin and clothing.

Dissolve 2 g of silver nitrate ($AgNO_3$) in 100 mL of distilled water. Place a small amount of the solution in a petri dish positioned on the stage of an overhead projector. Have the students observe as you add a few drops of a dilute solution of table salt (NaCl) to the petri dish. (A precipitate forms.) Have the students cite evidence that indicates a chemical reaction has taken place.

Chemical Reactions

SECTION 3

Jaime and Minh are walking home after seeing a chemistry show. They saw solutions that changed colors several times and chemicals that ignited when mixed. They even learned how to make a stink bomb.

"You know what I liked the most, Jaime?" says Minh. "The thermite reaction." "Yeah," replies Jaime, "that was a great ending. I liked it too." The chemist-magician had taken the audience outside. She had shown them a setup that held a paper cone filled with aluminum and iron(III) oxide over a sand box. Stuck in the cone was a strip of magnesium that she compared to a candle wick. When she lit the strip, it would burn like a wick and get the reaction started. The chemist lit the magnesium ribbon, which burned very brightly. All of a sudden the materials in the cone caught on fire. The mixture flared up, giving off huge amounts of heat, light, and smoke. Then a glowing molten ball of iron dropped from the cone into the sand below. WOW!

"Can you believe that the iron was between 2000°C and 2400°C?" says Minh. "I wish we could do reactions like that in school." "Are you crazy! That's too dangerous. But I sure would like to make a few stink bombs," adds Jaime. "Really," says Minh, "the best was understanding what was going on. The chemist told us the ingredients that were involved in each of the tricks. She even wrote the chemical reactions on the portable board. I wrote down the equation for the thermite reaction. I'll tell everyone about it in class tomorrow."

Objectives

Write and **balance** chemical equations.

Identify different types of reactions.

PROCESS SKILLS
- Inferring • Communicating
- Analyzing • Classifying/Ordering • Experimenting
- Observing

POSITIVE ATTITUDES
- Initiative and persistence
- Curiosity • Cooperativeness

TERMS
- chemical equation
- synthesis reaction
- decomposition reaction
- replacement reaction

PRINT MEDIA
A Consumer's Dictionary of Household, Yard and Office Supplies by Ruth Winter (see p. 513b)

ELECTRONIC MEDIA
Chemical Reactions Focus Media, Inc. (see p. 513c)

SCIENCE DISCOVERY
- Chemical reaction
- Candle extinguished by carbon dioxide

BLACKLINE MASTERS
Laboratory Investigation 20.2
Connecting Other Disciplines
Study and Review Guide

INTEGRATION– Industrial Arts

The thermite process can be used to produce molten iron hot enough to melt the broken ends of rails and heavy machine parts. Filings of other metals can be added to the thermite mixture so that the welding steel matches the steel in the broken part.

Chemical Goings-On

The thermite reaction that Minh and Jaime were talking about is really spectacular. However, it is usually not practical to describe a chemical reaction in a paragraph. Scientists generally describe chemical reactions in a shorter way called a *chemical equation*. A **chemical equation** lists the compounds or elements involved in a reaction and the new compounds or elements that form as a result.

Figure 20–6. The thermite reaction shown here gives off terrific amounts of heat and light.

● **Process Skill:** *Analyzing*

Have the students discuss how chemical equations are similar to recipes. Ask them to cite differences between the two. (A possible answer is that a recipe provides a specific procedure.)

● **Process Skills:** *Comparing, Inferring*

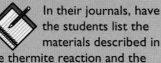

In their journals, have the students list the materials described in the thermite reaction and the substances found in the equation of the reaction. Ask if the equation indicates that magnesium is part of the reaction.

(no) Then ask the students to infer magnesium's role in the reaction. (The heat from the burning magnesium started the reaction; the chemical did not take part in the reaction.)

INTEGRATION–
Math

Challenge the students to draw an analogy between the yield sign in a chemical equation and a specific sign used in mathematical equations. (A yield sign in a chemical equation is equivalent to an equal sign in a mathematical equation.) Then point out that like a mathematical equation in which the numbers must be equal on both sides of the equal sign, the number and kinds of atoms on the left side of a chemical equation must be equal to the number and kinds of atoms on the right side of the equation.

THE NATURE OF SCIENCE

Chemist Jacqueline K. Barton has produced mirror-image molecules that bind to DNA—an accomplishment that has many applications in the sciences. Chemists use Dr. Barton's technique to examine DNA structure. Pharmaceutical companies might one day use her procedure to produce drugs that selectively affect parts of genes. In addition, biologists might utilize her research to pinpoint certain genetic structures.

The chemical equation for the thermite reaction is:

$$2Al + Fe_2O_3 \rightarrow 2Fe + Al_2O_3$$

A chemical equation is like a very condensed recipe. On the left side of the equation is a list of all the materials needed for the recipe. These ingredients are called the *reactants*. On the right side of the equation is a list of what is produced by the recipe. The new substances formed are called *products*.

Figure 20–7 Chemical equations are like recipes.

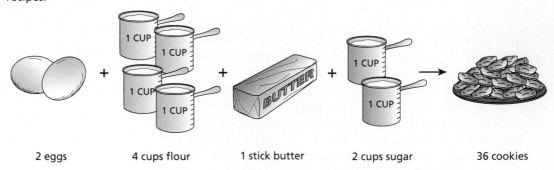

2 eggs   4 cups flour   1 stick butter   2 cups sugar   36 cookies

If you were making a stew, all the ingredients you added to the stew would still be there at the end. They would only be changed in flavor and appearance. The law of conservation of matter states that in a chemical reaction all matter involved in a reaction is conserved. Atoms are not lost while changing from reactants to products; they are only rearranged.

The arrow in a chemical equation is read "yields," which means turns into. The total mass of reactants used must equal the total mass of products formed. In order for this to be true, numbers are sometimes placed in front of formulas. These numbers are called *coefficients*. In the thermite reaction, the coefficients indicate that two atoms of Al react with one unit of Fe_2O_3 to form two atoms of Fe and one unit of Al_2O_3.

Do not confuse coefficients with subscripts. You can change coefficients in a chemical equation until the number of atoms of each element is balanced. **You cannot balance an equation by changing the subscripts.** If you change the subscripts, you change the composition of the substance to something totally different. Remember there is a great difference between H_2O (water) and H_2O_2 (hydrogen peroxide).

Table 20–2 is a checklist to help you balance a chemical reaction. The formation of hydrochloric acid (HCl) is used as an example.

● **Process Skill:**
Constructing/Interpreting Models

Have the students use two sets of two similar coins to repre-sent the hydrogen and chlorine molecules in the reaction $H_2 + Cl_2 \rightarrow 2HCl$. Ask them to construct the compound HCl from the coins and tell why the

coefficient 2 is needed to bal-ance the equation. (Each mole-cule of hydrogen and chlorine produced two molecules of HCl.)

Table 20–2 **Writing Chemical Equations**

Procedure	Result
Write the correct formulas for all reactants and products involved in the reaction. Separate each of the substances on the reactant side from each other by a *plus* sign. Separate reactants from products by an arrow. Remember that some gases are diatomic and must be written as such in the reaction.	$H_2 + Cl_2 \rightarrow HCl$
Balance the equation one element at a time. Use only coefficients to balance. Do not change the subscripts you have already written. Look at hydrogen first—there are two hydrogens on the left of the equation but only one on the right. Place the coefficient 2 before HCl.	$H_2 + Cl_2 \rightarrow 2HCl$
When you put the 2 in front of HCl, you have two hydrogens and two chlorines on the right side of the equation. Look back at the left side. Since chlorine is diatomic, there are two chlorines there also. The equation is balanced. Do not be afraid to go back and forth more than once.	
Do a final check. Make sure the chemical formulas are correct and the numbers of all the atoms are balanced.	

▼ **ASK YOURSELF**

Why are chemical equations useful?

Chemical Recipes

There are so many different types of chemical reac-tions it would be impossible to learn them all. The study of new reactions and the processes by which they occur is an ongoing area of research for many chemists. Most reactions, however, may be grouped into three basic types: synthesis, decomposition, and replacement reactions.

Synthesis Reactions When you make bread, you mix flour, water, yeast, and other ingredients and come out with a loaf of bread. This would be called a synthesis reaction by chemists. In a **synthesis reac-tion** two or more substances combine to form one new substance. The formation of water is an example.

$$2H_2 + O_2 \rightarrow 2H_2O$$

Figure 20–8. When iodine is added to zinc, zinc iodide is formed. This is an example of a synthesis reaction.

ONGOING ASSESSMENT
▼ **ASK YOURSELF**

Chemical equations list the com-pounds and elements involved in a chemical reaction and the new compounds and elements that are formed. These equations allow chemists to represent what actu-ally happens in a way that is easy for us to understand.

LASER DISC
46758–46877

Chemical reaction

SECTION 3 **555** ◀

● **Process Skill:** *Comparing*

Point out to the students that the corrosion of metals is a synthesis reaction. A metal combines with oxygen in the air to form an oxide. Explain that moisture is only a catalyst for this reaction but does not change the reaction. Then have volunteers describe real-life examples of metal corrosion. (Examples include corrosion of motor vehicles, cast iron furniture, and other metal objects.)

● **Process Skills:** *Comparing, Analyzing*

Explain to the students that a decomposition reaction is the reverse of a synthesis reaction. Many compounds break down, or decompose, when they are heated. Then challenge the students to describe real-life examples of decomposition reactions.

(For example, when wood is burned, it breaks down from a complex carbon compound into carbon, carbon dioxide gas, and water.)

INTEGRATION–
Language Arts

Have the students look up the definitions of *synthesize* and *decompose* and write them in their journals. Have them relate these definitions to the terms *synthesis reaction* and *decomposition reaction*.

✧ **Did You Know?**

Another name for a decomposition is analysis. A chemist might use a decomposition reaction to find out what elements make up a compound.

Figure 20–9. Decomposition of mercury (II) oxide; fire at fertilizer plant

Decomposition Reactions Sometimes people need to separate the atoms in a compound. Chemically, a **decomposition reaction** involves breaking a compound into two or more substances. For example, mercury(II) oxide can be decomposed into the elements mercury and oxygen gas by heating.

$$2HgO \rightarrow 2Hg + O_2$$

In Figure 20–9, you can see beads of mercury on the side of the test tube after this decomposition has taken place. Another example of a decomposition reaction is shown in the same figure. The burning factory was ignited by the spontaneous decomposition of fertilizer on a ship in the harbor. The factory became an uncontrollable decomposition reaction.

Some metals are purified from ore by decomposition reactions. Aluminum oxide is the ore from which we get aluminum foil. The reaction is shown below.

$$2Al_2O_3 \rightarrow 4Al + 3O_2$$

Figure 20–10. The production of aluminum from ore involves a decomposition reaction caused by electricity instead of heat.

● **Process Skill:**
Classifying/Ordering

Write the following equations on the chalkboard:

$2Al + Cr_2O_3 \rightarrow 2Cr + Al_2O_3$
$HCl + NaOH \rightarrow H_2O + NaCl$

In the first equation, circle the $2Al$ and the Cr_2 to point out that a single replacement took

place. In the second equation, circle the H and the OH and draw squares around the Cl and the Na to indicate that a double replacement occurred.

● **Process Skill:**
Classifying/Ordering

Cooperative Learning Have the students form small groups. Write the following equations on the chalkboard.

$2K + 2H_2O \rightarrow 2KOH + H_2$
$2NaOH + H_2SO_4 \rightarrow Na_2SO_4 + 2H_2O$

$Cl_2 + 2NaBr \rightarrow 2NaCl + Br_2$
$Zn + H_2SO_4 \rightarrow ZnSO_4 + H_2$
$FeS + 2HCl \rightarrow H_2S + FeCl_2$

Have the groups classify each reaction as a single or double replacement reaction. (1, 3, and 4 are single replacement reactions; 2 and 5 are double replacement reactions.)

Replacement Reactions

How would you like to make silver from a solution and copper wire? If a piece of copper is placed in a colorless silver nitrate solution, pieces of silver will be deposited on the copper and the solution will gradually turn blue. The blue solution is copper(II) nitrate. The equation for this reaction is:

$$Cu + 2AgNO_3 \rightarrow Cu(NO_3)_2 + 2Ag$$

Sometimes chemical reactions involve changing partners, like in a dance. These types of reactions are known as replacement reactions. During a **replacement reaction,** one atom or group of atoms in a compound is replaced with another atom or group of atoms. In Figure 20–11, copper from the wire replaced the silver in the silver nitrate solution. As the silver comes out of solution, it looks as if it "grows" on the copper wire. As more and more of the copper goes into the solution, the clear solution will become blue. The replacement of copper for silver in the silver nitrate solution is an example of a *single replacement reaction* because only one type of atom or group of atoms is being replaced. If two different types atoms or group of atoms exchange places in a reaction, the reaction is called a *double replacement reaction.*

Below is an example of a double replacement reaction. In this reaction, potassium (K) and lead (Pb) are changing partners in the compounds.

$$2KCl + Pb(NO_3)_2 \rightarrow PbCl_2 + 2KNO_3$$

Sometimes in a replacement reaction one of the compounds formed is not soluble in the reaction mixture. The compound forms a solid that comes out of solution. This solid is called a *precipitate.* This is just one way we know a reaction has taken place.

It is sometimes difficult to tell whether a reaction or chemical change has taken place when two substances are mixed. The formation of a precipitate is a sure sign that a chemical reaction has occurred. Other signs are a change in color, a temperature change, bubbles or a gas being released, or a buildup of the new substance (as in the copper and silver nitrate reaction).

Figure 20–11. When copper wire is placed in a solution of silver nitrate, the silver builds up on the wire. This is an example of a single replacement reaction.

Figure 20–12. In this double replacement reaction, sodium chromate (yellow solution) combines with silver nitrate (clear solution) to form sodium nitrate (clear solution) and silver chromate (pink precipitate).

MEETING SPECIAL NEEDS

Mainstreamed

Have mainstreamed students act out synthesis, decomposition, and replacement reactions. For example, to show a synthesis reaction of two elements joining to form a compound, have two students represent the reactants. Have them lock elbows to show they have joined to form a compound. Have them then unlock elbows to represent a decomposition reaction of a compound into two elements.

SCIENCE TECHNOLOGY SOCIETY If the students have not studied fire, discuss the three things that fire needs: fuel, oxygen, and heat to start the reaction. Have the students investigate fire prevention and fire control technologies, such as nonflammable materials or chemical extinguishers. Encourage interested students to present their findings to the class.

LASER DISC
49933–50332

Candle extinguished by carbon dioxide

GUIDED PRACTICE

Have the students write the equation for the decomposition of aluminum oxide.

$(2Al_2O_3 \rightarrow 4Al + 3O_2)$

For each element, have them count the number of atoms or ions in the reactants and in the product to show that the total atoms do not change in type or number during a reaction.

INDEPENDENT PRACTICE

Have the students provide written answers to the Section Review. In their journals, have them write balanced equations illustrating a synthesis reaction, decomposition reaction, single replacement reaction, and double replacement reaction.

EVALUATION

In their journals, have the students describe the three major types of reactions.

RETEACHING

Cut a large square (S), triangle (T), rectangle (R), and circle (C) from different colors of construction paper. Symbolically demonstrate synthesis reactions: $S + T \rightarrow ST$; $S + TR \rightarrow STR$; $ST + RC \rightarrow STRC$. Demonstrate decomposition and replacement reactions in a similar manner.

DISCOVER BY *Doing*

Baking soda and vinegar react, releasing water and carbon dioxide.

(a) $NaHCO_3 + HC_2H_3O_2 \rightarrow NaC_2H_3O_2 + H_2CO_3$

(b) $H_2CO_3 \rightarrow H_2O + CO_2$

The CO_2 is trapped in a foam formed by the alum and gelatin.

ONGOING ASSESSMENT

▼ **ASK YOURSELF**

In a synthesis reaction, two or more substances combine to form one new substance. In a decomposition reaction, one substance is broken down into two or more substances.

SECTION 3 REVIEW AND APPLICATION

Reading Critically

1. A chemical equation lists the substances involved in a chemical reaction and the products formed.

2. synthesis, decomposition, and replacement reactions

Thinking Critically

3. $AgNO_3$ and Ag_2SO_4 are ionic. H_2SO_4 and HNO_3 are covalent. $AgNO_3$ and H_2SO_4 are reactants. Ag_2SO_4 and HNO_3 are products.

$2AgNO_3 + H_2SO_4 \rightarrow Ag_2SO_4 + 2HNO_3$

4. $Ca(OH)_2 \rightarrow CaO + H_2O$

5. a. $C + H_2O \rightarrow H_2 + CO$
   b. $S + O_2 \rightarrow SO_2$
   c. $CuCO_3 \rightarrow CuO + CO_2$

Figure 20–13.
Electroplating is an example of a replacement reaction.

Replacement reactions are used in electroplating. You might have gold-plated or silver-plated jewelry. These items are not made out of pure gold or silver. Instead, a thin coating of these metals has been stuck on to an inexpensive metal. Beneath your silver-plated earrings might be steel or copper.

Sometimes more than one type of reaction occurs during a single process. Create your own replacement and decomposition reactions by trying the following activity.

DISCOVER BY *Doing*

 Some fire extinguishers contain CO_2, which puts out fires by smothering them. One of the older types of extinguishers used a compound called *foamite*. You can make foamite by adding two spoonfuls each of baking soda, alum, and powdered gelatin to a large beaker or glass. Stir well. Add about six spoonfuls of vinegar and stir again. What happens? Explain what has occurred using a chemical equation. ✎

▼ **ASK YOURSELF**

What is the difference between a synthesis and a decomposition reaction?

SECTION 3 REVIEW AND APPLICATION

Reading Critically

1. What information does a chemical equation give you?
2. What are three major types of reactions?

Thinking Critically

3. Look at the equation:
   $AgNO_3 + H_2SO_4 \rightarrow Ag_2 SO_4 + HNO_3$
   Which of the substances above are ionic? Which are reactants? Which are products? Balance the equation.

4. Hydroxide ions, OH^{-1}, may combine with metallic ions to form compounds. When heated, these metallic hydroxides decompose into metallic oxides and water. Write and balance the equation for the decomposition of calcium hydroxide.

5. Write the balanced equations for the following reactions:
   a. carbon + water → hydrogen gas + carbon monoxide
   b. sulfur + oxygen → sulfur dioxide
   c. copper(II) carbonate ($CuCO_3$) → copper(II) oxide + carbon dioxide

EXTENSION

 Have interested students research the terms *music synthesizer* and *biological decomposer*. Then have them write a brief paragraph in their journals relating these terms to the terms synthesis reaction and decomposition reaction.

CLOSURE

Have the students classify the type of reaction described by the Captain in the last paragraph of the literature passage on the first page of the chapter. (double replacement)

SKILL:

Drawing Conclusions

Process Skills: Analyzing, Classifying/Ordering

Grouping: Groups of 3 or 4

Objectives
● **Observe** the results of experiments.

● **Write** equations for chemical reactions.
● **Classify** reaction types.

Discussion

Some of the reactions might not be evident to students with limited science background. You might need to offer hints. In experiment 2, mention that if the gas were collected, it would flame a glowing splint. In experiment 3, indicate that the copper is Cu (I). Tell the students to write the formulas of the reactants before attempting to predict the products.

▶ **Application**

1. $Zn + 2HCl \rightarrow ZnCl_2 + H_2$
   single replacement reaction

2. $2KClO_3 \rightarrow 2KCl + 3O_2$
   decomposition reaction

3. $4Cu + O_2 \rightarrow 2Cu_2O$
   synthesis reaction

4. $NaCl + AgNO_3 \rightarrow NaNO_3 + AgCl$
   double replacement reaction

✳ **Using What You Have Learned**

The students' paragraphs should include a description of the situation, the information analyzed, and the conclusion formed. For example, a student might be considering whether to attend an outdoor event. The student would analyze the weather forecast and use the information to form a conclusion. Accept all reasonable paragraphs.

SKILL Drawing Conclusions

▼ PROCEDURE

Identifying Gases
Several tests can be used in order to identify common gases that may be given off in a reaction. Here are some examples.
1. Oxygen will cause a barely glowing wood splint to burst into flames.
2. Hydrogen will make a popping sound when brought into contact with a flaming splint.
3. Carbon dioxide will put out a flaming splint.

Solubility of Some Solids in Water
KEY: S = soluble, NS = not soluble
zinc chloride—S
potassium chloride—S
sodium nitrate—S
lead(II) iodide—NS
silver chloride—NS
silver nitrate—S
sodium chloride—S
copper(I) nitrate—S

▶ APPLICATION
Using the above information and information provided in this chapter, write the complete chemical equations for each experiment listed here. Balance the equations, and identify the type of reaction for each experiment.

Experiment 1
Richard dropped a piece of zinc (Zn) into 10 mL of hydrochloric acid (HCl) in a test tube. He observed a large number of bubbles forming around the piece of zinc and rising to the surface. When he brought a flaming splint close to the mouth of the test tube, he heard a loud pop. After a few minutes, all the zinc had disappeared.

Experiment 2
Liz heated a massed amount of potassium chlorate ($KClO_3$) in a special porcelain dish over very high heat. At the beginning, the solid was composed of white, translucent crystals. After heating for 30 minutes, the solid had changed to a white powder. She held a glowing splint over the dish, and it burst into flame. Liz massed the compound after heating and found it had less mass than before.

Experiment 3
Luis massed a few pieces of copper foil. He placed them in a test tube and began heating the test tube over a burner. After a few seconds, he noticed that the shiny copper began to turn black. Luis let the black substance cool and then massed it. Its mass was greater than the initial mass of copper.

Experiment 4
Leta had 20 mL of sodium chloride (NaCl) solution and 20 mL of silver nitrate ($AgNO_3$) solution in separate beakers. She slowly poured the silver nitrate solution into the sodium chloride. A white solid began to form as the silver nitrate was being added.

✳ Using What You Have Learned
Analyzing information is just as important in everyday life as it is in the laboratory. Write a short paragraph explaining a situation in which you analyzed information to form a conclusion.

⭐ **PERFORMANCE ASSESSMENT**

Cooperative Learning

 At the conclusion of this *Skill*, have each group of students describe an at-school situation in which they had to analyze information and form a conclusion. Accept all reasonable situations.

SECTION 3 **559** ◀

Observing Chemical Reactions

Process Skills:
Experimenting, Observing

Grouping: Groups of 3 or 4

Objectives
- **Observe** chemical reactions between solutions.
- **Classify** reactions between solutions.
- **Identify** halide ions in solution.

Hints
The equations and observations for this *Investigation* follow:

$2NaF + Ca(NO_3)_2 \rightarrow 2NaNO_3 + CaF_2$
white precipitate

$2NaCl + Ca(NO_3)_2 \rightarrow 2NaNO_3 + CaCl_2$
no precipitate

$2KBr + Ca(NO_3)_2 \rightarrow 2KNO_3 + CaBr_2$
no precipitate

$2KI + Ca(NO_3)_2 \rightarrow 2KNO_3 + CaI_2$
no precipitate

$NaF + AgNO_3 \rightarrow NaNO_3 + AgF$
no precipitate

$NaCl + AgNO_3 \rightarrow NaNO_3 + AgCl$
white precipitate

$KBr + AgNO_3 \rightarrow KNO_3 + AgBr$
yellow-white precipitate

$KI + AgNO_3 \rightarrow KNO_3 + AgI$
yellow-white precipitate

Label containers for the unknown halide solutions W, X, Y, and Z. Have each student test one or two solutions to determine the ion in each solution. Fluoride and chloride ions are readily identified, but there is little qualitative difference between the bromide and iodide ions.

Pre-Lab
Discuss the idea of tests with the students. Ask them why they take tests in school. Have the students relate tests they take to chemical tests.

Discussion
Before the students begin part B, be sure they understand that only the halide ions can be identified. The positive ions are used in more than one test with different results. Therefore, the observed changes could not be made from these tests.

▶Application
1. fluoride

2. Chloride, bromide, and iodide ions produce precipitates. Some students might feel that the color difference between AgBr and AgI helps to tell them apart. However, this is inconclusive.

3. The bubbles are caused by the release of CO_2 gas. This test could be used to identify CO_3^{-2}.

Post-Lab
Have the students complete the test analogy they began before the lab.

INVESTIGATION

Observing Chemical Reactions

▶ MATERIALS
- safety goggles
- laboratory apron ● test tubes (8) ● test-tube rack
- labels or wax pencil ● dropper bottles with the following solutions: sodium fluoride, sodium chloride, potassium bromide, potassium iodide
- graduate ● droppers ● calcium nitrate solution ● silver nitrate solution

CAUTION: Put on safety goggles and a laboratory apron, and leave them on throughout this investigation. Silver nitrate solution will stain your skin and your clothes.

▼ PROCEDURE

TABLE 1: OBSERVATIONS OF REACTIONS			
Reactants	**Products**	**Balanced equation**	**Observations**

1. Prepare a data table similar to the one shown. Leave plenty of room in each section. As you complete each reaction, fill in the reactant and observation sections.
2. Label four test tubes with the names of the solutions in the dropper bottles.
3. Place 5 mL of each of the solutions into the appropriate test tube.

4. Add 1 mL of calcium nitrate solution, $Ca(NO_3)_2$, to each of the four test tubes. One milliliter is approximately 20 drops. Record your observations.
5. Label four more test tubes as in step 2. Place 5 mL of each of the solutions into the test tubes.
6. Add 1 mL of silver nitrate solution ($AgNO_3$) to each

of the test tubes. Record your observations.
7. Before going farther, complete your data table and have your teacher check it.
8. Obtain an unknown solution from your teacher. Identify the halogen ions in your mixture. Record your procedure, observations, and conclusions.

▶ ANALYSES AND CONCLUSIONS
1. Which ion(s) produced a precipitate with calcium nitrate?
2. Which ion(s) produced a precipitate with silver nitrate? Does this test give conclusive evidence for all of the ions? Why or why not?
3. When hydrochloric acid, HCl, is added to limestone, $CaCO_3$, this reaction takes place.

$CaCO_3 + 2HCl \rightarrow CaCl_2 + CO_2 + H_2O$

You would observe bubbles fizzing at the spot where the acid touched the stone. Why? Could you use this as a test for an ion? Which ion? Explain.

✳ Discover More
Use the ion tests you have practiced in this investigation to test various liquids. Make a chart of your findings.

CHAPTER 20

*H*IGHLIGHTS

The Big Idea—Environmental Interactions

Lead the students to understand that reactions that tend to make the products more stable than the reactants often occur spontaneously or need little energy to start the reaction.

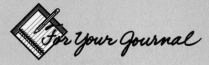

For Your Journal

The students' responses should indicate revisions of their initial answers. You might encourage the students to share their responses with classmates.

CONNECTING IDEAS

The concept map should be completed with words similar to those shown here.

1. change in color or release of gas

2. change in color or release of gas

3. synthesis or decomposition

4. synthesis or decomposition

The Big Idea

Everything in nature moves toward a lower energy state and becomes more stable. A ball rolls down hill. You get tired toward the end of the day and sleep. A hot solution gradually gives off heat and cools. Radioactive elements decay into more stable ones.

What about atoms? Atoms have a unique way of achieving chemical stability: they can exchange or share electrons in order to achieve a noble gas configuration. These movements of electrons are what we call chemical reactions. The reason iron rusts is because the product formed, iron oxide, is less reactive than the initial elements, iron and oxygen. It is amazing that so many different reactions occur spontaneously all around us. The only driving force is the need to achieve stability.

For Your Journal

What is the difference between an atom and an ion? How are atoms held together when matter interacts? Explain how chemical reactions can be written. Look back and review your original answers to these questions. How would you revise your ideas to show what you have learned?

Connecting Ideas

This concept map shows how the big ideas of this chapter are related. Copy the map into your journal, and fill in the blanks to complete it.

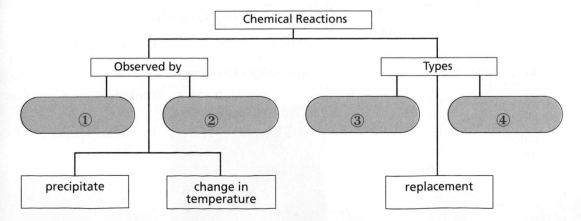

Understanding Vocabulary

1. (a) The terms refer to the force that holds atoms in a compound together. Chemical bond is a generic term that encompasses both covalent and ionic bonds. Covalent bonds hold atoms together as molecules through the sharing of valence electrons in covalent compounds. In ionic compounds, ionic bonds hold together ions, which are formed by atoms gaining or losing valence electrons.

(b) Each term describes a reaction in which new substances are formed. During a synthesis reaction, two or more substances combine to form another substance. During a decomposition reaction, a compound breaks into two or more substances. During a replacement reaction, one atom or group of atoms in a compound is replaced with another atom or group of atoms.

(c) Both are examples of codes used by chemists. A chemical formula contains the symbol for each atom in the compound. A chemical equation lists the compounds or elements involved in a reaction and the new compounds or elements that form as a result.

Understanding Concepts

Multiple Choice

2. a

3. b

4. c

5. c

6. c

7. c

8. b

Short Answer

9. Copper (II) oxide and hydrogen gas are produced in the reaction.

10. The formation of a precipitate, a change in color, a temperature change, bubbles or a gas being released, and a buildup of a new substance are signs that a chemical reaction has taken place.

11. Because sodium chloride is an ionic compound and only the smallest particle of a covalent compound is called a molecule, the term a molecule of sodium chloride is incorrect.

Interpreting Graphics

12. 4

13. 9

14. covalent

CHAPTER
20

REVIEW

Understanding Vocabulary

1. For each set of terms, compare and contrast their meanings.
   a) chemical bond (542), ionic bond (544), covalent bond (546)
   b) synthesis reaction (555), decomposition reaction (556), replacement reaction (557)
   c) chemical formula (547), chemical equation (553)

Understanding Concepts

MULTIPLE CHOICE

2. $A + B \rightarrow AB$ is an example of a
   a) synthesis reaction.
   b) decomposition reaction.
   c) single replacement reaction.
   d) double replacement reaction.

3. $AB \rightarrow A + B$ is an example of a
   a) synthesis reaction.
   b) decomposition reaction.
   c) single replacement reaction.
   d) double replacement reaction.

4. Which substance contains an ionic bond?
   a) CO **b)** CO_2
   c) KCl **d)** O_2

5. The smallest particle of a covalent compound is called a(n)
   a) atom. **b)** ion.
   c) molecule. **d)** reactant.

6. In any chemical equation, the arrow means
   a) "equals."
   b) "is greater than."
   c) "yields."
   d) "breaks down into."

7. The chemical formula for calcium chloride is
   a) $CaCl$. **b)** Ca_2Cl.
   c) $CaCl_2$. **d)** Ca_2Cl_2.

8. What is the coefficient needed in front of sodium bicarbonate, $NaHCO_3$, to balance the equation below?

   $$NaHCO_3 \rightarrow Na_2CO_3 + CO_2 + H_2O$$

   a) 1 **b)** 2
   c) 3 **d)** 4

SHORT ANSWER

9. What substances are the products in the chemical equation

   $$Cu + H_2O \rightarrow CuO + H_2?$$

10. List the signs that show that a chemical reaction has taken place.

11. Why is the phrase "a molecule of sodium chloride" inaccurate?

Interpreting Graphics

Study the computer-generated model of the compound C_2H_5OH shown below.

12. How many bonds are formed by each carbon atom?

13. How many bonds occur in the compound?

14. Are the bonds covalent or ionic?

Reviewing Themes

15. Valence electrons are either lost, gained, or shared by atoms until the atoms resemble the noble gas configuration closest to them in the periodic table. In gaining, losing, or sharing the electrons, chemical bonds are formed which produce products that are more stable than reactants.

16. Because of the law of conservation of mass, the number and type of each atom found in the reactants must be found in the products. In balancing a chemical equation, coefficients are placed in front of the formulas of the reactants and products so that the same number and type of atoms appear on both sides of the equation.

$$2H_2 + O_2 \rightarrow 2H_2O$$

Thinking Critically

17. Because polyatomic ions contain atoms which are covalently bonded, an ionic compound containing these polyatomic ions can contain covalent bonds.

18. Copper (II) ions in solution become copper atoms at the negative electrode. Chloride ions in the solution become chlorine atoms at the positive electrode, and combine to form molecules of chlorine gas.

Each copper (II) ion must gain two electrons. Only the negative electrode has excess electrons which can be gained by the copper (II) ions. Copper would form at the new negative electrode and chlorine gas would be produced at the new positive electrode.

19. Ammonium sulfate is an ionic compound containing two polyatomic ions. To show that it contains two ammonium ions, parentheses are placed around the chemical formula of the ammonium ion, NH_4 and a subscripted 2 is used to indicate two ammonium ions occur in the compound.

20. When electrons jump from one energy level to a lower energy level in an atom, they emit energy. Depending on the energy difference—which determines the frequency—the energy might be in the form of visible light rather than heat.

Reviewing Themes

15. *Environmental Interactions*
Describe the role of valence electrons in achieving stability through chemical bonding.

16. *Systems and Structures*
Oxygen gas and hydrogen gas react to form water. Use the law of conservation of matter to write a chemical equation for the interaction.

Thinking Critically

17. How can some ionic compounds contain covalent bonds?

18. If two carbon electrodes attached to a battery are placed in a beaker containing a solution of copper(II) chloride, two things happen. A layer of copper coats the electrode at the negative pole of the battery, and bubbles of chlorine gas form at the other electrode. Describe the chemical changes that occur. What must happen for Cu^{+2} to change to copper? Why does copper form only on the electrode at the negative pole of the battery? Predict what would happen if the poles of the battery were reversed.

19. Why is the formula of ammonium sulfate, a fertilizer, written as $(NH_4)_2SO_4$ rather than $N_2H_8SO_4$?

20. Some living organisms produce light through a chemical process called bioluminescence. Use your knowledge of chemical reactions to explain how a chemical reaction could occur that gives off light rather than producing or absorbing heat.

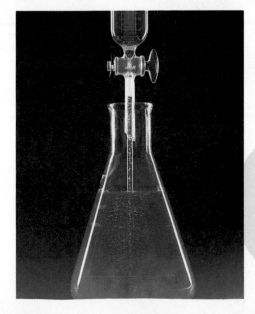

Discovery Through Reading

Uehling, Mark D. "Birth of a Molecule." *Popular Science* 240 (February 1992): 75–77, 88, 90. This article discusses use of computers to design new molecules with specific properties.

CHAPTER 21

ACIDS, BASES, and pH

PLANNING THE CHAPTER

B = Basic   **A** = Average   **H** = Honors
The coding Basic, Average, and Honors indicates subsections, features, and resources that might be appropriate for different levels of learners. For additional suggestions regarding choice of topic and depth of coverage, see the Pacing Chart on pages T28–T31.

*Frame numbers at point of use
(TR) Teaching Resources, Unit 6
(IT) Instructional Transparencies
(LI) Laboratory Investigations
(SD) *Science Discovery* Videodisc
     Correlations and Barcodes
(SRG) Study and Review Guide

CHAPTER MATERIALS

Title	Page	Materials
Teacher Demonstration	569	non-galvanized iron nail, can of soft drink
Discover by Doing	569	(per individual) blue and red litmus paper, journal
Discover by Doing	570	(per individual) baking soda, glass container, vinegar
Discover by Researching	570	(per individual) journal
Discover by Doing	572	(per individual) small cups (4), lemon juice, aluminum foil, iron nail, thumbtack, limestone
Discover by Doing	573	(per individual) strips of litmus paper (2) one blue and one red, ammonia
Activity: What does an antacid do?	574	(per individual) safety goggles, laboratory apron (per pair) beakers, dropper bottle with vinegar, blue and red litmus paper, spoon, wax paper, antacid tablet
Skill: Designing an Experiment	576	(per pair) to be determined by student-designed experiment
Discover by Doing	579	(per individual) pH paper, shampoo, detergent, ketchup
Investigation: Making Indicators	580	(per individual) safety goggles, laboratory apron (per group of 3 or 4) 50-mL beakers (8–10), mortar and pestle, flowers (4–5), 10-mL graduate, 10 mL of 95% ethanol, hot plate, stirring rod, funnel, filter paper, test tubes (5), test-tube rack, wax pencil, medicine dropper, 5 mL of hydrochloric acid, vinegar, distilled water, ammonium hydroxide, sodium hydroxide

ADVANCE PREPARATIONS

For the *Activity*, provide an assortment of brands of antacids. For the *Investigation*, keep the flowers fresh until they are used.

TEACHING SUGGESTIONS

Field Trip
Arrange for the class to visit a water purification facility to observe how water is treated before distribution to consumers. Ask the speaker to point out what is done to adjust the water's pH if the water is acidic or basic.

Outside Speaker
Invite an individual from your country's farm extension bureau to speak to the class about how the bureau works with farmers in relation to soil pH, fertilizers, and other issues related to acids, bases, and pH.

CHAPTER THEME—SYSTEMS AND STRUCTURES

This chapter defines acids and bases in terms of hydronium and hydroxide ions. The use of indicators and pH as a means to distinguish an acid from a base and a weak acid from a strong acid is also discussed. A discussion of the interaction of acids and bases to form salts is also included. A supporting theme in this chapter is **Environmental Interactions**.

MULTICULTURAL CONNECTION

Another name for sulfuric acid is "oil of vitriol." It was known to Arab scientists as far back as the eighth century. They wrote of sulfuric acid as a powerful "dissolving spirit."

MEETING SPECIAL NEEDS

Second Language Support

Due to the abstract nature of the material in this chapter, use visuals and concrete examples as much as possible. For example, have limited-English-proficient students list acids and bases they encounter on a regular basis, such as vinegar and baking soda. Then have the students write in their journals the name of each item in their native language opposite the English name.

CHAPTER
21 ACIDS, BASES, AND pH

It was the difficult Case of the Dying Lakes. *The patients were dying, but the doctors could not agree on what to do. Scientists knew that some lakes in the northeast United States were dying of acid poisoning, their fish completely gone. However, it took scientists and government officials a long time to diagnose and treat their sick patients.*

The first problem was finding the cause of the lakes' acidification. Many scientists and environmentalists blamed acid rain, the acid-laden rain and snow that fell on northeastern mountain ranges. Standing by a lake in upstate New York, it was easy to see the effects of acid rain. But scientists and officials found it harder to agree on the causes. Was it acidic pollution from power plants and automobiles in the Midwest United States, as many thought? How could this air pollution drift for thousands of miles and affect a handful of lakes in the Northeast? Could scientists pinpoint the sources of acidic pollution and advise governments about the most effective regulations?

While some scientists debated these difficult questions, others tried to give the patients some immediate relief. By the 1980s, up to 10 percent of lakes in some regions had become seriously acidified. Reducing air

CAUTION: Be sure that no student has food allergies to citrus fruits. If a student has such allergies, excuse the student from the activity or provide other sour foods, such as pickles.

Give each student a piece of citrus fruit to taste. Ask the students to describe how the fruit tastes. (tart, sour, not sweet) Explain to the students that certain molecules or atoms are responsible for the taste. Then introduce the term *acid* and explain to the students that they will learn that chemicals can be grouped into certain categories.

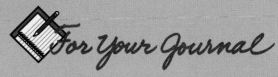

For Your Journal

The journal questions will help you identify misconceptions the students might have about acids, bases, and pH. In addition, these questions will help the students realize that they do possess knowledge about acids and bases. (The students will have a variety of responses to the journal questions. Accept all answers, and tell the students that at the end of this chapter they will have an opportunity to refer back to the questions.)

pollution would take 10 or 20 years. The *Case of the Dying Lakes* called for a short-term solution, something to restore the lakes temporarily until pollution controls took effect.

Scientists knew that the antidote for acid indigestion in humans was an antacid tablet, and to neutralize the acid on a car battery, you poured on a solution of baking soda. So scientists prescribed an antacid for the dying lakes. They poured the equivalent of baking soda into the lakes.

Scientists first attempted to neutralize acidic lakes by dumping carbonates from planes, helicopters, and snowmobiles. But they found that the carbonates didn't prevent fish kills. Acid rains and snowmelts brought acid "pulses" that killed eggs and young fish. So scientists tried a different prescription. They dumped carbonate pellets on the mountainsides around the lake and let the pellets neutralize the acid runoff before it reached the lake.

The acid lakes aren't out of the woods yet. Scientists still are not certain that neutralization and pollution controls will revive the dying lakes. The simple chemistry of acids and bases has created a very complicated environmental problem.

ABOUT THE PHOTO

The U.S. Department of Energy continues to support development of commercial technologies to fight acid rain. The atmospheric chemistry of acid rain is very complex, since it is influenced by temperature, solar radiation, and other atmospheric pollutants. Scientists are studying forests in many countries to determine the results of acid rain. Their goal is to ensure that Earth's forests will not be destroyed and that methods of preserving the forests will be compatible with the nations' economies and ecologies.

For Your Journal

- *What is an acid? What is a base?*
- *How can you determine if a substance is an acid or a base?*
- *How do acids and bases affect you every day?*

SECTION 1

Acids and Bases

Objectives

Distinguish between acids and bases.

Name some common acids and bases, and list their uses.

Compare the characteristics of strong and weak acids and strong and weak bases.

Hank is watching the lifeguard at the local pool take a sample of the water. The lifeguard adds some drops to the sample that make it turn pale yellow. To another sample of water she adds some red drops, and the solution turns violet. Hank is intrigued by what the lifeguard is doing. "Why do you add those drops to the samples of pool water?" he asks.

"I'm checking to see whether the amount of chlorine in the pool is right and what the pH is," Jessie, the lifeguard, responds. "Oh," says Hank. "I know it's important to have chlorine in the pool, but pH? What's pH?"

Jessie explains that the pool cannot be too acidic or too basic. If it is too basic she must add a substance to reduce the pH. "Wait, whoa, stop," Hank says. "I'm confused. I thought taking care of a pool was easy. What's an acid? What do you mean by basic and pH?"

Jessie smiles. "In order for you to understand pool chemistry, you will need to learn something about acids, bases, and salts. Why don't you come early tomorrow before the pool gets too crowded, and I can teach you."

● **Process Skill:** *Analyzing*

Remind the students that ions were discussed in Chapter 17. You might review the electron configuration of hydrogen at this time. Remind the students that hydrogen is a nonmetal.

● **Process Skill:** *Comparing*

Point out to the students that acids contain hydrogen and at least one other element. For example, sulfuric acid is made up of hydrogen, sulfur,

and oxygen. Then have the students identify the elements besides hydrogen in each chemical formula listed in Table 21–1 in their journals.

Figure 21–1. Citric acid got its name because it was first isolated from citrus fruits.

What's in an Acid?

Hank had heard of acids. In fact, he recently saw a horror movie that had an acid pit. It was full of a fuming, corrosive liquid that could "eat" through metal—and anything else. Surely Jessie could not be adding that kind of acid to the pool. Wouldn't that be too dangerous?

What Hank didn't know was that not all acids are as strong as the ones in the movie. In fact, Hank drinks and eats some acids every day. Acids are an important part of carbonated drinks, fruits, vegetables, meats, and even salad dressing!

Look at Table 21–1. Notice that each acid contains hydrogen. An **acid** is a compound that produces hydrogen ions (H^+) in solution. When an acid is dissolved in water, it separates into ions. Hydrochloric acid, for example, forms H^+ and Cl^- (chloride ions).

Table 21–1	Some Common Acids
Name	**Chemical Formula**
Acetic acid	$HC_2H_3O_2$
Citric acid	$HC_6H_7O_7$
Formic acid	$HCOOH$
Hydrochloric acid	HCl
Nitric acid	HNO_3
Phosphoric acid	H_3PO_4
Sulfuric acid	H_2SO_4

● **Process Skill:** *Analyzing*

Explain to the students that hydrogen ions (H⁺) are not stable in nature, because they are very reactive and combine quickly with other atoms or molecules. Hydronium ions (H₃O⁺) are the result of free H⁺ combining with water. The strength of an acid is determined by the concentration of H_3O^+ in solution for a given number of acid molecules.

● **Process Skill:**
Communicating

Display a poster that lists the safety precautions to be taken in a science laboratory. Have the students identify the precautions that particularly apply to the use of acids and bases.

✧ Did You Know?

While acids do react with metals and living tissue, they do not react with glass or most plastics. Therefore, strong acids are stored in glass containers or some types of plastic containers.

ONGOING ASSESSMENT
▼ ASK YOURSELF

A substance is probably an acid if its chemical formula contains hydrogen.

THE NATURE OF SCIENCE

In 1887 Swedish chemist Arrhenius gave a structural definition for acids and bases. Arrhenius stated that acids release hydrogen ions in a water solution, while bases form hydroxide ions in a water solution. In 1923 Bronsted and Lowry gave acids and bases a broader, more general definition: An acid is a proton donor; a base is a proton acceptor. Later, Lewis gave acids and bases even broader definitions: An acid accepts a pair of electrons; a base donates a pair of electrons.

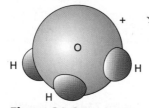

Figure 21–2.
Hydronium ions contain one oxygen atom and three hydrogen atoms. The number of hydronium ions in solution determines the strength of the acid.

What is H⁺? If you remove one electron from hydrogen you are left with a proton. Solitary protons do not wander around in solutions. Instead the H⁺ combines with a water molecule to form H₃O⁺, or a hydronium ion. Any solution that contains H₃O⁺ is an acidic solution. For simplicity, the symbol H⁺ will be used in this chapter.

▼ ASK YOURSELF
How can you tell from a chemical formula if a substance might be an acid?

Pucker Up

Jessie had told Hank that the safest way to taste an acid was to taste an unripe apple and compare it to a ripe apple. Hank went to the store and looked for the most unripe apple he could find. On the way home he bit into it. Wow! Was it ever sour!

One property of acids is that they taste sour. Most fruits have some acid content. Oranges and lemons contain citric acid; apples contain malic acid. Unripe fruit tastes sour because it has a greater concentration of acid than ripe fruit does. As the fruit ripens, the acids are converted into sugar.

Another common acid is vinegar. Vinegar is a solution of acetic acid in water. Vinegar and the acids in fruit are weak acids. Hank decided to look for the names of some more acids in the encyclopedia. Table 21–2 lists some weak acids and some strong acids listed in Hank's references. **CAUTION: Strong acids should never be touched or tasted.**

Table 21–2	Strengths of Some Acids	
Strong Acids		**Weak Acids**
Hydrochloric acid (HCl)		Acetic acid (HC₂H₃O₂)
Nitric acid (HNO₃)		Citric acid (HC₆H₇O₇)
Sulfuric acid (H₂SO₄)		Carbonic acid (H₂CO₃)
Hydrobromic acid (HBr)		

The strength of an acid is determined by the concentration of hydronium ions in solution. The more hydronium ions the acid produces in solution, the stronger the acid. Strong acids ionize completely in water, producing many hydronium ions. Weak acids do not ionize completely, so there are not as many hydronium ions in solution.

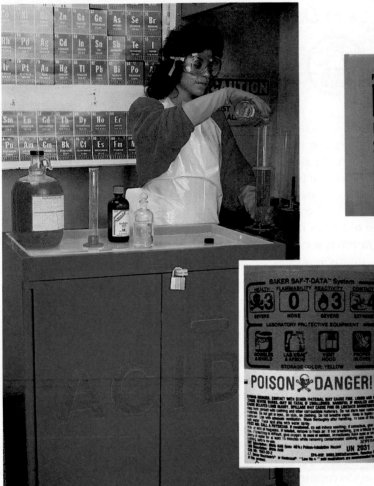

Figure 21–3. Strong acids must be handled with care. When using acids, you should always wear safety goggles and a laboratory apron and you should work in a well-ventilated area. The label on an acid bottle will provide any special instructions you might need.

DISCOVER BY *Doing*

 Use some blue and red litmus paper. What hap-pens when you dip them in vinegar or orange juice? Write the results in your journal. Can you think of any uses for what you have learned? ✐

Strong acids must be handled carefully because they can produce severe burns. Perhaps you have observed how metals react in strong acids. In addition to reacting with metals, acids also react with many other types of compounds. Try the next activity to find out about one kind of reaction.

SECTION 1 **569** ◀

TEACHING STRATEGIES, continued

● **Process Skill:**
Communicating

Point out to the students that the type of rock through which water travels often changes the acidity of the water. Water traveling through rocks that contain sulfur compounds form sulfuric acid and is, therefore, often quite acidic.

● **Process Skill:** *Applying*

Have volunteers name and locate on a map of the United States areas known for underground caves that contain stalactites and stalagmites. (Examples include Mammoth Cave in Kentucky and Blue Mounds Cave in Wisconsin.)

DISCOVER BY *Doing*

The mixture will bubble and release carbon dioxide gas (CO_2).

REINFORCING THEMES
Systems and Structures

Cooperative Learning Review with the students that limestone and marble—such as those found in the sculptures in Figure 21–5—contain carbonates and are, thus, susceptible to attack by acids. The air in industrial areas contains sulfur and nitrogen compounds that react with water in the air to form acids. These compounds combine with water in the air to form sulfuric or nitric acid. Then have the students work in pairs or small groups to design a diagram of how acid rain forms and damages the outside of buildings and other structures.

🔲 **LASER DISC**
619, 620, 621

Limestone cavern

DISCOVER BY *Researching*

Encourage the students to place their research on acid rain in their science portfolios.

ONGOING ASSESSMENT
▼ **ASK YOURSELF**

The strength of an acid is determined by the concentration of hydronium ions in solution. The greater the number of ions, the stronger the acid.

▶ **570** CHAPTER 21

Figure 21–4. Caves like this one are formed when acid reacts with limestone.

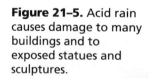 

Figure 21–5. Acid rain causes damage to many buildings and to exposed statues and sculptures.

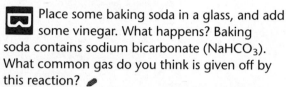

DISCOVER BY *Doing*

🔲 Place some baking soda in a glass, and add some vinegar. What happens? Baking soda contains sodium bicarbonate ($NaHCO_3$). What common gas do you think is given off by this reaction? ✏

A similar reaction to the one in the activity also takes place in nature. In areas where limestone is common, this reaction can cause the formation of caves and sinkholes.

Limestone is mainly calcium carbonate. When carbon dioxide in the air dissolves in rainwater to form a weak acid, the acid dissolves minerals in the cave rock. The dissolved minerals form stalactites and stalagmites.

Many building materials contain carbonates. The same reaction that occurs in caves can occur over time on concrete and marble.

DISCOVER BY *Researching*

Locate an article on acid rain in the library. Find the answers to the following questions about acid rain: What causes it? What are its effects? What are some of the acids found in it? What can be done to prevent it? ✏

▼ **ASK YOURSELF**
What determines the strength of an acid?

tion. Fatty acids— composed of carboxylic acid (a weak organic acid)—are essential components in organic cells.

Figure 21–6. Sulfuric acid has many uses in industry. One important use is in the manufacture of plastics.

INTEGRATION– Health

Have interested students use reference resources to find out how changes in the amount of stomach acid (hydrochloric acid) can cause ulcers. Have them write what they learn in their journals. (Ulcers can develop in the digestive tract when excessive secretion of stomach acid begins to "eat away" the digestive tract's lining. Once an ulcer forms, any factor that elevates the secretion of stomach acid will aggravate the ulcer. Aggravating factors include hasty, irregular eating habits and emotional strain.)

Uses of Acids

Hank was ready for another chat with Jessie. "I have learned a lot about acids," he says, "but I'm still not sure why it is important to add acid to the pool."

"One reason to add acid to the pool is to prevent deposits of carbonates," says Jessie. "I learned about how acids react with carbonates," says Hank. "Does that mean when you add acid the pool will foam?" "Usually not," Jessie assures him. "But we do use well water for our pool. Well water contains a lot of dissolved carbonates. If we don't do something about it, they will form a crust around the rim of the pool and clog the pipes. So I add just enough hydrochloric acid to react with the carbonates but not enough to harm the swimmers."

Hank went home and decided to look up more information on uses for acids. He found out that acids are very important to many industrial processes. Sulfuric acid, a strong acid, is used to make fertilizers and automobile batteries. It is also an important component in steel production and many other manufacturing processes. Sulfuric acid is used in tanning leather for shoes, in treating the paper

Figure 21–7.
Hydrofluoric acid is used to etch fine crystal. Because this acid "eats through" glass, it must be stored in plastic.

 **SCIENCE TECHNOLOGY SOCIETY** Sulfuric acid is the chemical produced in the largest amounts throughout the world. The production of sulfuric acid is so important that it serves as an indicator of how well the U.S. economy is doing.

● **Process Skill:** *Comparing*

Discuss with the students any situations in which they have used sulfuric, nitric, or hydrochloric acids. (For example, the students might be familiar with the warnings on car batteries that urge caution because the battery's cells contain sulfuric acid.)

● **Process Skills:** *Analyzing, Hypothesizing*

Ask the students to hypothesize whether water should be added to a concentrated acid, or vice versa. Have the students give reasons for their answers. (Acid is added very slowly to water so the heat generated can dissipate slowly. If water is added to concentrated acid, the water will hydrolyze on contact and splatter the acid.)

① Protective equipment is necessary to protect the worker from acid burns.

SCIENCE BACKGROUND

Nitric acid was one of the first acids known. Alchemists used a mixture of nitric and hydrochloric acids—called *aqua regia*— in their effort to change lead into gold. Sulfuric acid—which is the most widely used acid in industry—was discovered after the other two strong acids.

INTEGRATION–
Earth Science

Have interested students research techniques used by mining companies to protect the environment from acidic waste water. A local coal mining or power company are possible sources of information. Have the students record their findings in their journals.

DISCOVER BY *Doing*

The students should relate the rate of a reaction to the strength of the acid used (in this case, lemon juice). The reaction with the limestone will be the fastest; bubbles of gas should be visible on contact. The other reactions will be slower.

ONGOING ASSESSMENT
▼ ASK YOURSELF

Acids are important because they are used in the production of many products, such as chemical cleaners and fertilizers.

Figure 21–8.
Hydrochloric acid is used to clean buildings. Notice the protective gear the worker is wearing. Why do you think this equipment is necessary? ①

Figure 21–9. One use for nitric acid is to make dyes such as those shown here.

in this book, and even in making the vanilla flavoring for ice cream.

Nitric acid, another strong acid, is used to make explosives, such as nitroglycerin and dynamite. It is also used in the manufacture of ammonium nitrate (NH_4NO_3), an important ingredient in fertilizers.

Then there is the hydrochloric acid that Jessie uses in the pool. It is commonly called *muriatic acid*. Besides balancing the acid content of swimming pools, it is also used to clean concrete.

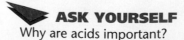

DISCOVER BY *Doing*

Test the effects of a common acid on different materials. Get four small paper cups, and fill them about half full of lemon juice. In the first cup, place a small piece of aluminum foil. In the next three cups, place an iron nail, a thumbtack, and a piece of limestone, respectively. Observe and describe the effect of the acid over a period of several days. ✐

▼ ASK YOURSELF
Why are acids important?

The Solution Is Basic

"Jessie, what would you do if the pool was too acidic?" asks Hank while sitting on the edge of the swimming pool splashing his feet. "I would add a base to neutralize the acid." "You mean like first base or home plate?" jokes Hank. "Very funny," Jessie answers. Then she proceeds to explain about bases.

Remember that an acid produces hydronium ions in solution. A **base** produces hydroxide ions (OH^-) in solution in the same way. When bases are added to water, they react to form hydroxide ions. The stronger the base is, the more hydroxide ions are released into solution.

Although many bases contain hydroxide ions, there are some bases that do not. Ammonia gas, for example, is a base. However, it produces hydroxide ions only when it is dissolved in water.

$$NH_3 + H_2O \rightarrow NH_4^+ + OH^-$$

The key to being a base, remember, is the formation of hydroxide ions. Since ammonia forms OH^- in solution, it is a base. Household ammonia is a solution of ammonia gas and water.

Solutions that contain bases are called basic solutions. However, the word *alkaline* is sometimes used instead of *basic*. Strong bases are just as dangerous as strong acids. They can cause serious burns and must be handled with care. **CAUTION: Never touch or taste a strong base.** Find out more about bases by trying the next activity.

Figure 21–10. Strong bases, such as the drain cleaner shown here, are as dangerous as strong acids.

 **DISCOVER BY** *Doing*

Take two strips of litmus paper, one red and one blue. Dip each into some household ammonia. What happens? How does this compare with the vinegar or orange juice test you did before? ✐

Soap solutions are mildly alkaline. If you ever had soap in your mouth, you are familiar with the bitter taste bases have. Bases usually have a slippery, soapy feel.

Bases react with acids to form salts. This process is called *neutralization*. Neutralization occurs when equal amounts of

ACTIVITY

What does an antacid do?

Process Skills: Measuring, Interpreting Data, Comparing

Grouping: Pairs

Hints

Different groups should use different brands of antacids for comparison. Instruct the students to gently swirl their beakers after adding each drop of vinegar.

Tell the students not to handle the litmus paper since perspiration can change the color of the paper.

▶ Application

1. An antacid dissolves in the stomach and neutralizes excess stomach acid. (Excess stomach acid can cause a burning feeling in the stomach.)

2. The antacid that worked best is the one that took the greatest amount of vinegar before the litmus paper turned red. These ions produce water, which is neutral.

GUIDED PRACTICE

Cooperative Learning Have the students work in pairs to develop a list of products that contain acids or bases. Encourage the students to compare their lists with the lists of their classmates.

INDEPENDENT PRACTICE

Have the students write a few paragraphs on how to maintain personal and environmental safety when working with acids and bases. Encourage the students to develop visuals to accompany their reports.

EVALUATION

Have the students use litmus paper to identify unlabeled samples of weak acids and bases. Ask them which type of litmus paper indicates acids and which indicates bases. (Blue litmus paper turns red in an acid; red litmus paper turns blue in a base.)

hydronium ions and hydroxide ions are in solution. These ions produce water.

$$H^+ + OH^- \rightarrow H_2O$$

Once the water is formed, the solution is no longer acidic or basic; it is neutral.

Neutralization occurs in the human body. When a person has an acid stomach, or heartburn, he or she may take an antacid tablet. The antacid tablet is a weak base that dissolves in the stomach and combines with the excess acid to neutralize it. Try the next activity to find out more about the neutralization process.

ACTIVITY   *What does an antacid do?*

MATERIALS

safety goggles, laboratory apron, beakers (2), dropper bottle with vinegar, blue and red litmus paper, spoon, wax paper, antacid tablet

CAUTION: Put on safety goggles and a laboratory apron, and keep them on throughout this activity.

PROCEDURE

1. Place 100 mL of water in a beaker, and add vinegar to it, one drop at a time. Test the solution with litmus paper after each drop is added. How many drops of vinegar did it take to make the solution very acidic (when the blue litmus paper turns bright red)? Record your results.

2. Place an antacid tablet on a small piece of wax paper. Crush the tablet with the back of a spoon, and dissolve it in about 100 mL of water.

3. Use litmus paper to find out whether the solution is acidic, basic, or neutral. Record your results.

4. Add vinegar, one drop at a time, to the antacid solution. Record the number of drops required to turn the litmus paper a bright pink. Compare your results with the solution that contained only vinegar and water. Compare your antacid with those of other groups.

APPLICATION

1. How does an antacid help with an acid stomach?

2. An antacid should neutralize excess acid in your stomach. Which brand worked best? Explain why you think so.

ONGOING ASSESSMENT
▼ ASK YOURSELF

Neutralization occurs when equal amounts of hydronium ions and hydroxide ions are in solution.

▼ ASK YOURSELF

Describe a neutralization reaction.

RETEACHING

Have the students practice reading, writing, and recognizing formulas for acids and bases by making a large table on the chalkboard. Label two columns "Acids" and "Bases." Provide cards with different ions printed on them, such as H^+, Na^+, Cl^-, OH^-, and so on. Then have volunteers select and arrange specific cards to show some of the chemical formulas discussed in this section. After confirming that a formula is correct, have the student write the formula under the correct column.

EXTENSION

Have interested students use reference resources to find out how pioneers used animal fat and fireplace ash to produce soap. Have them summarize their findings in their journals. (The animal fat was cooked with the ash, a base. The strength of the soap depended upon the amount of ash, or base, used.)

CLOSURE

Have the students use reference resources to identify each acid below as strong or weak.

$HClO_4$, perchloric acid (strong)
HCl, hydrochloric acid (strong)
HF, hydrofluoric acid (weak)
$HC_2H_3O_2$, acetic acid (weak)
H_3PO_4, phosphoric acid (weak)
H_2SO_4, sulfuric acid (strong)

Uses of Bases

Bases are as important as acids industrially. Bases are used in drain cleaners and soaps.

Sodium hydroxide is an important base. It has numerous uses that affect you. For instance, since it is used by the paper industry to remove pulp fibers from wood, it was probably used in producing this book.

Another commonly used base is ammonia. This compound is present in many cleaners because it can dissolve grease.

Many other bases are also used often. More than 16 billion kilograms of calcium hydroxide, $Ca(OH)_2$, are produced in the United States annually. This base is used to make cement, mortar, and plaster. Another base, magnesium hydroxide $Mg(OH)_2$, is used as an antacid. These are only a few examples of bases used every day.

Figure 21–11. Sodium hydroxide is a strong base used in the paper-making process.

▼ **ASK YOURSELF**

List two common bases and their uses.

Figure 21–12. Most households use many bases.

SECTION 1   REVIEW AND APPLICATION

Reading Critically

1. What makes an acid or a base strong or weak?

2. List three main differences between an acid and a base.

3. Why do unripe fruits taste sour?

Thinking Critically

4. An acid is sometimes called a proton donor. Why is an acid such as HCl thought of as a proton donor?

5. Would you use hydrochloric acid to clean silverware? Explain.

6. Why is magnesium hydroxide used as an antacid instead of sodium hydroxide?

Process Skills: Analyzing, Experimenting, Formulating Hypotheses, Observing, Predicting

Grouping: Pairs

Objective
● **Design** an experiment.

Hints

Be sure the students use only weak acids or bases in their experiments. Have the students take all necessary safety precautions when working with acids or bases. (use goggles, gloves, rubber apron, hood, conduct the experiment in a well-ventilated area, and so on)

Work with the students to develop hypotheses, if necessary. Be sure the students' experiments can be conducted in a safe manner.

Refer the students to Chapter 1 if they need to review the guidelines for a scientific method. Remind them to pay special attention to variables when designing their experiments; only one variable should be tested at a time.

▶ **Application**

Review the students' data and conclusions for logic and accuracy. Accept all reasonable conclusions based on the students' data.

✳ **Using What You Have Learned**

Review the students' proposed revisions for appropriateness. Accept all reasonable revisions that the students can justify.

PERFORMANCE ASSESSMENT

Cooperative Learning Have the students discuss their experiment designs and encourage classmates to suggest revisions, as applicable. Accept all revisions that the students can justify.

SKILL Designing an Experiment

▶ **MATERIALS**
● to be determined by student-designed experiment

▼ **PROCEDURE**

1. In earlier chapters you learned about the importance of controls and only changing one variable at a time when you perform an experiment. Suppose you wanted to find out more about the effect acid rain might have on different kinds of rocks. How would you go about designing an experiment that would provide that information?

2. Before doing an experiment, most scientists do library research to find out what is known about the subject. Check in the library for information on acid rain and the types of rocks that you might find in your area.

3. Write the hypothesis you are going to test. Have your teacher approve it before you go on. Scientists often have other researchers examine their work and offer advice.

4. Write a complete procedure for your experiment. Don't forget to include information on your variables. To complete this step in designing an experiment, you need to decide what acid you will use, what measurements you will make, and what types of rocks you want to test. Make a list of the materials you will need. Be sure to include any safety precautions you should use.

▶ **APPLICATION**
Collect your materials. Now there is only one thing left to do: conduct the experiment! Record your data as you proceed. Then draw conclusions based on your data.

✳ ***Using What You Have Learned***
After you have completed your experiment, go back and review your procedure. Would you change anything you did? How could you revise your procedure to improve the experiment?

Section 2:

pH AND INDICATORS

FOCUS

Section 2 discusses pH and the use of indicators. The students will also learn how to determine whether a substance is acidic, basic, or neutral by interpreting the color of an indicator.

MOTIVATING ACTIVITY

Demonstrate the use of an indicator by pouring 250 mL of alcohol into a flask. Then add 5–6 drops of bromothymol blue indicator solution. Stopper the flask tightly. The solution should be blue in color. If not, add a few drops of dilute NaOH. Be sure not to allow the solution in contact with the air before the

demonstration, or it will change color.

Explain to the students that this reaction will respond to certain people. Then have a volunteer remove the stopper and speak into the flask asking the solution to turn yellow. The volunteer should then stopper the flask and swirl the contents gently.

pH and Indicators

SECTION 2

Hank is still a bit puzzled about the way Jessie's pool test kit works. He knows that different drops produce different colored solutions. How does Jessie know whether to add acid or base to the pool? Hank decides to do some research for himself on this pH stuff Jessie keeps talking about.

Objectives

Classify various substances based on their pHs.

Define the term indicator, and give three examples of indicators.

Determine whether a substance is acidic, basic, or neutral by interpreting the color of an indicator.

Red or Blue—Acid or Base

The pH of a solution is a measure of the hydronium ions in the solution. The pH scale ranges from 0 to 14. Acids, bases, and neutral solutions are separated into regions on the scale. The middle point, pH = 7, is neutral. Thus, a solution with a pH of seven is neither an acid nor a base.

Acids have a pH of less than 7. The stronger the acid, the lower the number on the pH scale. For example, the pH of lemon juice is about 2.3, while tomato juice, which is less acidic, has a pH of 4.

Bases have a pH higher than 7. The higher the number above 7 on the pH scale, the stronger the base. The chart below shows the pH values for some common substances.

Figure 21–13. This chart shows the pH of some common substances.

PROCESS SKILLS
• Experimenting
• Interpreting data

POSITIVE ATTITUDES
• Scientific endeavor •
Initiative and persistence

TERM
• indicators

PRINT MEDIA
"Learning from an Acid Rain Program" in *Science* magazine by Leslie Roberts
(see p. 513b)

ELECTRONIC MEDIA
Acid, Bases and Salts. Coronet Film and Video.
(see p. 513c)

SCIENCE DISCOVERY
• Indicator solution

BLACKLINE MASTERS
Thinking Critically
Laboratory Investigation 21.2
Reading Skills
Study and Review Guide

INTEGRATION– Mathematics

The term *pH* literally means the "power of hydrogen." The pH scale is a logarithmic relationship. Each number of the scale differs from the next by a factor of ten. For example, an acid with a pH of 3 is ten times more acidic than an acid with a pH of 4. A base with a pH of 10 is ten times more basic than a base with a pH of 9.

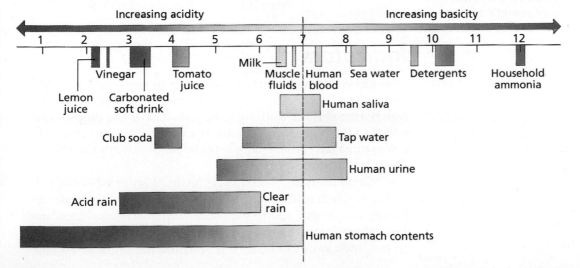

Increasing acidity | Increasing basicity

1 2 3 4 5 6 7 8 9 10 11 12

Lemon juice — Vinegar — Carbonated soft drink — Tomato juice — Milk — Muscle fluids — Human blood — Sea water — Detergents — Household ammonia — Human saliva — Club soda — Tap water — Human urine — Acid rain — Clear rain — Human stomach contents

Have other volunteers repeat the procedure until a color change is observed.

Ask the students to describe what happened. (The solution will turn yellow after approximately 10 students talk into the solution. Enough CO_2 will have dissolved in solution from the students' breath to form car- bonic acid: $CO_2 + H_2O \rightarrow H_2CO_3$.) Conclude by telling the students that bromothymol blue is an indicator.

GUIDED PRACTICE

Cooperative Learning Have the students work in pairs to identify and define the main concepts of this section.

INDEPENDENT PRACTICE

Have the students provide written answers to the Section Review. In their journals, have the students list several substances and their respective pH.

INTEGRATION–
Health

Have interested students use reference resources to find out what might cause a person's internal pH to change. (Possibilities include a blood disorder or a stomach disorder.)

ONGOING ASSESSMENT
▼ **ASK YOURSELF**

The pH range for acids is 0 to 7. The pH range for bases is 7 to 14.

✧ **Did You Know?**
In the human body, most fluids are slightly basic, except for gastric juices, which have a pH between 1 and 2.

MEETING SPECIAL NEEDS
Gifted

Have interested students use reference resources to find out how botulism sometimes occurs in improperly canned foods and what foods are less likely to promote the growth of botulism. (Acid foods—such as dill pickles—when properly canned, do not promote the growth of botulism.)

Figure 21–14. This scientist is testing lake water for acid rain contamination. She is using a pH meter.

Many plants and animals are very sensitive to changes in pH. Lake water usually has a pH between 6 and 7. However, due to acid rain, some lakes in the northeast United States have a pH as low as 3. This drastic change kills any plants and animals that are sensitive to pH changes.

Scientists are experimenting with methods to bring the pH of these acidic lakes back to normal levels. In order to do this, large quantities of weak bases, such as ground limestone or sodium bicarbonate, are dumped from planes into the lakes. Scientists hope that the addition of bases will neutralize the excess acid. The results of these efforts, however, will be only temporary if the sources of pollution are not stopped.

Your body is also very sensitive to changes in pH. The normal pH of human blood is between 7.38 and 7.42. This range is very narrow; any change in the concentration of hydronium ions has a marked effect on the body. If the pH is above 7.8 or below 7.0, the body cannot function normally. If this change is not corrected rapidly, it can be fatal.

▼ **ASK YOURSELF**
What is the pH range for acids? For bases?

Indicators

How can we know the pH of pool water or any solution if the hydronium ions and hydroxide ions cannot be seen? This question bothered Hank until he remembered that the drops Jessie added to the water sample made the water different colors. The colors must tell Jessie what the pH is. He decides to ask Jessie about it that afternoon.

"Good thinking, Hank," says Jessie. "I know what the pH is and how much acid or base to add by the color of the water solution. The drops I add to the water sample are called indicators. **Indicators** are substances that change color in an acid or base. In fact, indicators can be used to measure the pH of just about

RETEACHING

Have the students explain the pH scale in their own words. Encourage them to draw diagrams in their journals to accompany their explanations.

EXTENSION

Have interested students find out why shampoos should have a pH of about 5 to 6. (A pH of around 5 to 6 is close to the pH of human hair and skin. A harsh shampoo that has either a very high or a very low pH would make hair dull or even damage it.)

EVALUATION

Ask a volunteer to define the term *indicator* and explain how indicators differ. (An indicator can be a substance that changes color in an acid or base. Litmus paper is a commonly used indicator.)

CLOSURE

Cooperative Learning Have the students work in pairs or small groups to write a summary of this section. Ask each group to read aloud its summary and revise it, if necessary, based on classmates' feedback.

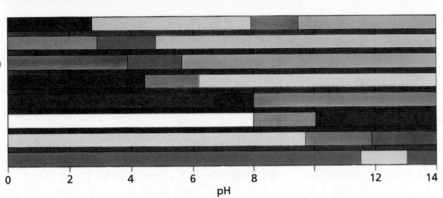

Figure 21–15. Indicators turn different colors to show acids or bases.

anything. My grandmother has a pH test kit for her garden so she can test the soil to see whether it is acidic or basic. Vegetables grow best within certain pH ranges. Grandma can add substances to the soil to get the pH she wants."

Some indicators show only two color changes—one for acids and another for bases. Other indicators, however, show a range of color changes depending on the pH. One type of indicator paper, called *pH paper*, is dyed with universal indicator. Universal indicator is a mixture of several different indicators, so the paper gives a different color for each pH. The advantage of this type of indicator is that it shows the exact pH of a solution rather than just whether the solution is an acid or a base.

DISCOVER BY Doing

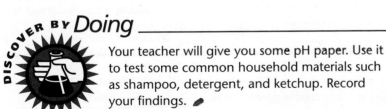

Your teacher will give you some pH paper. Use it to test some common household materials such as shampoo, detergent, and ketchup. Record your findings.

Figure 21–16. pH paper can be used to determine the pH of almost any liquid.

► ASK YOURSELF

How can scientists determine pH?

SECTION 2   REVIEW AND APPLICATION

Reading Critically
1. Define the term *pH*.
2. You have two bases, pH = 8 and pH = 12. Which is more alkaline? Why?

Thinking Critically
3. How could testing the pH of a lake be beneficial to people who fish?
4. What is the ideal pH of Jessie's pool? Why?

LASER DISC

50333–51120

Indicator solution

DISCOVER BY *Doing*

Encourage the students to test safe substances that have not been tested in class. Point out that some substances might need to be diluted because the color of the substance will mask the color of the pH paper. The students' findings should indicate that most soaps and cleaners are basic and most foods are acidic or neutral.

ONGOING ASSESSMENT
► ASK YOURSELF

Scientists can use pH paper, a pH meter, and other indicators to determine pH.

SECTION 2 REVIEW AND APPLICATION

Reading Critically

1. pH is the "power of hydrogen"; the pH scale, from 0 to 14, indicates how acidic or basic a substance is.

2. The base with a pH of 12 is more alkaline. The higher the pH, the more basic the substance.

Thinking Critically

3. Fish live in water with specific pH. Each type of fish has a preferred pH. Knowing the pH of the water would tell you what type of fish might be in an area.

4. The pool water should have a pH between 6 and 8 (neutral) so that the water is not irritating to swimmers.

SECTION 2   **579** ◄

Making Indicators

Process Skills:
Experimenting, Interpreting Data

Grouping: Groups of 3 or 4

Objective
● **Make** indicators from different flowers.

Pre-Lab
Review with the students why they have learned about indicators. Then explain to the students that in this investigation they will make their own indicators from common plants.

Hints
Ammonium hydroxide is available already prepared from several scientific supply houses. If you prepare the solution yourself, using ammonia and water, the solution will not have the pH required for this *Investigation*.

Almost any vegetable material can be used to make an indicator, but flowers usually produce the most brilliant colors. You might want to perform the extractions yourself before the *Investigation* is done in class. Have labeled test tubes of the extracts ready for the students to use.

Have the students wash the graduate between uses. Remind the students not to return any of the reagents to the reagent bottles.

This investigation will be unsuccessful for any student who brings a sample that won't act as an indicator. You might prepare some red cabbage juice indicator to have on hand for these students.

Analyses and Conclusions
1. The students should show an understanding that good indicators have a distinct color change over a narrow pH range. Accept all reasonable responses.

2. Yes. One of the indicators in each flower could be the same, or they could change colors close to the same pH range.

Application
Besides common solvents for chromatography, also try: n-butanol/concentrated NH_3 (4:1) and n-butanol/glacial acetic acid (4:1), but only under direct teacher supervision. Use a UV light, if possible, to develop your chromatogram. Many of these indicator dyes are fluorescent. Keep a record of what worked and under what conditions so they might be repeated the following year.

✳ Discover More
Test results will depend upon the natural substances chosen.

Post-Lab
Discuss with the students whether or not the extracts of other types of plant life could be used as indicators.

INVESTIGATION

Making Indicators

▶ MATERIALS

● safety goggles
● laboratory apron ● beakers, 50 mL (8–10) ● mortar and pestle ● flowers (4–5) ● graduate, 10 mL ● ethanol, 95%, 10 mL ● hot plate ● stirring rod ● funnel ● filter paper ● test tubes (5) ● test-tube rack ● wax pencil ● medicine dropper; ● prepared solutions of the following acids and bases, approximately 5 mL each:

HCl, hydrochloric acid (pH = 1) NH_4OH, ammonium hydroxide (pH = 11)
$HC_2H_3O_2$, vinegar (pH = 3)   NaOH, sodium hydroxide (pH = 13)
Distilled water (pH = 7)

CAUTION: Put on safety goggles and a laboratory apron, and keep them on throughout this investigation. Be sure your area is well ventilated.

▼ PROCEDURE

1. Prepare a table like the one shown here.

TABLE 1: COLOR CHANGES OF FLOWER INDICATORS

Solution	Flower Name	Flower Name	Flower Name	Flower Name
HCl				
$HC_2H_3O_2$				
Distilled water				
NH_4OH				
NaOH				

2. Using the mortar and pestle, crush the petals of one flower. Put them in a beaker.
3. CAUTION: Ethanol is flammable. Do not heat over an open flame. Add 10 mL of ethanol to the beaker,
and heat it gently in a hot-water bath or on a hot plate for 15 minutes. Stir often.
4. Allow the mixture to cool, and filter it into a clean beaker. Your indicator may not have any color.

5. Repeat steps 2–4 for each type of flower.
6. Label each test tube with the name of one of the prepared acid or base solutions. Place 5 mL of each solution into the correct test tube.
7. To each solution, add 1 or 2 drops of the indicator from one flower. Record any color changes you observe. If the color change is not clear, add more indicator—one drop at a time.
8. Clean the dropper, and repeat steps 6–7 for each indicator.

▶ ANALYSES AND CONCLUSIONS
1. Which flower made the best indicator? Give reasons for your answer.
2. Could two flowers of different colors be used to indicate the same pH? Explain your answer.

▶ APPLICATION
Find how many different pigments are in your flower indicator, by completing a paper chromatograph.

✳ Discover More
Test other natural substances for their suitability as indicators.

The Big Idea—Systems and Structures

Lead the students to an understanding that acids and bases affect and interact with both living organisms and nonliving substances. Also, point out to the students that an understanding of how acids and bases interact with each other—as well as the living and nonliving parts of the ecosystem—can help them be more environmentally conscious.

CHAPTER

21

HIGHLIGHTS

The Big Idea

Acids and bases affect natural systems as small as a swimming pool or as large as an ecosystem. Acids interact with nonliving substances such as building materials and rocks. Living organisms are also sensitive to changes in pH. In fact, most living systems are so sensitive to changes of this type that they cannot live if the pH changes drastically.

Acids and bases interact to form neutral substances in a process known as neutralization. By understanding the interactions of acids and bases with each other as well as with the living and nonliving parts of the ecosystem, you can make practical decisions about everyday occurrences. You will know what chemicals to add to your garden soil, how to counteract an acid spill, or what types of pollution would increase acid rain.

For Your Journal

At the beginning of the chapter, you wrote down your ideas about acids and bases. Look back at your ideas. How have your ideas changed? Revise your journal entries to show what you have learned.

For Your Journal

The students' ideas should reflect an understanding of acids, bases, pH, and indicators. You might have the students share their new ideas with classmates.

CONNECTING IDEAS

The students' concept maps will vary, but should include terms such as *base, hydroxide, slippery feel, high pH, ammonia,* and *soap,* as well as the appropriate safety precautions.

Connecting Ideas

Below is a concept map about acids. Study it carefully, and, in your journal, make a matching concept map about bases.

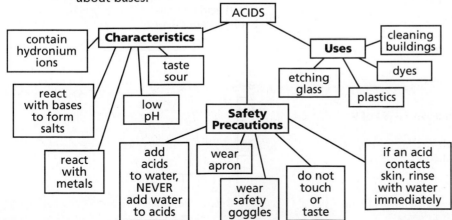

ACIDS

Characteristics
- contain hydronium ions
- react with bases to form salts
- react with metals
- taste sour
- low pH

Uses
- cleaning buildings
- dyes
- plastics
- etching glass

Safety Precautions
- add acids to water, NEVER add water to acids
- wear apron
- wear safety goggles
- do not touch or taste
- if an acid contacts skin, rinse with water immediately

Understanding Vocabulary

1. An acid is a substance that produces hydrogen ions in solution. A base is a substance which produces hydroxide ions in solution. pH is the measure of the acidity or basicity of a solution. An indicator is a material whose color shows the approximate pH of an acid or a base.

Understanding Concepts

Multiple Choice

2. b

3. c

4. c

5. a

6. c

Short Answer

7. Strong acids have a greater concentration of hydronium ions in solution than weak acids. Strong bases have a greater concentration of hydroxide ions in solution than weak bases.

8. Water and salt are formed when an acid and a base neutralize each other.

9. An acid turns blue litmus paper red. A base turns red litmus paper blue.

Interpreting Graphics

10. The solution is basic since pH increases as the solution is added.

11. Paper "a" indicates a base. Paper "b" indicates an acid.

CHAPTER

21

REVIEW

▶ Understanding Vocabulary

1. Explain how the following terms are related: acid (567), base (573), pH (577), indicator (578)

▶ Understanding Concepts

MULTIPLE CHOICE

2. An acid produces what type of ions in solution?
   a) oxygen
   b) hydronium
   c) hydroxide
   d) sulfur

3. One common acid is
   a) water.
   b) milk of magnesia.
   c) vinegar.
   d) bleach.

4. A base produces what type of ions in solution?
   a) oxygen
   b) hydronium
   c) hydroxide
   d) sulfur

5. A common base is
   a) ammonia.
   b) water.
   c) lemon juice.
   d) milk.

6. A substance with a pH of 9 would be
   a) neutral.
   b) an acid.
   c) a base.
   d) a salt.

SHORT ANSWER

7. Explain the differences between strong and weak acids and strong and weak bases.

8. Describe the chemical reaction that takes place when an acid and a base neutralize each other.

9. Explain how you would use litmus paper to determine if a substance were an acid or a base.

Interpreting Graphics

10. Study the graph below. Is the solution being added acidic or basic? Explain your answer.

11. Which litmus paper sample indicates an acid? a base?

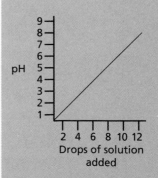

Reviewing Themes

12. Hydrochloric acid and sodium hydroxide would react to give sodium chloride (table salt) and water.

$$HCl + NaOH \rightarrow H_2O + NaCl$$

13. Chlorine reacts with water to form acidic compounds that can harm the skin and burn the eyes. Basic compounds will help neutralize the acidity of the water.

Thinking Critically

14. The ammonium hydroxide neutralizes the acid injected into the bite.

15. A base could be described as a proton acceptor since it contains many free hydroxide ions in solution. These hydroxide ions will accept free protons (hydrogen nuclei) to form water molecules.

16. Sulfur compounds which are emitted into the atmosphere can react with water in the atmosphere to form sulfuric acid, one cause of acid rain.

17. When acids and bases react, they cancel each other's properties.

18. Solution C is the strongest acid because it has the lowest pH. Solution A is the strongest base because it has the highest pH.

19. Magnesium metal reacts with hydrobromic acid to produce magnesium bromide and hydrogen gas.

20. neutral, pH = 7

Reviewing Themes

12. *Systems and Structures*
What products would be formed if you mixed hydrochloric acid (HCl) with sodium hydroxide (NaOH)? Write an equation for the reaction.

13. *Environmental Interactions*
Chlorine is added to swimming pools as a disinfectant. It is necessary to treat the resulting chlorinated water with basic compounds. Why must these bases be added?

Thinking Critically

14. Many insects bite. The reason the bite hurts is that some toxin is introduced into the victim. Often, the toxin is some form of acid. The centipede in the photo has been killed by a marauder ant, which injected it with formic acid.

Sometimes when people get an insect bite or sting, they treat it with a product similar to the one shown here. These products contain ammonium hydroxide. Using your knowledge of acids and bases, explain how ammonium hydroxide could stop a bite from itching or hurting.

15. Explain why a base can also be described as a proton acceptor.

16. Why is it important to control the emission of sulfur compounds into the atmosphere?

17. In what way might acids and bases be considered opposites?

18. Solutions A, B, C, and D have pH values of 12, 9, 1, and 5, respectively. The solutions' concentrations are equal. Identify the strongest acid and the strongest base in this group. Explain your answer.

19. Describe the reaction below in your own words.

$$Mg + 2HBr \longrightarrow MgBr_2 + H_2$$

20. Predict the approximate pH of a solution formed when a weak base reacts with a weak acid. Explain your prediction.

Discovery Through Reading

Thompson, Jon. "East Europe's Dark Dawn." *National Geographic* 179 (June 1991): 36–63. This extensive article focuses on the perils of acid rain and other environmental pollutants in eastern European countries.

PLANNING THE CHAPTER

B = Basic **A** = Average **H** = Honors
The coding Basic, Average,and Honors indicates subsections, features, and resources that might be appropriate for different levels of learners. For additional suggestions regarding choice of topic and depth of coverage, see the Pacing Chart on pages T28–T31.

*Frame numbers at point of use
(TR) Teaching Resources, Unit 6
(IT) Instructional Transparencies
(LI) Laboratory Investigations
(SD) *Science Discovery* Videodisc Correlations and Barcodes
(SRG) Study and Review Guide

CHAPTER MATERIALS

Title	Page	Materials
Discover by Doing	586	(per group of 3 or 4) soot coated beaker
Discover by Problem Solving	592	journal
Teacher Demonstration	593	safety goggles, laboratory apron, gloves, 10 mL of 5% aqueous hexamethylenediamine solution, 10 mL of 5% sebacoyl chloride in dry methylene chloride, 10-cm wire with a hook at the end, 50-mL beaker
Discover by Doing	593	(per group of 3 or 4) molecular model kit
Discover by Doing	594	(per group of 3 or 4) molecular model kit
Discover by Doing	596	(per individual) 20 mL of 4% polyvinyl alcohol, paper cup, several drops of food coloring, plastic stirrer, 3 mL of sodium tetraborate solution, plastic bag
Activity: How can you identify the odors of different natural oils?	601	(per group of 3 or 4) filter paper strips with different natural oils (5), strips of filter paper with unknown oils (3)
Skill: Using Models	603	(per group of 3 or 4) pencil, paper, mirror
Discover by Doing	607	(per individual) blocks (6 different colors)
Investigation: Testing for Nutrients	610	(per individual) safety goggles, laboratory apron, (per group of 3 or 4) wax pencil, test tubes (12), test tube rack, egg white solution, green onion shoots, saltine crackers, 10-mL graduate, stirring rod, glucose test strips (4), iodine solution, 10% sodium hydroxide solution, stoppers (4), 0.5% copper sulfate solution

ADVANCE PREPARATION

Prepare the soot covered beakers the day of *Discover by Doing* activity on page 586.

Prepare the 4% polyvinyl alcohol (40 g dissolved in 1 L of water) and the sodium tetraborate solution (4 g borax in 1 cup water) the day of the *Discover by Doing* activity on page 596.

Collect several weekly cafeteria menus or menus from nutrition and health magazines for the Motivating Activity on page 604.

Have students bring in several discarded magazines for the Extension Activity on page 609.

TEACHING SUGGESTIONS

Field Trip
Visit a paint or cleaning supplies store and find out about different organic solvents, pigments, or other compounds they use and safety precautions taken in their use.

Have a small group of students visit a laboratory (possibly in a hospital) that specializes in the separation of proteins. Have the group videotape the separation of proteins by centrifugation, chromatography, and electrophoresis and show the tape to classmates.

CHAPTER THEME—*SYSTEMS AND STRUCTURES*

Chapter 22 discusses how the properties of an element are related to its electron structure. The chapter also investigates how carbon can occur in several allotropes, which gives a clue as to why carbon compounds are so numerous.

A supporting theme in this chapter is **Technology.**

MULTICULTURAL CONNECTION

The chief crop of the Amerind people was a complex carbon compound—maize (which we now call corn). Evidence shows that Amerinds grew maize more than seven thousand years ago in both Peru and Mexico. By 1500 B.C. maize was the primary crop in Mexico.

The southwest Pueblo people grew many varieties of maize: white, yellow, black, blue, and even pink. Maize was not used in New England until about 1400. However, by that time, the Iroquois had more than 50 recipes for serving maize, such as corn on the cob, mush, cornbread, and several types of corn soup.

MEETING SPECIAL NEEDS

Gifted

Have gifted students investigate the properties of diamonds and how these properties affect their worth.

CHAPTER
22
CARBON COMPOUNDS

Carbon is one of the most ordinary and versatile elements; it is present in countless compounds. Yet, in its pure state, carbon can excite romance, intrigue, and violence. The rarest form of this ordinary element is the diamond, a stone of matchless hardness and beauty.

The history of diamonds is filled with legend and suspense. One legend surrounds the famous Koh-i-Noor, which passed among the Mogul, Persian, and Afghan conquerors of India.

Following the Mogul invasion, the diamond passed into the hands of Sultan Baber,...founder of the Mogul Empire in India. A much treasured possession of his, Baber refers to it in his diary in 1526 as "the famous diamond" of such value that it would pay "half the expenses of the world." It remained in the ownership of his descendants for the next two centuries, thus giving some substance to the legend that "he who owns the Koh-i-Noor rules the world." In 1739, however, it was lost to the Persians who, under their ruler Nadir Shah, invaded India and sacked Delhi. There is a story that for fifty-eight days the stone could not be found because the conquered Mogul emperor Mohammed Shah had hidden it in the folds of his turban. Told the secret by a member of the ex-emperor's harem, Nadir Shah invited him to a feast and, observing an ancient Oriental custom, proposed an exchange of turbans. Mohammed was in no position to refuse. Once he had the turban, Nadir Shah ran to his tent and on seeing the great diamond among the silk of the unrolled turban, cried "Koh-i-Noor," which means Mountain of Light, thus giving it the name it has borne ever since.

CHAPTER MOTIVATING ACTIVITY

Display a chunk of charcoal, a broken pencil lead (graphite), a stylus from a discarded record player (diamond), a few drops of lubricating oil floating on some water in a Petri dish, a plastic food storage bag, and some rayon cloth. Have the students develop one or more systems to classify these materials.

Then point out that these materials are all made of carbon compounds. Explain to the students that in this chapter they will learn about carbon compounds, where they are found, and how we use them.

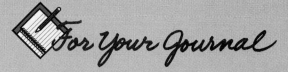

For Your Journal

The journal questions will help you identify any misconceptions the students might have about carbon compounds. In addition, these questions will help the students realize that they have prior knowledge about carbon and its many compounds. (Accept all reasonable answers. Tell the students that at the end of the chapter they will have an opportunity to refer back and reconsider their responses to these questions.)

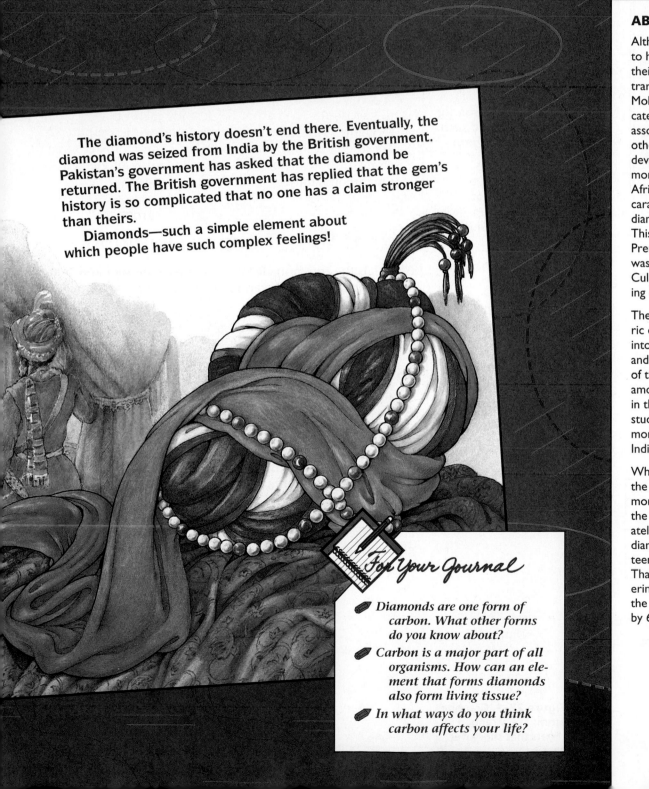

The diamond's history doesn't end there. Eventually, the diamond was seized from India by the British government. Pakistan's government has asked that the diamond be returned. The British government has replied that the gem's history is so complicated that no one has a claim stronger than theirs.

Diamonds—such a simple element about which people have such complex feelings!

For Your Journal

- Diamonds are one form of carbon. What other forms do you know about?
- Carbon is a major part of all organisms. How can an element that forms diamonds also form living tissue?
- In what ways do you think carbon affects your life?

ABOUT THE LITERATURE

Although diamonds were known to have great value because of their scarcity and hardness (arbitrarily given a value of 10 on the Mohs scale of hardness), the intricate cutting, which most people associate with diamonds and other gems, is a relatively recent development. The largest cut diamond in the world is the Star of Africa, weighing a little over 530 carats. It came from the biggest diamond ever found, the Cullinan. This stone, found in 1905 in the Premier mine of South Africa, was named after Thomas Cullinan, the chairman of the mining company.

The Cullinan weighed 3,106 metric carats (0.60 kg) but was cut into two large, seven medium, and 96 smaller stones. The largest of these, the Star of Africa, is now among the British Crown Jewels in the Tower of London. As the students have read, another diamond in this collection is the Indian diamond Koh-i-Noor.

When Queen Victoria accepted the confiscated Koh-i-Noor diamond in 1850 as reparation for the Indian mutiny, she immediately made plans to have the diamond—crudely cut by nineteenth-century standards—recut. That was quite a decision, considering that most cutting reduces the original weight of a diamond by 60 percent!

FOCUS

Section 1 explains why there are so many carbon compounds. It describes two allotropes of carbon—graphite and diamond. This section also discusses other elements in the carbon family.

MOTIVATING ACTIVITY

Cooperative Learning Have small groups of students organize the information they already know about carbon and tin using con-

cept maps. Using the maps, have the students discuss similarities and differences between the two elements.

POSITIVE ATTITUDES
- Cooperativeness
- Creativity

TERMS
- None

PRINT MEDIA
"Diamonds in the Rough" from *Discover* magazine by Ed Regis (see p. 513b)

ELECTRONIC MEDIA
How Atoms Combine Coronet Film and Video (see p. 513c)

SCIENCE DISCOVERY
- Diamond
- Graphite • Pencil

BLACKLINE MASTERS
Reading Skills
Study and Review Guide

MEETING SPECIAL NEEDS

Second Language Support

 In their journals, have limited-English-proficient students construct a table such as the one below to compare and contrast the properties of graphite and diamond.

Characteristics	Graphite	Diamond
Element		
Color		
Texture		
Hardness		
Structure		

DISCOVER BY *Doing*

Prepare soot samples by holding heatproof beakers with tongs over a candle flame. Invert the beakers, and let them cool. Answers should include similarities in color and differences in hardness and luster.

SECTION 1

From Soot to Computers

Objectives

Suggest a reason for the existence of such a variety of carbon compounds.

Summarize the uses for each of the elements in the carbon family.

Describe the problems associated with lead.

What came into your mind when you read the title of this section? "The author has lost it!" might have been a thought. What does soot have to do with computers? They don't seem to have anything in common. Yet some atoms in both belong to a prolific family of elements, the carbon family.

Soft as Soot, Hard as Diamond

Carbon is the common element that makes up soot and charcoal. What makes carbon and the carbon family so unique? Try the next activity to find out.

ER BY *Doing*

Compare the "lead" in your pencil with the soot collected on a beaker. Both are forms of carbon. How are they alike? How are they different? ✐

The different forms, or allotropes, of carbon have different properties. Graphite, at the tip of your pencil, feels slick and oily. Because of this, graphite is used as a lubricant in machines. Look at the diagram to see how the atoms of carbon are arranged in graphite. The "sheets" of carbon molecules can glide over each other, causing graphite's slipperiness.

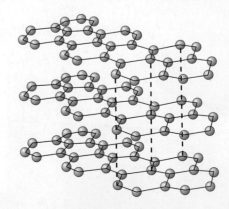

Figure 22–1. Graphite atoms are arranged in sheets that slide over each other.

TEACHING STRATEGIES

● **Process Skill:** *Inferring*

Have the students discuss how graphite's structure accounts for properties that make it usable as pencil "lead." (As the pencil lead moves across a sheet of paper, graphite is deposited on the paper because graphite layers are loosely held together and can easily slide past each other.)

● **Process Skill:** *Analyzing*

Have the students study the diagrams of graphite and diamond and discuss why a diamond is a single covalent molecule while graphite is composed of many covalent molecules. (In diamond, every carbon atom is covalently bonded to four carbon atoms, each of which is

bonded to four carbon atoms. Graphite does not have these tight bonds between sheets of atoms.)

● **Process Skills:**
Classifying/Ordering, Predicting

Have the students use the periodic table to locate the elements sodium, silicon, and chlorine. Then have them predict the order of increasing conductivity of these elements. (chlorine: nonmetal, silicon: metalloid, sodium: metal)

Another allotrope of carbon, with very different physical properties, is diamond. In a diamond the carbon atoms are bonded in a rigid crystalline structure. This structure is responsible for the extreme hardness of diamonds.

Scientists once thought the soot you observed in the activity was made of carbon that was not bonded in any particular pattern. Recent research, however, indicates that soot contains molecules made of 60 to 70 carbon atoms bonded into ball-shaped configurations. These molecules are called *Buckyballs* or *Buckminsterfullerenes* after Buckminster Fuller, the architect who invented geodesic domes.

The most important thing to remember about carbon is that it forms four bonds, either to other elements or to itself. This allows carbon to form long chains of carbon atoms or to combine in many different ways with other common elements, such as hydrogen, nitrogen, and oxygen.

There are so many carbon compounds that there is an entire field of chemistry dedicated to them. The area of chemistry in which carbon compounds are studied is called *organic chemistry*. Organic chemists study the properties of carbon compounds. They also make important molecules that are used in medicines, insecticides, plastics, and other synthetic materials. Why do you think the study of carbon is called organic chemistry? ①

 ASK YOURSELF

Why does carbon form different allotropes?

Silicon: The Element That Can't Make Up Its Mind

The second member of the carbon family is silicon. It combines with other elements to form a great number of compounds, but it does not make as many different types of compounds as carbon. Unlike carbon, pure silicon has a metallic shine; it is a metalloid. It may conduct electricity under certain conditions. Because of this property, silicon has been very valuable in making computer chips and semiconductors.

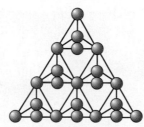

Figure 22–2. The crystalline structure of diamond is very rigid.

Figure 22–3. Carbon forms four bonds when it combines with itself or other elements.

Figure 22–4. Computer chips made of silicon

 LASER DISC
533

Diamond

538

Graphite

2349

Pencil

THE NATURE OF SCIENCE

Scientists have isolated another allotrope of carbon, fullerenes, which occur in soot. These hollow, soccer-ball-like molecules contain large numbers of carbon atoms. Fullerenes can withstand extreme compression, returning to their original configuration after the pressure has been removed. Current research indicates that fullerenes might be used as reaction chambers in which molecules trapped inside react when the fullerenes are subjected to great pressure.

① Carbon and carbon compounds are major parts of all organisms.

ONGOING ASSESSMENT
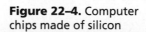 **ASK YOURSELF**

Carbon forms different allotropes because carbon atoms can combine with each other in different arrangements.

● **Process Skill:** *Applying*

Point out that silicon is never found naturally in a free state; it has to be refined from compounds. Silicon can be prepared from a halide, such as silicon tetrachloride ($SiCl_4$), which is reacted with hydrogen gas. Then the students write the equation for this reaction in their journals.
($SiCl_4 + 2H_2 \rightarrow Si + 4HCl$)

INTEGRATION–
Earth Science

Silicon is the second most abundant element in Earth's crust (approximately 27.7 percent). Silicon is found in various rocks in the form of silicates, or silicon-oxygen compounds. Free silicon is usually obtained by reducing silicon dioxide with carbon.

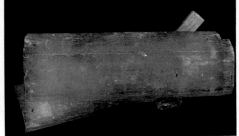

Figure 22–5. Many minerals contain silicon. The silicon-containing minerals shown here, clockwise from top left, are mica, amethyst, emerald, and beryl.

Even though silicon is a metalloid, its most common compounds are not metallic in character. It is hard to believe that the same substance that is used to make computer chips is found almost everywhere—as sand. Sand is a silicon compound called silicon dioxide (SiO_2). Other silicon-oxygen compounds are found in many minerals and gemstones.

Silicon, like carbon, can form four covalent bonds. Dr. Larry Hench of the University of Florida has used the stability of silicon and its similarity to carbon to make a unique kind of glass. This glass—made of silicon and salts of sodium, calcium, and phosphorus—can be cast and carved into the shapes of bones. The human body accepts the glass as if it were a natural bone transplant. Dr. Hench is currently studying many other applications of this glass in the body.

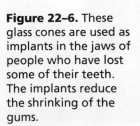

Figure 22–6. These glass cones are used as implants in the jaws of people who have lost some of their teeth. The implants reduce the shrinking of the gums.

● **Process Skill:** *Inferring*

Cooperative Learning Have small groups of students list characteristics of silicones, inferred from what the group knows about their uses. Then ask a member of each group to share the list with the class.

● **Process Skill:** *Comparing*

Explain to the students that a single silicon computer chip can perform one task or a variety of tasks. Then challenge the students to describe examples of how a silicon chip can perform multiple tasks. (For example, a single chip in a multifunctional

wristwatch might contain circuits for telling time, setting alarms, and providing the day and date.)

There are many silicon compounds in your home. The caulking used to seal windows and fill cracks is made from silicon compounds called *silicones*. Silicones are long chains of silicon and oxygen atoms with different carbon chains attached. They are also used as lubricants and as water-repellent coatings on masonry and cement.

ASK YOURSELF

How are carbon and silicon similar? How are they different?

Figure 22–7. Silicones are so heat resistant that this kitten can sit on a slab of silicone and not be burned by the flame below. This property of silicones makes them useful for heat-resistant tiles on spacecraft.

Cousins—Near and Far

What other elements in the carbon family are you familiar with? Do you know how they are used? How are they similar to or different from carbon and silicon? Other members of the carbon family might not be as abundant as carbon or silicon, but they are still important.

Germanium is a metalloid. It is a semiconductor used in some electric devices such as transistors. Small amounts of germanium can be added to glass to produce filters in optical instruments and in cameras.

The other two elements in the carbon family are tin and lead. They have metallic bonds rather than the covalent bonds found in the other members of the family. Since the bonds are different, their properties are also different. As metals, both tin and lead can be easily shaped.

Figure 22–8. Germanium is used to make camera filters like these.

ONGOING ASSESSMENT

ASK YOURSELF

Carbon and silicon are both Group 14 elements and can form four covalent bonds. Carbon is a nonmetal, and silicon is a metalloid. Carbon can form more compounds than silicon can.

GUIDED PRACTICE

Cooperative Learning Draw a single column of five rows on the chalkboard, and label it "Group 4." Have small groups of students complete the Group 4 column on a piece of paper, using the periodic table. Then have the students write in at least one use for each of the elements as well as the information they find in the periodic table.

INDEPENDENT PRACTICE

Have the students provide written answers to the Section Review. In their journals, have them write a paragraph summarizing similarities and differences among the Group 4 elements.

EVALUATION

Using blue chalk to represent nonmetals, yellow to represent metalloids, and red to represent metals, have volunteers list the Group 4 elements by writing on the chalkboard the chemical symbol of a member of that group in the appropriate color. Have other volunteers write the name of the element next to the symbol or state a use for the element.

INTEGRATION–
Language Arts

Have the students use dictionaries to find the etymology of the word *stannous* in stannous fluoride. (from the Latin word for tin, *stannum*)

◈ Did You Know?

Stannous fluoride limits the growth of bacteria that produce acids that cause tooth decay.

SCIENCE BACKGROUND

Many lead compounds taste sweet. One of the acetates of lead has the common name "sugar of lead." The sweet taste might be one reason children sometimes eat peeling lead paint. Caution the students that the ingestion of lead can result in poisoning, mental retardation, and, in extreme cases, death.

INTEGRATION–
Earth Science

Lead-detecting kits are available at some hardware stores. They enable you to test for the presence of lead in paints and glazes. Point out that any ceramic piece that might not be lead-free should not be used to cook, serve, or store foods.

Figure 22–9. These serving utensils are made of pewter, a metal alloy that contains tin.

Figure 22–10. Compounds that contain lead occur in many bright colors. For this reason, they were once used to color paints.

Tin is used to line the inside of metal cans and to make pewter. How would you like to clean your teeth with tin? What? Your teeth are not tough enough to stand a metal scouring. Nevertheless, you probably already use tin when you brush your teeth. It is found in the compound stannous fluoride. That's right, the compound that provides fluoride, which helps prevent tooth decay, contains tin.

Lead is a poisonous metal. At one time, most gasolines contained lead to prevent engine knock. Today, most cars use unleaded gasoline. Why the switch? Cars using leaded gasoline release small amounts of lead compounds into the air. These lead compounds can accumulate in the human body, eventually causing lead poisoning. Some of the results of lead poisoning are mental retardation and even death.

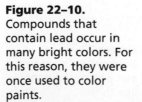

Some lead compounds are brightly colored. For this reason, they were used as pigments in house paints. The colors produced were attractive, but the lead caused problems. As the lead-containing paint peeled off walls, children were exposed to the lead. Small children are easily attracted to bright colors and some ate the paint

RETEACHING

Have the students construct a concept map using the following terms: *carbon family, silicon, tin, lead, carbon, diamond, germanium, graphite, semiconductors, toothpaste, some colored pigments.*

EXTENSION

Have the students research the effects of lead poisoning. Then have them hypothesize about lead as a cause of the downfall of the Roman Empire and report their findings to the class.

CLOSURE

 Have each student choose a Group 4 element that seems most representative of that group and write a paragraph in their journals explaining the choice. Encourage the students to share their responses with classmates.

because many lead compounds taste sweet. This gave them lead poisoning. Because of this, leaded paint is no longer sold. Some artists' paints still do contain lead, however, as does some colored clay pottery. You can check for lead by using an indicator stick that will change color if rubbed against a substance that contains lead compounds.

Some historians believe that the downfall of the Roman Empire might have been related to the use of lead. The early Romans used lead to line their aqueducts and seal their wine containers. Traces of lead have been found in human remains of that period. Imagine an element having such an impact on the course of history!

Figure 22–11. This Roman aqueduct contained pipes that were lined with lead.

▼ ASK YOURSELF

Describe one use each for germanium, tin, and lead.

SECTION 1 *REVIEW AND APPLICATION*

Reading Critically

1. What is the difference between silicon and silicones?
2. Name three things in your home that are made from silicon, germanium, tin, or lead.

Thinking Critically

3. How is the chemistry of carbon different from the chemistry of the other elements in its family?
4. Sugar, which is a typical organic compound, does not conduct electricity when dissolved in water. What can you conclude about the type of bonding in organic compounds from this information?

ONGOING ASSESSMENT
▼ ASK YOURSELF

Germanium is used in transistors and filters for optical instruments and cameras. Tin is used inside metal cans, in pewter, and in toothpaste (stannous fluoride). Lead is used in pewter and in some artists' paints and glazes for pottery.

SECTION 1 REVIEW AND APPLICATION

Reading Critically

1. Silicon is an element. Silicones are compounds containing silicon, oxygen, and carbon.

2. Silicon is used in computers and small electronics, gems, and caulking. Germanium is used in transistorized radios and TVs and in camera filters. Tin is used in pewter pieces, tin cans, and toothpaste. Lead is used in pewter, paintings, pottery glaze, storage batteries, and solder.

Thinking Critically

3. Carbon and silicon can both form four covalent bonds, but carbon forms more compounds than does silicon. Tin and lead form metallic bonds.

4. Organic compounds have covalent bonds. Since they do not form ions, they will not conduct electricity.

FOCUS

In this section the students learn about hydrocarbons, including structural isomers and polymers, and their uses.

MOTIVATING ACTIVITY

 Discuss with the students the meanings of the prefixes *mono-* and *poly-*. (*Mono-* means one; *poly-* means more than one.) Then have the students think of and list on the board words that begin with either prefix. (Examples include monorail, monograph, monologue, poly-ester, and polygon.) Explain to the students that in this section they will learn about the structures of certain organic compounds called monomers and polymers. Conclude by asking the students to hypothesize in their journals what monomers and polymers might be. (Accept all reasonable hypotheses.)

PROCESS SKILLS
- Observing • Constructing/Interpreting Models
- Comparing

POSITIVE ATTITUDES
- Caring for the environment
- Cooperativeness
- Initiative and persistence

TERMS
- structural formula
- hydrocarbons • isomer
- polymers • monomers

PRINT MEDIA
"Yews in Trouble" from *Discover* magazine by Betsy Hanson (see p. 513b)

ELECTRONIC MEDIA
Special Topics in Chemistry: Organic Chemistry Focus Media, Inc. (see p. 513c)

SCIENCE DISCOVERY
- Ball and socket joint; human
- Gasoline station; adding to tank
- Oil, crude; distillation
- Plastic, making; lab setup
- Yew

BLACKLINE MASTERS
Extending Science Concepts
Thinking Critically
Study and Review Guide

✧ Did You Know?
Organisms called methanogens can synthesize methane from carbon dioxide and hydrogen.

DISCOVER BY
Problem Solving

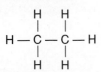

SECTION 2

Organic Compounds and Their Uses

Objectives

Sketch structural formulas for simple carbon compounds.

Explain what structural isomers are, and give two examples.

Define the term polymer, *and give an example of a polymer used in your home.*

Figure 22–12.
Structural formula for methane

Have you ever heard of swamp gas? If you have ever paddled a canoe through a marsh or bayou, you might have seen bubbles of gas rising to the surface. This swamp gas is called methane and is formed by decomposing organic matter at the bottom of the swamp.

Dot, Dot, Dash

Methane, the principal component in natural gas, can be used to illustrate the properties of many organic compounds. Methane has the molecular formula CH_4. How could you draw a molecule of methane? Look at the diagram. The dashes that connect the atoms represent covalent bonds. This type of formula is called a **structural formula**.

Methane

DISCOVER BY Problem Solving

Carbon can form only four bonds. Draw the structural formula for the molecule C_2H_6. ✎

TEACHING STRATEGIES

Figure 22–13. Propane (C_3H_8) is used as a fuel in torches, camping stoves, and gas grills.

The molecule you just drew is known as *ethane*. There are many compounds that are made of only carbon and hydrogen. These compounds are called **hydrocarbons.** When hydrocarbons form, long chains of carbons can bond to each other. You can find out more about how this happens by doing the next activity.

DISCOVER BY Doing

Using a molecular model kit, make models of methane and ethane. Then try to make two more hydrocarbons that have more than two carbon atoms. Use as many hydrogens as you need. In your journal, draw your compound and write the correct formula for it. Compare your models with those of other students. Are they all the same? Explain. ✎

Hydrocarbons are very common. You probably use some every day. One of their properties is that they are very flammable. Because of this, they are used as fuels. Methane, ethane, and propane are used to heat homes and to cook food.

The length of the carbon chain in a hydrocarbon determines the physical state of the compound. Hydrocarbons with less than five carbon atoms are gases at room temperature. Those with five to 20 carbon atoms are usually liquids. Gasoline is an example. Hydrocarbons containing more than 20 carbon atoms tend to be solids at room temperature. They are usually waxy to the touch.

Many organic compounds contain other elements besides carbon and hydrogen. Oxygen is a common element found in organic compounds. Find out more about these compounds by trying the next activity.

SECTION 2   **593** ◀

● **Process Skills:** *Comparing, Interpreting Data*

Point out that the two isomers for the hydrocarbon C_4H_{10} in Ask Yourself are called butane (the unbranched isomer) and isobutane (branched). Construct the following table on the chalkboard.

	Butane	Isobutane
Melting point	−138.3°C	−159.6°C
Boiling point	−0.5°C	−11.7°C

Have the students compare the melting and boiling points of the two compounds. Then ask them if this data indicates that branching in an isomer raises or low-ers its melting and boiling points. (It lowers the melting and boiling points. Some students may point out that one set of data is insufficient to draw a conclusion. Congratulate them on their scientific thinking and then tell them that this data does follow a proven trend.)

● **Process Skill:** *Comparing*

Some students might mistakenly think that since ethanol is consumed in alcoholic beverages, it is safe to consume in any amount. Point out that while ethanol is less toxic than dimethyl ether, a person can also die from overconsumption of ethanol, a condition called alcohol poisoning.

DISCOVER BY *Doing*

PORTFOLIO ASSESSMENT

The two possible structural formu-las are shown in Figure 22–14. After the students have com-pleted the *Discover by Doing*, have them place their diagrams of the isomers in their science portfolios.

MEETING SPECIAL NEEDS

Gifted

Have advanced students research the naming of classes of hydrocar-bons in which all the carbons in the molecule are joined by single, double, and triple bonds. (alkanes, alkenes, alkynes) The students should also find out the prefixes used to indicate the number of carbon atoms in the first ten members of each class. (1: meth-, 2: eth-, 3: prop-, 4: but-, 5: pent-, 6: hex-, 7: sept-, 8: oct-, 9: non-, and 10: dec-) Have the students present their findings to the class.

ONGOING ASSESSMENT
▼ ASK YOURSELF

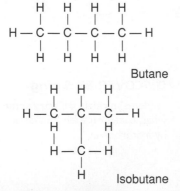

Butane

Isobutane

DISCOVER BY *Doing*

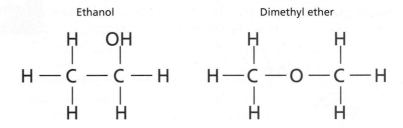

Using the same model kit as before, try to make different molecules that have the same formula—C_2H_6O. Remember that carbon forms four bonds and hydrogen forms one bond. Oxygen forms two bonds. How many different molecules could you make? ✎

There are two compounds that have the formula you worked with in the activity. Did you figure them both out? You can see them drawn below. Ethanol, also known as ethyl alcohol or grain alcohol, is shown on the left, and dimethyl ether is on the right.

Ethanol

Dimethyl ether

Figure 22–14. Structural formulas for ethanol and dimethyl ether

Think about the properties of both. Ethanol can be drunk; it is the alcohol found in alcoholic beverages. Dimethyl ether, on the other hand, can be used as an anesthetic and is poison-ous if drunk. Even though these two compounds have the same molecular formula, they have very different properties. Why do you think this is so? You guessed it—their structures are different.

These two compounds are known as structural isomers. An **isomer** is one of two or more compounds that have the same chemical formula but different structures. The structure refers to the arrangement of atoms. The number of isomers that a compound has increases with the number of carbon atoms in the molecule; more atoms available allow for more variations in how they are arranged.

▼ ASK YOURSELF
Diagram structural formulas for two isomers of C_4H_{10}.

● **Process Skill:** *Analyzing*

Point out to the students that not all polymers are synthetic. There are also natural polymers that are used for clothing. Then ask the students whether they know of any naturally occurring

polymers used for clothing. (Silk, cotton, and wool are all naturally occurring polymers that are used for clothing.)

● **Process Skill:** *Analyzing*

 Have the students examine the labels of five articles of clothing they own and make a list in their journals of the names of synthetic and natural fibers of which their clothes are made.

A Train of Carbon

The Riverfront Mall has a small shop that attracts a crowd. In it you can see a man dressed in a candy-striped jumpsuit making taffy. He kneads the taffy and pulls it into long strands almost a meter in length. The taffy seems to stretch forever.

Look at the beaker in the Figure 22–15. How is the substance inside the beaker similar to taffy? ①

The photograph shows two liquids in the beaker, one above the other. A long strand of nylon is being pulled out of the center. Yes, this stuff is the same material that is used to make tents and windbreakers!

Look around you, and make a list of all the things you see that are made of plastic or synthetic fibers. Did you know that many car parts are plastic? Your desk or counter top may be made from synthetic materials. Maybe the clothes you are wearing are synthetic. The pen you are using is probably made of plastic. Your world is filled with materials that did not exist 50 years ago. These synthetic materials are made of giant molecules called *polymers*. **Polymers** are large molecules made by "hooking" many identical small molecules together, much like railroad cars are linked to one another. The units in a single polymer might number in the thousands. The small units that make up the polymer are like the boxcars that make up a train. They are called **monomers.**

Figure 22–15. On the right, Dr. Hill, a co-inventor of nylon, is pulling a strand of nylon from the beaker. On the left is a close-up. How long do you think the strand can get? ②

① The substance stretches like taffy.

 Plastics used in containers and packaging are very difficult to destroy. As a result, plastic objects that are thrown out can remain intact for many years, taking up valuable landfill space. In recent years, researchers have been working to develop materials that change their properties over time in ways that are beneficial rather than harmful.

One approach to this problem has been to make new forms of plastic that become brittle and crumble into tiny pieces when exposed to sunlight. Industries are also helping out by recycling certain types of plastics.

LASER DISC
4156

Ball and socket joint; human

245–246

Plastic, making; lab setup

② The substance might be stretched for several hundred meters.

SECTION 2 **595** ◀

● **Process Skill:** *Inferring*

Point out that synthetic polymers are designed to have certain useful properties. For example, a polymer used in the tail sections of airplanes has a strength-to-weight ratio six times greater than steel. Discuss with the students the advantages of this polymer. (It is strong, but lightweight, an important characteristic of materials used in airplanes.)

● **Process Skill:** *Analyzing*

Point out to the students that monomers undergo chemical reactions to form polymers. Then have the students identify how the monomer ethylene $H_2C\!\!=\!\!CH_2$ is changed to the unit $-H_2C\!-\!CH_2-$ in the polymer polyethylene. (The double bond is broken, giving the unit a bond to join another monomer.) Point out that most molecular monomers contain a double bond.

DISCOVER BY *Doing*

Prepare 4 percent polyvinyl alcohol by dissolving 40 g of polyvinyl alcohol in 1 L of water. Prepare the solution of sodium tetraborate by dissolving 4 g of borax in 250 mL of water. Add 2–3 mL of the sodium tetraborate solution to each student's cup of polyvinyl alcohol solution. The students can store their slime in plastic bags for a day or two.

Answers should note that the solutions became more difficult to stir after the second solution was added. The change in properties indicates crosslinking occurred.

ONGOING ASSESSMENT
▼ ASK YOURSELF

A polymer is a large molecule made up of many identical small molecules, or monomers, hooked together.

Figure 22–16. Polymers form as long chains of molecules. Those polymers with no cross-bridges are flexible. Polymers with physical bridges or chemical crosslinks are more rigid. The greater the number of bridges or crosslinks, the harder the polymer.

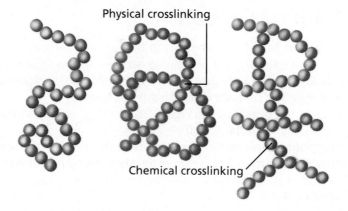
Physical crosslinking
Chemical crosslinking

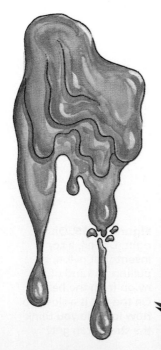

Why are some polymers elastic like rubber bands, while others are hard or brittle? In order to understand this, picture some cooked spaghetti (no sauce) on a plate. If you jiggle the plate, the strands shift every which way. Certain polymers are like that; they tend to be elastic.

In contrast, other polymers are hard. The seat you are sitting doesn't squirm like spaghetti. How can scientists change the properties of polymers so much? They hook the strands of polymer together with short bridges. This process is known as crosslinking. Crosslinked polymers have lost some degree of movement. These bridges make the polymers stiff so that they keep their shape. They tend to be less elastic and harder. You can make your own polymer in the next activity.

DISCOVER BY *Doing*

Your teacher will give you 20 mL of polyvinyl alcohol in a paper cup. Describe the liquid. Polyvinyl alcohol is a polymer. Add 2 to 3 drops of food coloring and stir. While you stir, your teacher will add a small amount of a second solution. Keep stirring until you can tell that some crosslinking has occurred. How would you know this? Describe what happens. **CAUTION: You can play with your slime, but don't eat it.** 🖉

Polyethylene is an example of one type of polymer that can be either elastic or rigid. You are probably familiar with many polyethylenes. Study Table 22–1 to find polymers you recognize.

▼ ASK YOURSELF
What is the difference between a polymer and a monomer?

Refer the students to Table 22–1, and discuss the many uses of polymers. Point out how many of our modern-day materials and conveniences are the result of polymers. You might also discuss some of the environmental problems that have developed as a result of polymers, such as nonbiodegradability.

Have the students investigate nonbiodegradable plastics and make notes in their journals. Ask them to come to class prepared to debate the pros and cons of an ever-increasing reliance on synthetic materials.

Table 22-1

Polyethylene and Its Thermoplastic Derivatives

Monomer	Polymer	Uses
$H_2C = CH_2$ ethylene	$-[H_2C—CH_2]_n$ polyethylene	packaging, bottles, toys
$H_2C = C{<}^H_{CH_3}$ propylene	$-[H_2C—CH(CH_3)]_n$ polypropylene	fibers, films, carpets, laboratory equipment, kitchenware
$H_2C = C—H$ (benzene ring) styrene	$-[H_2C—CH$ (benzene ring)$]_n$ polystyrene	Styrofoam insulation, packaging and packing material, household articles, toys
$H_2C = C{<}^H_{Cl}$ vinyl chloride	$-[H_2C—CH(Cl)]_n$ polyvinyl chloride	floor coverings, garden hoses, phonograph records, packaging
$H_2C = C{<}^H_{O—C(=O)—CH_3}$ vinyl acetate	$-[H_2C—CH(O—C(=O)—CH_3)]_n$ polyvinyl acetate	latex paints
$H_2C = C{<}^H_{C≡N}$ acrylonitrile	$-[H_2C—CH(CN)]_n$ polyacrylonitrile (Orlon or Acrilan)	textile fibers
$H_2C = CCl_2$ vinylidene chloride	$-[H_2C—CCl_2]_n$ polyvinylidene chloride (Saran)	self-adhering food wrap
$H_2C = C{<}^{CH_3}_{O=C—OCH_3}$ methylmethacrylate	$[CH_3—C(CH_2)(C(=O)—O—CH_3)]_n$ polymethylmethacrylate (acrylic, Lucite, Plexiglas)	unbreakable glass for windows, windshields; water-based latex paints
$F_2C = CF_2$ tetrafluoroethane	$-[F_2C—CF_2]_n$ polytetrafluoro-ethene (Teflon)	chemically inert items, gasket materials, cookware coatings

● **Process Skill:** *Inferring*

Point out that many soaps contain additives. When a scouring soap is made, sand and other abrasives are added. Ask the students to describe what they think manufacturers do if they want soap to float. (They inject air into the molten soap.)

INTEGRATION–
Language Arts

Ask the students to look up the terms *hydrophobic* and *hydrophilic* and write the definitions in their journals. Have them explain the structure of a soap molecule in writing using these terms. (A soap molecule has a hydrophobic, nonpolar end and a hydrophilic, polar end.)

⟡ **Did You Know?**
The water-soluble or polar groups of most synthetic detergents are sodium sulfonates and sodium sulfates. One serious disadvantage of some synthetic detergents is that they cannot be broken down by microorganisms. For many years, nonbiodegradable detergents accumulated in ground water and surface water supplies. This pollution is why these detergents have since been banned.

Figure 22–17. One end of a soap molecule bonds to dirt, and the other bonds to the water in which the items, such as your hands or dirty dishes, are being washed. These bonds pull the dirt off the items being washed.

ONGOING ASSESSMENT
▼ **ASK YOURSELF**

Water molecules cannot bond with oil molecules. Soap molecules can bond to water and to oil, allowing water to rinse the bonded oil and soap away.

Carbon Keeps You Clean?

Look at the ingredients listed on a bar of soap. There are probably many that you do not recognize. Today's soaps are made in laboratories using some synthetic substances. However, from Roman times to the late 1900s, soap was made by a different method. Large kettles containing animal fat, ashes, and water were heated and stirred over open fires. As the soap particles formed they were skimmed off the top of the liquid. Then the lumps were pressed together to form cakes.

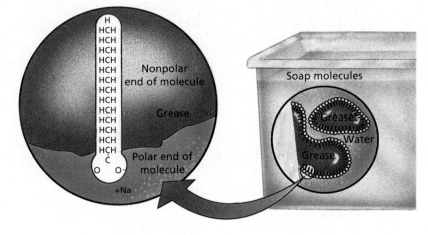

Soap molecules have two ends with different properties. One end of the molecule is a hydrocarbon; the other end is a salt. The hydrocarbon part bonds to fats (grease and dirt), while the salt end sticks to water. The grease is broken up into small droplets surrounded by soap molecules. These droplets can then be washed away with water.

▼ **ASK YOURSELF**
Why can soap remove oil stains that cannot be removed by water alone?

Fuel for Thought

You have probably realized how important organic chemistry is to you. Did you know that without hydrocarbons you would not be able to ride the school bus or drive to the mall? Vehicle fuels are hydrocarbons. Where do these fuels come from? The sources of many hydrocarbons in use today are

● **Process Skill:** *Inferring*

Ask the students to study Figure 22–18 to explain what function they think the wax on the apple serves. (The long-chain hydrocarbon is a protective covering.)

● **Process Skill:** *Interpreting Data*

Ask the students to describe the trend in the boiling points of the hydrocarbons listed in Table 22–2. (The hydrocarbons show an increase in the boiling point with an increase in numbers of carbons.)

● **Process Skill:** *Analyzing*

After the students have studied Table 22–2, have them list in their journals as many everyday uses as they can for the products listed in the table. For example, they are used as fuel in motor vehicles, lubricants in

cosmetics, and natural gas to heat their homes.

natural gas and petroleum. Petroleum is found trapped in pockets under the earth's surface. Natural gas deposits are usually found in the same places. Natural gas is a mixture of several hydrocarbons with very low boiling points.

You can learn more about different types of hydrocarbons by studying Table 22–2. It shows the different hydrocarbons that are present in petroleum.

Table 22-2 Hydrocarbons Separated from Petroleum

Fraction	Number of carbons	Boiling point (°C)	Common uses
Gas	1–5	Less than 30	Fuel
Petroleum ether	5–7	30–90	Solvent
Gasoline	5–12	40–200	Motor fuel
Kerosene	12–16	175–275	Fuel
Fuel oil Diesel oil	15–18	250–400	Fuel for furnaces and diesel engines, raw materials for petrochemicals
Lubricating oils Greases Petroleum jelly	More than 16	Higher than 350	Lubricants
Paraffin (wax)	More than 20	Melts at 50–55	Candles, matches, waterproofing

Figure 22–18. Wax, like that on the outside of these apples, is a long-chain hydrocarbon.

SCIENCE BACKGROUND

Petroleum is a complex mixture of hydrocarbons. It also contains small amounts of oxygen, nitrogen, water, and compounds containing sulfur. When petroleum was formed millions of years ago, organic compounds formed by organisms were covered by sediments. The sediments became rock as they were subjected to pressure and heat. The organic compounds were converted to petroleum.

LASER DISC
2633, 2634

Gasoline station; adding to tank

2663

Oil, crude; distillation

● **Process Skill:**
Communicating

Display a diagram of the substances produced when petroleum is separated using fractional distillation. Have the students identify the level at which each product listed in Table 22–2 is distilled in the tower.

MULTICULTURAL CONNECTION

As early as 900–300 B.C., the Olmecs—who lived in the area of what is now Mexico and neighboring Central American countries—used rubber, a natural polymer of 2-methylbutadiene from the havea tree, to make large balls for use in ritual games. Statuary also shows that the Olmecs wore what appears to be rubberlike protective headgear, garments, and footwear in these games. In the seventeenth century, Spanish explorers reported that descendants of the Olmecs made footwear and bottles from rubber molded on pottery forms.

THE NATURE OF SCIENCE

In 1960, in an effort to fight cancer, the National Cancer Institute established a natural products program. It worked with the U.S. Department of Agriculture to search the earth for substances that could fight cancer. Field workers and researchers tested everything available—from lichens, fungi, and mold, to insects and trees. From 1960 to 1981, over 130 000 plant and animal extracts were screened. Out of all those samples, only one safe, effective compound was found—taxol.

LASER DISC
2890

Yew

Figure 22–19.
Petroleum is separated using fractional distillation. The process takes place in fractionation towers such as those shown here.

Petroleum can be compared to a mixture of many different colored paints. Imagine trying to separate such a mess in order to see each color individually. It would be quite a job. Petroleum can be separated using a process called *fractional distillation,* which separates the hydrocarbons using the differences in their boiling points. The petroleum is heated at the bottom of a distillation column, and the compounds are turned into gases. Those with higher boiling points condense at the lowest part of the column. The substances with very low boiling points condense at the highest part.

Besides petroleum and natural gas, many other important organic compounds are found in nature. Compounds produced by living organisms are called *natural products*. These compounds are used in medicines, pesticides, perfumes, and food additives, among other things. Organic chemists try to isolate these natural products and to investigate their properties.

In the early 1990s, there was much medical interest in a substance found in a tree called a Pacific yew. The substance, called *taxol,* interested scientists because it can be used to fight certain types of cancer. The problem is that the Pacific yew is not very common, and thousands of trees must be cut in order to make just a few grams of taxol. When an important substance is found in too small a quantity in nature, scientists may try to synthesize it, or make it in the laboratory. By 1993, organic chemists had synthesized taxol inexpensively in the laboratory. These chemists made taxol using similar chemicals found in European yews. These trees are very common and in no danger of extinction.

Figure 22–20. The Pacific yew may have been saved from extinction now that the chemicals it contains can be made in the laboratory.

GUIDED PRACTICE

Have the students identify the double bond in several of the monomers listed in Table 22–1. Have volunteers show how the breaking of the carbon-carbon double bond in a monomer allows it to join other monomers to form a polymer.

INDEPENDENT PRACTICE

 Have the students provide written answers to the Section Review. In their journals, have them draw and label one monomer and its corresponding polymer.

EVALUATION

Ask the students to explain what a polymer is and provide some examples of practical uses.

What other useful things can organic chemists make? What about those wonderful scents that come from certain flowers and fruits? You might think that laboratories "stink," but scientists can prepare wonderful scents also. Many sweet, flowery scents are produced by organic compounds called *esters*. As you might expect, esters are very valuable to the perfume industry. What may surprise you is that esters are also used in prepared foods to enhance their aroma and flavor. You can experience some of these odors by trying the next activity.

Figure 22–21. This French scientist is testing the odors of various perfumes.

ACTIVITY

How can you identify the odors of different natural oils?

Process Skill: Observing

Grouping: Groups of 3 or 4

Hints

You will need several essential or natural oils. Banana oil (amyl acetate), clove, peppermint, cinnamon, and wintergreen oil (methyl salicylate) can be easily obtained. Label the known oils 1 through 5.

Prepare 5% (by volume) solutions of both the known and unknown solutions in dropper bottles. Label the unknown solutions A, B, C.

▶ **Application**

1. The types of oils will depend on the unknowns chosen.

2. Natural oils are produced in several ways, including distillation of plant matter. For specifics, the students should refer to a reference book.

3. Insects communicate by sensing various chemical substances known as pheromones. The ability of an olfactory organ to sense one or two pheromone molecules eliminates the need for other sensory organs to be large in order to have similar sensitivity.

ACTIVITY

How can you identify the odors of different natural oils?

MATERIALS
filter paper strips with different natural oils (5), strips of filter paper with unknown oils (3)

PROCEDURE
1. Make a table like the one shown.

TABLE 1: IDENTIFYING ODORS		
Strip	Odor	Matches unknown
1		
2		
3		
4		
5		

2. Your teacher will give you five pieces of filter paper with a different natural oil on each one. Do not let the strips touch each other. Smell and become familiar with the odor of each strip.

Do not smell any of the strips for too long because it may tire your sense of smell. Describe the odors on your data table.

3. Get one to three unknowns on strips of filter paper. Each unknown may have from one to four of the different oils you have already smelled. Again, don't tire your sense of smell. Try to identify the odors of the unknowns. Refer to the known samples if necessary.

APPLICATION
1. What were the oils in each of the unknowns? Refer to them by number if you do not know their names.
2. Find out how some of the natural oils you smelled are produced.
3. Insects can detect odors about 10^4 times better than humans. Why do you think a sense of smell is important to insects?

RETEACHING

Cooperative Learning Write the following analogy on the chalkboard: monomer: polymer, link: chain. Have small groups of students write similar analogies illustrating the relationship between monomers and polymers. Encourage the students to share their analogies with classmates.

EXTENSION

In their journals, have interested students list the names of fibers found on the labels of five articles of clothing they own. Have them classify the fibers as synthetic or natural polymers.

CLOSURE

Have the students discuss why clothing is often made of fabrics that are blends of synthetic and natural fibers. (For example, natural fibers allow the material to "breathe," while synthetic fibers resist wrinkling.)

Figure 22–22. *Kongfrontation* at Universal Studios in Orlando, Florida

ONGOING ASSESSMENT

▼ ASK YOURSELF

Natural products are compounds produced by living organisms.

SECTION 2 REVIEW AND APPLICATION

Reading Critically

1. To be a gas, propane must have between one and five carbons. Since methane has one carbon, and ethanol has two carbons, propane must have between three and five carbons. (To the teacher: Propane has three carbons; the formula is C_3H_8.)

2. A polymer is a large molecule made up of many small identical molecules hooked together.

Thinking Critically

3. Petroleum is a mixture. It can be separated into fractions through physical changes, such as melting and boiling.

4.

Have you ever been to a theme park? Universal Studios has a park in Orlando, Florida, in which one ride uses esters! Does that surprise you? It often surprises the people on the ride, too. The ride is called *Kongfrontation*. During one part of the ride, King Kong comes up to the cablecar you are riding in and breathes on you. His breath smells just like bananas. Now you know the odor is the result of synthetic esters!

▼ ASK YOURSELF

What is a natural product?

SECTION 2 *REVIEW AND APPLICATION*

Reading Critically

1. Bottled gas (propane) contains hydrocarbons. How many carbon atoms does each molecule of this hydrocarbon have?

2. What is a polymer?

Thinking Critically

3. Is petroleum a mixture or a complex compound? How can you tell? Be specific.

4. Orlon, a polymer used in synthetic fibers, has the structure shown. What is the monomer that makes up Orlon?

$$-CH_2-CH-CH_2-CH-CH_2-CH-CH_2-$$
$$\quad\quad CN \quad\quad\quad CN \quad\quad\quad CN$$

Process Skills: Observing, Constructing/Interpreting Models, Comparing

Grouping: Groups of 3 or 4

Objectives

● **Construct** models of tetrahedral molecules.
● **Observe** and compare mirror images of models.
● **Interpret** nonsuperimposable mirror-image models as enantiomers.

SKILL Using Models

▶ MATERIALS
● pencil ● paper ● mirror

▼ PROCEDURE

1. Print TOT on a sheet of paper. Place it in front of a mirror. On another sheet of paper draw the image you see. Is the mirror image the same as the original? How can you tell?

2. Now print POP on a sheet of paper, and repeat the process. Are the images the same? How are they different from the images in step 1?

3. Make a model of a molecule by taking a plastic-foam ball and inserting four different-colored straws in it. Place your model in front of the mirror, and make another model that is like the image in the mirror. Are the two models the same? Explain why or why not.

▶ APPLICATION

Your hands are mirror images of each other. You can place them facing each other and they match. However, if you put one on top of the other, they do not match. Some molecules are like your hands. They are alike but are mirror images of each other. These molecules are called *enantiomers* (ihn AN tee uh muhrz).

1. Which of the words you looked at in the mirror is an enantiomer?

2. Was the model you made an enantiomer?

3. Are the compounds shown here mirror images? Are they enantiomers? Explain.

4. Are three-dimensional models necessary in order to understand enantiomers? Why or why not?

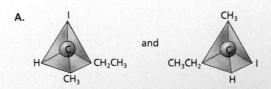

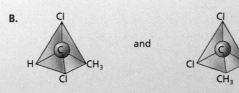

✴ Using What You Have Learned

How did using a model help you understand the structure of enantiomers? Describe another situation in which using a model could help you understand a new idea.

Application

1. POP

2. yes

3. Not as shown in the diagram. However, the model on the right in diagram A can be rotated around its center to produce a nonsuperimposable mirror image of the model on the left. The compounds in A are enantiomers. If the model on the right of B is rotated, its mirror image can be superimposed on the model to the left. They are the same compound and, therefore, not enantiomers.

4. Yes. Three-dimensional models are necessary to understand enantiomers because the models must be rotated and their mirror images compared.

✴ Using What You Have Learned

The students might note that the model helped them understand the structure of something that is difficult to visualize. A model can also help you understand the structure of something that is too small or too large to easily observe. Accept all reasonable responses.

FOCUS

Section 3 discusses carbohydrates, fats, and proteins. The students learn about the structure and use of these three types of polymers.

MOTIVATING ACTIVITY

Cooperative Learning Review with the students the basic food groups. Give small groups of students several weekly cafeteria menus or weekly nutrition menus provided by the school nurse or a dietitian. Have the groups use the menus to cite foods in each food group and discuss why each food group is important. Conclude by explaining to the students that they will learn more about what comprises each food group in this section.

PROCESS SKILLS
- Observing • Experimenting
- Communicating • Inferring

POSITIVE ATTITUDES
- Cooperativeness • Initiative and persistence

TERMS
- carbohydrates • lipids

PRINT MEDIA
"Aspirin" from *Chem Matters* magazine by Gail Marsella (see p. 513b)

ELECTRONIC MEDIA
Special Topics in Chemistry: Biochemistry Focus Media, Inc. (see p. 513c)

SCIENCE DISCOVERY
- Bear, black • Bear, grizzly • Bear, polar
- Pasta • Sugar

BLACKLINE MASTERS
Laboratory Investigations 22.1, 22.2
Connecting Other Disciplines
Study and Review Guide

 LASER DISC
2351

Sugar

2352

Pasta

✧ Did You Know?

Cellulose is referred to as fiber. Research indicates that the proper amount of fiber in a diet can reduce the risk of colon cancer and other intestinal problems. Because cellulose cannot be digested by humans, it is also added to reduced-calorie bread to provide calorie-free bulk. While humans cannot digest cellulose, special bacteria present in the digestive tracts of certain animals, such as cattle, break down plant cellulose into sugar.

SECTION 3

Molecules of Life

Objectives

Compare and ***contrast*** carbohydrates, proteins, and fats.

Differentiate between saturated and unsaturated fats.

So far in this chapter, we have discussed the importance of carbon in many manufactured goods. Your life is filled with carbon compounds; your life is also made up of them. You would not move, breathe, or exist if it were not for carbon compounds. The study of carbon compounds in living organisms is called *biochemistry*.

Fuel for Your Body

Jim is getting ready for a marathon in the morning. He must be in peak condition to win the race. Tonight he is meeting with other runners for a spaghetti feast. Jim does this before every marathon. He knows that spaghetti will provide the energy he needs to do well tomorrow. It is the fuel his body will burn as he runs. Bread, pasta, and rice are foods that contain carbohydrates. **Carbohydrates** are compounds made of carbon, hydrogen, and oxygen in a ratio of CH_2O. Figure 22–23 shows some common carbohydrates.

Figure 22–23. Carbohydrates are found in all the foods shown here.

TEACHING STRATEGIES

● Process Skill: *Applying*

Have a volunteer write the chemical equation for the breakdown of sucrose in living cells.
($C_{12}H_{22}O_{11} + H_2O \rightarrow 2C_6H_{12}O_6$)

● Process Skill: *Comparing*

Write the reaction of two glycine molecules on the chalkboard. Then have the students compare this reaction to the formation of sucrose shown on this page. (In both reactions, two monomers form a polymer by a hydration reaction.)

Sugars, starches, and cellulose are examples of carbohydrates. Glucose ($C_6H_{12}O_6$) is a simple sugar that makes up more complicated carbohydrates. All living cells use glucose as their "fuel." There are other simple sugars besides glucose. Fructose, for example, is commonly found in fruit. Simple sugars such as these are used as monomers in making carbohydrate polymers.

The sugar that you put in your lemonade is called *sucrose* ($C_{12}H_{22}O_{11}$). It is a compound formed by joining two simple sugars, glucose and fructose. Look closely at Figure 22–24. Find where the two molecules have joined and what is formed in the reaction. Notice that as the larger carbohydrate is formed, a molecule of water is released. This is how plants synthesize complex sugars and starches. In your body, the reverse reaction occurs. Water is used to break down the bigger molecules into simple sugars that can be used by your body to produce energy.

Figure 22–24. A dehydration reaction produces a large molecule from two smaller molecules. The reaction shown here is the formation of sucrose (table sugar) from glucose and fructose.

| Glucose $C_6H_{12}O_6$ | Fructose $C_6H_{12}O_6$ | Sucrose $C_{12}H_{22}O_{11}$ | Water |

ASK YOURSELF
Why are carbohydrates useful?

Muscle-Building Molecules

Other types of molecules important to your body are proteins. What kinds of food are rich in protein?

Like polymers, **proteins** are huge molecules made up of smaller ones. The building blocks of proteins are called *amino*

Figure 22–25. These foods are rich in protein.

REINFORCING THEMES
Systems and Structures

Using molecular model kits, construct ethanol and dimethyl ether as described in *Discover by Doing* on page 594. Review with the students the differences in the structures of the two molecules and the differences in their effect on the human body. Construct models of GLY-GLY-GLY, GLY-GLY-ALA, and GLY-ALA-GLY. Have the students compare the differences in their structures.

Remind the students that the structure of an enzyme determines how it affects the human body. Ask them whether the three enzymes would affect the body in the same way if these proteins were enzymes. (The structures are different, and therefore the ways they function are probably different.)

SCIENCE BACKGROUND

Proteins can be enzymes or structural molecules. Proteins are composed of amino acid monomers that each have a carboxyl group (–COOH) and an amino group (–NH$_2$) attached to the same carbon atom. Proteins are formed when a peptide bond forms between the carboxyl carbon of one amino acid and the amino nitrogen of another.

ONGOING ASSESSMENT
ASK YOURSELF

Carbohydrates supply the body with energy.

TEACHING STRATEGIES, continued

● **Process Skill:** *Inferring*

Point out that the essential amino acids are valine, leucine, isoleucine, phenylalanine, tryptophan, threonine, methionine, and lysine. Make sure the students understand that humans can get these amino acids only through eating animal and vegetable proteins. Ask them how the human body can contain proteins different from those taken in. (The ingested proteins are broken down into essential amino acids, which are rearranged to form other proteins.)

Demonstration

Combine any two of the amino acids listed in Table 22–3 via a condensation reaction. Point out to the students that condensation and hydrolysis reactions are very important to an understanding of biochemistry. If feasible, provide the students with opportunities to repeat this demonstration using amino acids other than the ones you used.

INTEGRATION–
Life Science

Encourage interested students to use reference resources to find out how vegans—vegetarians who eat no meat, fish, eggs, or dairy products—structure their diets to consume nutritionally complete proteins. Ask the students to share their findings with classmates.

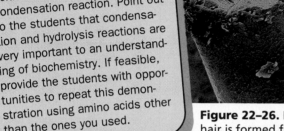

Figure 22–26. Human hair is formed from complex proteins. A magnified human hair is shown above. A computer-generated image of a complex protein is shown on the right.

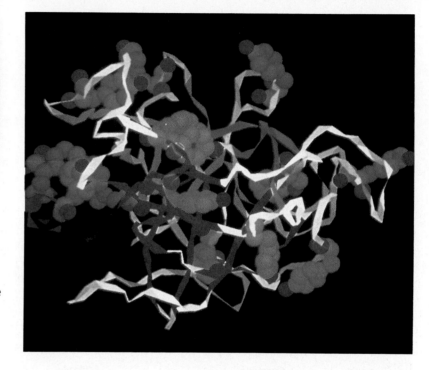

acids. There are two chemical groups by which you can recognize an amino acid. The -NH$_2$ part is an amino group, and the -COOH group is an organic acid. Put them together, and you have an amino acid. The other parts of this molecule can vary. These variable parts are called *R groups.* Table 22–3 shows the characteristic groups for some of the more common amino acids.

$$\begin{array}{c} H \\ | \\ R-C-COOH \\ | \\ NH_2 \end{array}$$

Figure 22–27. Amino acids are the building blocks of proteins.

Table 22-3

Some Common Amino Acids

Amino acid	Characteristic R group
Glycine	-H
Alanine	-CH$_3$
Serine	-CH$_2$OH
Valine	CH$_3$—CH—CH$_3$
Leucine	CH$_3$—CH—CH$_3$ CH$_2$

● **Process Skill:** *Comparing*

Ask the students to write a paragraph in their journals comparing amino acids and proteins with the alphabet and words. (There are only 26 letters in the alphabet. However, they combine to form thousands of words.)

There are 20 important amino acids. The number of different proteins, however, may be in the millions. If you had 20 different colored blocks and had to arrange them in a straight line, how many different combinations could you make? To give you an idea of how many combinations there could be, try the next activity.

DISCOVER BY Doing

Get six different-colored blocks from your teacher. Arrange them in as many straight-line patterns as you can. How many combinations did you find?

Amino acids link together just like boxcars in a train to make proteins. Your muscles, hair, tendons, and even parts of your blood are all made of proteins. Other proteins actually control all the chemical reactions that go on inside a cell. These proteins are called *enzymes*. These are of critical importance to living cells. Just one misplaced amino-acid boxcar in the protein train can sometimes mean the difference between life and death of a cell or even an entire organism.

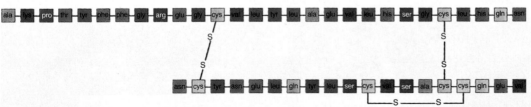

Figure 22–28. Proteins, like this one, are formed from long chains of amino acids.

In a protein, each boxcar represents one amino acid. Scientists use the first three letters of each amino acid to write a protein sequence. For example, glycine would be written as GLY and alanine as ALA. The diagram shown (Figure 22–28) illustrates an amino acid sequence in a protein. The -S-S- structure represents a sulfur bridge. These bridges increase the rigidity of the molecule. To what molecules do the sulfur bridges attach?

ASK YOURSELF

What are amino acids, and what function do they perform in the human body?

DISCOVER BY *Doing*

Have the students first look at combinations of two blocks, A and B. (AB and BA; 2 combinations: $2 \times 1 = 2$). Then have the students look at combinations of three blocks. (CAB, CBA, ACB, ABC, BCA, and BAC; 6 combinations: $3 \times 2 \times 1 = 6$) Then have the students continue with four blocks (DCAB, DCBA, DACB, DABC, DBCA, DBAC, CDAB, CDBA, CADB, CABD, CBDA, CBAD, ACDB, ACBD, ADCB, ADBC, ABCD, ABDC, BCAD, BCDA, BACD, BADC, BDCA, and BDAC; 24 combinations: $4 \times 3 \times 2 \times 1 = 24$). By induction, have the students arrive at 720 ($6 \times 5 \times 4 \times 3 \times 2 \times 1 = 720$) as the possible number of arrangements for six blocks.

ONGOING ASSESSMENT
ASK YOURSELF

Amino acids are the building blocks of proteins, which make muscles, hair, tendons, parts of your blood, and enzymes.

MEETING SPECIAL NEEDS

Gifted

Have advanced students use reference resources to research the induced-fit model of enzyme-substrate reactions in cells. Encourage them to report their findings to the class and be prepared to discuss questions. You may wish to have the students place their reports in their science portfolios.

GUIDED PRACTICE

Guide the students through the reaction shown in Figure 22–30. Have them identify it as a dehydration reaction. Then ask them to compare this reaction to the formation of sucrose in Figure 22–24.

(Both reactions are dehydration reactions in which monomers join to form polymers and water.)

INDEPENDENT PRACTICE

 Have the students provide written answers to the Section Review. Then have them cite in their journals an example of a carbohydrate, a protein, and a lipid.

EVALUATION

Ask the students to identify and describe at least one function each of a carbohydrate, a protein, and a lipid.

SCIENCE BACKGROUND

Lipids are defined by their solubility in nonpolar organic solvents, such as ether, chloroform, and benzene. Because lipids are insoluble in water, they are an important part of cell membranes. Cell membranes are made up of phospholipids, which are simple lipids combined with a phosphate group.

 LASER DISC

3996

Bear, black

4000

Bear, grizzly

4001

Bear, polar

Figure 22–29. When food supplies are low, some animals, such as these bears, hibernate and live off the fat stored in their bodies.

Figure 22–30. Fats (orange) are stored in the bodies of plants and animals. The reaction (inset) shows how fats are formed.

Slick Globs

What is the difference between vegetable shortening and vegetable oil? Why is one solid and the other liquid? These materials are all lipids. **Lipids** contain atoms of carbon, hydrogen, and oxygen—just as carbohydrates do. Lipids that are solids are called *fats*. Liquid lipids are called *oils*.

Most living organisms produce lipids as a storage fuel. Bears can hibernate through the winter without starving because they draw energy from the layer of fat they have built up over the summer. After your body burns all its sugar for energy, it will draw fat from its reserves.

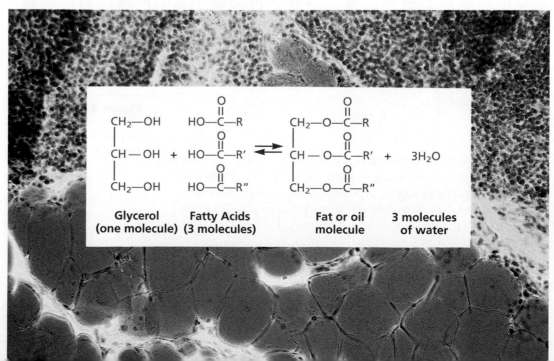

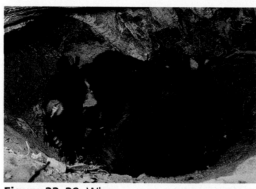

| Glycerol (one molecule) | Fatty Acids (3 molecules) | | Fat or oil molecule | 3 molecules of water |

You have probably heard commercials claiming that cooking with vegetable oil is better than using animal fat. The commercial might even have used the term *polyunsaturated fats*. What are polyunsaturated fats?

Remember that some lipids are solids and some are liquids. The difference is due to their chemical structure. Some lipids, called unsaturated or polyunsaturated oils, have many double bonds. They tend to be less dense and are usually liquids at room temperature. Olive oil, corn oil, safflower oil, and peanut oil are examples of unsaturated lipids.

Saturated lipids do not have double bonds. They have only single bonds and are, therefore, denser than unsaturated ones. Saturated lipids tend to be solids at room temperature. Research has shown that saturated fats contribute to heart disease. The American Heart Association suggests that eating a diet low in saturated fats is better for your health. This means a reduction of most animal fats in your diet.

Figure 22–31. Saturated and unsaturated fats, such as those shown here, are available for use in cooking.

 ASK YOURSELF

What is the difference between saturated and unsaturated fats?

SECTION 3 REVIEW AND APPLICATION

Reading Critically

1. What characteristics do carbohydrates, lipids, and proteins have in common?

2. What molecule is formed when sugars combine to form complex carbohydrates? When fats or oils are formed?

3. Carbohydrates and lipids both contain carbon, hydrogen, and oxygen. How are these molecules different?

Thinking Critically

4. Saturated fats can be responsible for clogging arteries. Based on what you know about the properties of saturated and unsaturated fats, why do you think this is true?

5. Different proteins may contain the same amino acids arranged in a different order. How many different combinations are possible with one unit of each of the following three amino acids: glycine, alanine, and serine? Write the different amino acid sequences.

Testing for Nutrients

Process Skills: Observing, Experimenting, Communicating, Inferring

Grouping: Groups of 3 or 4

Objective
● **Test** food samples for glucose, starch, and protein.

Pre-Lab
Ask the students how they might test for certain substances in food.

Hint
Make sure the students understand that they are performing three separate tests: one for glucose, one for starch, and one for protein.

Discussion
The color of the test strip in step 3 depends on the type of test strip used. Consult the manufacturer's instructions.

PERFORMANCE ASSESSMENT

Cooperative Learning
After this *Investigation,* have the students describe other situations in which an individual would need to know a food's chemical makeup for health reasons. (Examples include dietary restrictions and allergic reactions to certain foods, such as glucose and lactose.)

▶ **610** CHAPTER 22

▶ **Analyses and Conclusions**

1. Green onion should give a distinct positive test. Depending on the cracker used, a small amount of glucose might be detected.

2. cracker

3. Purple was a positive test for protein. The egg white tested positive.

▶ **Application**

The diabetic can accurately monitor his or her need for insulin.

✳ **Discover More**

The students might note that foods made from grains generally contain starch, fruits contain sugar, and most animal products contain protein. Accept all reasonable responses.

Post-Lab
Ask the students how their test results compared to the test procedures they proposed before the *Investigation.*

INVESTIGATION

Testing for Nutrients

▶ **MATERIALS**

● safety goggles
● laboratory apron ● wax pencil ● test tubes (12)
● test tube rack ● egg white solution ● green onion shoots ● saltine crackers ● graduate, 10 mL ● stirring rod ● glucose test strips (4)
● iodine solution ● sodium hydroxide, 10% ● stoppers (4) ● copper sulfate, 0.5%

▼ **PROCEDURE**

CAUTION: Put on safety goggles and a laboratory apron, and leave them on throughout this Investigation.

1. Copy the table shown, and record your observations on it.

TABLE 1: NUTRIENTS IN FOOD			
	Test results		
Substance Tested	**Glucose test strip**	**Iodine**	**NaOH/CuSO$_4$**
Egg white			
Green onion			
Cracker			
Control			

2. Using a wax pencil, label four test tubes 1–4. Put 5 mL of water in each. Place a small amount of egg white solution in #1, green onion shoots in #2, and a piece of cracker in #3. Leave only water in #4; this is your control.

3. Test for Sugar Add 4 mL of water to each test tube, and stir. Use a different glucose test strip for each test tube. Moisten the test strip in the liquid, remove it from the test tube, and note the color after 10 seconds. A violet to purple color indicates the presence of glucose. The darker the color, the more glucose in the mixture.

4. Test for Starch Repeat step 2, and test the mixtures with a few drops of iodine solution instead of the glucose strip.

5. Test for Protein Repeat step 2, and test the mixtures with 5 mL of sodium hydroxide solution. Stopper each solution, and shake well. Add several drops of copper sulfate solution, shaking well after each drop. Watch for a change of color.

▶ **ANALYSES AND CONCLUSIONS**

1. Explain the results of the glucose test strip test. Which material(s) tested contained glucose?

2. Iodine solution turns blue-black in the presence of starch. Which substance(s) gave a positive test for starch?

3. What was a positive test for protein? Which material(s) test positive for protein?

▶ **APPLICATION**

Glucose test strips are used to test urine for sugar. Many diabetics check themselves this way. What advantage would this method give a diabetic?

✳ **Discover More**

Choose other foods to test for sugar, starch, and protein. Make a chart to show your results. What conclusions can you draw about certain types of foods and their nutrients?

The Big Idea— Systems and Structures

Lead the students to realize that the structures of complex molecules such as carbohydrates, proteins, and lipids determine their use and functions within the cell.

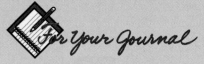

For Your Journal

Responses should reflect an understanding of how carbon can form so many different types of compounds, some of which are necessary for life, because of its valence electrons. You might ask the students to share their new ideas with the class.

CONNECTING IDEAS

Responses should reflect an understanding that carbon compounds occur as hydrocarbons in fossil fuels such as coal, carbohydrates such as cellulose in wood pulp, and synthetic polymers in fibers such as nylon.

CHAPTER 22 *H*IGHLIGHTS

The Big Idea

Differences in structure often mean differences in properties. Structural isomers such as ethyl alcohol and dimethyl ether have very different properties. Lipids can be harmful to your health depending on how many double bonds they contain. Combining amino acids in different orders produce different proteins. Crosslinking in polymers produces a stronger material.

The structure of a molecule affects the function of the molecule. Changing the shape of the molecule, even when the components remain the same, may change the function of the molecule drastically.

For Your Journal

You read about diamonds at the begining of the chapter. Now you have learned more about compounds of carbon. Revise your journal entry to include new ideas about these questions: How can an element that forms diamonds also form living tissue?

Connecting Ideas

Below are some photographs of compounds containing carbon. Look at the photographs, and write a paragraph in your journal that explains how each compound affects your lifestyle.

a. Coal

b. Wood chips

c. Nylon running clothes

Understanding Vocabulary

1. (a) An isomer is one of two or more compounds that have the same chemical formula but different structures. A polymer is a large molecule made by "hooking" many identical small molecules together. A monomer is one of the small units that make up a polymer. A structural formula shows the arrangement of atoms in a compound.

(b) Hydrocarbons are compounds that contain only the elements carbon and hydrogen.

2. Carbohydrates, proteins, and lipids are all carbon compounds necessary for human life. Carbohydrates supply the body with energy. Proteins are used to build muscle and produce enzymes that control cell functions. Lipids are fats and oils that are used to store energy.

Understanding Concepts

Multiple Choice

3. a
4. d
5. b
6. d
7. d
8. c
9. c

Short Answer

10. The chemical formula of a hydrocarbon shows the number of carbon and hydrogen atoms in the compound. The structural formula of a hydrocarbon shows the arrangement of the hydrogen and carbon atoms in the compound.

11. As the number of carbon atoms increases in a carbon compound, the possible arrangements of the chains increase, since each carbon atom can form four single covalent bonds.

12. fats: 570 kJ; carbohydrates and proteins: 255 kJ. You can gain weight by eating too much fat because you must expend more than twice the effort to use the energy in a gram of fat compared with a gram of protein or carbohydrate. If you do not expend the effort, the fat is stored as body fat.

CHAPTER
22
REVIEW

▶ Understanding Vocabulary

1. Explain how the terms in each set are related.
a) isomer (594), polymer (595), monomer (595), structural formula (592)
b) hydrogen (593), carbon (593), hydrocarbons (593)

2. Explain the similarities and differences among carbohydrates (604), proteins (605), and lipids (608).

▶ Understanding Concepts

MULTIPLE CHOICE

3. All of the following are members of the carbon family except
a) arsenic.
b) germanium.
c) lead.
d) tin.

4. The maximum number of covalent bonds a carbon atom can form is
a) one. **b)** two.
c) three. **d)** four.

5. The focus of most hydrocarbon reactions involves changes in bonds between
a) hydrogen atoms.
b) carbon atoms.
c) oxygen atoms.
d) nitrogen atoms.

6. The study of carbon compounds in living organisms is called
a) physics.
b) organic chemistry.
c) inorganic chemistry.
d) biochemistry.

7. Materials that have hydrocarbons and salts at opposite ends of their molecules are called
a) esters.
b) natural gas.
c) petroleum products.
d) soaps.

8. Which of the following characteristic ratios of elements represents a carbohydrate?
a) CHO
b) CHN
c) CH_2O
d) CNH_2

9. Silicones are long chains of what types of atoms with different carbon chains attached?
a) hydrogen and oxygen atoms
b) silicon and hydrogen atoms
c) silicon and oxygen atoms
d) silicon and nitrogen atoms

SHORT ANSWER

10. Explain the difference between the chemical formula and the structural formula of a hydrocarbon.

11. Why does the number of isomers of a carbon compound increase with the number of carbon atoms in the compound?

12. Every gram of fat provides 38 kJ (kilojoules) of energy, as compared to carbohydrates and proteins, which each provide 17 kJ of energy per gram. How many kJ of energy can be supplied by 15 grams of carbohydrates? 15 grams of protein? Why do you think eating too much fat would make you gain weight?

Interpreting Graphics

13. same substance

14. structural isomer

15. structural isomer

16. same substance

Reviewing Themes

17. The body does not have the ability to break down the mirror-image isomer and use the energy within it.

18. The energy absorbed from the sun restores the crosslinks between the polymers that give the band its shape.

Thinking Critically

19. Inorganic chemists study areas of chemistry that do not deal with carbon compounds.

20. Answers may include durability, nontoxicity, flexibility, and the ability to be accepted by the body's immune system.

Interpreting Graphics

Classify each pair below as structural isomers or the same substance.

13. CH_3—CH_2
  |
  CH_2—$CH3$

CH_3—CH_2—CH_2—CH_3

14. CH_3—CH_2—CH_2—CH_2—CH_3

CH_3
  \
  CH—CH_2—CH_3
  /
CH_3

15. CH_3—CH_2—$CH = CH_2$

CH_3—$CH = CH$—CH_3

16. CH_3—CH—CH_3
  |
  CH_2
  |
  CH_2
  |
  CH
  /   \
CH_3 CH_3

CH_3 CH_3
  \   /
   CH—CH_2—CH_2—CH
  /   \
CH_3 CH_3

Reviewing Themes

17. Systems and Structures
Glucose exists as two enantiomers. (See the Skill Activity on page 603.) One enantiomer can be used to supply energy by your body, but the other cannot. Why do you think this so?

18. Technology
If the polyethylene band that holds a six-pack of soda is stretched, it doesn't return to its original shape, as would a rubber band. However, if the polyethylene band is warmed in the sun, it begins to return to its original shape. Explain why this happens.

Thinking Critically

19. In this chapter you have studied organic chemistry. Another area of study is inorganic chemistry. Based on what you know about organic chemistry, what do you think an inorganic chemist studies?

20. One of the most interesting uses of polymers is in artificial body parts. Most artificial body parts, such as artificial knees or hips, contain polyurethane. What properties must a polymer have in order to function as a body part?

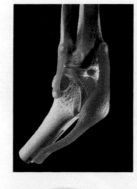

Discovery Through Reading

Edelson, E. "Buckyball: The Magic Molecule." *Popular Science* 239 (August 1991): 52–57+. The discovery, properties, and possible uses of buckminsterfullerenes, a new class of hydrocarbons, are discussed in the article.

Science
PARADE

SCIENCE APPLICATIONS

Discussion

● **Process Skills:** *Observing, Synthesizing*

This *Science Applications* is about the role of chemical communication in nature. Ask the students to describe instances of chemical communication **between animals that they have observed.** (Students' responses may include communication between animals they have observed in the field or communication between their own pets or farm animals.)

SCIENCE APPLICATIONS

Chemical Communication in Nature

Chemists are constantly trying to discover practical uses for chemicals, such as new medicines and insecticides. Where do chemists get their ideas? Sometimes they just look at nature.

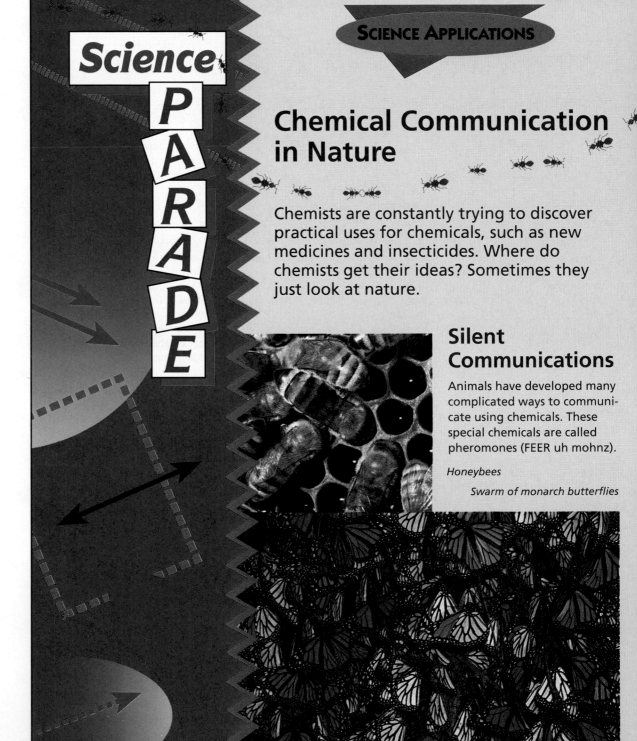

Silent Communications

Animals have developed many complicated ways to communicate using chemicals. These special chemicals are called pheromones (FEER uh mohnz).

Honeybees

Swarm of monarch butterflies

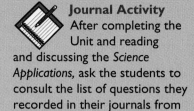

Pheromones can alert animals to danger or let them know where others of their species are located. Another important use of pheromones is to attract a mate. Some pheromones serve as a defense against predators, while other pheromones attract prey.

The pheromones used by honeybees have been widely studied. The queen bee produces a chemical that tells the worker bees there is a queen in the hive and another one is not needed at that time. It also tells the workers to feed substances to the eggs so that the eggs will produce only other workers.

Bees release a substance called *isoamyl acetate* when they sting. Isoamyl acetate is the same substance that gives bananas their characteristic odor. It produces an aggressive behavior in other bees. This explains why, when a person is stung by several bees, the bees tend to sting in the same area. The isoamyl acetate directs the bees to the spot.

Caribbean fruit flies attracted to artificial pheromones

Flies use their pheromones to attract mates. The female gives off the pheromone and the male fly comes to her.

Once the pheromone can be reproduced, it can be used to trap the insects. Sticky traps, sprayed with the pheromone and placed in groves, could attract and hold the flies without the use of insecticides.

Another insect that uses pheromones is the ant. Ants leave chemical trails so that other ants can follow them. If one ant finds a food source, its trail will lead other ants to the food. This is one way in which ants communicate.

Fly Away

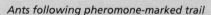

Ants following pheromone-marked trail

Pheromones are combinations of different chemicals in very specific amounts. If one chemical is missing, or if the amounts are not right, the pheromone does not work. Today there is much interest in insect pheromones. Researchers hope to use these pheromones to control insect growth and reproduction without pesticides. The Caribbean fruit fly, for example, is a pest that attacks orange trees. Chemists are trying to make the fly's pheromone in the laboratory.

Defend Yourself

Many animals use chemicals other than pheromones for defense. The bombardier beetle, for example, fires a chemical spray that quickly repels its enemies. Certain

EXTENSION

As a follow-up to the class discussion, you might have interested students prepare a report on chemical communication in plants. Encourage the students to share their findings with classmates.

Snail defending itself using a pheromone foam

types of millipedes store a compound in their bodies that can quickly produce hydrogen cyanide, a deadly poison. The compound, called *mandelonitrile,* is converted to hydrogen cyanide when the millipede is in danger. The hydrogen cyanide cannot be stored in the millipede's body because the chemical would kill the millipede.

Some animals do not make their own defensive chemicals but can obtain them from other sources. For example, the monarch butterfly eats milkweed plants. A chemical in the plant is stored in the body of the butterfly. When a bird attacks and eats a monarch butterfly, the chemical makes the bird sick. Even though the individual butterfly did not survive the attack, the bird may remember its bad experience. In the future, the bird may avoid eating another monarch butterfly. This

defense has worked so well that most birds avoid any type of butterfly that has the same wing colors as the monarch. This color copying, or mimicry, protects butterflies with similar coloration.

Nature still holds many secrets about chemistry and chemicals. Examples like these can help you understand how nature and chemistry are related. ◆

Bombardier beetle squirting a deadly pheromone at a predator

Background Information

Anabolic steroids are synthetic versions of the male sex hormone, testosterone. Anabolic steroids are complex chemicals that the human body cannot handle easily.

Some of the undesirable side effects of steroid use include: liver and kidney damage, decrease in ultimate height, increased risk of cancer, appearance of acne, loss of scalp hair, decrease in size of testicles and impotency in males, increased aggression and unpredictable mood changes, and decreased breast size in women. Point out to the students statistics that one survey determined one in ten high school seniors has used steroids. Discuss with the students that the temporary results from steroids do not outweigh the risks associated with their use.

READ ABOUT IT!

Steroids Build Muscles and Destroy Health

by A. T. McPhee from *CURRENT SCIENCE*

Eight runners settled into the starting blocks. The runners gazed toward the finish line, 100 meters (110 yards) away, and wondered who would cross the line first. The winner would bring home an Olympic gold medal.

The crowd fell silent. A referee pointed a starting pistol into the air, paused, and fired.

The men grunted. Bodies lurched forward. Feet pounded the track in a blur of blazing speed. A few blinks of an eye later, Canadian track star Ben Johnson raised his right hand in a winner's salute. Mr. Johnson had set a world's record, beaten an arch rival, and won an Olympic gold medal—all in just 9.79 seconds.

Yet, 18 hours later, Mr. Johnson lost all three prizes. Tests done after the race revealed that Mr. Johnson had been taking an anabolic steroid, a drug that might have improved his performance. Olympic rules forbid athletes to take such drugs. So that night, officials disqualified Mr. Johnson

How much extra power did steroids give Ben Johnson, shown here running in the Olympics? Maybe none. Many doctors say steroids don't increase power so much as they drive the athlete to train harder.

Discussion

● **Process Skills:** *Predicting, Analyzing*

Ask the students to discuss possible reasons that people take steroids. (Students might mention taking them to increase muscle size and strength as to enhance athletic performance or to look better.) You may wish to discuss with the students societal pressures some people might feel about the appearance of their bodies.

from further competition, canceled his world's record, and gave the gold medal to his rival, Carl Lewis.

Mr. Johnson's disqualification marked what many experts say is a major problem in sports: steroid abuse. Sports organizations such as the International Olympic Committee are playing tough with athletes who take steroids. These organizations say athletes will pay a price for steroid abuse—a price that for these athletes will mean losing their own good health.

Building Big Muscles, Hot Tempers

Why do athletes risk victory and good health by taking steroids? "Clearly," says Dr. Charles Yesalis, professor of health policy at Pennsylvania State University in University Park, "steroids give the athlete an advantage over people who aren't taking [the steroids]." That advantage may stem in part from the ability of anabolic steroids to increase muscle size and promote aggression.

Anabolic steroids are synthetic substances, usually taken in pill form, that resemble the natural male hormone testosterone. Testosterone promotes facial hair growth, deepens the voice, and increases muscle size.

Most athletes who take anabolic steroids do so for the drugs' effect on muscle size. These athletes insist that bigger muscles make them stronger.

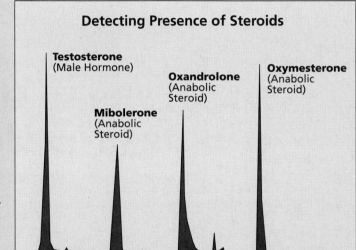

Detecting Presence of Steroids

Testosterone (Male Hormone)

Mibolerone (Anabolic Steroid)

Oxandrolone (Anabolic Steroid)

Oxymesterone (Anabolic Steroid)

This tracing from a urine test shows the presence of three anabolic steroids. A urine test taken from someone who has not used an anabolic steroid would show no such peaks. Ben Johnson's urine test showed he had taken a powerful steroid many times.

Many doctors doubt such claims. Steroids do make muscles larger, the doctors say, but not necessarily stronger.

Studies show that anabolic steroids, taken in moderate doses, increase strength only in women, children, and men with a low level of testosterone. Dr. Glenn Braunstein, professor of medicine at the University of California at Los Angeles, says, "The gains in muscle strength [from steroids] are minimal at best and do not occur in all individuals."

Even if steroids *do* increase strength somewhat, as some athletes and doctors claim, the drugs cause a far greater increase in aggressive behaviors. Dr. Robert Voy, chief medical officer for the U.S. Olympic Committee, says, "People don't realize that athletes on steroids become mean." The problem, say doc-

tors, is that athletes can't leave their aggression on the field.

One teen "strung out" on steroids had a friend videotape him while he slammed his car into a tree. Another steroid-pumping teen threw a wild temper tantrum at a friend's house. The teen smashed his friend's stereo, tossed a TV through a window, ripped a refrigerator door off its hinges, and was finally arrested when he beat up and severely injured a former teammate. Such aggression may help win football games, say doctors, but it wreaks havoc in the athletes' personal lives.

Steroids Halt Growth, Hurt Liver

Perhaps the greatest harm from anabolic steroids comes from their effects on the bones and the liver. Studies show that

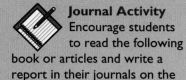

Journal Activity Encourage students to read the following book or articles and write a report in their journals on the information they learned.

"Clean and Slower." *Sports Illustrated* 75 (July 22, 1991): 26–29. This article discusses the

life of Ben Johnson, a Canadian track star who was banned from the Olympics after using steroids. Most recently, Johnson again tested positive for steroid use and has been banned from all track competition forever.

"The Devil's Juice." *Scholastic Update* 123 (May 1, 1991):

11–12. This article speaks to young adults about the death of a young football player who took steroids, the pressure that society places on young people to look good, and the side effects of steroids.

Dolan, Edward. *Drugs in Sports.* Watts, 1992. This book covers

the latest problems with drug use in sports. The composition and effects of each drug are clearly explained. The book concludes with a description of what professional sports leagues are doing to limit drug use among players.

Why People Use Steroids...

An athlete uses anabolic steroids in order to gain muscle mass to help him or her excel in a sport. A nonathlete uses anabolic steroids to look good either at the beach or in the gym. Steroids produce many harmful side effects in both males and females. Males may experience scarring acne, hair loss, and "roid rage"—an increase in anger and aggression. Females may notice a deepening of their voice, hair loss, and significant growth of facial hair. Anabolic steroids may also cause liver and heart damage, which can result in death.

long-term steroid use can stop a teen from growing as tall as he or she would grow without steroids.

Bones in the arms and legs grow by producing new bone tissue at areas near the ends of the bones. These areas are called *growth plates.* Growth plates normally stop producing bone tissue during the late teenage years. After the plates stop producing new bone, they close, or *fuse.* This fusion stops further growth at the plates.

Anabolic steroids can fuse growth plates much earlier than normal. Dr. Yesalis explains, "A kid who might grow to be 6 feet 2 inches tall [without steroids] could end up 5 feet 8 inches tall [with steroids]."

Besides stopping growth early, steroids also can cause extensive damage to the liver. Doctors say this damage may lead to liver cancer. "Teenagers don't understand [the dangers of steroid use]," says Armond

Colombo, football coach at Brockton (Mass.) High School. "Liver cancer really doesn't frighten them."

Nor do teens seem frightened by other possible side effects of steroid use—abnormal hair loss, kidney disease, high blood pressure, and an inability to have children. Some kids seem bent on taking steroids just to "look good at the beach."

Avoiding Steroid Abuse

World famous Olympic track star Edwin Moses says more frequent testing would help lower the number of athletes now using steroids. New lab tests done at some athletic events can determine whether an athlete took a certain drug days or even weeks before being tested.

Most experts agree that more frequent testing would help curb steroid abuse. They add that adults should be honest with kids. "You can't tell

kids that steroids don't work," says Pat Croce, a physical therapist in Philadelphia, Penn., "because they know they do."

Mr. Croce believes kids should take a healthier route to bigger muscles. "If you want to get big," Mr. Croce tells teens, "you have to eat right and sweat hard."

Former professional wrestling champion Ken Ventura understands the value of Mr. Croce's advice. Mr. Ventura took steroids until 1981, when he stopped because "I didn't want [the drugs] to tear my body apart."

Today, Mr. Ventura's message—and the message of experts the world over—is simple: "Don't pump trouble. Stay away from steroids." ◆

This teenager built up his muscles lifting weights. Some other teens take anabolic steroids to build muscles. One recent survey says that nearly one in ten high school seniors has used steroids.

• Gerty Theresa Radnitz Cori (1896–1957), an American biochemist who studied how carbohydrates are metabolized
• Dorothy Crowfoot Hodgkin (1910–), a British chemist who used X-ray crystallography to determine the structure of vitamin B_{12}

• Marjorie Caserio (1929–), an American chemist who specialized in studying organic reactions

EXTENSION

Students who are interested in Patrick Montoya's career, biomedical engineering, might want to read the following magazine article.
"Reshaping Our Lives—Advanced Materials." *National Geographic* 176 (December 1989): 746–781. This article discusses the wealth of new materials created by chemists that enhance peoples' lives.

THEN AND NOW

Gladys Emerson (1903 - 1984)

When Gladys Emerson was growing up in Caldwell, Kansas, she enjoyed listening to and playing music. As she grew older, she found it difficult to choose a career, because, in addition to enjoying the arts, she also liked science. When she went to college, she decided to take classes in both fields. Upon graduation from Oklahoma College for Women, she received two degrees—one in physics and chemistry and the other in English and history. She attended graduate school at Stanford University, where she completed a master's degree in history.

When she was in her twenties, Emerson taught nutrition at the University of California at Berkeley. That position allowed her to earn her Ph.D. in nutrition and biochemistry. It also started her on a road that would take her to the Garvan Medal, which is awarded for distinguished work in chemistry.

Among her many contributions were her work in helping to isolate vitamin E and in studying the chemistry and effects of vitamin B complex. She also studied the effect of diet on tumors. Gladys Emerson helped groups such as the World Health Organization and UNICEF work toward the goal of achieving good nutrition for all citizens. ◆

Patrick Montoya (1962-)

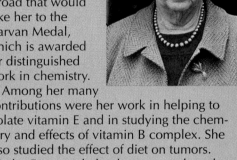

As a young adult, Patrick Montoya wanted to build and design automobiles. Once in college, however, he decided to combine his interest in engineering with classes in biology and chemistry. Now, Montoya's work as a biomedical engineer offers both challenges and rewards. It involves observing the processes of the human body, using computer models to study the way our lungs work, and designing artificial machines to temporarily take the place of organs.

Montoya designs machines that can perform vital functions. Sometimes, especially during an operation, a patient's lungs are not able to transfer oxygen to the blood. When this happens, physicians may use an oxygenator, better known as an artificial lung.

This oxygenator removes blood from the body, exchanges the carbon dioxide in the blood for oxygen, and returns the oxygen-rich blood to the body. Montoya's research has involved designing the pump that drives the blood through the oxygenator. This pump is a type of artificial heart, and together these machines form a heart-lung machine. He uses a computer model to understand better how the transport systems of the human body work and what improvements he should make to his designs. Although Montoya remains interested in automobile design, he has found fulfillment in using his engineering background to help people live longer and more healthful lives. ◆

Background Information

Dietitians must attend college for at least four years. They learn about nutrition and how to plan balanced diets for people in general and for people with special dietary needs, such as diabetics and coronary patients. Dietitians work in hospitals, nursing homes, schools, some restaurants and health clubs, and physicians' offices.

According to Annie Johnson, more and more dietitians are setting up private practice. They may work with several doctors to plan diets for patients.

Discussion

● **Process Skills:**
Classifying/Ordering, Inferring

Ask the students in what ways a dietitian could be helpful to a diabetic or a person who was recovering from a heart attack. (Responses might include that the dietitian could help a diabetic plan a menu of meals that the person would enjoy but that would limit sugar intake. For a heart patient, the dietitian could suggest low-fat, low-cholesterol foods to help the person reduce the risk of another heart attack.)

SCIENCE AT WORK

Annie Johnson, Dietitian

Annie Johnson knows about the nutritional value of school lunches. She is a dietitian, someone who chooses menus and supervises meals, often in hospitals, schools, and other institutions. In the following interview, she talks about what it means to be a dietitian.

What made you choose diet and nutrition as a career? I always enjoyed baking, and when I went to college, I decided to study food and nutrition. When I went to college, we studied everything about nutrition. Now students specialize in a particular field, such as therapeutic nutrition for hospitals. I graduated as a registered dietitian and started working in a hospital.

What does a dietitian do in a hospital? When I worked in the hospital, I helped teach patients about nutrition. I also helped individualize patients' diets according to their physicians' instructions. In the hospital, I always had the feeling I was helping someone.

What is your work like now? Now I supervise school lunch programs in an urban school district in Florida. Every day I'm out in the schools at lunch time. I check all the menus to see that they're balanced nutritionally. I allow cafeteria managers to set their own menus, so there is room for creativity.

How do school cafeteria programs teach students about nutrition? Cafeteria managers go into the classroom to help teach students about nutrition. They let students "taste test" unfamiliar foods. In the cafeteria, we encourage students to try new foods that they may never have at home. Some foods, like beets, aren't favorites, even among adults. I see the school lunch as a learning situation.

What has changed about school lunches since your career began? Years ago school cafeteria food was higher in fat. For example, all vegetables were cooked in butter. Now, vegetables are steamed, and meats are broiled instead of fried in fat. Instead of using fruit canned in heavy syrup, we buy fresh fruit. Now many school cafeterias offer salad bars and low-fat milk.◆

Discover More

For more information about careers for dietitians, write to the

American Dietetic Association
216 W. Jackson Blvd, Suite 800
Chicago, IL 60606

Discussion

● Process Skills: *Predicting, Classifying/Ordering*

Have the students read the article on artificial body parts and speculate on what kinds of arti- ficial body parts might be developed in the future and how those body parts would be useful. (Students' responses may include artificial body parts they have seen in science fiction movies or stories. Accept all reasonable responses.)

● Process Skills: *Formulating Hypotheses, Comparing*

Ask the students to hypothesize the advantages and disadvantages of having artificial replacement organs. (The advantages include living a more active life and being able to depend on oneself; disadvantages include the pain of surgery, the possibility of rejection of the replacement organ, and the surgical and hospital cost of the replacement organ.)

SCIENCE/TECHNOLOGY/SOCIETY

Artificial Body Parts

The human body is a biological machine whose parts may fail or simply wear out from use. Scientists and physicians are learning how to replace the body's parts with artificial organs that extend and improve human life.

Artificial Organs and Joints

People with cataracts have regained their sight through the implantation of lenses in the eye. Artificial knees and hips have enabled arthritis patients to walk. The hearing impaired may soon have their hearing restored by an electronic middle ear. Researchers continue to work on artificial replacements for the body's most complex organs, such as the heart and liver.

Even with artificial joints, the most common implant, scientists want a better match between the body and the artificial part. Artificial joints are usually made of space-age metals and plastics. However, the body's bone and muscle tissues do not easily bond to these foreign materials. Researchers have learned to coat the metal joint with "bioactive" ceramics (the material that clay pots are made of) that interact with body tissues. The ceramics actually stimulate bone growth and help cement the artificial joint in place.

Knee implants can now be custom designed for patients.

Journal Activity
Encourage interested students to research artificial organs and how they are being tested. Students might start their research with the following magazine article and record their research in their journals.

Yeaple, Judith Anne. "Winston's Wireless Heart." *Popular Science* 241 (November 1992): 30–31. This article briefly describes how researchers implanted a wireless artificial heart in a calf.

Researchers are also developing bioactive polymers (long chains of molecules) that are friendly to the body's tissues. Fibers coated with bioactive polymers serve as artificial ligaments to which muscle tissue will attach. Scientists are experimenting with a gel-like polymer that can be made to contract and relax like muscle.

Saving Lives with Artificial Organs

Newspaper headlines often tell of young patients waiting for kidney or liver transplants. Physicians hope they soon can routinely replace vital organs with artificial substitutes. Researchers continue to test several designs for an artificial heart. The most famous is the Jarvik-7, which has kept patients alive for several weeks.

Researchers are developing artificial cells that mimic the chemistry of the liver and kidneys. The artificial cells contain enzymes that convert the body's toxins into usable amino acids. Diabetics may soon benefit from an artificial pancreas. Scientists take cells that can manufacture insulin from a dog or calf and encase them in an artificial sac. The sac protects the animal cells from the body's rejection but allows glucose to reach the new cells and insulin to flow into the bloodstream.

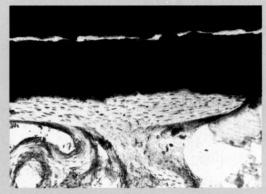

This micrograph shows a bioceramic coating (brown) that encourages bone cells (purple) to grow close to the surface of an artificial hip.

Many challenges remain in creating artificial body parts that effectively mimic the body's complex functions. Scientists are hampered especially by the difficulties of testing on humans. However, some researchers predict that technology may someday make replacement organs that are better than the originals. ◆

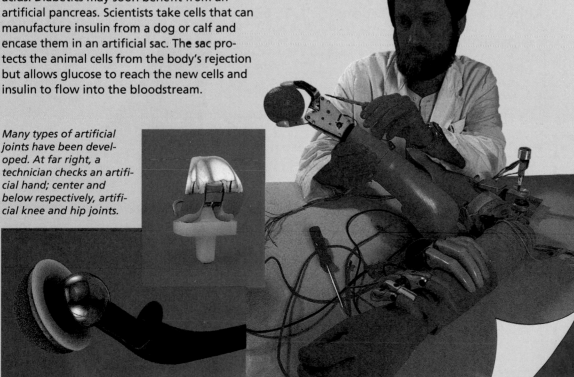

Many types of artificial joints have been developed. At far right, a technician checks an artificial hand; center and below respectively, artificial knee and hip joints.

REFERENCE SECTION

SAFETY GUIDELINES

Participating in laboratory investigations should be an enjoyable learning experience. You can ensure both learning and enjoyment from the experience by making the laboratory a safe place in which to work. Carelessness, lack of attention, and showing off are the major causes of laboratory accidents. It is, therefore, important that you follow safety guidelines at all times. If an accident should occur, you should know exactly where to locate emergency equipment. Practicing good safety procedures means being responsible for your classmates' safety as well as your own.

You will be expected to practice the following safety guidelines whenever you are in the laboratory.

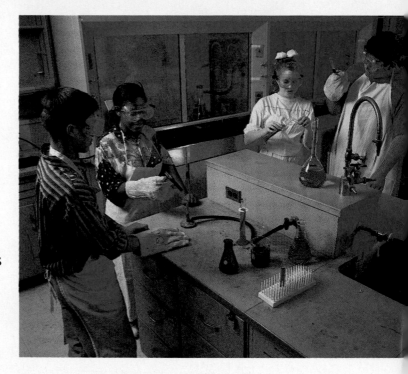

1. **Preparation** Study your laboratory assignment in advance. Before beginning your investigation, ask your teacher to explain any procedures you do not understand.
2. **Neatness** Keep work areas clean. Tie back long, loose hair and button or roll up long sleeves when working with chemicals or near an open flame.
3. **Eye Safety** Wear goggles when handling liquid chemicals, using an open flame, or performing any activity that could harm the eyes. If a solution is splashed into the eyes, wash the eyes with plenty of water and notify your teacher at once. Never use reflected sunlight to illuminate a microscope. This practice is dangerous to the eyes.

4. **Chemicals and Other Dangerous Substances** Some chemicals can be dangerous if they are handled carelessly. If any solution is spilled on a work surface, wash the solution off at once with plenty of water.
   - Never taste chemicals or place them near your eyes. Never eat in the laboratory. Counters and glassware may contain substances that can contaminate food. Handle toxic substances in a well-ventilated area or under a ventilation hood.
   - Never pour water into a strong acid or base. The mixture produces heat. Sometimes the heat causes splattering. To keep the mixture cool, pour the acid or base slowly into the water.
   - When noting the odor of chemical substances, wave the fumes

toward your nose with your hand rather than putting your nose close to the source of the odor.

• Do not use flammable substances near a flame.

5. **Safety Equipment** Know the location of all safety equipment, including fire extinguishers, fire blankets, first-aid kits, eyewash fountains, and emergency showers. Report all accidents and emergencies to your teacher immediately.

6. **Heat** Whenever possible, use an electric hot plate instead of an open flame. If you must use an open flame, shield the flame with a wire screen that has a ceramic center. When heating chemicals in a test tube, do not point the test tube toward anyone.

7. **Electricity** Be cautious around electrical wiring. Do not let cords hang loose over a table edge in a way that permits equipment to fall if the cord is tugged. Do not use equipment with frayed cords.

8. **Knives** Use knives, razor blades, and other sharp instruments with extreme care. Do not use double-edged razor blades in the laboratory.

9. **Glassware** Examine all glassware before heating. Glass containers for heating should be made of borosilicate glass or some other heat-resistant material. Never use cracked or chipped glassware.

• Never force glass tubing into rubber stoppers.

• Broken glassware should be swept up immediately, never picked up with the fingers. Broken glassware should be discarded in a special container, never into a sink.

10. **Unauthorized Experiments** Do not perform any experiment that has not been assigned or approved by your teacher. Never work alone in the laboratory.

11. **Cleanup** Wash your hands immediately after any laboratory activity. Before leaving the laboratory, clean up all work areas. Put away all equipment and supplies. Make sure water, gas, burners, and electric hot plates are turned off.

Remember at all times that a laboratory is a safe place only if you regard laboratory work as serious work.

The instructions for your laboratory investigations will include cautionary statements when necessary. In addition, you will find that the following safety symbols appear whenever a procedure requires extra caution:

 Wear safety goggles

 Wear laboratory apron

 Sharp/pointed object

 Biohazard/disease-causing organisms

 Flame/heat

 Dangerous chemical/poison

 Electrical hazard

 Gloves

 Radioactive material

LABORATORY PROCEDURES

READING A METRIC RULER

1. Examine your metric ruler. The numbers on it represent lengths in centimeters. The usual metric ruler is about 30 cm long. There are 10 marked spaces within each centimeter, which represent tenths of centimeters (0.1 cm).

2. To measure the width of a piece of paper, place the ruler on the paper. The zero end of the ruler must line up exactly with one edge of the paper. Look at the other edge of the paper to see which of the marks on the ruler is closest to that edge. In Figure A, for example, the edge of the paper is nearest to the second line beyond the 7. Therefore, the width of the paper is 7.2 cm.

3. The edge of the paper might fall exactly on one of the centimeter marks. In Figure B, the edge is just on the 5-cm mark. The width of this paper is 5.0 cm. You must write in the .0 to indicate that the measurement is accurate to the nearest tenth of a centimeter; that is, it is more than 4.9 cm and less than 5.1 cm.

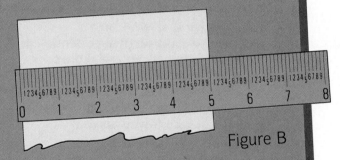

Figure B

4. Sometimes you may want to make a reading with more accuracy. It is possible to estimate readings to the nearest hundredth of a centimeter, but you must be very careful. Look at Figure A again. You can guess the number of tenths in the distance between the marks. The edge of the paper is about 3 tenths of the space between 7.2 and 7.3. The best estimate, then, is that the width of the paper is 7.23 cm.

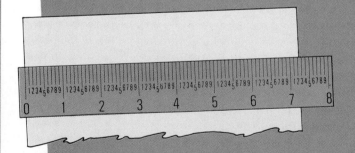

Figure A

5. In Figure C, the edge of the paper falls exactly on the 8.6 mark. If you are taking careful readings, accurate to the nearest hundredth of a centimeter, you must record the width as 8.60 cm.

6. Note the general rule: You can estimate scale readings to the nearest tenth of a scale division. If the scale is marked in tenths, you can estimate the hundredths place but never more than that.

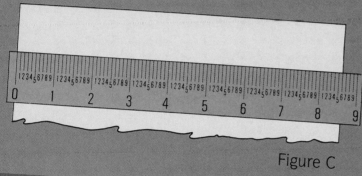

Figure C

CONVERTING SI UNITS

In SI, it is easy to convert from unit to unit. To convert from a larger unit to a smaller unit, move the decimal to the right. To convert from a smaller unit to a larger unit, move the decimal to the left. Figure D shows you how to move the decimals to convert in SI.

Figure D

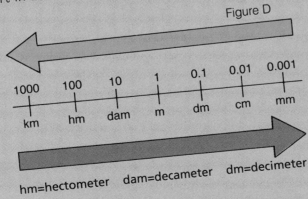

hm=hectometer   dam=decameter   dm=decimeter

SI Conversion Table

SI Units		Converting SI to Customary		Converting Customary to SI	
Length		1 km	= 0.62 mile	1 mile	= 1.609 km
kilometer (km)	= 1000 m	1 m	= 1.09 yards	1 yard	= 0.914 m
meter (m)	= 100 cm		= 3.28 feet	1 foot	= 0.305 m
		1 cm	= 0.394 inch	1 foot	= 30.5 cm
centimeter (cm)	= 0.01 m	1 mm	= 0.039 inch	1 inch	= 2.54 cm
millimeter (mm)	= 0.001 m				
micrometer (µm)	= 0.000 001 m				
nanometer (nm)	= 0.000 000 001 m				
Area		$1 km^2$	= 0.3861 square mile	1 square mile	= $2.590 km^2$
square kilometer (km^2)	= 100 hectares	1 ha	= 2.471 acres	1 acre	= 0.4047 ha
hectare (ha)	= 10 000 m^2	$1 m^2$	= 1.1960 square yards	1 square yard	= $0.8361 m^2$
square meter (m^2)	= 10 000 cm^2			1 square foot	= $0.0929 m^2$
		$1 cm^2$	= 0.155 square inch	1 square inch	= $6.4516 cm^2$
square centimeter (cm^2)	= 100 mm^2				
Mass		1 kg	= 2.205 pounds	1 pound	= 0.4536 kg
kilogram (kg)	= 1000 g	1 g	= 0.0353 ounce	1 ounce	= 28.35 g
gram (g)	= 1000 mg				
milligram (mg)	= 0.001 g				
microgram (µg)	= 0.000 001 g				
Volume of Solids		$1 m^3$	= 1.3080 cubic yards	1 cubic yard	= $0.7646 m^3$
1 cubic meter (m^3)	= 1 000 000 cm^3		= 35.315 cubic feet	1 cubic foot	= $0.0283 m^3$
		$1 cm^3$	= 0.0610 cubic inch	1 cubic inch	= $16.387 cm^3$
1 cubic centimeter (cm^3)	= 1000 mm^3				
Volume of Liquids		1 kL	= 264.17 gallons	1 gallon	= 3.785 L
kiloliter (kL)	= 1000 L	1 L	= 1.06 quarts	1 quart	= 0.94 L
liter (L)	= 1000 mL	1 mL	= 0.034 fluid ounce	1 pint	= 0.47 L
milliliter (mL)	= 0.001 L			1 fluid ounce	= 29.57 mL
microliter (µL)	= 0.000 001 L				

READING A GRADUATE

1. Examine the graduate and note how the scale is marked. The units are milliliters (mL). A milliliter is a thousandth of a liter and is equal to a cubic centimeter. Note carefully how many milliliters are represented by each scale division on the graduate.

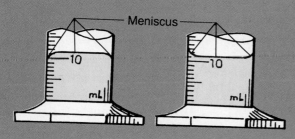

2. Pour some liquid into the cylinder and set the cylinder on a level surface. Notice that the upper surface of the liquid is flat in the center and curved at the edges. This curve is called the *meniscus* and may be either upward or downward. In reading the volume, you must ignore the curvature and read the scale at the flat part of the surface.

3. Bring your eye to the level of the surface and read the scale at the level of the flat surface of the liquid.

USING A LABORATORY BALANCE

1. Make sure the balance is on a level surface. Use the leveling screws at the bottom of the balance to make any necessary adjustments.

2. Place all the countermasses at zero. The pointer should be at zero. If it is not, adjust the balancing knob until the pointer rests at zero.

3. Place the object you wish to mass on the pan. **CAUTION: Do not place hot objects or chemicals directly on the balance pan, because they can damage its surface.**

4. Move the largest countermass along the beam to the right until it is at the last notch that does not tip the balance. Follow the same procedure with the next largest countermass. Then move the smallest countermass until the pointer rests at zero.

5. Determine the readings on all beams and add them together to determine the mass of the object.

6. When massing crystals or powders, use a piece of filter paper. First, mass the paper; then add the crystals or powders and remass. The actual mass is the total minus the mass of the paper. When massing liquids, first mass the empty container, then mass the liquid and container. Finally, subtract the mass of the container from the mass of the liquid and the container to get the mass of the liquid.

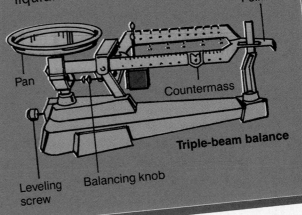

Triple-beam balance

USING A BUNSEN BURNER

1. Before lighting the burner, observe the locations of fire extinguishers, fire blankets, and sand buckets. Wear safety goggles and an apron. Tie back long hair and roll up long sleeves.

2. Close the air ports of the burner and turn the gas full on by using the valve at the laboratory outlet.

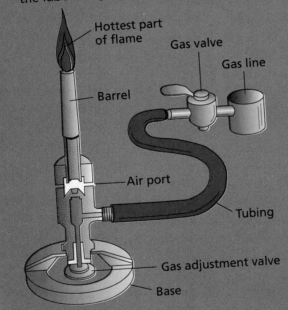

Hottest part of flame

Gas valve

Gas line

Barrel

Air port

Tubing

Gas adjustment valve

Base

3. Hold the striker in such a position that the spark will be just above the rim of the burner. Strike a spark.

4. Open the air ports until you can see a blue cone inside the flame. If you hear a roaring sound, the ports are open too wide.

5. CAUTION: If the burner is not operating properly, the flame may burn inside the base of the barrel. Carbon monoxide, an odorless gas, is released from this type of flame. Should this situation occur, immediately turn off the gas at the laboratory gas valve. Do not touch the barrel of the burner. After the barrel has cooled, partially close the air ports before relighting the burner.

6. Adjust the gas-flow valve and the air ports on the burner until you get a flame of the desired size with a blue cone. The hottest part of the flame is just above the tip of the blue cone.

FILTERING TECHNIQUES

1. To separate a precipitate from a solution, pass the mixture through filter paper. To do this, first obtain a glass funnel and a piece of filter paper.
2. Fold the filter paper in fourths as shown. Then open up one fourth of the folded paper. Put the paper, pointed end down, into the funnel.

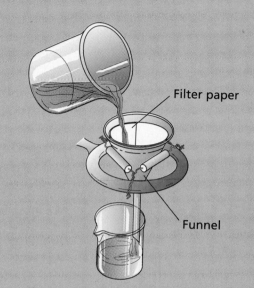

Filter paper

Funnel

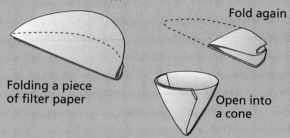

Folding a piece of filter paper

Fold again

Open into a cone

3. Support the funnel, and insert its stem into a beaker.
4. Stir the mixture to be filtered and pour it quickly into the filter paper within the funnel. Wait until all the liquid has flowed through.

5. You may wish to wash the solid that is left in the filter paper. If this is the case, pour some distilled water into the filter paper. Your teacher will tell you how much water to use.

USING REAGENTS

1. For safety reasons, it is important to learn how to pour a reagent from a bottle into a flask or a beaker. Begin with the reagent bottle on the table.
2. While holding the bottle steady with your left hand, grasp the stopper of the bottle between the first and second fingers of your right hand. Remove the stopper from the bottle. **DO NOT** put the stopper down on the table top.
3. While still holding the stopper, lift the bottle with your right hand and pour the reagent into your container.
4. Replace the bottle on the table top and replace the stopper.

If you are left handed, reverse the instructions.

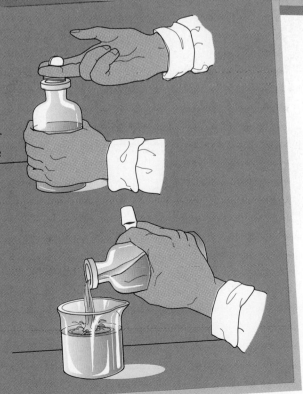

THE PERIODIC TABLE

1	G
H	
Hydrogen	
1.01	

Atomic number 92 | S |
Symbol **U**
Uranium
Atomic mass 238.03

State: S Solid
L Liquid
G Gas
X Not found in nature

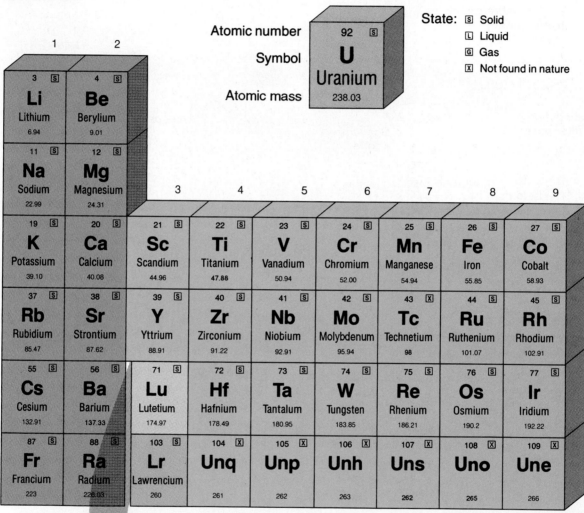

	1			2	

3	S	4	S
Li		**Be**	
Lithium		Berylium	
6.94		9.01	

11	S	12	S
Na		**Mg**	
Sodium		Magnesium	
22.99		24.31	

| | 3 | | 4 | | 5 | | 6 | | 7 | | 8 | | 9 |

19	S	20	S	21	S	22	S	23	S	24	S	25	S	26	S	27	S
K		**Ca**		**Sc**		**Ti**		**V**		**Cr**		**Mn**		**Fe**		**Co**	
Potassium		Calcium		Scandium		Titanium		Vanadium		Chromium		Manganese		Iron		Cobalt	
39.10		40.08		44.96		47.88		50.94		52.00		54.94		55.85		58.93	

37	S	38	S	39	S	40	S	41	S	42	S	43	X	44	S	45	S
Rb		**Sr**		**Y**		**Zr**		**Nb**		**Mo**		**Tc**		**Ru**		**Rh**	
Rubidium		Strontium		Yttrium		Zirconium		Niobium		Molybdenum		Technetium		Ruthenium		Rhodium	
85.47		87.62		88.91		91.22		92.91		95.94		98		101.07		102.91	

55	S	56	S	71	S	72	S	73	S	74	S	75	S	76	S	77	S
Cs		**Ba**		**Lu**		**Hf**		**Ta**		**W**		**Re**		**Os**		**Ir**	
Cesium		Barium		Lutetium		Hafnium		Tantalum		Tungsten		Rhenium		Osmium		Iridium	
132.91		137.33		174.97		178.49		180.95		183.85		186.21		190.2		192.22	

87	S	88	S	103	S	104	X	105	X	106	X	107	X	108	X	109	X
Fr		**Ra**		**Lr**		**Unq**		**Unp**		**Unh**		**Uns**		**Uno**		**Une**	
Francium		Radium		Lawrencium													
223		226.03		260		261		262		263		262		265		266	

57	S	58	S	59	S	60	S	61	X	62	S
La		**Ce**		**Pr**		**Nd**		**Pm**		**Sm**	
Lanthanum		Cerium		Praseodymium		Neodymium		Promethium		Samarium	
138.91		140.12		140.91		144.24		145		150.4	

89	S	90	S	91	S	92	S	93	X	94	X
Ac		**Th**		**Pa**		**U**		**Np**		**Pu**	
Actinium		Thorium		Protactinium		Uranium		Neptunium		Plutonium	
227.03		232.04		231.04		238.03		237.05		244	

Metals
Transition Metals
Nonmetals
Noble gases
Lanthanide series
Actinide series

18				

2 ⒢
He
Helium
4.00

13	14	15	16	17

5 ⓢ	6 ⓢ	7 ⒢	8 ⒢	9 ⒢	10 ⒢
B	**C**	**N**	**O**	**F**	**Ne**
Boron	Carbon	Nitrogen	Oxygen	Fluorine	Neon
10.81	12.01	14.01	16.00	19.00	20.18

13 ⓢ	14 ⓢ	15 ⓢ	16 ⓢ	17 ⒢	18 ⒢
Al	**Si**	**P**	**S**	**Cl**	**Ar**
Aluminum	Silicon	Phosphorus	Sulfur	Chlorine	Argon
26.98	28.09	30.97	32.07	35.45	39.95

10	11	12

28 ⓢ	29 ⓢ	30 ⓢ	31 ⓢ	32 ⓢ	33 ⓢ	34 ⓢ	35 ⒧	36 ⒢
Ni	**Cu**	**Zn**	**Ga**	**Ge**	**As**	**Se**	**Br**	**Kr**
Nickel	Copper	Zinc	Gallium	Germanium	Arsenic	Selenium	Bromine	Krypton
58.69	63.55	65.39	69.72	72.61	74.92	78.96	79.90	83.80

46 ⓢ	47 ⓢ	48 ⓢ	49 ⓢ	50 ⓢ	51 ⓢ	52 ⓢ	53 ⓢ	54 ⒢
Pd	**Ag**	**Cd**	**In**	**Sn**	**Sb**	**Te**	**I**	**Xe**
Palladium	Silver	Cadmium	Indium	Tin	Antimony	Tellurium	Iodine	Xenon
106.42	107.87	112.41	114.82	118.71	121.75	127.60	126.90	131.29

78 ⓢ	79 ⓢ	80 ⒧	81 ⓢ	82 ⓢ	83 ⓢ	84 ⓢ	85 ⓢ	86 ⒢
Pt	**Au**	**Hg**	**Tl**	**Pb**	**Bi**	**Po**	**At**	**Rn**
Platinum	Gold	Mercury	Thallium	Lead	Bismuth	Polonium	Astatine	Radon
195.08	196.97	200.59	204.38	207.2	208.98	209	210	222

63 ⓢ	64 ⓢ	65 ⓢ	66 ⓢ	67 ⓢ	68 ⓢ	69 ⓢ	70 ⓢ
Eu	**Gd**	**Tb**	**Dy**	**Ho**	**Er**	**Tm**	**Yb**
Europium	Gadolinium	Terbium	Dysprosium	Holmium	Erbium	Thulium	Ytterbium
151.96	157.25	158.93	162.50	164.93	167.26	168.93	173.04

95 ⓧ	96 ⓧ	97 ⓧ	98 ⓧ	99 ⓧ	100 ⓧ	101 ⓧ	102 ⓧ
Am	**Cm**	**Bk**	**Cf**	**Es**	**Fm**	**Md**	**No**
Americium	Curium	Berkelium	Californium	Einsteinium	Fermium	Mendelevium	Nobelium
243	247	247	251	252	257	258	259

ELECTRIC CIRCUIT SYMBOLS

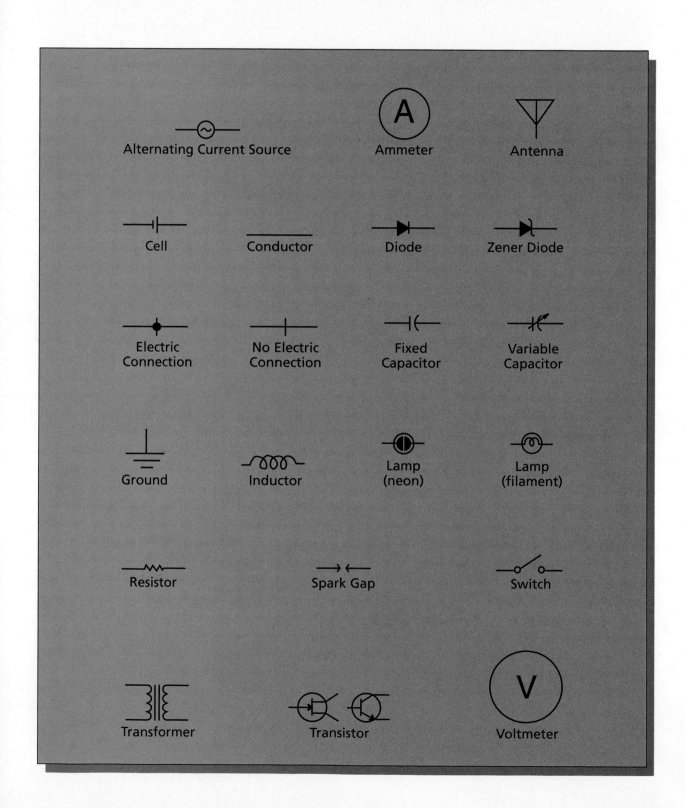

Alternating Current Source

Ammeter

Antenna

Cell

Conductor

Diode

Zener Diode

Electric Connection

No Electric Connection

Fixed Capacitor

Variable Capacitor

Ground

Inductor

Lamp (neon)

Lamp (filament)

Resistor

Spark Gap

Switch

Transformer

Transistor

Voltmeter

PHYSICAL SCIENCE EQUATIONS

$$\text{density} = \frac{\text{mass}}{\text{volume}}$$

$$d = \frac{m}{V}$$

$$\text{mechanical advantage} = \frac{\text{force output}}{\text{force input}}$$

$$MA = \frac{F_o}{F_i}$$

$$\text{acceleration} = \frac{\text{distance}}{\text{time}^2}$$

$$a = \frac{d}{t^2}$$

$$\text{kinetic energy} = \frac{\text{mass} \times \text{speed}^2}{2}$$

$$KE = \frac{1}{2}mv^2$$

$$\text{acceleration} = \frac{\text{force}}{\text{mass}}$$

$$a = \frac{F}{m}$$

$$\text{average speed} = \frac{\text{total distance}}{\text{total time}}$$

$$v = \frac{d}{t}$$

$$\text{acceleration} = \frac{\text{final velocity} - \text{initial velocity}}{\text{time}}$$

$$a = \frac{v_f - v_i}{t}$$

$$\text{power} = \frac{\text{work}}{\text{time to do work}}$$

$$P = \frac{W}{t}$$

$$\text{work} = \text{force} \times \text{distance}$$
$$W = Fd$$

$$\text{power} = \text{current} \times \text{potential difference}$$
$$P = IV$$

$$\text{wave speed} = \text{wavelength} \times \text{frequency}$$
$$v = \lambda f$$

GLOSSARY

Pronunciation Key					
Symbol	**As In**	**Phonetic Respelling**	**Symbol**	**As In**	**Phonetic Respelling**
a	b<u>a</u>t	a (bat)	ô	d<u>o</u>g	aw (dawg)
ā	f<u>a</u>ce	ay (fays)	oi	f<u>oi</u>l	oy (foyl)
â	c<u>a</u>reful	ai (CAIR fuhl)	ou	m<u>ou</u>ntain	ow (MOWN tuhn)
ä	<u>a</u>rgue	ah (AHR gyoo)	s	<u>s</u>it	s (siht)
ch	<u>ch</u>apel	ch (CHAP uhl)	sh	<u>sh</u>eep	sh (sheep)
e	t<u>e</u>st	eh (tehst)	u	l<u>o</u>ve	uh (luhv)
ē	<u>ea</u>t	ee (eet)	u̇	p<u>u</u>ll	u (pul)
	sk<u>i</u>	ee (skee)	ü	m<u>u</u>le	oo (myool)
ėr	f<u>er</u>n	ur (furn)	zh	trea<u>s</u>ure	zh (TREH zhuhr)
i	b<u>i</u>t	ih (biht)	ə	med<u>a</u>l	uh (MEHD uhl)
ī	r<u>i</u>pe	y (ryp)		<u>e</u>ffect	uh (uh FEHKT)
	<u>i</u>dea	eye (eye DEE uh)		seri<u>ou</u>s	uh (SIHR ee uhs)
k	<u>c</u>ard	k (kahrd)		<u>o</u>ni<u>o</u>n	uh (UHN yuhn)
o	l<u>o</u>ck	ah (lahk)		tal<u>e</u>nt	uh (TAL uhnt)
ō	<u>o</u>ver	oh (OH vuhr)			

A

acceleration the rate at which velocity changes **(75)**

acid a compound that produces hydrogen ions (H^+) in solution **(567)**

acid rain rain formed when air pollutants (sulfur oxides and nitrogen oxides) combine with oxygen and water vapor in the atmosphere to form weak sulfuric and nitric acids **(182)**

acoustics the branch of physics that deals with the transmission of sound **(252)**

active solar heating complex solar heating systems; method of heating that uses mechanical devices such as fans to move heat from one place to another **(161)**

Acceleration

allotropes one of two or more molecular forms of the same element; graphite and diamond are allotropes of carbon **(472)**

alloy mixture of two or more metals, or a mixture of metals and nonmetals **(459)**

alpha decay type of nuclear decay in which an atom emits an alpha particle (a helium nucleus), which is made up of two protons and two neutrons **(485)**

alternating current (AC) electric current in which the electrons repeatedly change direction rather than flowing in a single direction **(350)**

amplitude the height of a wave, or the distance a wave moves from a rest position **(220)**

analog something that represents something else; an audio signal is a series of electric impulses and is an analog of a sound wave **(408)**

atomic mass the sum of the relative masses of the protons and neutrons in an atom **(439)**

atomic number the number of protons in the nucleus of an element **(440)**

Allotropes

B

base a compound that produces hydroxide ions (OH⁻) in solution **(573)**

beta decay type of nuclear decay in which an atom emits a beta particle, which is formed when a neutron is converted into a proton and an electron **(485)**

C

carbohydrates compounds made of carbon, hydrogen, and oxygen in a ratio of CH_2O **(604)**

chain reaction a reaction that occurs over and over as the result of a single beginning action **(497)**

characteristic property a property that always stays the same and is characteristic of a particular kind of matter **(41)**

chemical bond the force that holds the atoms in a compound together **(542)**

chemical equation a list of the compounds or elements involved in a reaction and the new compounds or elements that form as a result **(553)**

Base

Circuit breaker

circuit breaker a device that protects circuits from overloads **(405)**

coherent light the type of light created when all the light waves in a beam are in step **(313)**

colloid a mixture in which the particles are so small that they stay in solution **(533)**

compound a pure substance that has molecules made of two or more different kinds of elements that have been combined chemically **(35; 520)**

compound machines two or more simple machines combined to do work **(115)**

compressions the areas where the particles in a longitudinal wave are close together **(219)**

concave lens a lens that is thinner in the middle than it is at the edges; it causes light rays to diverge **(268)**

conclusion a judgment based on the analysis of data gathered from experiments or field studies **(8)**

conduction the transfer of heat energy from one substance to another by direct contact **(142)**

conductor a material that readily allows the transfer of heat or the flow of electric current **(143; 355)**

conservation the careful and efficient use of resources **(188)**

convection the transfer of heat in liquids and gases as groups of molecules move in currents **(144)**

convex lens a lens that is thicker in the middle than it is at the edges; it causes light rays to converge **(268)**

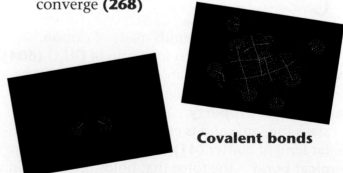

Covalent bonds

covalent bond a bond formed between two or more atoms in which the valence electrons are shared **(546)**

crests the high points of a transverse wave **(217)**

D

decibels (dB) unit used to measure the volume of sound **(246)**

decomposition reaction a reaction in which a compound is broken down into two or more substances **(556)**

density the ratio of the mass of an object to its volume **(24)**

diatomic molecules molecules composed of two atoms **(470)**

digital signals a series of electric pulses that stand for two digits, 0 and 1 **(409)**

direct current (DC) electric current in which the electrons flow steadily in one direction **(350)**

Doppler effect a change in the frequency of waves caused by a moving wave source or a moving observer **(252)**

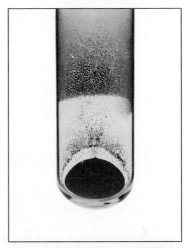

Decomposition reaction

E

electric circuit the path electricity follows **(355)**

electric current flow of electrons through a conductor **(348)**

electric motor a device that converts electrical energy into mechanical energy **(394)**

electromagnet a strong magnet created when a solenoid with an iron core has electric current passed through it **(381)**

electromagnetic induction the process by which a moving magnetic field produces an electric current in a conductor **(383)**

electromagnetic spectrum all types of electromagnetic waves **(285)**

electromagnetic waves transverse waves that carry both electric and magnetic energy **(284)**

electron the negatively charged particle found in atoms **(432)**

element a pure substance that contains only one kind of atom **(34)**

energy levels orbits in which electrons circle the nucleus **(435)**

excited state the state of an atom in which it contains stored energy **(314)**

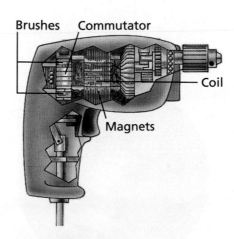

Brushes   Commutator
Coil
Magnets

Electric motor

F

focal length   the distance from the lens to the focus **(268)**

force   a push or a pull **(78)**

fossil fuels   fuels formed from plant and animal material that was buried in sediment millions of years ago; coal and petroleum **(153)**

frequency   the number of vibrations a wave produces each second **(221)**

friction   the force that opposes motion between surfaces that touch **(81)**

fuse   an electric safety device containing a short strip of metal with a low melting point **(404)**

G

gamma decay   type of nuclear decay in which an atom emits high-energy radiation similar to X-rays **(486)**

Geiger counter   a hand-held instrument that measures the level of radiation **(490)**

geothermal energy   natural heat within the earth **(166)**

greenhouse effect   the increase in the temperature of Earth due to the increase of carbon dioxide in the atmosphere **(179)**

ground state   the state an atom returns to as it gives off the energy gained to reach an excited state **(314)**

groups   sets of elements arranged in vertical columns on the periodic table **(449)**

H

heat   the measure of the total kinetic energy of the random motion of the atoms and molecules of a substance **(139)**

hologram   an interference pattern created by a laser on a piece of film **(321)**

hydrocarbons   compounds that are made of only carbon and hydrogen **(593)**

hypothesis   an educated guess about what might happen **(5)**

Geothermal Energy

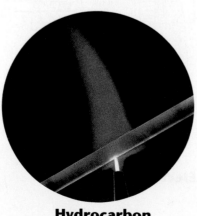

Hydrocarbon

inclined plane   a simple machine consisting of a flat, sloping surface; a ramp **(114)**

indicator   a substance that changes color in an acid or a base **(578)**

inertia   the resistance any object has to a change in velocity **(88)**

insulator   a material that is a poor conductor of heat or electric current **(143; 355)**

ion   an atom or group of atoms that has an electric charge as the result of losing or gaining electrons **(459)**

ionic bond   a chemical bond resulting from the attraction between a positive and a negative ion **(544)**

isomer   one of two or more compounds that have the same chemical formula but different structures **(594)**

isotopes   atoms of the same element that have different numbers of neutrons **(441)**

joule (J)   the amount of work equal to the force of 1 N moving an object a distance of 1 m **(104)**

kinetic energy   the energy an object has due to its mass and its motion **(127)**

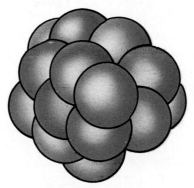

Isotope, C–14

larynx   the organ that produces the human voice; the small box made of cartilage located in the front of your neck **(254)**

law   a summary of many experimental results and observations **(9)**

law of conservation of momentum   the law that states that momentum can be transferred from one object to another but cannot change in total amount **(92)**

lever   a simple machine; a bar used for prying or dislodging something **(110)**

light pipe   a clear plastic rod that can be bent into a curved shape to carry light into otherwise unreachable places **(322)**

lipid   solid fat containing atoms of carbon, hydrogen, and oxygen **(608)**

longitudinal wave   a wave in which the particles move parallel to the path of the wave **(219)**

lux   a unit of measure for the brightness of light **(263)**

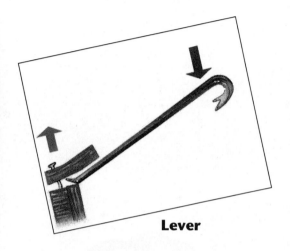

Lever

machine   a device that helps to do work by changing the size or direction of a force **(101)**

magnetic field   any region in which magnetic forces are present **(375)**

magnetism   a force of repulsion or attraction between like or unlike poles **(368)**

mass   the measure of the amount of matter in an object **(21)**

mass number   the whole number nearest the atomic mass of an atom; its total protons + neutrons **(439)**

mechanical advantage (MA)   the number of times a machine multiplies force; the ratio of the force that comes out of a machine to the force that is put into the same machine **(106)**

metalloids   substances that exhibit some, but not all, of the properties of metals **(462)**

mixture a combination of two or more different substances that are not combined chemically **(36)**

momentum an object's mass multiplied by its velocity **(77)**

monomers the small chemical units that make up a polymer **(595)**

N

neutron a particle found in the nucleus of an atom that has no charge **(434)**

newton the force required to move 1 kg of mass a distance of one meter per second each second **(79)**

noise pollution noise that is harmful to human health **(246)**

nonrenewable resources resources that cannot be replaced after they are used **(153)**

nuclear reactors complex devices that are used to convert nuclear energy into heat energy **(170)**

nuclear waste spent fuel and other radioactive products produced by nuclear reactions **(185)**

O

optical fiber a long, thin light pipe that uses total internal reflection to carry light **(323)**

P

parallel circuit an electric circuit that has two or more separate paths through which electricity can flow **(359)**

passive solar heating simple solar heating systems; heating as a result of the simple absorption of solar energy **(160)**

period a row of elements in the periodic table **(445)**

periodic table the organization of elements by atomic number into a table; similar elements are grouped in rows (periods) and columns (groups) **(445)**

phosphors the chemicals that are used to coat the inside of TV screens; when struck by electrons they glow to produce a picture **(411)**

physical change a change in matter that does not produce a new kind of molecule **(43)**

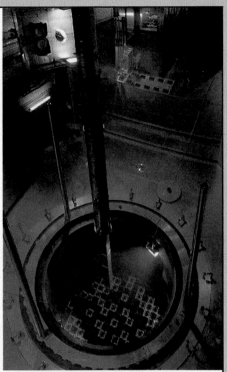

Nuclear reactor

Optical fibers

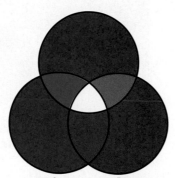

Primary colors

Protein

physical property   a property of matter that can be observed or measured without changing the composition of the matter **(42)**

pitch   the highness or lowness of a sound; determined by frequency **(245)**

plasma   a phase of matter consisting of a low-density, super-heated gas in which the electrons and nuclei are free to move at random **(499)**

polymers   large molecules that are made by hooking many identical small molecules together **(595)**

potential difference   a value expressed in volts related to electrical potential energy that is measured by a voltmeter **(352)**

potential energy   the energy stored in an object due to its position **(125)**

power   the measure of the amount of work done in a certain period of time **(102)**

power grid   a network of electric transmission lines that provide electricity to homes and other buildings **(391)**

primary colors   red, green, and blue; when combined these three colors can produce any other color **(302)**

protein   huge molecules made up of amino acids; needed for tissue growth and repair **(605)**

proton   the positive particle found in the nucleus of an atom **(434)**

pulley   a simple machine that consists of a wheel that is free to spin on an axle **(112)**

pumping   the process of raising atoms to an excited state during the production of a laser beam **(314)**

pure substance   matter that contains only one kind of molecule **(34)**

R

radiation   the transfer of energy by electro-magnetic waves **(145)**

radioactive   the condition in which the nuclei of unstable atoms spontaneously break apart, giving off particles or high-energy radiation **(483)**

rarefactions   the areas where the particles in a longitudinal wave are far apart **(219)**

real image   the image created when a collection of points of light is focused by a lens **(271)**

refinery a large industrial plant that separates crude oil into products such as gasoline, diesel fuel, heating fuel, petroleum jelly, and asphalt **(155)**

reflection the bouncing back of a wave after it strikes a barrier **(226)**

refraction the change of direction when a wave enters a different medium **(228)**

replacement reaction a reaction in which one atom or group of atoms in a compound is replaced with another atom or group of atoms **(557)**

resistance a measure of how much a substance opposes the flow of electricity; the ratio of potential difference to electric current **(356)**

reverberation the combination of many small echoes occurring very close together; caused when reflected sound waves meet at a single point, one right after another **(252)**

Refinery

saturated solution a solution in which the maximum amount of solute is dissolved in a given amount of solvent **(526)**

screw a simple machine composed of an inclined plane wrapped around a cylinder **(115)**

secondary colors yellow, cyan, and magenta; colors formed by combining two primary colors **(302)**

series circuit an electric circuit in which there is only one path for the current to follow **(357)**

short circuit any accidental connection that allows electric current to take an unintended path instead of passing through an appliance **(402)**

simple machines machines that have only one or two parts; the lever, the wheel and axle, the pulley, the inclined plane, the wedge, and the screw **(109)**

solar energy energy from the sun **(159)**

solenoid a wire coil through which electric current can flow **(381)**

solubility the maximum amount of solute that can be dissolved in 100 g of solvent at a specific temperature and pressure **(527)**

solute the substance in a solution that is dissolved in the solvent **(523)**

solution a homogeneous mixture **(523)**

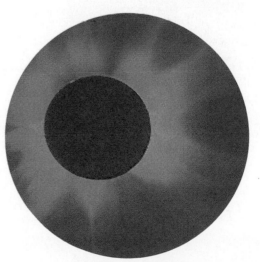

Solar energy

Static electricity

Telecommunications

solvent   the material in a solution in which the solute is dissolved **(523)**

sonar   a navigation system that uses echoes of ultrasonic waves to find the depth of water; an acronym for **so**und **na**vigation **r**anging **(247)**

speed   the measure of how fast an object moves over a certain distance in a specific time **(72)**

standard units   units that are agreed upon and used by large numbers of people **(15)**

standing waves   waves that form a pattern where portions of the waves do not move and other portions move with increased amplitude **(232)**

static electricity   a buildup of an electric charge on an object **(343)**

stimulated emission   the process of forcing identical atoms into step **(315)**

structural formula   a simplified drawing that shows the pattern in which atoms are bonded in a molecule **(592)**

sublimation   a phase change in which a solid changes directly into a gas **(45)**

suspension   a heterogeneous mixture in which one of the parts is a liquid **(531)**

synthesis reaction   a chemical reaction in which two or more substances combine to form one new substance **(555)**

telecommunications   the science and technology of sending messages over long distances **(297)**

temperature   the measure of the average kinetic energy of the moving atoms and molecules of a substance **(140)**

theory   an explanation of why things work the way they do **(9)**

thermal expansion   a physical change that occurs when the volume of a substance increases as the temperature increases **(49)**

total internal reflection   in an optical fiber, the complete reflection of light from the inside surface; in this way, light waves travel long distances through the fibers **(322)**

transformer a device that uses electromagnetic induction to change the electric potential (voltage) of alternating current **(397)**

transverse wave the wave created when the particles within the wave move perpendicular to the path of the wave **(218)**

troughs depressions in a transverse wave **(217)**

velocity the speed of an object in a particular direction **(74)**

virtual image an image that forms only because the light seems to diverge from it **(272)**

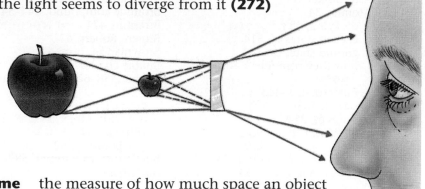

volume the measure of how much space an object takes up; the softness or loudness of a sound **(20; 244)**

wave a disturbance that travels through matter or space **(216)**

wavelength the distance between one point on one wave and the identical point on the next wave **(221)**

wedge a simple machine consisting of an inclined plane; used to pry objects apart **(114)**

weight the force of gravity pulling one object toward the center of another object **(85)**

wheel and axle a simple machine consisting of a lever connected to a shaft **(111)**

wind turbines machines that convert the energy of wind into electricity **(163)**

work the measure of the force required to make something move **(101)**

Wind turbines

INDEX

Boldface numbers refer to an illustration on that page.

Magnetic field(s), 374–379, **374, 376**
  around electric current, 380
  and solenoids, 381, **381**
  turning on and off, 381, **382**
Magnetic lines of force, 377–378, **377, 378, 380**
Magnetic poles, 372, **372**
Magnetic Resonance Imaging (MRI), 513
Magnetism, **369**
  defined, 368
  and electricity, 380–383
  properties of, 369
Magnetite. *See* Lodestone
Magnets
  fields around, 376–377, **376, 377**
  inside, 371–372, **371**
  in use, 372–373, **372**
Mandelonitrile, 616
Maniet, Monique, 334
Mason, Jane, 334
Mass, 21, **21**
  measuring, 20–22
  and weight, 85, **85**
Mass number, 439
Matter, 32–55,
  ancient Greeks on nature of, 428–429, **429**
  characteristics of, 40–45, **40, 42, 43, 44, 45**
  mixtures, 36–38, *table* 37
  model of, 46–51, **46, 47, 48, 49, 50, 51**
  physical properties of, *table* 43
  properties of, 40
  and purity, 34, 36
  states of, 32–34, **32, 33, 34**
  structure of, 428–449
  types of, 518–521, **519, 520, 521**
Measurement, invention of, 15, **15**
Mechanical advantage (MA), 106–107, **107**
Mechanical energy, 124, **124**
Medium, 216
Meltdown, 171
Melting, as phase change, 51, **51**
Melting point, 44
Mendeleev, Dmitri, 445–448, **445, 448**
Meniscus, 14, **20**
Menzel, Jack, 419
Metal atoms, 458
  arrangement of, **459**
Metallic families, reactivity of, 461–462, **461, 462**
Metalloid(s), 462–463, **463**
  silicon as, 587
Metals, properties of, 456–462, **456, 458, 459, 460, 461, 462**

Meter, 19
Meter stick, **104**
Meter technician, 422
Methane, 592, 593
  structural formula for, **592**
Microcrystalline polymer (MCP) suspensions, 532, **532**
Microgravity, 57
Microwaves, 298–299, **298**
  radiation from, 299
Milliliter, 20
Millirems, 500
Mimicry, 616
Mirage, 144
Mixture, 36–37, 518–521
  vs. compounds, 521
  physical methods of separating parts of, *table* 37
  types of, 519, **519**
Model of matter, 46–51, **46, 47, 48, 49, 51**
Moderator, **170**, 171, **171**
Molecules, 34, 546
  of life, 604–610
  and physical changes, 49–50
  and physical properties, 48–49, **48–49**
Momentum, 76–77, **76, 77**
  law of conservation of, 92–93, **92, 93**
Monarch butterflies, **614,** 616
Mono–, 549
Monomers, 595, **595**
Montoya, Patrick, 620, **620**
Moses, Edwin, 619
Motion
  describing, 72–77, **72, 73, 74, 75, 76, 77**
  first law of, 87–89, **87, 88, 89**
  forces and, 78–86, **78, 79, 80, 81, 82, 83, 84, 85**
  perpetual, 135, **135**
  predicting and explaining, 87–93
  second law of, 90–91, **90, 91**
  third law of, 91–92, **91**
Muriatic acid, 572
Mutagenics, 493

Natural gas, 156, **156**
Natural products, 600
Nearsightedness, 275, **276**
Neon, 469
  and laser production, 314–316, **314, 316**
Neutral atom, 438
Neutralization, 573–574
Neutrons, 434, 438
Newton, Isaac, 87, 265, 286–287

Newton (N), 79, **104**
Nitric acid, 572, **572**
Nitrogen, 471
Nitrogen, liquid, 423
NMR Spectroscopy, 512–513, **512, 513**
Noble gases, 466
  as inert, 466, **466**
Nodes, 232, 233
Noise pollution, 246, **246**
Nonliving, defined, 495
Nonmetal families, 470–475, **470, 471, 472, 473, 475**
Nonmetals, 465–473, **465, 466, 468**
  characteristics of, 459
Nonrenewable sources, 153
Nonvisible spectrum, 289–298, **290, 292, 293, 294, 295, 296, 297, 298**
Normal, of wave, 227, **227**
North Pole
  geographic, 374–375, **376**
  magnetic, 374–375, **375, 376**
Nuclear energy, 170–172, **170, 171**
  as not waste free, 185–186, **185, 186**
Nuclear fission, 497
  See also Fission
Nuclear fusion, 337, **337**
Nuclear magnetic resonance (NMR), 511, **511**
Nuclear reactions, 486, **486, 497–501**
Nuclear reactors, 170–172, **170, 171**
Nuclear waste, 185–186, **185, 186**

Ocean thermal energy conversion (OTEC), 169, **169**
Oersted, Hans Christian, 380
Ohm, Georg, 356
Ohmmeters, 356, **356**
Ohms, 356, **356**
Oils, 608, **609**
Optical fiber, 323–324, **324**
Orbitals, 436
Organic chemistry, 587
Organic compounds, 592–601, **592, 593, 594, 595, 596, 598, 600, 601, 602,** *tables* 597, 599
Organization for Petroleum Exporting Countries (OPEC), 197
Oscilloscope, **350**
Overloads, 403–405, **404, 405**
Oxygen, 593
  and ozone, 473, **473**
Ozone, 473–475, **473**
Ozone layer, 180–181, **181**

CREDITS

PHOTOS

Abbreviations used: (t) top, (c) center, (b) bottom, (l) left, (r) right, (bkgrd) background.

Page: ii, Romilly Lockyer/The Image Bank; vi(bl), Park Street; vi(br), HRW Photo by Dennis Fagan; vii, HBC Photo by Rodney Jones; viii(t), Stephen Wade/AllSport; viii(b), Frank Whitney/The Image Bank; ix(t), J. Novak/SuperStock; ix(c), Dr. Don James/Adventure Photo; ix(b), Park Street; x(t), HRW Photo by Eric Beggs; x(c), J. Leifer/SuperStock; x(b), Steve Weber/Picturesque; xi, E. R. Degginger; xii, William E. Ferguson; xiii(t), E. R. Degginger; xiii(b), NASA; xiv(t), Richard A. Nelridge/Phototake; xiv(b), Yoav Levy/Phototake; xv(t), E. R. Degginger; xv(br), HRW Photo by Stanley Livingston; xv(br), Walter Bibikow/The Image Bank; xvi(t), Ken Eward/Science Source/Photo Researchers; xvi(b), Culver Pictures, Inc.; xvii(r), Phil Degginger; xvii(b), HBC Photo; xviii(tr), Audrey Gibson/The Stock Market; xviii(b), Lee Foster/Bruce Coleman, Inc.; xix(t), Alfred Pasieka/Peter Arnold, Inc.; xix(b), Ned Haines/Photo Researchers; xx(t) HRW photo by John Langford; xx(b) HRW photo by Felicia Martinez/PhotoEdit; xxi, Dan R. Boyd; 1-2, Douglas B. Nelson/Peter Arnold, Inc.; 4, David R. Frazier Photolibrary; 5, Will & Deni McIntyre/Science Source/Photo Researchers; 6(all), with permission of CURRENT SCIENCE Magazine; 7, Dan McCoy/Rainbow; 10(t), Archive Photos; 11(tr), HBC Photo by Rodney Jones; 11(bl), Howard Dratch/The Image Works; 12, Gary Halpern/Sygma; 15, C. Alexander Marshack; 19, HBC Photo by Rodney Jones; 21(b), HBC Photo by Jerry White; 23(br), HBC Photo by Rodney Jones; 24, Chip Clark; 29(r), Audrey Gibson/The Stock Market; 29(c), Peter Steiner/The Stock Market; 29(l), Park Street; 30, Didier Givois/Agence Vandystadt/Allsport; 32, E. R. Degginger; 33, Rudy Muller/Envision; 34(r), Dan McCoy/Rainbow; 34(l), Dan McCoy/Rainbow; 35(r), Warren Jacobi/Picture Perfect USA; 35(l), E. R. Degginger; 36(r), HBC Photo by Rodney Jones; 36(l), Ron Dahlquist/Tony Stone Images; 37, HRW Photos by Dennis Fagan; 38, Island Water Association; 40(r),A. S./Duomo; 40(l), Focus on Sports; 42, Focus on Sports; 44(r), H. Gritscher/Peter Arnold, Inc.; 44(c), David Barnes/The Stock Market; 44(l), HBC Photo by Rodney Jones; 45,

HBC Photo; 46(t), Tom Stack/Tom Stack & Associates; 46(b), HBC Photo; 48, Sigrid Owen/International Stock Photography; 49, Comstock; 50, Richard Choy/Peter Arnold, Inc.; 51(t), HBC Photo by Rodney Jones; 51(b), Chip Clark; 53, Didier Givois/Agence Vandystadt/Allsport; 55(t), Jay Freis/The Image Bank; 55(b), Chuck O'Rear/Westlight; 56(both), Lehigh University; 57(all), Marshall Space Flight Center/NASA; 58(b), Richard Faverty; 59(cr), Paul Silverman/Fundamental Photographs; 59(bl), The Bettmann Archive; 59(tr), Richard Parker/Photo Researchers, Inc.; 59(cl), The Bettmann Archive; 59(br), The Bettmann Archive; 60(bc), Guido A. Rossi/The Image Bank; 60(tr), George S. Bolster Collection, Historical Society of Saratoga Springs, New York; 60(cl), courtesy Saratoga County Chamber of Commerce; 60(bl), The Bettmann Archive; 60(br), Lawrence Ruggeri/The Stockhouse; 61(tr), John Bova/Photo Researchers, Inc.; 61(tl,c,bl,br), HRW photos by Park Street; 62(c), J. Koivula/Photo Researchers, Inc.; 62(tl), Ludek Pesek/Science Photo Library/Photo Researchers, Inc.; 62(br), Mark Tuschman; 62(bl), Cazenave/Vandystadt/Photo Researchers, Inc.; 62(tr), Patrice Loiez, CERN/Science Photo Library/Photo Researchers, Inc.; 63(bl), Culver Pictures, Inc.; 63(tr), AT&T Bell Labs; 64-65, HBC Photos; 66(inset), Bruce Iverson/Photomicrographs; 66(br), HRW Photo by Eric Beggs, product courtesy of American Excelsior Company; 67, Rafael Macia/Photo Researchers, Inc.; 68-69, J. Messerschmidt/H. Armstrong Roberts; 71, National Archives; 72, Gerard Vandystadt/Allsport; 73(tl), Arthur Hustwitt/Leo de Wys, Inc.; 73(tc), Bill Staley/West Stock, Inc.; 73(tr), Al Satterwhite/The Image Bank; 74(tl), Ken Biggs/The Stock Market; 74(bl), Focus on Sports; 75(tr), Ron McQueeney/Sportschrome; 75(tl), NASA; 75(b), Steve Swope/Indy 500 Photos; 76, Craig Melvin/Sports Photo Masters; 77(t), Tim Bieber/The Image Bank; 77(b), Willy Ostgathe/Leo de Wys, Inc.; 78, B. de Lis/Gamma-Liaison; 79(r), Tom Grill/Comstock; 79(l), HBC Photo by Rodney Jones; 80, Paul Kennedy/Leo de Wys, Inc.; 82(b), E. R. Degginger; 82(t), Russ Kinne/Comstock; 83, David Madison/Duomo Photos; 85, NASA; 87, General Motors Co.; 88, Loomis Dean/LIFE Magazine © 1954 Time Warner, Inc.; 89, NASA; 90(r), Tony Duffy/Allsport; 90(l), David

Madison; 92, Nathan Bilow/Allsport; 93, Roger Ressmeyer/Starlight; 97(l), M. Timothy O'Keefe/Bruce Coleman, Inc.; 97(br), A. Hubrich/H. Armstrong Roberts, Inc.; 98(bkgrd), Ralph Clevenger/Westlight; 98, The Butler Institute of American Art; 101, HBC Photo by Rodney Jones; 102, A. Hubrich/H. Armstrong Roberts; 103(c), David Madison/Bruce Coleman, Inc.; 103(l), Mitchell B. Reibel/Sports Photo Masters; 103(r), Todd Powell/Adventure Photo; 105, HBC Photo by Beverly Brosius; 106, David Madison; 107, Hanson Carroll/Peter Arnold, Inc.; 110(r), Sportschrome; 110(l), Richard Hutchings/Photo Researchers; 110(c), Whitney L. Lane/The Image Bank; 111-115, HBC Photos by Rodney Jones; 116, © 1968 Reprinted by special permission of King Features Syndicate; 117(l), Richard Megna/Fundamental Photographs; 117(r), David Young-Wolff/PhotoEdit; 118, From THE WORLD BOOK ENCYCLOPEDIA, © 1993, World Book, Inc., by permission of the publisher; 119(tl), The Butler Institute of American Art; 121(bl), HBC Photo by Rodney King; 121(r), HRW Photos by Lisa Davis; 122-123, NASA; 124, Stephen Wade/Allsport; 125, Ben Rose/The Image Bank; 126(all), Rich Clarkson; 127(r), David W. Hamilton/The Image Bank; 127(l), G. Zimmerman/Sports Illustrated/Time Warner, Inc.; 129, Bradley Leverett/Tony Stone Images; 130, HRW Photo by Michelle Bridwell; 133, Paul J. Sutton/Duomo; 134, HBC Photo by Rodney Jones; 135(l), Hank Morgan/Rainbow; 135(r), Culver Pictures, Inc.; 136, G. C. Kelley/Photo Researchers; 139(b), NASA/Lewis Research Center; 139(tl), Jerry Sieve/FPG; 139(tr), Culver Pictures, Inc.; 139(cr), Culver Pictures, Inc.; 142, Frank Siteman/Stock, Boston; 143(tr), Nick Nicholson/The Image Bank; 143(bl), HBC Photo by Rodney Jones; 144(tl), HBC Photo; 144(bl), Jack Ryan/Photo Researchers; 144(br), Frank Siteman/Stock, Boston; 147, NASA; 149, Gregory Dale; 150, United Mine Workers of America Archives; 151, Lewis Hine, United Mine Workers Journal; 153(b), Farrell Grehan/Photo Researchers; 153(c), John D. Cunningham/Visuals Unlimited; 153(t), Kage/Peter Arnold, Inc.; 154, Visuals Unlimited; 155, Sam C. Pierson/Photo Researchers; 157, David Hundley/The Stock Market; 159, Thayer Department of Engineering/Dartmouth University Solar Team; 160, Frank Whitney/The

ILLUSTRATIONS

For permission to reprint copyrighted material, grateful acknowledgment is made to the following sources:

George G. Blakey: From *The Diamond* by George G. Blakey. Copyright © 1977 by George G. Blakey.

Butler Institute of American Art: Illustration by Albert Bierstadt from the front and back cover of *The Oregon Trail* by Leonard Fisher.

Children's Television Workshop: From "CAT Scans: High-Tech Medicine Can Save Animals, Too" from *3-2-1 Contact*, May 1991. Copyright © 1991 by Children's Television Workshop. All rights reserved.

Cobblestone Publishing, Inc: From "The Trouble with CFCs" from *Odyssey*, vol. 11, no. 1, January 1989. Copyright © 1989 by Cobblestone Publishing, Inc., Peterborough NH 03458.

The Continuum Publishing Company: From "Elective Affinities" from *The Sufferings of Young Werther and Elective Affinities* by Johann Wolfgang von Goeth, edited by Victor Lange. Copyright © 1990 by the Continuum Publishing Company.

The Crown Publishing Group: From "Child Labor in the Mines" from *A Pictorial History of American Mining* by Howard N. Sloane and Lucille L. Sloane. Copyright © 1970 by Howard N. Sloane and Lucille L. Sloane.

Current Science®: "The Flashy Science of Fireworks" from *Current Science®*, vol. 76, no. 18, May 10, 1991. Copyright © 1991 by The Weekly Reader Corporation. "How Safe is Irradiated Food?" from *Current Science®*, vol. 77, no. 16, April 10, 1992. Copyright © 1992 by The Weekly Reader Corporation. From "Sneaker Science: How to Choose your Shoes" from *Current Science®*, vol. 77, no. 15, March 27, 1992. Copyright © 1992 by The Weekly Reader Corporation. "Steroids Build Muscles and Destroy Health" from *Current Science®*, vol. 74, no. 9, January 6, 1989. Copyright © 1989 by The Weekly Reader Corporation. "You Are There: Riding a Roller Coaster-Thrills, Chills, and Fun Physics" from *Current Science®*, vol. 76, no.18, May 10, 1991. Copyright © 1991 by The Weekly Reader Corporation.

J. M. Dent & Sons, Ltd: From "Communications: Messages by Wire" from *A History of Invention* by Egon Larsen, with illustrations by George Land. Copyright © 1961, 1969 by Egon Larsen. Published by Everyman's Library.

Doubleday, a division of Bantam Doubleday Dell Publishing Group, Inc.: From "Dreaming is a Private Thing" from *The Complete Stories*, Volume 1, by Isaac Asimov. Copyright © 1990 by Nightfall, Inc.

Down Beat Magazine: From "Wynton Marsalis: 1987" from *Down Beat Magazine*, vol. 54, no. 11, November 1987. Copyright © 1987 by Down Beat Magazine.

Holiday House: From *The Oregon Trail* by Leonard Everett Fisher. Copyright © 1990 by Leonard Everett Fisher.

Kids Discover: From "Bubbles" from *Kids Discover*, April 1992. Copyright © 1992 by Kids Discover.

Lansing Lamont: From *Day of Trinity* by Lansing Lamont. Copyright © 1965 by Lansing Lamont.

Los Alamos National Laboratory: Illustrations of sectional view of atomic bomb and six photographs of test explosion of 1st atomic bomb. Copyright © 1945 by Los Alamos National Laboratory.

Lothrop, Lee & Shepard Books, a division of William Morrow & Co., Inc.: Excerpt and 1 illustration from *Near The Sea: A Portfolio of Paintings* by Jim Arnosky. Copyright © 1990 by Jim Arnosky. From *To Space and Back* by Sally Ride and Susan Okie. Copyright © 1986 by Sally Ride and Susan Okie.

National Geographic Society: From "Chernobyl-One Year After" by Mike Edwards from *National Geographic*, vol. 171, no. 5, May 1987. Copyright © 1987 by National Geographic Society.

National Geographic WORLD: From "A New World At Your Fingertips" from *National Geographic WORLD*, no. 192, August 1991. Copyright © 1991 by National Geographic Society. Art by Dave Jonason. Copyright © 1991 by National Geographic WORLD.

Newsweek: From "The Longest Night" from *Newsweek*, November 22, 1965. Photograph by Bernard Gotfryd. Copyright © 1965 by Newsweek, Inc. All rights reserved. Reprinted by permission.

Noble, M. Lee: "Walk on a Winter's Day" by M. Lee Noble. Copyright © 1994 by M. Lee Noble.

Random House, Inc: From *The Wright Brothers: Pioneers of American Aviation* by Quentin Reynolds. Copyright © 1950 by Random House, Inc.; renewed copyright © 1978 by James J. Reynolds & Frederick H. Rohlfs, Esq.

Sierra Club Books: From *Ascent: The Mountaineering Experience in Word and Image* by Allen Steck and Steve Roper. Copyright © 1989 by Sierra Club Books. From "The Yosemite" from *The American Wilderness in the Words of John Muir* by the Editors of Country Beautiful. Copyright © 1973 by Country Beautiful Corporation.

CREDITS

Annotated Teacher's Edition

PHOTO

Abbreviations used:(t) top,(c) center,(b) bottom,(l) left,(r) right.

Page T2-3, Romilly Lockyer/The Image Bank; T6, Ginger Chih/Peter Arnold; T7, Rudy Muller/Envision; T22-T27 borders, Lowell Georgia/Science Source/Photo Researchers; T25, HRW photo by Lisa Davis; T26, Tim Bieber/The Image Bank; T27, John Zimmerman; T28-T31 borders, E. R. Degginger; T29, G. C. Kelley/Photo Researchers; T31, John Gerlach/Earth Scenes; T32 border, Sam C. Pierson/Photo Researchers; T33(c), Jim Whitmer; T33 border, Gregory Dale; T34-T35 borders, Tom Branch/Photo Researchers; T34, David E. Kennedy/TexaStock; T36-T37 borders, Jeffrey Blackman/Index Stock International; T37, Steve Chenn/Westlight; T38 border, Comstock; T38(tr), Mary Messenger; T38(b), HBC photo by Richard Haynes; T39 border, SEUL/Science Source/Photo Researchers; T40 border, Steve Weber/Picturesque; T40, HRW photo by Richard Haynes; T41 border, H. Gritscher/Peter Arnold; T42 border, Merrel Wood/The Image Bank; T42, Will & Deni McIntyre/Science Source/Photo Researchers; T43, HRW photo by John Langford; T43 border, Dan McCoy/Rainbow; T44-T45 borders, Manfred Kage/Peter Arnold; T44, HBC photo by Rodney Jones; T45, HRW photo by John Langford; T46-T47 borders, Richard Megna/Fundamental Photographs; T47(tr), Phil Degginger; T47(b), E. R. Degginger; T48-T51 borders, Ron Dahlquist/Tony Stone Images; T48(tr), Chuck O'Rear/Westlight; T48(b), John Madere/The Stock Market; T49, Tom Grill/Comstock; T50(l), Alfred Pasieka/Peter Arnold; T50(r), Willy Ostgathe/Leo de Wys; T51, D. B. Owen/Black Star.

ART

Icons: Rosario Cosgrove, Joan Rivers
Page T7, David Griffin; **T24(t),** Holly Cooper; **T24(b),** Holly Cooper; **T35,** Holly Cooper; **T46,** Don Collins